COMPREHENSIVE ANALYTICAL CHEMISTRY

ELSEVIER SCIENCE B.V.
Sara Burgerhartstraat 25
P.O. Box 211, 1000 AE Amsterdam, The Netherlands



First edition 2003

Library of Congress Cataloging in Publication Data
A catalog record from the Library of Congress has been applied for.

British Library Cataloguing in Publication Data
A catalogue record from the British Library has been applied for.

ISBN: 0-444-51037-0
ISSN: 0166-526X

⊗ The paper used in this publication meets the requirements of ANSI/NISO Z39.48-1992 (Permanence of Paper).
Printed in The Netherlands.

**Wilson & Wilson's**

**COMPREHENSIVE ANALYTICAL CHEMISTRY**

*Edited by*

D. BARCELÓ

*Research Professor*
*Department of Environmental Chemistry*
*IIUQAB-CSIC*
*Jordi Girona 18-26*
*08034 Barcelona*
*Spain*

**Wilson & Wilson's**

**COMPREHENSIVE ANALYTICAL CHEMISTRY**

VOLUME XXXIX

# INTEGRATED ANALYTICAL SYSTEMS

*Edited by*

SALVADOR ALEGRET

2003

ELSEVIER
Amsterdam – Boston – Heidelberg – London – New York – Oxford
Paris – San Diego – San Francisco – Singapore – Sydney – Tokyo

# CONTRIBUTORS TO VOLUME XXXIX

S. Alegret
*Department of Chemistry, Autonomous University of Barcelona, 08193 Bellaterra, Catalonia, Spain*

A. van der Berg
*MESA + Research Institute, University of Twente, PO Box 217, 7500 AE Enschede, The Netherlands*

P. Bergveld
*MESA + Research Institute, University of Twente, PO Box 217, 7500 AE Enschede, The Netherlands*

S. Böhm
*Coventor B.V., Kabelweg 37, 1014 BA Amsterdam, The Netherlands*

E.J. Calvo
*INQUIMAE, Departamento de Química Inorgánica, Analítica y Química Física, Facultad de Ciencias Exactas y Naturales, Universidad de Buenos Aires, 1428 Buenos Aires, Argentina*

L.F. Capitán-Vallvey
*Department of Analytical Chemistry, University of Granada, Fuentenueva, s/n, 18071 Granada, Spain*

S. Casans Berga
*Electronics Engineering Department, University of Valencia, Doctor Moliner 50, 46100 Burjassot, València, Spain*

C. Danilowicz
*INQUIMAE, Departamento de Química Inorgánica, Analítica y Química Física, Facultad de Ciencias Exactas y Naturales, Universidad de Buenos Aires, 1428 Buenos Aires, Argentina*

C. Domínguez
*Centre Nacional de Microelectrònica, CSIC, Campus UAB, 08193 Bellaterra, Catalonia, Spain*

M.D. Fernández Ramos
*Department of Analytical Chemistry, University of Granada, Fuentenueva, s/n, 18071 Granada, Spain*

E. Forzani
*INQUIMAE, Departamento de Química Inorgánica, Analítica y Química Física, Facultad de Ciencias Exactas y Naturales, Universidad de Buenos Aires, 1428 Buenos Aires, Argentina*

L.M. Lechuga
*Centro Nacional de Microelectrónica,CSIC, Isaac Newton 8, 28760 Tres Cantos, Madrid, Spain*

A. Legin
*Laboratory of Chemical Sensors, Department of Chemistry, St. Petersburg University, St. Petersburg 199034, Russia*

H.J. van der Linden
*MESA + Research Institute, University of Twente, PO Box 217, 7500 AE Enschede, The Netherlands*

J.L. Luque-García
*Department of Analytical Chemistry, University of Cordoba, Campus of Rabanales, 14071 Córdoba, Spain*

M.D. Luque de Castro
*Department of Analytical Chemistry, University of Cordoba, Campus of Rabanales, 14071 Córdoba, Spain*

R.D. McDowall
*McDowall Consulting, 73 Murray Av., Bromley, Kent BR1 3DJ, United Kingdom*

A. Merkoçi
*SensoChip Lab, Department of Chemistry and Biochemistry, New Mexico State University, MSC 3C, Las Cruces NM 88003, USA*

D.N. Muraviev
*Department of Chemistry, Autonomous University of Barcelona, 08193 Bellaterra, Catalonia, Spain*

R. Narayanaswamy
*Department of Instrumentation & Analytical Science, University of Manchester Institute of Science and Technology, PO Box 68, Manchester M60 1QD, United Kingdom*

A. E. Navarro Antón
*Electronics Engineering Department, University of Valencia, Doctor Moliner 50, 46100 Burjassot, València, Spain*

W. Olthuis
*MESA + Research Institute, University of Twente, PO Box 217, 7500 AE Enschede, The Netherlands*

M. Otero
*INQUIMAE, Departamento de Química Inorgánica, Analítica y Química Física, Facultad de Ciencias Exactas y Naturales, Universidad de Buenos Aires, 1428 Buenos Aires, Argentina*

D.G. Pijanowska
*Polska Akademia Nauk, Institute of Biocybernetics and Biomedical Engineering, 02-109 Warszawa, Poland*

A. Poghossian
*Institute of Thin Films and Interfaces, Research Centre Jülich GmbH, D-52425 Jülich, Germany*

D. Ramírez
*Electronics Engineering Department, University of Valencia, Doctor Moliner 50, 46100 Burjassot, València, Spain*

J.A. Rodríguez
*Physics Faculty, Havana University, PO Box 6426, 10600 Habana 6, Cuba*

A. Rudnitskaya
*Laboratory of Chemical Sensors, Department of Chemistry, St. Petersburg University, St. Petersburg 199034, Russia*

M. J. Schöning
*Institute of Thin Films and Interfaces, Research Centre Jülich GmbH, D-52425 Jülich, and University of Applied Sciences Aachen, Ginsterweg 1, D-52428 Jülich, Germany*

F. Sevilla III
*Research Centre for the Natural Sciences, University of Santo Tomas, Manila 1008, Philippines*

A.J. Sprenkels
*MESA + Research Institute, University of Twente, PO Box 217, 7500 AE Enschede, The Netherlands*

R. Stuetz
*School of Water Sciences, Cranfield University, Cranfield, MK43 0AL, United Kingdom*

K. Toko
*Department of Electronics, Graduate School of Information Science and Electrical Engineering, Kyushu University, Fukuoka 812-8581, Japan*

Yu. Vlasov
*Laboratory of Chemical Sensors, Department of Chemistry, St. Petersburg University, St. Petersburg 199034, Russia*

T.T. Veenstra
*MESA+ Research Institute, University of Twente, PO Box 217, 7500 AE Enschede, The Netherlands*

T. Vo-Dinh
*Advanced Biomedical Science and Technology Group, Oak Ridge National Laboratory, PO Box 2008, Oak Ridge TN 37831-6101, USA*

A. Wolosiuk
*INQUIMAE, Departamento de Química Inorgánica, Analítica y Química Física, Facultad de Ciencias Exactas y Naturales, Universidad de Buenos Aires, 1428 Buenos Aires, Argentina*

WILSON AND WILSON'S

# COMPREHENSIVE ANALYTICAL CHEMISTRY

## VOLUMES IN THE SERIES


| | |
|---|---|
| Vol. IA | Analytical Processes |
| | Gas Analysis |
| | Inorganic Qualitative Analysis |
| | Organic Qualitative Analysis |
| | Inorganic Gravimetric Analysis |
| Vol. IB | Inorganic Titrimetric Analysis |
| | Organic Quantitative Analysis |
| Vol. IC | Analytical Chemistry of the Elements |
| Vol. IIA | Electrochemical Analysis |
| | Electrodeposition |
| | Potentiometric Titrations |
| | Conductometric Titrations |
| | High-Frequency Titrations |
| Vol. IIB | Liquid Chromatography in Columns |
| | Gas Chromatography |
| | Ion Exchangers |
| | Distillation |
| Vol. IIC | Paper and Thin Layer Chromatography |
| | Radiochemical Methods |
| | Nuclear Magnetic Resonance and Electron Spin Resonance Methods |
| | X-Ray Spectrometry |
| Vol. IID | Coulometric Analysis |
| Vol. III | Elemental Analysis with Minute Sample |
| | Standards and Standardization |
| | Separation by Liquid Amalgams |
| | Vacuum Fusion Analysis of Gases in Metals |
| | Electroanalysis in Molten Salts |
| Vol. IV | Instrumentation for Spectroscopy |
| | Atomic Absorption and Fluorescence Spectroscopy |
| | Diffuse Reflectance Spectroscopy |
| Vol. V | Emission Spectroscopy |
| | Analytical Microwave Spectroscopy |
| | Analytical Applications of Electron Microscopy |
| Vol. VI | Analytical Infrared Spectroscopy |
| Vol. VII | Thermal Methods in Analytical Chemistry |
| | Substoichiometric Analytical Methods |
| Vol. VIII | Enzyme Electrodes in Analytical Chemistry |
| | Molecular Fluorescence Spectroscopy |
| | Photometric Titrations |
| | Analytical Applications of Interferometry |
| Vol. IX | Ultraviolet Photoelectron and Photon Spectroscopy |
| | Auger Electron Spectroscopy |

| | |
|---|---|
| | Plasma Excitation in Spectrochemical Analysis |
| Vol. X | Organic Spot Tests Analysis |
| | The History of Analytical Chemistry |
| Vol. XI | The Application of Mathematical Statistics in Analytical Chemistry |
| | Mass Spectrometry |
| | Ion Selective Electrodes |
| Vol. XII | Thermal Analysis |
| | Part A. Simultaneous Thermoanalytical Examination by Means of the Derivatograph |
| | Part B. Biochemical and Clinical Applications of Thermometric and Thermal Analysis |
| | Part C. Emanation Thermal Analysis and other Radiometric Emanation Methods |
| | Part D. Thermophysical Properties of Solids |
| | Part E. Pulse Method of Measuring Thermophysical Parameters |
| Vol. XIII | Analysis of Complex Hydrocarbons |
| | Part A. Separation Methods |
| | Part B. Group Analysis and Detailed Analysis |
| Vol. XIV | Ion-Exchangers in Analytical Chemistry |
| Vol. XV | Methods of Organic Analysis |
| Vol. XVI | Chemical Microscopy |
| | Thermomicroscopy of Organic Compounds |
| Vol. XVII | Gas and Liquid Analysers |
| Vol. XVIII | Kinetic Methods in Chemical Analysis |
| | Application of Computers in Analytical Chemistry |
| Vol. XIX | Analytical Visible and Ultraviolet Spectrometry |
| Vol. XX | Photometric Methods in Inorganic Trace Analysis |
| Vol. XXI | New Developments in Conductometric and Oscillometric Analysis |
| Vol. XXII | Titrimetric Analysis in Organic Solvents |
| Vol. XXIII | Analytical and Biomedical Applications of Ion-Selective Field-Effect Transistors |
| Vol. XXIV | Energy Dispersive A-Ray Fluorescence Analysis |
| Vol. XXV | Preconcentration of Trace Elements |
| Vol. XXVI | Radionucleide X-Ray Fluorescence Analysis |
| Vol. XXVII | Voltammetry |
| Vol. XXVIII | Analysis of Substances in the Gaseous Phase |
| Vol. XXIX | Chemiluminescence Immunoassay |
| Vol. XXX | Spectrochemical Trace Analysis for Metals and Metalloids |
| Vol. XXXI | Surfactants in Analytical Chemistry |
| Vol. XXXII | Environmental Analytical Chemistry |
| Vol. XXXIII | Elemental Speciation – New Approaches for Trace Element Analysis |
| Vol. XXXIV | Discrete Sample Introduction Techniques for Inductively Coupled Plasma Mass Spectrometry |
| Vol. XXXV | Modern Fourier Transform Infrared Spectroscopy |
| Vol. XXXVI | Chemical Test Methods of Analysis |
| Vol. XXXVII | Sampling and Sample Preparation for Field and Laboratory |
| Vol. XXXVIII | Countercurrent Chromatography: The Support-Free Iiquid Stationary Phase |
| Vol. XXXIX | Integrated Analytical Systems |

# Contents

Series Editor's Preface . . . . . . . . . . . . . . . . . . . . . . . . . . . . . . xxi
Editor's Preface . . . . . . . . . . . . . . . . . . . . . . . . . . . . . . . . . . xxiii

*Chapter 1. Integration, a new paradigm in analytical chemistry* . . . . . . . . . 1
S. Alegret
1.1 Integration in science and technology . . . . . . . . . . . . . . 1
1.1.1 Integration in analytical chemistry . . . . . . . . . . . 3
1.2 Parts and components . . . . . . . . . . . . . . . . . . . . . . . . 7
1.2.1 Supported reagents . . . . . . . . . . . . . . . . . . . . . . 7
1.2.2 Separation membranes . . . . . . . . . . . . . . . . . . . 9
1.3 Systems . . . . . . . . . . . . . . . . . . . . . . . . . . . . . . . . . 10
1.3.1 Total analysis systems . . . . . . . . . . . . . . . . . . . 10
1.3.2 Miniaturised systems . . . . . . . . . . . . . . . . . . . . 13
1.3.3 Networked systems . . . . . . . . . . . . . . . . . . . . . 14
1.4 Sensors . . . . . . . . . . . . . . . . . . . . . . . . . . . . . . . . . 15
1.4.1 Electrochemical sensors . . . . . . . . . . . . . . . . . . 17
1.4.2 Optochemical sensors . . . . . . . . . . . . . . . . . . . 19
1.5 Array systems . . . . . . . . . . . . . . . . . . . . . . . . . . . . . 19
1.5.1 Redundant-sensor array systems . . . . . . . . . . . . 20
1.5.2 Selective-sensor array systems . . . . . . . . . . . . . 20
1.5.3 Cross-selective sensor array systems . . . . . . . . . 21
1.6 Microsystems . . . . . . . . . . . . . . . . . . . . . . . . . . . . . 22
1.6.1 Microsensors . . . . . . . . . . . . . . . . . . . . . . . . . 23
1.6.2 Analytical microsystems . . . . . . . . . . . . . . . . . 25
1.6.3 Array microsystems . . . . . . . . . . . . . . . . . . . . 28
1.7 Nanosystems . . . . . . . . . . . . . . . . . . . . . . . . . . . . . . 30
1.8 Conclusions and perspectives . . . . . . . . . . . . . . . . . . . 32
Acknowledgements . . . . . . . . . . . . . . . . . . . . . . . . . . . . 33
References . . . . . . . . . . . . . . . . . . . . . . . . . . . . . . . . . 33

*SYSTEMS*

*Chapter 2. Integrated separation systems* . . . . . . . . . . . . . . . . . . . . 37
D. N. Muraviev
2.1 Introduction . . . . . . . . . . . . . . . . . . . . . . . . . . . . . . 37
2.2 General principles of bi-phase separation . . . . . . . . . . . . 38
2.2.1 Thermodynamics of bi-phase equilibrium . . . . . . . 40
2.2.2 Integration concepts in bi-phase separation . . . . . . 40
2.2.3 Integration of uptake and stripping steps . . . . . . . 42

2.3 Multiplication of single separation effect . . . . . . . . . . . . . 48
2.3.1 Frontal ion exchange chromatography . . . . . . . . . 50
2.3.2 Reverse frontal ion exchange chromatography . . . . 52
2.3.3 Displacement chromatography . . . . . . . . . . . . 54
2.3.4 Tandem ion exchange fractionation . . . . . . . . . . 55
2.4 Combined separation techniques . . . . . . . . . . . . . . . . 57
2.4.1 Solvent extraction-ion exchange. Aqua impregnated resins . . . . . . . . . . . . . . . . . . . . . . . . . 58
2.4.2 Ion exchange-crystallisation. Ion exchange isothermal supersaturation . . . . . . . . . . . . . . 65
2.4.3 Ion exchange supersaturation of zwitterlites . . . . . 67
2.4.4 Ion exchange supersaturation of electrolytes . . . . . 71
Acknowledgements . . . . . . . . . . . . . . . . . . . . . . . . . 73
References . . . . . . . . . . . . . . . . . . . . . . . . . . . . 73

*Chapter 3. Solid-phase spectrometric assays* . . . . . . . . . . . . . . . . . . . . 81
L.F. Capitán-Vallvey and M.D. Fernández Ramos
3.1 Introduction . . . . . . . . . . . . . . . . . . . . . . . . . . 81
3.2 Integration of processes in solid-phase spectrometric assays . 83
3.3 Types of solid-phase spectrometric assays . . . . . . . . . . . 84
3.4 Features of solid-phase spectrometric assays . . . . . . . . . . 85
3.5 Particulated solid-phase spectrometric assays . . . . . . . . . 86
3.5.1 Fixation process . . . . . . . . . . . . . . . . . . . . 87
3.5.2 Operational aspects . . . . . . . . . . . . . . . . . . . 96
3.5.3 Analytical characteristics . . . . . . . . . . . . . . . 112
3.5.4 Mixtures resolution . . . . . . . . . . . . . . . . . . 118
3.5.5 Analytical applications . . . . . . . . . . . . . . . . . 121
3.6 Membrane solid-phase spectrometric assays . . . . . . . . . . 122
3.6.1 Membrane filtration systems . . . . . . . . . . . . . 122
3.6.2 Membrane 'problem' equilibration systems . . . . . . 125
3.6.3 Membrane 'problem' deposit systems . . . . . . . . . 129
Acronyms . . . . . . . . . . . . . . . . . . . . . . . . . . 147
References . . . . . . . . . . . . . . . . . . . . . . . . . 149

*Chapter 4. Continuous flow analytical systems* . . . . . . . . . . . . . . . . . 161
M.D. Luque de Castro and J.L. Luque-García
4.1 Introduction . . . . . . . . . . . . . . . . . . . . . . . . . . 161
4.2 Reverse flow injection . . . . . . . . . . . . . . . . . . . . . 164
4.3 Integrating effect of conventional flow injection units . . . . . 164
4.3.1 Confluence points . . . . . . . . . . . . . . . . . . . 164
4.3.2 Exchanged units . . . . . . . . . . . . . . . . . . . . 166
4.3.3 Modified units . . . . . . . . . . . . . . . . . . . . . 166
4.3.4 Duplicate units . . . . . . . . . . . . . . . . . . . . . 171
4.4 Derivatisation reactions in flow injection systems . . . . . . . 174
4.4.1 Redox reactions involving solid reagents . . . . . . . 174

4.4.2 Micellar media . . . . . . . . . . . . . . . . . . . . . 174
4.4.3 Photoinduced reactions . . . . . . . . . . . . . . . . 175
4.4.4 Electrogenerated reagents . . . . . . . . . . . . . . . 175
4.4.5 Catalytic reactions . . . . . . . . . . . . . . . . . . 176
4.5 External energy sources integrated with flow injection . . . . 176
4.5.1 Conventional heat sources . . . . . . . . . . . . . . . 176
4.5.2 Ultrasound energy sources . . . . . . . . . . . . . . 177
4.5.3 Use of electrical energy . . . . . . . . . . . . . . . . 180
4.5.4 Microwave energy assistance . . . . . . . . . . . . . 180
4.6 In-line coupling of simple non-chromatographic continuous separation units and flow injection manifolds . . . . . . . . . 182
4.6.1 Couplings with techniques involving gas-separation: gas-diffusers, pervaporators and others . . . . . . . 184
4.6.2 Couplings with liquid–liquid separators: dialysers and liquid–liquid extractors . . . . . . . . 190
4.6.3 Couplings with liquid–solid separators and solid phase formation . . . . . . . . . . . . . . 196
4.6.4 On-line separation equipment and flow injection manifolds . . . . . . . . . . . . . . . . . . . . . . . 205
4.7 On-line coupling of robotics and flow injection manifolds . . . 214
4.8 Detection in flow injection . . . . . . . . . . . . . . . . . . . 217
4.8.1 Flow injection–detector interfaces . . . . . . . . . . 219
4.8.2 Automatic calibration . . . . . . . . . . . . . . . . . 220
4.8.3 Special uses of conventional detectors coupled to FI . . . . . . . . . . . . . . . . . . . . . . . 221
4.8.4 Three-dimensional and complex detectors coupled to FI . . . . . . . . . . . . . . . . . . . . . . . 224
4.8.5 Screening and flow injection . . . . . . . . . . . . . . 226
4.9 Integration and flow injection . . . . . . . . . . . . . . . . . 226
References . . . . . . . . . . . . . . . . . . . . . . . . . . . . 227

*Chapter 5 Distributed analytical instrumentation systems* . . . . . . . . . . . 245
D. Ramírez, A. E. Navarro Antón and S. Casans Berga
5.1 The remote concept . . . . . . . . . . . . . . . . . . . . . . 245
5.1.1 Elements in a measurement system . . . . . . . . . 245
5.1.2 Distributed systems topologies . . . . . . . . . . . . 247
5.2 The remote place . . . . . . . . . . . . . . . . . . . . . . . . 256
5.2.1 The benefits of distributed intelligence . . . . . . . . 256
5.2.2 The computer-controlling function . . . . . . . . . . 259
5.2.3 Virtual instruments . . . . . . . . . . . . . . . . . . 262
5.2.4 Smart/intelligent sensors . . . . . . . . . . . . . . . 265
5.3 The link . . . . . . . . . . . . . . . . . . . . . . . . . . . . . 269
5.3.1 Industrial networks . . . . . . . . . . . . . . . . . . 270
5.3.2 Ethernet . . . . . . . . . . . . . . . . . . . . . . . . 271
5.3.3 Wireless links . . . . . . . . . . . . . . . . . . . . . 274

5.4 The local place . . . 276
5.5 Remote analytical instruments/systems: application examples . . . 279
5.6 Conclusion . . . 281
References . . . 283

*Chapter 6. Laboratory information management systems* . . . 287
R.D. McDowall
6.1 The analytical laboratory . . . 287
6.1.1 Role of an analytical laboratory . . . 287
6.1.2 Need to increase productivity . . . 288
6.1.3 The aims of laboratory automation . . . 288
6.1.4 Problems with laboratory automation . . . 289
6.1.5 Solutions for laboratory automation . . . 290
6.2 What is laboratory automation? . . . 291
6.2.1 A definition of laboratory automation . . . 291
6.2.2 Laboratory automation constituent groups . . . 291
6.2.3 Instrument automation . . . 292
6.2.4 Communications . . . 294
6.2.5 Data to information conversion . . . 295
6.2.6 Information management . . . 296
6.2.7 A laboratory automation strategy in practice . . . 296
6.3 Laboratory Information Management Systems . . . 298
6.3.1 What is a LIMS? . . . 298
6.3.2 A LIMS has two targets . . . 299
6.3.3 Construction of the LIMS matrix . . . 301
6.3.4 LIMS matrix views . . . 304
6.3.5 Organisational integration and LIMS . . . 310
6.4 LIMS and the system development life cycle . . . 311
6.4.1 System development life cycle . . . 311
6.4.2 Project proposal . . . 313
6.4.3 The LIMS project team . . . 314
6.4.4 User requirements specification and system selection . . . 315
6.4.5 Functional specification . . . 320
6.4.6 Qualification of the system . . . 321
6.4.7 User training and roll-out strategies . . . 322
6.4.8 Project close-out . . . 324
6.4.9 Post-implementation review . . . 324
6.4.10 Enhancement of the system and controlling change . . . 324
References . . . 325

*SENSOR SYSTEMS*

*Chapter 7. Chemically modified electrodes with integrated biomolecules and molecular wires* . . . . . 327
E.J. Calvo, C. Danilowicz, E. Forzani, A. Wolosiuk and M. Otero
7.1 Introduction . . . . . 327
7.2 Enzyme redox catalysis . . . . . 329
7.3 Redox hydrogels . . . . . 333
7.4 Self-assembled polyelectrolyte and protein films . . . . . 339
7.5 Self-assembled enzyme films . . . . . 340
7.6 Electrocatalysis . . . . . 345
7.6.1 Electron hopping . . . . . 346
7.6.2 Different molecular architectures . . . . . 349
7.7 Structure of self-assembled enzyme films . . . . . 357
7.7.1 Atomic force microscopy . . . . . 357
7.7.2 Ellipsometry . . . . . 360
7.7.3 Combination of QCM and ellipsometric measurements . . . . . 361
7.7.4 Infrared spectroscopy (FTIR) . . . . . 364
7.8 Conclusions . . . . . 367
Acknowledgements . . . . . 369
References . . . . . 369

*Chapter 8. Composite and biocomposite materials for electrochemical sensing* . . . . . 377
S. Alegret and A. Merkoçi
8.1 Introduction . . . . . 377
8.2 Composite electrode materials . . . . . 379
8.2.1 Conducting composites . . . . . 380
8.2.2 Conducting biocomposites . . . . . 388
8.3 Composite- and biocomposite-based electrochemical sensors . . . . . 394
8.3.1 Conductometric sensors . . . . . 394
8.3.2 Potentiometric sensors . . . . . 395
8.3.3 Amperometric sensors . . . . . 397
8.3.4 Thick-film sensors . . . . . 402
8.3.5 Sensors for voltammetric stripping techniques . . . . . 405
8.4 Conclusions . . . . . 406
Acknowledgements . . . . . 407
References . . . . . 407

*Chapter 9. Optical chemical sensors and biosensors* . . . . . 413
F. Sevilla III and R. Narayanaswamy
9.1 Introduction . . . . . 413
9.2 Sensor structure . . . . . 414
9.2.1 Optical fibers . . . . . 415
9.2.2 Optoelectronic instrumentation . . . . . 418

9.2.3 Molecular recognition element . . . . . . . . . . . . 421
9.2.4 Sensor designs . . . . . . . . . . . . . . . . . . . . 425
9.3 Modes of optical signal measurements . . . . . . . . . . . . . 427
9.3.1 Absorbance measurement . . . . . . . . . . . . . . . 427
9.3.2 Reflectance measurement . . . . . . . . . . . . . . . 429
9.3.3 Fluorescence measurement . . . . . . . . . . . . . . 430
9.3.4 Chemiluminescence measurement . . . . . . . . . . 432
9.4 Conclusion . . . . . . . . . . . . . . . . . . . . . . . . . 434
References . . . . . . . . . . . . . . . . . . . . . . . . . . 435

*ARRAY SYSTEMS*

*Chapter 10. Electronic tongues: new analytical perspective of chemical sensors* . . . . . . . . . . . . . . . . . . . . . . . . . 437
A. Legin, A. Rudnitskaya, Yu. Vlasov
10.1 General approach to the application of sensor arrays . . . . 437
10.1.1 Why use sensor systems? . . . . . . . . . . . . . . . 437
10.1.2 Inspirations from chemometrics and biology . . . . 439
10.1.3 Advantages of sensor systems in comparison with discrete sensors . . . . . . . . . . . . . . . . . 444
10.1.4 Specific features of the sensors for the electronic tongue . . . . . . . . . . . . . . . . . . . 446
10.2 Electronic tongue systems . . . . . . . . . . . . . . . . . 449
10.2.1 Sensors . . . . . . . . . . . . . . . . . . . . . . . . 449
10.2.2 System designs . . . . . . . . . . . . . . . . . . . . 453
10.2.3 Hybrid systems . . . . . . . . . . . . . . . . . . . . 457
10.2.4 Data processing . . . . . . . . . . . . . . . . . . . . 458
10.3 Selected applications of the electronic tongue . . . . . . . . 458
10.3.1 Application areas and analytes . . . . . . . . . . . . 459
10.3.2 Quantitative analysis . . . . . . . . . . . . . . . . . 464
10.3.3 Qualitative analysis, recognition, identification and classification . . . . . . . . . . . . . . . . . . . 467
10.3.4 Comparison with human perception of flavours . . . 473
10.3.5 Taste quantification . . . . . . . . . . . . . . . . . . 475
10.3.6 Application of hybrid systems . . . . . . . . . . . . 478
10.4 Problems and perspective . . . . . . . . . . . . . . . . . . 482
References . . . . . . . . . . . . . . . . . . . . . . . . . . 483

*Chapter 11. A Taste sensor* . . . . . . . . . . . . . . . . . . . . . 487
K. Toko
11.1 Introduction . . . . . . . . . . . . . . . . . . . . . . . . . 487
11.2 Structure of the taste sensor . . . . . . . . . . . . . . . . 488
11.3 Response characteristics . . . . . . . . . . . . . . . . . . 490
11.4 Amino acids . . . . . . . . . . . . . . . . . . . . . . . . . 490
11.4.1 Classification of taste of amino acids . . . . . . . . . 490
11.4.2 Discrimination of D-amino acids from L-amino acids . . . . . . . . . . . . . . . . . . . . . 496

11.5 Quantification of the taste of foods . . . . . . . . . . . . . . 499
11.6 Interaction between taste qualities . . . . . . . . . . . . . . 502
11.6.1 Suppression of bitterness due to phospholipids . . . 502
11.6.2 Scale of bitterness . . . . . . . . . . . . . . . . . . . . 502
11.6.3 Suppression of bitterness due to taste substances . . . . . . . . . . . . . . . . . . . . . . . . 506
11.7 Detection of wine flavor using taste sensor and electronic nose . . . . . . . . . . . . . . . . . . . . . . . . 507
11.8 Perspective . . . . . . . . . . . . . . . . . . . . . . . . . . 509
References . . . . . . . . . . . . . . . . . . . . . . . . . . . . . 511

*Chapter 12. Application of electronic nose technology for monitoring water and wastewater* . . . . . . . . . . . . . . . . . . . . . . . . 513
R. Stuetz
12.1 Introduction . . . . . . . . . . . . . . . . . . . . . . . . . . 513
12.2 Electronic nose technology . . . . . . . . . . . . . . . . . . 514
12.2.1 Sensor types . . . . . . . . . . . . . . . . . . . . . . . 515
12.2.2 Analysis of electronic nose data . . . . . . . . . . . 516
12.3 Electronic nose instrumentation . . . . . . . . . . . . . . . 520
12.3.1 Sensor array components . . . . . . . . . . . . . . . 520
12.3.2 Commercial systems . . . . . . . . . . . . . . . . . . 522
12.4 Application to water and wastewater monitoring . . . . . . 524
12.4.1 Laboratory-based systems . . . . . . . . . . . . . . . 526
12.4.2 On-line monitoring systems . . . . . . . . . . . . . 528
References . . . . . . . . . . . . . . . . . . . . . . . . . . . . . 537

*MICROSYSTEMS*

*Chapter 13. Integrated optical transducers for (bio)chemical sensing* . . . . . 541
C. Domínguez, J.A. Rodríguez and L. M. Lechuga
13.1 Introduction . . . . . . . . . . . . . . . . . . . . . . . . . . 541
13.2 Basic concepts . . . . . . . . . . . . . . . . . . . . . . . . . 542
13.3 Fundamentals of optical waveguides . . . . . . . . . . . . 545
13.4 Detection principles: Types of devices . . . . . . . . . . . 549
13.5 Technologies for integrated optical transducer fabrication . . . . . . . . . . . . . . . . . . . . . . . . . . 553
13.5.1 Substrate materials and specific processes . . . . . 554
13.5.2 Basic technological processes . . . . . . . . . . . . . 557
13.6 Integrated optical sensors . . . . . . . . . . . . . . . . . . 574
13.6.1 Absorbance sensor . . . . . . . . . . . . . . . . . . . 574
13.6.2 Grating coupler . . . . . . . . . . . . . . . . . . . . . 577
13.6.3 Resonant mirror . . . . . . . . . . . . . . . . . . . . . 579
13.6.4 Mach–Zehnder interferometer . . . . . . . . . . . . . 580
13.7 Towards a total integrated system . . . . . . . . . . . . . . 582
References . . . . . . . . . . . . . . . . . . . . . . . . . . . . . 584

*Chapter 14. 'High order' hybrid FET module for (bio)chemical and physical sensing* . . . . 587
A. Poghossian and M. J. Schöning
14.1 Introduction . . . . 587
14.1.1 Design concepts of (bio)chemical sensor arrays . . . 587
14.1.2 'High order' sensor module based on an identical transducer principle . . . . 589
14.2 Experimental . . . . 591
14.2.1 Hybrid module design . . . . 591
14.2.2 ISFET fabrication . . . . 591
14.2.3 Measuring system and sensor configurations . . . . 592
14.3 Multi-parameter detection of both (bio)chemical and physical quantities using the same transducer principle . . 594
14.3.1 ISFET-based pH sensor . . . . 594
14.3.2 ISFET-based penicillin sensor . . . . 596
14.3.3 ISFET-based temperature sensor . . . . 601
14.3.4 ISFET-based flow-velocity sensor . . . . 605
14.3.5 ISFET-based flow-direction sensor . . . . 609
14.3.6 ISFET-based diffusion-coefficient sensor . . . . 610
14.3.7 ISFET-based bioelectronic sensor . . . . 612
14.4 Applications of the hybrid sensor module . . . . 615
14.4.1 pH determination in human urine . . . . 615
14.4.2 pH measurement in rain droplets . . . . 616
14.5 Summary and conclusion . . . . 619
Acknowledgements . . . . 620
References . . . . 620

*Chapter 15. Microdialysis based lab-on-a-chip, applying a generic MEMS technology* . . . . 625
P. Bergveld, W. Olthuis, A.J. Sprenkels, D.G. Pijanowska, H.J. van der Linden and S. Böhm
15.1 Introduction . . . . 625
15.1.1 The need for in vivo monitoring . . . . 626
15.1.2 Microdialysis . . . . 627
15.1.3 The microdialysis lab-on-a-chip . . . . 628
15.2 The micromachined double lumen microdialysis probe connector . . . . 631
15.2.1 The conventional microdialysis probe . . . . 631
15.2.2 Theory . . . . 632
15.2.3 Experimental . . . . 634
15.2.4 Results and discussion . . . . 636
15.3 The passive and the active calibration system . . . . 639
15.3.1 Passive control of a calibration plug . . . . 639
15.3.2 Active control of a calibration plug . . . . 641
15.3.3 Closed-loop controlled electrochemically actuated microdosing system . . . . 643

15.4 The flow-through potentiometric and amperometric sensor array . . . . 646
15.4.1 The flow-through potentiometric sensor array . . . . 646
15.4.2 The flow-through reference electrode . . . . 649
15.4.3 The flow-through amperometric sensor . . . . 649
15.5 The integrated microdialysis-based lab-on-a-chip . . . . 652
15.5.1 The complete integrated microdialysis lab-on-a-chip . . . . 653
15.5.2 Experimental . . . . 657
15.5.3 Measurements . . . . 660
15.6 General conclusions . . . . 662
References . . . . 662

*Chapter 16. Design methodology for a lab-on-a-chip for chemical analysis: the MAFIAS chip* . . . . 665
T.T. Veenstra and A. van den Berg
16.1 Introduction . . . . 665
16.2 The design path . . . . 666
16.3 The design . . . . 667
16.3.1 Chemistry . . . . 667
16.3.2 System schematics . . . . 669
16.3.3 Channel geometry . . . . 669
16.3.4 Specifications for the components . . . . 670
16.3.5 The components . . . . 671
16.3.6 System . . . . 676
16.4 Conclusions . . . . 682
References . . . . 683

*NANOSYSTEMS*

*Chapter 17. Nanosensor and nanoprobe systems for in vivo bioanalysis* . . . . 685
T. Vo-Dinh
17.1 Introduction . . . . 685
17.2 Background on biosensors and bioreceptors . . . . 686
17.2.1 Biosensing systems . . . . 686
17.2.2 Bioreceptor probes . . . . 687
17.3 Fiberoptics nanosensor system . . . . 689
17.3.1 Fabrication of the fiberoptic nanoprobe . . . . 689
17.3.2 Immobilization of receptors onto fiber nanoprobes . . . . 691
17.3.3 Experimental system and protocol for nanoprobe investigation of single cells . . . . 693
17.3.4 Optical measurement system . . . . 694

17.4 Applications in bioanalysis . . . . . . . . . . . . . . . . . . . . 695
17.4.1 Optical nanofiber probes for fluorescence measurements . . . . . . . . . . . . . . . . . . . . . . 695
17.4.2 Single-cell measurements using antibody-based nanoprobes . . . . . . . . . . . . . . . . . . . . . . . 696
17.5 Conclusion . . . . . . . . . . . . . . . . . . . . . . . . . 698
Acknowledgements . . . . . . . . . . . . . . . . . . . . . . . . 699
References . . . . . . . . . . . . . . . . . . . . . . . . . . 699

Index . . . . . . . . . . . . . . . . . . . . . . . . . . . . . . . 701

## *Series Editor's Preface*

It is a great pleasure for me to introduce this new book, *Integrated Analytical Systems*, edited by Salvador Alegret, in the Comprehensive Analytical Chemistry series. When I took on the editorship of this series, one of my first ideas was to introduce modern analytical chemistry topics. At a certain moment I met my old friend and colleague from Barcelona, Salvador Alegret, a well-known expert in the field of sensors and biosensors, who had the proposal of this current book in mind. It was quite easy for me to convince him to work out the proposal and submit it. The outcome is now in your hands.

Screening techniques are an emerging issue in modern analytical chemistry due to the need to obtain direct analytical information in the field. The importance of this area has been already recognised by us with the recent publication in this series of Volume XXXVI, *Chemical Test Methods of Analysis*, by Yu.A. Zolotov, V.M. Ivanov and V.G. Amelin. The present book complements the latter very well and it is an additional step in the area of rapid analysis. It offers a comprehensive survey on automation, miniaturisation and communication systems using more sophisticated and advanced technological approaches.

The book contains 17 chapters and summarises the main integration methodologies in the field of analytical chemistry. The first chapters are devoted to different conceptual aspects of integration, including examples on solid-phase spectrometric analysis, continuous-flow analytical systems and laboratory information management systems (LIMS). The second set of chapters is devoted to various sensor systems such as electrochemical and optical transducers, providing excellent examples of the integration of the recognition element with the measurement devices. The book also contains topics on array systems – including the fashionable analytical devices of electronic tongues and the electronic nose. Modern microfabrication techniques are described in various chapters, with examples comprising chemical, physical and biological sensing, and design methodologies for lab-on-a-chip. In the closing chapter, the possibilities of nanosystems are explored.

Overall this book covers many of the technological advances applied to analytical chemistry that are the focus of attention today. I am convinced that this book opens a new window in analytical chemistry. It brings interdisciplinary experts together to cover a broad range of expertise comprising physical chemistry, microelectronics, biological sciences, analytical chemistry and automation. In conclusion, I would like to thank the editor and the various authors for their contributions in compiling such a unique book on modern technologies, the integrated analytical systems.

D. Barceló

## *Editor's Preface*

Since the 1950 s there has been immense change in the design of analytical instrumentation. In part this change has been brought about by the steady increase in demands for more and better chemical information. At first, analytical instrumentation, based mainly on simple electrochemical and optical techniques, took the guise of metrological instruments housed in a laboratory for the measurement of a specific physical or chemical parameter of a sample or for detecting the end-point of a titration. Prior to reaching this stage of analysis the sample had to pass through a sequence of manual operations or steps.

Little by little a broad range of electrochemical and spectroscopic principles were implemented and applied in qualitative, quantitative and structural analysis, at the expense of complex and sophisticated instrumental design. Meanwhile, the demand for the number of samples being analyzed in the laboratory increased, requiring the redesign of classical instruments. As a result of this redesigning, total or partial automation of the analyzer occurred. In some cases, instrumental setups were 'connected' to the sampling site allowing for on line monitoring of variable material systems. A good example of this evolution can be seen in chromatographic techniques, that progressed from the classic preparative techniques to current advanced analytical versions where the separation step is coupled to a sophisticated detection system.

Additionally, while the number of samples is continuously increasing, preliminary screening techniques have become a priority. Also, the demand for direct analytical information from places outside the laboratory ("on site analysis") is growing. Frequently, these analyses are implemented by untrained personnel using rapid, simple and cost-effective procedures, requiring simple instrumentation designs.

The aforementioned trends have resulted in the development of new instrumentation systems or updated versions of old analytical instrumentation that are sometimes an alternative but also a complement to traditional analytical designs. However, while a broad range of designs are being implemented, common traits exist between the different types of instruments. An important group of these are based in different strategies to 'connect' (mechanically, physically or chemically) the steps of the analytical process, particularly separation and reaction with the measurement step. The term 'integrated analytical systems' (IAS) illustrates the convergence of these strategies in the analytical process, including simplification, miniaturization, automation and communication. The convergence of these 'integration' strategies confers new characteristics to instrumental designs, especially in terms of information, quality, reliability, speed, robustness, portability and cost of the chemical analyses.

This book summarizes the main integration strategies currently found in analytical instrumentation or laboratories, giving a wide and updated panorama of Analytical Chemistry. The objective of this book is to highlight the different integration solutions by treating them in groups of one or several chapters. A general introduction to integrated analytical systems (IAS) ("Integration, a new paradigm in analytical chemistry") is followed by a group of reports (Chapters 2–6) describing selected examples of systems in the analytical laboratory using different methodological or conceptual solutions to the integration problem. These solutions revolve around connection and communication between different subsystems or between the stages or steps that form the analytical process. Chapters 2–6 are dedicated to "Integrated separation systems", "Solid-phase spectrometric assays", "Continuous flow analytical systems", "Distributed analytical instrumentation systems" and "Laboratory information management systems". Chromatographic techniques are not discussed specifically in this book, as they have been reviewed extensively elsewhere in the literature. Nevertheless, various aspects of chromatography are incorporated in Chapter 2.

A follow on from these chapters deals with distinct cases of analytical integration represented by *sensor systems* (Chapters 7–9), using selected reports on chemosensors and biosensors based on either electrochemical transducers ("Chemically modified electrodes with integrated biomolecules and molecular wires" and "Composite and biocomposite materials for electrochemical sensing") or optical transducers ("Optical sensors and biosensors"). The main emphasis of these chapters being the simplification of the analytical process by integrating the recognition reaction with the measuring step or, in other words, by connecting both elements without break of continuity.

*Array systems* are dealt with in Chapters 10–12 where analytical systems based on low selectivity sensors with cross sensitivities, representing a biomimetic approach to the generation of analytical information, are presented. Integration is achieved precisely on the one hand by the array platform itself and on the other by the processing and analysis of multicomponent signals produced by the sensor array ("Electronic tongues: new analytical prospective of chemical sensors", "A taste sensor", and "Applications of electronic nose technology for monitoring water and waste water").

Also present are analytical microsystems where the concept of integration has a strong technological component and some of the IAS described in earlier chapters are now confined to a small space using modern microfabrication techniques. These spatial integration techniques together with some examples of the microinstruments developed are described in chapters 13–16 ("Integrated optical transducers for (bio)chemical sensing", "High order" hybrid FET module for (bio-) chemical and physical sensing", "Microdialysis based lab-on-a-chip, applying a generic MEMS technology" and "Design methodology for a lab-on-a-chip for chemical analysis: the MAFIAS chip").

Finally the book is brought to a close by analysing the potential held by analytical nanosystems ("Nanosensor and nanoprobe systems for in vivo bioanalysis").

This book is aimed at graduate students and researchers in industry and academy to help them acquire a critical vision of some of the current tendencies in Analytical Chemistry. It is written by an international team of researchers active within the fields reported here. The authors come from a wide diversity of technological and scientific fields including chemistry, electrochemistry, physics, materials science, microelectronics, electronic engineering, and the instrumentation industry highlighting the multidisciplinary nature of Analytical Chemistry in today's world. I wish to express my gratitude for the time and the effort they put into the preparation of excellent manuscripts with a common but diverse focus on the questions involved in the integration from the point of view of their respective areas of research, producing a report that is the first of its kind in the specialized literature on Analytical Chemistry.

Salvador Alegret
August 2002

*Chapter 1*

# *Integration, a new paradigm in analytical chemistry*

***Salvador Alegret****

## 1.1 INTEGRATION IN SCIENCE AND TECHNOLOGY

Integration is defined as the action or the effect of joining parts to form a whole resulting in unique features and functions not present in the original constituting parts. Integration is a far-reaching concept in technology but it is less used in science.

Transistors, integrated circuits (IC) and digital computers can be seen as a paradigm of integration in microelectronics, illustrating the hierarchical levels produced by integration. In fact, the concept of integration in microelectronics is related to the confinement of a great number of interconnected electronic components in a reduced space. The electronics industry underwent a significant revolution with the advent of miniaturised electronic components, produced at high volume and low-cost. Most electronic systems in use today, especially those pertaining to information and communication technologies, owe their existence to devices that show a high degree of miniaturisation. These devices are known as ICs. Millions of transistors are built and interconnected in a small silicon area in order to attain processing power in a computer. The relationship between computing power and transistor size is illustrated by the fifty-fold reduction in the size of the transistor in the last 25 years. The smallest feature of the transistor has

* Tel.: +34-93-5811976; fax: +34-93-5812379. *E-mail address:* salvador.alegret@uab.es.

*Comprehensive Analytical Chemistry XXXIX, pages 1–36*


ISSN: 0166-526X

shrunk from ten micron to sub-micron dimensions, allowing progression from a large scale integration (LSI), comprising between $10^3$ and $10^5$ transistors in a single IC, to an ultra large scale integration level (ULSI), with between $10^6$ and $10^7$ transistors per IC. The immediate future is giga scale integration (GSI) comprising between $10^7$ and $10^8$ transistors per IC. All this has been possible due to microfabrication techniques, especially microphotolithography, a key tool for the mass production of ICs.

The current progression towards GSI will lead to integrated systems made of nanometric electronic components. This will greatly impact on the approach to designing devices and circuits. In fact, it has been stated that the capacity and speed of computer chips doubles every 18 months. This is known as Moore's law, after Gordon Moore—one of the founders of Intel Corporation—who observed this trend in 1965. However, the increase in speed is directly related to the reduction in size, and at some point the size will reach atomic level, and a natural limit to miniaturisation will be reached. This limit or frontier has brought the concept of 'quantum computing' to the fore. In passing from a microscopic world to a nanoscopic one, the discrete nature of matter and electric charge and also the existence of quantum effects need to be considered. A radical concept shift is needed to produce designs that profit from quantum effects instead of designs that work despite quantum effects. However, these levels of miniaturisation and integration imply high production costs and also pose new production challenges that have not yet been fully resolved.

In biology, the term integration is used informally to explain the complexity of biological systems where an initial integration results in cells that are the building blocks for more complex structures (tissues) that are used to construct edifices with a wide variety of form and function (organisms).

The concept of integration is more often used in materials science and in chemical technology, but in comparison to these disciplines it is sparsely used in chemistry. Composites are a clear example of integration where materials do not lose their original properties upon mixing; rather the new material presents features different from those of the separate components. In photoelectrochemical systems radiant energy (solar light) is converted to electric energy using a semiconductor. This energy is used in redox reactions to generate certain products, as is the case in the transformation of environmental toxicants to innocuous products. These two examples, composites and photoelectrochemical systems, are known as *integrated chemical systems* (ICS) with a low level of integration. An instant colour photographic film represents an example of an ICS with a high level of integration. When it is compared with modern photo processing techniques, a clear conceptual and methodological distance can be established between the two. The latter uses large and complicated machines, involves working with wet chemistries and requires specialised personnel, while the former is based on dry chemical processes and is resolved in situ by the user. The instant film is made of a large number of integrated components and interfaces.

The above examples are taken from the only book dedicated to the subject, carrying the explicit title 'Integrated Chemical Systems', by Professor Allen J. Bard, of the University Texas at Austin [1]. This book has a nanotechnological focus, making its object of study the confinement of chemical systems at a molecular or atomic level.

### 1.1.1 Integration in analytical chemistry

Within chemistry, the term integration is more widely used in analytical chemistry with special reference to analytical systems. In analytical chemistry, the concept of integration has a richer meaning than in general chemistry or biology as the integrated parts are the steps or operations of the analytical process. Additionally, it may also refer to the analytical devices comprising integrated steps—especially when the reaction step is integrated with the detection step. Integration in analytical chemistry refers more to the design of the analytical procedure, i.e. the way in which the analytical process is done, and not just the miniaturisation of the procedure.

Within the context of instrumentation, a 'system' is defined [2] as a group of devices and instruments forming a whole. This group forms a network for a common purpose or for a common distribution method.

Analytical instruments are systems that produce chemical information. In this type of system, information is obtained following a well-established sequence (the analytical process). The main steps of this sequence are: sampling, sample transport and processing, separation, reaction, transduction, signal acquisition and processing (see Fig. 1.1). Most analytical systems are designed following this sequence in full or in part. The steps of the analytical process either can be carried out discretely, using machines, device or instrument systems whose

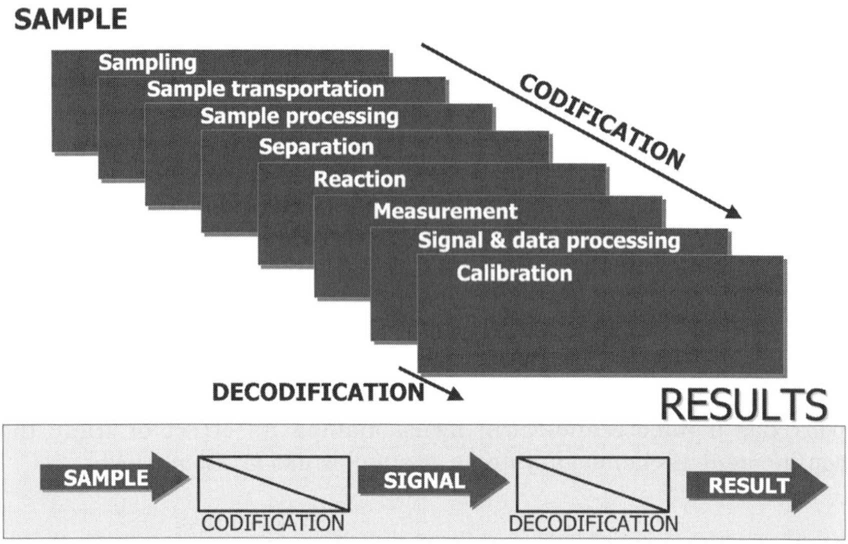


Fig. 1.1. Main steps of the classical analytical process for generating analytical information. The information contained in the sample is coded into an analytical signal which is subsequently uncoded as useful analytical data through the calibration of the analytical system.

connectivity calls for human intervention, or may be realised in an *integrated* fashion where connectivity between the parts of the system is achieved without human intervention.

Different attributes are associated with an 'integrated analytical system'. These attributes include: the number of units that make up the whole, the type of integration (relationship, interconnection, communication, spatial confinement) involved, the new functions obtained and the aim or goal of the system. Thanks to integration, new analytical procedures and instruments can be implemented providing new perspectives in simplification, miniaturisation, automation, communication, information, quality, speed, portability and cost of chemical analyses. Integrated analytical systems (IAS) show a wide typology with several levels of integration (see Table 1.1). One of the positive attributes of analytical systems is their adaptability to changes in the environment, e.g. automatic temperature compensation in optical or electrochemical measurements. Additionally, 'smart' IAS show improved abilities to process analytical information

TABLE 1.1

Different attributes of IASs

| System | Main integrated materials, components, devices or operations | Interconnection or relationship between integrated parts | Examples |
|---|---|---|---|
| Supported reagents | Reagents and supports | Without break in continuity | Spot test papers and analytical strips; dip-and-read sticks; reactors; chromatographic packaging and columns; SLMs |
| Chemical sensors | Receptors and transducers | Without break in continuity | Chemosensors and biosensors; electrochemical sensors; ion-selective electrodes (ISE); optochemical sensors |
| Flow analytical systems | Machines, devices and analytical operations | Flow conduits | Gas or liquid chromatographs; continuous flow analysers; automated analysers; multiparametric analysers; total analysis systems (TAS) |
| Laboratory systems | Analytical modules, workstations and instruments | Communication interfaces | Laboratory information management systems (LIMS); paperless laboratory; automated laboratory; roboted analytical workstations |
| Distributed analytical systems | Distributed analytical instruments | Communication interfaces | Networked analytical instruments or laboratories; cybernetic systems |
| Array analytical systems | Selective sensor array | Sensor platform/ sensor signal processing | Sensor array; array sensor; multiparametric sensor system; multiparametric analysers |
| | Cross-selective sensor array | Multicomponent sensor signal processing | Electronic noses; electronic tongues; taste sensors |

(*continued on next page*)

TABLE 1.1 (*continued*)

| System | Main integrated materials, components, devices or operations | Interconnection or relationship between integrated parts | Examples |
|---|---|---|---|
| Chemical microsensors | Receptors and transducer structures confined in a micrometric space | Without break in continuity/sensor platform | Ion-sensitive field effect transistors (ISFET); amperometric microsensors; integrated optical sensors; mass-sensitive acoustic-wave devices (BAW and SAW); microsensor arrays |
| Analytical microsystems | Analytical operations and devices confined in a micrometric space | Monolithic flow microconduits and platform | MEMS and BioMEMS; micro total analysis system (μTAS); laboratory-on-a-chip (LOC); sensor microarrays; biochips |
| Nanosystems | Atomic and molecular devices confined in a nanometric space | Atomic and molecular forces or bonds | Nanoprobes; nanosensors |

intelligently, including self-diagnosis, automatic calibration or compensation of baseline drift and communication with other systems, instruments, devices or machines.

It is possible to predict the different integrated levels required to form a given analytical system. The system may be integrated horizontally by groups of analytical subsystems of a different class (designed with a different number of integrated steps) and use several connectivity forms (mechanical, electronic, photonic, fluidic, etc.). Alternatively, it is possible to integrate the system vertically, such that each subsystem can group structures of different sizes (millimetric, micrometric or nanometric). Additionally, it is possible that a system may be a part of a set of subsystems located either close to or remote from each other and forming a laboratory or a node in a network of analytical stations (see Fig. 1.2).

Although some authors within the analytical chemistry field use the concept of integration solely for analytical devices fabricated using micromachining techniques, creating an analogy with ICs (see Section

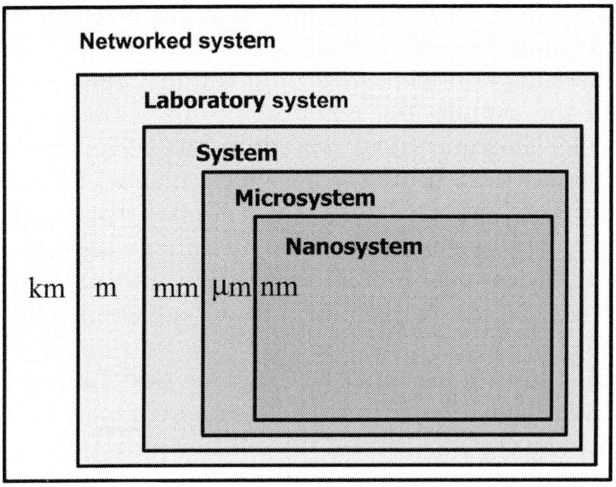


Fig. 1.2. Hierarchical levels found in the integration of analytical systems.

1.6), there is a growing body of authors that link integration to a more general implementation of the analytical process [3,4].

In the following sections we review several IAS, involving a variety of both conceptual and methodological approaches, levels and purposes (see Table 1.1). The IAS described are working examples but by no means a comprehensive list.

## 1.2 PARTS AND COMPONENTS

Two types of analytical systems will be described as examples of systems that can be considered simple IAS or building blocks for more complex IAS.

### 1.2.1 Supported reagents

The simplest IAS is an analytical reagent, synthetic or natural, confined to an insoluble solid support (active or inert), and immobilised by chemical or physical procedures [5,6].

Many commercial companies involved with qualitative and semi-quantitative control tests have developed *spot tests* or plastic *strip tests* with a reactive zone consisting of an adsorbent material impregnated with reagents, buffers, sequestering agents and other additives. Either a drop of the sample is deposited onto the test zone, or the strip is immersed in the sample; the resulting colour is then compared to a reference scale. This analytical procedure has been greatly simplified due to the ability to integrate the analytical reagent to a solid support [7,8]. Chapter 3 of this book deals with solid-phase spectrometric assays where one or more reagents are immobilised on a solid phase. As will be seen, such devices offer simple assays that integrate several analytical operations, i.e. derivatisation, preconcentration, separation and measurement.

In general, the use of supported reagents (see Table 1.2) presents several analytical advantages: (A) The microenvironment created at the interface between the support and the solution provides a better control of the generation of the product as the reagent remains immobilised; this is a very interesting feature for continuous analytical processes. (B) Reagent use and the associated cost of the analysis are reduced as immobilisation allows either reuse of the reagent or more rational use. (C) The analytical procedure is simplified as the reagent is confined

TABLE 1.2

Immobilisation of reagents in analytical chemistry. Different aspects related to the integration of reagents with supports

| | |
|---|---|
| Materials (chemical and biological) | Reagents and ionophores; enzymes and cells; antibodies and antigens; chemoreceptors and agonists DNA and RNA |
| Supports (inorganic and organic; natural and synthetic) | Beads and particles; fibres and hollow fibres or columns; films, layers and membranes; strips; transducing materials |
| Immobilisation procedures (surface; bulk) | Adsorption; entrapment; covalent bonding |
| Analytical techniques or systems | Supported reagents and analytical kits; reactors; chromatographic packaging; separation membranes; chemosensors and biosensors |

to a defined area facilitating its manipulation as stated above. This simplification may also facilitate the automation of the analysis with both wet and dry chemistry methods. (D) It provides new surface properties to the supporting materials, modifying them chemically or biologically and producing either new materials (for instance chromatographic packagings, reactor fillings or membranes (see Section 1.2.2)) or devices (chemosensors or biosensors (see Section 1.4)) with a greater stability, selectivity or sensitivity.

### 1.2.2 Separation membranes

A membrane is a barrier between two phases. Depending on the material used, the membrane can act as a simple, permselective diffusional barrier discriminating solutes passively according to size, charge or polarity. More complex membranes are based on active transport mechanisms that are more selective. *Supported liquid membranes* (SLM) [9,10] immobilise a selective carrier molecule (recognising element) which is dissolved in an organic solvent within the support membrane. With membranes of this class, transport is not determined in any way by the membrane or membrane material but by the carrier molecule. In response to a driving force applied to these membranes (see Fig. 1.3), the analyte (gas, ionic or non-ionic compounds) is transported selectively from one phase (feed phase) to the other (receiving phase) through a carrier-mediated mechanism. This membrane action is to a certain extent similar to that of cell membranes [11].

In addition to supported reagents, SLMs are a good example of an analytical system that integrates two separation processes that are commonly performed separately in laboratories and in industrial plants. The normal procedure involves separation of the analyte to an organic phase, containing the dissolved selective reagent (extracting phase), followed by the transfer of the analyte to a new and more adequate aqueous phase (stripping phase). Use of an SLM simplifies the whole separation process. Fig. 1.3 shows how separation with SLM is comparable to, but simpler than, a double separation liquid–liquid of the analyte ($M^+$).

Membranes of the same class as SLM are used as the transducing element in ion-selective electrodes (see Section 1.4.1 and Fig. 1.6).

The next chapter of this book is focused on the application of the integration concept within the field of separation sciences. Membrane

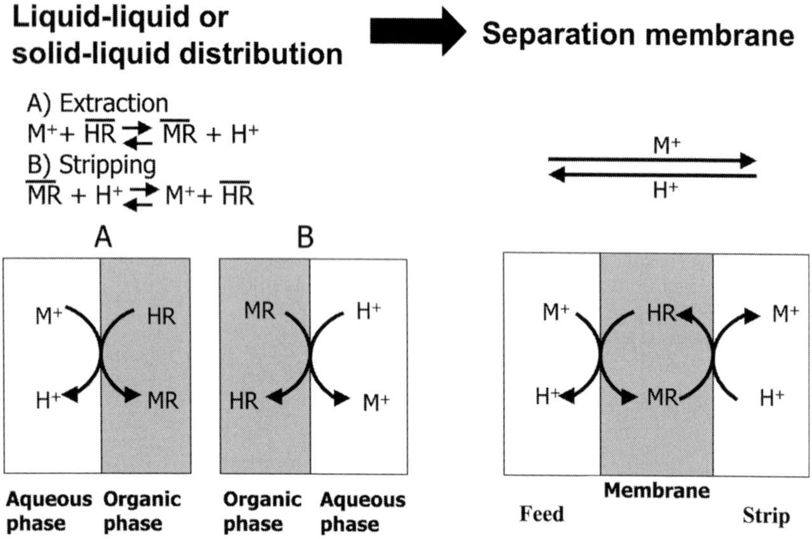


Fig. 1.3. Comparison between a common separation process through two consecutive liquid–liquid or solid–liquid distributions and the same separation process but using an integrated separation system such as an SLM.

separations such as those based on SLM represent a first level of integration. Additionally, other systems of a higher order of integration, based on several separation stages or different separation methods combining different separation techniques, are reviewed.

## 1.3 SYSTEMS

### 1.3.1 Total analysis systems

Traditionally, all the steps of an analytical process were performed manually with different instruments. Latest trends indicate that a more automated approach is required to monitor or control a process (e.g. in biomedical, environmental or industrial fields), inside or outside (in-field analysis) the laboratory. Sample handling, separation, reaction, and detection steps are integrated into a single instrument, with most steps automated and controlled by a computer. An IAS that groups conventional machines, devices and instruments to carry out an analytical process (partially or totally) is known simply as an 'analytical system' [2]. If it performs the process in full, it is called a 'total analysis system' (TAS). Examples of TAS are systems based on continuous flow analysis, electrophoresis, chromatography or mass spectrometry [4]. In fact, these systems integrate separation and detection techniques and can perform a total analysis on a wide range of samples. In general, these systems offer a high degree of automation but a pre-treatment step prior to the conventional analysis is often needed, and can sometimes involve large quantities of reagents. In some configurations, the sample transport and the separation speeds are slow, even though they are bench-top instruments.

The solutions adopted to attain a good analytical connection (integration) of the various parts have produced 'automated analytical systems'. The large automated analysers used today are either batch or flow analysers. In a batch analyser, the connection among its parts is mechanical, with each discrete sample assigned to a container. The sample remains in the same container throughout the analysis. This type of integration is very complex as it has many moving parts with limited lifetimes.

In continuous flow analysers, the analyte is continuously measured in a stream of liquid that may or may not be air-segmented. Successive samples pass through the same tube, in which a carrier solution is flowing and reagents merge at strategic points. The mixing of the sample and the resulting reaction take place while the sample plug travels to a flow-through detector, where the signal is continuously recorded. The continuous flow approach is a very flexible way to perform the operations needed for a total analysis. In addition to sample dispensing, dilution, heating, mixing and reagent addition—as in any batch analyser—the continuous flow mode may perform dialysis, gas diffusion, evaporation, distillation, solvent extraction, and other types of separations, and can be integrated to various detection techniques. Flowing streams interconnect the main components of continuous flow analysers with few moving parts. This type of integration is less complex mechanically and it is easier to build than a batch analyser.

*Flow-injection analysis* (FIA) systems [12,13] based on the injection of a liquid sample into a non-segmented carrier stream are used more frequently now than air-segmented continuous flow systems. This is due to a better reproducibility and a greater simplicity since careful control of air bubbles is critical, requiring considerable time for start-up and stabilisation. However, the multichannel air-segmented Technicon

AutoAnalyzer is still used in many centralised hospital laboratories. To date it is still the most popular automatic analyser ever marketed.

In the techniques discussed so far, the samples are injected or aspirated directly into a continuous carrier solution flow. However, in *sequential injection analysis* (SIA) systems [14] the sample and reagent plugs are sequentially aspirated into a channel and then the flow is reversed and the stacked plugs are transported to the detector while mutual penetration of stacked zones and dispersion of the sample take place. The SIA approach has important advantages over conventional FIA including robustness, easy sample handling, a versatile manifold, lower reagent and sample use and multiparametric configuration.

On the other hand, flow systems where the detection is integrated to other analytical steps (chemical reactions, separations or both) forming a differentiated subsystem are known as 'flow-through chemical sensor systems' [3].

All the above-mentioned aspects of continuous flow analytical systems as an example of IAS are covered in depth in Chapter 4.

*High-performance liquid chromatography* (HPLC) resembles FIA, except that a column replaces the reaction coil, while the rest of the system (i.e. pump, injection port, tubes, flow-through detector, and recorder) is basically the same. These two examples of TAS are very different since their principles and purposes diverge. The purpose of HPLC is to measure several sample components while FIA aims to analyse a large number of samples.

The integration of two or more analytical techniques produces a more efficient analytical tool. This concept is studied in greater detail in Chapter 2. There are examples of sophisticated TAS capable of processing complex samples by integrating a separation technique with a spectroscopic detection technique. Such integraton allows for an efficient and rapid on-line identification of sample components, and usually results from the combination of two spectroscopic techniques (e.g. MS and MS) or two separation techniques (GC and LC), or the coupling of two different techniques (e.g. GC-MS, GC-FTIR or LC-NMR) (hyphenated systems). Hyphenation of HPLC and MS or NMR—the most useful spectroscopic techniques for elucidation of structures of unknown compounds—offer extremely valuable information [15]. Besides HPLC, electrically driven separations such as CE and CEC have been coupled with increasing success to spectroscopic techniques [16].

Most hyphenated systems require an interface between the separation technique and the detector. The main advantage of these TAS is the additional dimensionality of the data obtained.

### 1.3.2 Miniaturised systems

The miniaturisation of IAS is a dominant trend in current analytical chemistry. The mere reduction in size of an analytical system can be seen as a step towards spatial integration. Miniaturisation provides a large number of additional advantages.

In principle, miniaturisation in analytical chemistry is linked to the possibility of placing the instrument in reduced spaces, allowing in situ measurements while preserving the identity of the sample. In classical analytical methods based on the use of reactions, miniaturisation is also related to the use of small samples. However, in some cases the size of the instrument does not inevitably relate to the size of the sample. Currently, we speak of a miniaturised analytical system when it is implemented with 'ministructures', that is, according to the decimal submultiples of SI units, structures in the milli range (i.e. dimensions of $10^{-3}$ to $< 10^{-2}$ m) (see Table 1.4).

The up-down scaling of IAS is aimed at producing systems with new improvements. Research and development in this field are directed at improving performance by increasing speed, raising throughputs, reducing costs associated with reagent use and automating different steps and processes. Miniaturised analytical systems in general offer a greater response stability and a lower consumption of power augmenting their autonomy and, additionally, they produce less waste and weigh little, making them portable and usable far from the laboratory [17].

Sometimes, solutions for the miniaturisation of an analytical process are not simply a geometric reduction but allow for the implementation of new analytical concepts. A new analytical domain has recently emerged in medicine: *point-of-care diagnostics*. Clinicians demand on-site diagnostic tests to assess the state of their patients quickly in emergency units, under the assumption that 'faster is better'. However, it is imperative that rapid testing does not provide inaccurate results. All features found in traditional analytical systems must also be provided in small portable instrument based on miniaturised disposable cartridge systems incorporating either electrochemical or optical chemo/biosensing [18].

simple IAS has brought simplification, miniaturisation, robustness, speed, portability and low cost to the analytical process [20].

There are two basic aspects for the development of this type of device. First, the availability of more selective receptors permits direct (and eventually continuous) measurements without the need for a previous separation step. Receptors are usually highly selective substances, both synthetic (acyclic and macrocyclic ligands, molecular imprinted polymers, etc.) and natural (enzymes, microorganisms, plant and animal tissues, antibodies, DNA, etc.). Sensors incorporating chemical recognition elements are known as *chemosensors* while those using biological elements are known as *biosensors* [21] (see Table 1.3). Biosensors have paved the way for bioelectronics [22].

Secondly, integration (i.e. immobilisation [5,6]) of the receptor and the transducer is a crucial aspect for the development of these devices. The main objective of this integration is to preserve the functional features of both parts while achieving a stable coupling. In this incidence, stability means that the information obtained from the sample is transmitted from the recognition element to the transducer with little distortion and as far as is possible.

Due to their spatial and temporal integration, chemical sensors represent an ideal approach for portable, handheld field instruments thanks to their ability to operate at low-power, small-size and relatively low cost [23].

TABLE 1.3

Different receptor materials and transducer devices used in the development of chemical sensors

| Transducers | Receptors | Chemical sensors |
|---|---|---|
| Potentiometric electrodes; ISFET; amperometric electrodes; optical transducers; bulk acoustic wave (BAW) devices; surface acoustic wave (SAW) devices | Ionophores; ion carriers; modifiers; mediators; molecular imprinted polymers; chromophores | Chemosensors |
| | Enzymes; microorganisms; plant and animal tissues; antibodies; cell receptors; DNA/RNA | Biosensors |

### 1.4.1 Electrochemical sensors

*Modified electrodes* enhance the power and scope of electrochemical transducers. The chemical or biological modification of a conventional electrode surface (platinum, gold or carbon) is very attractive as it combines the high sensitivity of amperometric transducers with the improved selectivity and stability provided by the surface modifier. Many of the current limitations of electrochemical sensors can be overcome with surface microstructures based on molecular architecture concepts that meet specific sensing needs.

There are different ways by which chemically and biologically modified electrodes (known also as *amperometric (bio)sensors*) may benefit analytical applications. These benefits include the acceleration of electron transfer reactions (electrocatalysis or bioelectrocatalysis), prevention of fouling, permselective transport (discrimination based on analyte size, charge or polarity) and preferential accumulation (coupled to stripping techniques) [24].

The most common way to modify the surface of amperomeric sensors is to attach the modifier directly (using surface-modification procedures) [5,6] or doping a carbon–polymer composite paste (bulk-modification procedures) [25,26]. The latter produces *composites* and *biocomposites*. These materials can be hard, soft or fluid and represent a very attractive option for sensor technology as they integrate the receptor and the electrochemical transducer in the same material (see Fig. 1.5).

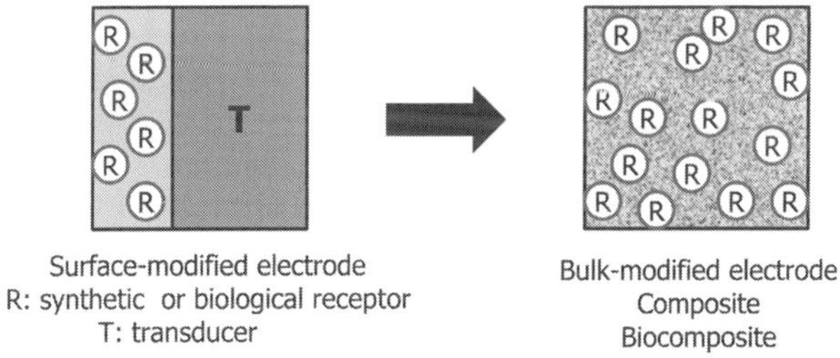


Fig. 1.5. Composite materials used as chemically or biologically bulk-modified electrodes (compared with surface-modified electrodes) are an attractive and useful concept in electrochemical sensing, since they integrate in a single phase both recognition and transducer elements (cf. Fig. 1.4).

Chemically and biologically surface-modified electrodes are described in depth in Chapter 7, while Chapter 8 reviews composite and biocomposite materials for electrochemical sensing.

Fluid composites are suitable for the preparation of sensing inks that are employed for single or multiple layer depositions using thick-film technologies such as screen or ink-jet printing [27]. This is the methodology behind disposable *sensor strips*, such as the well-known glucose biosensor that allows diabetics to measure their own blood sugar levels at home. Obviously, the promise of low-cost, large-scale manufacturing of identical devices has been the driving force behind research on the integration of different (bio)electroactive layers on an inert support (*planar sensors*) [28].

Asymmetrical configurations such as 'all-solid-state' versions of *ion-selective electrodes* (ISE) where the traditional internal solution has been substituted by direct contact of the membrane with a solid conductor or with an FET (*field-effect transistor*) are good examples of a more integrated (and simplified) design for sensing technologies [29] (see Fig. 1.6 and Section 1.6.1. See also Chapter 10 for ISEs).

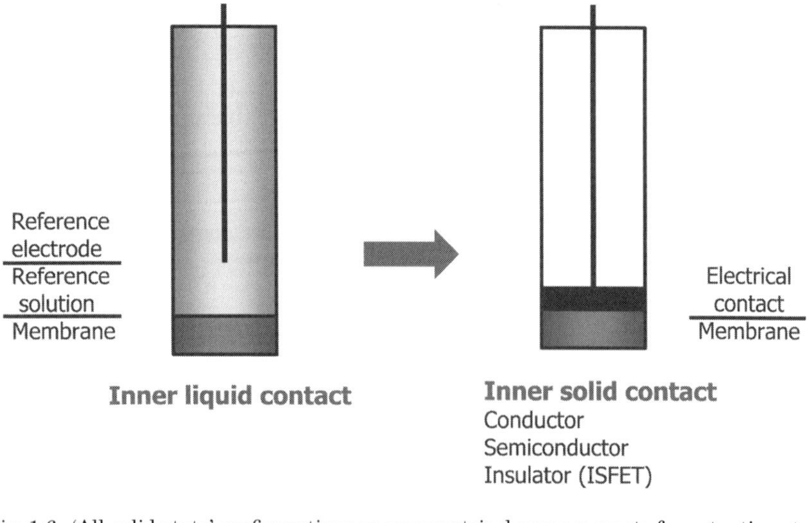


Fig. 1.6. 'All-solid-state' configurations or asymmetrical arrangements for potentiometric membrane sensors represent a more simple and integrated solution than the inner liquid contacts found in the symmetrical arrangements for classical potentiometric electrodes.

### 1.4.2 Optochemical sensors

Optical techniques for chemical analysis are well established. Most optochemical sensing systems rely on conventional analytical configurations, where light directly strikes a layer of reagent immobilised on a support located in free space or on the distal end of an optical fibre. The integration of optical fibres and reagents that respond optically (e.g. by absorption or fluorescence) to the analyte is attractive as a transduction mechanism. The terms *optrode* or *optode* are often used to describe this type of reagent-mediated optical sensor. Significant advantages can be obtained by adopting an alternative approach based on evanescent wave interactions. In intrinsic optical sensors based on this approach the fibre cladding is chemically or biologically sensitive and is interrogated via the evanescent field of the guided light.

*Fibre-optic chemical sensors* (FOCS) [30,31] can benefit from the geometric flexibility and miniaturisation of optical fibres (the core diameter of a fibre ranges from a few micrometers in the case of single-mode fibres to over one millimetre in multimode fibres). Also, two key features of optical fibres, low attenuation and electrical noise immunity, are suitable for remote sensing. Work in this field has benefited from access to high-quality fibre-optic telecommunication equipment and to a variety of optical techniques already developed for routine chemical analysis.

Optical chemo- and biosensors using optical fibres as the integrating medium are reviewed in Chapter 9.

## 1.5 ARRAY SYSTEMS

The sensing systems described in Section 1.4 are designed explicitly as individual sensing units for a given analyte. From a technological point of view it is easy to imagine, and implement, the integration of several sensors in a single platform using the same technology or different (hybrid) technological processes. This involves the organisation of a group of devices, *sensor arrays* or *array sensors*, ranging from a bundle of conventional ion-selective electrodes or fibre-optic chemical sensors to a group of sensors integrated in a microfabricated device. These arrangements represent a new category of IAS that creates great interest and potential for analytical chemistry applications due to the additional dimensionality of the data produced by such devices.

### 1.5.1 Redundant-sensor array systems

These systems integrate a group of sensors for the same analyte with identical response characteristics. In principle, these systems provide an average signal with a greater precision and stability than the signal of an individual sensor. In fact, the added value of this type of integrated configuration is as a self-diagnosing system. The malfunction of an individual sensor (faulty electrical connection, degradation, air bubbles, abnormal noise, etc.) can be detected and the related response can be automatically discarded from the pattern by an intelligent subroutine within a control loop. Furthermore, when the array sensor is integrated into a system (e.g. FIA), additional information on the system is also obtained (leakage, no sample injection, poisoning due to the sample, etc.), producing an instrument capable of real-time self-diagnosis and analytical interpretation leading to more reliable results [20].

Additionally there are two-dimensional array systems formed by identical individually addressable chemical sensors. This type of design can obtain a 2D chemical image by analysing the spatial distribution of an analyte [32]. It is possible to imagine that each element of the array can be a microfabricated needle with regularly spaced redundant sensing surfaces (e.g. for pH, oxygen, potassium, etc.). In this fashion information on local concentrations can be obtained in 3D chemical images. This type of redundant multisensing microneedle array offers, for example, the potential to monitor tissues in donated organs preserved for transplant.

### 1.5.2 Selective-sensor array systems

There is considerable interest in the development of multianalyte sensor systems capable of detecting a range of chemical and biological substances simultaneously. These systems are formed by a set of sensors sensitive to different analytes, i.e. non-redundant sensors. Each sensor responds to a single component in a multi-component mixture.

This is also the case with multi-channel analysers or multiple parameter analysers that show additional advantages over single parameter analysers, including reduced use of reagents, power and space, and lower cost.

Gene micro arrays also have to be considered as an example of this class of array system (see Section 1.6.3).

### 1.5.3 Cross-selective sensor array systems

These systems are a group of similar sensors with low selectivity showing a cross-selective response pattern. Such a design represents a new approach in measurement technology and is used to measure complex phenomena such as quality, odour or flavour, rather than just a single parameter.

There is growing interest in using multivariate signal processing of sensor signals from complex media to obtain information. Gas-sensor arrays have proven to be a practical way to measure concentrations in multicomponent gas samples. Gas sensors used in an array can be based almost on any detection principle but metal oxide or conductive polymer-based chemoresistors have been largely used. In this way, intelligent sensor array systems integrating sensors with different cross-selectivity patterns, a signal-collecting unit and a pattern recognition computer engine have been implemented. It is possible to retrieve, with some accuracy, the desired qualitative or quantitative information from the sample. In fact, a complex sample (a vapour, an odour, a headspace of complex liquids, etc.) presented to a gas-sensor array produces a characteristic signature or pattern. By presenting many different samples, a database of signatures is built. This database is used to train and configure a pattern recognition system to produce a unique classification or quantification of the components, providing capabilities for automated qualitative or quantitative analysis.

These intelligent chemical sensor arrays are similar to the human olfactory sense and are known as 'electronic noses' [21,33]. Indeed, in the human olfactory sense, the receptor cells in the olfactory bulb generate a signal pattern. The mitral cells carry out the first treatment of the data and the resulting pattern is further processed by the olfactory cortex for recognition and learning. Although the selectivity of each receptor cell may be low, the combination of several selectivity patterns produces a large information content.

Chapter 12 reviews electronic nose sensor technology and data processing and discusses the current status of integrating this non-selective sensing technology into water and wastewater monitoring systems.

Similar concepts, applied to the analysis of liquids (wine, beer, milk, water), have been reported more recently. These systems are related in similar ways to the tasting sense. Thus, for these systems the terms 'electronic tongue' and 'taste sensor' have been coined [34,35]. Chapters 10 and 11 describe both of these integrated systems as new analytical prospects in the chemical sensor field.

Artificial sensor systems are also known as 'biomimetic systems', as they mimic an approach seen in nature. When molecular imprinted polymers are used as recognition material in the sensor array, the sensing mechanism is also biomimetic as it resembles antibody–antigen interactions [36]. Furthermore, the chemometric treatment of the data may also be biomimetic if artificial neural networks (ANN) are used.

## 1.6 MICROSYSTEMS

Microsystems are microstructured devices, integrated as structures in the micrometric range. The prefix *micro* is often confusing. In strict sense and according to the decimal submultiples of SI units, a 'microstructure' presents dimensions in the range $10^{-6}$ to $<10^{-5}$ m. The integration of structures and devices of a given geometric range normally produces bigger analytical systems. It should not come as a surprise that the up-down scaling of devices in general uses prefixes that are loosely defined, specially for micro ($<1$ mm) and nano ($<1$ μm) (see Table 1.4).

Sometimes the decimal prefix is related to the volume of the processed sample and not to the size of the structure. Each ten-fold reduction in sample holder diameter reduces the volume by a thousand fold.

TABLE 1.4

Different aspects related to instrument size and sample volume of IASs

| Term | Feature size | Term | Typical volume | Fabrication technology |
|---|---|---|---|---|
| Mini | $\gg 1$ mm | Micro | μL | Integration of discrete components |
| Micro | 1 mm–50 μm | Nano | $>10$ nL | Micromachining of components in silicon and plastics |
| | 50–1 μm | | 10 nL–1 pL | Up-down scaled micromachining |
| Nano | $\ll 1$ μm | | fL–aL | Proprietary; down-up scaled chemical synthesis |

For example, minititer plates that process the sample at the millilitre level become microtiter plates for microlitre sample scale (see Table 1.4).

In some microanalytical techniques (for samples between 1 and 10 mg) such as microgravimetry or microtitration, the size of the instrument does not inevitably follow the size of the sample. This is why 'microanalytical system' and 'analytical microsystem' are completely different concepts.

The prefix *micro* is associated with microstructures and microdevices developed using microfabrication techniques whose origins are in the microelectronic industry. There is a range of techniques and materials available for three-dimensional micromachining using the controlled removal or deposition of material at a micrometer scale. Generally, features of approximately 1 μm are possible using conventional photolithography and wet etching techniques on silicon. Other fabrication methods can produce even smaller structures. Due to the optical properties of glass there is growing interest in utilising it for micromachining. Several machining and fabrication methods have also been developed for plastic microchips. Additionally, moving parts can be fabricated in microactuators by sacrificial layer techniques, where an intermediate sacrificial layer is selectively removed to release an overlying structure.

An advantage of microfabrication is that it offers a systematic approach to miniaturisation profiting from flexible designs and easy fabrication, comparable to the processes used to build macro scale devices [37].

### 1.6.1 Microsensors

The term *microsensor* describes a sensor whose active sensing area is in the micro range, whereas the sensing device itself is much larger for easier handling. *Microelectrode* and *ultramicroelectrode* (UME) are terms used in electrochemistry instead of microsensor, although electrodes with smaller radii (nanometric size) are currently used. When UMEs such as a platinum wire or carbon fibre embedded within a solid frame or micromachined electrode structures are fabricated with a dimension smaller than the diffusion length of the analyte, they have several desirable attributes in addition to steady-state responses and short response times. Small faradaic currents (pA–nA) observed in these microelectrodes often completely eliminate the ohmic drop.

Consequently, the quantification of analytes in previously inaccessible media is feasible including non-polar solvents, supercritical fluids, and even solid media. The small size of these electrodes makes diffusional mass transport extremely efficient and double layer capacitance very small, attaining greater signal-to-noise ratio and lower detection limits than those of macroelectrodes [38,39].

Advanced microelectrode arrangements are those based on either sensor array or array sensor systems that follow three basic integration strategies: (a) an array of microsensors operating in parallel to exploit the microelectrode quality response, being the current level obtained that is measurable using conventional instrumentation [40]; (b) interdigitated microsensor arrangements to enable redox recycling between the two comb-shaped microsensors (acting as anode and cathode) resulting in a dramatic increase of sensitivity [41]; and (c) an independent addressable microsensor array used to construct multianalyte sensor systems [42] (see also Section 1.6.3).

Much effort was made to develop silicon-based microsensors known as *ion-sensitive field-effect transistors* (ISFET) [43,44] mainly used for pH determinations. Sensitivity to other ions is achieved by covering FET microstructures with ISE-PVC membranes in an all-solid-state configuration (ChemFET). Direct integration of a sensor membrane and a signal processing microstructure within a single chip offers the potential for a more sensitive detection since this integration offers a much higher rejection of electromagnetic interferences, thus improving the signal-to-noise ratio.

Potentiometric ion-selective electrodes are devices with a high output impedance. Their responses are prone to corruption as their leads are affected by stray capacitance and electrical noise. These effects mean that the sensor outputs cannot be transmitted over long distances or they may require special instrumentation techniques. However, by integrating the sensor membrane with a FET structure that converts the high-impedance sensor response to a low-impedance electrical signal results in enhanced stability, sensitivity, and resolution of the sensor [45]. Chapter 14 is devoted to this type of microsensor. The authors propose these devices for the development of 'high order' sensing microsystems, i.e. systems capable of multiparameter detection of both (bio-)chemical and physical quantities using the same transducer principle [46,47].

Integrated optical devices combine microelectronic production technology with the inherent advantages of optical sensing. Chemical sensors have also been implemented on integrated optical transducers. Optical microdevices based on planar wave-guides, surface plasmon resonance structures, grating couplers or interferometers are turned into sensors with chemical or biological recognition layers [48]. Chapter 13 reports on integrated optical transducers for chemical sensing based on planar waveguides fabricated with conventional microelectronic technology.

Analogously, mass-sensitive devices fabricated on piezoelectric silicon and based on surface acoustic waves are turned into gas or liquid sensors with adequate surface-modification [49].

### 1.6.2 Analytical microsystems

The technology developed in the microelectronics industry has evolved considerably and now pervades several fields, including chemistry and the life sciences, striving to miniaturise common laboratory instruments. Besides microsensors, different machines, devices and other instruments have been developed down to a size as small as a credit card [50]. Microelectromechanical systems (MEMS) are examples of these miniaturised devices which integrate electrical and mechanical components within micrometric dimensions using glass, quartz or plastic substrates [51].

Liquids may be transported within microstructures etched in a substrate with standard lithographic and wet-etching techniques. Separation, reaction and detection can occur in these microstructures. Channels in microfluidic structures are typically 5 μm wide and 200 μm deep. Fluid control operations include sampling from a specimen container, transport of the sample to different regions of the microchip for processing, and distribution to multiple test sites. Off-chip pumps are inconvenient for integration. For this reason, the design of micromachined pumps is an active field of research within the field of MEMS. A range of pump and valve (or valveless) technologies has been developed to fulfil the basic requirements of pulse-free liquid handling in separation and detection systems. For example, fluid transportation can be carried out by electroosmotic flow. Flow can also be established by thermal capillary pumping using addressable heating devices or with pumping using piezoelectric membrane actuators placed on the raised flat roof of the microchannels guiding the flow, as in a peristaltic pump.

The low internal volume of a microchip analyser (pL–μL) produces an economic benefit as reagent consumption (and waste fluid) per analysis is correspondingly very low. Sample volume is also very small (pL–nL), offering a clear advantage in clinical applications.

Liquids are manipulated in some designs through a micro channel network and detected in microsensor structures. The increased throughput is achieved faster not only by individual runs but also by parallel processing. The ability to handle multiple samples (multiplexing) using microchip devices, either simultaneously or sequentially, is one of the most significant advantages offered by micro fluidic networks. Eventually, simultaneous multireplicated assays in one microfabricated device provide analytical redundancy and robustness, a safeguard that is not easily attained in conventional analysers.

The integration of these fluidic microstructures with other analytical microstructures (e.g. microreactors, microsensors and microactuators) is the ultimate goal of the development of 'micro total analysis systems' (μTAS) [52,53] that function as large analytical instruments. Sampling, pre-treatment steps (enrichment, separation, and reaction) and detection can all be done in these integrated devices. A lack of selectivity often prevents the use of sensors in complex samples. This is solved using a μTAS.

The term 'lab-on-a-chip' has been coined to describe a wide variety of μTAS. This term comprises any type of laboratory operation, be it analytical or not, including those related to liquid handling, synthetic and combinatorial chemistry, reactor technology and biotechnology, drug screening, PCR amplification, etc.

The main advantages of these micro laboratories are improved analytical performance through increased speed and higher throughputs due to multifunctional design, i.e. interconnected channel networks of many parallel subsystems with negligible dead volumes. Additionally, analytical microsystems offer increased mechanical stability due to (monolithic) integration, reduced resource consumption (sample, reagents, and mobile phase) and waste production, and are suitable for inexpensive mass fabrication methods.

Therefore, these analytical microsystems will be used in the near future as stand-alone units directly at the point of sampling (in situ or point-of-care testing), to perform (automated) analysis without the need for a conventional laboratory. Also, due to their uncomplicated use and operation, they will run unattended for longer periods and offer the advantage of being operated by non-skilled personnel. However, the full integration of these devices has not yet been realised and one or more steps are still performed off-chip.

Chapters 15 and 16 describe in detail two different designs of lab-on-a-chip. Each microlab integrates various subsystems relevant for different chemical analysis techniques and procedures.

*Chromatographic and electrophoretic microsystems*

Complex samples, especially clinical samples, need separation units as part of the analytical microsystem. It is well known that miniaturisation improves the performance of chromatographic separations. This has lead to the development of small-diameter packing materials and open-tubular micro column separation techniques such as capillary gas chromatography, micro bore HPLC and capillary electrophoresis (typically ∅ range between 25 and 100 μm). However, micromachining technology has more to offer for analytical separations than just miniaturised columns. It can provide, with negligible additional costs, networks of coupled separation channels including frits, mixing chambers, flow restrictors and reagent compartments, of virtually any shape. All these microstructures can be integrated in a single chip. Additionally, these systems can integrate sample injectors or on-line (pre- or post-column) derivatisation subsystems. Moreover, these operations can be done without introducing significant dead volumes to the system, which would be detrimental for precise handling of pico- to nanolitre sample volumes. Most chromatographic separation methods implemented on μTAS use an open channel approach where packing materials are avoided and surface-modification or in situ polymerisation schemes are used to generate a stationary phase inside the microchannel directly. Complete gas and liquid chromatographs with all microstructures integrated in a monolithic platform have been demonstrated [54].

Among different chip-based separation methods, electrophoresis is preferred to chromatography. The application of a voltage across the terminals of microchannels is much easier than the application of a differential pressure. Understanding of the electrokinetic phenomena (electroosmotic flow and electrophoretic migration) makes it easier to manipulate molecules and liquids. One effect of miniaturisation in chip-based electrophoretic separations is the decrease in heating (Joule effect) as power is dissipated more efficiently. This leads to a reduction in analysis times with a high resolution due to the higher voltages attainable in electrophoretic microsystems. Additionally, these microsystems display unique features with respect to separation speed, sample injection, and the consumption of sample and buffer solutions. Analysis time is reduced by a factor of between 10 and 100 compared to conventional electrophoresis, without a significant decrease in separation efficiency [55].

The analysis of biomolecules, namely DNA, oligonucleotides, proteins and peptides, is the main field of application for electrophoretic microsystems. Research has focused on exploring ways to combine the separation step with techniques preceding or following it, within a highly integrated system. Some examples of existing techniques involving electrophoretic microsystems are cell lysis, polymerase chain reaction (DNA amplification) and hybridisation [56]. The resulting microchips can be fabricated in thousands of addressable parallel test sites. Alternatively, these electrophoretic microsystems can be coupled to mass spectroscopy.

### 1.6.3 Array microsystems

Instead of thinking about individual (bio)chemical reactions as common analytical systems, researchers are adopting a new paradigm whereby many reactions or analytes are monitored or measured simultaneously, hence enabling the study of complex processes or complex samples. This is done through massive parallel processing that allows for a high output of data per unit time. As seen in Section 1.6.2, parallel processing is easily achievable using microsystems.

Array microsystems permit simultaneous detection and can be seen as an example of simple parallel processing. These systems can be formed with sensor arrays similar to those described in Section 1.5, but developed with micromachination techniques in a more integrated fashion. Small multiparametric devices have been developed integrating microsensors, either selective or showing a cross-sensitivity, both optical or electrochemical. An interesting multiparametric sensing microsystem based on ISFET technology for the measurement of six physicochemical parameters is explained in detail in Chapter 14.

Microsensor arrays (see also Section 1.6.1) can also be classified within this section. Advanced fibre-optic imaging technology has been used to develop microsensor arrays. An example of this is the site-selective photopolymerisation of a monomer doped with fluorescent dyes through distinct optical pathways at the distal end of an imaging fibre bundle ($\varnothing < 1$ mm), comprising several thousand individual fibres. Simultaneous monitoring of several chemical parameters (pH, $CO_2$ and $O_2$) using CCD video acquisition of fluorescent temporal responses observed at one end of the bundle has been demonstrated using this technique [57].

Other types of array microsystems are those based on microarraying. Microarraying techniques can accommodate a large number of reactions within a very small area without requiring large amounts of material or robotic assemblies. Microarray technology represents the most recent, and exciting, advance in DNA analysis. Information obtained from the human genome project opens the door to tremendous analytical opportunities, particularly for massive DNA analysis that is faster, simpler and cheaper than traditional hybridisation assays.

DNA microarrays, gene chips or biochips are often intermixed to describe this new type of device. Most of these microarrays use the binding of labelled-target ssDNA to complementary DNA probes immobilised on the chip. Two approaches are currently used to create a DNA probe microarray (see Table 1.5): (A) Light-directed synthesis uses precision photolithographic masks to define the positions where single, specific nucleotides are added to a growing in-situ single-stranded nucleic acid chain. As there are four nucleotides that make up DNA, synthesis is quite efficient and a set of $4^n$ different polynucleotides ($n$-mers) can be synthesised in $4n$ photolithographic masks. Currently

TABLE 1.5

Typical parameters for current DNA biochip microarrays

| | Typical parameter | Trend |
|---|---|---|
| *Light-directed synthesis* | | |
| Probe length | 4–8 | 12 |
| Chemical steps | 16–32 | 48 |
| Number of possible probes per chip | 256–65,536 | 16,777,216 |
| *Microfluidic delivery* | | |
| Number of spots per array | 100–10,000 | 50,000 |
| Number of simultaneous deposits | 1–32 | |
| Volume deposited per spot | 50–500 pL | 10 pL |
| Diameter of a spot | 100–250 μm | 50–1000 nm |

the state of the art is at $n = 8$ level. (B) Microfluidic delivery techniques allow for reproducible arrays of micro depositions (spots) of solutions and suspensions on suitable substrates.

Chips are fabricated using glass, silicon or plastic supports with sizes of 1–2 $cm^2$ comprising thousands of reaction micro zones where each individual oligonucleotide has been deposited or synthesised, resulting in high-density arrays (up to $10^6$ probes/$cm^2$). Arraying technologies often require the supply of pL volumes to discrete locations on the chip. Microarray manufacture is only a first step in the experimental cycle. Additional downstream requirements include hybridisation with labelled targets (usually with fluorescent labels), data acquisition and analysis, and the reconstruction of the target sequence. Data are usually acquired through an imaging system (a microarray scanning system) and offer the ability to view the entire array in a single image.

The use of DNA microarrays is revolutionising many aspects of genetic analysis, including the diagnosis of genetic diseases, the detection of infectious agents, drug screening, forensic analysis and gene expression [58].

## 1.7 NANOSYSTEMS

Nanoscale chemical analysis refers to analysis done in sub-microlitre volumes of liquid [37] and usually using instruments of micrometric or higher dimensions (see Table 1.4). Commercially available microtiter plates are established as reactors usually coupled to robotic feed systems that fit perfectly to their dimensions. A drastic reduction in the reaction volume (picoliter) and an increase in the number of reaction sites ('wells') has lead to the so-called nanotiter plates (e.g. with $36 \times 96 = 3456$ wells per $cm^2$). Confinement of a reaction to a picoliter volume produces peculiar phenomena. A well filled up with 1 pM solution will be occupied by only one molecule on average and, for example, it would be possible to amplify only one DNA target molecule in mixtures containing many kinds of molecule (i.e. a DNA library).

In the strict sense, the word 'nanosystem' refers to a system that has the main structures in the range from $10^{-9}$ to $10^{-7}$ m. In fact, when we speak of nanosystems we refer to molecular-sized systems, built with atomic precision.

Experimental confirmation that atoms could be treated as building blocks for fabricating a variety of hypothetical nanodevices has propelled nanotechnology forward as an independent field of research. Currently, it is known that atoms are robust enough to have an independent existence and that they can be isolated and counted. Atoms are also strong enough to form reliable parts of working nanodevices. Today, individual atoms can be moved and their positions manipulated on suitable substrates.

Useful materials, devices and systems in the nanometer range may be obtained by two different approaches. Firstly, the 'top-down' approach to nanodevices involves making existing devices so small that they eventually finish up as nano-sized objects ($<1\ \mu m$). This is typified by the manufacture of increasingly smaller transistors on a chip (see Section 1.1). Bulk materials or thin films can be shaped into structures in the 10–100 nm range through lithography and etching or very precise cutting tools.

The 'bottom-up' approach involves building nanodevices from their constituent parts: atoms or small molecules. Fabrication is achieved by one of two possible routes, namely, the physical or chemical manipulation of the atoms or molecules into their pre-assigned positions. The physical route involves manoeuvring the atoms or molecules mechanically with laser tweezers or with the tip of an atomic force microscope. Chemical manipulation relies on instructions programmed into the system to determine the ultimate location of each atom (e.g. self-organising molecular processes).

For comparison, the resolution of a light microscope is $10^{-6}$ m, implying that working with nanostructures requires special visualising tools. Progress in this area has been rapid due to chemically sensitive nanometer-scale probes such as scanning tunnelling microscopy (STM) and atomic-force microscopy (AFM), or scanning electrochemical microscopy (SECM), which can provide information on the distribution of chemical species on suitable surfaces on a micron or sub-micron scale.

Although the devices produced to date (batteries, switches, wires, tubes, rotors or sensors [59]) are rudimentary, they do demonstrate the feasibility of constructing nanomachines and nanodevices. Already, the application of submicron fibreoptic chemical sensors has been reported [60,61]. The usefulness and potential of nanotechnological-based biosensor systems in biological research and applications in single-cell analysis are discussed in Chapter 17.

The future indicates a definite trend towards more sophisticated nanosystems, including those integrating several devices in supramolecular assemblies. Probably, the method of choice for fabricating both the devices and the systems (the assemblies) is likely to be self-assembly from a solution, because any method based on the physical manipulation of atoms would be too time-consuming.

## 1.8 CONCLUSIONS AND PERSPECTIVES

An intrinsic trend in analytical chemistry is the constant improvement in the quality of the analytical information. During the latter half of the past century, this trend has coexisted with the development of conceptual, methodological and instrumental tools to build analytical equipment that requires minimum human intervention.

All this has led to different strategies to 'connect' (mechanically, physically or chemically) the steps of the analytical process. The term 'IAS' illustrates these strategies, including simplification, miniaturisation, automation and communication. In fact, the term illustrates the convergence of these strategies rather than the pre-eminence of one in particular. The 'integration' of an analytical process implies some degree of simplification, miniaturisation, automation and communication.

Although these strategies or trends are found independently in other scientific and technological fields, within the field of analytical chemistry taken together they have a special meaning, producing instruments with additional advantages with respect to information, speed, robustness, portability and cost. Because of this, integration in general and its application to analytical systems in particular can be seen as a new paradigm in scientific instrumentation.

At the turn of the century, thanks to microfabrication techniques, we have at our disposal techniques to produce analytical devices with a high degree of integration in a systematic way. Additionally, this constitutes a new paradigm in the field of analytical instrumentation that, it is predicted, will spawn a revolution in analytical chemistry in the coming future. This revolution will be embodied by 'integrated analytical microsystems' that will provide new features and will be designed ex profeso and fabricated on a massive scale for a wide range of analytical problems.

Today, analytical microsystems are being tested and gradually applied to real problems, mainly in the life sciences. As this takes place, the appearance of a new way of producing analytical information can be envisaged using highly IAS produced with nanotechnological processes that will permit access to new analytical properties in the quantum domain, and this will perhaps happen in the next two decades.

## ACKNOWLEDGEMENTS

This chapter is indebted to discussions that arose during seminars for graduate students given recently by the author in different academic and research centers such as Universitat Autònoma de Barcelona, Spain; Universidad Autónoma Nacional de México, Universidad Autónoma Metropolitana, and CINVESTAV-Instituto Politécnico Nacional, México; Universidad Técnica Federico Santa María, Chile; and University of Santo Tomas and University of San Agustín, Philippines.

## REFERENCES

1 A.J. Bard, *Integrated Chemical Systems. A Chemical Approach to Nanotechnology*, Wiley, New York, 1994.
2 J. Inczédy, T. Lengyel, A.M. Ure, A. Gelencsér and A. Hulanicki, *Compendium of Analytical Nomenclature. Definitive Rules 1997*, International Union of Pure and Applied Chemistry, 3rd ed, Blackwell, Oxford, 1998.
3 M. Valcárcel and M.D. Luque de Castro, *Flow-through (Bio)Chemical Sensors*, Elsevier, Amsterdam, 1994.
4 R. Kellner, J.-M. Mermet, M. Otto and H.M. Widmer (Eds.), *Analytical Chemistry*, Ch. 15. Wiley/VCH, Weinheim, 1998.
5 S. Lam and G. Malikin (Eds.), *Analytical Applications of Immobilized Enzyme Reactors*, Blackie Academic and Professional, London, 1994.
6 T. Cass and F.S. Ligler (Eds.), *Immobilized Biomolecules in Analysis. A Practical Approach*, Oxford University Press, Oxford, 1998.
7 F. Feigl and V. Anger, *Spot Test in Inorganic Analysis*, 6th ed, Elsevier, Amsterdam, 1972.
8 Y.A. Zolotov, V.M. Ivanov and V.G. Amelin, *Chemical Test Methods of Analysis*, Elsevier, Amsterdam, 2002.
9 M. Mulder, *Basic Principles of Membrane Technology*, Kluwer, Dordrecht, 1992.

10 T. Araki and H. Tsukube (Eds.), *Liquid Membranes: Chemical Applications*, CRC Press, Boca Raton, 1990.
11 J. Koryta, *Ions, Electrodes and Membranes*, 2nd ed, Wiley, Chichester, 1992.
12 J. Ruzicka and E.H. Hansen, *Flow Injection Analysis*, 2nd ed, Wiley, New York, 1988.
13 M. Valcarcel and M.D. Luque de Castro, *Flow Injection Analysis. Principles and Applications*, Ellis Horwood, Chichester, 1987.
14 J. Ruzicka and G.D. Marshall, *Anal. Chim. Acta*, 237 (1990) 329.
15 I.D. Wilson, E.D. Morgan, R. Lafont, J.P. Shockcor, J.C. Lindon, J.K. Nicholson and B. Wright, *Chromatographia*, 49 (1999) 374.
16 K. Pusecker, J. Schewitz, P. Gförer, L.H. Tseng, K. Albert and E. Bayer, *Anal. Chem.*, 70 (1998) 3280.
17 G.W.H. Sanders and A. Manz, *Trends Anal. Chem.*, 19 (2000) 364.
18 (a) i-STAT Corporation, Kanata, Canada. (b) AVL OPTI, AVL Biosense Corp, Atlanta, USA.
19 W.A. Warr and C. Suhr, *Chemical Information Management*, VCH, Weinheim, 1992.
20 D. Diamond, (Ed.), *Principles of Chemical and Biological Sensors*, Wiley, New York, 1998.
21 E. Kress-Rogers, (Ed.), *Handbook of Biosensors and Electronic Noses. Medicine, Food and the Environment*, CRC Press, Boca Raton, 1997.
22 I. Willner and B. Willner, *Trends Biotechnol.*, 19 (2001) 222.
23 D.M. Wilson, S. Hoyt, J. Janata, K. Booksh and L. Obando, *IEEE Sensor J.*, 1 (2001) 256.
24 M. Murray, (Ed.), *Molecular Design of Electrode Surfaces*, Wiley, New York, 1992.
25 F. Céspedes, E. Martínez-Fàbregas and S. Alegret, *Trends Anal. Chem.*, 15 (1996) 296.
26 F. Céspedes and S. Alegret, *Trends Anal. Chem.*, 19 (2000) 276.
27 M. Prudenziati, (Ed.), *Thick Film Sensors*, Elsevier, Amsterdam, 1994.
28 M. Alvarez-Icaza and U. Bilitewski, *Anal. Chem.*, 65 (1993) 525A.
29 J. Janata, *Principles of Chemical Sensors*, Plenum Press, New York, 1989.
30 O.S. Wolfbeis, (Ed.), *Fibre Optic Sensors and Biosensors*, 2 vols, CRC Press, Boca Raton, 1991.
31 B. Kuswandi, R. Anders and R. Narayanaswamy, *Analyst*, 126 (2001) 1469.
32 H. Meyer, H. Drewer, J. Krause, K. Cammann, R. Kakerow, Y. Manoli, W. Mokwa and M. Rospert, *Sens. Actuat. B*, 18–19 (1994) 229.
33 J.W. Gardner and Ph.N. Bartlett, *Electronic Noses. Principles and Applications*, Oxford University Press, Oxford, 1999.
34 Y. Vlasov and A. Legin, *Fresenius J. Anal. Chem.*, 361 (1998) 255.
35 K. Toko, *Biomimetic Sensor Technology*, Cambridge University Press, Cambridge, 2000.

36 B. Sellergren, (Ed.), *Molecularly Imprinted Polymers. Man-made Mimics of Antibodies and their Applications in Analytical Chemistry*, Elsevier, Amsterdam, 2001.
37 S. Verpoorte, ed. *Nanoscale Chemical Analysis.* Speical issue. *Trends Anal. Chem.*, 19(6), 2000.
38 A.J. Bard and L. Faulkner, *Electrochemical Methods. Fundamentals and Applications*, 2nd ed, Wiley, New York, 2001.
39 R.J. Forster, *Chem. Soc. Rev.*, (1994) 289.
40 G. Wittstock, B. Gründig, B. Strehlitz and K. Zimmer, *Electroanalysis*, 10 (1998) 526.
41 R. Hintsche, J. Alberts, H. Bernt and A. Eder, *Electroanalysis*, 12 (2000) 660.
42 R.C. Matos, M.A. Augelli, C.L. Lago and L. Angnes, *Anal. Chim. Acta*, 404 (2000) 151.
43 R.E.G. van Hal, J.C.T. Eijkel and P. Bergveld, *Sens. Actuat. B*, 24/25 (1995) 201.
44 R.E.G. van Hal, J.C.T. Eijkel and P. Bergveld, *Adv. Col. Interf. Sci.*, 69 (1996) 31.
45 P. Bergveld and A. Sibbald, *Analytical and Biomedical Applications of Ion-selective Field-Effect Transistors*, Elsevier, Amsterdam, 1998.
46 A. Poghossian, H. Lüth, J.W. Schultze and M.J. Schöning, *Electrochim. Acta*, 47 (2001) 243.
47 A. Poghossian, L. Berndsen, J.W. Schultze, H. Lüth and M.J. Schöning, In: M. Butler, P. Vanysek and N. Yamazoe (Eds.), *Chemical and Biological Sensors and Analytical Methods*, The Electrochemical Society, Pennington, 2001, p. 143.
48 P.V. Lambeck, *Sens. Actuat. B*, 8 (1992) 103.
49 D.S. Ballantine, R.M. White, S.J. Martin, A.J. Ricco, E.T. Zellers, G.C. Frye and H. Wohltjen, *Acoustic Wave Sensors: Theory, Design and Physico-Chemical Applications*, Academic Press, San Diego, 1997.
50 A. Manz and H. Becker (Eds.), *Microsystem Technology in Chemistry and Life Sciences*, Springer, Berlin, 1999.
51 D. Figeys and D. Pinto, *Anal. Chem.*, 330A(May) (2000).
52 A. Manz, N. Graber and H.M. Widmer, *Sens. Actuat. B*, 1 (1990) 224.
53 A. van der Berg and P. Bergveld (Eds.), *Micro Total Analysis Systems*, Kluwer, Boston, 1994.
54 C.S. Effenhauser, In: A. Manz and H. Becker (Eds.), *Analytical and Biomedical Applications of Ion-selective Field-Effect Transistors*, Elsevier, Amsterdam, 1998, p. 51.
55 J.P. Kutter, *Trends Anal. Chem.*, 19 (2000) 352.
56 E.T. Lagally, P.C. Simpson and R.A. Mathies, *Sens. Actuat. B*, 63 (2000) 138.
57 J.A. Ferguson, B.G. Healey, K.S. Bronk, S.M. Barnard and D.R. Walt, *Anal. Chim. Acta*, 340 (1997) 123.

58 M. Schena, (Ed.), *Microarray Biochip Technology*, Eaton Publishing, Natick, 2000.
59 T. Vo-Dinh, B.M. Cullum and D.L. Stokes, *Sens. Actuat. B*, 74 (2000) 2.
60 E. Betzig, J.K. Trautman, T.D. Harris, J.S. Weiner and R.L. Kostelak, *Science*, 251 (1991) 1468.
61 E. Betzig and R.J. Chichester, *Science*, 262 (1993) 1422.

*Chapter 2*

# *Integrated separation systems*

**Dmitri N. Muraviev***

## 2.1 INTRODUCTION

Separation is one of the basic concepts in chemistry and chemical technology. The appearance of highly efficient separation and purification techniques has permitted the purity of substances to be substantially increased; this is particularly important for the intensive development of a number of science-based industries, such as microelectronics, fiber- and IR-optics, thermonuclear energy and pharmacology [1]. The choice of separation technique and the desired degree of its efficiency usually depend on the complexity of the separation task and the scale of separation. For example, fractionation of stable isotopes can be considered to be the most complicated task in separation science due to (1) small values of the equilibrium separation coefficient ($\alpha$, see below Eq. (2.1)), which is very close to one for all elements except hydrogen, and (2) low initial concentration of the target component to be separated and recovered (desired isotope). Hence, to achieve the desired separation degree (i.e. the desired concentration of the target isotope) one needs to solve the problem of multiplication of a single separation effect (see below) through combination or *integration* of single separation stages into a cascade. The separation of isotopes of light elements is carried out by applying physico-chemical methods based upon the distribution of isotopes between either two phases or two different compounds located in different phases. These methods include

* *E-mail address:* dimitri@icmab.es.

*Comprehensive Analytical Chemistry XXXIX, pages 37–79* 

ISSN: 0166-526X

rectification and isotope exchange, as well as other techniques. Fractionation of isotopes of intermediate and heavy elements is mainly carried out using physical methods, which are based on distribution of isotopes under separation within one phase under the action of a gradient of some force. The methods of this type include centrifugation, thermodiffusion, electromagnetic, and mass-diffusion techniques [2].

Separation is also particularly important in Analytical Chemistry, as the mixtures of substances must be in many instances separated prior or in the course of their analytical determination. The techniques, which are usually applied to separate substance mixtures, are based on a non-equal distribution of the mixture components between two contacting phases (solid–liquid, liquid–liquid or solid–gas) composing a separation system. The higher the non-equality of distribution of the mixture components between two phases, the more easily the desired degree of separation is achieved. The second phase can either (1) be introduced artificially into a solution of substance mixtures (liquid or gas), or (2) appear simultaneously due to the modulation of some extensive physico-chemical parameters of the system, such as temperature and pressure. The separation methods of the first group include ion exchange, extraction, adsorption and different chromatographic techniques based upon these methods. The second group includes crystallisation, distillation and rectification [3–5].

## 2.2 GENERAL PRINCIPLES OF BI-PHASE SEPARATION

The use of separation methods of both groups can be based on different modes of operation, such as (1) single-stage operations, (2) step-wise or semi-continuous operations and (3) continuous operations. The single-stage operation can be qualified as the most simple and traditional one. It is based on the use of single-stage separation units, which can be either real (such as ion exchange or extraction batch reactors) or hypothetical (a plate of rectification column or theoretical plate of chromatographic column). A single-stage batch operation includes (in the case of separation methods of the first group) the following main sequential steps: (1) uptake (adsorption, absorption, extraction, etc.) of substances by one of the contacting phases; (2) separation of phases; (3) stripping (desorption, evaporation, re-extraction, etc.) of substances from the second phase (other than liquid or gas solution), and (4) isolation of the desired component for analytical purposes or recovery. In this case the fractionation degree is determined by a *single-stage equilibrium separation effect*, which can be quantitatively characterised by a number of equilibrium parameters of the system, such as equilibrium separation coefficient $\alpha$ or selectivity coefficient $K$. For a binary mixture of components A and B (e.g. ions), $\alpha$ and $K$ values can be expressed as follows:

$$\alpha_{\mathrm{B}}^{\mathrm{A}} = \frac{Y_{\mathrm{A}}}{Y_{\mathrm{B}}} \times \frac{X_{\mathrm{B}}}{X_{\mathrm{A}}} = \frac{Y_{\mathrm{A}}}{(1 - Y_{\mathrm{A}})} \times \frac{(1 - X_{\mathrm{A}})}{X_{\mathrm{A}}} \tag{2.1}$$

$$K_{\mathrm{B}}^{\mathrm{A}} = \frac{Y_{\mathrm{A}}^{1/z_{\mathrm{A}}}}{Y_{\mathrm{B}}^{1/z_{\mathrm{B}}}} \times \frac{X_{\mathrm{B}}^{1/z_{\mathrm{B}}}}{X_{\mathrm{A}}^{1/z_{\mathrm{A}}}} \tag{2.2}$$

Here $Y$ and $X$ are the equivalent fractions of components in contacting phases at equilibrium (for example, $X_{\mathrm{A}} = z_{\mathrm{A}}C_{\mathrm{A}}/(z_{\mathrm{B}}C_{\mathrm{B}} + z_{\mathrm{A}}C_{\mathrm{A}})$, where $C$ is the concentration, equiv./dm$^3$); $z_{\mathrm{A}}$ and $z_{\mathrm{B}}$ are the charges of ions; and indices A and B denoting mixture components are usually chosen so that $\alpha > 1$ and $K > 1$.

Equation (2.1) is widely used in separation science to describe the selectivity of a certain separation system [2]. For example, in distillation the $\alpha$ value is called relative fugitiveness and $X$ and $Y$ denote the mole fractions of components of a binary mixture in the vapour and liquid phases, respectively. Quite obviously, Eqs. (2.1) and (2.2) characterise the equilibrium of a bi-phase exchange reaction (proceeding in, for example, ion exchange or extraction system) or a bi-phase distribution (in a liquid–gas or solid–liquid system), which can be described by a stoichiometrically strict equation (characterising the mass and charge balance in the system). For example, Eqs. (2.1) and (2.2) can refer to the following ion-exchange reaction, which proceeds in a bi-phase system including a solid (e.g. ion-exchange resin) or a liquid (e.g. ion-exchange extractant) exchanger (S) bearing mono-functional, in this case carboxylic, groups:

$$\frac{1}{z_{\mathrm{B}}}(\mathrm{S} - \mathrm{COO}^{-})_{z_{\mathrm{B}}}B^{+z_{\mathrm{B}}} + \frac{1}{z_{\mathrm{A}}}A^{+z_{\mathrm{A}}} \Leftrightarrow \frac{1}{z_{\mathrm{A}}}(\mathrm{S} - \mathrm{COO}^{-})_{z_{\mathrm{A}}}A^{+z_{\mathrm{A}}} + \frac{1}{z_{\mathrm{B}}}B^{+z_{\mathrm{B}}} \tag{2.3}$$

### 2.2.1 Thermodynamics of bi-phase equilibrium

The thermodynamics of reversible exchange reaction (2.3) under isothermal conditions is described by the isotherm equation:

$$\Delta G_{P,T} = \Delta G_T^o + RT \ln\left( \frac{a_{\mathrm{AR}}^{-1/z_\mathrm{A}} \; a_{\mathrm{BX}}^{1/z_\mathrm{B}z_\mathrm{X}}}{a_{\mathrm{BR}}^{-1/z_\mathrm{B}} \; a_{\mathrm{AX}}^{1/z_\mathrm{A}z_X}} \right) \tag{2.4}$$

where $\Delta G^o = -RT \ln K$; $K$ is the thermodynamic equilibrium constant of reaction (2.3) corresponding to the condition $\Delta G_{P,T} = 0$; $a$ stands for the activity of the respective component; X is the co-ion; and a bar over symbols denotes the exchanger phase (e.g. resin or extractant).

Although the equilibrium and selectivity coefficients ($\alpha$ and $K$) are thermodynamically meaningless parameters, it is possible to correlate them with the thermodynamic equilibrium constant ($K$) and, hence, with the Gibbs energy $\Delta G$ of the system. In the simplest case (for equally charged ions disregarding the redistribution of water between contacting phases) this can be done through the use of the following equation [6–10]:

$$\Delta G^o = -RT \ln K = -RT \int_0^1 \ln \alpha \, \mathrm{d}Y \tag{2.5}$$

where $Y$ has the same meaning as in Eqs. (2.1) and (2.2).

Determination of $\alpha$ versus composition ($Y$ or $X$) dependencies at different temperatures allows for evaluation of the enthalpy term of Gibbs energy ($\Delta H$) from the temperature dependence of $K$ by using van't Hoff's equation:

$$\left( \frac{\partial \ln K}{\partial T} \right)_P = \frac{\Delta H^o}{RT^2} \tag{2.6}$$

From the known $\Delta G$ and $\Delta H$ values, the entropy term can also be calculated, i.e. the system can be thermodynamically described.

### 2.2.2 Integration concepts in bi-phase separation

The integration concept is widely applied in Separation Science both to improve the efficiency of existing fractionation techniques and to develop new separation methods. One can distinguish three main types (or levels) of integration, which are used in bi-phase separation:

(1) integration of uptake and stripping steps (see above) within one separation stage,
(2) integration of several single separation stages (real or hypothetical) based upon the same fractionation technique (e.g. ion exchange) within one more efficient separation method (e.g. ion exchange chromatographic methods), and
(3) integration of different separation methods (ion exchange-extraction, ion exchange-crystallisation, etc.) within combined fractionation techniques (see below).

In all the above cases the main result of integration is a substantial improvement of separation efficiency. In certain instances it gives a number of additional advantages, which cannot be achieved without integration.

The choice of desired level of integration depends on the type of separation task. For example, the following main types of separation can be distinguished in 'aquatic chemistry' systems:

1. separation of non-electrolyte mixtures;
2. separation of weak or non-electrolytes from strong electrolyte impurities;
3. separation and conversion of electrolytes from one ionic form to another;
4. separation of electrolyte mixtures;
5. separation accompanied by selective concentration of the target substance;
6. selective extraction and concentration of micro components, followed by their analysis.

The degree of complexity of the separation task and the level of integration which is required for its successful solution primarily depend on the difference in properties of the substances to be separated. Indeed, a dramatic difference in the chemical properties of non-electrolytes (e.g. water) and electrolytes (e.g. salts) substantially simplifies the problem of their separation so that the first type of integration and a single stage operation appear in many instances sufficient to achieve the desired separation degree. In the case of complex electrolyte mixtures such as mixtures of rare earth elements, only a combination of the first and the second levels of integration (e.g. within ion-exchange displacement chromatography techniques) permits their complete separation (see Ref. [1], and below). Let us now consider all types of integration by using ion exchange, extraction and membrane separations as the most typical examples.

### 2.2.3 Integration of uptake and stripping steps

Examples of the first level of integration (integration of the uptake and stripping steps) can be found in many separation systems where the second step of single-stage operation (separation of phases, see above) is not required. This situation occurs, for example, in some ion-exchange and extraction systems (e.g. in the case of polymer immobilised extractants, known also as extraction chromatographic materials [10] or solvent impregnated resins [11–16] permitting the column mode of operation to be used). In the majority of membrane systems the phases are separated *a priori* which also creates the conditions for integration of the first type.

*Membrane separations*
Membrane systems represent the most typical example of integrated separation systems of the first type as membranes are primarily used for separation, and membrane processes are generally separation processes. A membrane process requires two bulk phases, *feed* (f) and *stripping* (s), physically separated by a third phase, the *membrane* (m) (see Fig. 2.1). The membrane is a semi-permeable barrier between two bulk phases representing either a homogeneous phase or a combination of heterogeneous phases [17–21]. The membrane separations proceed in fact due to integration of the uptake and the stripping processes. Indeed, the uptake of a substance by the membrane surface in the feed compartment of the cell (see Fig. 2.1) is followed by its simultaneous stripping from the membrane surface in the receiver compartment. This integration creates conditions for the transport of a substance through the membrane due to the existence of the inter-membrane flux ($J$).

Membranes can be classified according to different criteria, e.g. their origin, either synthetic or biological, their separation mechanism, their structure or their geometry. One or more driving forces cause the movement of species across the membrane. These driving forces arise from a gradient of pressure, chemical potential, electrochemical potential or temperature. In the simplest case of separation of a solvent (e.g. water) from electrolyte or non-electrolyte impurities (membrane

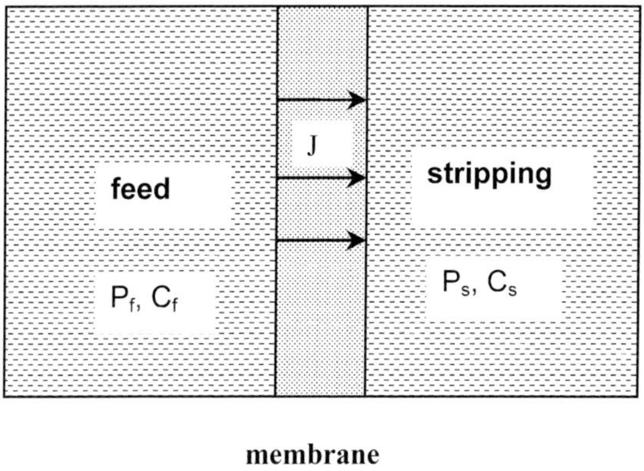


Fig. 2.1. Schematic diagram of membrane cell (see text).

concentration and desalination processes) the driving force ($F_d$) is one of the main factors determining the separation degree. The use of only an 'internal' driving force such as gradient of concentration (dialysis) or osmotic pressure (osmosis) in many instances does not permit the desired separation degree to be reached as the driving force gradually diminishes when the system tends to equilibrium ($J \to 0$ when the gradient $\to 0$). To maintain $J$ value at a sufficiently high level an external driving force is required to increase the separation efficiency. Some examples of membrane processes with internal and external driving forces are given in Table 2.1.

By following the general concept of this book, a sort of *integrated description* of membrane separation processes can be proposed. This description can be based on the use of an Ohm-like equation connecting $J$ and $F_d$ values with each other in many membrane systems as follows:

$$J = \frac{F_d}{R_m} = F_d P_m \tag{2.7}$$

where $R_m$ and $P_m$ are the membrane resistance (similar to electrical resistance in Ohm's law) and the membrane permeability (similar to electroconductivity), respectively.

To illustrate this conclusion let us consider as an example a pressure-induced transport in the reverse osmosis membrane system. The feed phase contains an aqueous solution of a non-electrolyte. This system is characterised by two driving forces: concentration and pressure

TABLE 2.1

Examples of membrane separation processes

| Process | Dialysis | Electrodialysis | Osmosis | Reverse osmosis |
|---|---|---|---|---|
| Driving force | Concentration gradient | External electrical field | Osmotic pressure gradient | External pressure |
| Membrane type | Conventional | Ion exchange | Conventional | Asymmetric |
| Flux direction | High C ⇒ Low C | Low C ⇔ High C | Low C ⇒ High C | High C ⇒ Low C |
| Separation degree | Determined by equilibrium | Desired degree is achievable | Determined by equilibrium | Desired degree is achievable |

gradients. The general equation for the trans-membrane flux can be written as follows [22,23]:

$$J = \frac{DA}{L}\left[C_{\mathrm{f}} - C_{\mathrm{s}} \exp\left(\frac{-\bar{\nu}\Delta\pi}{RT}\right)\right] \tag{2.8}$$

where $D$ is the diffusion coefficient; $L$ is the membrane thickness; $\Delta\pi = P_{\mathrm{f}} - P_{\mathrm{s}}$; $C$ and $P$ are the concentration and the pressure, respectively; indices f and s denote the feed and the stripping phases, respectively, and $A$ is a constant.

The equation for water flux, $J_{\mathrm{w}}$, in the reverse osmosis cell with the use of external pressure $\Delta P \gg \Delta\pi$ can be easily derived from Eq. (2.8) by assuming that water concentration in the feed remains almost constant $C_{\mathrm{f}} = C_{\mathrm{f_0}} \approx C_{\mathrm{s}}$. Taking into account that $\exp(x) \approx 1 + x$ for low $x$, one obtains:

$$J_{\mathrm{w}} = \frac{D_{\mathrm{w}}A_{\mathrm{w}}}{RTL} C_{\mathrm{f_0}} \bar{\nu}_{\mathrm{w}} (\Delta P - \Delta\pi) \tag{2.9}$$

Equation (2.9) connects water flux $J_{\mathrm{w}}$ with the gradients of external pressure and osmotic pressure created by the dissolved substance. At $\Delta\pi = 0$ Eq. (2.9) transforms to the Darcy's law equation:

$$J_{\mathrm{w}} = \frac{D_{\mathrm{w}}A_{\mathrm{w}}}{RTL} C_{\mathrm{f_0}} \bar{\nu}_{\mathrm{w}} \, \Delta P \tag{2.10a}$$

As follows from the comparison of Eqs. (2.7) and (2.10a), $F_{\mathrm{d}} = \Delta P$ and $P_{\mathrm{mw}} = (D_{\mathrm{w}}A_{\mathrm{w}}/RTL)C_{\mathrm{f_0}}\bar{\nu}_{\mathrm{w}}$, which is known as the water permeability constant of the membrane.

An equation describing the flux of non-electrolyte, $J_{ne}$, in the system under consideration can be also derived from Eq. (2.8) by assuming that $C_f \gg C_s$. If $\Delta P$ is not too high (not higher than 100 atm) the term $[\exp(-\bar{\nu}\,\Delta\pi/RT)] \approx 1$, and we obtain the Fick's law equation:

$$J_{ne} = \frac{DA}{L}(C_f - C_s) = \frac{DA}{L}\Delta C \qquad (2.10b)$$

Comparison of Eqs. (2.10b) and (2.7) shows in this case $F_d = \Delta C$, and $P_m = DA/L$ the solute permeability constant. Note that, from Eqs. (2.10a) and (2.10b), the flux of the solute does not depend on pressure gradient. The proposed integrated description of different membrane separation systems allows for estimation of their separation efficiency through the comparison of the desired $F_d$ values permitting a given separation degree at a given $P_m$ or $R_m$ to be achieved. This approach also allows the inverse problem to be solved, i.e. the comparison of $P_m$ (or $R_m$) values of different membrane systems used in a certain separation process (e.g. desalination) at a given $F_d$ value.

The membrane itself is another principal factor (other than the driving force) determining the selectivity of separation and the flux. In fact, the nature of the membrane (its chemical properties, physical structure and the type of membrane material) determines the type of application, ranging from the separation of microscopic particles to the separation of molecules of almost identical size or shape. Modern industrial membrane processes are based on the use of solid membranes such as polymeric membranes, or in some cases ceramic or glass membranes. The choice of a given polymer as a membrane material is not arbitrary but is based on its specific properties, originating from a number of structural factors and the separation task to be solved. For example, polymeric ion-exchange (sulphonated) membranes were successfully used both to carry out the H–D isotope exchange reaction (deuteration) and to quantitatively separate the deuterated product from the undesired side products [24,25].

*Liquid membranes*

Separation can also occur through a liquid film (liquid membrane) separating the feed and the stripping phases, in which a component is being transported by diffusion [20,21]. This process is often enhanced by the addition of a carrier facilitating the transport of a specific solute. Table 2.2 shows examples of the most typical carriers which are used in

TABLE 2.2

Examples of carriers (extractants) used for separation of metal ions in liquid membrane systems

| | |
|---|---|
| *Basic extractants* | |
| Prime | $(CH_3)_3C(CH_2C(CH_3)_2)_4NH_2$ |
| Aliquat 336 | $R_3N(CH_3)^+Cl^-$ |
| *Acidic extractants* | |
| Di-2-ethylhexylphosphoric acid | $(C_4H_9CH(C_2H_5)CH_2O)_2PO_2H$ |
| Versatic 10 | $R_3CCO_2H$ |
| Synex 1051 | 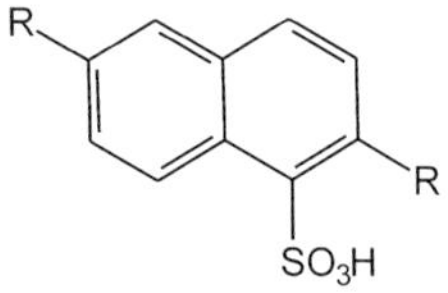 |
| *Solvating extractant* | |
| Tributyl phosphate | $(RO)_3PO$ |
| Trioctylphosphine oxide | $R_3PO$ |
| *Chelating extractants* | |
| Kelex 100 | 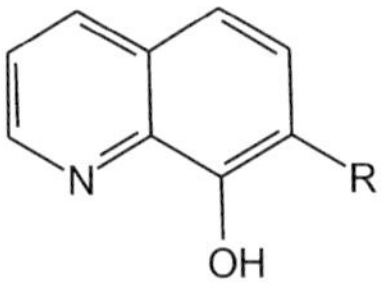 |
| LIX 65 N | 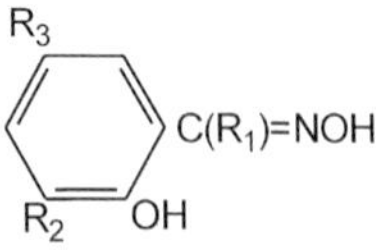 |

different membrane separation processes. The addition of a carrier to a liquid membrane system (facilitated transport), which rapidly and reversibly forms complexes with the desired guest species, can dramatically improve the membrane permeability and selectivity (which primarily depend on the carrier–guest interaction). A proper carrier selectively binds one of the components in the feed phase and transports it across the membrane from the feed to the stripping phase. This process is driven by a gradient of electrochemical potential of the component over both sides of the liquid membrane. The component (guest) to be separated first reacts with the carrier to form a complex which is soluble in the membrane but not in the adjacent solutions. After diffusion across the membrane, this complex decays into the free carrier and the original guest species. The transport of a component by a carrier is mostly coupled and can be either co-coupled or counter-coupled. In co-coupled transport processes, both components of the mixture (A and B) move in the same direction, whereas in counter-coupled transport processes they move in opposite directions, as shown in Fig. 2.2. As seen, both extraction and stripping processes proceed simultaneously, i.e. are integrated.

Most liquid membranes are composed of a carrier dissolved in an organic solvent. Numerous kinds of liquid membrane systems have been applied to a wide variety of uses, such as gas separation, removal of

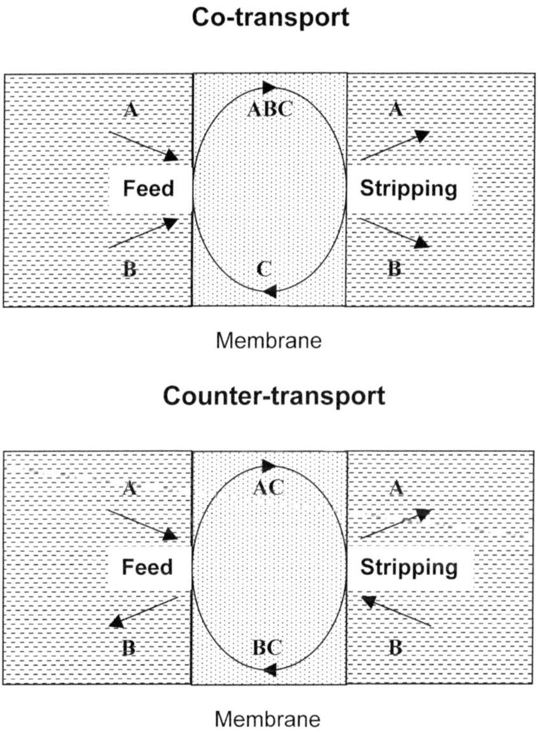


Fig. 2.2. Schematic illustration of co- and counter-coupled transport: A = component to be transported; B = coupled species; C = carrier species.

organic compounds, separation and recovery of metal ions, removal of toxic wastes from industrial effluents, development of selective sensing devices, enzyme reaction and recovery of fermentation products [21].

Separation based on the use of liquid membranes is the most typical example of integration of the first type as it combines the conventional solvent extraction and the stripping processes within one separation stage. The use of a liquid film offers a possibility of much higher separation rates than in polymeric membranes because of the higher diffusion rates in the former [21]. Much attention is paid nowadays to the investigation and development of liquid membranes (supported liquid membranes in particular) as they can be successfully applied to solve complex and specific separation problems including enrichment, clean-up, separation and determination of organic and inorganic contaminants in air and water samples [26–31]. For example, liquid membranes with enhanced stability (activated composite membranes, ACM) recently developed by Oleinikova et al. [32–36] have been shown to be applicable for efficient separation of metal ions from multicomponent mixtures.

## 2.3 MULTIPLICATION OF SINGLE SEPARATION EFFECT

In many instances a single equilibrium effect is not high enough to achieve a significant fractionation degree in a single-stage operation. To achieve the desired separation degree one needs to 'multiply' a single stage separation effect. Multiplication of this effect can be achieved through *integration* of single-stage separation units into a cascade or by using a separation column. The column must be of sufficient separation efficacy, which is characterised by a number of theoretical plates. The separation degree achieved at each plate equals a single separation effect (characterised by $\alpha$ value, see Eq. (2.1)), i.e. in other words a separation column can also be imagined to be a sort of a cascade of *hypothetical* single-stage separation units. The number of single-stage separation units to be cascaded to achieve the desired separation degree depends on the $\alpha$ value as it is shown schematically in Fig. 2.3, where $Y$ versus $X$ diagrams (isotherms) for two bi-phase separation systems are shown. As seen in Fig. 2.3, in the case of a lower $\alpha$ value (see Fig. 2.3(a)) more separation stages are required to achieve the same separation degree, which can be obtained in just two stages at a higher $\alpha$ (see Fig. 2.3(b)). Let us consider separation techniques permitting

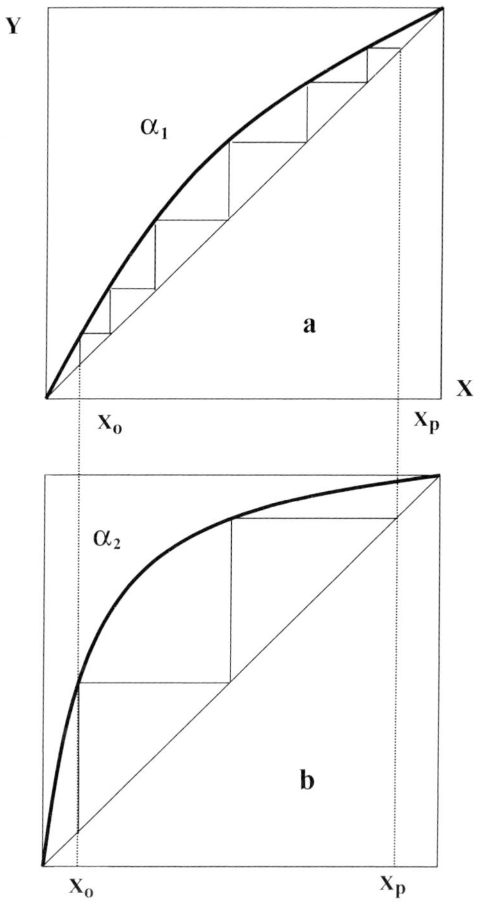


Fig. 2.3 *Y* versus *X* diagrams for two bi-component mixtures with different $\alpha$ values: $\alpha(b) > \alpha(a)$. In distillation vertical and horizontal lines correspond to evaporation and condensation, respectively.

multiplication of a single separation effect by using ion exchange as an example. Attention will mainly be paid to preparative separation methods as analytical separations are out of the scope of this chapter. Note that all the techniques which are considered below are based on *integration* of the sorption and desorption or stripping stages (see above) within one separation stage. For example, sorption and desorption (at least partial) processes proceed simultaneously in the course of percolation of the mixture to be separated through the bed of ion exchanger (ion exchange column). Due to the different affinity of the ion exchanger towards mixture components they are characterised by different mobilities inside the column. Numerous repetitions of sorption–desorption acts lead to redistribution of components along the column so that they form under certain conditions a zone of a pure substance.

Two main routes of multiplying a single separation effect are known in ion exchange separation and purification processes:

(1) reversing the flow (or reflux, similar to distillation) of the mixture under separation at the edge of the separating system;
(2) altering the direction of the interphase transfer by changing the selectivity of the ion exchanger towards the ions to be separated.

The first route is widely used in ion exchange separation and purification processes in both fixed bed and counter-current modes of operation (see e.g. Refs. [1] and [37] in Chapters 5, 7 and 8). The separation techniques to be applied in this case are the following:

1. frontal separation (frontal ion exchange chromatography);
2. reverse frontal separation (reverse frontal ion exchange chromatography);
3. displacement chromatography, and
4. tandem ion-exchange fractionation.

The above separation methods will be discussed and illustrated primarily for the fixed bed mode of operation.

### 2.3.1 Frontal ion exchange chromatography

This separation procedure allows the separation of components *which are sorbed better* than the target substance. Let us consider a binary mixture, for example an electrolyte AX containing an impurity BX. In this case $\alpha_{\mathrm{B}}^{\mathrm{A}} < 1$ (see Eq. (2.1)). In the frontal separation process a solution containing the AX and BX mixture is passed through a column with an ion exchanger bearing a counter-ion C, which must be bound to the resin weaker than A and B ($\alpha_{\mathrm{C}}^{\mathrm{A}} > 1$ and $\alpha_{\mathrm{C}}^{\mathrm{B}} > 1$). A and B ions displace C from the resin phase, resulting in the formation of a stationary sorption front between the mixture under separation and the C ions. This sorption front moves along the column without changing its shape or length. The difference between the sorbabilities of A and B results in the accumulation of the weaker sorbed component A in the head portions of the solution collected from the column. This is shown in Fig. 2.4(a), where the schematic diagram of a typical breakthrough curve for the frontal separation process is presented. As seen in Fig. 2.4(a), only a certain amount of substance A appears to be completely separated from component B.

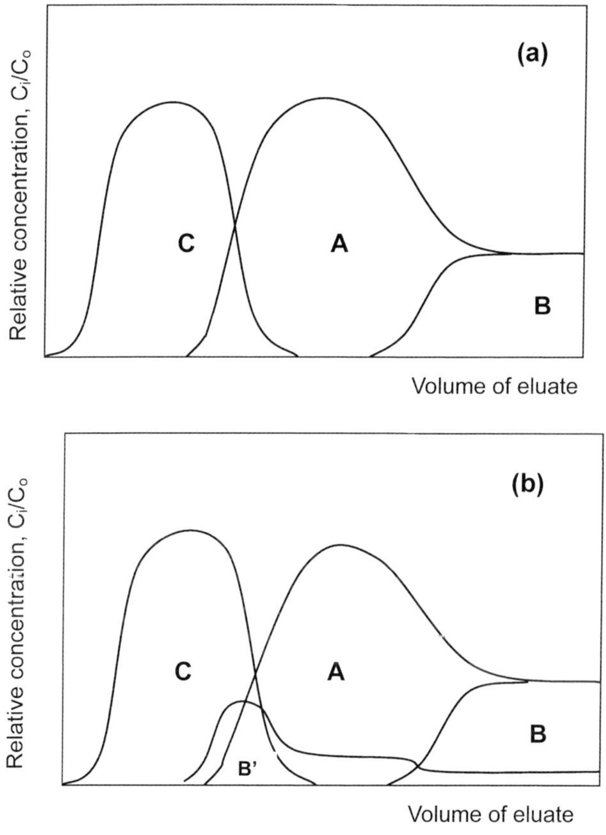


Fig. 2.4. Concentration–volume histories of separation by frontal ion exchange chromatography of component A from (a) stronger sorbed impurity B, and from (b) both stronger (B) and weaker (B′) sorbed impurities.

If the initial solution of A also contains a related B′ impurity, which is sorbed weaker than A, frontal separation is not applicable. This can be seen in Fig. 2.4(b). If the ion exchanger contains ions other than C that are sorbed stronger than A, the purified product will be recontaminated. A necessary condition for the ion exchanger when used in the frontal separation process is the complete absence of ions that are more preferably sorbed than A. Frontal ion exchange chromatography is widely used to solve different separation tasks, including high purification of acids, bases and salts, as illustrated by the following examples.

For the purification of quaternary ammonium hydroxides from $Cl^-$ impurities, a 5% aqueous solution of $Me_4NOH$ was passed through a bed of strong base anion exchanger in the OH-form [38]. The level of $Cl^-$ ions decreased 200 times to 0.2 ppm. Aqueous HF solutions (75%) were purified from arsenic impurities by being passed through a thiophenol resin in the H-form. The As content decreased from 3 ppm to 30 ppb [39]. To obtain highly pure $H_2IrCl_6$, the iridium was first reduced to Ir(III), then treated with a Wofatit KPS-200 cation exchanger for complete elimination of the ionic impurities, and subsequently reoxidized in 6 M HCl media [40]. Highly pure barium acetate (used in the production of $BaF_2$) was obtained by treating acetate salt solution on the Ba-form of two resins: amino carboxylic ion exchanger ANKB-35 or phosphoacidic cation exchanger KF-7. The residual content of Fe, Co, Ni and Cu impurities was reduced to less than $10^{-4}$ mass % [41]. Ultra purification of concentrated NaCl solutions (used in electrolytic membrane process for NaOH production) from $Ca^{2+}$ impurity was effectively carried out on the Na-form of pyridine carboxylic polyampholite VPK [42]. Concentrated NaCl solutions used in mercury-alkali cells were purified from Fe, Co, Mo and V admixtures by being passed through a bed of a granulated titanium polyphosphate ion exchanger. The same solution was purified from Ca, Mg, Sr, Ba and Ni impurities on the Na-form of polyvinylpyridine resin [43]. Purification of NaF solutions from iron down to $1.5 \times 10^{-4}$ mass % was carried out on Dowex-1A iminodiacetic resin, a Warion KS(II) sulphonate cation exchanger and Dowex 2 anion exchange resin [44].

### 2.3.2 Reverse frontal ion exchange chromatography

This separation technique allows for the separation of substances from components *with weaker sorbability* than that of the substance being purified. The procedure is based on the preliminary equilibration (loading) of the ion exchanger with the ion mixture (for example, A and B′, see above) and subsequent displacement of the mixture with an ionic reagent D (displacer), which displaces A and B′, since it is better sorbed by the ion exchanger than either of them. The better sorbed ion, A, accumulates at the end of the zone, as shown in Fig. 2.5. The necessary condition in the reverse frontal separation is the absence of weaker sorbed ions (in comparison with A and B′) in the displacer solution. The reverse frontal separation allows purification of only part of the substance entering the process. The quantity of substance purified is related to the $\alpha_{B}^{A}$, and is higher for high $\alpha$ values.

For instance, reverse frontal separation was used to obtain highly pure $Pd(NH_3)_4(HCO_3)_2$ [45]: palladium, as a complex cation, was initially sorbed from a $Pd(NH_3)_4Cl_2$ solution on a Wofatit KPS cation exchanger in the $NH_4$-form, then, after the resin phase was rinsed with water, it was selectively stripped with a 10% solution of $(NH_4)_2CO_3$. Note that the frontal separation technique is more widely used in ion exchange purification processes than reverse ion exchange chromatography. Nevertheless, as both frontal and reverse frontal separation methods are based on the first type of integration, their application is limited by the possibility of separating only one type of undesired component (impurity) from the target substance.

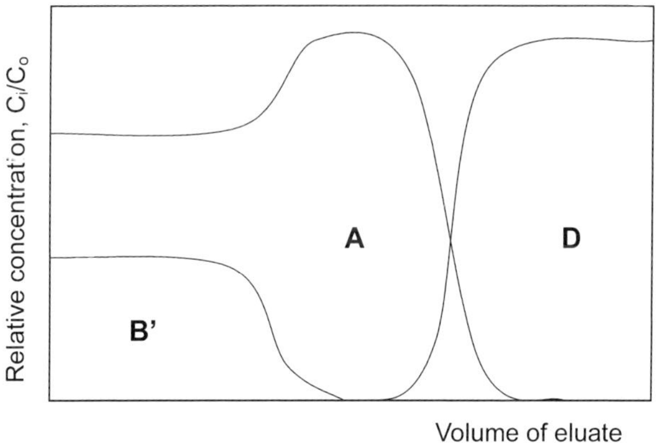


Fig. 2.5. Concentration–volume histories of separation by reverse frontal ion exchange chromatography of component A from weaker sorbed impurity B′.

### 2.3.3 Displacement chromatography

This separation technique represents an example of the second level of integration (see above) as it is based on the *simultaneous combination* of frontal and reverse frontal separations. In this regard it can be qualified as being based upon *parallel integration* of these two ion exchange separation methods. As a result, displacement chromatography permits purification of the target substance from *both better and weaker sorbed impurities*. The purification procedure in this case includes loading the bottom part of the column with an ion exchanger in the initial C-form, followed by loading a portion of the mixture being separated (AX, BX and B′X) on to the upper part of the resin bed. Then the components of the mixture under separation are displaced (desorbed) from the ion exchanger with the auxiliary ion D (displacer). In some cases, complexing agents are used as the displacer. Conditions for both steps of the fractionation process, in terms of corresponding $\alpha$ values for couples of the components of the system, are as follows:

$$\text{loading}: \quad \alpha_{\mathrm{C}}^{\mathrm{A}} > 1;\ \alpha_{\mathrm{C}}^{\mathrm{B}} > 1;\ \alpha_{\mathrm{C}}^{\mathrm{B'}} > 1,\ \text{(FS) and}$$

$$\text{displacing}: \quad \alpha_{\mathrm{A}}^{\mathrm{D}} > 1;\ \alpha_{\mathrm{B}}^{\mathrm{D}} > 1;\ \alpha_{\mathrm{B'}}^{\mathrm{D}} > 1\ \text{(RFS)}.$$

Here C is the initial counter-ion, and D is the displacer. These conditions correspond to those for proceeding with frontal (FS) and reverse frontal separation (RFS) processes (see above).

The displacement leads to the migration of the zone containing the ion mixture along the resin bed, resulting in redistribution of the components inside this zone, so that each tends to occupy the position which corresponds to its sorbability. The resulting situation, shown in Fig. 2.6, can be achieved when both the migrating zone and the resin bed have sufficient lengths. Note that this migrating zone is equivalent to a hypothetical counter-current separation column where, instead of the motion of both phases, one uses the motion of one phase (solution) and that of the zone itself.

One of the first and most impressive applications of displacement chromatography was the early work of Spedding et al. [46–48] on the separation of rare earth elements (REE) on a sulphonate cation exchanger with 0.1% citrate displacer solution. Since then, displacement chromatography has been widely used for obtaining highly pure REE compounds [49]. This method appears to be the most convenient for separation of multi-component mixtures of REE, which is known to

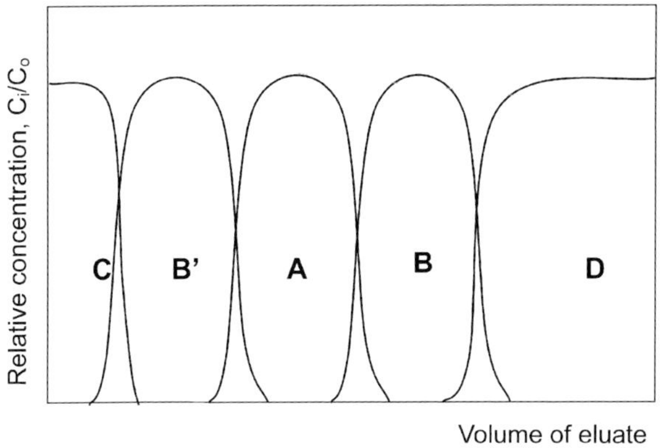


Fig. 2.6. Concentration–volume histories of separation by displacement ion exchange chromatography of tri-component mixture (A, B and B′).

be one of the most difficult tasks in Separation Science (similar to isotope separation, see above). Different complexing agents were applied as displacers. In this case, an ion exchanger is usually preloaded with ions (initial ionic form or C-form, see above) forming more stable complexes with the complexing agent than with the ions being separated. The displacer contains cations with less complexing abilities than the RE. Application of displacement chromatography for purification of REE compounds is demonstrated in the following examples. A mixture of RE chlorides (60% Gd; 33% Sm; 2.3% Eu; 1.4% Dy; 3.3% Tb; 3.3% Y and 3.3% Nd) was sorbed on a Wofatit KPS × 8 sulfonate cation exchanger in the $NH_4$-form, and then displaced with a mixture of 0.26 M lactic acid and 0.01 M EDTA at pH 3.4. Spectroscopically pure Gd was obtained from the middle fraction of the solution collected [50]. A similar procedure was used for producing spectroscopically pure Y, Eu [51] and highly pure La [52].

### 2.3.4 Tandem ion-exchange fractionation

This novel ion exchange separation technique has been recently developed by Muraviev et al. [1,53] and is also based on the combination (integration) of frontal and reverse frontal separations, but unlike displacement chromatography, can be defined as their *sequential integration*. The Tandem Ion-Exchange Fractionation (TIEF) technique also allows the purification of substances from impurities of either sorbability (in comparison with that of the target component). Separation of $n$ ($n > 2$) components by TIEF is carried out on ($n - 1$) columns sequentially connected to each other, as shown in Fig. 2.7(a) for the quadri-component mixture of A, B, C and D cations with equivalent ratio A:B:C:D = 1:1:1:1. The initial mixture of AX, BX, CX and DX (here X denotes the mutual anion) is passed through three columns preloaded with cation exchangers, which preferentially adsorb one of the mixture components (B, C or D) and *manifest the weakest affinity towards the same cation*, e.g. A. The last condition is the necessary one when applying TIEF technique. The loading of each column with the mixture under separation is carried out until the breakthrough of the better sorbed component is observed (see Fig. 2.7(a)). Hence, as seen in Fig. 2.7(b), the sequential repetition of the frontal separation leads to

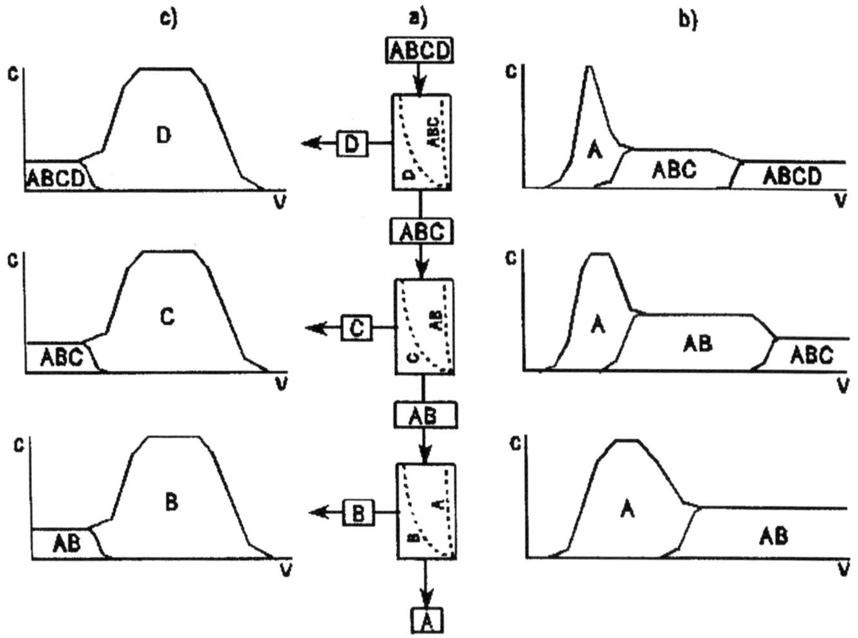


Fig. 2.7. Schematic diagram of separation of quadri-component mixture by tandem ion exchange fractionation technique (see text). Reprinted with permission from [D. Muraviev, J. Noguerol, M. Valiente, *Anal. Chem.*, 69 (1997) 4234]. Copyright (1997) American Chemical Society.

the step-wise elimination of the number of components in the solution entering the next column, and to the accumulation of A ion in the head portions of the eluate. Highly pure component A can be recovered from the solution collected from the last column. Components D, C and B are accumulated in the resin phase in columns 1, 2 and 3, respectively (as shown in Fig. 2.7(a) by dotted lines) and can be obtained in a highly pure state during the stripping stage by using the RFS technique (see above). The distribution of the components in the stripping solution is shown in Fig. 2.7(c).

A mixture of $Cu^{2+}$, $Al^{3+}$, $Mg^{2+}$ and $Zn^{2+}$ sulphates (37.5% Cu; 9.8% Al; 14.4% Mg and 38.3% Zn) was separated on three columns loaded with a Lewatit TP-207 iminodiacetic ion exchanger (columns 1 and 3) and Lewatit R-250K carboxylic resin (column 2). Highly pure magnesium, copper and aluminium, containing less than 0.001% of impurities, were obtained as the products [53].

## 2.4 COMBINED SEPARATION TECHNIQUES

The separation methods based upon integration of the third level (see above), i.e. combining different fractionation techniques within one separation method, give a number of advantages when compared to conventional techniques. As with integration of the second level, at the third level one can also distinguish two types of integration: parallel (e.g. when two separation processes proceed simultaneously) and sequential. One of the first communications reporting the use of integrated separation techniques of the first type dates back to 1917 when Kober published his paper entitled 'Pervaporation, perstillation and percrystallization' [54]. The results reported by Kober were the first attempt to combine membrane permeation with evaporation (pervaporation), distillation (perstillation, known now as membrane distillation [55]) and crystallisation (percrystallization). The first two integrated separation techniques have found wide application in modern Chemical Technology (see, e.g. [56–58]). For example, pervaporation that uses non-porous membranes in certain applications benefits from enhanced separation efficiency, selectivity, and flux rates compared to other membrane based separation techniques. Pervaporation, the most well-known membrane process for the separation of liquid/liquid mixtures, allows for a variety of possible application areas:

(1) dewatering of organic fluids like alcohols, ketones and ethers,
(2) separation of organic fluid mixtures from narrow boiling temperatures to constant (azeotrop) boiling temperatures,
(3) removal of organic pollutants from water like hydrocarbons, halogenated hydrocarbons, ketones and alcohols,
(4) separation of fermentation products,
(5) separation of organic products in low concentration from aqueous media, e.g. aromas and scents.

Presently this technique shows promise as a significantly lower energy consuming and environmentally friendly substitute for distillation processes commonly used to 'break' or separate azeotropic mixtures, a requirement in many petrochemical processes, and in the dehydration and purification of various organic solvents.

One of the brightest examples of integrated separation methods of the second type (based upon *sequential integration* concept) is the combination of chromatographic techniques with mass spectroscopy. Nowadays the breadth of practical applications of LC-MS and GC-MS techniques is indefinitely wide, from the solution of different environmental problems to analysis of complex biological samples. Let us consider in detail below two examples of less-known combined separation techniques based upon parallel and sequential integration concepts.

### 2.4.1 Solvent extraction-ion exchange. Aqua impregnated resins

A substantial increase in separation efficiency of ion exchange or extraction can be achieved through their *parallel integration*, i.e. by using combined ion exchange-liquid extraction systems. These systems are known in the literature as combined ion exchange–solvent extraction (CIESE) and aqua-impregnated resins (AIR). The idea of combined ion-exchange-solvent extraction separation (CIESE) was suggested by Korkisch [59–63], who applied mixed water-organic solvents for selective separation of ions on ion-exchange resins. The CIESE concept is based on simultaneous ion exchange and liquid extraction by a bifunctional separation mechanism which occurs in applying CIESE-active organic compounds such as *water-soluble* ethers and ketones in the presence of mineral acids to selectively elute ions

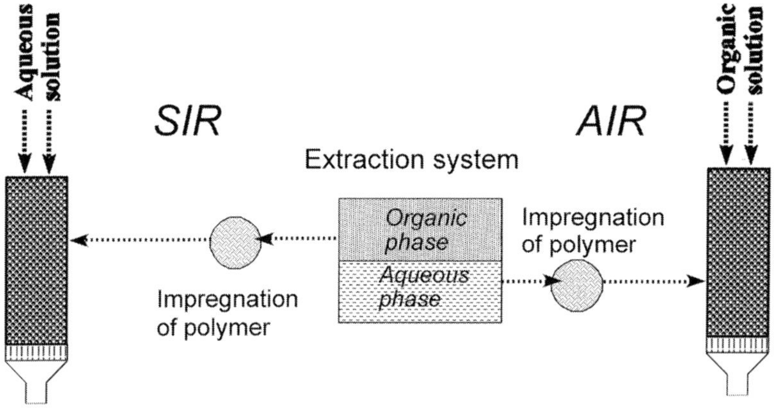


Fig. 2.8. Comparison of SIR and AIR concepts. Reprinted from [D. Muraviev, *Solv. Extr. Ion Exch.*, 18 (2000) 753 with permission from Marcel Deker].

from ion exchange resins. Since then, many CIESE systems involving both cation and anion exchangers have been studied [64–71].

Unlike CIESE, the AIR concept (introduced for the first time by Muraviev et al. [72]) refers to biphasic systems involving *water immiscible solvents* and a water swollen ion exchange resin. From this viewpoint AIR systems can be considered as a combination of liquid extraction and ion exchange in a true sense. The term Aqua-Impregnated Resins actually reflects an apparent symmetry between these ion exchange systems and those involving solvent impregnated resins (SIR) [72–74] (see Fig. 2.8).

The higher efficiency of separation in AIR systems (in comparison with ion exchange and extraction ones) can be understood through consideration of the ion exchange equilibria characterising the redistribution of ions between the resin and extractant phases. Let us consider for this purpose a hypothetical three phase system composed of an aqueous solution (AS) containing a couple of divalent metal ions $M_1^{2+}$ and $M_2^{2+}$, an ion exchange resin (IER) and an organic solution of an extractant (OSE) as shown schematically in Fig. 2.9. The equilibrium in this system can be described by the following ion exchange reactions associated with the corresponding $\alpha$ values (rewritten, unlike Eq. (2.1) in absolute concentrations) [74,75]:

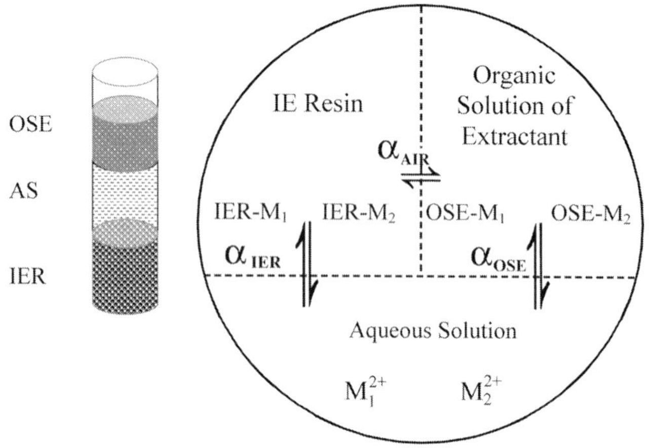


Fig. 2.9. Schematic diagram of a three phase system involving solid ion exchange resin, organic solution of extractant, and aqueous solution of metal ions. Reprinted from [D. Muraviev, *Solv. Extr. Ion Exch.*, 18 (2000) 753] with permission from Marcel Deker.

(1) IER-AS equilibrium

$$\mathrm{IER-M_2 + M_1^{2+} \Leftrightarrow IER-M_1 + M_2^{2+}} \tag{2.11}$$

$$\alpha_{\mathrm{IER}} = \frac{[\mathrm{M_1^{2+}}]_{\mathrm{IER}}}{[\mathrm{M_2^{2+}}]_{\mathrm{IER}}} \frac{[\mathrm{M_2^{2+}}]_{\mathrm{AS}}}{[\mathrm{M_1^{2+}}]_{\mathrm{AS}}} \tag{2.12}$$

(2) OSE-AS equilibrium

$$\mathrm{OSE-M_2 + M_1^{2+} \Leftrightarrow OSE-M_1 + M_2^{2+}} \tag{2.13}$$

$$\alpha_{\mathrm{OSE}} = \frac{\left[\mathrm{M_1^{2+}}\right]_{\mathrm{OSE}}}{[\mathrm{M_2^{2+}}]_{\mathrm{OSE}}} \frac{\left[\mathrm{M_2^{2+}}\right]_{\mathrm{AS}}}{[\mathrm{M_1^{2+}}]_{\mathrm{AS}}} \tag{2.14}$$

(3) IER-OSE equilibrium

$$\mathrm{OSE-M_2 + IER-M_1 \Leftrightarrow OSE-M_1 + IER-M_2} \tag{2.15}$$

$$\alpha_{\mathrm{IER}}^{\mathrm{OSE}} = \frac{\left[\mathrm{M_1^{2+}}\right]_{\mathrm{OSE}}}{[\mathrm{M_2^{2+}}]_{\mathrm{OSE}}} \frac{\left[\mathrm{M_2^{2+}}\right]_{\mathrm{IER}}}{[\mathrm{M_1^{2+}}]_{\mathrm{IER}}} \tag{2.16}$$

The aqueous solution in contact with IER and OSE is the same (see Fig. 2.10), therefore Eq. (2.15) can be obtained by subtracting Eq. (2.11) from Eq. (2.13). Hence, the aqueous phase can be eliminated from the three phase system under consideration which reduces it to the respective AIR (IER-OSE) one. Then Eq. (2.16) can be rewritten in the following form [74,75]:

$$\alpha_{\mathrm{AIR}} = \frac{\alpha_{\mathrm{OSE}}}{\alpha_{\mathrm{IER}}} \tag{2.17}$$

Relationship (2.17) is the fundamental equation describing the shift of ion exchange equilibrium in AIR systems to the desired direction in comparison with the respective equilibria in a single IER-AS or OSE-AS system. Indeed, as can be seen from Eq. (2.17) the necessary condition for $\alpha_{\mathrm{AIR}} > \alpha_{\mathrm{OSE}}$ is $\alpha_{\mathrm{IER}} < 1$. Eq. (2.17) can also be applied to pre-select the conditions for the most efficient separation of a given ion mixture by using the AIR technique. This pre-selection is based on the data in the literature on ion exchange and liquid extraction equilibrium, from which one can estimate the efficiency of separation ($\alpha_{\mathrm{AIR}}$) through the use of respective $\alpha_{\mathrm{OSE}}$ and $\alpha_{\mathrm{IER}}$ values.

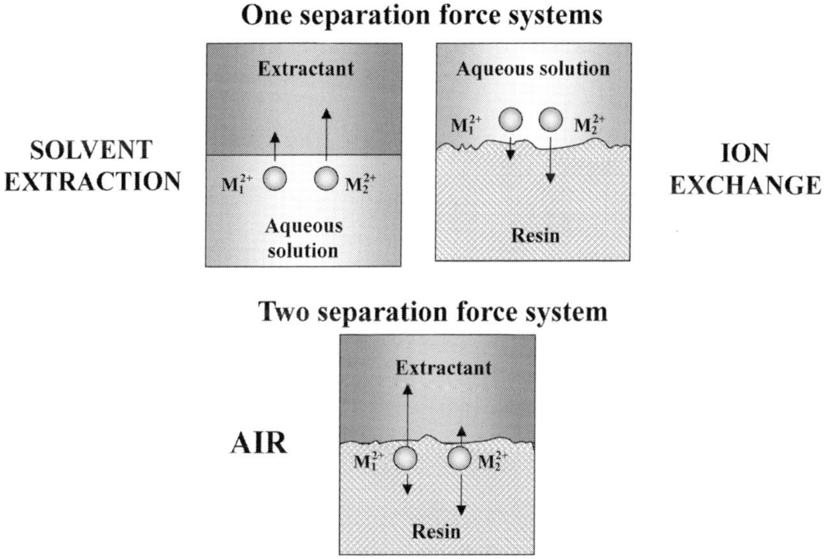


Fig. 2.10. Comparison of separation efficiency of liquid extraction, ion exchange and AIR systems. Reprinted from [D. Muraviev, *Solv. Extr. Ion Exch.*, 18 (2000) 753] with permission from Marcel Deker.

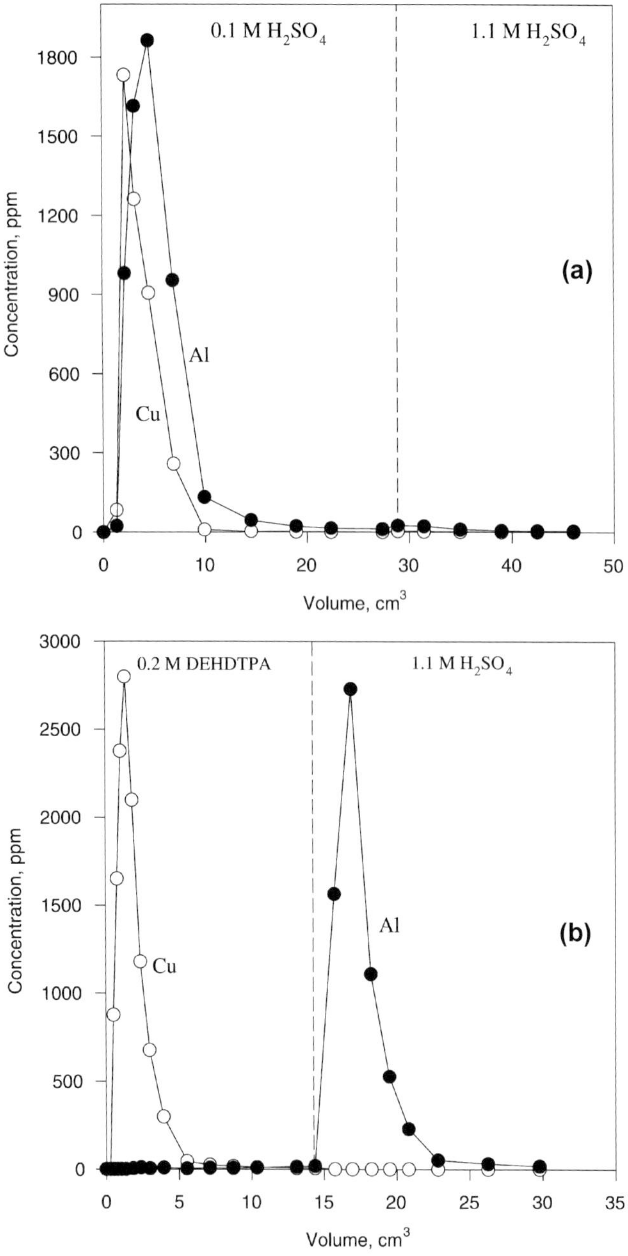

0.1 M $H_2SO_4$
1.1 M $H_2SO_4$
Concentration, ppm
Al
Cu
(a)
Volume, $cm^3$
0.2 M DEHDTPA
1.1 M $H_2SO_4$
Concentration, ppm
Al
Cu
(b)
Volume, $cm^3$

The separation mechanism in AIR systems can also be viewed as follows: the separation in conventional ion exchange or liquid extraction system proceeds due to the action of 'one separation force' determined by the different affinity of the ion exchange resin or extractant toward the ions under consideration (see Fig. 2.10). In AIR systems, the separation of ions can be defined as proceeding due to a 'couple of separation forces' acting in opposite directions (similar to the action of two forces in physics). This conclusion illustrates the main result of the parallel integration of ion exchange and extraction within the combined AIR technique.

The confirmation of the above conclusions is clearly seen in Fig. 2.11 [75], where the separation efficiency in a conventional ion exchange system (see Fig. 2.11(a)) is compared with that of an AIR system (Fig. 2.11(b)). Although the example shown in Fig. 2.11 (separation of $Cu^{2+}$ and $Al^{3+}$ mixture) may seem to be not so impressive, it illustrates an extremely high efficiency of metal separation in the AIR system. Both metals are quantitatively separated and can be recovered from the respective stripping solutions in a high pure state (>95% purity) with a sufficiently high yield (>95%). This is the result of the combined action of two 'separation forces' (see Fig. 2.10), one of which is determined by the higher selectivity of the resin towards $Al^{3+}$ (separation force directed to the resin phase) and the other one is due to the far higher selectivity of the extractant for $Cu^{2+}$ (separation force directed out of the resin), leading to complete separation of $Cu^{2+}$ and $Al^{3+}$ in the AIR system (see Fig. 2.11(b)). Indeed, the selectivity of the resin is higher for $Al^{3+}$ than for $Cu^{2+}$ ($\alpha_{Cu}^{Al} \approx 1.8$ [75]). This means that the former is retained by the resin phase more strongly than the latter. At the same time, the selectivity of di-2-ethylhexyl dithiophosphoric acid (DEHDTPA) towards $Cu^{2+}$ is much higher than that for $Al^{3+}$ ion [76,77]. In fact, the extraction constant, $K_{ex}$, for $Cu^{2+}$ has been reported by Levin et al. [76] to be $\log K_{ex} = 12.3$ while, according to Alimarin et al. [77], $Al^{3+}$ is not extracted by DEHDTPA. Quantitative comparison of

---

Fig. 2.11. Concentration–volume histories obtained by elution of $Cu^{2+}$ and $Al^{3+}$ from polyacrylic resin Lewatit R-250K with (a) 0.1 M $H_2SO_4$ (conventional IE system) and (b) 0.2 M DEHDTPA in heptane (AIR system). Conditions: column of 0.5 cm i.d., 10 cm of length filled with 0.5 g of resin; flow rate = 0.5 $cm^3$/min. For other details see Ref. [75]. Reprinted with permission from [M. Oleinikova, D. Muraviev and M. Valiente, *Anal. Chem.*, 71 (1999) 4866]. Copyright (1999) American Chemical Society.

the separation efficiency of the AIR technique versus conventional ion exchange fractionation is given in Table 2.3.

Another paramount feature of metal separation in AIR systems is also clearly seen in Fig. 2.11(b). This feature clearly follows from the comparison of the shape of the $Cu^{2+}$ peaks obtained in the elution of this metal ion from the resin with $H_2SO_4$ (in IE system, see Fig. 2.11(a)) and DEHDTPA (in AIR system, see Fig. 2.11(b)). Indeed, although the concentration of both eluents is identical (0.2 equiv./$dm^3$), the width of the $Cu^{2+}$ elution peak in the AIR system appears to be approximately half of that observed in the ion exchange system, while the height of the peak obtained in the first case differs from that in the second by a factor of 1.5. This difference in the elution efficiency is attributed to the specific interaction mechanism of OSE with the water swollen (aqua-impregnated) IER. This type of ion exchange mechanism was studied for the first time by Muraviev and Omarova [78] and is known as 'contact granulo-micellar exchange' [9,78], which can be considered

TABLE 2.3

Comparison of separation efficiency of ion exchange (IX) and aqua impregnated resins (AIR) separation techniques. Adapted from Ref. [75]

| System type | Ion mixture | Resin type | Striping solution | Separation process | IPP (%) | AFPP (%) | Yield of product with PFP > 95% |
|---|---|---|---|---|---|---|---|
| IX | Cu, Al | PAR | $H_2SO_4$[a] | Al from Cu | Al (29.8) | 68 | 0 |
| AIR | Cu, Al | PAR | DEHDTPA[b] and $H_2SO_4$ | Al from Cu, Cu from Al | Al (29.8), Cu (70.2) | 99.2, 98.6 | 98.5, 96 |
| IX | Cu, Zn | IDA | $H_2SO_4$ | Cu from Zn | Cu (49.3) | 95.2 | 72 |
| AIR | Cu, Zn | IDA | DEHPA[c] and $H_2SO_4$ | Cu from Zn | Cu (49.3) | 97.2 | 97 |
| IX | Cu, Al, Zn | PAR | $H_2SO_4$ | Al from Cu | Al (17.3) | 77 | 55.4 |
| AIR | Cu, Al, Zn | PAR | DEHDTPA, DEHPA and $H_2SO_4$ | Cu from Al, Zn, Al from Cu, Zn | Cu (40.8), Al (17.3) | 97.3, 98.1 | 96, 98.7 |

PAR, polyacrylic resin; IDA, iminodiacetic resin; IPP, initial purity of product; AFPP, average final purity of product.

(a) Aqueous 0.1 and/or 1.1 M solution.

(b) 0.2 M heptane solution.

(c) 0.2 M heptane solution.

a particular version of 'contact ion exchange' (that is, proceeding through the contact of resin beads in different ionic forms) investigated by Nikolaev and Bogatyrev [79–81] and later by Bunzl and Schultz [82].

The extractant (DEHDTPA) exists in the organic (heptane) solution in the form of reverse micelles [9,83] (with the hydrophilic >PSSH functional groups facing towards the hydrated micelle nuclei) as its concentration (0.2 mol/dm$^3$) substantially exceeds the cmc (critical micelle concentration). Hence, the interaction between the metal bearing resin and the extractant can only occur through collisions between the resin beads and the extractant micelles. The collisions must be accompanied by the 'partial opening' of the micelles when approaching the hydrophilic surface of the resin bead as this is, in fact, the necessary condition for such an interaction to proceed. A schematic diagram of the contact granulo-micellar exchange in an AIR system is shown in Fig. 2.12. As seen, the interaction of the extractant micelles with the resin beads proceeds through the following steps: (1) diffusion of reverse micelles to the surface of the resin bead, (2) opening of the micelles at the bead surface followed by interaction of extractant with metal ions (denoted by $M^{2+}$ in Fig. 2.12), and (3) closing of the micelles containing captured metal ions followed by their diffusion from the bead surface.

According to the mechanism presented in Fig. 2.12, metal ions desorbed from the resin by the extractant appear to be 'entrapped' by the reverse micelles. The interaction among extractant micelles in a non-polar solvent is negligibly weaker in comparison with the interionic interactions in aqueous solutions [79]. Hence, the broadening of the peaks in the course of the elution of stripped metals along the column in AIR systems appears to be far less than in their ion exchange analogues. It seems quite obvious that this feature of AIR systems makes this separation mode of great interest for application in AIR-HPLC and we intend to continue our investigations of this new version of combined ion exchange-liquid extraction chromatography.

### 2.4.2 Ion exchange-crystallisation. Ion exchange isothermal supersaturation

Ion exchange is widely used not only for separation (or purification) but also for the concentration and recovery of some ionic species from dilute solutions (see below and, e.g. [81]). In many cases, the ion exchange

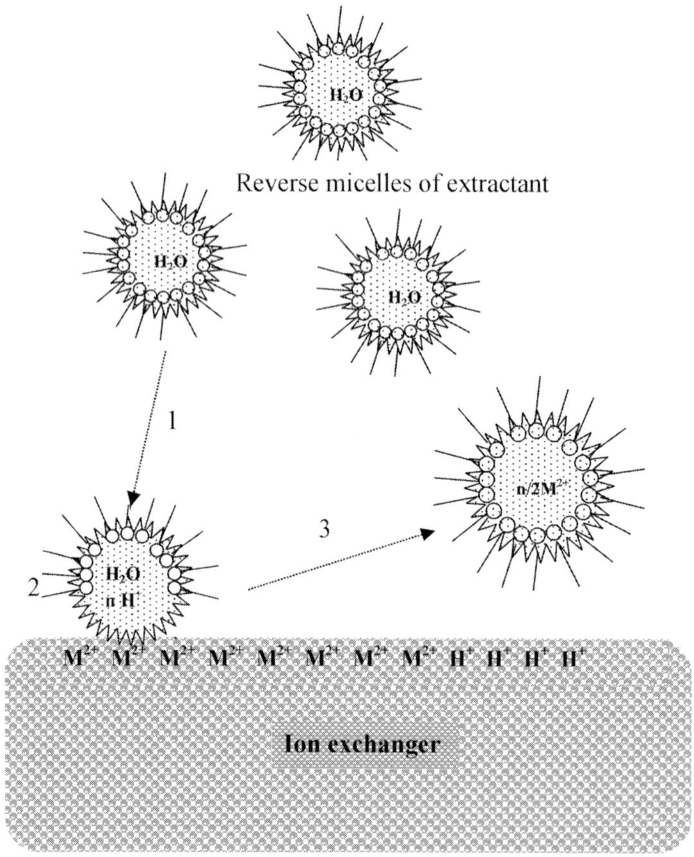


Fig. 2.12. Schematic diagram of contact granulo-micellar exchange in AIR systems. Reprinted with permission from [M. Oleinikova, D. Muraviev and M. Valiente, *Anal. Chem.*, 71(1999) 4866]. Copyright (1999) American Chemical Society.

purification is accompanied by a simultaneous concentration process. This can be clearly seen in Figs. 2.4 and 2.5, where the purification of component A by frontal (Fig. 2.4) or reverse frontal (Fig. 2.5) separation eventually leads to an increase of A concentration to the overall concentration of the initial mixture. In certain cases, both purification techniques allow a concentration of the purified substance to be reached, up to and beyond the level exceeding the solubility of the product at a given temperature. Moreover, this supersaturated solution remains stable within the column interbed space for a long period [84]. This phenomenon was discovered for the first time by Muraviev [85] and is known as 'Ion-Exchange Isothermal Supersaturation' (IXISS).

Frontal or reverse frontal separation integrated with a supersaturation process provides the possibility of purifying the target substance (supersaturated) from both better and weaker sorbed impurities. The first is removed during the ion-exchange purification step (see Fig. 2.4(a)), while the last (see Fig. 2.4(b)) can be separated from the product during crystallisation of the collected supersaturated solution. In this case, the impurity B′ (Fig. 2.4(b)) remains in the supernatant.

### 2.4.3 Ion exchange supersaturation of zwitterlites

Frontal separation can be effectively used for manufacturing highly pure zwitterlites such as amino acids and proteins, by applying either (1) cation or (2) anion exchangers in the H- or OH-form, respectively. In the first case, the zwitterlite (e.g. amino acid, AA) under purification must first be converted into the anionic form (at pH > pI, where pI is the iso-electric point of a given amino acid), e.g. used as a monosodium salt. In the second case, it must be converted into cationic form (at pH < pI, e.g. as a hydrochloride).

For instance, when carrying out the frontal separation of a monosodium amino acid salt (NaAA) on a sulfonate cation exchanger in the H-form, the interaction of $Na^+$ with the ion exchanger results in protons being released from the resin phase, followed by their interaction with amino acid anions ($AA^-$), leading to the conversion of the latter into zwitterions ($AA^{\pm}$), i.e. to final pure amino acids:

$$R{-}SO_3^-H^+ + Na^+ \rightarrow R{-}SO_3^-Na^+ + H^+ \qquad (2.18)$$

$$AA^- + H^+ \rightarrow AA^{\pm} \qquad (2.19)$$

where $AA^+$ is the cation of the amino acid.

The result of reactions (2.18) and (2.19) when proceeding in an ion exchange column in the course of frontal separation of $Na^+$ and $AA^-$ ions is the formation of a pure amino acid zone (zone of component A in Fig. 2.4(a)).

Similar interactions occur when frontal separation is carried out on an anion exchanger in the OH-form, e.g.

$$R{-}N(CH_3)_3^+OH^- + Cl^- \rightarrow R{-}N(CH_3)_3^+AA^- + OH^- \qquad (2.20)$$

$$AA^+ + OH^- \rightarrow AA^\pm \tag{2.21}$$

In both cases, the concentration of the purified amino acid becomes equal to twice the concentration of the initial salt (usually far more soluble than the amino acid) and, under certain experimental conditions, exceeds its solubility.

Amino acids with low solubility, such as glutamic [84–88]; aspartic [85,89] and tyrosine [90–92], were shown to form very stable supersaturated solutions during ion exchange purification on both cation (see e.g. [84, 85]) and anion exchangers (see e.g. [90, 91]) in the interbed space of the columns. It has been suggested that serine, asparagine and threonine behave similarly [93–95]. The mechanism of stabilisation of supersaturated zwitterlite solution in the interstitial space of an ion exchange column investigated in detail by Muraviev et al. [85,93–97] can be interpreted as follows:

(1) after formation of a pure amino acid zone (see above and Fig. 2.4(a)), amino acid zwitterions interact with a cation exchanger (e.g. sulphonate) in the H-form according to the following reaction:

$$R{-}SO_3^-H^+ + {}^+NH_3{-}R'{-}COO^- \rightarrow R{-}SO_3^- + NH_3{-}R'{-}COO^-H^+ \tag{2.22}$$

Protonated carboxylic groups of the adsorbed amino acid cations are known to dissociate in the resin phase [98,99], i.e. sulfonate cation exchanger in the amino acid form may be considered as a somewhat carboxylic cation exchanger (see below). This means that reaction 2.22 can be continued due to the further aminecarboxylate interaction of amino acid molecules according to the following scheme [85]:

$$\begin{aligned}
&R{-}SO_3^-{}^+NH_3{-}R'{-}COO^-H^+ + {}^+NH_3{-}R'{-}COO^- \\
&\quad\rightarrow R{-}SO_3^-{}^+NH_3{-}R'{-}COO^-{}^+NH_3{-}R'{-}COO^-H^+ \\
&\qquad + R{-}SO_3^-{}^+NH_3{-}R'{-}COO^-{}^+NH_3{-}R'{-}COO^-H^+ \\
&\qquad + {}^+NH_3{-}R'{-}COO^- \\
&\quad\rightarrow R{-}SO_3^-{}^+NH_3{-}R'{-}COO^-{}^+NH_3{-}R'{-}COO^-{}^+NH_3{-}R'{-}COO^-H^+ \text{ etc.}
\end{aligned} \tag{2.23}$$

The first molecule of the amino acid chain is fixed on the resin phase while the rest of the chain may be located in the solution film surrounding the resin bead.

Stabilization of amino acid supersaturated solutions by the resin or by any other granulated phase bearing charges on the surface may also be explained by sorption of the amino acid pre-crystalline associates which are formed in the interbed space. Hence, the second version of the stabilization mechanism can be described as follows [94]:

$$m^{+}NH_3{-}R'{-}COO^{-} \rightarrow (^{+}NH_3{-}R'{-}COO^{-})_m \tag{2.24}$$

$$R{-}SO_3^{-}H^{+} + (^{+}NH_3{-}R'{-}COO^{-})_m \rightarrow R{-}SO_3^{-}(^{+}NH_3{-}R'{-}COO^{-}H^{+})_m \tag{2.25}$$

According to this version of the stabilisation mechanism, formation of the pre-crystalline associates, representing in fact polymolecular zwitterions (see Eq. (2.24)), is followed by their sorption on a cation exchanger (Eq. (2.25)) and is accompanied by the conversion of polymolecular zwitterions into cations. The mechanism of this conversion is identical to that of the monomeric zwitterions, which is well known from the literature (see e.g. [100,101]). The second version of the proposed mechanism does not contradict the first one, since both versions can be considered to complete each other and help to explain some other features of amino acid uptake by ion exchangers such as the superequivalent sorption [94,96,97]. An experimental confirmation of the aminecarboxylate interaction of adsorbed amino acid molecules was reported by Muraviev et al. [85,93–97]. An additional confirmation of this mechanism was obtained by Selemenev et al. [91,102–105] and later by Zuyi Tao et al. [106,107].

The amino acid supersaturated solutions coexist with granulated ion exchangers for several hours without any alteration of their concentrations, as shown in Fig. 2.13, but spontaneously crystallise following removal of the solution from the column (see curve 5 in Fig. 2.13). As clearly follows from Fig. 2.13, a decrease of the supersaturation degree ($\gamma$) substantially increases stability of the supersaturated solution. The crystalline product obtained by using the IXISS technique was, as a rule, highly pure and well shaped, and could be easily separated from the supernatant containing soluble impurities. The IXISS technique was used for purification of amino acids from mineral salt admixtures [108]. It was also successfully applied for purification of L-glutamic acid from racemate admixture (DL-glutamic acid) [84,98].

An additional advantage of the IXISS technique in separation of impurities from low soluble amino acids must be emphasised.

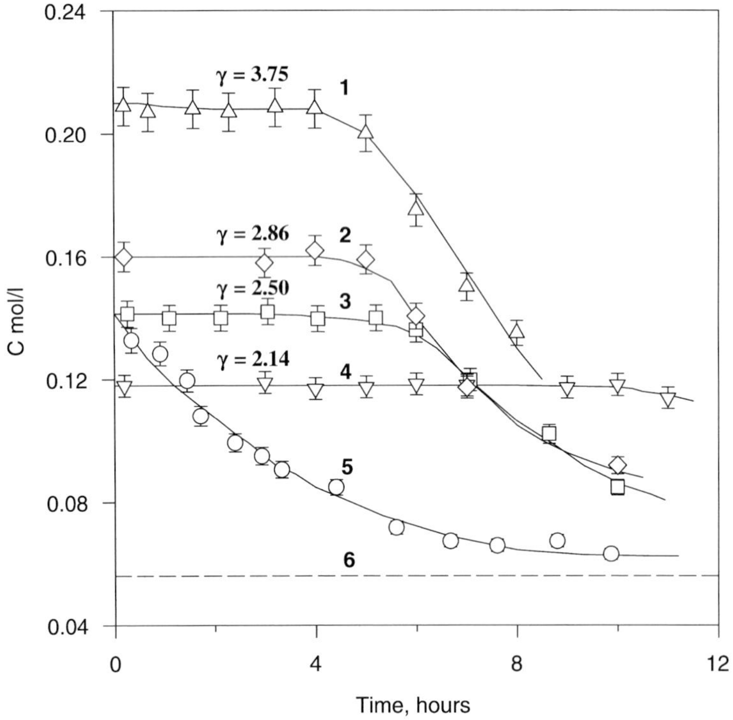


Fig. 2.13. Concentration-time histories (stabilities) of L-glutamic acid (L-Glu) solutions with different supersaturation degrees $\gamma$ (curves 1–4) in contact in the interstitial space of the column with an equivalent mixture (mixed bed) of strong acid and strong base ion exchangers in H- and OH-forms, respectively (1); sulfonate cation exchanger KU-2 × 8 in H-form (2,4), and carboxylic resin KB-4P2 in H-form (3) in comparison with L-Glu supersaturated solution removed from column (curve 5). $T = 293$ K. Dotted line (6) denotes solubility of L-Glu at $T = 293$ K. For other details see Ref. 96. Reprinted with permission from [D. Muraviev, *Langmuir*, 14 (1998) 4169]. Copyright (1998) American Chemical Society.

Their purification is usually complicated by the need to subsequently concentrate their purified solutions. The concentration of these solutions by evaporation at increased temperature may, in some cases, lead to side-reactions, resulting in recontamination of the purified product. For instance, evaporation of glutamic acid solutions at $T > 373$ K results in the formation of the respective lactame (pyrrolidon carboxylic acid) [109].

Since IXISS is a combination of ion exchange separation and a 'mild' concentration technique and can be applied even at low temperatures (around 273 K), it avoids these and similar complications and is recommended for separation, purification, concentration and recovery of temperature-sensitive compounds. Sequential integration of ion exchange separation and crystallisation within the IXISS method permits the recovery of highly pure crystalline products right after the ion exchange separation cycle.

### 2.4.4 Ion exchange supersaturation of electrolytes

Unlike zwitterlites, the ion-exchange supersaturation of electrolytes (e.g. inorganic) is observed by using the reverse frontal separation. In this case, the tailored application of the IXISS effect is based on the use of the IXISS-active stripping agents, which must meet the following requirements:

(1) The IXISS-active eluent must be bearing, on one hand, the desired counterion (displacer) to combine the desorption of the target component with the regeneration of the ion exchanger. On the other hand, it must also contain an appropriate co-ion to provide the formation of a stable supersaturated solution of a low-solubility compound (the product) and to shift the ion exchange equilibrium in the system to the desired direction. For example, the ion exchange reaction of displacement of a divalent metal ion, $M_1^{2+}$, by a monovalent one, $M_2^+$, from a carboxylic cation exchanger can be written as follows:

$$(\text{R–COO}^-)_2\text{M}_1^{2+} + 2\text{M}_2^+ \Leftrightarrow 2\text{R–COO}^-\text{M}_2^+ + \text{M}_1^{2+} \tag{2.26}$$

The equilibrium in reaction 2.26 is characterised by the equilibrium coefficient, $K_{M_2}^{M_1}$, which unlike Eq. (2.2) can be rewritten in absolute concentrations as follows:

$$K_{M_2}^{M_1} = \frac{Q_{M_2}(C_{M_1})^{1/2}}{(Q_{M_1})^{1/2}C_{M_2}} \tag{2.27}$$

where $C$ and $Q$ are the concentrations of ions in the solution and resin phases, respectively. For nitrate or chloride media, $K_{M_2}^{M_1}$ is usually $\ll 1$. If reaction 2.26 proceeds in a carbonate media, for example, when the resin in the $M_1$-form is treated with $M_2$ carbonate solution ($C_o$, mol/dm$^3$), it is coupled with the formation of $M_1$ carbonate, which can be described by the solubility product of $M_1CO_3$, $L_{M_1CO_3}$ :

$$L_{M_1CO_3} = C_{M_1^{2+}}C_{CO_3^{2-}} \tag{2.28}$$

If $M_1CO_3$ forms a stable supersaturated solution, where it exists in an associated (molecular) form at a concentration $C_M$ mol/dm$^3$ exceeding $\gamma$ times its solubility, $C_s$, at a given temperature, Eq. (2.28) takes the following form:

$$L_{M_1CO_3} = C_{M_1^{2+}}(C_o - C_M) \tag{2.29}$$

By introducing $C_M = \gamma C_S$, and after substitution of $C_{M_1^{2+}}$ from (2.29) into (2.27), one obtains:

$$K_{M_2}^{M_1} = \frac{Q_{M_2}(L_{M_1CO_3})^{1/2}}{(Q_{M_1})^{1/2}C_o(C_o - \gamma C_S)} \tag{2.30}$$

As follows from Eq. (2.30), at constant $C_o$, $C_S$, and $L_{M_1}CO_3$, $K_{M_2}{}^{M_1}$ increases with $\gamma$ and may reach a sufficiently high value ($\gg 1$), as $\gamma \rightarrow C_o/C_S$. Equation (2.30) can be rewritten in a more general form for the displacement of the divalent metal ion from the resin with an IXISS-active stripping agent bearing a monovalent counterion as follows [110]:

$$K_{Sss}^{Dis} = \frac{Q_{Dis}L_{Sss}^{1/2}}{(Q_{Sss})^{1/2}C_o(C_o - \gamma C_S)} \tag{2.31}$$

Here 'Dis' and 'Sss' superscript and subscript denote the displacer and the substance under supersaturation (the product), respectively; $C_o$ is the concentration of the displacer solution, $L$ is the solubility product of the target compound, $C_S$ is the solubility of the product at a given temperature, and $\gamma$ is the supersaturation degree of the product solution. Relationship 2.31 is the fundamental equation describing the shift of ion exchange equilibrium in IXISS systems of different types. On the other hand Eq. (2.31) helps one to understand that this shift of the equilibrium is the result of integration of ion exchange separation and supersaturation in a given IXISS system.

(2) The successful application of the IXISS effect requires, on one hand, maximum stability of the supersaturated solution in the interstitial space of the column during the ion exchange separation or purification cycle and, on the other hand, fast decomposition (crystallisation) of this solution after its removal from the column. In the case of inorganic substances a unified interpretation of the IXISS phenomenon must be based on general principles of the aggregative stability of dispersion systems, adapted to the particular ion exchange system.

(3) The following main factors may influence the stability of dispersions of precrystalline molecular aggregates in the interstitial space of a column: (1) an effective charge of the polymolecular aggregate (micelle), which is due to the sorption of either counter- or co-ions on the particle surface, and (2) the ionic strength of the media, which may strongly influence the coagulation (crystallisation) conditions. If, for example, an excess of co-ions exists in the interstitial space, the charge of micellae will be the same as that of the functional groups of ion exchanger. In this case the sorption of the micellae on the surface of the ion-exchanger beads becomes impossible and a stabilising action of the resin bed towards supersaturated solution can be expected. In contrast, in the presence of an excess of counterions the charge sign of the precrystalline aggregates will be opposite to that of the functional groups and fast decomposition of supersaturated solution will be observed due to the sorption of micellae on the surface of the ion exchanger followed by crystallisation of the component under supersaturation. The formation of precrystalline molecular aggregates in the interstitial space of an ion exchange column was indirectly confirmed by studying the dynamics of ion exchange in supersaturated solutions [111].

The high efficiency of separation, which is observed in a number of known IXISS systems, permits this integrated separation technique to be successfully used in different fields of science and technology [1,110,112–117].

## ACKNOWLEDGEMENTS

The author would like to thank all his co-workers mentioned throughout the text and cited in the various references for their assistance and efforts in making this publication possible. The Universitat Autònoma de Barcelona and Institut de Ciència de Materials de Barcelona are acknowledged with thanks for the financial support to the author during preparation of this manuscript.

## REFERENCES

1 V.I. Gorshkov, D.N. Muraviev and A. Warshawsky, Ion-exchange methods for ultra purification of inorganic, organic and biological substances, In: D. Muraviev, V.I. Gorshkov and A. Warshawsky (Eds.), *Ion Exchange, Highlights of Russian Science*, Vol. 1. Ch. 1. Marcel Dekker, New York, 2000.

2 D. Muraviev, V.I. Gorshkov and A. Warshawsky (Eds.), Special Issue on Stable Isotopes and Isotope Exchange. Highlights of Russian Science, *Sep. Sci. Technol.*, 36 (2001), 1596–2119.
3 I.D. Wilson, C.F. Poole, T.R. Adlard and M. Cooke (Eds.), *Encyclopedia of Separation Science*, Academic Press, London, 2000.
4 J.S. Watson, *Separation Methods for Waste and Environmental Applications*, Marcel Dekker, New York, 1999.
5 D.G. Peters, J.M. Hayes and G.M. Hieftje, *Chemical Separations and Measurements. Theory and Practice of Analytical Chemistry*, Khimia, Moscow, 1978 (Russian Edition).
6 W.J. Argensinger Jr., A.W. Davidson and O.D. Bonner, *Trans. Kansas Acad. Sci.*, 53 (1950) 404.
7 E. Ekedahl, E. Högfeldt and L.G. Sillén, *Acta Chem. Scand.*, 4 (1950) 556.
8 V.S. Soldatov, Thermodynamics of ion exchange, In: K. Dorfner (Ed.), *Ion Exchangers*, Ch. 6.1. Walter de Gruyter, Berlin, 1991.
9 D. Muraviev, *Solv. Extr. Ion Exch.*, 16 (1998) 381.
10 D. Muraviev, *Solv. Extr. Ion Exch.*, 18 (2000) 753.
11 D. Muraviev, Surface impregnated sulfonate ion exchangers: preparation, properties and application, In: D. Muraviev, V.I. Gorshkov and A. Warshawsky (Eds.), *Ion Exchange. Highlights of Russian Science*, Vol. 1. Ch. 7. Marcel Dekker, New York, 2000.
12 A. Warshawsky, Extraction with Solvent Impregnated Resins, In: J.A. Marinsky and Y. Marcus (Eds.), *Ion Exchange and Solvent Extraction*, Vol. 8. Ch. 3. Marcel Dekker, New York, 1981.
13 J.L. Cortina and A. Warshawsky, Developments in Solid–Liquid Extraction by Solvent Impregnated Resins, In: J.A. Marinsky and Y. Marcus (Eds.), *Ion Exchange and Solvent Extraction*. Vol. 13. Ch. 5. Marcel Dekker, New York, 1997.
14 A. Warshawsky, *Talanta*, 21 (1974) 962.
15 D. Muraviev, *Chemica Scripta*, 29 (1989) 9.
16 D. Muraviev, L. Ghantous and M. Valiente, *Reactive & Functional Polymers*, 38 (1998) 259.
17 W.S.W. Ho and K. Sirkar (Eds.), *Membrane Handbook*, Van Nostrand Reinhold, New York, 1992.
18 K. Scott, *Handbook of Industrial Membranes*, Elsevier, Oxford, 1995.
19 M. Mulder, *Basic Principles of Membrane Technology*, Kluwer Academic Publishers, Dordrecht, 1992.
20 T. Araki and H. Tsukube (Eds.), *Liquid Membranes: Chemical Applications*, CRC Press, Boca Raton, Florida, 1990.
21 R.D. Noble and J.D. Way, *Liquid Membranes: Theory and Applications*, ACS Symposium Series 347, ACS, Washington, 1987.
22 J.G. Wijmans and R.W. Baker, *J. Membrane Sci.*, 107 (1995) 1.

23 N.M. Kocherginsky, personal communication, National University of Singapore, 1998.
24 L. Bromberg, D. Muraviev and A. Warshawsky, *J. Phys. Chem.*, 97 (1993) 967.
25 D. Muraviev and A. Warshawsky, *Sep. Sci. Technol.*, 36 (2001) 2087.
26 B. Lindegard, L. Mathiasson, J.A. Jonson and B. Akesson, *J. Chromatogr.*, 293 (1990) 514.
27 N. Partharathy and J. Buffe, *Anal. Chim. Acta*, 254 (1991) 1.
28 N. Partharathy and J. Buffe, *Anal. Chim. Acta*, 284 (1994) 649.
29 G. Audunsson, *Anal. Chem.*, 58 (1986) 2714.
30 G. Nilve and R. Stebbins, *Chromatographia*, 32 (1991) 269.
31 M. Knutsson, G. Nilve, L. Mathiasson and J. Jonsson, *J. Agric. Food Chem.*, 40 (1992) 2413.
32 M. Oleinikova, Preparation and characterization of activated composite membranes containing organophosphorous compounds for separation of polyvalent metal ions, PhD Thesis, Universitat Autònoma de Barcelona, 2000.
33 M.J. Ariza, E. Rodríguez-Castellón, R. Rico, J. Benavente, M. Muñoz and M. Oleinikova, *J. Coll. Interface Sci.*, 226 (2000) 151.
34 M. Oleinikova, M. Muñoz, J. Benavente and M. Valiente, *Anal. Chim Acta*, 403 (2000) 91.
35 M. Oleinikova, M. Muñoz, J. Benavente and M. Valiente, *Langmuir*, 15 (2000) 430.
36 M. Oleinikova and M. Muñoz, *Solv. Extr. Ion Exch.*, 18 (2000) 401.
37 B. Tremillon, *Les Separations par les Resines Echangeuses d'Ions*, Ganthiers–Villar, Paris, 1965.
38 Y. Kofuchi and T. Deo, Electrochemical Manufacture of High Purity Ammonium Hydroxide Aqueous Solutions, Jap. Patent 62139890 (1987).
39 I. Lehms, R. Kaden, D. Mross, M. Knuepfer, R. Walter, L. Feistel and G. Schwachula, Method for Manufacture of High Purity Hydrofluoric Acid, Patent DDR 254373 (1988).
40 M. Knotze, P. Schlesier and E.V. Vebel, Manufacture of Hexachloroiridium (IV) Acid from Iridium-Containing Solutions, Patent DDR 274142 (1989).
41 A.M. Garbar, A.V. Loginov, G.L. Murskii, V.I. Rodchenko and V.G. Pimenov, *Vysokochistye Veshchestva (High Purity Substances)*, 2 (1989) 212 (Russian).
42 N.P. Nikolaev, V.A. Ivanov, V.I. Gorshkov, N.B. Ferapontov, T.E. Mitchenko, L.F. Postolov, L.N. Vishnetskaja, *Vysokochistye Veshchestva (High Purity Substances)*, 1 (1988) 75 (Russian). *High Purity Substances* (Engl. Transl.), 6 (1989) 1014.
43 T.E. Mitchenko, L.E. Postolov, I.A. Skripnik, V.G. Ovchinnikov and N.V. Makarova, *Khim. Tekhnol.* (Kiev), 1 (1992) 19 (Russian).
44 R. Voszka, K. Raksanyi and I. Foldari, *Kr. und Techn.*, 8 (1973) 1347.

45 M. Knotze, P. Schlesier and E.V. Vebel, Manufacture of High Purity Palladiumtetramine-di(hydrocarbonate), Patent DDR 288065 (1991).
46 F.H. Spedding, J.E. Powell, E.I. Fulmer and T.A. Butler, *J. Am. Chem. Soc.*, 72 (1950) 2354.
47 F.H. Spedding and J.E. Powell, *J. Am. Chem. Soc.*, 76 (1954) 2546.
48 F.H. Spedding, J.E. Powell and E. Wheelwright, *J. Am. Chem. Soc.*, 76 (1954) 2557.
49 L. Chen, W. Xin, C. Dong, W. Wu and S. Yue, High Pressure Ion Exchange Separation of Rare Earths, In: J.A. Marinsky and Y. Marcus (Eds.), l *Ion Exchange and Solvent Extraction*, Vol. 12. Ch. 1. Marcel Dekker, New York, 1995.
50 W. Brzyska, I. Soltysiak and J. Cydan, Preparation of High Purity Rare Earth Oxides by Ion-Exchange Chromatography, Pol. Patent 147137 (1989).
51 W. Brzyska, I. Soltysiak and J. Cydan, *Przem. Chem.*, 68 (1989) 510.
52 B. Brandel and W. Brzyska, *Przem. Chem.*, 70 (1991) 39.
53 D. Muraviev, J. Noguero and M. Valiente, *Anal. Chem.*, 69 (1997) 4234.
54 P.A. Kober, *J. Am. Chem. Soc.*, 39 (1917) 944. Reprinted in *J. Membr. Sci.*, 100 (1995) 61.
55 K.W. Lawson and D.R. Lloyd, *J. Membr. Sci.*, 124 (1997) 1.
56 S. Hwang and K. Kammermyer, *Membranes in Separation*, John Wiley & Sons, New York, 1975, pp. 99–123.
57 R. Singh, *Chemtech.*, April (1998) 33.
58 E. Favre, T.N. Quang and S. Bruneau, *J. Chem. Technol. Biotechnol.*, 65 (1996) 221.
59 J. Korkisch, *Sep. Sci.*, 1 (1966) 159.
60 J. Korkisch, *Nature*, 210 (1966) 626.
61 J. Korkisch, *Österr. Chem. Ztg.*, 67 (1966) 309.
62 J. Korkisch, *Microchim. Acta*, 4–5 (1966) 634.
63 J. Korkisch, *Handbook of Ion Exchange Resins: Their Application to Inorganic Analytical Chemistry*, CRC Press, Boca Raton, FL, 1989; Vol. 1, p. 88.
64 R. Kuroda and N. Hasoi, *Chromatographia*, 14 (1981) 359.
65 A.K. De and C.R. Bhattacharyya, *Anal. Chem.*, 44 (1972) 1686.
66 D. Henriet and P. de Gelis, *Chim. Anal.*, 50 (1968) 519.
67 U. Gerwien and R. Oberhauser, *Materialpruefung*, 20 (1978) 233.
68 K.A. Orlandini and J. Korkisch, *J. Sep. Sci.*, 3 (1968) 255.
69 T.A. Belyavskaya, I.P. Alimarin and G.D. Brykina, *Vestn, Mosk. Univ. Ser. Khim.*, 1 (1967) 53 (Russian).
70 J. Sherma and F. Van Lentan, *J. Sep. Sci.*, 6 (1971) 199.
71 F.W.E. Strelow, M.D. Hanekom, A.H. Victor and C. Eloff, *Anal. Chim. Acta*, 76 (1975) 377.

72 D. Muraviev, I. Rogachev, L. Bromberg and A. Warshawsky, *J. Phys. Chem.*, 98 (1994) 718.
73 D. Muraviev and A. Warshawsky, *React. Polym.*, 22 (1994) 55.
74 D. Muraviev, M. Oleinikova and M. Valiente, *Langmuir*, 13 (1997) 4915.
75 M. Oleinikova, D. Muraviev and M. Valiente, *Anal. Chem.*, 71 (1999) 4866.
76 I.S. Levin and V.V. Sergeeva, *Izv. Sib. Otd. Acad. Nauk SSSR, Ser. Khim. Nauk*, No. 3 (1974) 53 (Russian).
77 I.P. Alimarin, T.V. Rodionova and V.M. Ivanov, *Russ. Chem. Rev.*, 58 (1989) 863.
78 D. Muraviev and F. Omarova, In: *Applications of Ion-Exchange Materials in Industry and Analytical Chemistry, Proceedings of the National Conference*, VGU, Voronezh, Russia, 1986; Vol. 1, p. 21 (Russian).
79 A.V. Nikolaev, V.L. Bogatyrev and F.V. Zhurko, *Dokl. AN SSSR*, 200 (1971) 886 (Russian).
80 A.A. Tobolov, V.L. Bogatyrev and F.V. Zhurko, *Izv. Sib. Otd. AN SSSR, Ser. Khim. Nauk*, 12 (1976) 20 (Russian).
81 A.A. Tobolov and V.L. Bogatyrev, *Izv. Sib. Otd. AN SSSR, Ser. Khim. Nauk*, 12 (1976) 14 (Russian).
82 K. Bunzl and W. Schultz, *J. Inorg. Nucl. Chem.*, 43 (1976) 791.
83 D.A. Friedrichsberg, *Colloid Chemistry*, 2nd edn, Khimia, Leningrad, 1984, p. 325 (Russian).
84 D. Muraviev, S.A. Fesenko and V.I. Gorshkov, *Zh. Fiz. Khim.*, 54 (1982) 1567 (Russian).
85 D. Muraviev, *Zh. Fiz. Khim.*, 53 (1979) 438 (Russian).
86 D. Muraviev and V.I. Gorshkov, *Zh. Fiz. Khim.*, 56 (1982) 1560 (Russian).
87 V.F. Selemenev, L.A. Ogneva, G.V. Trubetskih and G.A. Chikin, *Zh. Fiz. Khim.*, 58 (1984) 2525 (Russian).
88 V.F. Selemenev, Z.P. Miroshnikova, L.A. Ogneva, I.I. Ermakova, D.L. Kotova and G.Yu. Oros, *Zh. Fiz. Khim.*, 59 (1985) 1992 (Russian).
89 D. Muraviev, A.D. Saurin, *Zh. Fiz. Khim.*, 54 (1980) 1271 (Russian).
90 V.F. Selemenev, A.A. Zagorodny, N.V. Polupanov and L.A. Ogneva, *Zh. Fiz. Khim.*, 60 (1986) 1461 (Russian).
91 V.F. Selemenev, D.L. Kotova, A.N. Amelina and A.A. Zagorodny, *Zh. Fiz. Khim.*, 65 (1991) 996 (Russian).
92 V.F. Selemenev, A.A. Zagorodny, V.A. Uglianskaya, T.A. Zavyalova and G.A. Chikin, *Zh. Fiz. Khim.*, 66 (1992) 1555 (Russian).
93 D. Muraviev, S.A. Fesenko, *Zh. Fiz. Khim.*, 56 (1982) 1960 (Russian).
94 D. Muraviev and O.N. Obrezkov, *Zh. Fiz. Khim.*, 60 (1986) 396 (Russian).
95 D. Muraviev, R.Kh. Khamizov and N.A. Tikhonov, *Solvent Extr, Ion Exch.*, 16 (1998) 151.
96 D. Muraviev, *Langmuir*, 14 (1998) 4169.
97 D. Muraviev, N.V. Drozdova, N.B. Dolgina and A.A. Karabutov, *Langmuir*, 14 (1998) 1822.

98 E.M. Savitskaja, P.S. Nys and M.S. Bulycheva, *Khim. Farm. Zh.*, 7 (1969) 32 (Russian).
99 E.M. Savitskaja, P.S. Nys and M.S. Bulycheva, *Khim. Farm. Zh.*, 7 (1969) 32 (Russian).
100 D.J. Greenland, R.H. Laby and J.P. Quirk, *Trans. Faraday Soc.*, 61 (1965) 2013–2024.
101 O. Talibudeen, *Trans. Faraday Soc.*, 51 (1955) 582.
102 V.F. Selemenev, L.B. Antakanova and D.L. Kotova, *Zh. Fiz. Khim*, 64 (1990) 1405 (Russian).
103 V.F. Selemenev, G.A. Chikanov, P. Fryolih, A.A. Zagorodny and A.F. Sholin, *Zh. Fiz. Khim*, 64 (1990) 3330 (Russian).
104 V.F. Selemenev, G.A. Chikin and V.Ju. Khokhlov, *Solv. Extr. Ion Exch.*, 17 (1999) 851.
105 V.F. Selemenev, G.A. Chikin and V. Ju. Khokhlov, Interionic and Intermolecular Interactions in Ion Exchange and Sorption Systems Involving Physiologically Active Substances, In: D. Muraviev, V.I. Gorshkov and A. Warshawsky (Eds.), *Ion Exchange. Highlights of Russian Science*, Vol. 1. Ch. 10. Marcel Dekker, New York, 2000.
106 Z. Baolin, S. Fenling and T. Zuy, *Chem. J. Chinese Univ.*, 13 (1992) 840.
107 T. Zuyi, Ion Exchange Equilibria of Amino Acids, In: J. Marinsky and Y. Marcus (Eds.), *Ion Exchange and Solvent Extraction*, Vol. 12, Marcel Dekker, New York, 1995, p. 353.
108 D. Muraviev, R.Kh. Khamizov and N.A. Tikhonov, Ion-Exchange Isothermal Supersaturation, In: D. Muraviev, V.I. Gorshkov and A. Warshawsky (Eds.), *Ion Exchange. Highlights of Russian Science*, Vol. 1. Ch. 3. Marcel Dekker, New York, 2000.
109 R. Powell, *Monosodium Glutamate and Glutamic Acid*, Noyes Development Corp., New-Jersey-London, 1968, p.153.
110 D. Muraviev, Ecologically Safe Ion Exchange Technologies, In: I.D. Wilson, C.F. Poole, T.R. Adlard and M. Cooke (Eds.), *Encyclopedia of Separation Science*, Academic Press, London, 2000.
111 D. Muraviev, R. Khamizov, N.A. Tikhonov and V.V. Kirshin, *Langmuir*, 13 (1997) 7186.
112 R. Khamizov, D. Muraviev and A. Warshawsky, Recovery of Valuable Mineral Components from Sea Water by Ion Exchange and Sorption Methods, In: J. Marinsky and Y. Marcus (Eds.), *Ion Exchange and Solvent Extraction. Series of Advance*, Vol. 12. Ch. 3. Marcel Dekker, New York, 1995.
113 R. Khamizov, L.I. Mironova, N.A. Tikhonov, A.V. Bychkov and A.D. Poezd, *Sep. Sci. Technol.*, 31 (1996) 1.
114 V.I. Gorshkov, D.N. Muraviev and A. Warshawsky, *Solv. Extr. Ion Exch.*, 16 (1998) 1.

115 D. Muraviev, R. Khamizov, N. Tikhonov, A. Krachak, T. Zhiguleva and O. Fokina, *Ind. Eng. Chem. Res.*, 37 (1998) 1950.

116 R. Khamizov, D. Muraviev, N. Tikhonov, A. Krachak, T. Zhiguleva and O. Fokina, *Ind. Eng. Chem. Res.*, 37 (1998) 2496.

117 D. Muraviev, J. Noguerol, J. Gaona and M. Valiente, *Ind. Eng. Chem. Res.*, 38 (1999) 4409.

*Chapter 3*

# *Solid-phase spectrometric assays*

***Luis Fermín Capitán-Vallvey*** *
***and María Dolores Fernández Ramos***

## 3.1 INTRODUCTION

Solid-phase spectrometric assays are a group of optical batch analytical procedures based on the integration of four analytical operations: derivation, preconcentration, separation and measurement. These procedures are carried out using an insoluble solid phase that may contain one or more reagents immobilized by chemical or physical procedures and which retains the analyte or a derivative product from a solution. Subsequent measurement of the appropriate optical property conducted over this same solid support makes it possible to determine the concentration. Thus, the need to desorb the analyte or its derivative before measurement is eliminated and the procedure is simplified.

Within solid-phase spectrometric assays, two different strategies have been employed. The first consists of the use of powdered solid supports with small-sized particles which are placed in contact with the 'problem', usually in solution, and to which are added the reagents needed for derivation, elimination of interferences, adjustment of conditions (pH, ionic strength, solvent, etc.) and retention on the solid phase. The collection of this solid and the direct measurement of an

* Corresponding author. Tel.: +34-958248436; fax: +34-958243328. *E-mail address:* lcapitan@ugr.es.

*Comprehensive Analytical Chemistry XXXIX, pages 81–159* 

ISSN: 0166-526X

optical property, usually absorbance, diffuse reflectance or fluorescence, make it possible to then quantify the analyte. These assays show good preconcentration factors and, consequently, good sensitivity and detection limits, with an acceptable reproducibility and equilibration times that are generally not very long. They use cheap and conventional instrumentation for analysis at a μg- or sub-μg/L level and do not use solvents. Additionally, the sensitivity of the procedures can be adjusted by modifying the volume of the sample being used. Compared to conventional methods they can be slower and require, on occasion, specifically prepared solid phases.

The second approach uses membranes on which the analyte or a derivative is retained and which may contain none, some or all of the necessary reagents. When placed in contact with the 'problem' via filtration, immersion with or without agitation, moistening or the deposit of a volume of the problem, a set of reactions and processes is unleashed that generates an optical response that is measured on the membrane itself. This response can manifest itself in diffuse reflectance, absorption or luminescent emission.

The complexity of the system differs depending on the degree of simplification of the analytical procedure that it contains. With the most basic systems, the membrane is simply made up of a polymer, such as PVC or cellulose, capable of retaining the analyte or a derivative. This requires all of the necessary reagents to be added to the problem solution. It is possible to achieve a greater degree of integration by preparing membranes that contain only some of the reagents incorporated in them, which at the same time simplifies the operative procedure. However, it is only when all of the necessary reagents are integrated in one or more layers in a reactive zone that it is possible to develop rapid analysis methods for specific problems. Assays of this nature are called strip tests.

With this second type of assay, the simplification that the procedure offers is usually more important than the preconcentration that it achieves, which is to say, its sensitivity. This occasionally results in the inconveniences of providing only semi-quantitative information and a short shelf life.

The two strategies—powdered supports and membranes—share the same basic concept and both are capable of being automated, though in different forms. In the first case, the solid supports are packed into flow-through cells and use FIA-type systems. The flow-through sensors that result are treated in the next chapter of this book. Membranes, on the other hand, have been used with automated systems that make it possible to deposit samples of a fluid on multi-parametric test strips and measure the diffuse reflectance.

## 3.2 INTEGRATION OF PROCESSES IN SOLID-PHASE SPECTROMETRIC ASSAYS

Solid phase spectrometric assays use a solid phase as an axis that makes it possible to integrate different analytical operations. Though the forms they take may differ significantly, with all of these assays, the measurement is carried out on a solid phase.

The degree of integration differs among the various types of existing solid phase assays, which results in different general schemes. The most basic is that of specific retention and is based on an analytical property intrinsic to the analyte. The solid phase can take the form of a finely divided solid, like a filter or a polymeric membrane, which makes it possible to separate this analyte under a determined chemical form, such as a molecular or ionic form of an organic compound or a metallic ion-like hydroxo- or aquocomplex. It may be necessary to adjust the conditions that make retention possible, such as pH, solvent, or surfactant agent. The retention can be carried out by equilibration or filtration of the problem with or through the solid phase, respectively.

Specific retention of a derivative of an analyte generated in solution on the solid phase is the scheme most commonly used to produce an analytical property. In this case, the derivative and auxiliary reagents are added to the solution, with several different operational forms being possible (see below).

The use of modified solid phases, either by covalent immobilization of the reagents, by adsorption or by physical trapping, makes it possible to integrate the derivation and separation stages. As above, it is possible to use particulated solid phases such as chelating resins or modified resins, filters like modified cellulose, or polymeric membranes like PVC which contain a recognition element such as ligands or ion carriers. From here, it is only necessary to adjust the experimental conditions of the problem solution, especially the pH, so that, by equilibration or filtration, it produces the derivative in the solid phase where it has been retained.

One last scheme integrates not only the derivatizating reagents but also all of the auxiliary reagents needed to produce the analytical signal, such as buffers or surfactants, all of which are usually integrated into a mono- or multi-layered membrane. In addition to making the chemical or biochemical reaction possible, the membranes may incorporate elements to separate interferents, like a dialysis membrane, or masking agents or some means of improving measurement, like a reflectant surface. In this way it is only necessary to place the problem in contact with the analytical element, by immersion, passing it through or deposit. The most common form for this last type of system are test strips, which are analytical formulations for solution analysis, in which all the reagents needed for analysis are present in a dry state on or in absorptive or swelling carriers or films and are mounted in the form of devices, so-called test carriers such as test strips or slides [1].

Nevertheless, other types of device do exist, such as indicator tubes filled with indicator powders based on modified hydrophobic silica, through which the problem passes; these are used for gasses, aerosols or solutions [2].

Test strips can be considered as single-use sensors or drop planar sensors. They are miniature analytical devices that respond to the presence or concentration of an analyte. However, unlike most sensors, they do not work in a fully reversible manner or operate in a continuous fashion [3].

The degree of integration of the measurement stage in these batch assays is different. In the procedures that use particulated supports, the solid must be collected after the reaction and retention in a form that makes it possible to measure the optical property in a conventional spectrometer. With more simple membrane procedures, the optical measurement is measured after the reaction either with the help of conventional systems or with hand-held dedicated devices. However, in other cases it is possible to integrate the measurement stage with the help of an automatic system that makes it possible to situate the disposable test strips, usually multi-analyte, add the problem, for example, urine, and after the reaction raise the strip to the measurement compartment. This is the type of system used in clinical analysis.

## 3.3 TYPES OF SOLID-PHASE SPECTROMETRIC ASSAYS

Given the great variety of existing assays, it is difficult to cover them with a single classification; thus this section presents several classifications according to various criteria related to solid phases, the reaction that takes place and the measurement.

(1) Type of solid phase: (a) ion exchangers; (b) sorbents, both hydrophilic and hydrophobic; (c) modified supports.
(2) Form of the solid phase: (a) unstructured (powdered); (b) structured (monolayer membranes, multilayer membranes).
(3) Form of placing the sample in contact with the solid phase: (a) by immersion, equilibration by agitation (vibration, rotation) or without agitation; (b) by filtration; and (c) by deposit of a small volume of the problem on the solid phase.
(4) Process that takes place: (a) selective separation; (b) chemical or biochemical reaction. This is related to the property of the measured analyte: intrinsic or extrinsic.
(5) Type of optical property measured: (a) UV–Vis molecular absorption; (b) diffuse reflectance; (c) photoacoustic absorption; (d) luminescent emission (fluorescence, phosphorescence); (e) atomic absorption; (f) atomic emission; (g) visual comparison; (h) other.
(6) Number of determined species: monoparametric, multianalyte.

## 3.4 FEATURES OF SOLID-PHASE SPECTROMETRIC ASSAYS

The properties of solid phase assays relate to the quality of results, the assay operation and the reliability of the assays. The quality of the results refers to their accuracy, precision, sensitivity and selectivity. The precision is usually limited by the measurement of the analytical property on the solid phase. The sensitivity is usually high for the assays that use a particulated solid phase due to the preconcentration process that they use and is less for those that use a membrane. The selectivity depends as much on the solid phase as on the derivatizating reactions used; this can be improved with chemometric techniques.

The second group of properties relates to the assay operation. The analysis time, typically between 15 and 60 min, is usually better with assays that use particulated solid phases because of the need to reach equilibrium in the reactions and implied processes and thus attain a good preconcentration factor. With test strip assays, the response time may be very short, in all cases less than 5 min and typically just a few seconds; if this were not the case, their applicability would be greatly compromised. The stability of the analyte or of the derivative in the solid phase is usually adequate (hours, days) and on occasions very high (months), which makes its measurement easier. The stability of the system before its use for the test depends on the level of integration and the type of reagents that it uses. This property—shelf life—is very important, especially with test strips, and is key in determining their introduction into the marketplace.

The third group of properties relates to reliability, especially simplicity, flexibility, ruggedness and cost. The simplicity of the assay depends on the degree of integration, with greater integration meaning greater simplicity. In the case of the assays that use a particulated solid phase, simplicity refers primarily to the instrumentation that is used, while with the test strips it can even refer to the simplicity of sample treatment, since this tends to be null. Flexibility, understood to be the capacity to adjust the assay to the problem, is greater in the particulated solid phase assays than in the test strips. Depending on the concentration level of the problem, it is possible to modify the volume of the problem or the mass of the solid phase being used. Ruggedness and low cost are very important in dedicated systems such as test strips.

## 3.5 PARTICULATED SOLID-PHASE SPECTROMETRIC ASSAYS

The existence of a linear relationship between absorbance of a particulated solid phase and concentration of an analyte in the solution problem was suggested by Yoshimura et al. in 1976, who called this ion-exchanger phase absorptiometry [4], though its origins go back to the 1920s with the so-called spot tests. From the beginning, the most important studies about spot tests on gels were performed by Fujimoto, Kakihana, Kato, Murase and Koblyanski, who on several occasions started with studies carried out by Feigl and employed the traditional methodology described for these techniques. The well-known books of Feigl [5,6] review the most useful applications of resin spot tests.

Fundamentally, these spot tests use a small amount of colorless ion-exchanger or adsorbent upon which a drop of the problem is placed, along with the appropriate reagents needed to adjust the experimental conditions. The appearance or enhancement of color in the solid phase is the most common indicator of the presence of the species being tested for. Although visual observation of color in the solid phase is most common, measurement of the intensity of the color can provide, in some cases, a semiquantitative way to determine the analyte.

These resin spot tests have several advantages with respect to tests in solution: (1) a substantial increase in sensitivity as a consequence of the concentration process; (2) high selectivity from selective fixation of the analyte or a derivative in the resin; (3) occasional high stability of the color produced; and (4) the possibility of using derivatives that are only formed in the solid support.

Solid phase techniques can be performed by batch or in-flow methods [7–9]. Initially, solid phase spectrometry was devised as a batch technique, but its confluence with present-day flow injection techniques creates a fruitful integration of pre-concentration, reaction and detection steps. Here only solid phase spectrometry batch techniques are considered.

In the case of batch solid phase techniques, the analytical process is determined by the following aspects: fixation of the analyte or analytes on the solid support by means of an equilibration process, measurement of the optical property directly on the solid phase, and the treatment of data in order to obtain the best calibration set. Obviously, all the experimental parameters that have a role in the fixation process can potentially influence the value of the analytical signal and, consequently, must be controlled and optimized before proceeding to the metrological step.

### 3.5.1 Fixation process

Three aspects should be considered in this process: the nature of the solid phase, the nature of the analyte/s and the interaction between analyte, solid support and reagents, if applicable.

*Solid phases*

Different inorganic and organic solid particles of diverse granulometry and composition have been used as solid support. These could be classified as ion exchangers, sorbents and modified sorbents.

A large variety of synthetic polymers have been used as ion exchangers. Though not common, inorganic ion-exchangers have been used; as with tin(IV) antimonophosphate in $H^+$ form for the retention and determination of Cu(II) [10]. Organic polystyrene type ion exchangers cross-linked with divinylbenzene hold a hydrophobic matrix and ionic groups: cationic ($-SO_3^-$, $-COO^-$ or $PO_3^{2-}$) or anionic ($-NH_3^+$, $-NRH_2^+$ or $-NR_3^+$). Among exchangers of this commonly used styrene type are the anionic resins Dowex 1-X8 and Dowex 1-X2 of 50–100, 100–200 and 200–400 mesh in cycle $Cl^-$, $NO_3^-$ or $SO_4^{2-}$ or the macroreticular Amberlyst A-27 and A-26 all in cycle $Cl^-$; and among the cationic ion exchangers Dowex 50W-X2 or -X4 with 100–200 or 200–400 mesh in $H^+$ or $Li^+$ form, or the macroreticular Amberlyst 15 in $H^+$ or $Na^+$ form.

The above exchangers were once used often, but currently dextran type gels are the most commonly used because their optical properties are better. These exchangers are formed by chains of polydextran cross-linked with epichlorohydrin (Sephadex), containing both anionic diethylaminoethyl (DEAE) and diethyl[2-hydroxy-propyl] aminoethyl (QAE) groups in chloride form, or cationic carboxymethyl (CM) and sulfopropyl (SP) groups in $H^+$ form. In all cases the particles may be of different sizes as indicated by the numbers 15, 25 or 50. Other ion exchanger materials, such as protonated chitin, have also been used [11].

The most common hydrophilic sorbents are polydextran type, which contain a large number of hydroxyl groups that give the gels hydrophilic properties and come in several particle sizes (A-15, A-25 and A-50). In the case of the hydrophobic sorbents, C-18 silica (typically 40–63 μm, usually < 100 μm) is the most common, but a less hydrophilic hydroxypropyl derivative of dextran gel (Sephadex LH) and even paraffin have been used.

The most common modified sorbents are chelating sorbents either with grafted groups or loaded with chelating ligands [12]. Among those included in the covalently bound group are styrene polymers with iminodiacetic groups used for Cu(II), hydroxamate acrylic polymer [13] or silica [14] both used for Fe(III), *N*-2-aminoethyl-3-aminopropyl-trimethoxysilano immobilized on silica gel for Cu(II) and Fe(III) [15]. An inclusion sorbent containing β-cyclodextrin units has been used for retention and direct determination of vanillin [16]. Much more common has been the sorption of chelating ligands on ion exchanger materials or on adsorbents. Usually the level of modifier Q must be low ($10^{-5}$–$10^{-7}$ mol/g) to avoid a significant increase in the background signal. The preparation of modified sorbents is carried out starting with diluted solutions under static procedure and in defined conditions. Conventional ion exchange materials are used for preparing modified sorbents [17], e.g. EBT (0.23 mg/g sorbent) [18], TAN3,6S ($5 \times 10^{-7}$ mol/g) [19], XO ($1 \times 10^{-6}$ mol/g) [20] all on AV-17X8, or Zincon on Dowex 1-X2 [4]. Here, the immobilization of organic reagents takes place by both ion exchange and hydrophobic interaction of condensed aromatic rings of reagents with the matrix [12]. Fixation of the inorganic reagents on resins such as Dowex 1-X8 in a molybdate cycle has also been used, which makes it possible to form heteropoly compounds on the resin [21,22]. Another method is the immobilization of chelating reagents by irreversible inclusion in β-cyclodextrin polymers, as with 5BrPADAP [23].

Adsorption on silica gel also makes it possible to prepare modified sorbents as with macroporous silica gel (200–350 μm) PAR ($6 \times 10^{-6}$ mol/g) [24]. Another method is the incorporation of reagents such as PAN (0.42–16.7 μmol/g) into silicic acid xerogels [25]. Adsorption of chelating reagents on hydrophobic sorbents is also possible, and PAN ($1 \times 10^{-6}$ mol/g), CV ($17 \times 10^{-6}$ mol/g) or Cadion ($2.5 \times 10^{-6}$ mol/g) have been retained on C-18 silica in this manner [2].

*Sorption mechanisms*
Interaction between the analyte and particulated solid phase can be accomplished in several ways.

*a. Retention of the analyte itself.* Retention by ion exchange has been used for organic compounds that intrinsically display an adequate analytical property. The presence of ionizable groups in the molecule and the adjustment of experimental conditions make it possible to fix them on ion exchangers and to attain the separation of other compounds in solution. Thus it is possible to retain the colorant SYFCF on the anionic exchanger SDEAE A-25 in the presence of its non-sulfonated homologe Sudan I which accompanies it as an impurity [26]. Compounds of a basic character can be retained on cation exchangers, as is the case with sulphatiazole on SSP C-25, which does not retain other sulphonamides like sulphanilamide, sulphametoxazole or sulphaquinoxaline [27].

Fixation of molecular species by adsorption can be achieved using both hydrophilic and hydrophobic sorbents. Depending on the sorbent selected, it is possible to retain groups of PAH. PAH of lesser molecular weight are retained better on SG-15, while starting with anthracene they are retained on C-18 silica, with the exception of crisene and benzo[a]anthracene. The retention of aromatic species on dextran gels probably involves the π-electron system of the solute and a deficient electron or electronegative portion of the gel. Additionally, when the crosslinking of the dextran gel (as with SG-25) is reduced, the affinity for the aromatic species decreases; this is justified by an increase in the residual water in the polymeric matrix [28].

*b. Sorption of a derivative of the analyte.* The formation of complexes, colorants, ion-pairs or other derivatives with measurable optical properties can be carried out in various ways.

*b.1. Formation of complexes*

*b.1.1. Derivatization in solution.* This procedure is useful when the derivatizating reaction is specific enough to determine the analyte of interest and the reaction product is retained in the solid phase.

Retention of addition complexes is one of the ways used to derivatize metallic ions. In general, in the presence of ion exchangers there is a displacement in the equilibrium towards highly coordinated and charged complexes because of the peculiar nature of the internal solution of the ion exchanger [29] (see Table 3.1). This usually means an improvement in the optical characteristics of the complexes that are formed.

The fixation of chelate complexes formed in solution is much more frequently used. Fixation in these cases depends on the charge, hydrophobicity and configuration of the compounds. These chelates are retained because: (a) the complex itself is charged, i.e. Fe(II)–Phen in Dowex 50W-X2 [4] or Cu(I)-BC in Diaion PK228 [42]; (b) the ligand has charged auxiliary groups, as with carboxylate, i.e. CA [43] or BCA [44], or sulfonic, i.e. FZ [35] or NCPA [45]; (c) the complex becomes charged upon coordination of an addition ligand present in the solution, i.e. Cu(II)-PAR-$Cl^-$ [46] or Be(II)-2-hydroxy-3-naftoic acid-$OH^-$ [31]. At times it is necessary to adjust the state of oxidation in the metallic ion, by reductions, i.e. iron [35] or copper [44] or oxidations, i.e. manganese [47]. Table 3.1 presents some examples of systems that have been studied, as well as their stoichiometry in resin phase.

In addition to electrostatic interactions, it is necessary to consider hydrophobic interactions between the complex and the resin matrix. On certain occasions, even the retention of charged complexes in resins of ionic exchange can be exclusively due to hydrophobic interactions. This is the case with the 1:2 complex of Cu(I) with BC which is retained in strongly and weakly basic anion exchange resins despite having a positive charge. The 1:3 Cu(II) complex with the same ligand is also retained, although more slowly, due to its octahedral configuration with two positive charges and a less favorable hydrophobic interaction [42].

TABLE 3.1

Derivatization of metallic ions

| Analyte | Reagent | Derivative | Solid phase | pH | $D$ (L/kg) | Type | Reference |
|---|---|---|---|---|---|---|---|
| Co(II) | $NH_4SCN$ | $Co(SCN)_{x>4}^{(x-2)-}$ | Dowex 1-X2 Amberlyst A-27 | 4 | – | AC | [4,30] |
| Bi(III) | HCl | $BiCl_5^{2-}/BiCl_6^{3-}$ | Sephadex QAE A-25 | 0.2 M HCl | $2.8 \times 10^3$ | AC | [31] |
| U(VI) | $NH_4SCN$ | – | Sephadex QAE A-25 | 0.1 M HCl | $4 \times 10^3$ | AC | [32] |
| Al(III) | CA S | $AlL_2^{x-}$ | QAE A-25 | 4.6 | $2.34(\pm 0.06) \times 10^4$ | CC | [33] |
| V(V) | 5BrSHA | $VO_2L_2^-$ | QAE A-25 | 7.0 | $3.8(\pm 0.7) \times 10^5$ | CC | [34] |
| Fe(III) | FZ | $Fe(II)L_3^{4-}$ | QAE A-25 | 3.0 | $5.9(\pm 0.1) \times 10^4$ | CC | [35] |
| Co(II) | NRS | $CoL_3^{6-}$ | QAE A-25 | 7.7 | $3 \times 10^5$ | CC | [36] |
| W(VI) | PV | $WO_2(HL)_2^{4-}$ | QAE A-25 | 2.1 | $1.15(\pm 0.07) \times 10^6$ | CC | [37] |
| Be(II) | ATCA/EDTA | $Be_2YL^{x-}$ | DEAE A-25 | 5.5 | – | TC | [38] |
| Zr(IV) | $XO/F^-$ | $ZrFL_4^{x-}$ $ZrFL_2^{x-}$ | Dowex 1-X8 | 1.9 | – | TC | [39] |
| Ti(IV) | $CRT/C_2O_4^{2-}$ | $TiOC_2O_4L^{4-}$ | Dowex 1-X8 | 2.0 | $1.12 \times 10^6$ | TC | [40] |
| U(VI) | $PAR/F^-$ | $UO_2FL^-$ | Dowex 1-X8 | 8.0 | – | TC | [41] |

AC: addition complex; CC: chelate complex; TC: ternary complex.

For a given ligand, the actuation form is different depending on the experimental conditions and on the metallic ion. Thus, aluminum is fixed on cationic SSP C-25 as a 1:1 cationic complex with morin at pH 5.3, while morin, which acts as an anion at this pH, is not retained due to Donnan exclusion [48]. On the other hand, beryllium at pH 11.5 is fixed as a 1:1 complex on anionic gel SQAE, through a different mechanism that involves the previous immobilization of anionic morin and subsequent preconcentration of beryllium [49].

The formation and fixation of ternary species on ion exchangers makes it possible to increase the sensitivity of the determinations by a hyperchromic effect, and the selectivity by masking agents, thus offering a way to determine auxiliary ligands. Table 3.1 presents some examples.

The use of hydrophilic sorbents can offer good results on certain occasions. It is, for example, possible to jointly fix Al and Be complexes with morin on SG-15, since they are retained in cationic and anionic exchangers, respectively, as shown above [50]. Another example is found in the blue Al–ECR complex, which is strongly retained on cellulose while the reagent ECR remains in solution [51], or the retention of the Ni–DMG complex on cellulose or silica gel [52]. Hydrophobic sorbents have been used for the retention of some complexes, as with the Be–ECR complex retained on C-18 silica [53] or the sorption of the ionic pair formed by $AuCl_4^-$ and Rhodamine S in paraffin granules [54].

*b.1.2. Preseparation and derivatization on solid phase.* For reactions with poor selectivity, selective retention of the analyte on the solid phase and subsequent derivatization with an adequate reagent can be performed. This happens with cadmium preconcentrated as the iodide complex $CdI_4^-$ in anionic Dowex 1-X2 and then derivatized with PAR, while Fe(III), Mn(II), Co(II), Ni(II) and others do not react in this way [55].

*b.1.3. Derivatization in solid phase.* This entails the previous immobilization of the reagent by ion exchange, adsorption or covalent immobilization in the solid, and the subsequent sorption and reaction (usually complexation) of the analyte in the solid phase (Table 3.2). The solid phase here is a chelating sorbent and has several advantages. It can be used when the complex cannot be sorbed from the solution and the reagent is readily fixed, the reagent consumed is less than that in solution ($10^{-5}$–$10^{-7}$ mol/g) and the influence of ionic strength of the sample is lower in solid phase than in solution [56].

TABLE 3.2

Derivatization in solid phase

| Sorbed reagent | Solid phase | Fixation mechanism | Particle size | Analyte | $Q$ | Reference |
|---|---|---|---|---|---|---|
| AEAPS | Silica gel | Covalent | 60–200 mesh | Cu(II) Fe(III) | 0.75 mmol/g | [58] |
| 5BrPADAP | β-cyclodextrin polymer | Inclusion | 60–80 mesh | Mn(II) | 17.5 μmol/g | [23] |
| DMG | Microcrystalline cellulose | Adsorption | 100–250 μm | Ni(II) | – | [59] |
| EBT | AV-17X8 | Ion exchange | 0.10–0.25 mm | Zr(IV) | 0.23 mg/g | [18] |
| NRS | Amberlite CG-400 | Ion exchange | 100–200 mesh | Co(II) | 1.54 mmol/g | [60] |
| PAN | Dowex 50W-X2 | Ion exchange | 100–200 mesh | Ni(II) | – | [61] |
| PAN | Xerogel silicic acid | Adsorption | 0.25–0.50 mm | Co(II) | 16.7 μmol/g | [25] |
| TAAP | AV-17X8 | Ion exchange | 0.10–0.25 mm | U(VI) | 1 μmol/g | [62] |
| TAN3,6S | AV-17X8 | Ion exchange | 0.10–0.25 mm | Cu(II), Pd(II) | 0.5 μmol/g | [19] |
| XO | AV-17X8 | Ion exchange | 0.10–0.25 mm | Zr(IV) | 1 μmol/g | [20] |

$Q$: quantity included in solid phase.

Usually, the composition of species formed in solid phase is the same as in solution, but the stability of complexes in the sorbent phase is lower than in solution due to their deformation [56–58].

The fixation and retention of metal ions by modified ion exchangers depends on the composition of the complex being formed and its stability. The formation of 1:1 complexes in solid phase does not require any additional expenditure of energy to deform the polymer matrix, and the sorption of the metal is therefore high. Sorption is considerably lower when two or more ligand groups intervene in the formation of the complex, since this entails additional energy costs to deform the matrix in solid phase. For this reason, the formation of weak complexes is not used with modified ion exchangers [63]. On certain occasions, the retention of metallic ion on a modified sorbent is not due to the reaction with the complex agent retained, as with the retention of uranium as a carbonate complex on anion exchanger AB-17 X8 modified with TAAP, where only the subsequent adjustment of pH to 6 makes it possible to form the complex with the immobilized reagent [62].

The addition of colloids during the retention process increases the analytical signal, as when the fixation of Cu(II) on the anionic exchanger AV-17 modified with TAN3,6S increases the signal by 20% in the presence of pentone or gelatin, which is explained by the more complete sorption of the metallic ion [19].

*b.2. Formation of heteropoly compounds*

The formation of heteropoly compounds and their subsequent reduction to molybdenum blue species has been used for the derivation of phosphate and silicate. Anion exchangers such as Amberlyst A-27 [64] or SDEAE have been employed as solid phase, using molybdate in solution alongside a reductor, usually antimonyl potassium tartrate, $SnCl_2$ or ascorbic acid; ion exchangers have also been used in molybdate cycle [21,22]. The retention of reduced blue species in hydrophilic adsorbents like SG-25 is also possible; here it should be observed that the distribution data fit a Langmuir-type isotherm for both phosphate and silicate [65,66].

*b.3. Redox reactions*

The redox nature of an analyte has been used to generate species with analytical properties adequate to fix or form them on the solid phase. The formation of intensely colored complexes with metallic ions in a low state of oxidation offers a way to determine ascorbic acid based on the reduction of Fe(III) in the presence of FZ and the subsequent fixation of the ferrous complex on SQAE A-25 [67]. Similarly, stable and reversible xerogels modified with a mixture of series 12 and 18 phosphomolybdic heteropoly acids (Vavale reagent) (60 mg/g) have been used to derivatize ascorbic acid and hydrazines to molybdenum blue species [68]. The use of redox indicators, as with tetrabenzotetraaza-cyclohexadecine as a copper complex adsorbed on silica gel (5 μmol/g) for ascorbic acid and analgin, has also been proposed [69]. Another method is the fixation of the generated redox product as an ion pair in solid phase, i.e. selenium (IV) derivatization through oxidation of iodide and fixation of triiodide as ion pair with Rhodamine B in a dextran type lypophilic gel, with a $D$ value of $2.00(\pm 0.26) \times 10^5$ [70].

*b.4. Organic reactions*

Different types of reactions have been employed. Most common is the formation of colored compounds, i.e. hydrogen sulfide derivatization through methylene blue formation from *N,N*-dimethyl-*p*-phenylenedia-mine and Fe(III) and fixation on a mixture of Amberlyst A-27 and 15 [71] or sulfur dioxide or sulfite through pararosaniline and formal-dehyde reaction and sorption of pararosanilinemethylsulfonic on Dowex 1-X8 [72,73].

The formation of azo dyes has been used for different ends, as with the determination of nitrites with sulfanilamide as a substratum for diazotation and *N*-(1-naphtyl)ethylendiamine as a coupling agent and Dowex 50W-X2 as a solid phase [74] or a mixture of anionic and cationic resins (Amberlyst A-27 and 15) [75]. Xerogel modified with 1-naphthol has also been used for aromatic amines [76]. At times, it is necessary to couple a redox reaction with the organic reaction to attain the species of interest; this occurs with the derivatization of formaldehyde through reaction with 4-amino-3-hydrazino-5-mercapto-1,2,4-triazole and sub-sequent oxidation with potassium periodate and fixation on a mixture of Amberlyst A-27 and 15 [77]. Another possibility is the formation of complexes with metallic ions. At times, a given reagent that participates in the reaction is already a complex and via a multi-step process generates a colored product that is retained on the solid phase, e.g. acetoacetate transformation with nitroprusside and glycine and sorption on Dowex 1-X8 [78]. In other situations, the process is carried out in various stages that involve the formation of a ligand, subsequent complexation and retention on the solid phase. This is the case with aromatic esters that generate the corresponding hydroxamic acid and then react with V(V) to form a complex that is fixed on Dowex 1-X8 [79,80].

The hydrolysis of organic compounds makes it possible to attain products that can be fixed on the solid phase and display good analytical properties. This is the case with the pesticide carbaryl hydrolyzed in an alkaline medium to fluorescent 1-naphthol that is fixed on SQAE A-25 [81,82].

*b.5. Ion pair formation*

The formation and fixation of ion pairs can be used for derivatization. Surfactants such as lauryl sufate can be efficiently extracted to lipophilic Sephadex LH in the presence of ferroin complex [83] and polyethers, such as dicyclohexyl-18-crown-6, can be preconcentrated on silica gel with Orange II as a counter ion [84].

### 3.5.2 Operational aspects

Several systems have been used to concentrate the analyte in the solid phase and subsequently measure the optical property.

*Equilibration and collection of the solid phase*

Usually, the transference of an analyte or its derivative to a solid phase is done under static conditions using equilibration with the solution containing the analyte and all of the necessary reagents. To speed the process up, the solution is usually agitated either by magnetic stirring in a glass or tube or by shaking in a rotating agitator in polyethylene or glass bottles. The time needed to achieve equilibration depends on several factors: agitation rate, solution volume, and particle and pore size of the solid phase, though the usual time ranges between 5 and 60 min. A different kinetic is observed depending on the type of process that takes place, such that when there is no derivation reaction, the kinetic is usually quite fast, while in the opposite case, it is slower. This could be because the controlling step is the reaction in solution or, in other cases, because of the slowness of the internal diffusion of the derivative, usually a complex, in the particles (Fig. 3.1).

In general, a reduction in the degree of crosslinking is observed with ionic exchanger materials or with polydextran type sorbents, which increases the process rate. With silica gel modified with reagents, it has been suggested that the reaction be diffusion-controlled by a mixed diffusion mechanism with a greater contribution of inside diffusion. The increase in pore size increases rate constants of both outside and inside diffusion, that is, it increases the reaction rate [85].

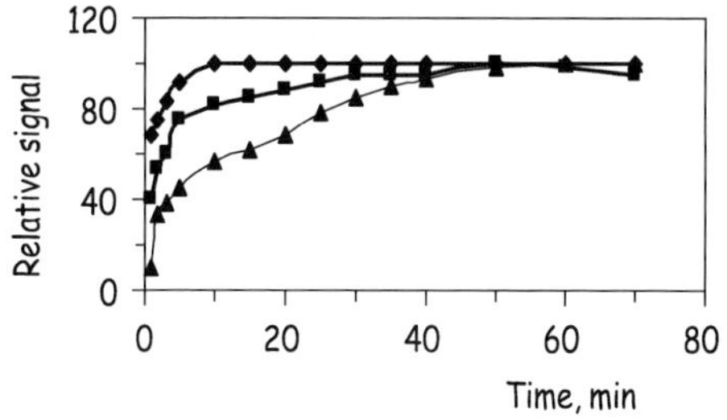


Fig. 3.1. Equilibration time. (a) *o*-phenylphenol in SQAE; (b) Co(II)–PAR in Dowex 1-X8; (c) Mn(II)–3BrBHA in SQAE.

With modified ion exchangers, diffusion in the gel is not the rate-determining step because of the surface distribution of the modifier. When 1:1 complexes are formed in this solid phase, the limiting step is diffusion of the metallic ions in the film covering the ion exchanger particles. The formation of a 1:2 complex, which is associated with the deformation of the ion exchanger framework, requires a longer time, which probably occurs via a sterically controlled mechanism [12].

The ion exchange reaction rate is proportional to the square of the radius of the resin particles. For this reason, particles of various sizes have been used to fix the analytes or derivatives formed in the solution with the only condition being uniformity of particle size, the most common being 100–200 and 200–400 mesh [86]. The use of smaller sizes makes it necessary to use another form of equilibration and to carry out the measurement as shown below. Suspensions of macro-reticular ion exchangers, such as Amberlyst (A-27, 15) or Diaion (PA318, WA30, PK228), have been used this way with particles ranging from 20 to 30 μm and capacities from 7 to 15 μequiv./mL [42,87].

An alternative method is to use a coagulated mixture of finely divided macroreticular ion exchangers with a particle size less than 30 μm, which collect the analytical species more rapidly than conventionally sized beads. When one ion exchanger is added to an oppositely charged particle, they coagulate to form a bulk material, which is easily sedimented and collected. Given that the coagulation of an exchanger system by the addition of an oppositely charged ion exchanger only consumes around 5% of the total ion exchange sites, the coagulated resin continues to have considerable exchange capacity [77,88].

Measurement preparation involves attaining an evenly covered layer with the solid phase. Three procedures can be used to prepare the solid phase with the analyte retained in order to measure it. In the first, the solid is packed into an optical glass or into quartz cells with parallel walls. To do this, the slurry is filtered under suction through a plate of sintered glass filter or through a small column fitted with a disc, and the solid phase is collected, with the aid of a pipette, and packed into the cell. Alternatively it is possible to omit filtration, and simply decant the particles, discard the liquid and withdraw the solid with a dropper. If the solid support is a hydrophilic solid, packing into the cell is performed with a small volume of filtrate, but if it is hydrophobic, it is first dried by passing an air current through the filter itself before the packing operation. The optical length of the cell ranges from the more common 1 mm up to 10 mm, and at times even 20 mm.

In order to attain dense packing of the solid phase (related to the reproducibility of the measurements), it is usual after filling the cell to wait for a while, at times as long as 15 min, depending on the type of solid phase and the particle size. To hasten sedimentation, it has been proposed that the cells should be centrifuged with the solid phase inside usually at 25 g, either in a modified desk centrifuge [12,64] or using cells with one [89] or two [66] small holes at the bottom, because the solution flows out through the holes and the beads settle down more quickly.

A second system used to collect the equilibrated solid when working with solid phases of small particle size or coagulated ion exchangers is by means of filtration under suction through an inert membrane (cellulose acetate, cellulose nitrate, polyamide, teflon, etc.). A disc with a thin layer of preconcentrated solid (15–20 mm in diameter and 0.05–0.30 mm thick) supported by the membrane is placed in a filter holder. For absorbance measurements, a dipping solution is used to increase the color, to stabilize it, and to keep the membrane wet before the optical measurement in order to increase its transparency [30,87]. When working with finely divided ion exchangers, the addition of colloids improves the process characteristics, as with the retention of phosphate as molybdenum blues species in the presence of gelatin on anionic Amberlyst A-27 [87]. This occurs in the same way with the addition of salts, like $NaClO_4$, which facilitates filtration, lowers the blank in antipiryne dye formation from phenols and is attributed to possible coagulation of resin particles [90]. These techniques are called ion-exchange resin thin-layer spectrophotometry (TLS) [77,91] and coagulated ion-exchanger spectrophotometry (CIS) [75,92]. A similar system, filtration through paper and measurement in wet state, is used for diffuse reflectance measurements [51,59] and photoacoustic measurements [93], although with the latter, a dry concentrated solid phase is also used [94].

A third system used to prepare the equilibrated solid phase for measurement consists of preparing a membrane by fusion. This is done when solid phase paraffin in granules is used. After filtration, the granules are placed in a container and melted in a 50°C water bath. The measurement cell is filled with the paraffin that is measured after cooling [95] (Fig. 3.2).

*Measurement of the analytical property*

Only the most common techniques are going to be discussed here, and not those less frequently used which also measure directly over the solid phase like suspension-introduction graphite-furnace atomic absorption spectrometry [96,97], laser ablation sampling inductively coupled plasma atomic emission spectrometry [98] or radiochemical measurements [99,100].

*UV–Vis spectrophotometric absorption*. The principal drawbacks in the measurement of the analytical signal directly on solid phase particles packed in an optical cell are the strong scattering of radiation,

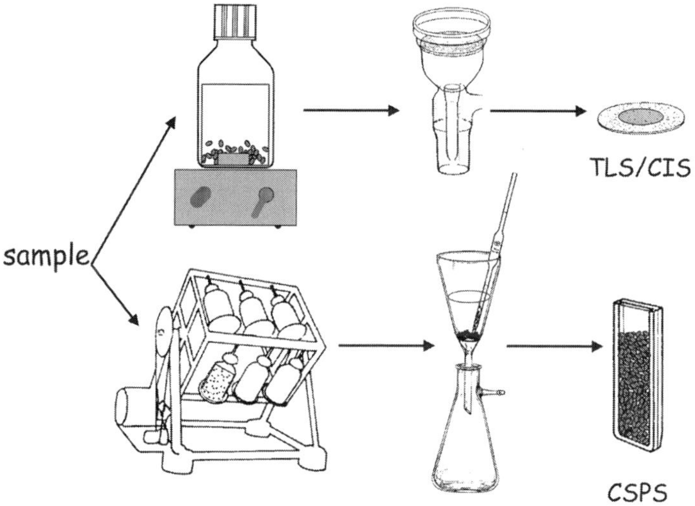


Fig. 3.2. Operational procedures in particulated solid phase assays.

depending on the particle size and the wavelength [12], and poor reproducibility in the absorbance value due to variability when packing the solid into the cell.

To increase the S/N ratio measurements can be performed: (a) using a solid phase blank equilibrated with the same reagents as the sample [101]; (b) against frosted glass [102]; (c) against a perforated metal plate having a constant absorbance over a wide range of wavelengths (1.0 or 2.0) [103]; or (d) using derivative spectrophotometry [104].

To obtain good measurements, the absorbance—which is really attenuation, because the light is scattered as well as absorbed by the system—is obtained at two wavelengths, namely heterochromatic extrapolation. According to Yoshimura [4] the measured absorbance can be due to several contributions, $A = A_a + A_{sol} + A_S + A_{SL}$, where $A_a$ is the absorbance of the analyte or derivative, $A_{sol}$ is the absorbance of the solution present between the solid particles (which can usually be disregarded if the distribution coefficient $D$ is high), $A_{SL}$ is the absorbance due to the free reagent, if necessary, in the solid phase and $A_S$ is the background arising from light scattering and absorption by the solid phase. In an ideal case, $A_a$ could be obtained by direct measurement at the wavelength of the compound compared with a reference blank, but in practice, it is not possible to cancel the background attenuance. Since solid phase background spectra are parallel under similar packing conditions, $A_a$ can be obtained by measuring both sample ($\Delta A_{\lambda_1,\lambda_2}$) and reference ($\Delta A_{\lambda_{1b},\lambda_{2b}}$) at two wavelengths, $\lambda_1$, the maximum absorption wavelength, and $\lambda_2$, a wavelength at which the analyte or its derivative practically does not absorb, typically between 700 and 800 nm. Then the difference gives the net absorbance due to analyte: $A_a = \Delta A_{\lambda_1,\lambda_2} - \Delta A_{\lambda_{1b},\lambda_{2b}}$.

An alternative way to measure at two wavelengths is to place frosted glass, such as Opal glass [101], dull Dacron films [105] or a sheet of Lavsan [20] in both the sample and reference compartment. The frosted glass scatters the transmitted and scattered radiation equally, and an averaged, constant portion of the total amount of the light flux passing through the cell reaches the detector with the scattering due to suppression of the solid phase particles. In this way, it is possible to perform measurements at a single wavelength. The problem with this technique is that the decrease in the radiation reaching the detector may increase the measurement error in absorbance.

As shown above, it is also possible to correct variations in the background of the concentrated solid phase using derivative spectrophotometry, taking into consideration the form of the background signal of solid phase. Ishii [104] has shown that using the net derivative value corresponding to the analyte, obtained by measuring the derivative spectrum against a solid phase blank, it is not necessary to use heterochromatic extrapolation.

The use of a membrane to collect the analyte retained on the solid phase, in the two variants TLS and CIS, offers the advantage of not needing heterochromatic extrapolation because the reproducibility of the solid phase over the membrane is higher than in the cell [106]. However, in some occasions, a slightly convex shape in the calibration function is observed [87]. Consequently, it is possible to measure the absorbance at a characteristic wavelength against a membrane, as long as both membranes are wet.

A strong absorption showed by the frequently used styrene based ion exchangers in the UV region prevents the use of UV measurements over a solid phase. Here, solid phases with low UV background have been used, as with dextran type polymers [32,107].

The UV–Vis absorption spectra of analytes or their derivatives are usually influenced by the type of solid phase on which they are retained. In general, hyperchromic effects responsible for enhanced sensitivity of the reaction in solid phase and also batochromic shifts from the solution to solid phase are observed. This is what occurs with the reaction of Be(II) with ECR and EDTA, which shows in solution a maximum at 512 nm when fixed on SQAE 570 and 580 nm when fixed on SDEAE [108]. The same occurs with more simple systems like the retention of copper on the tin(IV) antimonophosphate solid phase where the maximum is displaced from 770 to 820 nm [10].

*Spectrofluorimetry.* One of the aspects to bear in mind in the measurement of fluorescence is the position of the measurement cell. For this reason, a variable angle device has been used (Fig. 3.3) which makes it possible to situate the measurement cell at the intersection of the incident beam and the emergent beam.

Fig. 3.4 shows typical behavior. There are four maximums of different intensity of relative fluorescence (RFI) corresponding to the four positions in which the cell plane forms angles of $+45^\circ$ to $-45^\circ$, with the optical line directed from the center of the cell to the center of the detector. Those which show higher RFI values correspond to $+45^\circ$ and the transmitted fluorescence is measured in this position, since the excitation beam intersects one side of the cell, measuring the RFI on the other. The other two maximums of lesser value correspond to

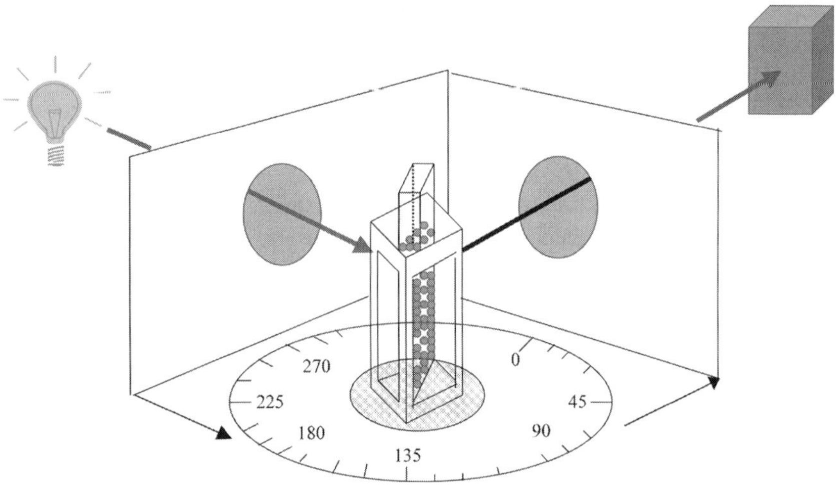


Fig. 3.3. Variable angle device for the measurement of fluorescence in solid phase.

the angle $-45°$, with the reflected fluorescence being measured here, and both the excitation and the measurement are carried out on the same side [136]. With solid phase spectrofluorimetry (SPF), the intensity of the transmitted fluorescence is usually measured, for which a cell holder fixed in a position at a 45° angle with the excitation and emission beam is used. When the reagent itself also shows fluorescence emission at the same wavelength, the difference between fluorescence intensities of the solid phase measured with and without analyte should be taken.

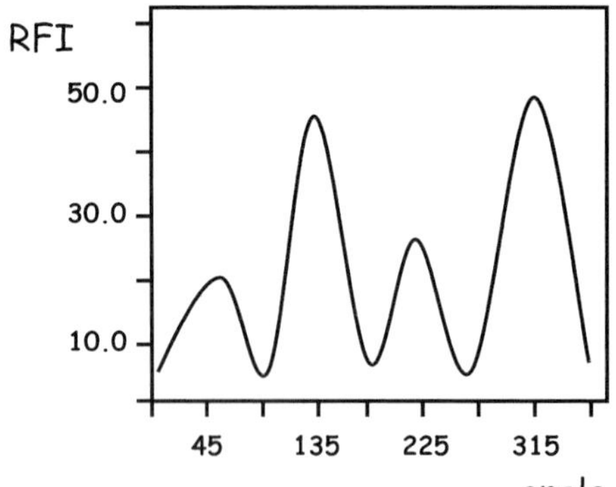


Fig. 3.4. Dependency of the fluorescence signal with the measurement angle.

With solid phase spectrofluorimetry measurements, the most common lamp is the xenon discharge lamp. However, the use of a laser brings about a significant improvement in sensitivity, as with the determination of carbaryl using SQAE where the ratio of slopes of the calibration graphs using the two source systems is 9.4 [82].

The fluorescence spectra features of the species fixed on solid phase can be modified considerably on certain occasions with respect to the solution by (a) increasing the rigidity due to fixation on the solid phase which then decreases the deactivation processes; and, in general, (b) the modification of the surrounding environment of the species. For example, the herbicide bentazone shows an intense fluorescence in SQAE A-25 gel phase with 436/338 nm and a very low one in water at 430/353 nm [109] while the 1:1 complex molybdenum-CA shows 523/583 nm in SQAE and 530, 560/590 nm in solution [110]. In other instances, the fixed species are relatively insensitive to their environment as indicated by the absence of modification of their characteristics peaks. This is the case with 1:1 gallium complex of SOAPh fixed on SSP [111] or aluminum complex with 2,2′,4-trihydroxyazobenzene covalently immobilized on silica gel [112].

*Diffuse reflectance spectroscopy.* Generally, diffuse reflection coefficients of the sample and reference are measured and, from these measurements, the Kubelka–Munk function $F(R)$ or its variations $\Delta F(R)$ are calculated according to:

$$\Delta F = \left[(1 - R)^2/2R\right] - \left[(1 - R_0)^2/2R_0\right]$$

Various transformations have been used to obtain linear relationships with concentration such as the Williams–Clapper transformation [113].

The use of chromaticity characteristics of the preconcentration solid support as analytical parameters has been suggested as an alternative [51,59]; these would be color lightness ($L$), chromaticity coordinates ($A$, $B$), saturation ($S$), hue ($T$), brightness ($Y$), whiteness ($W$), yellowness ($G$) and the total color difference ($\Delta E$) of the sample and the reference ($R_0, L_0, A_0, B_0, S_0, T_0, Y_0, W, G_0$). The values used as possible parameters are $\Delta L$ ($L_0 - L$), $\Delta A$ ($A - A_0$), $\Delta B$ ($B - B_0$), $\Delta S$ ($S - S_0$), $\Delta Y$ ($Y_0 - Y$), $\Delta G$ ($G - G_0$), $\Delta E$ ($100 - W$) and $\Delta T$ calculated according to $[(\Delta E)^2 - (\Delta L)^2 - (\Delta S)^2]^{1/2}$.

*Photoacoustic spectrometry.* The measurement of the photoacoustic response of particulated solids is strongly influenced by scattering light. The use of lasers as excitation light sources has greatly improved sensitivity and extended its applicability. Since the cell volume must be as small as possible, Yoshimura has designed a cell to collect and easily remove the solid phase [114]. To perform the measurements, the height of the collected beads in the cell—which is filled with water for the measurements—and the distance between the center of the laser beam and that of the piezoelectrical transducer must be fixed and maintained. The analytical parameter used is the longest peak-to-peak height intensities ($\Delta h$) of positive and negative excursions observed in the photoacoustic signal waveform. Alternatively, the solid phase has been collected by filtration through a cellulose acetate filter and then, after drying, placed on a flat glass plate in a brass photoacoustic cell [93]. A function that combines photoacoustic amplitude and phase into a single response function has also been used as the analytical parameter [58,115], taking advantage of the greater sensitivity of the phase measurements at large absorptivities.

*Variables and their quantitative relationships*

Given a mass $m$ (grams) of solid support balanced with a volume $V$ (mL) of solution whose initial analyte concentration is $C_0$ (mol/L), of the millimoles initially present ($C_0V$), it is supposed that $M_r$ is retained on the solid support and the rest ($M_D$) remains in the solution, which can be written as:

$$C_0V = M_r + M_D \tag{3.1}$$

If the distribution coefficient of the analyte between the solid phase and the solution is $D = (M_S/m)/(M_D/V)$ and the volume of the solid support $v_s$ and its density $\rho$, substituting this expression in Eq. (3.1) and finding the value of $M_r$ leaves:

$$M_r = \frac{C_0VmD}{mD + V} \tag{3.2}$$

The concentration of the analyte, fixed on the solid support, would finally be:

$$C_r = \frac{M_r}{\nu_r} = \frac{C_0VD\rho}{mD + V} \tag{3.3}$$

an equation that makes it possible to establish the relationship between the concentration of the analyte on the solid phase and the analytical parameter.

When using molecular absorption spectrophotometry (CSPS, TLS, CIS), the absorbance due to the analyte $A_a$, supposing that $D$ has a very large value, is given by:

$$A_a = \varepsilon_a l_r C_S = \varepsilon_a l_r \frac{C_0 V D \rho}{mD + V} \tag{3.4}$$

where $\varepsilon_a$ is the molar absorptivity of the analyte or its derivative in the solid phase and $l_r$ the mean light-path length through the solid phase, which is different from the cell length. For diffuse reflectance measurements, the relationships are very similar. If $D$ is sufficiently high, as is usual (see Table 1), the above Eq. (3.4) could be simplified to:

$$A_a = \varepsilon_a l_r \frac{C_0 V \rho}{m} \tag{3.5}$$

When the measurement performed over the solid phase is fluorescent emission, the intensity of the emitted fluorescence is called $I_F$, the intensity of the excitation beam $I_0$ and the quantum yield of the fluorescent emission process $\Phi$. Using Zweidinger and Winerfordner's expressions [116] to calculate the intensity of the fluorescence emitted in optically heterogeneous, non-transparent media, simplified for the case where the concentration of analyte is small, gives:

$$I_F = 2\phi I_0 \frac{2.3\varepsilon_a l_r}{1 + 2S l_r} \frac{C_0 V D \rho}{mD + V} \tag{3.6}$$

where $\varepsilon_a$ and $l_r$ are known and $S$ is the radiation dispersion index per unit of solid support thickness, whose value here is influenced not only by the dispersant capacity of the solid but also by the degree of packing.

With photoacoustic spectrometry and for low absorption conditions where the absorbed power is small, the intensity of photoacoustic signal $I_{PAS}$ could be approximately [114,117] expressed as:

$$I_{PAS} = k I_0 \varepsilon_a l_r \beta C \tag{3.7}$$

where $k$ is an instrumental proportionality constant, $I_0$ is the excitation intensity, $\varepsilon_a$, $l_r$ and $C_0$ are known and $\beta$ is a measure of the conversion efficiency of absorbed power to heat by radiationless processes. Substituting the value of $C_0$:

$$I_{PAS} = k l_0 C_0 \varepsilon_a l_r \beta \frac{C_0 V D \rho}{mD + V} \tag{3.8}$$

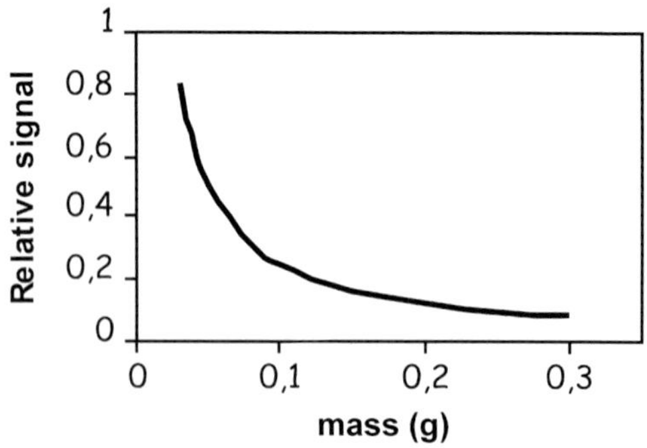

Fig. 3.5. Dependency of the signal on the mass of the solid phase.

The expressions given by Eqs. (3.4), (3.6) and (3.8) show a clear division between the variables that influence the measurement stage and those that do so in the stage of fixing the analyte on the solid support.

Lessening of solid phase mass, $m$, generates an increase in the analytical signal as the earlier equations indicate (Fig. 3.5). Eq. (3.5) indicates that the dependence is a branch of an equilateral hyperbole ($A_a = K/m$). Table 3.3 presents the values for the $K$ experimentals and theoreticals for some systems.

The amount of resin used must be the least possible in a form that gives the highest signal and the greatest ease of handling, with amounts typically ranging between 30 and 200 mg, depending on the type of solid phase, although in some cases it has been possible to use amounts as large as 800 mg [122]. The amount must be the minimum to fill the cell and conduct the measurement of the analytical signal. The most common is the use of 1 or 10 mm cells with a transversal quartz spacer of variable dimensions along with a frontal spacer in such a way that only the path-length is occupied by the solid phase [66].

TABLE 3.3

Influence of mass of solid phase

| Analyte | Reagents | Solid phase | $K_{exp}$ | $K_{teor}$ | Reference |
|---|---|---|---|---|---|
| Hydrazine | pDMAB | Dowex 50W-X8 | $6.7 \times 10^{-2}$ | $6.4 \times 10^{-2}$ | [118] |
| Thiamin | – | S CM C-25 | $2.5 \times 10^{-3}$ | $2.4 \times 10^{-3}$ | [119] |
| Ascorbic acid | Fe(III)–FZ | S QAE A-25 | $2.5 \times 10^{-2}$ | $2.6 \times 10^{-2}$ | [35] |
| V(V) | PAR | Dowex 1-X8 | $4.4 \times 10^{-2}$ | $4.6 \times 10^{-2}$ | [120] |
| Cu(II) | PAR | Dowex 1-X8 | $8.4 \times 10^{-2}$ | $8.4 \times 10^{-2}$ | [46] |
| Co(II) | PAR | Dowex 1-X8 | $1.07 \times 10^{-1}$ | $1.06 \times 10^{-1}$ | [121] |

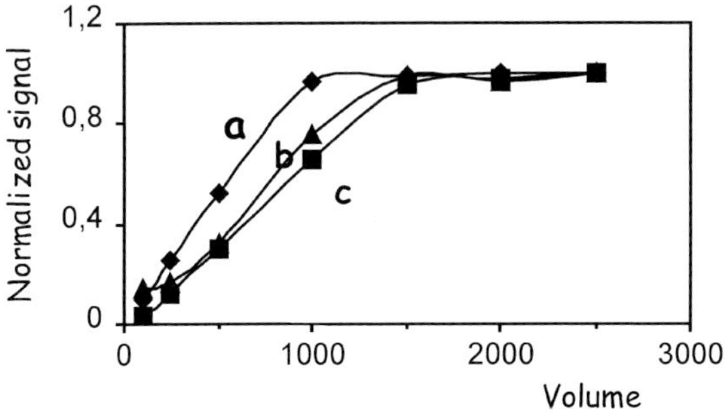


Fig. 3.6. Influence of sample volume on analytical signal. (a) Hydrazine–pDMAB on Dowex 50W-X8 ($D$: 117,600); (b) Mn(II)–3BrPADAP on SQAE ($D$: 67,000); (c) Co(II)–PAR on Dowex 1-X8 ($D$: 28,500).

A fused-silica tube apparatus to collect and transfer solid phase to a small-volume flow cell has been designed to make it possible to use a smaller quantity of resin and attain reproducible measurements (0.035 mL). Here, amounts as small as 0.1 mL of solid phase slurry are workable [123].

The influence of the volume of the solution equilibrated with the solid phase on the analytical signal is seen clearly in Fig. 3.6, from which one of the advantages of the analytical methods based on this technique can be deduced: sensitivity increases when the initial volume of the solution used for equilibration with the solid is also increased. This increase in sensitivity, which is limited by the value of $D$ in each concrete case, makes it possible to construct methods appropriate to the measurement of the problem. The practical limit of sample volume may be a few liters, 2 or 3 at the most, due to the increase in the equilibration times.

An increase in the path-length of the solid phase layer by use of longer cells should cause an increase in analytical signal, as seen earlier. However, the use of a very long cell results in radiation loss by scattering or absorption to such an extent that the radiation intensity to be measured may be below the detection limit of the detector being used. Yoshimura [124] demonstrated that the attenuance of the background generated by solid phase is a linear function of the logarithm of the cell length, according to $A = (2 - a)\log l + k$ where $a$ is a directionality parameter usually larger than zero, $l$ is the path length and $k$ is a constant, while there is a direct proportionality between measured net absorbance and cell length. This shows that up to 10 mm it is advantageous to increase the cell length for the determination of sample components present at low levels. An example of this is the procedure for phosphate determination by molybdenum blue sorption in SG-25 where the slope of the calibration function is made 1.88 times greater when passed through a cell between 1 and 2 mm, 2.47 times greater when passed through one between 2 and 5 mm and 1.78 when passed through one between 5 and 10 mm [65].

When the cell length is increased, it is necessary to minimize the radiation loss from scattering in a special way. Two systems have been used to collect the maximum quantity of radiation possible: (1) locating the cell as close as possible to the detector window and (2) placing a slotted cylinder mirror between the cell and the detector [124].

The dependence of the analytical signal on $D$ is very important, as can be seen in Fig. 3.7. The $D$ value is conditioned by the nature of the analyte and support, and particularly by the reactions that take place. For this reason, it is especially influenced by pH, ionic strength, reagent concentration, percentage of cosolvent and temperature.

The most critical factor in arriving at an adequate retention in the solid phase is pH, which is a condition as much for the formation of derivatives as for the fixation of analytes or derivatives. Additionally, in the case of ion exchangers, both protons and hydroxyls can compete for the retention of the species of interest. Thus, different behaviors are observed (Fig. 3.8). The most common is to show a maximum in the intermediate pH zone, with the signal lowering towards both sides. This drop may be due to three main causes. (a) Competition of ions of water by ionic sites of the gel, as is the case with the complex Fe(II)–FZ on SQAE A-25 [35]. (b) Formation of the retainable species in solution, which depends on the p$K_a$ of the groups involved and on the retention

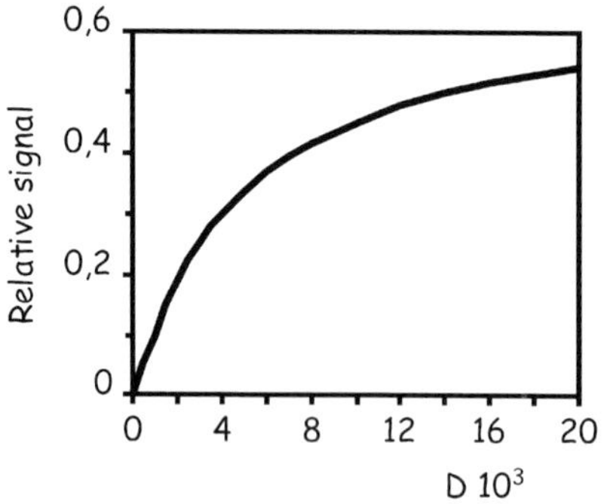


Fig. 3.7. Influence of $D$ value on the signal.

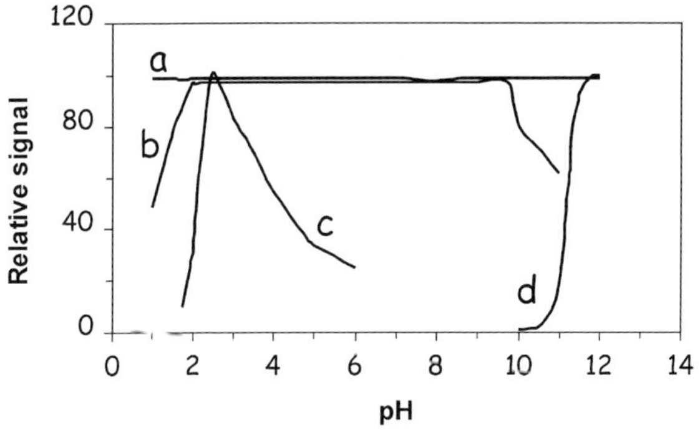


Fig. 3.8. pH dependence on signal. (a) anthracene on C-18 silica; (b) Fe(II)–FZ in SQAE; (c) sulphatiazol in SSP; (d) morin–Be(II) in SQAE.

mechanism. For sulphatiazol retention on SSP, this is due to the combination of both factors, which generates a narrow dependency [27]. In the case of the retention of morin–Be(II) complex, it only starts to grow from pH 11.5, and is inconsiderable at lower pH values [49]. At times, the pH margin can be narrower than in the solution because of the need to conjoin different processes. Thus the retention of (1,1′-biphenyl)-2-ol on SQAE occurs at pH 10 because of its necessary retention as an anion, while in solution the fluorescence is optimal from pH 2 [125]. On some occasions, the signal is independent of the pH, as occurs with anthracene fixed on C-18 silica [126]. (c) Protonation of weak acid or base exchanger groups of the resin.

At times, a high concentration of strong acid is required, due to the need for reaction protons, as with the determination of nitrite by formation and fixation of an azo dye on Dowex 50W which uses 0.8 M HCl [74] or the determination of Se(IV) based on the oxidation of iodide and retention as an ionic pair on lipophilic Sephadex, which must operate at 0.16 M HCl [70]. Another characteristic process is the fixation of the blue molybdenum species, where its adsorbability on both Sephadex adsorbent and ion exchangers may depend on the acidity of the solution [64,66]. In other situations, the anion of the acid justifies the necessary acidity, as with vanadium complex with BHA fixed on anionic Dowex 1-X8 between 2 and 4 M HCl as a ternary species with chloride [86]. In every case, the adsorbability of the derivative must be very high.

The influence of electrolytes—usually inorganic salts coming from the sample itself or from the analytical treatment—on the analytical signal can be considerable. When ion exchanger solid phases are used, an increase in the contents of salts lowers the distribution ratio and the sorption rate of the species, and consequently lowers the signal; but it also generates a volume contraction of the resin beads, which favors packing and increases the signal. Depending on the system, the result can be different, but in general it is not usual that more than 1 M as sodium chloride is tolerated. In general, more is tolerated when the degree of crosslinking of the ion exchanger is lower, since it is possible for the matrix to contract. Thus the retention of molybdenum as a Tiron complex on macroreticular Amberlyst A-27 has no influence up to 0.75 M [127]. On the other hand, many systems cannot tolerate concentrations of salts higher than around 0.01 M since the competition between the ion of salt or buffer for the ionic sites of the solid phase predominates, i.e. with Co(II)-PAR on Dowex 1-X8 quantities higher than $4 \times 10^{-2}$ M are not possible [121] or in organic compounds such as sulphatiazole in SSP where the limit is $9 \times 10^{-3}$ M [128]. At times, the influence of the salt content is due to an increase in the blank by adsorption of the reagent as an ionic pair, as with the reaction of sulfite with pararosaniline on Dowex 1-X8 [73]. This decrease in signal occurs especially when fluorescence is measured, possibly due to the deactivation process, where concentrations are not usually tolerated above $10^{-3}$ M. One example of this is the Al-morin system fixed on SSP C-25 whose fluorescence does not depend on ionic strength ($I$) up to $4 \times 10^{-3}$ and decreases at higher levels according to $\mathrm{RFI} = 2.71 I^{-1/3}$ [48].

In the case of hydrophobic adsorbents, the signal usually increases with the saline concentration of the medium, a salting out effect, as with the retention of morestan on C-18 silica, which tolerates up to 1 M in salts [129] or with PAH again over C-18 where the signal increases lineally with saline concentration at least up to 1.5 M [130]. This effect has been used to reach complete retention of an antipyrine dye formed from phenols by adsorption on anionic Diaion WA30 using 20 M $Na_2SO_4$ solution [90].

The concentration of the reagent used when the reaction takes place in solution cannot be very high, just enough to displace the reaction sufficiently, since reagents usually fix themselves on the solid phase in competition with the derivative and increase the background signal if the signals are close, or when measuring fluorescence, generate a quenching effect owing to the re-absorption effect of the reagent fixed on the solid phase. This is why using a reagent/analyte ratio from 2 to 50 is common [44,131]. On the occasions when it is necessary to use a greater excess of reagent to displace the equilibrium, but this is retained and causes interference, it has been proposed that the excess be eliminated, as when Co(II) with NRS on SQAE is eliminated by washing with 0.5 M NaCl [36]. There are occasions when the reagent does not interfere because it is excluded from the solid phase, as with the complex Al–morin on SSP C-25, since at pH 5.3 morin has an anionic nature [48]. High ratios are only used when the reaction is slightly displaced and the reagent is not affected, as with bismuth in Dowex 1-X2 (ratio Bi(III)/$I^-$: 3000) [132] or with molybdenum in Dowex 1-X8 (ratio $MoO_4^{2-}$/$SCN^-$: 64,000) [133].

Sometimes the use of organic solvents is necessary to dissolve analytes or reagents, though the proportion is generally small, usually less than 10%, since an increase in the proportion of the organic solvent reduces the sorption rate on solid phase [16], while it may increase an analytical signal like fluorescence [48]. On occasions, it is necessary to use high proportions of solvent so that the derivatization reaction grows adequately. This occurs in the determination of aromatic esters through hydroxamation and complexation steps, where the proportion of methanol increases the signal in the first stage and decreases it in the second [79,80].

Temperature is a variable that can influence both the process of reaction and fixation on the solid support and measurement, especially with fluorescence. With some systems, temperature affects not only the reaction itself, but also the value of the blank and for this reason it must be kept constant within $\pm 1°C$, as with the formation of blue molybdenum from phosphate [65,134]. In other situations, the reaction kinetic is slow, as when Co(II) is retained on an ion exchanger modified with NRS in which it operates at near boiling temperatures [60]. However, it is not usually necessary to work at high temperatures when carrying out the derivatization reaction in solution.

A slow sorption kinetic is responsible for the need to work above room temperature in some cases, as happens to the cationic copper complex with T4MPP retained on Dowex 50W-X2 which must be worked at temperatures above 35°C [135]. In other situations, it is necessary to work at temperatures that are as low as possible—i.e. work at room temperature—since the fixation process is exotermic, as has been described for inclusion of vanillin in β-cylodextrin polymers [16]. However, in the majority of cases it is not necessary to adjust the temperature during either the reaction or retention stage. Temperature has an affect during measurement in the case of fluorescence, which generally decreases when the temperature increases, a totally reversible effect due to higher internal conversion as the temperature increases, facilitating non-radiative deactivation of the excited singlet state [43,111].

The stability of the solid phase with the analyte or its retained derivative, defined by the variation in the analytical signal over time, is an aspect that must be kept in mind. In the case of complexes, stability does not usually last very long, generally between 3 and 5 h after preparation [38,72]. In the case of organic compounds the situation is more varied, from thiabendazole on SG-15 (3 h) [136], to PAH on C-18 silica (5 h) [137], to sulphonamides on SSP (24 h) [128,138] to thiamin on SCM (13 days) [119].

### 3.5.3 Analytical characteristics

Solid phase spectrometric assays have been used for the analysis of a considerable number of inorganic and organic analytes. They have been used for the determination of 27 different metallic ions in different states of oxidation. Tables 3.4–3.6 present some examples of procedures that have been proposed using different measurement techniques.

*Calibration*

The treatment of experimental data for analytical purposes is usually performed by univariate calibration. In some instances, derivative measurements are used as the analytical signal not only to make measurement in a single wavelength possible (in order to solve the problem of the irreproducibility of packing shown above), but also to eliminate the necessity for a blank, if wavelength measurements are chosen where the signal derived from this blank is null. Thus in the spectrophotometric determination of W(VI) [37] or Mo(VI) [139] using PV, the first derivative signal used is at 674.0 or 716.0 nm, respectively. In the case of observing matrix effects, calibration is usually carried out by the standard addition procedure.

*Sensitivity*

As Eq. (3.3) and related equations show, it is possible to attain a high sensitivity by using large $V$ and/or $D$ or small $m$, as adequate values for

TABLE 3.4

Analytical characteristics of particulated solid phase spectrophotometric procedures

| Analyte | Volume (mL) | LDR | DL | RSD[a] (%) | Procedure | Reference |
|---|---|---|---|---|---|---|
| Be(II) | 1000 | 0.03–1.0 μg/L | 0.006 μg/L | 1.6–2.0 | CSPS | [38] |
| Co(II) | 40 | 1–15 μg | 25 μg/L | 0.89 | CIS | [30] |
| Cu(II) | 1000 | 0.1–1.2 μg/L | 0.072 μg/L | 3.5–6.0 | CSPS | [135] |
| Mo(VI) | 1000 | 2–8 μg/L | 0.024 μg/L | 4.3 | CSPS | [139] |
| U(VI) | 1000 | 1–10 μg/L | 0.28 μg/L | 4.1 | CSPS | [41] |
| $F^-$ | 1000 | 0.8–16.0 μg/L | 0.5 μg/L | 2.8 | CSPS | [140] |
| $PO_4^{3-}$ | 1000 | 0.9–13.0 μg/L | 0.4 μg/L | 4.3 | CSPS | [64] |
| $SO_2$ | 100 | 5–70 μg/L | 1.2 μg/L | 2.1 | CSPS | [72] |
| Ascorbic acid | 20 | 0.3–5.0 mg/L | 0.053 mg/L | 0.74 | CSPS | [141] |
| Phenols | 50 | up to 1.0 μg | 0.5 μg/L | 0.2 | TLS | [90] |
| Quinoline yellow | 500 | 1.5–15 μg/L | 0.4 μg/L | 5.7 | CSPS | [142] |
| Thiamin | 1000 | 15–170 μg/L | 3.0 μg/L | 2.6 | CSPS | [119] |

[a]Expressed as repeatability.

TABLE 3.5

Analytical characteristics of particulated solid phase photoacoustic and diffuse reflection spectrophotometric procedures

| Analyte | Volume (mL) | LDR | DL | RSD[a] (%) | Procedure | Reference |
|---|---|---|---|---|---|---|
| Cr(VI) | 20 | 0.04–2 μg | 0.58 μg/L | – | PAS | [143] |
| Fe(II) | 30 | 0.05–0.8 μg | 20 ng/L | 5.4 | PAS | [93] |
| V(V) | 20 | up to 0.5 μg | – | 1.0 | DRS | [144] |
| Mo(VI) | 20 | up to 5 μg ? | 0.037 μg | 0.8 | DRS | [127] |
| Al(III) | 25 | 0.2–2.0 μg | 0.004 μg/L | 0.04–3.7 | CFDRS | [51] |
| Ni(II) | 25 | 5–20 μ | 0.1 μg/mL | 0.3–1.3 | CFDRS | [59] |
| In(III) | 25 | 0.2–16 μg | – | 2.0–2.4 | CFDRS | [24] |
| Ascorbic acid | 50 | 0.01–0.23 mM | 0.9 mg/L | <10 | CFDRS | [69] |

[a]Expressed as repeatability.

the characteristic parameters of each measurement technique, such as $\varepsilon_a$ and l in UV–Vis spectrophotometry. From all of these factors, the one that is easiest to modify is the volume of the sample to be equilibrated. For this reason it is common with particulated solid phase assays to prepare procedures for different sample volumes, typically between 100 and 1000 mL [145]. To do this, it is only necessary to increase the quantity of the necessary reagents proportionally and adjust the equilibration time. Yoshimura [89] calculates the gains in sensitivity when the sample volume is increased by using the mass ratio, which can also be defined as a sensitivity ratio, $S_{V_2}/S_{V_1}$. This is calculated from Eq. (3.3):

$$\frac{M_{r_2}}{M_{r_1}} = \frac{S_{V_2}}{S_{V_1}} = \frac{mD + V_1}{mD + V_2}\,\frac{V_2}{V_1}$$

The theoretical increase in sensitivity with the sample volume with respect to a reference volume of 200 mL and for different $D$ values is shown in Table 3.7. The values may be somewhat different in practice, due to kinetic problems or non-linear retention isotherm.

Generally, it can be said that the solid phase methods are between 10 and 1000 times more sensitive than their corresponding methods in solution; see for example, the procedure for nitrites based on azo dye formation which is 9.5 times more sensitive [74], that of cadmium with PAR which is 85 times [55], that of copper with BCA which is 317 times [44], or that of aluminum with CAS which is 1300 times [33].

TABLE 3.6

Analytical characteristics of particulated solid phase spectrofluorimetric procedures

| Analyte | Volume (mL) | LDR | DL | RSD[a] (%) | Procedure | Reference |
|---|---|---|---|---|---|---|
| Be(II) | 1000 | 0.10–0.70 μg/L | 0.02 μg/L | 1.2 | SPF | [49] |
| Al(III) | 1000 | 0.2–1.2 μg/L | 0.03 μg/L | 2.7 | SPF | [48] |
| Ga(III) | 500 | 2.0–10.0 μg/L | 0.3 μg/L | 1.3 | SPF | [111] |
| Mo(VI) | 1000 | 1.0–15.0 μg/L | 0.3 μg/L | 1.3 | SPF | [110] |
| $Br^-$ | 1000 | 0.05–0.3 μg | 0.02 μg/L | 1.4 | SPF | [146] |
| Anthracene | 1000 | 30–500 ng/L | 14 ng/L | 0.8 | SPF | [126] |
| Benzo[a]pyrene | 500 | 0.52–3.45 μg/L | 0.07 μg/L | 1.2 | SSPF | [147] |
| Fluoranthene | 500 | 0.3–5.0 μg/L | 0.21 μg/L | 1.2 | SSPF | [130] |
| Carbaryl | 500 | 0.07–40.0 μg/L | 0.02 μg/L | 2.4 | SPF | [82] |
| 1-Naphthylacetic acid | 500 | 0.3–18 μg/L | 0.9 μg/L | 1.4 | SPF | [148] |
| Thiabendazole | 500 | 1.5–10 μg/L | 0.1 μg/L | 1.4 | SPF | [136] |
| Ciprofloxacin | 1000 | 0.3–10.0 μg/L | 0.1 μg/L | 1.1 | SPF | [149] |

[a]Expressed as repeatability.

TABLE 3.7

Effect of volume on sensitivity

| Sensibilities ratio | Thiamin–SCMC[a] | | Fe–FZ–SQAE[b] | |
|---|---|---|---|---|
| | Theor. | Exp. | Theor. | Exp. |
| $S_{1000}/S_{10}$ | 42.2 | 39.1 | 40.9 | 46.4 |
| $S_{500}/S_{10}$ | 29.8 | 29.8 | 29.1 | 30.9 |
| $S_{250}/S_{10}$ | 18.8 | 18.8 | – | – |
| $S_{100}/S_{10}$ | 8.9 | 8.9 | 8.8 | 8.3 |

[a]*D*: 23,800 L/kg; *m*: 0.03 g [119].
[b]*D*: 16,850 L/kg; *m*: 004 g [35].

*Precision*

The precision of solid phase methods is good, as can be seen in the examples gathered in Tables 3.4–3.6. With CSPS, values of relative standard deviation are found that are between 1 and 6%, though the most common are between 2 and 4%. With SPF the precision is better, usually being between 1 and 2%. On some occasions, an increase in the RSD value is seen when the sample volume is increased (Table 3.8); however, the controlling factor for precision is more commonly the packing of the solid phase. Thus in SPF determination of anthracene, the reproducibility of the packing is found to be around 0.7–0.8% [126], which is also indicated for the determination of formaldehyde where the RSD for 10 mL is from 1.62% and for the blank 1.75% [91].

In general, centrifuging of the solid phase within the cell improves the reproducibility of the procedures, as with the determination of aluminum with CAS in SQAE, where the RSD is 4.4%, while centrifuging at 15*g* for 1 min lowers it to 3% and the absorbance values increase 12% [33].

TABLE 3.8

Effect of volume on precision

| System[a] | 10 | 100 | 250 | 500 | 1000 | Reference |
|---|---|---|---|---|---|---|
| Thiamin–CMC | 1.5 | 1.7 | 2.1 | 2.3 | 2.6 | [119] |
| V-PAR-D.1-X8 | – | 1.4 | – | 1.7 | 1.5 | [120] |
| Anthracene-C-18 | – | 0.6 | – | 0.7 | 0.7 | [126] |

[a]*n*: 10, precision expressed as repeatability.

The robustness of these procedures has been established several times, checking that it is good, although strict control of the stirring time and final volume is necessary [150].

*Selectivity*

The use of a solid phase usually offers advantages in the realm of selectivity. Various strategies have been used to attain retention or selective response for a given analyte. (1) The choice of which solid phase to use is the most common. Thus the interference of Be(II) in the determination of Al(III) with ECR can be eliminated using cellulose as a support, since neither ECR nor the Be(II) complex is retained on it [151]; another example is the UV determination of ascorbic acid in pharmaceutical preparations based on its selective retention in SQAE, since other vitamins, common tablet fillers, and other drugs, such as paracetamol, codeine or caffeine, are not retained [141]. (2) The combination of an adequate support and a reaction can improve selectivity, as is the case with the above-mentioned retention of Cd(II) as iodide complex on anionic Dowex 1-X2 followed by reaction with PAR in the same support, since it eliminates the usual interferences in solution [55]. (3) The modification of experimental conditions, usually pH, gives good results on some occasions. The interference of Fe(III) in the determination of Co(II) with NRS is reduced when the reaction occurs in solution at pH 7.7 and the subsequent retention of the complex on SQAE at pH 2.7 [36]. (4) The addition of masking agents is a common resource for avoiding the retention of an interferent, like the interference of Fe(III) in the determination of Mo(VI) with PV on anionic SQAE which is eliminated with Phen, since the resulting cationic complex is not fixed [139]. (5) The selective elimination of interferent fixed species is another way to attain the necessary selectivity for application; thus treatment with nitric acid destroys copper and iron complexes of NRS on Amberlite CG-400, leaving the cobalt complex unaltered [60]. This decomposition can also make it possible to modify experimental conditions, in this way taking advantage of differences in behavior according to temperature. The tolerance of Al(III) in Ga(III) determination can be increased fivefold by heating the packed solid phase in the cell at 70°C before measuring the fluorescence at 20°C, because the aluminum–SOAPh complex is irreversibly destroyed, while that of Ga(III) is stable [111]. (6) Treatment prior to the reaction and fixation of the analyte on solid phase makes it possible to increase tolerance, as with silicate interference in phosphate determination in waters like in the case of molybdenum blue sorbed in SDEAE. The previous treatment of the water sample with 0.1 M $H_2SO_4$ for 5 min raises the tolerance level up to 20,000 μg/L, probably due to adsorption of hydrogen ions by colloidal silicic acid, usually negatively charged, which could hinder the uptake of silicate by the anion exchange gel [64]. (7) Modification of spectral properties on the solid phase has been used to improve selectivity. This occurs with the determination of zirconium using an anion exchanger modified with ECT where it is possible to avoid the usual interference of hafnium in solution by using the first derivative signals, while in solid phase the absorption maximum is displaced 20 nm with respect to zirconium [18]. (8) The use of more selective signals makes it possible to increase tolerance, as with synchronous SPF which reduces the overlapping of fluorescent signals from other PAH, especially naphthalene and acenaphtene, on those of fluorene [152].

### 3.5.4 Mixtures resolution

The advantages of the SPS methodology can be used to resolve mixtures of compounds (Table 3.9). Several possible strategies have been devised. The first is to secure selective signals by adequate selection of the solid phase. This is the case with mixtures of colorant matters and their subsidiary unsulphonated products in foods, drugs and cosmetic products that have been spectrophotometrically determined by means of the fixation of each one of the colorants on a specific solid support. SYFCF has been fixed on SDEAE A-25 gel in this way, while Sudan I was retained in C-18 silica gel [153]. The fungicide mixture 1-naphthylacetic acid and 1-naphthalenacetamide was determined spectrofluorimetrically in solid phase by fixation, the first element on SQAE and the second in solution [154].

The second strategy attains selective signals by measuring the optical signals under determined parameters. In some cases, it is possible to use absorption spectra directly if their spectral peaks are sufficiently separate, as with nickel and cobalt, both of which react with PAN, forming complexes that are strongly retained on Dowex 50W-X2 and showing absorption maxima at 566 and 628 nm [61]. Similarly, the use of signals of synchronous fluorescence, intrinsically narrower than fluorescent ones, makes it possible to resolve the mixture

TABLE 3.9

Resolution of mixtures by particulated solid phase spectrometric assays

| Analyte 1 | Analyte 2 | Technique | LDR 1[a] | LDR 2[a] | DL 1[a] | DL 2[a] | RSD 1 (%) | RSD 2 (%) | Reference |
|---|---|---|---|---|---|---|---|---|---|
| SYFCF | Sudan I | CSPS | 15–500 | 20–200 | 3.5 | 4.4 | 2.8 | 3.3 | [26] |
| Al(III) | Be(II) | DCSPS | 10–40 | 0.50–4.0 | 2.25 | 0.04 | 3.5 | 3.8 | [156] |
| NAA | NAD | SPF | 16–60 | 6–120 | 3 | 2 | 4.5 | 1.2 | [154] |
| Benomyl | Morestan | SSPF | 0.50–15.0 | 0.60–15.0 | 0.15 | 0.18 | 1.2 | 1.5 | [155] |
| Pyrene | BaP | DSSPF | 0.70–4.5 | 0.40–2.5 | 0.1 | 0.04 | 1.4 | 1.1 | [157] |
| Fe(III) | Cu(II) | PAS | >160 | 300–2200 | – | – | <5 | <10 | [58] |

[a]Concentrations expressed as μg/L.

of the pesticides benomyl and morestan [155]. The use of variable-angle scanning spectrofluorimetry signals makes it possible to increase selectivity in the case of strong overlapping spectra, as with the PAHs fluorene, acenaphtene and phenanthrene mixture (DL 70, 40 and 50 ng/L, RSD 0.94, 0.95 and 0.74%, respectively) [137].

The use of derived signals is more useful with analytes that show some amount of overlapped spectra on solid phase. This includes everything from relatively simple cases such as the determination of the colorants, Quinoline Yellow and Brilliant Blue FCF, in cosmetics where the absorption spectrum of the second overlaps the first. The use of signals derived from the first order and zero order, respectively, makes it possible to resolve the mixture [158]. Mixtures of Be(II) and Al(III) in natural waters, orange juice, and biological and geological samples have been determined using the formation of the respective complexes with ECR in the presence of EDTA, as a masking agent, fixed in an anionic dextran gel and using the first derivative spectra [156].

A combination of selective retention and derived techniques makes it possible to resolve more complex mixtures, such as a mixture of four colorants, namely Sunset Yellow, Sudan I, Quinoline Yellow, and spirit soluble Quinoline Yellow which was resolved after fixation of the water soluble colorants in SDEAE A-25 gel using the second derivative, and the unsulphonated colorants in C-18 silica gel using the first derivative [153]. Signals derived from synchronous fluorescence are very useful for resolving mixtures of PAH [157], pesticides [159] or metallic ions [50,160].

In cases where it is not possible to attain selective signals by any of the above methods, the use of equations has been proposed, as is the case with the determination of the PAH benzo[a]pyrene, benzo[a]-anthracene and pyrene fixed on SG-25 using synchronous spectra [161]. Alternatively, full-spectrum multivariate calibration methods have been used. Here, a mixture of the sulphonated colorants Tartrazine, Ponceau 4R and SYFCF fixed in SDEAE gel can be determined by means of a partial least squares (PLS-1) multivariate calibration (TT: RRMSD 4.00%, $R^2$: 0.9978; SY: RRMSD 6.07%, $R^2$: 0.9977; P4R: RRMSD 4.19%, $R^2$: 0.9954; number of factors: 3 in all cases) [162].

The advantages of solid phase procedures have been used for the speciation of various ions, such as Fe(II) [163,164] Se(IV) [70], V(V) and V(IV) [165] or Sn (IV) where it does not interfere with Sn(II) or organotin compounds, since although the latter react with the reagent (PV) they do not become fixed on the anionic gel [131].

### 3.5.5 Analytical applications

A considerable number of methods have been developed for the analysis of a variety of inorganic and organic analytes in a wide range of real samples as seen above. Fig. 3.9 presents the fields where these procedures have been applied, as a percentage. With notable differences, procedures have been most often applied to the analysis of trace elements, pesticides and PAH, as well as other elements like phenols in waters of various provenance, e.g. natural waters including river, spring, hot spring, rainwater and seawater, and additionally, tap water, distilled water, industrial water, mine drainage water, and wastewater. The procedures have also been applied to the determination of minority constituents of minerals, rocks, crude petroleum, metals and alloys. In food and beverages they have determined traces of metals, both additives and colorants, preservatives and anti-oxidants, and flavor intensifiers. The analysis of trace elements in plant and animal tissues has been handled by solid phase procedures. Sulfamides, vitamins and metals have been determined in pharmaceutical preparations, as well as drugs in biological fluids like urine and serum.

Many of the procedures have been well validated against reference methods for specific applications, as for ciprofloxacin in urine [149] or

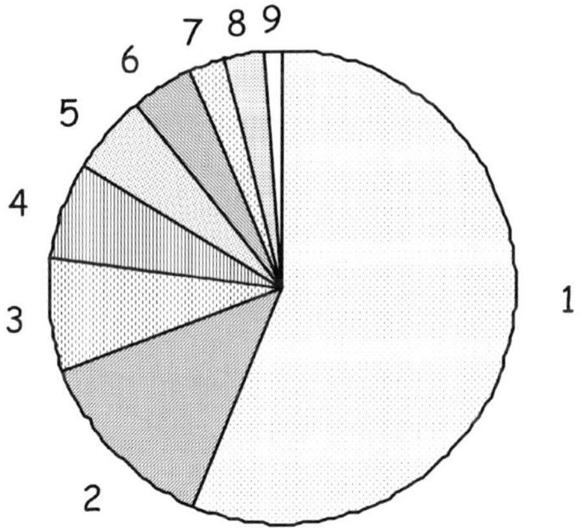


Fig. 3.9. Applications of particulated solid phase spectrometric assays. (1) Water 56.4%; (2) minerals, rocks, petroleum, soils, and metals 12.9%; (3) beverages 7.8%; (4) pharmaceutical formulations and drugs 6.2%; (5) food products 5.7%; (6) vegetable and animal tissues 4.1%; (7) bodily fluids 2.6%; (8) various 2.6%; (9) gases in the atmosphere 1.5%.

thiamin in pharmaceuticals [119], either using certified materials like beryllium in water [53] or aluminum in botanical samples [156] or using chemometric techniques, as for the determination of zinc in water [150] or naphthylacetic acid in commercial formulations [148].

## 3.6 MEMBRANE SOLID-PHASE SPECTROMETRIC ASSAYS

As indicated above, these assays use a membrane in which the analyte or a derivative is retained and which may contain none, some or all of the necessary reagents. When the membrane is placed in contact with the solution via filtration, immersion, with or without agitation, wetting or deposition of a volume of the solution, it unleashes a set of reactions and processes that generate an optical response, typically diffuse reflectance, absorption or luminescent emission, which is measured over the membrane itself.

It is possible to classify membranes into three groups, depending on the form in which the membrane is put into contact with the problem: (a) systems that filter the problem through the membrane; (b) systems where the membrane is immersed in the problem; and (c) systems where the problem is deposited onto the membrane (Fig. 3.10).

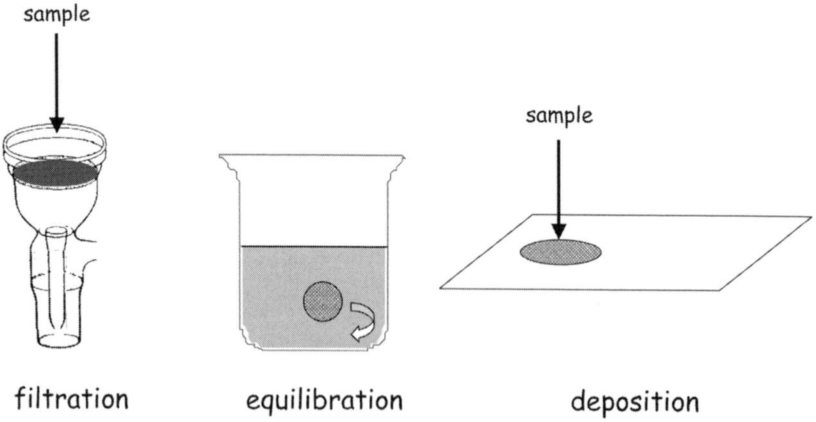


Fig. 3.10. Membrane solid-phase spectrometric assay types.

### 3.6.1 Membrane filtration systems

These systems use a porous membrane through which the problem can be filtered. The filters can be unmodified or modified depending on the type of derivatization reaction that takes place. Among the unmodified membranes, the most common are cellulose nitrate, cellulose acetate, ion exchanger cellulose, polyamide, polyethersulphone, polytetrafluoroethylene, polyacrylonitrile fibers filled with ion exchangers and borosilicate glass filter, all with pore sizes typically between 0.1 and 1 μm. Modified membranes achieve immobilization of the reagents through various procedures: (1) covalent immobilization, as with polydentate cellulose containing formazane or hydrazone groups in position 6 of α-cellulose chains [166]; (2) immobilization by ion-exchange as with PAR on polyacrylonitrile fiber filled with anion exchanger AV-17 membrane [167]; (3) adsorption on the membrane, as is used to immobilize PVP on polystyrene–divinylbenzene disk [168].

The sorption of a derivative of the analyte on the membrane can be accomplished by three mechanisms: ion exchange, adsorption and complexation, depending on where the derivation reaction takes place.

It is quite common for the derivatization reaction to take place in solution, after which it can be retained on the filter by ion exchange, as with the cationic ferrous complex of Phen on cellulose phosphate cation exchanger paper [169]. However, it is more common to retain the hydrophobic complex by adsorption in the presence of surfactants. The role of the surfactant must differ according to the situation, so that some: (a) act as an agent for forming ionic pairs with the derivative in a manner that assures a high hydrophobicity of the derivative, as with the Cu(II) complex with TPPS which is extracted in the presence of cationic Zephiramine as $CuL^{4-}$–$(Zephiramine)_4$ in cellulose nitrate [97]; (b) form ternary complexes, as for the retention of aluminum on cellulose nitrate membranes such as 1:2:4 Al(III)–CAS–Zephiramine complex [170]; (c) make the reaction between the analyte and the reagent possible as with the complex forming reaction in solution of niobium with PNF that takes place in Zephiramine micellar phase, from which the complex is extracted on the cellulose nitrate membrane [171]; (d) suppress the retention of the reagent in the membrane, as with the retention of the Co(III) complex with 5BrPAPS in the presence of the surfactant DBS, due to repulsion between the 5BrPAPS and DBS anions [172].

Derivatization in solid phase can be done by previous immobilization of the reagent in the membrane by adsorption, as with the retention of iodine as a PVP complex on a polystyrene membrane [168], or by covalent immobilization, as when using polydentate cellulose to selectively retain different metals with a definite electron structure and size capable of forcing the formation of 1:1 complexes [166].

Preseparation and derivatization in solid phase has been used to increase the reactionability and selectivity as with lead preconcentrated as anionic thiosulfate or iodide complexes on polyacrylonitrile fiber filled with anion exchanger AV-17 membrane before a subsequent reaction in the membrane with PAR [167].

Another type of derivatization reaction is the retention of ionic pairs, for example for the retention of cationic surfactants with the anionic 1:2 Co(III) complex with HQNS in a PTFE membrane [173]. Redox reactions have also been used, as with the oxidation of TMB by $KIO_4$ catalyzed by Mn(II) for the determination of the same upon retention by filtration in a DETATA paper-based chelating membrane and subsequent addition and reaction with the other reagents [174].

Passing the solution containing the analyte or its derivative through the membrane is done either by filtration under suction using a filter holder with a glass filter base [175], by pumping through the filter holder using a peristaltic pump [176] or manually by a syringe provided with a plastic support holder [168].

The most commonly measured analytical property in assays based on filtration through a membrane is diffuse reflectance [172], as with absorbance either over opaque membranes of different types, either dry or wet [169,177], or using transparent membranes such as a hydrophilized PTFE membrane [173]. The absorbance measurement is analogous to that carried out with the TLS or CIS procedures previously discussed.

The influence of different factors (pH, reagent concentration, surfactants, sample volume, temperature, etc.) on the derivation and fixation of analytes in the membranes depends on the derivation and fixation mechanism, in a way similar to that discussed in the section on particulated assays. Two factors specific to this type of assay are the pore size of the membrane and the filtration rate. In general, a decrease in pore size, typically from 1.0 to 0.1 μm, which is to say an increase in specific surface, means an increase in the analytical signal, suggesting that sorption on the membrane is a surface phenomenon [172,178]. The filtration rate of the solution through the membrane must be adequate to attain equilibrium in those dynamic conditions, a rate usually between 10 and 20 mL/min, independent of the filtration form and the retention mechanism [168,169,176,179].

The analytical characteristics of some procedures for filtration through the membrane are shown in Table 3.10, highlighting the sensitivity that they offer and the small volumes of the problem. Frequently, the calibration function presents a curvature, possibly due to the absorbance measurement that is simply done against a blank. For this reason, it is common to give two calibration intervals.

These procedures have been applied fundamentally to the determination of traces of metallic ions in liquid samples, such as waters of various provenance and beverages. It is important to note the need for some preconcentration procedures to attain the required sensitivity. Examples include the determination of Co(II) in tap water and beverages with previous preconcentration as a PAN complex on cellulose nitrate membrane and after wet ashing, the determination of Co(II) as 5BrPAPS complex on the same kind of membrane [172] or the preconcentration of Be(II) from river and hot spring water on glass-fiber membrane and subsequent determination as CAB beryllium complex on cellulose nitrate membrane [177].

### 3.6.2 Membrane 'problem' equilibration systems

These systems use membranes that are non-porous, transparent and generally hydrophobic, are capable of retaining the analyte or its derivative, and have the optical measurement carried out over them. Several alternatives have been used: (a) systems that do not use a reagent, except for adjustments to the pH, retaining the analyte in the membrane; (b) systems that are derivatized in solution and extract the derivative to the membrane; (c) systems that use membranes containing reagents and extract the analyte that reacts in the membrane. In all of these cases it is necessary to insert the membrane into the problem until it reaches equilibrium with or without shaking. The most common membranes are plasticized PVC, polyurethane foams and ion exchange films.

Procedures that use a membrane to retain the analyte must measure a sufficiently selective property, as with transparent cation exchange films which have been used to preconcentrate copper in water, making it

TABLE 3.10

Analytical characteristics of membrane filtration solid phase spectrometric procedures

| Analyte | Volume (mL) | LDR | DL | RSD[a] (%) | Procedure | Reference |
|---|---|---|---|---|---|---|
| Be(II) | 30 | up to 1.3 μg/L | 0.013 μg/L | 1.3 | TLS[b] | [177] |
| Mn(II) | – | 0.005–2.5 μg/L | <0.005 μg/L | ≤5 | TLS | [174] |
| Fe(III) | 100 | 1.0–10.0 μg/L | 0.2 μg/L | 4.9–5.3 | TLS | [169] |
| Co(II) | 20 | up to 1.5 μg/L 1.5–3.0 μg/L | 0.018 μg/L | 0.2–0.4 | DRS | [172] |
| Cu(II) | 30 | up to 1 μg/L | 0.007 μg/L | 1.8 | TLS | [97] |
| Ge(IV) | 40 | up to 2.5 μg/L 2.5–6.5 μg/L | 0.047 μg/L | 0.76–1.6 | TLS | [180] |
| Nb(V) | 30 | up to 5.3 μg/L 6.7–12.0 μg/L | 0.13 μg/L | 0.9–3.1 | TLS | [171] |
| Mo(VI) | 60 | up to 1.0 μg/L 1.3–2.3 μg/L | 0.13 μg/L | 1.8 | TLS | [175] |
| Pb(II) | 25 | 0.02–0.2 μg/L | 0.01 μg/L | <15 | DRS | [176] |
| $PO_4^{3-}$ | 25 | 22–220 μg/L | 0.9 μg/L | 11 | TLS | [181] |
| $I_2/I^-$ | 10 | 0.1–5.0 mg/L | 0.05 mg/L | 6.0–9.7 | DRS | [168] |
| Cationic surfactants | 20 | $4 \times 10^{-6}$–$8 \times 10^{-7}$ M[c] | – | 1.7–4.2 | TLS | [173] |

[a]Expressed as repeatability.
[b]In all cases the measurement technique is similar to TLS.
[c]In Zephiramine.

necessary only to adjust the pH to 5.0 [182,183]. The use of plasticized PVC makes it possible to retain the germicide nalidixic acid using tributyl phosphate as a plasticizer [184].

Derivatization of the analyte in solution and its retention in a membrane has been achieved by both adsorption and ion exchange. In the first case, the formation of ionic pairs with an highly absorbent species has been quite common, as with the fixation of alkylbenzene-sulfonates as an ion pair with CV [185] or with Fe(III) as 1:2 2,2′-dihydroxyazobenzene complex and an ion pair with CV [186], both in a PVC membrane plasticized with DOP.

The use of polyurethene foam loaded with a liquid anion exchange and TBP as plasticizer makes it possible to use ion exchange to preconcentrate phosphate as molybdenum blue species formed in solution [187,188].

In the last group, the derivatization reaction takes place in the membrane itself where some or all of the necessary reagents are found. One example is the use of a membrane plasticized with NPOE containing bathophenanthroline for the determination of Fe(III) which requires hydroxylammonium sulfate as a reducer, potassium iodide as a counterion and pH 5 acetate buffer, all in solution [189]. With another example, all of the reagents are incorporated in the membrane, making it necessary only to adjust the pH, if that. In this way plasticized PVC membranes or ethyl cellulose containing chelating agents such as ditizone or thenoyltrifluoroacetone have been used for the determination of several metal ions in solution [190,191]. The use of D4 polyurethane hydrogel membranes containing FZ, ascorbic acid and the lipophilic anion TCB makes the determination of Fe(III) possible upon reduction, complexation and retention as an ionic pair in the membrane [192]. In the same way, the use of ion exchanger membranes constructed of Nafion makes it possible to retain cationic species, as with BC impregnated membranes for Cu(I) [193] or membranes with BC and *p*-anisidine as a reducer for Cu(II) [194]. With plasticized PVC membranes containing 18-crown-6-ether as an ionophore, and lipophilised Nile Blue as a chromoionophore, it is possible to determine calcium by an ion-exchange mechanism [195]. In the last three cases, the only reagent needed is the pH buffer.

On certain occasions, the membranes do not need to be prepared, as with transparent cation exchange films [182], though preparation is usually necessary. Membrane preparation can be done in various ways: (a) by impregnation, such as with Nafion self-supporting membrane film in ethanolic solution of BC for 24 h [193]; (b) by slow evaporation of a solution containing the polymer, plasticizer and reagents, if necessary, in a Petri dish, glass slides or stainless steel vat and then cutting the membrane into pieces, for example $5 \times 30\ mm^2$ and 0.150 mm thick [186]; (c) by deposit and evaporation of a small volume of solution on a plastic sheet, typically polyester and usually ∅6 mm and between 15 and 25 μm thick [194].

The equilibration of the membrane with the solution containing the analyte or its derivative can be carried out by adding the membrane to the solution and shaking it magnetically or with a vibrating agitator for times of up to 60 min [189,194]. Alternatively, it is possible to introduce the membrane hanging from a support into a magnetically stirred solution, making certain that the membrane makes perpendicular contact with the surface of the sensing zone for times of up to 60 min [196]. When the reaction is very fast, agitation is not necessary, just a short period of contact [190].

The most commonly controlled variables—in addition to any necessary reagent concentration—are pH, temperature and equilibration time. With derivation in solution, it is usually necessary to work at high temperatures, up to 70°C, because of the slow extraction kinetic of the derivatives formed in solution [197].

The most commonly used analytical property with this type of assay is spectrophotometry, either placing the membrane on a glass plate [197] or in a cell holder prepared for membranes [198], though reflectance spectrophotometry has also been used [199]. Room temperature phosphorescence in the membrane offers interesting possibilities because of the preconcentration factors that it achieves and the lack of a need for heavy atom due to the system's greater rigidity [184,196] (Fig. 3.11).

Table 3.11 presents the analytical characteristics of some assays based on membrane. In general, good sensitivity and reproducibility are attained using small sample volumes, though the contact time with the problem is usually critical. As expected, the precision decreases a little when using membranes with the reagents incorporated.

These procedures have been applied to the determination of metallic ions [186,189,193,200], anions like nitrites [198] and organic compounds [196] in waters of various provenance, as well as metallic ions and organic compounds in milk, wine and body fluids, such as serum, plasma and urine [186,194]. The interference of proteins in the determination of Cu(II) in human plasma has been eliminated using

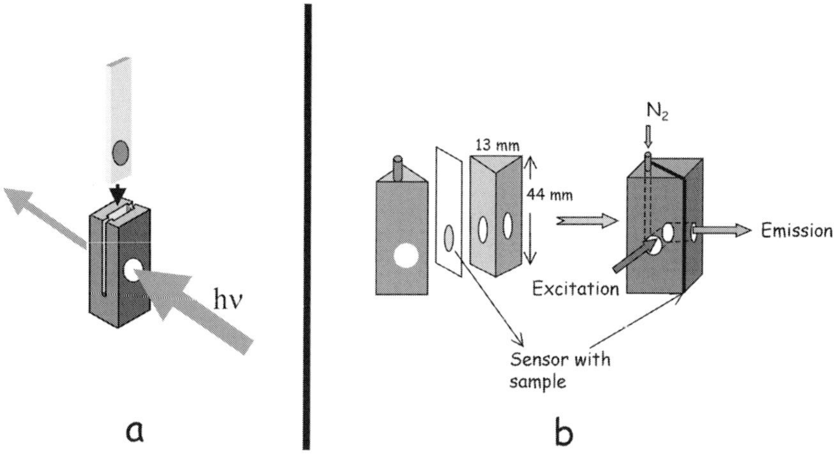


Fig. 3.11. Cell holders for membrane assays. (a) UV–Vis spectrophotometry; (b) room temperature phosphorimetry.

a two-layer design containing a copper-sensitive layer and a dialysis layer made with D-6 polyurethane hydrogel [194]. Validation of the results has been made regularly against reference methods.

### 3.6.3 Membrane 'problem' deposit systems

These systems are the most integrated solid phase spectrometric assays and could be considered as self-contained analytical devices. More precisely, they can be described as integral analytical elements which typically take the form of thin pads or film, and contain all of the reagents required for an assay distributed in a dry (dry-to-the-touch) form within the pad or film [201]. The reaction zones are fastened to the lower region of small plastic bars. To use the systems, a drop or precise volume of a fluid problem is placed on one of the surfaces of the solid phase element by dipping or addition, where it diffuses into the reaction zone and dissolves the reagents. After the reaction, an optical property of the generated derivative is measured on the device itself.

As seen above, the necessary reagents and the precise processes needed to make the performance of certain assays possible have been integrated into a single element in a form which makes it unnecessary to prepare, dispense or mix the needed reagents, or perform operations like reaction, extraction, precipitation or dialysis.

TABLE 3.11

Analytical characteristics of membrane immersion solid phase spectrometric procedures

| Analyte | Volume (mL) | LDR | DL | RSD (%) | Procedure | Reference |
|---|---|---|---|---|---|---|
| Cu(II) | 2000 | up to 2 mg/L | – | – | S, A | [182] |
| Nalidixic acid | 50 | 0.06–1.5 mg/L | 0.02 mg/L | 2.0 | P, A | [184] |
| Morestan | 50 | 1.0–200.0 μg/L | 0.28 μg/L | 2.9 | P, A | [196] |
| Fe(III) | 20 | up to 130 μg/L | 0.6 μg/L | 3 | S, B | [186] |
| Al(III) | 20 | up to 50 μg/L | 3 μg/L | 4.3 | S, B | [200] |
| Alkylbenzene sulfonate | 100 | 0.2–10.0 mg/L | – | – | S, B | [185] |
| Cu(II) | 5 | up to 2.7 mg/L | 9.5 μg/L | 4.9 | S, C | [197] |
| Fe(III) | 5 | up to 2.4 mg/L | 37.9 μg/L | 4.3 | S, C | [189] |
| Fe(III) | 100 | 4–50 μg/L | 1 μg/L | 7.6 | S, C | [192] |
| Cu(II) | 50 | 0.03–2.0 mg/L | 9 μg/L | 4.2 | S, C | [194] |
| $NO_2^-$ | 25 | 2.6–100 μg/L | 1.4 μg/L | 8.8 | S, C | [198] |
| Cd(II) | – | up to 5.0 μg/L | 0.015 μg/L | 1.4 | R, C | [199] |

S: spectrophotometry; P: phosphorimetry; R: reflectance spectrophotometry; A: without reagents membrane assays; B: derivatization in solution membrane assays; C: Reagents in membrane assays.

These assays have their origin in test papers, i.e. paper strips impregnated with reagents, which have long been used in analytical chemistry for the detection of inorganic ions, organic substances and especially pH (Prout reports the use of litmus paper for testing alkalinity in urine in 1817). A milestone in the use of test strips came with the appearance of Ames Clinistix for testing urinary glucose in the 1950s, and since then, clinical chemistry has been the field where strip tests have been most highly developed. They have been used to monitor numerous physiological functions by merely dipping a test strip into a sample of bodily fluid, such as urine or blood, and observing a detectable response, and have become popular because of their relatively low cost, ease of usability, and speed in obtaining results. The first formats for dry test strips relied on visual comparison using various procedures, and the results taken from a color chart were qualitative or semi-quantitative. However, while these tests continue to be used, many others have been developed that make it possible to measure various optical or electrical properties. Here, only quantitative test strips that use optical measurements are discussed.

These membrane assays must be based on reactions of sufficient selectivity in conjunction with different analytical operations to achieve results for real problems without previous treatment of the sample or addition of reagents. Additionally, they must be designed in a form that is easy to use, robust, stable during storage and with which measurements can be carried out safely using, usually, dedicated instruments. Reactions that are used range from complexation reactions to enzymatic reactions or antigen–antibody reactions.

*Classification*

It is possible to classify these membranes by how they relate the problem containing the analyte to the test strip, that is, into vertical and horizontal systems. With vertical systems, the fluid goes through the layer or layers that make up the reagent zone in a perpendicular way, undergoing a series of reactions and processes that make it possible to develop an analytical property. In the horizontal systems, the sample is deposited on a reception zone and is moved by fluid flow through the tests strip, producing chromatographic separations, reactions and/or analyte retentions or interferents that generate the analytical property (Fig. 3.12).

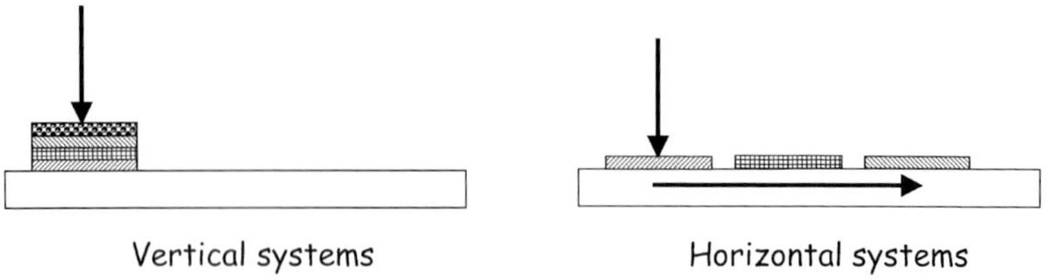


Fig. 3.12. Membrane 'problem' deposit systems.

*Vertical systems*

Vertical systems seek to integrate all the necessary analytical operations into one single set above which the sample is applied, which brings about, after dissolution of the reagents, different reactions and processes that conclude when the measurement is made. The types of operations that must be carried out depend on the type of problem and the chemistries used for its determination.

With these systems, the reagents are in the solid phase and may either be distributed throughout the matrix of the test strip or be compartmentalized into specific domains within the infrastructure of this matrix. The previous distribution of the reagents and necessary components can be achieved in two ways: (1) using a single layer, the so-called monolayer systems; or (2) using a series of differentiated layers that are piled upon each other, the so-called multilayer systems.

The greater or lesser complexity of the system depends on the analyte itself, on the complexity of the matrix that contains it and on the selectivity of the recognition reaction.

*Monolayer systems*. Monolayer systems use membranes made up of a single layer, which can be homogeneous or heterogeneous.

The homogenous monolayer membrane systems are the simplest. They achieve a homogenous distribution of components through: (1) impregnation processes of an adsorbent support (fiber-impregnated systems); (2) deposition of a solution of reagents on a non-absorbent support, (3) fabrication of a film using a film-forming material along with the reagents on a plastic support. This can be achieved through the techniques of immersion, painting, spraying, applying or casting.

Fiber-impregnated systems are the oldest and have been used as a base for the preparation of a large number of commercially available diagnostic products. In general, fiber-impregnated systems consist of a sheet of cellulose matrix which is either porous or semi-permeable with respect to solute diffusion. The chemistries needed for this type of system are distributed homogeneously in a dry form both in the interstices of the cellulose matrix and on the surface of its integral structural fibers [202]. Their widespread use is explained by their low cost, ease of function, and good reactivity on cellulose fibers. However, they present some problems, principally due to non-uniformity of paper, which leads to semi-quantitative results in determinations, but also due to their instability in the face of strong acids, bases or oxidizing substances. Another problem with paper is a chromatographic effect (often called ringing, targeting, doughnuting or banding) due to the tendency of the applied sample to exert a washing effect on the incorporated reagent. Alternatively, several synthetic fibers have been used instead of cellulose, such as, for example, felts or fleeces of polyamides impregnated for testing urobilinogen or ketonic body [202].

Reagent films were introduced to solve the problems of the fiber-impregnated systems. Their principal advantage is a high degree of uniformity with the reagent layer. By using reagent films and optical measurements, especially reflectance, it is possible to achieve measurement results with test strips that have the quality of those obtained with wet chemical methods.

The reagent film either is applied directly to a synthetic polymer film that serves as a handle or is present on an additional carrier film that gives it the stability necessary for production and processing.

In comparison with most papers and cloth-like supports, reagent films have a smooth surface and only a small proportion of hollow spaces, which, apart from the advantages already cited, gives rise to various disadvantages. Among those cited are a low reaction rate or low rate of the entrance of oxygen during enzymatic reactions, for example in the determination of glucose based on glucose oxidase [203].

Generally speaking, the use of homogenous monolayer systems is restricted to specific tests that are simpler or have a lesser degree of integration because of certain inherent problems like: (1) the reagents may be mutually incompatible; (2) some of the necessary reagents may be soluble in water and others in organic solvents; (3) some reagents may be leached during their use; (4) it may be necessary to modify the experimental conditions, such as pH, during the different reactions that take place; (5) it may be necessary to carry out separation processes, like extraction or dialysis, incorporated in the analytical scheme; (6) the equilibration times may be very high, which may disqualify their use.

There are several solutions to these problems, one of which is the use of heterogeneous monolayer membraness. With these, the membrane is compartmentalized either to segregate mutually incompatible reagents or to separate particular sequential steps of the overall analytical reaction.

The strategies used to prepare the monolayer heterogeneous membranes depend on the type of problem that they seek to resolve. Some of these are discussed below, classifying them according to their objective:

(1) To create hydrophilic and hydrophobic zones in the membrane. This method seeks to incorporate soluble reagents in one of the two zones, which makes possible reactions that involve the transference of material from one to the other. Among the solutions used are: (a) use of a hydrophilic carrier matrix (gelatin, agarose) incorporated, by emulsion, with finely divided globules of a hydrophobic liquid, or plasticizer, such as the high boiling point oxygen donor NPOE, with the emulsion then coated onto a support member. This system has been used for the determination of potassium ion in serum through an ion-exchange mechanism between a potassium-selective ionophore, namely valinomycin, and a fluorescent reporter substance, typically a liphophilic pH indicator, both placed in the hydrophobic compound. The buffer needed to adjust pH is placed in the hydrophilic carrier [204]; (b) a porous carrier matrix (paper) containing, by successive impregnations, the buffer and a homogeneous hydrophobic solution of reagents in an organic liquid that usually includes a polymer. These have been used for alkaline and alkaline-earth ion determination in serum based on the same principle as above [205]; (c) design of core/shell polymer beads comprising a hydrophobic and unswellable core and a hydrophilic outer shell. The individual particles are connected to one another by adhesion of the hydrophilic outer layers and placed onto a support. This design has a good retention capacity for hydrophobic or high-molecular weight compounds but shows problems of particle adhesion [206].

(2) To increase the porosity of the membrane. This seeks to resolve problems that reagent films present, namely to increase the reaction rate, increasing the area of contact with the problem and permitting access by molecules of high molecular weight, such as enzymes, with the goal of being able to analyze them. Among the solutions that have been tested are as follows. (a) The production of porous membranes by introducing film openers, i.e. non-reactive, fine and insoluble solid organic or inorganic particles such as $TiO_2$ or cellulose, in the hydrophobic film. These have been used, for example, for cholesterol determination in serum using a membrane of polyvinyl propionate,

$TiO_2$ and cholesterol oxidase, cholesterol esterase, peroxidase and other reagents [207]. This strategy has also been used in test strips for alkaline and alkaline-earth ion determination in serum based in ionophores. Hydrophobic components (ionophore, lipophilic pH indicator, organic acid, plasticizer) and hydrophilic (buffer, others) are both placed on the surface of porous diatomaceous earth particles. The preparation form assumes the recovery of particles with the aqueous solution, eliminating the solvent, and later treating it with organic solution or vice-versa [204]. (b) The use of an hydrophobic film-forming polymer (i.e. polycarbonate) with a hydrophilic organic polymer (i.e. polyacrylamide), that contains the reagents, dispersed as discrete particles prepared by W/O emulsion polymerization. Such systems have been used for glucose determination in urine by glucose oxidase/ peroxidase reaction [208]. (c) The use of a water insoluble hydrophilic polymer (i.e. cellulose triacetate) in the form of microspheres ($< 100$ μm) containing the reagents as microvacuoles. Preparation uses a complicated W/O/W emulsion procedure and has been used, for example, to determine catalase in cow's milk [209].

(3) To achieve the separation of constituents with a high molecular weight. The presence of macromolecules and, especially, of particles such as cellular and particulated matter in the problem impedes direct determination of analytes. A characteristic example is interference in the determination of analytes such as glucose, uric acid or enzymes in whole blood caused by the presence of erythrocytes or free hemoglobin. Various systems have been proposed to integrate elements that make determination possible without prior separation. Among these are the following. (a) The use of a coated matrix or one impregnated first with reagents and subsequently in separating components (i.e. water-soluble salts or carbohydrates like mannitol). The whole blood first contacts the separating components that retain the particles, generating the serum fraction migration and reaction, with which it is possible to read the response on the other side of the sample deposit. This can generate errors due to the separating components which can produce hemolysis [210]. (b) The use of a porous matrix with a small pore size (i.e. polyamide 0.1–150 μm) to which separating components (polyethylene glycol, polystyrene sulfonic acids) and testing reagents are covalently bound or impregnated. The matrix allows the sample to flow, retarding the cells, and the separating component acts as a surface tension modifier capable of releasing analytes from the separated component fluid. If the matrix has a sufficient thickness, formation of a derivative

on the opposite side of the sampling is possible [210,211]. (c) The use of anisotropic membranes, i.e. with a density gradient from one planar surface to the other, composed usually of cellulose nitrate and cellulose nitrate–cellulose acetate mixed ester polymers, with reagents incorporated in different ways. It is possible to prepare these with the settling from a slurry during evaporation [212] or by a coagulation process [213], although with the second procedure it is possible to attain higher anisotropy, that is, a higher ratio of the mean pore diameter between the surfaces. The sample receptive surface of the membrane is relatively dense, to avoid the penetration of cells, particles and/or macromolecules, whereas the more porous surface allows for infusion of the reagent system during manufacture, and the formation, diffusion and measurement of the derivative of the analyte of interest.

(4) To attain stabilization of reagents or isolation of incompatible reagents. This is one of the most common problems, reagents that react among themselves, which reduces—at times drastically—the shelf life of these assays or makes the reagents unstable. Among the various strategies used for monolayer assays are the following. (a) Microencapsulation of reagents that are freed in contact with the problem, such as those that have been used for hydroperoxides, by means of gelatin or carboxyvinyl polymers, in a test strip for hidden blood in feces. This is based in a hydroperoxide as oxidant, in addition to an oxidation indicator, for example *o*-tolidine. The reaction of these two components, which is catalyzed by blood in the aqueous phase, cannot be fully prevented on the test strip [214]. (b) Preparation of synthetic polymers that contain some of the reagents incorporated during preparation in the matrix, while the rest are placed, via impregnation of the polymer, in the pores [215]. (c) Incorporation of incompatible reagents in different batches of the same hydrophilic absorbent, such as powdered cellulose, aluminum oxide or silica gel. After this, a mixture of the two dried impregnated sorbents may be secured to a support with an adhesive and/or using pressure [216]. (d) Preparation of two types of polymeric hydrophilic microspheres containing the incompatible reagents, each one dissolved or dispersed as small vacuoles. Different kinds of beads (100 μm in size) were applied as a monolayer on a strip using an adhesive. In this case, the products of the reaction remain within the microspheres. This has been used for the preparation of a test strip to identify catalase in cow's milk using the β-galactosidase/β-galactose oxidase system [209].

(5) To improve optical measurement. Most solid phase test strips are monitored by reflectance methods, so the inclusion of a reflective component in the membrane minimizes the absorptive and screening properties of the matrix, improving diffuse reflectance measurement. The reflective particles ($TiO_2$, ZnO, $SiO_2$, $BaSO_4$) must be optically opaque granular or particulated matter, with no substantial internal void volume, and chemically inert to reagents and sample. The combination of a high refractive index (around 1.6) and irregular particle shape (0.1–200 μm) enhances the scattering of incident radiation. The reflective particles can be dispersed along with the reagents in a binder material. A test strip along these lines has been proposed for potassium determination in serum based on ionophore and a reporter substance dissolved in a plasticizer and dispersed with the $TiO_2$ reflective particles in cellulose acetate [217]. Alternatively, the reagents can be deposited on the surface of the reflective particles and dispersed uniformly in the binder. This disposition has been used as well for the determination of alkaline ions in serum by means of ionophores [218].

With monolayer systems, both homogeneous and heterogeneous, materials and reagents in addition to those discussed above have been used to carry out a series of auxiliary functions. The addition of emulsifiers or surfactants, especially anionic, zwitterionic or non-ionic, in analytical elements can enhance solvent extraction and coatability of composition formulations and also enhance the extent and range of wetting in elements that are not easily wetted by samples. The use of thickeners, like ethylcellulose, improves the consistency of films. The addition of a dye color additive can improve the test strip measurement [219]. Other components include masking agents for removal of interferents, stabilizers, photochemical stabilizers, preservatives and the like [220].

*Multilayer systems.* The use of monolayer formats has certain advantages as far as manufacture. However, they do not efficiently resolve all problems, nor can they be applied to all sets of reactions or processes. Some compounds in aqueous solution can only rarely be determined with just one reagent and, consequently, a combination of several reagents is often used for multi-stage reactions. For this reason, multilayer systems have been developed. These contain a set of layers of different natures through which the sample runs vertically, producing a sequential series of analytical operations, such as dissolution, filtration, dialysis or extraction, along with chemical reactions of different types, such as neutralization, complexation, redox reaction, formation of insoluble compounds and stereospecific reactions, all of which are integrated in a film with a cross-sectional thickness of less than 100 μm.

The number of layers may vary depending on the problem to be solved (analyte and matrix), on the chemistries and on the measurement system being used. Usually whole layers are built on a support material that consists of a thin, rigid plastic or plastic-like material that may be transparent or reflective. Where the support is hydrophobic and the adhesion between the support and the hydrophilic binder of the reagent layer is insufficient, it may be subjected to processes that render the surface hydrophilic, such as ultraviolet ray irradiation or electron beam irradiation. It is also possible to subject it to an auxiliary process such as (1) the formation of an undercoating layer made up of a substance with appropriate adhesion to both the support and the hydrophilic binder of the reagent layer on the surface of the support or (2) the formation of minute irregularities on the surface of the support (e.g. by brushing and electrolytic etching), but only to the extent that the light transmission property is not greatly reduced.

Different types of layer that can be structured in different ways for each case exist. The following are the most common.

(a) *Reflecting layer or radiation-blocking layer.* The function here is to reflect to a detector any radiation not absorbed by the chemistry of the carrier and that can be permeable to a predetermined analyte or to a reaction product dissolved or dispersed in a liquid. The porous layer is usually made up of pigments such as $TiO_2$ or $BaSO_4$ included in a hydrophilic substance, such as gelatin or polyvinyl alcohol, and additional wetting agents and suspending agents [221]. It is also possible to use reflective material such as metal foils and, if the reagent zone is made up of paper, the paper matrix itself can act as a reflective layer. Such reflectance can also be provided by a spreading layer.

(b) *Opacifying layer.* This is an optional layer deposited on top of the reflecting layer used to impart an opaque appearance, improve the optical measurement, prolong the stability of the generated color and, along with the reflecting layer, screen compounds separated from the sample. Thus, in whole blood determinations, the purpose of these layers is to screen the color of erythrocytes [222]. Additionally, these two layers effectively seal the reagent layer from the outside environment.

The reflecting and opacifying layers can also contain optional components such as buffers, interferant removal substances, i.e. typically polyamine polycarboxilic compounds, or decreasing hematocrit dependence substances, i.e. sodium chloride. Positioning these components in the initial layers assures their redissolution in the first stages of the entire process.

(c) *Spreading layer or metering layer.* Used within the layer for absorption and distribution or metering of some component of a liquid sample applied to the element or a derivative to provide, at any given time, a uniform concentration of the substance at the surface of the spreading layer facing the reagent layer. Usually this layer is isotropically porous; that is, it is porous in every direction within the layer. The spreading is generated principally from a combination of hydrostatic pressure of the liquid sample, capillary action within the spreading layer, surface tension of the sample, and wicking action of layers in fluid contact with the spreading layer. Uniform permeability ensures that undesirable concentration gradients within reagent layers are avoided due to a chromatographic effect. Such layers can be prepared using a variety of components, such as particulated material with interconnected spaces between the particles, i.e. pigments, diatomaceous earth and microcrystalline colloidal materials derived from natural or synthetic polymers. Alternatively isotropically porous polymers, i.e. blush polymers, made with polycarbonates, polyamides, polyurethanes or cellulose esters, can be used [221,223]. The spreading layer is usually the first layer and may be comprised of a pigment acting as a reflection layer, though the reflecting layer can be situated later.

(d) *Reagent layer.* The reagent layer can consist of a single layer, several consecutive layers or layers separated by auxiliary layers. One example of a single reagent layer is that used for the determination of urea in serum composed of a hydrophilic binder (gelatin) with urease and a pH buffer, and fine hydrophobic particles dispersed in the binder containing the pH indicator bromothymol blue. The hydrophobic particles are permeable to $NH_3$ gas but not to an aqueous solution [223]. A two-layer system for blood urea analysis uses a porous film containing urease and a pH buffer separated by a semi-permeable membrane that makes the diffusion of the ammonia generated possible from a second layer containing a pH indicator [224]. Different multi-stage reactions can be accomplished in an element having a plurality of reagent layers, each of which may be adapted to enhance or effect particular reaction stages. For example, in the determination of the enzyme glutamic-oxalacetic transaminase in serum, sequential reactions can be used. This enzyme catalyzes the conversion at pH 7.4 of α-ketoglutarate and aspartate to the corresponding oxalacetate and glutamate. The oxalacetate can be measured by its reaction with NADH catalyzed by maleic dihydrogenase wherein the oxalacetate is converted to the corresponding malate and the NADH is oxidized to $NAD^+$. The course of the reaction is observed by decreases in fluorescence or absorption of NADH. To facilitate the reactions, the reagents can be separated into two, preferably discrete, layers. Thus, α-ketoglutarate and aspartate can be incorporated in a first reagent layer that is coated over a second reagent layer that contains the maleic dehydrogenase and NADH [221].

Homogeneous specific binding immunoassays can be mono- or multiphasic in format, containing a labeled conjugate and a binding partner for the ligand, e.g. an antibody or fragment thereof or a natural receptor protein, along with the ancillary reagents needed to measure the labeling substance in the labeled conjugate. For example, phenytoin can be determined by a homogeneous prosthetic group-labeled immunoassay, because the carrier matrix format makes simultaneous competition between prosthetic group-labeled phenytoin and the analyte for binding to an antibody for phenytoin possible [225].

A combination of reagent layers can be used to isolate incompatible reagents. This is the case with the determination of uric acid based on its reaction with iodine in alkaline media, where a test strip with two reagent layers is used. The first contains a mixture of the alkalines iodide and iodate, starch and a solid and less soluble base, either solid, such as alkaline-earth carbonate, or a base encapsulated in gelatin, such as alkaline carbonate. The acid needed to generate iodine from iodide and iodate (oxalic acid) is placed in the other layer. A calculated amount of all reagents makes it possible to carry out a first reaction in an acidic medium and, by neutralization, to have the basic medium available for the next reaction with uric acid [226]. Another example is the determination of ketone bodies using sodium nitroprusside in a strongly alkaline solution, in which sodium nitroprusside is very unstable. One of the many solutions available is a test strip with two reagent layers, one of which contains phosphate buffer and glycine while the other is a water-soluble paper (carboxymethyl cellulose) impregnated in nitroprusside [227].

Typically, the reagent and support layers are radiation-transmissive, so that a spectrophotometric measurement of the optical density of the dye formed in the reagent layer can be made with the element remaining intact.

(e) *Registration layer.* Some of the colored derivatives formed in the reagent layer may migrate or wander into the opaque spreading layer where they would not be detected during measurement, thereby reducing the sensitivity and the accuracy of the analysis.

To solve this problem, a layer can be included to receive any reaction products or other materials whose relative presence or absence is related to the detection of an analytical result. Such a layer is in fluid contact with the reagent layer and may be separated from this reagent layer by a reflecting and/or opacifying layer to facilitate the result measurement. Such registration layers can include hydrophilic colloids, such as those useful in reagent layers.

Additionally, when dyestuffs are produced in the element, the registration layer may contain mordant materials, such as antibodies or polymers containing quaternary ammonium units [228], to fix them in this layer. It has been found that if the mordant is placed directly in the reagent layer, it often interferes with the reactions initiated by the presence of the analyte and prevents, or significantly inhibits, the formation or release of the detectable species [229,230]. Further, the registration layer can also contain, if desired, interactive materials such as leuco dyes or other materials appropriate to form materials initially that contribute to the analytical signal in the registration layer.

(f) *Interference removal layer.* The elimination of interferences can be accomplished by various mechanisms, including complexation, precipitation and oxidization. At times, the agent for eliminating interferences is not located in a differentiated layer, but incorporated in other layers, such as a spreading layer or reagent layer. This is the case with complexones, which are used to eliminate the interference of alkaline-earths in the determination of alkalines in serum [205,231] or with oxine, to eliminate magnesium interference in the determination of calcium [232]. As an example, ascorbate oxidase may be incorporated in an element to remove ascorbate ion, which can interfere with analysis for glucose. [221].

(g) *Separation layer*. These layers are used for the separation of sample substances based on size or polarity. Thus a filter or dialysis layer may be included at an appropriate location in the element. A typical problem is the determination of an analyte in blood. In the first systems, erythrocytes must be eliminated by an extra manipulative step, such as wiping, blotting or rinsing with water after contact with the test, which lessens accuracy and efficiency. A large number of possible solutions separating cellular blood components have been proposed, such as using dialysis layers made on cellulose [221], polymethacrylate [233] or dextran [234] or the use of asymmetrically porous membranes, i.e. polysulfone membranes, having pores of progressively decreasing diameter from the dosing surface to the determining surface [235].

A phase composed of a dispersion of a hydrophobic solvent in a hydrophilic phase, or vice versa, can serve to separate hydrophobic or hydrophilic components [236].

(h) *Unstable reagents layer*. One of the problems with test strip elements is the need to incorporate very reactive, unstable chemical products such as highly alkaline or acidic compounds. Various solutions, such as using compound solids such as carbonates or phosphates for the bases whenever possible, have been proposed [237]. Another solution is using rupturable pod-like members containing a liquid reagent for the assay. Rupture of the member can be affected by diverse means including pressure, solvents, or heat. Such containers are prepared from a rectangular sheet of fluid- and air-impervious material folded longitudinally upon itself to form two walls which are sealed in such a way that a liquid reagent is contained. Then, this liquid layer is sandwiched between porous spacer layers. For example, for creatinine assay, pressure-rupturable pods are composed of strips of polyethylene by folding the strip and sealing the edges by heat, and contain 0.1 mL of 1.0N NaOH solution and hydroxyethyl cellulose as a thickener [238].

(i) *Photoresponsive layer*. These have been proposed for chemiluminescent analytical devices and usually are a photoresponsive imaging layer of the type which is permanently transformed by exposure to a light response produced by the system in proportion to the amount of radiation emitted. These include photographic film, particularly self-developing photographic black and white film and color film [239].

(j) *Other optional layers* include: (1) subbing layers that may be included between other layers to provide the required adhesion and fluid contact between such layers; (2) protective layers made of some inert material and usually mounted over the spreading layer to prevent damage to the set of layers during sample application [240]; (3) swellable interlayers used to enhance the permeation of analyte through the spreading layer into the reagent layer [228].

(k) *Additional reagents.* It can be advantageous to incorporate one or more surfactant materials such as anionic and nonionic materials in the layers of the element. These can enhance the coatability of layer formulations and enhance the extent and rate of spreading in spreading layers that are not easily wetted by liquid samples in the absence of surfactants. Interactive materials can also be present in the spreading layer, if desirable, for a particular analysis. As an example, proteins or other higher molecular weight materials can be divided into more easily spreadable and analyzable components of lower molecular weight by having an interactive material such as an enzyme, e.g. a protease or esterase, in the spreading layer. Other common reagents include preservatives, thickeners and stabilizers.

*Horizontal systems*

With horizontal systems, a lateral sample movement by fluid flow is imposed that makes it possible to take up the sample while avoiding cross contaminations, arrange analytical operations sequentially and carry out chromotographic separations that cannot be performed with vertical systems.

With these test strips, a capillary transfer plane member with capillary transfer passages is placed above a solid support to control sample flow rate, above which are consecutively different areas for reaction, separation and measurement as well as other areas obtained by impregnation or deposited in the form of one or more layers.

There is usually a metering zone to deposit or absorb the sample, one or more reagent zones and a measurement zone. From the metering zone the sample moves towards the reagent zone, passing first through areas that can contain oxidizers, or precipitants to eliminate interferents, or buffers to adjust the pH. There may be different reagent zones if different sequential reactions that require different conditions exist or if it is necessary to collect and separate some reaction product, as with ammonia in the determination of urea with urease, where a reaction area is needed to free gas and act as a barrier [241]. The measurement zone usually has a mordant to avoid the dispersion of the reaction product [242]. Usually the systems are based on unilateral directional flow capable of transporting reagents, products, and/or sample [243], although bilateral flow systems that avoid the accumulation of solids in the reaction zone have been proposed [244].

Horizontal systems have been used to implement heterogeneous specific binding immunoassays in which the separation of the free- and bound-phases are accomplished by capillary migrations along the carrier element [243,245,246–248].

*Influencing factors*

Among the various factors that influence response with this type of solid phase assay, it is important to mention sample volume, contact time and variability of the test strips.

The way that the test strip is put in contact with the sample can be by deposition of a given volume or by immersion of either the entire strip or simply the sample area in the problem. If the test strips are used for quantitative analyses, exact metering of the sample is necessary, as done with multi-analyte dedicated instrumentation, but not with self-diagnosis applications. Different systems have been devised so that the test strip takes up only the given volume of the sample by immersion and to avoid non-uniform moistening, such as using a depression of the carrier, whose size determines the volume of sample which can be absorbed, having on its edge an absorbent material which takes up the excess of sample [249]. Another proposed solution is to include an additional dosaging step in the test strip, such as in [250] where a distribution membrane is situated over a set of layers that impedes lateral distribution of the problem, thus regulating the quantity of sample taken up; alternatively [251] a slowly-absorbing layer can be placed directly over the support and on it, a water insoluble film, a liquid-impermeable reagent layer and, finally, a mesh layer covering the entire thing. After a brief immersion in the problem, the mesh serves to conduct the problem towards the slowly-absorbing layer where it penetrates laterally to the reagent layer.

The analyte reaction time on the test strip is a very important variable since it is only by measuring the response after a predetermined time, the normal read time, that it is possible to obtain quantitative information.

With the use of automated systems this is not a problem since it is easy to control the reaction time after the addition of the problem. Some strategies for precluding the need to time the reading have been developed in the form of 'dip-and-read' strips. Similarly, an inhibitor system has been proposed which, upon contact with the sample, prevents the reactant system from interacting further with the component after a predetermined time has elapsed. One example is heat-generating agents to deactivate an enzyme or chemical inhibitors such as microencapsulated metallic ions [252]. The use of a chemical timer; i.e. an element on the strip which changes color to indicate that the time needed for the reaction has passed, has been proposed, as with the enzymatic oxidization of glucose in the presence of a defined quantity of ascorbic acid as an inhibitor [253]. The use of measured quantities of a chemical antagonist of the derivative being measured that impedes its appearance until it runs out is one method for test strips with a very slight dependency on time and for directly readable systems [253–255].

Variability in test strips is usually controlled externally, but also by adding a control zone to the test strip that contains an analyte or a derivative which frees the analyte in contact with the sample [256]. This can control correct operation of the device, correct for strip-to-strip variability, and even minimize lot-to-lot and run-to-run variations [244].

*Analytical characteristics and use*

Given the structure of different types of test strips, they are usually opaque, which is why the usual measurement method is by diffuse reflectance spectroscopy, measuring at one or two wavelengths [257]. Measurement can be performed in two ways. After reaction in the surface of the analytical element, the reflectance of the derivative in the sampling surface is determined. This approach has been the one adopted so far for monolayer fiber-impregnated test strips. With the second method, the sample is applied to the top surface and the color is measured by monitoring reflectance in the reverse surface of the element. This approach is common with multilayer test strips. Chromaticity functions can in some instances improve results [258]. The measurement of absorbance by transmission is not very common, although some systems that use water swellable transparent membranes have been proposed [259]. The measurement of luminescence (fluorescence, phosphorescence, chemiluminescence) in its diverse variations makes it possible to increase sensitivity and selectivity in these determinations [245,260].

TABLE 3.12

Analytical characteristics of test strip procedures

| Analyte | Sample | LDR | DL | RSD (%) | Procedure | Reference |
|---|---|---|---|---|---|---|
| K(I) | Blood | up to 1.75 μM | – | – | IN, FFF | [261] |
| Ca(II) | Serum | up to 10 mM | 1 mM | – | IN, DRSCF | [205] |
| Chlorine | 10 | 0.5–20 mg/L | – | – | R, DRS | [262] |
| Trinitrotoluene | Water | 1–10,000 μg/L | 0.7 μg/L | – | I, DRS | [263] |
| Putrescine | Tuna | up to 75 μM | 0.5 μM | <20 | E, DRS | [264] |
| Uric acid | Urine, serum, saliva | up to 10 mg/dL | – | <6 | R, DRS | [265] |
| BUN | 20 | 1–150 mg/dL | – | – | E, DRS | [224] |
| 2,4-D | Water, fruit, urine | 0.5–100 μg/L | 0.02 μg/L | – | I, DRS | [266] |
| Glucose | Blood | up to 90 mg/dL | – | <5 | E, DRS | [241] |
| Bilirubin | Urine | 0.5–3 mg/dL | – | – | D, DRS | [267] |
| Cholesterol | Blood plasma | up to 300 mg/dL | – | <6 | E, FFF | [268] |
| Protein | Urine | up to 100 mg/dL | 5 mg/dL | – | PE, DRSCF | [258] |

IN: Ionophore reaction; R: redox reaction; I: immunoassay; E: enzymatic; D: dye formation; PE: protein error phenomena; BUN: blood urea nitrogen.

Table 3.12 presents the analytical characteristics of different types of test strip procedures. One of the more interesting characteristics is the small sample size needed, since most of the dry reagent chemistries do not require more than 10 μL of serum or 30 μL of whole blood. In general, the performance of the results is good, with RSD values $<2\%$ in some cases and good correlation with reference methods.

Test strips have particularly been used in clinical chemistry, where they have been one of the motors for change in using the clinical laboratory to analyze blood metabolites, enzymes, therapeutic drugs and electrolytes and have been commercialized by different companies (Bayer (Ames), Roche (Boehringer Mannheim), Johnson & Johnson (Eastman Kodak), Menarini), along with hand-held dedicated devices or large freestanding instruments. Although to a lesser extent, test strips have also been used for analysis of food [269,270,271], and for industrial [272] and environmental samples [273,274].

The advantages of the independent usage and immediacy that this type of integrated system presents for obtaining chemical information are so significant that there is great expectation regarding their widespread development in different fields of the clinical laboratory in the future.

## ACRONYMS

| | |
|---|---|
| AEAPS | *N*-2-Aminoethyl-3-aminopropyltrimethoxisilane |
| ATCA | Aurintricarboxlic acid |
| BCA | Bicinchoninic acid |
| BaP | Benzo[a]pyrene |
| BC | Bathocuproine; 2,9-dimethyl-4,7-diphenhyl-1, 10-phenantroline |
| BHA | Benzohydroxamic acid |
| 3BrBHA | 3-Bromobenzohydroxamic acid |
| 5BrPADAP | 2-(5-Bromo-2-pyridylazo)-5-diethylaminophenol |
| 5BrPAPS | 2-(5-Bromo-2-pyridylazo)-5-[*N*-propyl-*N*-(3-sulfopropyl)amino]phenol |
| 5BrSHA | 5-Bromosalcylhydroxamic acid |
| CA | Carminic acid |
| CAS | Chromazurol S |
| CV | Cristal Violet |

| | |
|---|---|
| CIS | Coagulated ion-exchanger spectrophotometry |
| CRT | Chromotropic acid |
| CSPS | Conventional solid phase spectrophotometry |
| 2,4-D | 2,4-Dichlorophenoxyacetic acid |
| DBS | Sodium dodecylbenzene sulfonate |
| DCSPS | Derivative conventional solid phase spectrophotometry |
| DETATA | Diethylenetriaminetetraacetate |
| DMG | Dimethylglyoxime |
| DOP | Dioctylphthalate |
| DRS | Diffuse reflection spectroscopy |
| DRSCF | Chromaticity functions diffuse reflection spectroscopy |
| DSSPF | Derivative synchronous solid phase spectrofluorimetry |
| EBT | Eriochrome Black T |
| ECR | Eriochrome Cyanine R |
| EDTA | Ethylenediaminetetraacetic acid |
| FFF | Front-face fluorescence |
| FZ | Ferrozine; 3-(2-pyridyl)-5,6-diphenyl-1,2,4,-triazine |
| HQNS | Sodium 4-hydroxy-3-(8-quinoyl)azo-1-naphthalene sulfonate |
| NAA | 1-Naphthylacetic acid |
| NAD | 1-Naphthalenacetamide |
| NCPA | *p*-Nitro-chlorophosphonazo |
| NPOE | *o*-Nitrophenyl octyl ether |
| NPOE | 2-Nitrophenyl octyl ether |
| NRS | Nitroso R Salt |
| PAH | Polycyclic aromatic hydrocarbons |
| PAN | 1-(2-Pyridylazo)-2-naphthol |
| PAR | 4-(2-Pyridylazo)-resorcinol |
| PAS | Photoacustic spectrometry |
| pDMAB | *p*-Dimethylamine benzaldehyde |
| Phen | 1,10-Phenanthroline |
| PNF | 2,3,7-Trihydroxy-9-phenyl-6-fluorone |
| PTFE | Polytetrafluoroethylene |
| PV | Pyrocatechol Violet |
| PVC | Polyvinylchloride |
| PVP | Polyvinylpyrrolidone |
| RRMSD | Relative residual mean standard deviation |
| SCPAR | Sulfochlorophenolazorhodanine |
| SDEAE | Sephadex DEAE |
| SG-15 | Sephadex G-15 |

| | |
|---|---|
| SG-25 | Sephadex G-25 |
| SOAPh | Salicylidene-*o*-aminophenol |
| SPF | Solid phase spectrofluorimetry |
| SQAE | Sephadex QAE |
| SSP | Sephadex SP |
| SSPF | Synchronous solid phase spectrofluorimetry |
| SYFCF | Sunset yellow FCF |
| TAAP | 2-(2-Thiazolylazo)-5-diethyl-*m*-aminophenol |
| TAN3,6S | 1-(2-Thiazolylazo)-2-naphthol-3,6-disulfonic acid |
| TCB | Potassium tetrakis-(4-chlorophenyl) borate |
| TLS | Ion-exchange resin thin-layer spectrophotometry |
| TMB | 3,3′,5,5′-Tetramethylbenzidine |
| T4MPP | α,β,γ,δ-Tetrakis-(4-*N*-methylpyridyl)porphine |
| TPPS | 5,10,15,20-Tetraphenyl-21*H*,23*H*-porphinetetra sulfonic acid |
| XO | Xylenol orange |
| Zephiramine | Benzyldimethyltetradecylammonium chloride |

## REFERENCES


1 M. Fritz, D. Thym, P. Vogel and D. Mosoiu, US Patent 5,211,914, 1993.
2 E. Morosanova, A. Velikorodnyi and Yu. Zolotov, *Fresenius, J. Anal. Chem.*, 361 (1998) 305.
3 A. Molina-Diaz, M.J. Ayora Cañada and M.I. Pascual-Reguera, *Spectrosc. Lett.*, 31 (1998) 503.
4 K. Yoshimura, H. Waki and S. Ohashi, *Talanta*, 23 (1976) 449.
5 F. Feigl and V. Anger, *Spot Test in Inorganic Analysis*, 6th ed, Elsevier, Amsterdam, 1972.
6 F. Feigl, *Spot Test*, Elsevier, Amsterdam, 1954.
7 M. Valcárcel and M.D. Luque de Castro, *Flow-through (Bio) Chemical Sensors*, Elsevier, Amsterdam, 1994, pp. 20–21.
8 K. Yoshimura, S. Nawate and G. Kura, *Analyst*, 115 (1990) 772.
9 L.F. Capitán-Vallvey, M.C. Valencia and G. Mirón, *Anal. Chim. Acta*, 289 (1994) 365.
10 S.K. Mittal, *Indian J. Chem.*, 32A (1993) 1106.
11 S. Hoshi, K. Konuma, K. Chiba, K. Sugawara, K. Akatsuka and M. Uto, *Bunseki Kagaku*, 48 (1999) 1129.

12 G.D. Brykina, L.S. Krysina and V.M. Ivanov, *J. Anal. Chem.*, 43 (1988) 1247.
13 C.Y. Liu, H.T. Chang and L.J. Uang, *Fresenius J. Anal. Chem.*, 328 (1987) 61.
14 Y.V. Kholin and I.V. Khristenko, *Zh. Prikl. Khim.*, 70 (1997) 939.
15 R.S. Field, D.E. Leyden and R.S. Shereedhara Murthy, *Anal. Chim. Acta*, 186 (1986) 123.
16 R. Li, Z. Jiang, L. Mao and H. Shen, *Talanta*, 47 (1998) 1121.
17 P. Burba, *Fresenius J. Anal. Chem.*, 341 (1991) 709.
18 M.I. Shtokalo, E.E. Kostenko and I.Z. Zhuk, *Zh. Anal. Khim.*, 46 (1991) 1093.
19 G.D. Brykina, T.E. Didenko and G. Agapova, *Zh. Anal. Khim.*, 43 (1988) 2195.
20 G.D. Brykina, G.G. Lebedeva and G. Agapova, *Zh. Anal. Khim.*, 45 (1990) 1838.
21 T. Tanaka, K. Hiiro and A. Kawahara, *Bunseki Kagaku*, 28 (1979) 43.
22 T. Tanaka, K. Hiiro and A. Kawahara, *Bunseki Kagaku*, 30 (1981) 131.
23 Z.T. Jiang, *Anal. Chim. Acta*, 392 (1999) 243.
24 N.I. Ershova and V.M. Ivanov, *Anal. Chim. Acta*, 364 (1998) 235.
25 E. Morosanova, A. Velikorodnyi, Yu. Zolotov and V. Skornyakov, *J. Anal. Chem.*, 55 (2000) 1136.
26 L.F. Capitán-Vallvey, M.D. Fernández-Ramos, I. de Orbe Payá and R. Avidad-Castañeda, *Analyst*, 120 (1995) 2421.
27 L.F. Capitán-Vallvey, I. de Orbe, M.C. Valencia and J.J. Berzas-Nevado, *Talanta*, 41 (1994) 1327.
28 H. Determan and I. Walter, *Nature*, 219 (1968) 604.
29 H. Waki, S. Takahasi and S. Ohashi, *J. Inorg. Nucl. Chem.*, 35 (1973) 1259.
30 K. Suzuki, K. Ohzeki and T. Kambara, *Bull. Chem. Soc. Jpn*, 56 (1983) 2624.
31 H. Waki, S. Noda and M. Yamashita, *Reactive Polym.*, 7 (1988) 277.
32 H. Waki and J. Korkich, *Talanta*, 30 (1983) 95.
33 A. Molina-Diaz, J.M. Herrador-Mariscal and M.I. Pascual-Reguera, *Talanta*, 40 (1993) 1059.
34 M.I. Pascual-Reguera, A. Molina-Diaz and N. Ramos-Martos, *Anal. Lett.*, 24 (1991) 2245.
35 M.L. Fernández de Córdova, A. Ruiz-Medina and A. Molina-Diaz, *Fresenius J. Anal. Chem.*, 357 (1997) 44.
36 T. Nakashima, K. Yoshimura and H. Waki, *Talanta*, 37 (1990) 735.
37 M.I. Pascual-Reguera, A. Molina-Diaz, M.C. Pacheco-Castillo, M.C. Anguita-Fernández, M.L. Fernández de Cordova and L.F. Capitán-Vallvey, *Mikrochim. Acta*, 112 (1994) 225.
38 A. Amin, *Anal. Chim. Acta*, 437 (2001) 265.

39 L.F. Capitán-Vallvey, J.M. Bosque-Sendra and M.C. Valencia, *Analusis*, 17 (1989) 601.
40 L.F. Capitán-Vallvey, M.C. Valencia and I. de Orbe Payá, *Microchem. J.*, 166 (1989) 166.
41 J.M. Bosque-Sendra, M.C. Valencia and L.F. Capitán-Vallvey, *Int. J. Environ. Anal. Chem.*, 38 (1990) 551.
42 K. Ohzeki, M. Minorikawa, F. Yokota, I. Nukatsuka and R. Ishida, *Analyst*, 115 (1990) 23.
43 F. Capitán, J.P. de Gracia, A. Navalón, L.F. Capitán-Vallvey and J.L. Vilchez, *Analyst*, 115 (1990) 849.
44 F. Capitán, J.M. Roman Navarro and L.F. Capitán-Vallvey, *Anal. Lett.*, 24 (1991) 1201.
45 C.G. Hsu, Q. Xu and J.M. Pan, *Mikrochim. Acta*, 126 (2001) 83.
46 M.L. Fernández de Córdova, A. Molina-Diaz, M.I. Pascual-Reguera and L.F. Capitán-Vallvey, *Fresenius J. Anal. Chem.*, 349 (1994) 722.
47 A. Molina-Diaz, J.J. Vida-Sagrista and M.I. Pascual-Reguera, *Int. J. Environ. Anal. Chem.*, 45 (1991) 219.
48 F. Capitán, E. Manzano, J.L. Vilchez and L.F. Capitán-Vallvey, *Anal. Sci.*, 5 (1989) 549.
49 F. Capitán, E. Manzano, A. Navalón, J.L. Vilchez and L.F. Capitán-Vallvey, *Analyst*, 114 (1989) 969.
50 F. Capitán, E. Manzano, A. Navalón, J.L. Vilchez and L.F. Capitán-Vallvey, *Talanta*, 39 (1992) 21.
51 N.I. Ershova and V.M. Ivanov, *Anal. Chim. Acta*, 408 (2000) 145.
52 N.I. Ershova and V.M. Ivanov, *Fresenius J. Anal. Chem.*, 367 (2000) 210.
53 N.I. Ershova and V.M. Ivanov, *Fresenius J. Anal. Chem.*, 371 (2001) 556.
54 Y. Wan, K. Liao and T. Gong, *Fenxi Huaxue*, 28 (2000) 1563.
55 K. Yoshimura and S. Ohashi, *Mem. Fac. Sci. Kyushu Univ. Ser. C*, 11 (1978) 181.
56 G.D. Brykina, L.S. Krysina and V.M. Ivanov, *Zh. Anal. Khim.*, 43 (1988) 1547.
57 G.D. Brykina, T.V. Marchak and T.A. Belyavskaya, *Zh. Anal. Khim.*, 54 (1980) 2094.
58 G.D. Brykina, T.V. Marchak, L.S. Krysina and T.A. Belyavskaya, *Zh. Anal. Khim.*, 35 (1980) 2294.
59 N.I. Ershova and V.M. Ivanov, *Fresenius J. Anal. Chem.*, 363 (1999) 641.
60 M.D. Petit Domingez, M.T. Sevilla Escribano, J.M. Pinilla Macias and L. Hernández, *Microchem. J.*, 42 (1990) 323.
61 K. Yoshimura, Y. Toshimitsu and S. Ohashi, *Talanta*, 27 (1980) 693.
62 G.D. Brykina, G. Agapova, V.F. Kalinina and L.S. Krysina, *Zh. Anal. Khim.*, 43 (1988) 1461.
63 G.D. Brykina, T.V. Marchak, L.S. Krysina and T.A. Belyavskaya, *Zh. Anal. Khim.*, 38 (1983) 1463.

64 J.M. Bosque-Sendra, C. Mérida and L.F. Capitán-Vallvey, *Analusis*, 18 (1990) 531.
65 K. Yoshimura, M. Ishii and T. Tarutani, *Anal. Chem.*, 58 (1986) 591.
66 K. Yoshimura, M. Motomura and T. Tarutani, *Anal. Chem.*, 56 (1984) 2342.
67 P. Ortega-Barrales, M.L. Fernández de Córdova and A. Molina-Diaz, *Anal. Chim. Acta*, 360 (1998) 143.
68 E. Morosanova, E. Reznikova and A. Velikorodnyi, *J. Anal. Chem.*, 56 (2001) 173.
69 O. Zaporozhets, E.A. Krushinskaya, N.A. Lipkvskaya and V. Sukhan, *J. Anal. Chem.*, 56 (2001) 524.
70 M.C. Valencia, E. Arana Nicolas and L.F. Capitán-Vallvey, *Talanta*, 49 (1999) 915.
71 Y. Ohno, T. Nakumara and K. Matsuhisa, *Bull. Chem. Soc. Jpn*, 23 (1984) 117.
72 J.M. Bosque-Sendra, M.F. Molina and E. López, *Analyst*, 116 (1991) 871.
73 J.M. Bosque-Sendra, M.F. Molina and M. Nechar, *J. AOAC Int.*, 77 (1994) 1224.
74 F. Capitán, M.C. Valencia and L.F. Capitán-Vallvey, *Mikrochim. Acta*, III (1984) 303.
75 K. Matsuhisa and K. Ohzeki, *Nippon Kagaku Kaishi*, 11 (1983) 1593.
76 A. Velikorodnyi and E. Morosanova, *J. Anal. Chem.*, 55 (2000) 994.
77 K. Matsuhisa and K. Ohzeki, *Analyst*, 111 (1986) 1175.
78 M.F. Molina, M.C. Valencia, J. de la Torre and L.F. Capitán-Vallvey, *J. Pharm. Biomed. Anal.*, 7 (1989) 843.
79 L.F. Capitán-Vallvey, M.C. Valencia and J. García, *Quim. Anal.*, 6 (1987) 445.
80 M.C. Valencia, J. García and L.F. Capitán-Vallvey, *Anal. Lett.*, 23 (1990) 1095.
81 J.L. Vilchez, R. Avidad-Castañeda, A. Navalón, J. Rohand and L.F. Capitán-Vallvey, *Int. J. Environ. Anal. Chem.*, 53 (1993) 139.
82 M. del Olmo, J. Laserna, D. Romero, J. Rohand and J.L. Vilchez, *Talanta*, 44 (1997) 443.
83 L.F. Capitan-Vallvey, M.C. Valencia and E. Arana, unpublished results.
84 N.A. Lipkovakaya, V.V. Sukhan, A. Yu. Nazarenko and A.A. Chuiko, *Zh. Anal. Khim.*, 47 (1990) 1410.
85 E. Morosanova, A. Velikorodnyi, A.A. Kuz'min and Yu. Zolotov, RF Patent 98,113,445, 1998, *Buyll. Izobret.*, 28 (1999).
86 F. Capitán, M.C. Valencia, L. Cuadros and L.F. Capitán-Vallvey, *Microchem. J.*, 31 (1985) 396.
87 K. Matsuhisa and K. Ohzeki, *Analyst*, 111 (1986) 685.
88 M. Abe, K. Ohzeki and T. Kambara, *Bull. Chem. Soc. Jpn*, 51 (1978) 1090.
89 K. Yoshimura and S. Ohashi, *Talanta*, 25 (1978) 103.

90 I. Nukatsuka, S. Nakamura, K. Watanabe and K. Ohzeki, *Anal. Sci.*, 16 (2000) 269.
91 K. Matsuhisa, K. Ohzeki and T. Kambara, *Bull. Chem. Soc. Jpn.*, 56 (1983) 3847.
92 K. Matsuhisa, K. Ohzeki and T. Kambara, *Bull. Chem. Soc. Jpn.*, 55 (1982) 3335.
93 J. Shida and I. Masuda, *Anal. Sci.*, 14 (1998) 333.
94 J. Shida, K. Kasama, H. Honda and K. Oikawa, *Anal. Chim. Acta*, 233 (1990) 135.
95 Y. Wan, L. Guo and C. Dai, *Fenxi Huaxue*, 28 (2000) 606.
96 K. Anezaki, X. Chen, T. Ogasawara, I. Nukatsuka and K. Ohzeki, *Anal. Sci.*, 14 (1998) 523.
97 T. Shimizu, I. Nukatsuka and K. Ohzeki, *Anal. Sci.*, 15 (1999) 153.
98 W.T. Chan and H.H.C. Yip, *Anal. Chem.*, 69 (1997) 4872.
99 D.M. Beals, B.S. Crandall and P.D. Fledderman, *J. Radioanal. Nucl. Chem.*, 243 (2000) 495.
100 S.K. Fiskum, R.G. Riley and C.J. Thompson, *J. Radioanal. Nucl. Chem.*, 245 (2000) 261.
101 K. Ohzeki, C. Toki, R. Ishida and T. Saitoh, *Analyst*, 112 (1987) 1689.
102 Y. Tagaki and M. Yoshida, *Anal. Chim. Acta*, 165 (1984) 195.
103 K. Yoshimura and H. Waki, *Talanta*, 32 (1985) 345.
104 H. Ishii, *Fresenius Z. Anal. Chem.*, 319 (1984) 23.
105 T.M. Nikolaeva and A.I. Lazarev, *Zavod. Lab.*, 58 (1992) 10.
106 K. Suzuki, G.D. Park, K. Ohzeki and T. Kambara, *Bull. Chem. Soc. Jpn.*, 55 (1982) 2851.
107 L.F. Capitán-Vallvey, I. de Orbe, M.C. Valencia and J.J. Berzas Nevado, *Quim. Anal.*, 12 (1993) 141.
108 M.C. Valencia, S. Boudra and J.M. Bosque-Sendra, *Analyst*, 118 (1993) 1333.
109 J.L. Vilchez, D. Torres-Bustos, R. Blanc and A. Navalón, *J. AOAC Int.*, 79 (1996) 567.
110 J.L. Vilchez, G. Sánchez-Palencia, R. Blanc, R. Avidad-Castañeda and A. Navalón, *Anal. Lett.*, 27 (1994) 2355.
111 F. Capitán, A. Navalón, J.L. Vilchez and L.F. Capitán-Vallvey, *Talanta*, 37 (1990) 193.
112 M.A. Ditzler, G. Doherty, S. Sieber and C. Allston, *Anal. Chim. Acta*, 142 (1982) 305.
113 F.C. Williams and F.R. Clapper, *J. Opt. Soc. Am.*, 43 (1953) 595.
114 S. Yamada and K. Yoshimura, *Talanta*, 39 (1992) 1013.
115 L.W. Burggraf and D.E. Leyden, *Anal. Chem.*, 53 (1981) 759.
116 R. Zweidinger and J.D. Winefordner, *Anal. Chem.*, 42 (1970) 639.
117 M.J. Adams, A.A. King and G.F. Kirkbright, *Analyst*, 101 (1976) 73.

118 P. Ortega-Barrales, A. Molina-Diaz, M.I. Pascual-Reguera and L.F. Capitán-Vallvey, *Anal. Chim. Acta*, 353 (1997) 115.
119 P. Ortega-Barrales, M.L. Fernández de Córdova and A. Molina-Diaz, *Anal. Chem.*, 70 (1998) 271.
120 M.L. Fernández de Córdova, A. Molina-Diaz, M.I. Pascual-Reguera and L.F. Capitán-Vallvey, *Talanta*, 42 (1995) 1057.
121 M.L. Fernández de Córdova, A. Molina-Diaz and M.I. Pascual-Reguera, *Anal. Lett.*, 25 (1992) 1961.
122 Y.L. Yang, C.G. Hsu and J.M. Pan, *Microchem. J.*, 48 (1993) 178.
123 T. Nakashima, K. Yoshimura and T. Taketatsu, *Talanta*, 59 (1992) 523.
124 K. Yoshimura and H. Waki, *Talanta*, 34 (1987) 239.
125 J.L. Vilchez, A. Navalón, R. Avidad-Castañeda, J. Rohand and L.F. Capitán-Vallvey, *Fresenius J. Anal. Chem.*, 345 (1993) 716.
126 L.F. Capitán-Vallvey, E.J. Alonso, R. Avidad-Castañeda, M. del Olmo and J.L. Vilchez, *Anal. Sci.*, 9 (1993) 117.
127 M.M.A. Shriadah, M. Kataoka and K. Ohzeki, *Analyst*, 110 (1985) 125.
128 L.F. Capitán-Vallvey, I. de Orbe Payá, M.C. Valencia and J.J. Berzas-Nevado, *Anal. Chim. Acta.*, 282 (1993) 75.
129 J.L. Vilchez, R. Avidad-Castañeda, J. Rohand, A. Navalón and L.F. Capitán-Vallvey, *Anal. Chim. Acta*, 282 (1993) 445.
130 J.L. Vilchez, M. del Olmo, R. Avidad-Castañeda and L.F. Capitán-Vallvey, *Quim. Anal.*, 13 (1994) 22.
131 M.C. Valencia, D. Gimeno and L.F. Capitán-Vallvey, *Anal. Lett.*, 26 (1993) 1211.
132 Y. Toshimitsu, K. Yoshimura and S. Ohashi, *Talanta*, 26 (1979) 273.
133 F. Capitán, L.F. Capitán-Vallvey and M.C. Gómez, *Quim. Anal.*, 6 (1978) 343.
134 K. Matsuhisa, K. Ohzeki and T. Kambara, *Bull. Chem. Soc. Jpn*, 54 (1981) 2675.
135 K. Yoshimura, S. Nigo and T. Tarutani, *Talanta*, 29 (1981) 173.
136 F. Capitán, E.J. Alonso, R. Avidad-Castañeda, L.F. Capitán-Vallvey and J.L. Vilchez, *Anal. Chem.*, 65 (1993) 1336.
137 L.F. Capitán-Vallvey, M. del Olmo, R. Avidad-Castañeda, A. Navalón, I. de Orbe Payá and J.L. Vilchez, *Anal. Chim. Acta*, 302 (1995) 193.
138 L.F. Capitán-Vallvey, I. de Orbe Payá, M.C. Valencia and J.J. Berzas-Nevado, *Mikrochim. Acta*, 111 (1993) 223.
139 A. Molina-Diaz, M.I. Pascual-Reguera, E. Liñán-Veganzones, M.L. Fernández de Córdova and L.F. Capitán-Vallvey, *Talanta*, 43 (1996) 185.
140 L.F. Capitán-Vallvey, M.C. Valencia and J.M. Bosque-Sendra, *Analyst*, 113 (1988) 419.
141 A. Ruiz-Medina, M.L. Fernández de Córdova and A. Molina-Diaz, *J. Pharm. Biomed. Anal.*, 20 (1999) 247.

142 L.F. Capitán-Vallvey, I. de Orbe Payá, S. López-Rodriguez and R. Avidad-Castañeda, *Mikrochim. Acta*, 117 (1995) 129.
143 S. Yamada, *Talanta*, 39 (1992) 1013.
144 M.M.A. Shriadah and K. Ohzeki, *Analyst*, 110 (1985) 677.
145 J.L. Vilchez, L.F. Capitán-Vallvey, J. Rohand, A. Navalón and R. Avidad-Castañeda, *Analyst*, 120 (1995) 1609.
146 F. Capitán, L.F. Capitán-Vallvey and J.L. Vilchez, *Quím. Anal.*, 10 (1991) 21.
147 L.F. Capitán-Vallvey, M. del Olmo, R. Avidad-Castañeda and J.L. Vilchez, *Anal. Lett.*, 26 (1993) 2243.
148 A. Navalón, R. Blanc and J.L. Vilchez, *Mikrochim. Acta*, 126 (1997) 33.
149 A. Navalón, O. Ballesteros, R. Blanc and J.L. Vilchez, *Talanta*, 52 (2000) 845.
150 M.F. Molina, M. Nechar and J.M. Bosque-Sendra, *Anal. Sci.*, 14 (1998) 791.
151 V.M. Ivanov and N.I. Ershova, *J. Anal. Chem.*, 56 (2001) 1089.
152 L.F. Capitán-Vallvey, M. del Olmo, R. Avidad-Castañeda and J.L. Vilchez, *Anal. Quim.*, 90 (1984) 125.
153 F. Capitán, L.F. Capitán-Vallvey, M.D. Fernández-Ramos, I. de Orbe Payá and R. Avidad-Castañeda, *Anal. Chim. Acta*, 331 (1996) 141.
154 J.L. Vilchez, R. Blanc and A. Navalón, *Anal. Lett.*, 27 (1994) 2355.
155 J.L. Vilchez, A. Navalón, J. Rohand, R. Avidad-Castañeda and L.F. Capitán-Vallvey, *J. Fluor.*, 5 (1995) 225.
156 M.C. Valencia, S. Boudra and J.M. Bosque-Sendra, *Anal. Chim. Acta*, 327 (1996) 73.
157 L.F. Capitán-Vallvey, R. Avidad-Castañeda, M. del Olmo and J.L. Vilchez, *Mikrochim. Acta*, 112 (1993) 55.
158 L.F. Capitán-Vallvey, N. Navas Iglesias, I. de Orbe Payá and R. Avidad-Castañeda, *Talanta*, 43 (1996) 1457.
159 L.F. Capitán-Vallvey, J. Rohand, A. Navalón, R. Avidad-Castañeda and J.L. Vilchez, *Talanta*, 40 (1993) 1695.
160 F. Capitán, G. Sánchez-Palencia, A. Navalón, L.F. Capitán-Vallvey and J.L. Vilchez, *Anal. Chim. Acta*, 259 (1992) 345.
161 J.L. Vilchez, M. del Olmo, R. Avidad-Castañeda and L.F. Capitán-Vallvey, *Analyst*, 119 (1994) 1211.
162 L.F. Capitán-Vallvey, M.L. Fernández de Córdova, I. de Orbe Payá and R. Avidad-Castañeda, *Talanta*, 47 (1998) 861.
163 S. Nigo, K. Yoshimura and T. Tarutani, *Talanta*, 28 (1981) 669.
164 F. Yokota and S. Abe, *Anal. Commun.*, 34 (1997) 111.
165 R.F. Gur'eva and S.B. Savvin, *J. Anal. Chem.*, 56 (2001) 901.
166 V.M. Ostrovskaya, *Fresenius J. Anal. Chem.*, 361 (1998) 303.
167 O.P. Shvoeva, V.P. Dedkova and S.B. Savvin, *J. Anal. Chem.*, 56 (2001) 376.
168 M.P. Arena, M.D. Porter and J.S. Fritz, *Anal. Chem.*, 74 (2002) 185.

169 F. Capitán, R. Checa-Moreno, R. Avidad-Castañeda and L.F. Capitán-Vallvey, *Talanta*, 42 (1995) 711.
170 K. Ohzeki, T. Uno, I. Nukatsuka and R. Ishida, *Analyst*, 113 (1988) 1545.
171 I. Nukatsuka, T. Miura, K. Ohzeki and R. Ishida, *Anal. Chim. Acta*, 248 (1991) 529.
172 C. Shima, I. Nukatuka and K. Ohzeki, *Anal. Sci.*, 14 (1998) 337.
173 S. Taguchi, K. Morisaku, Y. Sengoku and I. Kasahara, *Analyst*, 124 (1999) 1489.
174 M.K. Bklemishev, T.A. Stoyan and I.F. Dolmanova, *Analyst*, 122 (1997) 1161.
175 I. Nukatsuka, A. Nishimura and K. Ohzeki, *Anal. Chim. Acta*, 304 (1995) 243.
176 O.P. Shvoeva, V.P. Dedkova and S.B. Savvin, *J. Anal. Chem.*, 56 (2001) 1080.
177 I. Nukatsuka, T. Ohba, H. Ishida, H. Satoh, K. Ohzeki and R. Ishida, *Analyst*, 117 (1992) 1513.
178 S. Taguchi, T. Inaba, M. Nishio, N. Hata, I. Kasahara and K. Goto, *Analyst*, 114 (1989) 489.
179 R.F. Gur'eva and S.B. Savvin, *J. Anal. Chem.*, 55 (2000) 249.
180 I. Nukatsuka, K. Takahashi, K. Ohzeki and H. Ishida, *Analyst*, 114 (1989) 1473.
181 J. Shida, H. Takahashi and Y. Sakamoto, *Talanta*, 41 (1994) 143.
182 C.M. Hill, K.W. Street, W.H. Philipp and S.P. Tanner, *Anal. Lett.*, 27 (1994) 2589.
183 C.M. Hill, K.W. Street, S.P. Tanner and W.H. Philipp, *Anal. Lett.*, 33 (2000) 2779.
184 L.F. Capitán-Vallvey, O.M.A. Al-Barbarawi, M.D. Fernández-Ramos, R. Avidad-Castañeda and V. Ramirez, Gonzalez, *Analyst*, 125 (2000) 2000.
185 T. Tanaka, K. Hiiro and A. Kawahara, *Bunseki Kagaku*, 23 (1974) 650.
186 M. Kaneko, H. Tanno and Y. Takao, *Mikrochim. Acta*, 1 (1991) 37.
187 T. Tanaka, K. Hiiro and A. Kawahara, *Bunseki Kagaku*, 31 (1982) E123.
188 S.B. Savvin, V.P. Dedkova and O.P. Shvoeva, *Russ. Chem. Rev.*, 69 (2000) 187.
189 T. Saito, *Anal. Chim. Acta*, 268 (1992) 351.
190 R. Bloch, R. Bauer and B.F. Phillips, US Patent 3,635,679, 1972.
191 L.F. Capitan-Vallvey, R. Avidad-Castañeda, M.D. Fernández-Ramos, P. Alvarez de Cienfuegos and A. Ariza-Avidad, SP Patent 9,900,718, 1999.
192 L.F. Capitán-Vallvey, E. Arroyo, C. Berenguer, M.D. Fernández-Ramos and R. Avidad-Castañeda, *Fresenius J. Anal. Chem.*, 369 (2001) 139.
193 J.W. Foerster, K. Ewing, A. Ervin and R. Lamontagne, WO Patent 99/47917, 1999.

194 L.F. Capitán-Vallvey, C. Cano-Raya, C. Esparza del Valle, M.D. Fernández-Ramos, I. de Orbe Payá and R. Avidad-Castañeda, *Anal. Lett.*, 35 (2002) 615.
195 L.F. Capitán-Vallvey, M.D. Fernández-Ramos and P. Alvarez de Cienfuegos, *Anal. Chim. Acta*, 451 (2002) 231.
196 L.F. Capitán-Vallvey, M.D. Fernández-Ramos, R. Avidad-Castañeda and M. Deheidel, *Anal. Chim. Acta*, 440 (2001) 131.
197 T. Saito, *Talanta*, 41 (1994) 811.
198 L.F. Capitán-Vallvey, A. Ariza-Avidad, M.D. Fernández-Ramos, E. Arroyo and R. Avidad, *Anal. Bioanal. Chem.*, 373 (2002) 289.
199 J. Wu, *Fenxi Shiyanshi*, 19 (2000) 36.
200 M. Kaneko, H. Tanno and T. Yotsuyanagi, *Mikrochim. Acta*, III (1988) 333.
201 J.Y. Wang, US Patent 4,618,475, 1986.
202 H. Lange, W. Rittersdorf and H.G. Rey, US Patent 3,897,214, 2001.
203 A. Rothe, W.R. Knappe and H.F. Trasch, US Patent 4,604,264, 1986.
204 S.C. Charlton, P. Hemmes and A.L.Y. Lau, US Patent 4,649,123, 1987.
205 M.L. Gantzer, P.R. Hemmes and D. Wong, US Patent 4,670,218, 1987.
206 D.V. Mehta, FR Patent 2,303,290, 1976.
207 P. Vogel, H.P. Braun, D. Berger and W. Werner, US Patent 4,312,834, 1982.
208 K.E. Piejko, B. Boemer, H. Bartl and G. Frank, US Patent 4,780,411, 1998.
209 I. Siddiqi, US Patent 4,438,067, 2001.
210 E.J. Kiser, E.G. Rice and M.F. Tomasco, US Patent 5,306,623, 1994.
211 E.J. Kiser, E.G. Rice and M.F. Tomasco, US Patent 5,418,142, 1995.
212 L. Terminiello and J.L. Aronowitz, US Patent 4,774,192, 1988.
213 K. Hildenbrand, H. H. von Dohren, H. Perrey, G. Frank and R. Dhein, US Patent 4,824,639, 1989.
214 E.C. Adams and J.A. Peterson, US Patent 3,092,463, 1963.
215 M. Koyama and K. Okinawa, US Patent 4,486,537, 1984.
216 B.J. Verbeck, US Patent 3,993,451, 1976.
217 A. Kumar and K.W. Leong, EP Patent 141,647, 1984.
218 K. Anand and L. Koon-Wah, EP Patent 0,141,647, 1985.
219 K.J. Phillips, US Patent 5,556,761, 1996.
220 A.L.Y. Lau, J.H. Pedergrass and C.A. Ritucci, US Patent 5,096,833 (1992).
221 E.P. Przybylowicz and A.G. Millikan, US Patent 3,992,158, 1976.
222 S.C. Charlton, R.L. Fleming, P. Hemmes and A.L. Lau, US Patent 4,645, 744, 1987.
223 M. Kitajima, F. Arai and A. Kondo, US Patent 4,356,149, 1982.
224 B.J. Bruschi, US Patent 4,066,403, 1978.
225 A.C. Greenquist, T.M. Li, P.A. Rupchock, R.J. Tyhach and B. Walter, US Patent 4,668,619, 1987.
226 V.H. Esch and R.M. Fristrom, US Patent 4,205,043, 1980.
227 H. Lange, W. Rittersdorf, H.G. Rey, W. Werner and P. Rieckmann, US Patent 4,061,468, 1977.

228 T.W. Wu and G.M. Dappen, US Patent 4,069,017, 1978.
229 M.J. Smith-Lewis and J. Figueras, US Patent 4,166,093, 1979.
230 A.C. Greenquist, US Patent 4,806,312, 1989.
231 M.L. Gantzer, P. Hemmes and D. Wong, US Appl. 583,127, 1984.
232 K. Arai, M. Koyama and K. Okinawa, US Patent 4,594,225, 1986.
233 R.C. Gross and S.C. Gross, US Patent 4,478,944, 1984.
234 C.B. Maddox, US Patent 5,212,060, 1993.
235 G.M. Daffern and T.N. Thompson, US Patent 4,994,238, 1991.
236 P. Vogel, D. Thym, M. Fritz and D. Mosoiu, US Patent 5,211,914, 1993.
237 V.H. Esch and R.M. Fristrom, US Patent 4,205,043, 1980.
238 J.R. Schaeffer, L.M. Minsk and R.E. Stevens, US Patent 4,110,079, 1978.
239 P.O. Vogelhut, US Patent 4,231,754, 1980.
240 P. Vogel, D. Thym, M. Fritz and D. Mosoiu, US Patent 5,302,346, 1994.
241 S.M.H. Sherman, R.K. Nakahiro and M.P. Okamoto, *Am. J. Health-Syst. Pharm.*, 54 (1997) 1058.
242 K.H. Kallis, US Patent 4,281,062, 1994.
243 J. Gordon, M.E. McMahon and S. Ching, US Patent 4,960,691, 1990.
244 R.M. Thayer, A.J. Polito, R.K. Dinello, G.H. Sierra and H.J. Wieck, WO Patent 00/77524, 2000.
245 F.C. Horstman and J. Whiteside, US Patent 5,006,474, 1991.
246 D. Bernstein, US Patent 4,770,853, 2002.
247 R.W. Eisinger, M.H. Khalil, D.H. Katz and R.B. Sargeant, US Patent 4, 943,522, 1990.
248 M. Owens, WO Patent 00/42430, 2000.
249 H. Engelmann, US Patent 4,738,823, 1988.
250 E.P. Przybylowicz and A.G. Millikan, GB Patent 1,440,464, 1973.
251 A. Rothe, A.K. Selle and B. Sojka, US Patent 4,587,099, 1986.
252 F.R. Jerome, US Patent 3,995,019, 1976.
253 E.J. Kiser, M.F. Tomasco, E.G. Rice and Y.S. Yu, US Patent 5,719,034, 1998.
254 H.T. Hochstrasser, US Patent 3,964,871, 1976.
255 H.T. Hochstrasser, US Patent 4,042,329, 1977.
256 S.C. Charlton, R.P. Hatch and P.R. Hemmes, US Patent 5,520,883, 1996.
257 R. Phillips, G. McGarraugh, F.A. Jurik and R.D. Underwood, US Patent 5,563,042, 1996.
258 A.L. Y. Lau and I.N. Granger, US Patent 5,096,833, 1992.
259 K. Hildenbrand, H.H. von Döhren, H. Perrey and K. Wehling, US Patent 4,824,640, 1989.
260 J.D. Olson, US Patent 4,959,307, 1990.
261 S.C. Charlton, EP Patent 0,475,045, 1991.
262 C.O. Rupe and R. Bauer, US Patent 4,092,115, 1978.
263 C. Heiss, M.G. Weller and R. Niessner, *Anal. Chim. Acta*, 396 (1999) 309.

264 M. Hall, P.A. Sykes, D.L. Fairclough, L.J. Lucchese, P. Rogers, W. Staruszkiewicz and R.C. Bateman, *J. AOAC Int.*, 82 (2000) 1102.
265 T.Y. Lee, Y.C. Lei, S.Y. Sheu, Y.F. Tsai, T.G. Wu, H.H. Ho, M.Y. Kuo, J.J. Liu, T.H. Chang, H.F. Chen and L.Y. Liu, EP Patent 1,160,571, 2001.
266 N.V. Cuong, T.T. Bachmann and R.D. Schmid, *Fresenius J. Anal. Chem.*, 363 (1999) 584.
267 Super Aution EX SA-4250 brochure. Menarini Diagnostics.
268 K. Iwata, K. Kawahara and H. Nakajima, US Patent 5,912,139, 1999.
269 A. Boettger, K. Hagmann and H.J. Pieper, *Branntweinwirtschaft.*, 136 (1987) 34.
270 D.D. Rawles, G.J. Flick and R.E. Martin, *Adv. Food Nutr. Res.*, 39 (1996) 329.
271 H. Rehbein and T. Schmidt, *Informat. Fischwirtschaft.*, 43 (1997) 37.
272 Q.H. de Campos, A.N. Lizama and B.M.G. Marquez, *Agricultura Técnica.*, 54 (1994) 318.
273 F.W. Scheller, F.F. Bier and N. Gajovic, *Lebensmittel Biotechnol.*, 14 (1998) 95.
274 C. Wittmann, U. Bilitewski, T. Giersch, U. Kettling and R.D. Schmid, *Analyst*, 121 (1996) 863.

*Chapter 4*

# Continuous-flow analytical systems

**M.D. Luque de Castro* and J.L. Luque-García**

## 4.1 INTRODUCTION

Since Skeggs described continuous segmented systems in 1957 [1], so many different researchers have investigated dynamic systems in such a way that at present, the pioneer design is almost forgotten and others, requiring simpler, cheaper and more versatile equipment, are in common use. The two most outstanding features of dynamic systems are decreased sample manipulation and easy automation. The most popular among present dynamic approaches is Flow Injection (FI). After appearing in 1974–1975 with the name of Flow Injection Analysis (FIA) [2,3], the acronym has been shortened because of both its similarity with that of Fluoro Immuno Assay and the fact that total analysis is developed in the dynamic system in very few FI methods. Despite the importance of the two above-mentioned features of dynamic systems— namely reduced manipulation and ease of automation, both greatly fulfilled by FI— the most outstanding characteristic of FI is versatility.

As FI is at present a well-known technique, readers are referred to monographs [4–6] for generalities and only a basic scheme and short comments are presented in this introduction as a starting point for the development of the chapter. So a basic FI manifold (Fig. 4.1) consists of (1) a propulsion unit (usually a peristaltic pump) which establishes a

* Corresponding author. Tel./fax: +34-957218615. *E-mail address:* qa1lucam@uco.es

*Comprehensive Analytical Chemistry XXXIX, pages 161–243* 

ISSN: 0166-526X

flow-rate of either carrier or reagents along the system that is as constant as possible; (2) an injection system (usually a six-way low-pressure valve) which inserts in the carrier or reagent stream a precisely measured volume of the order of μL without disturbing the flow; (3) transport tubing, sometimes improperly named a 'reactor'—this part of the FI manifold is more easily handled as a series of units that can be in-line coupled with it; and (4) a flow-cell located in a detector for monitoring the passage of the reactant—or not a reactant—plug, which provides a transient signal collected by a computer or shown by a recorder. From this basic manifold more complex ones can be designed and easily built depending on the type of integration required, which can range from dilution of the sample for suiting the concentration to the linear range of the calibration curve to complex open-closed systems. Miniaturisation, a growing trend in analytical chemistry, has also reached FI systems in the form of microcircuits [7], miniaturised total chemical analysis systems [8] and ultra-micro flow injection analysis [9], thus providing advantages additional to those of each approach—miniaturisation and FI—when used separately. Because microsystems are the subject matter of subsequent chapters (chapters 13–16), they are not included in the present one. Flow-through chemical sensors [10], biosensors [11] and immunoassay systems [12], even those developed in FI approaches, are not the subject of the present chapter but of chapters 7–9.

Integration in FI can be considered from several points of view.

(A) The simplest integration in FI is achieved by changing its conventional *modus operandi* by injecting the reagent into a sample stream, thus causing the reverse mode [4–6].
(B) Another simple integration is achieved in a basic FI manifold by including in it typical units of this technique, thus obtaining the

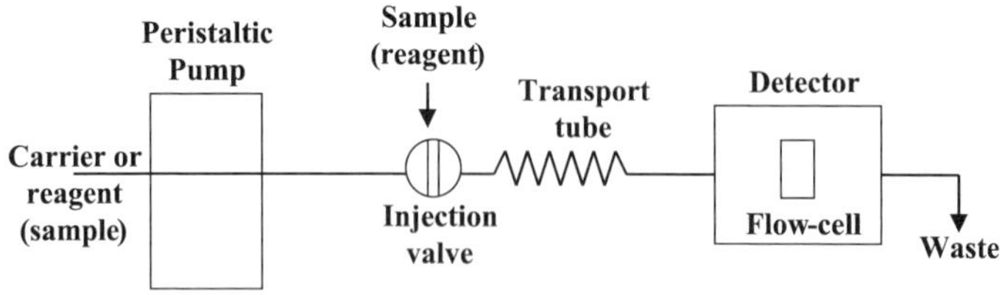


Fig. 4.1. Scheme of a basic FI manifold (in brackets the sample–carrier exchange characteristic of the reverse mode).

desired effect. Such is the case of the inclusion of: (b.1) a confluence point, thus allowing the injected sample to merge with a liquid stream for dilution, with a buffer solution for conditioning, and a reagent solution for derivatisation purposes; (b.2) an exchange of a common FI unit for a more sophisticated unit from other dynamic systems (e.g. use of an HPLC injection valve for inserting into the FI volumes smaller than those allowed by a conventional low-pressure FI valve [13]); (b.3) a modified unit (e.g. a programmable impulsion unit for better manipulation of either the flow-rate or the sense of the stream it provides); (b.4) a duplicate unit (e.g. an additional injection valve for simultaneous sequential introduction—related to the main valve—of either sample or reagent).

(C) Derivatisation reactions also constitute a simple and versatile way of improving the features of FI methods. They can be implemented very easily, using confluence points, but there are a number of other ways of implementing them, thus creating additional advantages.

(D) A slight complication as compared with previous integration modes is the use of external energy sources such as heat, ultrasound or microwaves, whose effect can drastically modify (both in expeditiousness and nature) the steps to be developed in an FI manifold.

(E) Integration of FI manifolds with either commercial or simple laboratory-made units specially designed for in-line coupling. Such is the case with modules for development of non-chromatographic continuous separation processes (e.g. dialysers, gas-diffusers, pervaporators, gas-liquid separators and sorbent minicolumns) and dilution or titration chambers.

(F) On-line coupling of commercial equipment, usually bigger than FI manifolds, as is the case with dynamic extractors using sub- or supercritical fluids as extractants, chromatographs, capillary electrophoresis equipment or robotic stations. In these cases the FI manifold either can work as a simple slave of the main equipment (e.g. for sample introduction or derivatisation) or it can develop a key role in the overall process. The interface required for these couplings is usually very simple and affordable.

(G) The on-line connection of the FI manifold with a detector is a basic necessity for obtaining proper information about the steps developed in the former. Due to this fact, conventional detectors will be discussed below and an in-depth emphasis will be made on both special designs and high-resolution detectors, because of the considerable benefits obtained by these couplings.

## 4.2 REVERSE FLOW INJECTION

The reverse mode of FI—that is, injection of the reagent into a sample stream, as shown in Fig. 4.1—produces the following effects: decrease of reagent consumption, which can be of enormous interest when the reagent is expensive and the sample abundant and not valuable, particularly in routine analysis; and increase of sensitivity [14,15]. Typical examples of the use of this FI mode are for monitoring both pollutants in natural water (e.g. manganese [16], cobalt [17], residual chlorine [18], total dissolved iron [19]) and industrial processes (e.g. thermal power plants [20], fluoride and monofluorophosphate ion in toothpastes using a single calibration for both analytes [21] and fructose in syrups [22]). An on-line dilution effect can also be achieved with a reverse use of flow injection by decreasing the volume of the injected reagent, as is the case for isotope-dilution ICP-MS methods [23].

## 4.3 INTEGRATING EFFECT OF CONVENTIONAL FLOW INJECTION UNITS

The integrating effect of some conventional and very simple FI units is so well-known that the users do not realise the benefits they report. To the simple flow injection manifold in Fig. 4.1 the basic components and modifications introduced are as follows.

### 4.3.1 Confluence points

The presence of a confluent point in a flow manifold permits the main channel which transports the injected sample to merge with a stream whose nature depends on the pursued objective, that can be one of those detailed below.

*Dilution of the injected sample*
Dilution is one of the most common steps of the analytical process which is a significant source of error, particularly for non-skilled operators, but it is also time-consuming when performed manually. The inclusion of a

confluence point with the main channel of transport tubing for the proper diluent and connection of this tubing with a pump tubing with the adequate diameter for the required dilution constitutes the simplest, fastest and most reproducible—providing the impulsion system is also reproducible—way of achieving the concentration of a given analyte in a sample to be lowered as required. Usual dilutions up to 10 times can be easily achieved by T, Y or W confluence points [4] and a convenient length of the subsequent transport tubing for proper homogenisation of the resulting stream. Dilutions of 1:1000 or higher require the inclusion in thc FI manifold of a mixing chamber (e.g. a 10 cm × 5.0 mm i.d. tube packed with 3–4 mm glass beads, i.e. a single bead string reactor, SBSR [24]) located after the confluence point [13,25], and cascade dilution or other similar devices for achieving the appropriate degree of homogenisation. Sample dilution, as required most often in ICP-MS for multielement determination [26], and on-line cascade dilution systems are commonly implemented with simple flow-injection approaches.

*pH conditioning and masking effects*

Another common aim of the inclusion of a confluence point is conditioning, other than dilution, of the sample after injection; that is, adjustment of the sample conditions for subsequent proper detection or derivatisation. A mixture of the sample with a solution of a buffer or complexing agent can be the key for avoiding interferences from other sample components. The relative flow-rate of these channels with respect to the main channel where the sample is injected depends on the necessity or not of sample dilution, which can also be achieved in this way using more or less diluted reagents.

*Derivatisation reactions*

One immediate consequence of integrating either chemical or biochemical reactions with FI systems is the kinetic character of the methods thus developed. Fixed-time measurements in non-equilibrium (neither chemical nor physical) conditions, typical of FI methods, endow them with selectivity, which is even increased when the kinetic measurements are of the reaction-rate type. The introduction of a reagent solution is one of the main uses of confluence points in FI manifolds. Practically all types of reaction have been developed in these dynamic systems.

Taking into account that the most usual detector connected to an FI manifold has been a conventional photometer, coloured reactions have been the commonest in FI, particularly those based on *complex formation* with metal cations as the target species [27–29].

*Redox reactions*
Redox reactions have been mainly used for the determination of anions such as nitrate and nitrite [30,31], phosphate and silicate [32,33] and thiocyanate and iodide [34], but also for organic compounds such as ascorbic acid [37] and promazine [38], and cations [35,36] including ammonium formed after Kjeldahl digestion, which can be determined at a rate of 20–100 samples/h [39,40].

Special methods developed in these very simple FI manifolds subjected to very short reaction times are very selective for the determination of analytes in very complex samples, as is the case with the promoting effect of oxalate on the oxidation of ferroin by Cr(VI), which enables the determination of this anion in urine [41].

It is worth emphasising that the majority of these steps can be developed in a flow manifold such as that in Fig. 4.1 by injection of the sample in the diluent, buffer, masking solution or derivatising solution. Nevertheless, the merging point provides a more efficient, homogeneous sample–other stream mixture in all instances.

### 4.3.2 Exchanged units

A dilution effect can also be achieved by substituting the common low-pressure injection valve—usually dead volume of more than 20 μL—by an HPLC injection valve, thus allowing injection volumes of a few μL or less [42]. In-line dilution is also an excellent way to extend the calibration range in atomic spectrometric methods [43], easily achieved by injecting a submicrolitre volume of the sample into the dynamic system. Also, the use of electroosmotic pumps enables the in-line dilution effect in atomic spectrometric methods, as proved by Gan et al. [44].

### 4.3.3 Modified units

The units of a flow injection manifold that can be easily modified are the inner diameter of the dynamic system and the impulsion unit.

*Modification of the inner diameter of the transporting system*

An alternative for reducing the injected volume other than exchanging the common FI valves for others from more sophisticated dynamic approaches is to decrease the dead-volume of the former by shortening the connection tubes and reducing their diameter.

The usual inner diameter of the transporting tubing (between 0.3 and 0.8 mm) can either be decreased, thus creating miniaturised flow injection systems which allow smaller sample and reagent consumption—as mentioned in the introduction—or increased, for different purposes. Thus, when FI is used for on-line monitoring of evolving processes [45], free-from-particle solutions to be introduced into the dynamic manifold can be achieved either because the monitored system is cleaned enough by itself or through a series of filters—either located in a cascade mode in the FI manifold or before it—that retain the suspended particles. Alternatively, the larger inner diameter of all the units involved in the dynamic approach enables samples within suspension particles to be directly monitored, provided the detector is not affected by the heterogeneous system and the heterogeneity itself is not the parameter to be monitored. Different flow injection methods for on-line monitoring of penicillin V (phenoxymethylpenicillin) using a rapid response glass electrode—on which β-lactamase was immobilized—which measures the pH change caused by the hydrolysis of penicillin V to penicilloic acid [46], or through glucose and ethanol using a split-flow modified thermal biosensor [47], have been used. On-line monitoring of glucose in industrial bioprocess engineering [48], and that of recombinant protein production [49], but particularly the automation of dissolution tests of solid pharmaceuticals [50] are key applications of these systems requiring straightforward modifications of the conventional units of an FI manifold.

Slurries is the simplest mode of working with solid samples in atomic techniques and flow injection, the most convenient mode for introducing these heterogeneous systems into the spectrometer, thus causing a crucial integration effect. Depending on the sample-particle size, the dynamic manifold has been modified accordingly. Thus, the determination of aluminium in juices by GF-AAS [51], that of potassium and magnesium [52], and manganese, lead, zinc, calcium, magnesium, iron, sodium and potassium, both in cements by AAS [53], and that of iron, calcium and magnesium [54] in samples with high silica content also by AAS, are some of the examples that show the usefulness of this integration, which can also be helped by some types of auxiliary energy, as discussed later.

*Modification of the impulsion system—peristaltic pump*

The simplest and most versatile modification to a flow injection manifold is related to the impulsion unit, which is often a peristaltic pump. The changes in the normal functioning of the pump can range from halting the flow one of several times during the development of a given method to changing the flow direction or changing the flow-rate by modifying the velocity of the rollers.

*Stopped-flow modes*. The flow of a dynamic system can be halted when the injected plug is at different points of the manifold—namely, in a reactor or at the detection point—depending on the objective, and thus giving place to the non-kinetic and kinetic stopped-flow modes, respectively [55]. The former is used when a longer residence time of the reacting plug in the system is required without increasing dispersion, as this phenomenon increases with the two usual modes of obtaining long residence times (namely, decreasing the flow-rate or increasing the reactor length). Non-kinetic stopped-flow is a way of achieving a higher development of a given reaction without appreciable dispersion of the reactant plug [56].

The most usual time for halting the flow is when the reactant plug is at the detector, thus giving place to kinetic stopped-flow methods, or reaction-rate kinetic methods. In this way, instrumental signal-time data are obtained, thus avoiding contributions from the sample matrix and from both faster and slower reactions than that in which the target analyte is involved (e.g. colour in the case of absorptiometric measurements). Usually, both chemical and biochemical catalytic effects are the basis of the methods proposed. Thus, the catalytic effect of vanadium on the eriochrome blue black B/potassium bromate system [57] and those of iron, titanium and vanadium on starch–$I_2$ complex formation [58] have been used for the individual and simultaneous photometric determination, respectively, of the analytes. In the latter case, differential rate of catalysis allowed multidetermination. Non-metal analytes, such as iodine in urine samples [59] and that of ethylenethiourea [60], and substrates of biocatalysed reactions, such as L-tartrate [61] and L-phenylalanine [62], have been selectively determined by taking advantage of the stopped-flow mode.

Dual stopped-flow constitutes a more complex FI mode but it provides additional advantages when compared with the simple mode.

Thus, both non-kinetic/kinetic and doubly kinetic methods have been developed. An example of the former is a method for the determination of both free and bounded $SO_2$ in wines in which a dual injection of two plugs of sample into two channels of different lengths allows the shorter one to drive the corresponding plug to merge with the derivatisation reagents (namely, *p*-rosaniline and formaldehyde) and then to the flow cell of a photometric detector, the flow being halted when this plug reaches the detection point where absorbance-time data are collected, later related with the concentration of free $SO_2$. Meanwhile, the other plug is merged with a basic solution for breaking the bond of the analyte to organic molecules during the stop-time. Then, the flow is re-established and the second plug is also stopped at the detection point after merging with the derivatisation reagents. The data now collected are a combination of both the free and bounded $SO_2$ originally present in the wine [63]. A doubly kinetic stopped-flow was used both for the enzymic determination of oxalate in urine for proper contact with each of the two biocatalysts involved in the overall process [64] and also for the photometric determination of amiloride hydrochloride in a multi-commutated flow system [65].

*Programmed change of the roller's velocity.* Conventional FI methods are based on concentration gradients: the sample plug or that of the sample-reagent mixture is transported along the manifold, where the concentration gradient is changed as a function of the dispersion undergone by the plug. Modes as isotopic or electronic dilution [66–68] are based on the concentration gradient existing in the plug involving the sample. A major degree of dispersion is created in 'gradient chambers', thus causing FI-titrations [69]. None of these approaches requires special functioning of the propulsion system, so they have been implemented from the very beginning of FI. Temporal variation of the flow-rate also allows gradient profiles to be created along the dynamic manifold, whose precedent is the controlled variation of a hydrostatic head [70], in which the reservoir is raised or lowered to change the flow velocity, thus creating a flow gradient. However, nowadays more sophisticated experimental set-ups have been proposed mainly based on the use of a programmable peristaltic pump.

The most general working mode resulting from application of temporal variation of the flow-rate uses two pumps: in one of them the flow-rate is programmed according to a preset pattern and the other works at a constant flow-rate. The variable and constant flow-rate channels merge and, after a mixing or reaction coil, the mixture reaches the detector. The pumps are controlled by a microcomputer through an electronic interface. Both quantitative applications—namely, acid–base titrations [71], complexometric and redox [72]—and also calculation of equilibrium constants—namely, critical micelle concentrations of both anionic and cationic surfactants, acid–base constants, solubility products, stoichiometries of complexes and Michaelis-Menten constants [73–76]—have been successfully developed.

*Programmed change of the flow direction*. The iterative change of the flow direction has also been named flow-reversal, but this name lead to misunderstanding because of its similarity to reverse-flow, used for designing the concept of FI systems where the reagent is introduced into the sample which acts as a carrier. The iterative change of the flow direction is a strategy by which a plug injected into an FI system might go forward and backward through it by appropriate programming of the performance of the, usually, peristaltic pump in the two possible directions, thus giving rise to, for example, the iterative passage of the plug through the detector.

The iterative change of the flow direction in an FI system uses a single detector and allows the dispersion of the process to be manipulated over a wide range [77,78]; it also allows any signal within the dynamic detection range to be accommodated. Thus, the iterative change of the flow direction shares with open-closed FI [79] the capability for multidetection using a conventional, single detector, but also differs in some respects, namely: (a) the volume in which the sample plug is diluted can be indefinitely large and the final equilibrium reached in open-closed systems is unattainable here; (b) multidetection is performed on a small zone of the sample plug so that an effective sample concentration is set by choosing the start time for the cycles from the injection instant; (c) the multidetection frequency (the chief weakness of open-closed systems) is considerably increased in the iterative approach. These advantageous features are of special interest for multidetection by introducing new elements in the configuration and using the kinetic information it provides [78].

The iterative change of the flow direction in FI systems has been used for purposes other than multidetection, such as for increasing the sensitivity in acid–base titrations [78,80] and minimising the consumption of sample and reagents as well as for shortening the analysis time [81,82]. Wang et al. have used the iterative change of the flow direction for improving stripping voltammetry [83] by repeated changes of the flow direction during deposition in FI-stripping voltammetry, thus promoting improved plating efficiency and hence better detection limits. This FI mode has also been used for sample incubation in the determination of olive-oil iodine-value, dramatically shortening the determination time [84].

Maybe one of the most salient applications of iterative change of the flow direction is for liquid–liquid extraction without phase separation, which will be discussed in Section 4.6. The iterative change of the flow direction is also a useful tool for developing ultrasound-assisted leaching, discussed in Section 4.5.

### 4.3.4 Duplicate units

Conventional units of an FI manifold are frequently duplicated in a single design, thus achieving additional benefits which can either improve the method thus developed or make key steps which would be unaffordable without duplication of these units possible.

*Duplicate confluence points*

The most common duplicate unit in FI manifolds is the confluence point. Two, three, or even more of these devices are necessary most often for merging, with the main channel where the sample is inserted; all the necessary reagents in a sequential manner: diluents, pH conditioners, masking reagents and derivatising reagents are thus put into contact with the sample at the appropriate time along the transport system.

*Duplicate injection valves*

Another common duplicate unit is an injection valve, the location of which related to the other can be in-parallel, in-series or even in inner coupling (Fig. 4.2).

*In-parallel valve arrangements* make two main channels into which the contents of these valves are injected mandatory. These contents can be the sample, as is the case for simultaneous determinations, either using two detectors—one for each main channel—(Fig. 4.2(A1)) or a single detector after a confluence point (Fig. 4.2(A2)). In the latter case, delay of one of the injected plugs is mandatory by using a long reactor in such a way that its residence time is long enough to avoid overlapping while passing through the detection point. This step can be implemented by using a single flow-cell [85]—a previous confluence point is required for the two previous main channels—or two cells—each

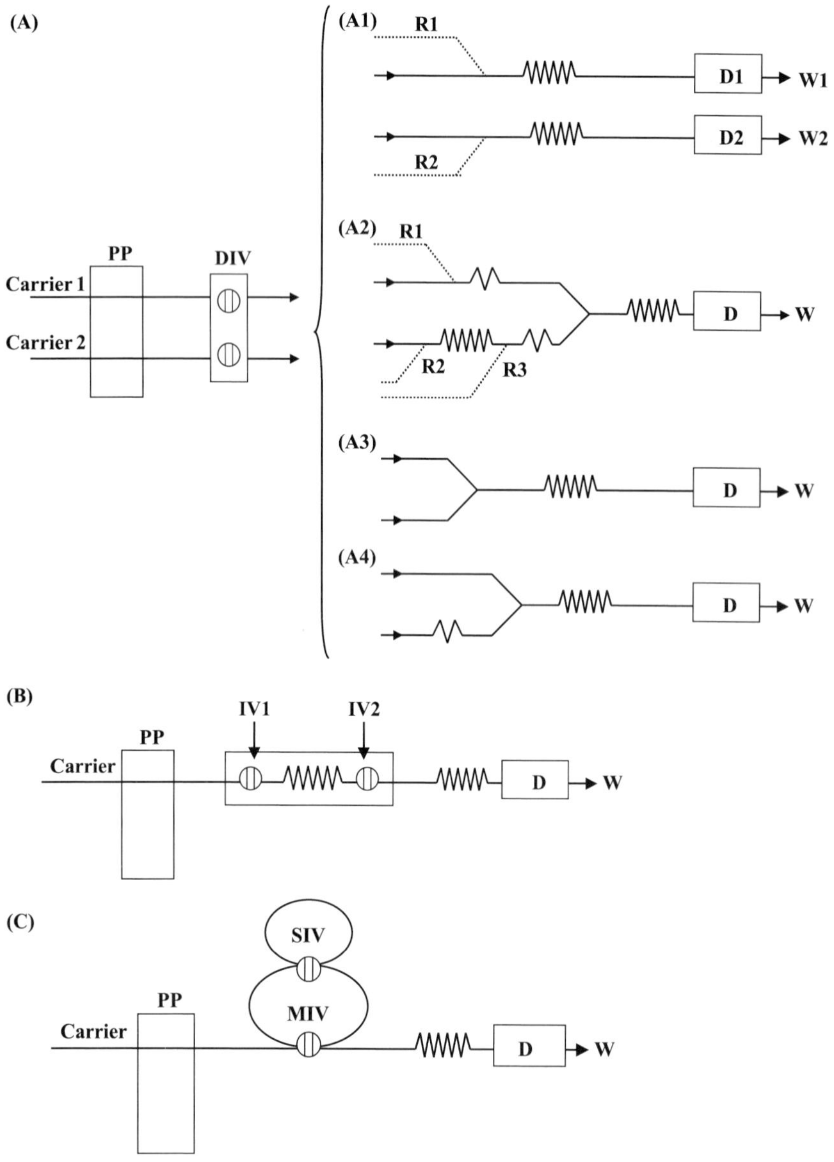

(A)
PP
DIV
Carrier 1
Carrier 2
(A1)
R1
D1
W1
D2
W2
R2
(A2)
R1
D
W
R2
R3
(A3)
D
W
(A4)
D
W
(B)
IV1
IV2
PP
Carrier
D
W
(C)
SIV
PP
MIV
Carrier
D
W

receiving a plug from one channel—can be located in series in the detector.

The two in-parallel valves can simultaneously inject sample and reagent into two main channels when both are expensive or valuable; thus causing the *symmetrical merging-zone mode* when the two main channels have the same length before reaching a point where both merge [86,87] (Fig. 4.2(A3)). A more versatile version of this approach is the *asymmetrical merging-zone mode* in which the two channels merge at the confluence point with partial overlapping of the injected plugs after the simultaneous injection of sample and reagent [88]—or simultaneous injection of two reagents in a reverse FI mode [89]. A richer understanding about the system under study is thus achieved (Fig. 4.2(A4)).

*In series-valve arrangements* (Fig. 4.2(B)). This can be implemented with very different aims. The simplest objective can be injection of two sample plugs for achieving two different residence times with minimal complication of the experimental set-up [90]. Such is the case for the speciation analysis of nitrogen as ammonia and urea. A urease reactor located in the transporting tubing between the two valves with subsequent merging of this main channel with reagents for the determination of ammonia enables the determination of the ammonia initially present in the sample using the plug closer to the detector, and the determination of the original ammonia plus that formed in the enzymic reaction in the plug from the other valve.

Other functions of the second valve—the one closest to the detector—can be housing in its loop a separation unit, which will be more widely discussed in Section 4.6. The enormous advantages of locating the separator in this position deserve a detailed explanation.

*Innerly coupled valves or nested loop valves*. This simple and versatile arrangement (Fig. 4.2(C)) has provided support for the development of interesting and useful methods. Multidetermination [91], calibration

---

Fig. 4.2. Duplicate injection valve arrangements. (A) In-parallel coupling without plugs-merging and either different (A1) or a common detector (A2); and with symmetrical (A3) or asymmetrical merging of the two plugs (A4). (B) In-series coupled valves. (C) Innerly coupled valves. PP, peristaltic pump; DIV, dual injection valve; MIV, main injection valve; SIV, secondary injection valve R, reactor; D, detector; W, waste.

[92], speciation studies [93], construction of pH gradients [94] and blank-sample measurements [95] are some of the more remarkable applications of these configurations.

In addition, inner coupled valves are used for locating a separation unit in the loop of the secondary valve.

## 4.4 DERIVATISATION REACTIONS IN FLOW INJECTION SYSTEMS

Either reactions developed with the aim of suiting the target analyte to the features of the detector can be of the general types of the previous subsection, easily implemented with simple FI manifolds involving confluence points, or a more complex aim can be pursued, thus requiring more sophisticated designs as discussed below.

### 4.4.1 Redox reactions involving solid reagents

The use of solid reagents is one of the best ways to decrease sample dilution in an FI method. The reactors designed with this objective are small columns connected in-line in the dynamic manifold after the injection unit. One of these approaches containing a solid-phase lead(II) chromate reactor has been used for the determination of sulphide in effluent streams with a detection limit of 3.7 mg/L and between days relative standard deviation $<3\%$ [96]. Meanwhile, for the in-line removal of copper and iron, and reduction of Se(IV) prior to hydride formation for subsequent determination by ICP-AES, the limit of detection obtained was 0.4 $\mu$g/L and the relative standard deviation 2.3% [97]. Even a cadmium microcolumn has been used in an on-chip flow injection device for the determination of nitrate [98].

### 4.4.2 Micellar media

Surfactants provide a favourable medium for improving the features of the species to be involved in any kind of detection. These organised media are easily introduced into FI manifolds, this binomial causing a synergistic effect which has made possible the development of very sensitive and selective methods not achievable by the separate use of dynamic and micellar media. Thus, photometric methods for the determination of gadolinium with 1-(2-pyridylazo)-2-naphthol in the presence of Triton X-100 [99], and of nicotinic acid in the presence of *N*-cetylpyridinium [100], and spectrofluorimetric methods such as that for selenium(IV) in a cationic micellar medium [101] have been reported. Chemiluminescence detection and reverse micellar media have been combined with on-line solvent extraction using quinolin-8-ol for the determination of iron(II) and (III) [102]. Also a reverse micellar medium has been used for the determination of the antioxidant *t*-butylhydroxyanisole using an amperometric enzyme biosensor [103]. Finally, micellar-catalysed reactions [104] have been the basis for the determination of pyridoxal-5-phosphate [105].

### 4.4.3 Photoinduced reactions

Flow injection manifolds constitute an excellent support for subjecting the sample–reagent mixture to an appropriate light source, thus causing the required development of the photoinduced reaction. Several types of photoinduced reactions such as digestion [106], degradation [107], redox [108,109] and chemiluminescent reactions [110,111] have been carried out using flow injection manifolds, thus providing methods for the on-line monitoring of inorganic [106,108], organic [107,110,111] or even organometallic compounds [109]. These approaches have the advantages of high sample-throughput rate, simple operation, low detection limits and excellent precision.

### 4.4.4 Electrogenerated reagents

For economy of both material and human resources, avoiding reagent preparation would be advantageous, particularly with respect to flow analysis methodologies involving chemiluminescent reactions where the reagents usually are very expensive and unstable [112–117]. The use of electrogenerated reagents offers all the same analytical advantages as conventional chemiluminescent procedures, with the added benefits of increased control over the rate and course of the reactions in which the electrogenerated reagents are involved and the ability to use *in situ* unstable reactive species when required and formed at the electrode from passive precursors in the carrier stream. Aromatic tertiary amines [112], quinine [113], ethanol [114], citarabini hydrochloridum [115] and inorganic compounds such as ammonium [116] and sulphite [117] have been determined using electrogenerated reagents in a chemiluminescent FI system.

### 4.4.5 Catalytic reactions

The binomial chemical catalysis-flow injection has been used with different objectives and always produced the desired integrating effect. Thus, the simplest and most common use of catalysis is for the determination of the catalyst, a metal cation most often, as is the case with the determination of manganese [118] or cobalt [119]. Not so frequent is the catalysis by non-metal as has been used for bromide, which can be determined at the trace level by its catalytic effect on the 4,4′-bis(dimethylamino)diphenylmethane-chloramine T reaction [120].

Catalysis has been used in FI not only for either making a method possible or improving it, but also for expediting other analytical steps, as with the catalytic digestion of urine prior to the spectrophotometric determination of iodine [121].

Biocatalysed reactions very frequently developed in FI approaches using either in solution or immobilized biocatalysts [122] are not dealt with in this chapter, in order to avoid overlap with chapters 4–6. Readers are referred to references [123,124] dealing with the present situation with this matter.

## 4.5 EXTERNAL ENERGY SOURCES INTEGRATED WITH FLOW INJECTION

There are various energy sources such as heat, ultrasound, electrical current or microwave that, when integrated with FI approaches, cause singular results.

### 4.5.1 Conventional heat sources

Heat is the most usual and simple energy source used with FI from the beginning of the technique. The most general aim of this coupling has been to accelerate endothermal reactions, thus making their appropriate development possible in the usual short residence time of a sample in an FI manifold—typically between 15 s and 2 min. It is worth emphasising the key role of temperature in enzymatic and immuno reactions [125,126].

### 4.5.2 Ultrasound energy sources

Ultrasound is an almost forgotten source of energy in analytical chemistry in general and in flow injection in particular [127]. It is worth noting that the sparse use of ultrasound—either in the bath or probe versions—has always provided excellent results with liquids, but even more so with solid samples. One of the first ultrasound–FI arrangements used a bath and it was applied to liquid samples. Although baths are more widely used, probes are more efficient as a result of (a) the uniformity in the transmission of ultrasounds (in baths, only a small fraction of the total liquid volume in the immediate vicinity of the ultrasound source experiences the effects of cavitation) and (b) the constancy of power with time (the decline of which leads to exhaustion of the energy applied to baths). Both phenomena result in substantially decreased experimental repeatability and reproducibility. By contrast, ultrasonic probes (also known as 'sonotrodes') focus their energy on a specific sample zone, cavitation in which is thus enormously boosted. Also, they are not subject to any exhaustion restrictions, so they are much more suitable for use in analytical chemistry than ultrasonic baths.

Ultrasound has been used in FI systems with different purposes, mainly for acceleration of leaching steps and for improvement of filtration steps.

The usefulness of continuous ultrasound-assisted leaching using an FI manifold has been demonstrated with applications such as the determination of iron in plant material [128], boron in soil [129] and hexavalent chromium in sediments [130]. There are three different dynamic approaches to ultrasound-assisted leaching in continuous-flow systems.

(a) The simplest approach, based on the manifold in Fig. 4.3(A), can only be used when the leached species respond to a specific type of detector placed in-line with the close circuit. The leacher is circulated in one direction, the leaching step can be continuously monitored and a rising signal is obtained, the slope of which reflects the kinetics of the leaching process. Attainment of the extraction

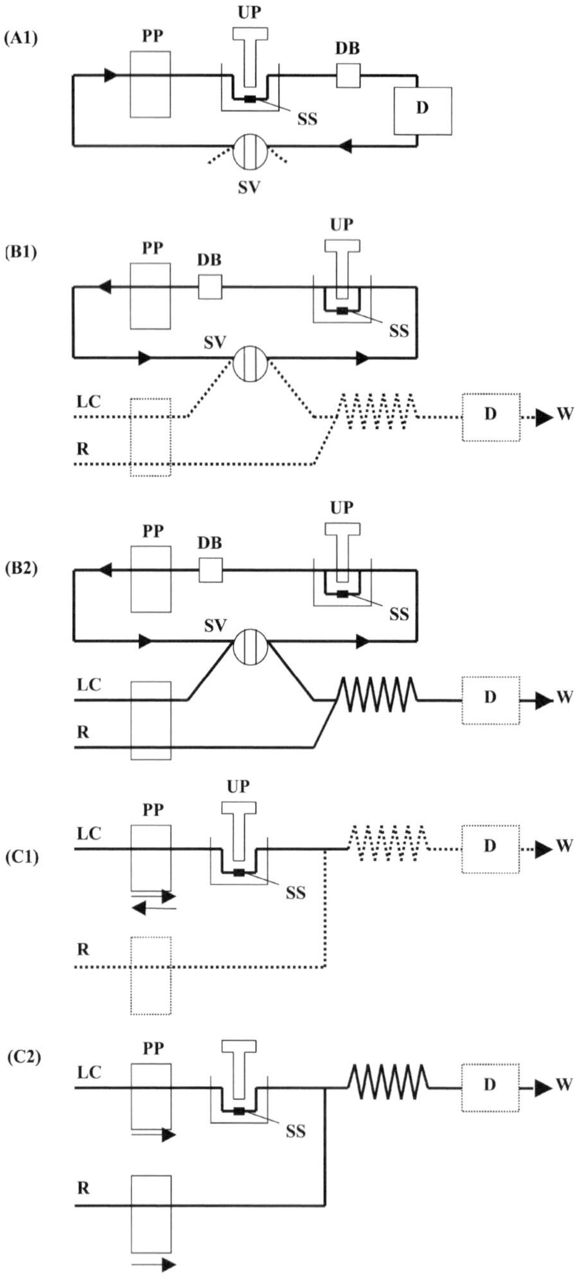
(A1)
PP
UP
DB
SS
D
SV
(B1)
PP
DB
UP
SS
SV
LC
D
W
R
(B2)
PP
DB
UP
SS
SV
LC
D
W
R
(C1)
UP
PP
LC
SS
D
W
R
(C2)
PP
LC
SS
D
W
R

equilibrium is shown by a plateau in the detector response, after which SV in the figure is switched to its open position and the extract driven to waste.

(b) The manifold in Fig. 4.3(B) is used when the leached species must be derivatised in order to obtain a response at the detector. In this case, the leacher is recycled as in the previous approach and, after an optimal leaching interval, SV is switched to its open position and the carrier drives the circuit contents to a point where they merge with a reagent stream, the derivatisation reaction taking place between this point and the detector. Valve SV is in fact an injection valve in whose loop the recycling system is built; it operates as such in the filling position (Fig. 4.3(B1)). The contents of this macro-loop are injected into the carrier solution in the derivatisation-detection step (Fig. 4.3(B2)).

(c) When the derivatisation reaction is required after leaching, the dynamic manifold can be simplified by programming the functioning of the peristaltic pump in such a way that the leacher is driven front and back through the sample container at preset intervals (Fig. 4.3(C1)). After a number of cycles, the extract is merged with the derivatisation reagent and driven to the detector, as in the previous approach (Fig. 4.3(C2)).

When necessary, a debubbler is connected in-line with the closed circuit to remove occluded air in the solid, which can give rise to parasitic signals at the detector.

Particles (either solids or cells) suspended in an ultrasound standing wave can be rapidly aggregated and separated in the pressure nodes located at half-wavelength intervals along the wave [131]. Recently, this idea was used to design a filter for the separation of particles from various media by combining ultrasound aggregation and gravity sedimentation. Such ultrasound filtration provides a new technology for use in flow injection on-line sampling. The fouling problems associated with other techniques are avoided because ultrasound separation does not rely on a physical barrier to particle separation; this makes the devices ideal for

---

Fig. 4.3. Approaches for implementing continuous ultrasound-assisted leaching. (A) With direct monitoring of the process. (B) and (C) with a derivatisation step before detection. (B1) with leaching in the recirculation mode; (C1) with leaching in the go-and-back mode. (B2) and (C2) derivatisation-detection steps. PP, peristaltic pump; UP, ultrasonic probe; SS, solid sample; DB, debblubler; SV, selecting valve; D, detector; LC, leaching carrier; R, reagent.

use in a process environment. The system is very robust, with no moving parts, and can be stream-sterilised (which makes it highly suitable for fermentation and clinical applications).

Only three types of ultrasound filters have so far been designed for as many types of sample that can be coupled to flow injection systems [132]. Each type of filter uses an ultrasound resonance field generated by a solid-state transducer to cause particle aggregation without any physical filtering surface.

### 4.5.3 Use of electrical energy

Few ways have been published of using electrical energy for helping FI systems because of its limited applicability. This joint use is mainly applicable to metal, solid samples, usable as the anode of an electrical system in which the sample and the cathode are in a flow-cell and between both electrodes an appropriate current is applied in such a way that the anode is dissolved. By keeping the dissolution step during the required time, sufficient amount of the target analyte can be obtained for its subsequent determination by either an atomic or a molecular detector. In the latter case a previous derivatisation reaction is mandatory. One of the first uses of this approach was for the determination of aluminium in alloys. After dissolution, the solution was merged for derivatisation with Eriochrome cyanine R solution containing HCl, ascorbic acid and hexamine for pH adjustment, and photometric detection [133]. Multielemental determinations have also been developed using an appropriate potential for the oxidation of iron, tungsten, molybdenum, vanadium and chromium in tool steels followed by ICP-AES measurements [134]. The main problem of this approach is the shape and characteristics of the sample, which significantly restrict its applicability, and the difficulty for calibration because of the absence of standards for running the calibration curve. A dual use of electrical energy was proposed by Cardwell et al. [135] for electrolytic clean-up and reductive amperometric detection of sulphur dioxide in wines and beverages.

### 4.5.4 Microwave energy assistance

Microwave energy is one of the most useful sources of additional energy for a series of steps in the analytical process, from sample treatment (e.g. leaching, dissolution, digestion) to detection. Flow injection has also taken advantage of using this type of energy. Since Burguera et al. reported the first instance the use of a microwave oven and a flow injection system for solid sample treatment in 1988 [136], a host of applications have been developed to exploit the many advantages of these on-line systems over off-line methods; such advantages include automatability, reduced delays between sample delivery and analysis, protection of samples from ambient agents and improved personal safety. Basically, most applications such as digestion, leaching, total nitrogen (Kjeldahl) determination and distillation use an FI system where the sample is introduced as a stabilised slurry; the system includes a reaction coil that is inserted in a microwave oven to perform a sample treatment which is usually a digestion [137–145] but can also be a leaching operation [146,147]. The FI system allows the solution or the resulting extract to be transferred to an appropriate detector in order to determine the target species after the microwave treatment. The sample digestion throughput can be improved by using several coils simultaneously in the oven [143]. Microwave ovens have so far been coupled to detectors based on spectrophotometry [141,143,148], flame atomic absorption spectroscopy (FAAS) [149,150], electrothermal atomic absorption spectroscopy (ETA-AAS) [151], inductively coupled plasma-atomic emission spectroscopy (ICP-AES) [138,144] and inductively coupled plasma-mass spectrometry (ICP-MS) [142]; and with derivatisation reactors in techniques such as hydride generation—atomic absorption spectrometry (HG-AAS) [140], hydride generation-atomic fluorescence spectroscopy (HG-AFS) [139] and cold vapour-atomic fluorescence spectroscopy (CV-AFS) [145]. In each coupled system, an FI manifold was used to merge the sample with the required reagents (usually acids) prior to the microwave treatment or to add the reagents, separation devices or components needed to derivatise the sample prior to analysis.

The main problem with these systems is the inherent inhomogeneity of the microwave power distribution within the oven cavity: with only a small area occupied by the reaction coil, a high proportion of the microwave power within the cavity is not absorbed. Thus, the positioning and distribution of the reaction coil within the cavity has been identified as an important parameter in achieving uniform heating of the sample and reagents with a view to ensuring efficiently controlled digestion. In addition to being inefficient, the high proportion of unabsorbed power probably reduces the lifetime of the oven.

Any resulting heating of the magnetron may in fact reduce the microwave output.

Some methods use a domestic or commercially manufactured microwave oven into which the sample is not introduced as a slurry but rather placed in an appropriate vessel that is then introduced in the oven [136,146,147,149,152].

In some commercially available microwave systems, the sample is placed directly in the vessel [152] or in an extraction cartridge that is in turn placed in the vessel [147]. The use of a pump as the interface between the extractor and an FI manifold (Fig. 4.4) allows microwave-assisted extraction to be coupled to filtration, preconcentration, individual chromatographic separation and UV detection steps with a view to automating the whole analytical process [147].

## 4.6 IN-LINE COUPLING OF SIMPLE NON-CHROMATOGRAPHIC CONTINUOUS SEPARATION UNITS AND FLOW INJECTION MANIFOLDS

One of the main uses of FI manifolds is as supports of small, simple separation units which can thus work in a continuous way and are located in the dynamic systems between the propulsion unit and the detector [153,154]. All types of interface (namely, gas–liquid, liquid–liquid and liquid–solid) can be involved in the separation step developed with the help of these units which can be commercial, but a number of experienced users prefer to design their own modules which are thus better adapted to the necessities of the target method. Membrane-based separation techniques (namely, gas-diffusion, pervaporation and dialysis) and liquid–liquid extraction require very similar or identical modules, as shown in Fig. 4.5, the mean difference between techniques being the type of membrane used. Meanwhile solid-phase extraction uses a different module (a minicolumn, which can adopt

---

Fig. 4.4. Experimental set-up used for the automation of the whole process. (A) Leaching step: CT, controller; MD, microwave digestor; S, sample; R, refrigerant; WR, water reservoir; TCPP, two-channel piston pump; ER, extract reservoir. (B) Clean up-preconcentration step: M, methanol; A, air; B, buffer; PP, peristaltic pump; F, filter; EL, elution loop; MC, minicolumn; R, retention sense; E, elution sense; IV, injection valve; SV, selecting valve, W, waste. (C) Individual separation-detection step: HPIV, high pressure injection valve; AC, analytical column; DAD, diode array detector; SR, solvent reservoir.

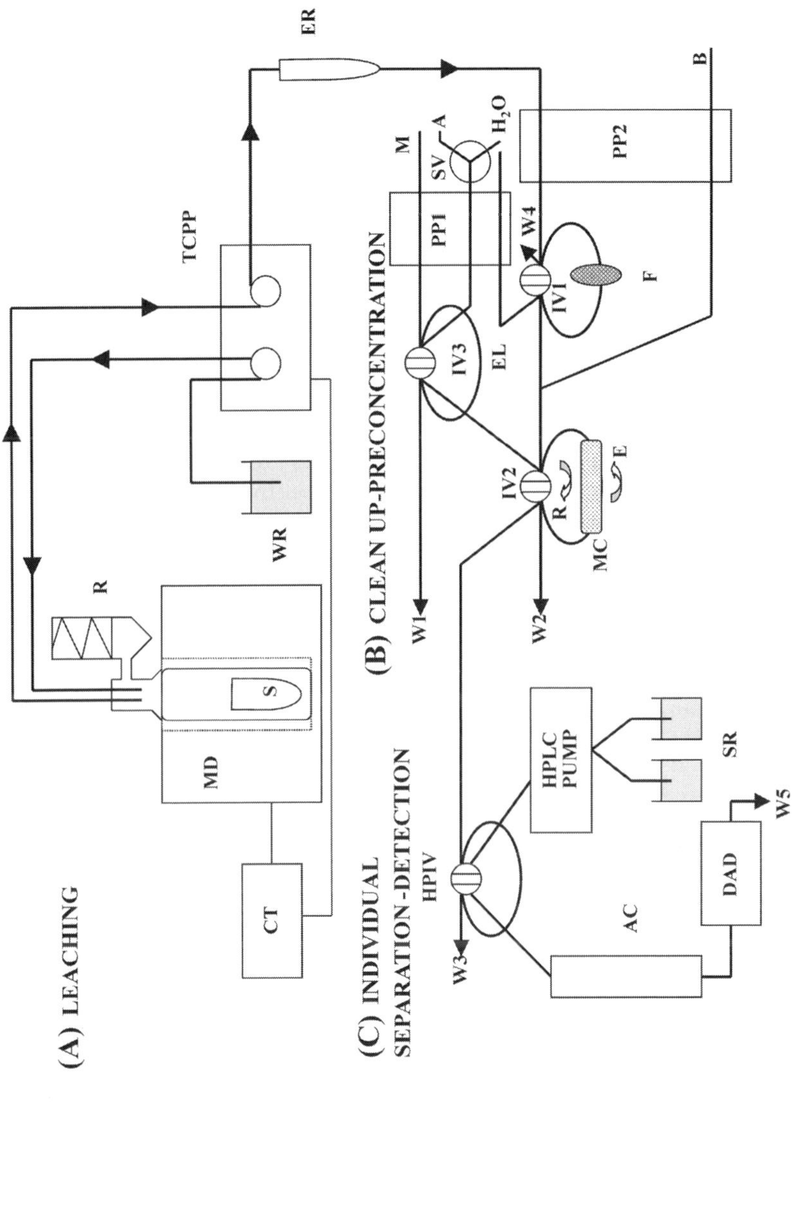
(A) LEACHING
CT
MD
S
R
WR
TCPP
ER
(B) CLEAN UP-PRECONCENTRATION
M
SV
A
$H_2O$
PP1
PP2
B
W4
IV1
F
IV3
EL
IV2
R
E
MC
W1
W2
(C) INDIVIDUAL
SEPARATION-DETECTION
HPIV
W3
HPLC
PUMP
SR
W5
DAD
AC

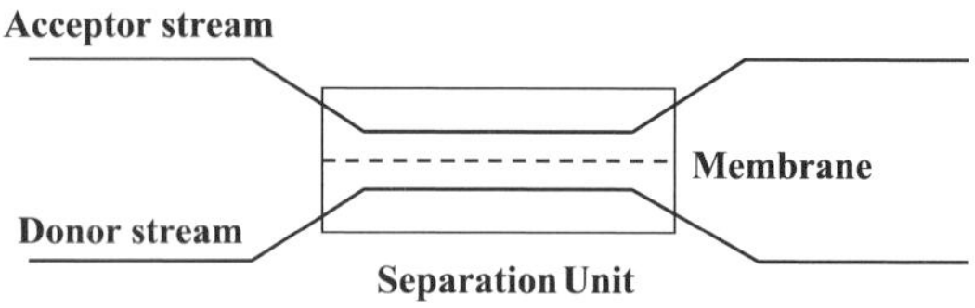


Fig. 4.5. Membrane-based separation module.

various shapes) which is also very easily designed and constructed in the laboratory.

### 4.6.1 Couplings with techniques involving gas-separation: gas-diffusers, pervaporators and others

When a non-chromatographic separation technique is implemented in a continuous mode with the help of an FI manifold the immediate, general benefits of this in-line coupling are easy automation, increased selectivity as a consequence of the non-equilibrium conditions under which separation is performed and rapidity as a result of the time saved which would otherwise be required for complete removal of the target species. In turn, the efficiency of separation decreases with respect to separation-equilibrium conditions and additional strategies must be designed in order to increase the removal yield from samples with low concentrations of the target compounds. The integrating effect of FI manifolds which incorporate separation units for gaseous analyte removal are individually discussed.

*Gas-diffusion*

One of the oldest membrane-separation techniques integrated with FI was gas-diffusion [155]. Despite the fact that most authors do not give data about efficiency, the flow rate used and the small surface for mass-transfer allow a poor removal to be assumed. Two strategies—both shown in Fig. 4.6—are used at present for improving the efficiency of gas-diffusion; they are also used with other membrane-based separation techniques such as pervaporation and dialysis.

*Halting the flow of the acceptor solution for a higher enrichment* (Fig. 4.6(A)). This can be done either by stopping the impulsion system of the corresponding channel—producing a backward effect of the flow and distortion of the baseline at the detector—or by locating the acceptor

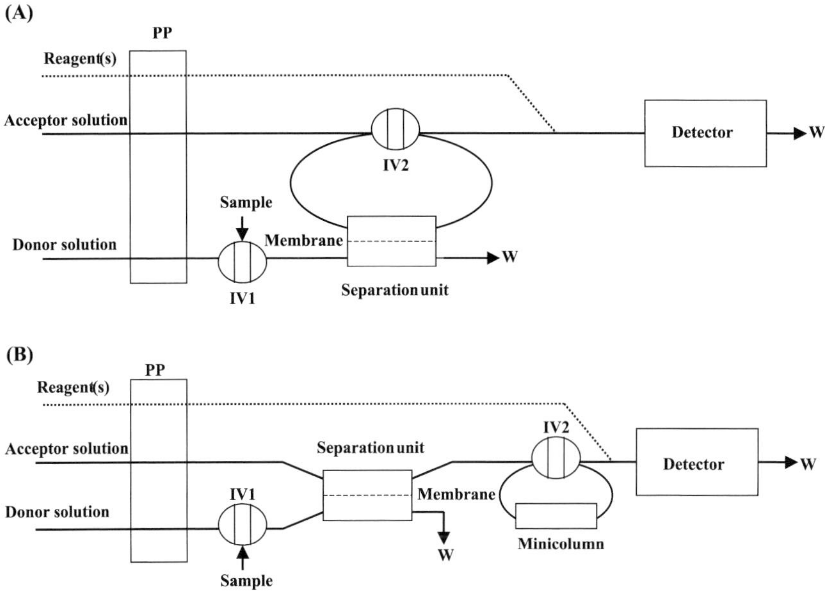


Fig. 4.6. Strategies for improving the efficiency in gas-removal processes. (A) With the acceptor chamber in the loop of an auxiliary injection valve. (B) With an on-line sorbent minicolumn located in the loop of an auxiliary injection valve for circumventing associated shortcomings. Abbreviations as in previous figures.

chamber in the loop of an auxiliary injection valve that, in its filling position, keeps the content of the loop static while the flowing solution is driven to the detector, thus keeping the baseline stable. When the valve is switched to the inject position, the content of the loop is driven to the detector, with or without prior merging with reagents. When the sample is scant the flow in the donor, sample chamber can also be stopped, but the efficiency is thus lower than with continuous introduction as the sample was impoverishing in the analyte.

*Locating a sorption minicolumn after the acceptor chamber* (Fig. 4.6(B)). In this way, the analyte crossing the membrane is always collected into a fresh acceptor; thus the amount retained in the solid-phase can be as high as required, provided enough sample is available.

The integrating effect of gas-diffusion and flow injection has been widely demonstrated. One of the analytes more frequently removed from the sample matrix has been ammonia using either the conventional approach in Fig. 4.5 [156] or that with the gas-diffusion cell in the loop of an additional injection valve (Fig. 4.6(A)) [157]. Other analytes such as fluoride [158], chlorine dioxide [159], and oxides of nitrogen, carbon and sulphur [160] have been individually determined in a variety of complex matrices after gas-diffusion removal. Methods for simultaneous determinations have also been reported with previous separation of the analytes based on their volatile character (or that of their reaction products), as is the case with the method for carbon and sulphur dioxides in wines based on a pH-measurement for the joint determination of both and then derivatisation of the sulphur dioxide by the *p*-rosaniline/formaldehyde reaction followed by photometric detection [161]. Two chemical forms of cyanide (free and that forming labile complexes) have also been determined using photodissociation as the step previous to gas-diffusion and fluorimetric detection [162].

*Analytical pervaporation*

Pervaporation is a relatively new technique in the analytical field [163, 164] which is characterised with respect to both gas-diffusion and industrial pervaporation by the presence of a head-space between the sample surface and the membrane that preserves it from clogging and deterioration and allows work with solid and slurry samples. The conventional pervaporator is quite similar to dialysis and gas-diffusion modules (Fig. 4.5), but a higher membrane surface is available to both volatile species and the acceptor solution, thus favouring mass-transfer as compared with other types of separation modules.

In addition to its use for the separation of analytes from liquid samples—with [165,166] or without a reaction for producing the volatile species [167,168]—the highest versatility of pervaporation is clearly demonstrated by its usage as an interface between fermentors and monitoring [169,170], for proposing methods for direct measurement on solid samples [171,172], for integration of separation and detection [173, 174] and on slurries [175,176]. One of the most significant uses of pervaporation is as an alternative to head-space in both the static and dynamic modes (or purge and trap). This approach has been proposed for on-line coupling particularly to gas chromatographs, but also to HPLC equipment, as commented in Section 4.7.

One very interesting and practical integrating effect of the FI-pervaporation binomial is availability for multideterminations [177] that can be based either (a) on manipulation of parts or parameters of the overall approach or (b) on coupling of the dynamic manifold of the upper chamber to a high discriminating separation technique such as gas chromatography.

*Multideterminations involving liquid samples* are the easiest and most frequently carried out in pervaporation-based approaches. These multideterminations have involved either speciation analysis or not. Speciation analysis in liquid samples based on pervaporation has provided methods for two volatile species, or one volatile (determined after passing to the acceptor chamber) and one non-volatile species determined either in the effluent from the donor chamber or using other sample aliquot injected simultaneously with that driven to the pervaporator. The methods for free and bound $SO_2$ in wine [178] and that of ammonia and urea in serum and urine [179] are typical examples of sequential speciation analyses of two volatile species, while that of chemical oxygen demand and inorganic carbon in bleaching liquors and domestic sewage [180] constitutes an example for the case of volatile and non-volatile speciation analysis.

Both wine fermentation monitoring and analysis of the final product constitute liquid samples with a number of volatile compounds and enough complex matrix to make the use of pervaporation convenient. Thus, methods for the determination of two volatile species using either a single photometric detector (free and bound $SO_2$ [178]) or two potentiometric and photometric in-series detectors ($CO_2$ and $SO_2$ [178]) and the determination of one volatile and one non-volatile species based on the use of either a single photometric detector (volatile and total acidity [181]) or two, photometric and fluorimetric, in-parallel detectors (ethanol and glycerol [182]) are representative examples of the wide range of possibilities offered by pervaporation.

*Multideterminations in FI-pervaporation involving solid samples* have been focused on speciation analysis [183]. The two more representative examples developed so far are the speciation analysis of mercury (as inorganic and organic forms), and that of nitrogen (as ammonia and urea), using the approaches in Fig. 4.7(A) and (B), respectively.

The speciation analysis of mercury in soil was based on FI-microwave-assisted pervaporation/atomic fluorescence detection in the simple, hybrid discontinuous–continuous manifold shown in Fig. 4.7(A). Focused microwaves proved to be highly efficient when analytes are sufficient polar to interact with this type of energy (e.g. $HgNO_3$ and PhHgAc). In the proposed manifold the sample is weighed in the donor chamber of the pervaporation cell which is tightly closed

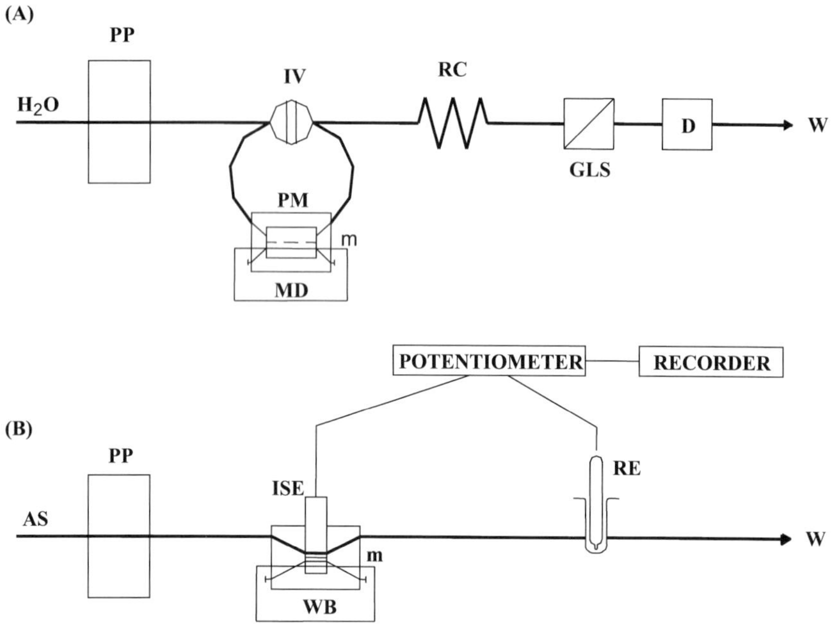


Fig. 4.7. Multideterminations in FI-pervaporation involving solid samples. Speciation analysis of mercury, (A), and nitrogen, (B). PM, pervaporation module; m, membrane, RC, reaction coil; MD, microwave digestor; GLS, gas-liquid separator; AS, acceptor stream; ISE, ion-selective electrode; RE, reference electrode; WB, thermostatic water bath.

and connected to the injection valve in the loop of which the upper chamber of the pervaporator is located. This allows the carrier stream (water or argon) to pass either through the acceptor chamber and then to the detector, or via a bypass to the detector while the flow in the acceptor chamber is stopped. Once the sample is in the donor chamber, the reagents are injected via the inlet of the donor chamber. One minute after the reagent ($SnCl_2$ acid solution) is injected, the microwave device is switched on at a preset power for a given time, after which the cell is allowed to cool for 1 min before the valve is activated to divert the carrier through the acceptor chamber, collecting one of the mercury compounds to the gas–liquid separator if the acceptor carrier is liquid, or directly to the atomic fluorescence detector if the carrier is gaseous. When the detector baseline is restored, the valve is switched again to isolate the pervaporator, the reagents (oxidising $-KBr/KBrO_3-$ solution and then the oxidant $-SnCl_2$ in this case), and the microwave digester is switched on again for another preset time and at a suitable power setting for the other mercury, now organic, compound [184]. The method was validated using a certified reference material (CRM 145R from the BCR). The speciation analysis of nitrogen (as ammonia and urea) in soils is based on enzymic reaction and integrated pervaporation/potentiometric detection in an FI manifold [172] (Fig. 4.7(B)).

*Other separation techniques for volatile species coupled to FI*
There was a net division between separator devices for the removal of a gas from either a liquid or solid sample depending on whether the separation was based on gas diffusion or gas expansion. The main difference between the two types of separator was that a membrane was used almost always for gas diffusion from one channel to another, whereas gas expansion separators were characterised by separation chambers with significantly larger capacities, but usually without a membrane. However, presently there is an increasing tendency to implement membranes in expansion chambers, leaving the geometrical aspect as the only difference between the two species of separators. Both *hydride and cold vapour mercury generation* belong to the gas expansion category, most often developed without the presence of separation membrane.

The species typically determined prior hydride formation are Se, As, Sb, Bi, Pb, Te, Sn and Ge [153,154]. Despite Hg being considered as a hydride-forming element, some authors have doubts about the nature of the volatile species formed and they consider the possible partial formation of a hydride at the beginning of the derivatisation step, rapidly evolving to the elemental, also volatile form [185]. Between these analytes, Se, As and Sb readily form hydrides only at lower oxidation states, whereas higher oxidation states need to be reduced. All inorganic species of these elements eventually form the same hydride ($SeH_2$, $AsH_3$, and $SbH_3$, respectively), thus making possible speciation analysis based on developing the formation of the hydride in two different steps with an intermediate reduction of the most oxidised forms. These steps are easily developed in an automated fashion in an FI manifold, then connected to a gas–liquid (in some cases to a gas–solid) separator for removal of the gas formed with subsequent transport to the atomic spectrometer.

In addition to hydride generation and cold mercury vapour formation in a dynamic fashion, the FI manifold enables other steps of the analytical process to be developed in an unattended manner. Such is the case with the method for the determination of As in a nickel-based alloy and other matrices in which the sample dissolution leads the analyte to its highest oxidation state and the mandatory reduction to As(III) is performed in-line helped by halting the flow. Then, hydride generation and transportation to an atomic absorption spectrometer is also performed in the FI manifold [153,154]. In-line cryogenic trapping with microwave-assisted removal, together with microwave-assisted hydride generation, have been arranged in the FI approach for the speciation analysis of arsenic [186]. Also microwave-assistance provides a way for accelerating and improving the speciation analysis of selenium with separation based on hydride generation and AA [187] or ICP-MS detection [188].

Cold vapour mercury (CV) can be formed in an FI system by reduction—with either sodium tetrahydroborate or stannous ion—from any of its oxidation states. The assistance of microwaves significantly accelerates the pretreatment of complex samples such as biological fluids (namely, blood or urine) for the subsequent determination of total mercury [189,190] or speciation analysis [191] with atomic absorption detection in all instances.

Other, less common gas-separator devices integrated with flow injection systems are as follows. First there is the *microdistillation-FI* approach designed and applied by Lane et al. for the determination of ammonia [192] using a flow-through micro-pH electrode instead of an $NH_3$ gas sensing probe [193,194]. The use of computerised control and a data acquisition system gave good control over the system, improved data acquisition rates, good graphics display and various data processing options. On the other hand, do Nascimento et al. have demonstrated the usefulness of a *microdiffusion-FI* method followed by polarographic detection of the determination of cyanide in biological fluids [195].

### 4.6.2 Couplings with liquid–liquid separators: dialysers and liquid–liquid extractors

The involvement of two liquid phases in a continuous separation technique makes the presence of a membrane as an interface mandatory when they are miscible—as in the case of dialysis in the normal and micro format—meanwhile the separation unit may or may not be provided with a membrane when the two involved phases are immiscible, as in liquid–liquid extraction.

*Dialysis*

Despite the fact that dialysis provides a less selective separation than the above-mentioned gas-phase separation techniques, its applicability is more general, as it can be used for any kind of analyte and provides a sample free from high-weight molecules. In its simplest version (namely, using a dialyser such as that in Fig. 4.5 directly [196] or with the acceptor chamber—or even both the donor and the acceptor chambers—located in the loop of an auxiliary injection valve [197]) it allows the characteristic clean up effect, but also a dilution effect, which can be modified as required through the flow-rate of the acceptor stream in the former case for subsequent determination of a single analyte. In addition, multidetermination methods are very common in dialysis-FI approaches when coupled to additional devices. Thus, a stream splitting and a dialysis unit in an FI system before the corresponding detectors makes the determination in soil of K and Na by FES possible and also that of Ca and Mg by AAS [196]; two in-series dialysers also permit the simultaneous determination by FES of Na and K in vegetables [197], meanwhile three of these units, also in-series, were used for the simultaneous determination of Na, K and Cl in blood serum [198]. A single unit can also enable the dual enzymic determination of malic and lactic acid in wines using a single dialysate by connecting an auxiliary injection valve in-line in the outlet of the acceptor chamber for trapping a part of the dialysate in the loop of this valve while the rest is driven to a bioreactor where malate dehydrogenase was immobilised and finally to the detector; then, the trapped dialysate was driven to other bioreactor with immobilised lactate dehydrogenase for the subsequent monitoring of the reduced form of the coenzyme [199].

A higher degree of selectivity enables the use of an electrodialyser coupled to the FI system, as proposed by van Staden for the determination of Zn in pharmaceuticals [200], which also provided enhanced selectivity for the analysis of L-amino acids after amino acid oxidation [201].

*Flow injection-Donnan dialysis*—namely, an ion exchange membrane process that can be used for the purification and concentration from diluted solutions—coupled to ICP-MS constitutes an excellent approach for preconcentration-determination of metal cations [202]; meanwhile, for speciation analysis of cadmium the FI-Donnan dialysis binomial was coupled to a differential-pulse anodic-stripping voltammetry equipment [203].

*Microdialysis (MD)*, the miniaturised version of dialysis, when coupled to flow injection manifolds offers the best way for continuous development and quasi real-time monitoring, which is of enormous interest for *in-vivo* analysis. Thus, FI-MD, combined with either chemiluminescence or electrochemical detection, has been used for studying drug–protein interactions [204,205]. *In-vivo* microdialysis-based sampling has been coupled to FI and then to chemiluminescence detection for the study of nitric oxide release in rat brain following traumatic injury [206] and monitoring of glucose in intravenous and subcutaneous tissue [207]; to fluorescence detection for the determination of calcium in blood [208]; and to tandem MS for R-84760 in blood [209,210]. Following microdialysis sampling, glucose, L-glutamate and acetylcholine have been determined after enzymic reactions on the corresponding immobilised biocatalysts and with subsequent amperometric detection [211]. Also L-lactate has been monitored by an electrochemically synthesised bilayer membrane-based biosensor in a microdialysate [212]. The wide field of application and possibilities of microdialysis when coupled to FI manifolds is also shown by the stopped-flow-FI-MD approach for automated multivessel high-resolution drug dissolution testing [213]. In this system, the perfusion solution ($H_2O$) was delivered (2.8 mL/min) to microdialysis probes immersed in the drug dissolution media. Stopped-flow dialysis was performed for 50 s, after which the dialysate was delivered (2.8 mL/min) to a UV detector and the absorbance measured at 254 nm. A six-vessel drug dissolution testing system was used, providing an overall sampling frequency of 360 samples/h (60 samples/h for each test vessel).

*Liquid–liquid extraction*

Mass transfer between two immiscible phases when developed in a continuous fashion helped by an FI approach may or may not involve separation of the two phases prior to detection.

*Liquid–liquid extraction with phase separation* in FI systems was widely developed by B. Karlberg in the very early 1980s [214–216] using both density-based and membrane-based separators, thus demonstrating the integrating effect of developing this separation process in an FI system [217]. New phase-separators continue to be proposed by FI researchers [218–220], and are used and compared with the commercial units, which are those commonly used by the majority of

FI practitioners. The arrangement of the extraction step in the dynamic manifold can be implemented in two main configurations depending on the relative positions of the extractor and the injection valve. As shown in Fig. 4.8 for aqueous samples, the separation step can be developed after injection of the sample into a carrier stream with a derivatising reaction before extraction, if required (Fig. 4.8(A)); alternatively, the

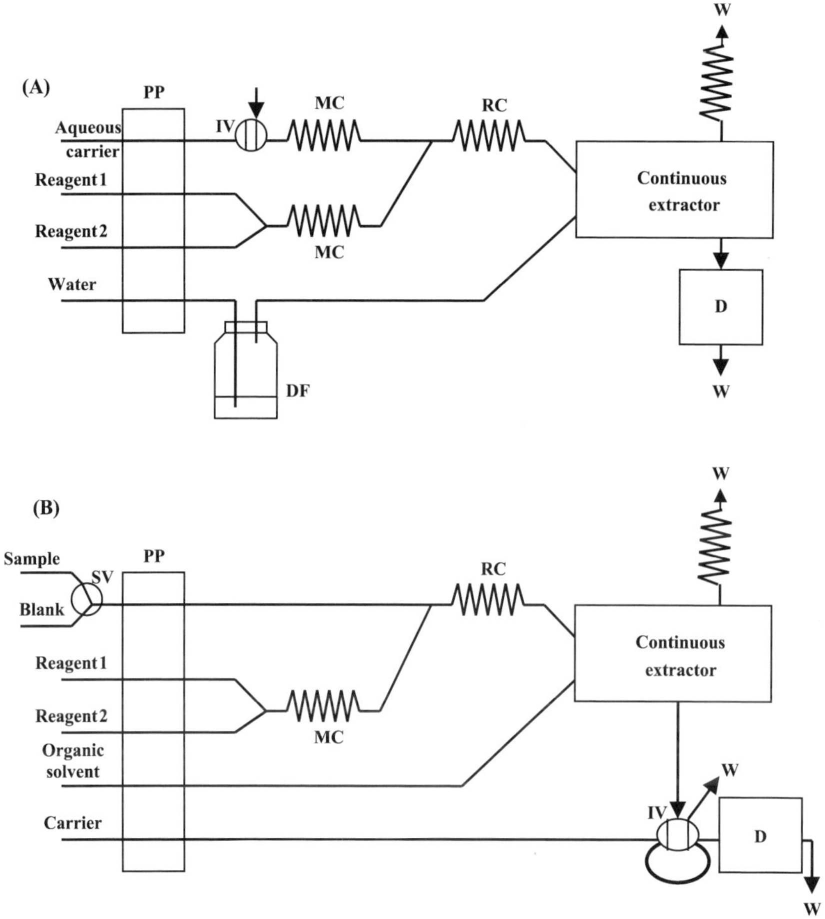


Fig. 4.8. Locations of the liquid–liquid step in FI manifolds. (A) After, and (B) before the injection valve. MC, mixing coil; DF, displacement flask, other abbreviations, as in previous figures.

extract can fill the injection valve if the separation unit is located before the injection valve (Fig. 4.8(B)).

Flow injection liquid–liquid (LL) extraction with phase separation has been proposed before determination of bismuth in pharmaceutical products [221] and for that of anionic surfactants using a special microporous membrane [222]. More complex designs involve a double membrane separator to ensure separation of the aqueous phase and the organic phase so that only the organic phase is transferred to the flow cell in the detector [223]. The capacity and versatility of FI manifolds for coupling other steps and/or units have also been exploited for solving problems appeared when more complex samples and/or analytes are involved. The simplest coupling of FI–LL extraction with other steps involves derivatisation, which can be developed in different points of the dynamic system depending on the objective: before the segmenter for analyte conditioning, in the segmenter itself or along the extraction coil depending on the kinetics of the reaction which facilitates extraction, or after the phase separator, for proper detection. An example of the last type is the use of a reverse micellar mediated chemiluminescence detection tool used after LL extraction for trace level determination of atropine [224]. More complex is the method based on FI-HG-AAS following LL extraction preconcentration and back-extraction for the determination of As in environmental samples [225]. Other two-separation steps based on dialysis sampling-LL extraction-FI have been reported for automation of continuous monitoring of drug dissolution processes [226].

Progress in *liquid–liquid extraction without phase separation and event without segmentation* have been made during the last 20 years. The simplest but least effective way of performing this type of extraction is by using a module similar to a dialyser and circulating the immiscible phases, one through the donor and the other through the acceptor chamber, depending on the relative density of the phases [227] and with supported liquid membranes [228]. Continuous LL extraction with on-line monitoring and non-phase separation for the determination of anionic surfactants in water was reported more than six years ago [229]. In this case, the sample stream was merged with the reagent and the resulting stream was passed through a mixing coil and then through a plug (200 μL) of $CHCl_3/CCl_4$ (3:1) situated in the cuvette of a spectrophotometer where the absorbance of the ion-pair complex formed was measured at 650 nm as a function of time. None of the units in an extractor are necessary in approaches based on the programmed iterative reversal of the flow direction and the location of the detector flow-cell in the loop of the injection valve, as proposed by Cañete et al. [230]. As shown in Fig. 4.9(A1), in the load position, the organic phase fills the loop of the injection valve, thereby establishing the baseline, and the sample is continuously introduced and merged with a reagent stream, if required, and sent to waste. In the inject

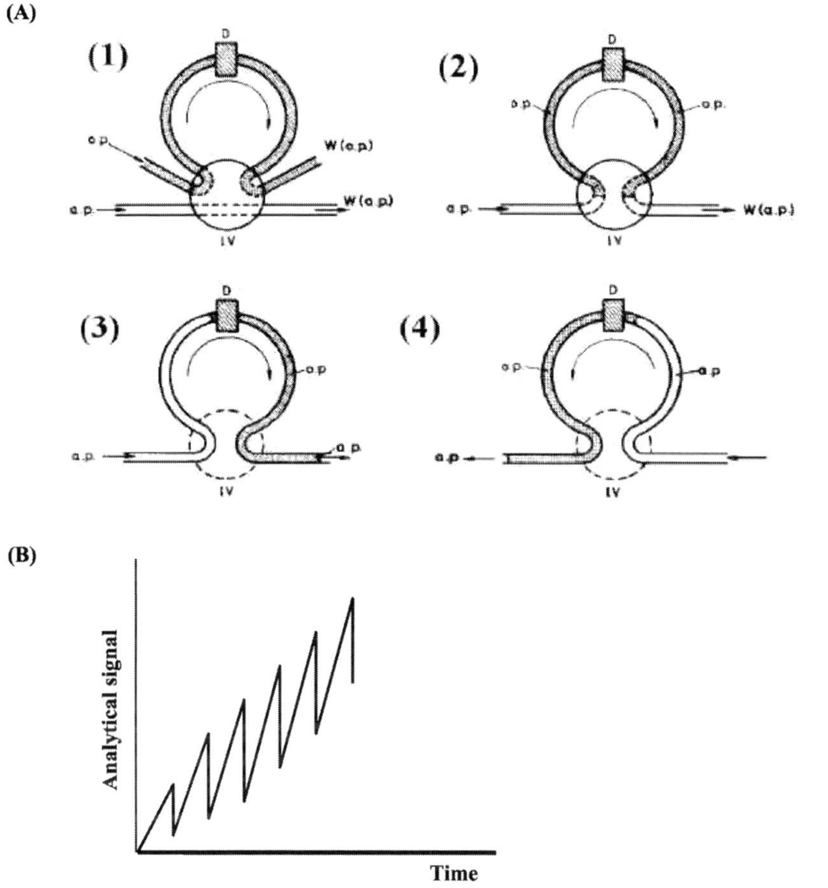


Fig. 4.9. (A) Performance of liquid–liquid extraction without phase separation. (A1) Baseline establishment (injection valve in the filling position). (A2) Switching of the valve to the injection position. (A2) and (A3) Go-and-back movement of the flow. o.p., organic phase; a.p., aqueous phase; IV, injection valve, D, detector. (B) Multipeak recording provided by the approach.

position, two liquid interfaces are established within the valve (Fig. 4.9(A2)). The flow circulates left to right for a preset time, and then back, as shown in Fig. 4.9(A3) and (A4). Only one of the halves of the organic plug is monitored, and the interface is never allowed to pass through the flow-cell, in order to avoid parasitic signals. The multipeak recording thus obtained (Fig. 4.9(B)) provides a wealth of information that can be exploited both for theoretical studies and for the implementation of analytical methods both for extraction into the organic phase—for the determination of surfactants [230] or anions in waters [231]—and for that of polyphenols in olive oil by exchanging the aqueous and organic phases [232].

The phase separator can also be deleted in an FI-segmented-LL extraction system if a diode array spectrophotometer is used and an expansion chamber is located after the extraction coil, thus giving place to the formation of organic and aqueous segments long enough for each to fill the path-length for a short time, thus allowing the detector to monitor signals from each at the required wavelengths. Another wavelength far from those of the two phases is used for subtracting noise signals [233]. Continuous monitoring of the extraction process without the typical units is performed by leaving the aqueous samples to a flow-cell containing an organic extractant denser than water and bubbling the sample at the flow-cell bottom. The mass transfer is produced during the travelling of the aqueous drops through the organic phase to its surface, from where it is sent to waste. The process is simple, but not efficient, as a consequence of the short contact time and small mass-transfer surface between the two immiscible phases [234].

The wide diversity of possibilities provided by the wealth of FI-LL extraction approaches and the number of other separation modules and FI modes which can be implemented allow integrated and very profitable methods which always replace the manual approaches with easy automatable alternatives.

### 4.6.3 Couplings with liquid–solid separators and solid-phase formation

There are three most common ways to remove an analyte in liquid samples to either a solid support or in a solid form, namely: (1) in a solid material (either by sorption, ion-exchange, etc.) packed in a microcolumn which can adopt different forms; (2) by precipitation/filtration using an appropriate in-line filter; and (3) by a redox reaction at the working electrode with either deposition on its surface or amalgam formation, depending on the nature of the electrode.

*Solid-phase extraction in flow-injection approaches*

Removal in a solid support of some species from a liquid sample can be done with one or two aims: to eliminate the interferent effect of the removed species or, more commonly, to preconcentrate analytes as they are eluted with a liquid volume smaller than that of the initial sample. The latter is different from the most common preconcentration mode, which is easily developed in an FI manifold as all the steps—time consuming steps when developed manually—involved in the process can be carried out in an automated way and at the microscale level, minimal reagent—(and time)—consumption and without contact of the chemical or biochemical system with the atmosphere.

The *type of retention material or solid support* depends on the features of the species to be removed. Ion exchangers are the usual supports for preconcentrating both cations and anions [235–240], but fullerenes [241–243] and carbon [244] are also used prior to determination by an atomic technique; meanwhile, for subsequent speciation analysis of mercury by capillary gas chromatography with a microwave-induced plasma emission detector, a dithiocarbamate resin has been used [245] and different types of coatings for solid-phase microextraction (SPME) coupled to flow injection have been reported for demonstrating the integrating effect of SPME-FI prior to gas chromatography [246].

The *shape and size of the column* used for the separation step can be diverse. Despite the most common shape being tubular, conical columns provide higher concentrated analytes when used properly—that is, when sample loading takes place from the narrower to the broader end and elution in the opposite direction, which produces a very short plug of the eluted species, provided the elution kinetics is fast enough. The size of the column—in both length and diameter—is a crucial aspect which depends on the amount of analyte(s) to be retained. For large amounts of species, it is preferable to increase the length of the column rather than the diameter as a higher dispersion of the eluted species is foreseeable by increasing the diameter than when the length is increased in such a way that the same amount of support is packed in both cases. Connection of the columns in the dynamic system can be made, either by using customised small connecting tubes of an intermediate size or—using a more elaborate and efficient method—by commercially available connectors, as shown in Fig. 4.10, where the different shapes of these devices are also found. Either, a small plug of glass wool or cotton—or better a porous disc of PVC or glass—must be

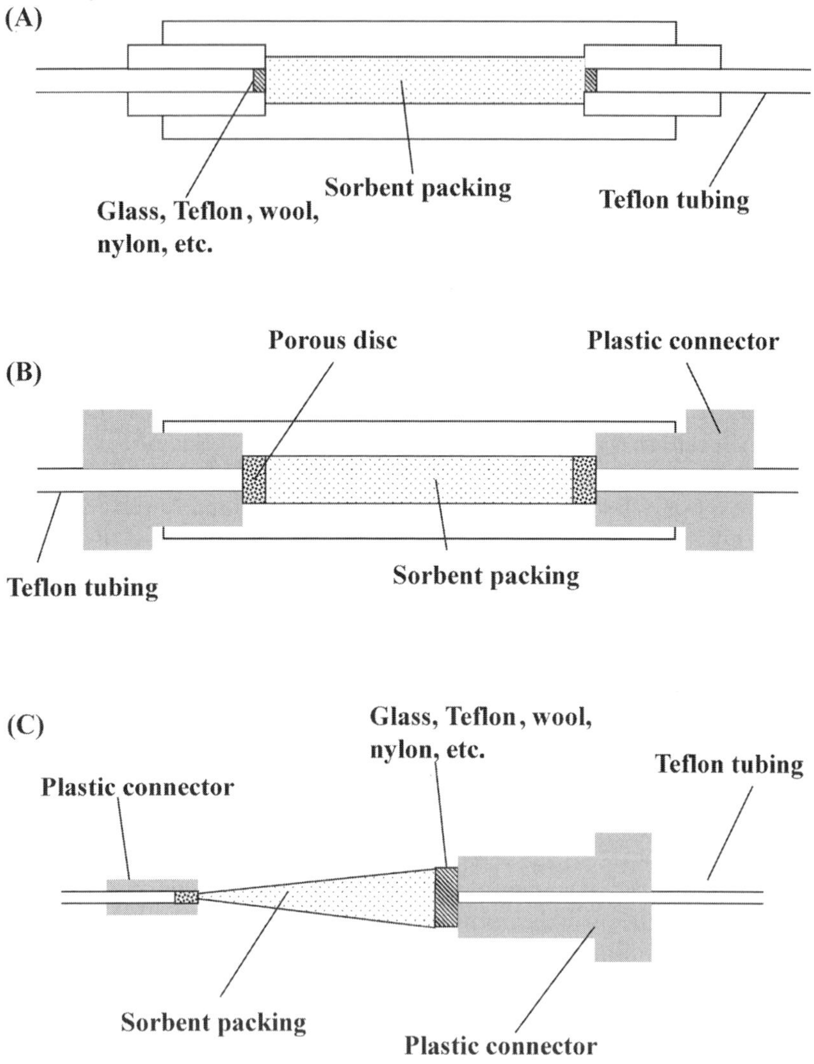


Fig. 4.10. Types of sorption columns. (A) and (B) tubular. (C) Conical.

located at both ends of the tubes to avoid loss of the solid material, which can clog the dynamic system.

The *location of the solid-phase microextraction minicolumn in the FI manifold* depends on the aim of the unit and this location can be established with respect to (a) sample channel and injection system, and (b) use of some auxiliary unit for more effective performance.

(a) The removal of either interferent species in the sample or impurities in the reagents calls for a position before sample injection in the sample channel or in the reactor after sample injection. For analyte preconcentration the column can be positioned either before injection—in the sample stream—or after injection.

(b) The column can be directly connected in-line in the flow stream in the selected point. This very simple approach has a number of shortcomings, the most serious being: (1) uneven mixing of sample and carrier (generally a buffer), which results in pH changes along the sample plug and in partial separation of the analyte in the column; (2) gradual increase in the column compactness at the exit by effect of the continuous-flow in the same direction and the sequential swelling and contraction of the exchange material; (3) the production of parasitic signals and the risk of clogging the burner outlet (atomic detectors) and poisoning or fouling the sensor (electrochemical detectors). Despite the number of complex manifolds with either multiple valves or pumps, the simplest and most effective way of overcoming the above problems is to locate the minicolumn in the loop of an auxiliary injection valve, whose filling position is used for preconcentration by passage of the sample in one direction, meanwhile the injection position allows the eluent to pass in the opposite, detector direction. Most of the present FI-preconcentration minicolumn couplings are constructed with this criterion.

The shortcoming derived from the long time required to achieve a large enrichment factor (2–3 min or even 20 min preconcentration times have been reported) is avoided by using several columns arranged in parallel which operate simultaneously and alternatively to deliver the eluates sequentially to the flow manifold.

One other constraint of the methods developed with these approaches is the relatively large sample volumes used in each determination.

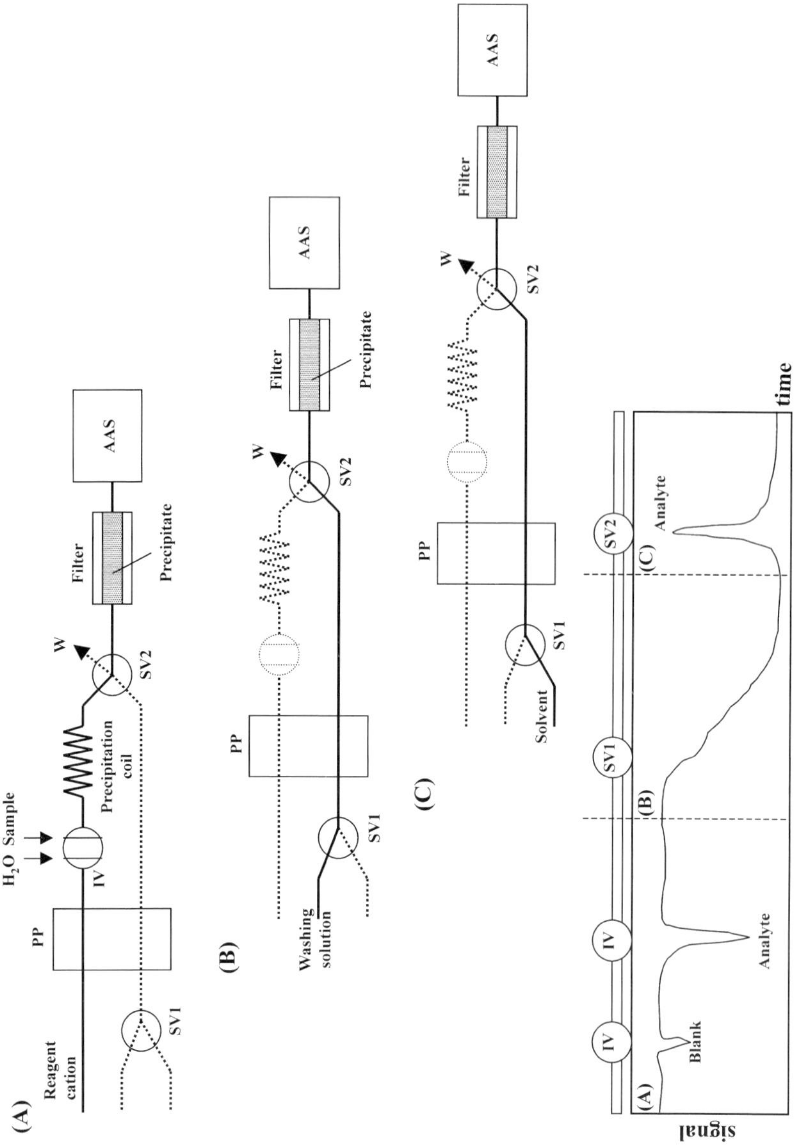
(A)
Reagent cation
PP
$H_2O$ Sample
IV
Precipitation coil
SV1
SV2
W
Filter
Precipitate
AAS
(B)
Washing solution
PP
SV1
SV2
W
Filter
Precipitate
AAS
(C)
Solvent
PP
SV1
SV2
W
Filter
AAS
signal
time
(A)
(B)
(C)
IV
IV
SV1
SV2
Blank
Analyte
Analyte

Depending on both the analyte concentration and enrichment factor to be achieved, the sample volume can range from 2 to 100 mL, which is far from the microlitre volume range typical of FI. In general, the efficiency of the system can be improved through more careful control of the dispersion and reduction of the sample loading period. Extensive efforts aimed at reducing the eluate volume for improving the enrichment factors have resulted in isolation of the eluant zones with air, thus decreasing dispersion [247,248].

*Separation by solid-phase formation*

The popularity of conventional precipitation reactions has declined in the last few decades mainly because of the slowness and intensive involvement of the operator in the process and the difficulty of implementing precipitation and dissolution reactions in automated systems, whether discrete or continuous.

*Filtration*, the immediate step following precipitation for separation purposes in a continuous system—which is not the case in turbidimetry or nefelometry where solid-phase formation facilitates detection—had been previously used most often in automatic analysers for the removal of suspended particles from sample or reagents. The solid-phase formed in a liquid, continuous system for analytical purposes can be used either as such or after dissolution.

The most significant integrating effects achieved by flow injection–filtration/precipitation coupling, with or without dissolution, are preconcentration, and automation of precipitation, filtration, washing of the precipitate, dissolution—if required—and transportation to the detector [249].

Precipitation—usually of the analyte, except when the objective of the step is interference removal—developed in a flow injection manifold is generally followed by atomic detection, either after filtration of the precipitate or after filtration/dissolution of the solid: PF and PFD, respectively (it is worth noting that dissolution is always necessary, but in PF it is performed after detection in order to prepare the system for introduction of a new sample). In any case, the key unit of these systems is a filter, the most common being a cylindrical filter of the type shown in Fig. 4.11.

---

Fig. 4.11. Scheme of a continuous precipitation-washing-dissolution system with sample injection for the determination of anionic species also showing the FI peaks obtained in each operation. AAS, atomic absorption spectrometer.

A series of flow injection configurations—either normal and reverse—have been reported depending on whether or not the precipitate is dissolved, and the objective (indirect determinations, preconcentration-dissolution, etc.). As an example of the versatility of these approaches, Fig. 4.11 shows the scheme of an FI with sample injection involving the sequential formation and filtration of a precipitate (A), its washing (B), and its dissolution and transport of the dissolved species to an atomic spectroscopic detector (C). The figure also shows the typical signal obtained in each of the above stages, as well as the time programming of the valves used.

The preconcentration of metal traces by precipitation followed by filtration/dissolution/atomic detection is the immediate application of FI-PFD. Thus, methods for the determination of lead in tap water [250], that of copper [251] and cobalt [252] in silicate rocks, and manganese [253], and cadmium in mussels [254], have been reported. This type of approach has also been used after on-line microwave assisted processes for the determination of iron in geothermal fluids containing high concentrations of sulphur compounds [255] and for that of bismuth in biological samples [256], followed by atomic detection in both cases. Co-precipitation without filtration has been used by Fang et al. for the determination of silver in geological materials [257], and for that of lead in biological samples [258], with simplification of the operational setup. Also the use of a filter can be avoided by in-line reductive precipitation and collection with magnetic particles, as proposed by Sella et al. [259].

Typical applications of FI-PF are indirect determinations of anions such as chloride [260], chloride and iodide in foodstuffs [261], sulphate [262], phosphorous [263], and citric acid [264] by atomic techniques, meanwhile FI-PFD has been used for the molecular absorption determination of ultratrace of phosphate [265].

Other methods based on FI-PF with or without dissolution prior to detection involve the determination of drugs such as sulphonamides [266] and chlorhexidine [267] in pharmaceuticals and saccharin in mixtures of sweeteners [268].

Less common, but also interesting, is the use of continuous precipitation for eliminating interferences posed by substances accompanying the analyte in the sample. Such is the case with the precipitation of calcium with oxalate to isolate this alkaline earth from synthetic rock-sample solutions containing aluminium, silicon and phosphorus as interferents. In addition to the increased selectivity of the method, an enrichment factor of 650 was obtained [269].

*Electrolytic or adsorptive preconcentration prior to stripping coupled to FI*

Electrolytic deposition is one of the major alternatives for preconcentrating metal traces, meanwhile adsorptive preconcentration is the most suitable for organic and non-electroactive species. Independently of the nature of the subsequent stripping step, the development of both types of preconcentration in a dynamic fashion using an FI manifold confers to the method, thus a series of advantages developed resulting from the integrating effect of coupling the two approaches. These advantages are even more outstanding when complex samples requiring special treatment are dealt with. A fully automated approach for implementing such a method is as simple as that shown in Fig. 4.12, which consists of: (a) an autosampler which, in addition to the samples, contains the standards, washing solution and solution for regenerating the electrode, if necessary. A lower automation degree is achieved by substituting the autosampler with a selecting valve. An optional injection valve would allow sample injection instead of continuous aspiration. (b) A programmable propulsion system—usually a peristaltic pump—whose functioning can be stopped at preset intervals, as the most reproducible monitoring is obtained by keeping the solution into

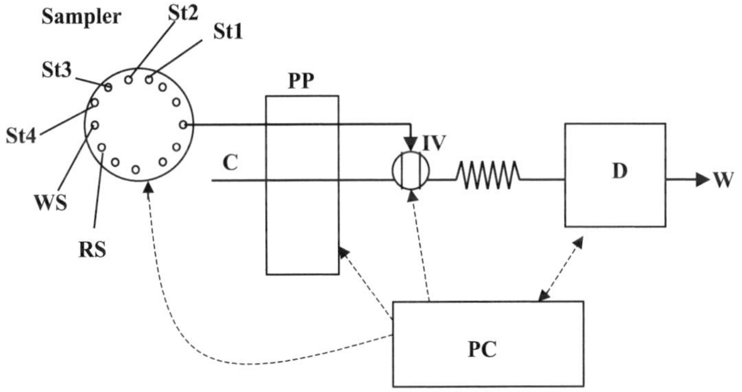


Fig. 4.12. FI automated system for implementing stripping methods. St, standard; WS, washing solution; RS, regenerating solution Hg(II); C, carrier; PC, personal computer. Other abbreviations as in previous figures.

which the analytes are stripped static. (c) A potentiostat, that in voltammetric both conventional and adsorptive stripping has the dual function of applying the constant potential for deposition and the scan potential for stripping, meanwhile in potentiometric both conventional and adsorptive stripping the potentiostat is only used in the deposition step and a potentiometer is mandatory for the stripping, determination step. (d) A cell and electrode system. The most suitable cell for stripping methods is of a wall-jet type [270], meanwhile the most convenient working electrode is a film mercury deposited on a glassy carbon, reticulated vitreous carbon or carbon fibre electrode. Metals (gold, platinum) are scarcely used as working electrodes. The nature and position of both reference and auxiliary electrodes are not crucial. (e) A computer for proper control of the overall equipment via active and passive interfaces.

When the typical steps of the stripping method (namely, electrode pretreatment, analyte preconcentration and stripping of the deposited substances) are developed in a continuous, dynamic fashion, other steps can also be included in the overall process if required. These include the following. (a) The introduction of a complexing agent via a merging point can be useful for shifting the redox potential of interferents far from that of the target analyte. (b) The development of the deposition step in one base electrolyte and its replacement by a secondary electrolyte can endow the stripping step with higher selectivity. This step is rarely developed manually because of the contact of the working electrode with the atmosphere during electrolyte change. This problem is fully overcome by continuous systems, which eliminate one of the electrolytes by flushing with the next one. (c) The inclusion of some other separation step by in-line location of a separation unit before the electrode cell.

All stripping techniques have been successfully implemented in FI manifolds. Thus, both voltammetric modes—anodic [271] and cathodic [272]—have been used for the determination of selenium with excellent results; meanwhile the simultaneous determination of copper, lead, cadmium and zinc in previously lyophilised biological tissues has been carried out by anodic stripping [273].

Potentiometric stripping and flow injection are contemporary and they were coupled very early. One example is the method for the determination of Cd and Pb in whole blood using carbon fibre electrodes, which required only a single $1 + 19$ dilution of 0.2–0.4 mL sample aliquots with 0.5 M HCl, and a standard addition to one of the aliquots.

The computer-controlled flow system used automatically executed a pre-programmed number of cycles on each sample pair before delivering the final result [274]. The increased sensitivity obtained by halting the flow was demonstrated in the determination of lead and tin in fruit juices and soft drinks: the Pb and Sn signals increased by a factor of about 10 with respect to those obtained when the stripping flow-rate was maintained at that of the deposition when the flow was stopped 20 s before stripping [275]. The characteristic versatility of FI is shown in the special methodology combining voltammetric and potentiometric stripping for the automated determination of iodine species in natural water. Two flow-cells were used for analyte preconcentration and detection. Iodide was electrochemically oxidised to iodine, which was concentrated quantitatively at a carbon wool electrode housed in the preconcentration cell. After interfering ions had been removed from the cell, iodine was eluted with a reductant and subsequently determined at a polished $Ag_3SI$ electrode in the detection cell. Iodate was also determined after chemical reduction to iodide [126]. Chronopotentiometric stripping [276–278] and coulometric stripping [279] have also been implemented in FI manifolds with the added integrating effect in all instances.

The most versatile stripping technique—namely, adsorptive stripping—has been widely coupled to FI for the determination of analytes which range from metal cations adsorbed as complex—e.g. Ni(II) and Co(II) as dimethylglyoximates [280], speciation analysis of Fe(II)–Fe(III) in wines by complexation of the latter with Solochrome Violet Red [281]—to chloride and cyanide adsorbed on pH electrodes [282] and organic compounds such as chlorpromazine [283] and doxorubicin [284].

### 4.6.4 On-line separation equipment and flow injection manifolds

Separation units of dimensions bigger than that described in Section 4.6. (namely, sub- and supercritical fluid extractors, liquid and gas chromatographs and capillary electrophoresis devices) coupled to flow injection manifolds are dealt with in this section. The arrangements have been on-line most often; nevertheless, some cases exist in which the big unit has been inserted inbetween two sub-manifolds, and the detector located after the last one.

*Coupling sub- and supercritical fluid extractors with fluid injection manifolds*

Taking into account that the most common fluids used as extractants in the subcritical state are liquid—under both normal conditions and those required during extraction—and that the almost unique fluid used as extractant in the supercritical state is $CO_2$, the interface for a continuous extractor to be coupled to a flow injection manifold is simpler when the extractant is in the subcritical state. Depressurisation and cooling are the only steps prior to interfacing the extractor and the FI system.

*Subcritical extractors and flow injection manifolds.* The adjective subcritical is ambiguous as a liquid in the subcritical state can, in principle, be at any temperature. Nevertheless, the term is commonly used for superheated liquids; that is, those at temperatures above their boiling point which remain as liquids because the appropriate pressure is applied. Both static and dynamic subcritical solvent extractors have been coupled on-line to an FI manifold in order to carry out several steps of the analytical process such as filtration [247,248], preconcentration [247,248,285], derivatisation [286,287] and continuous monitoring of the extracted analytes [288].

*Static subcritical solvent extraction* as implemented in commercial equipment is basically discrete in nature, so it is rarely coupled to other operations of the analytical process. In fact, only in one reported application was the static mode coupled on-line to an FI continuous fluorescence monitoring system [288]. This coupling was primarily used to develop an independent matrix extraction method allowing the extraction kinetics to be monitored with a view to determining the number of extraction cycles required to quantitatively extract the analytes irrespective of the kind of matrix concerned and of whether the species concentrations in it were known. The extract obtained after each cycle was driven to a graduated reservoir from which it was aspirated into a flow-injection manifold connected to a detector. This system was used for the extraction and fluorimetric monitoring of PAHs in soil, sediment and food [288].

The *dynamic mode of subcritical solvent extraction* usually requires implementation of a preconcentration step prior to the determinative step and, because the extracted analytes are dissolved in a liquid—usually aqueous—phase, SPE is a highly useful tool for avoiding the dilution effect. For this purpose, SPE columns packed with appropriate sorbents and coupled on-line to the extractor outlet [247,248] have been employed. A solid-phase trap fitted on-line with the extractor provides an excellent means for concentration prior to detection. The mini-column is placed either in the transport tube or in the loop of an auxiliary injection valve of a flow-injection system, thus enabling elution in the opposite direction to retention. The analytes are eluted using a second injection valve to control the eluted volume to ensure reproducibility.

One very common, undesirable occurrence in dynamic subcritical solvent extraction is the presence of solid particles in the liquid extract due to oversaturation during cooling. This shortcoming can be circumvented by fitting a filtration unit in-line with the extractor; the unit can also be placed either in the transport tube or in the loop of the injection valve—the latter choice allows the filter to be cleaned between samples by passing water at a high flow-rate in the opposite direction to filtration [247,248].

When the extracted analytes provide no response on passing through a flow-cell located in a detector connected on-line to the continuous extractor via a dynamic manifold, the extractor outlet can also be connected to a manifold where the extract is merged with a stream of an appropriate reagent to derivatise the analytes as they are extracted, thereby enabling their subsequent on-line determination. This approach has been used to extract selenium from sand and sludge, and selenium, arsenic and mercury from coal. In the latter application, on-line derivatisation with sodium tetrahydroborate (for selenium and arsenic) and tin chloride (for mercury) allowed the analytes to be determined in a direct manner using atomic fluorescence detection [286,287].

*Continuous supercritical fluid extractors and flow injection manifolds.* While continuous monitoring of the extract in the supercritical state invariably requires a windowed high-pressure cell as the interface with the detector [289], monitoring after depressurisation can be carried out with a wide variety of low-pressure flow injection (extractor–detector) interfaces each reporting a series of additional advantages. The most representative of these approaches are shown in Fig. 4.13. In the first approach (Fig. 4.13(A)) a membrane phase separator was used to remove $CO_2$ from the extract. In this way, interferences were suppressed while ensuring quantitative transfer of the solutes (*viz* chloramphenicol and penicillin G) to the flow injection system [290].

The interface between the SF extractor and the photometric flow-through sensor for sulfaquinoxaline (Fig. 4.13(B)) is more complicated since, following expansion of $CO_2$ emerging from the extractor, the

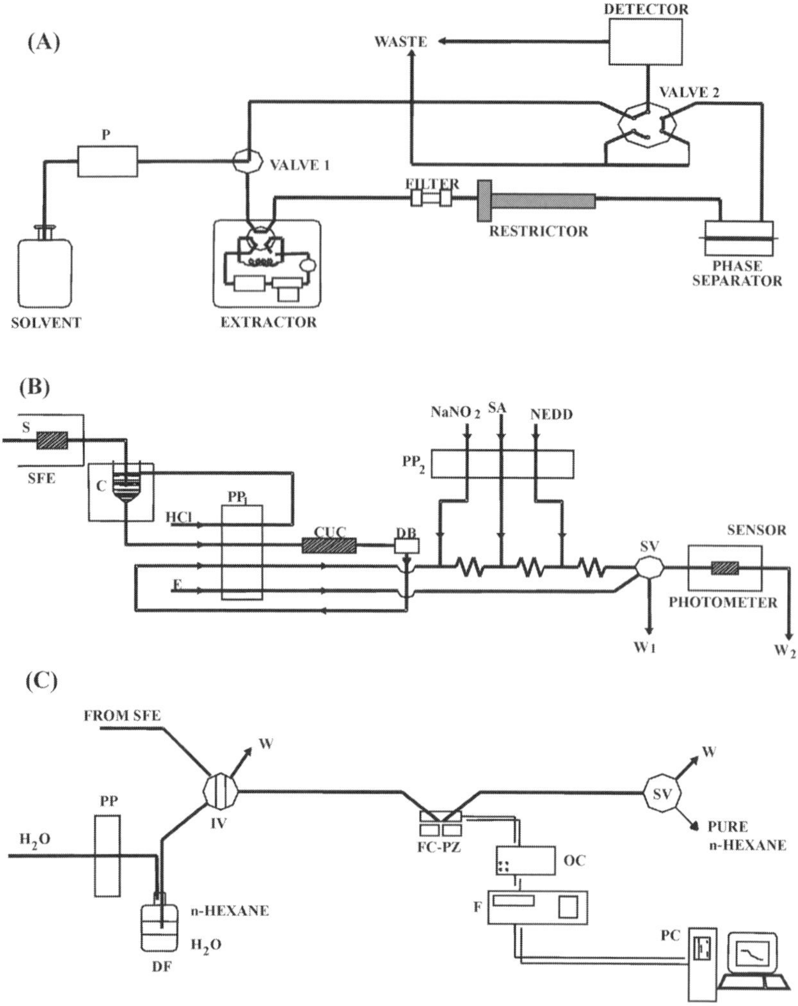


Fig. 4.13. Flow injection manifolds for interfacing a supercritical fluid extractor (SFE) to a detector. (A) SFE interface to a photometric detector across a membrane. (B) Interface to a flow-through photometric sensor with prior derivatisation by the modified Griess reaction. (C) Interface to a piezoelectric detector. P, peristaltic pump; C, collector; CUC, clean-up column, DB, debubbler; SA, sulfamic acid; NEDD, *N*-(1-naphthyl)ethylenediamine dihydrochloride; SV, switching valve; W, waste; DF, displacement flask; IV, injection valve, FC-PZ, flow-cell-piezoelectric crystal; OC, oscillator circuitry; F, frequency counter, PC, personal computer.

analyte must be transferred to a continuously renewed 0.1 M HCl solution in a collector. This carrier solution is passed through a silica clean-up column that retains water-insoluble extracted compounds so as to avoid turbidity in the circulating stream. A debubbler is used to remove the remaining $CO_2$ prior to derivatising the analyte by the Griess reaction, the coloured product formed being retained in the flow-cell (packed with $C_{18}$ bonded silica) to enable photometric monitoring of the extraction kinetics [291].

The on-line piezoelectric system in Fig. 4.13(C) was used for the quantitative gravimetric determination of total fat in foods. The extract obtained by rinsing the cryogenic trap with $n$-hexane was aspirated to the low-pressure injection valve of the FI system, which injected the extract into an $n$-hexane carrier that drove the injected plug to the flow-cell for monitoring with the aid of an At-Cut piezoelectric quartz crystal coated with a gold plated electrode [292].

*Coupling gas and liquid chromatographs to flow injection manifolds*

A key difference in coupling a flow injection system to a chromatograph is that pre- and/or post-column couplings can be arranged with liquid chromatographs, meanwhile only pre-column coupling is, obviously, possible in the case of gas chromatographs.

*Coupling a liquid chromatograph to a flow injection manifold.* The pre- or post-column coupling of FI to LC (usually HPLC) depends on the pursued objective: a pre-column position is mainly used for implementing a continuous separation step prior to chromatographic individual separation and, in a smaller extension, for developing pre-column derivatising reactions; on the other hand, a post-column position is most often used for derivatising purposes.

*Precolumn (FI-HPLC) assemblies* in which the FI manifold includes a separation unit [293,294] have been mainly devoted to the use of microcolumns with the following objectives: (1) trace preconcentration, using different materials such as ion-exchangers, ion-pairing reagents, ligand-exchangers or size-exclusion gels which have provided preconcentration factors up to 10 000. (2) Sample clean-up, by taking advantage of the differences in the interaction between the components of a given sample and the sorbent (thus, weakly retained phenol compounds or acid substances can be readily separated from phthalate esters strongly adsorbed on a non-polar sorbent). (3) Sample storage, by taking advantage of the relatively inert character of many sorbent materials. This is of special interest when samples have to be collected in remote places. (4) Protection of the analytical system, as the solid-phase microcolumn acts as a protective filter, lengthening the useful lifetime of the separation unit. (5) Pre-column derivatisation, by using a sorbent impregnated with the reagent, a solid redox agent, or a support for retention of a (bio)catalyst. A number of devices for automated development of one or several of these steps are commercially available. In any case, the eluate containing the target analytes is transported directly either to the injection valve of the chromatograph or to the top of the analytical column.

Also other separation steps such as liquid–liquid extraction [295], membrane extraction [296] and dialysis [297] have been coupled to HPLC through an FI manifold. The solution containing the analytes once separated from the matrix is injected into the chromatograph.

*Post-column coupling of FI to HPLC* has been mainly used for developing a derivatising reaction prior to detection, as is the case with the speciation analysis of antimony [298] and arsenic [299] where the hydride is generated before the eluted species reach the AA spectrometer. In the latter case, the versatility of FI also makes in-line photooxidation or microwave-assisted oxidation of the target analytes possible, thus increasing the integration effect.

A continuous liquid–liquid extractor has been coupled to a chromatograph in a post-column position with two main purposes: (a) to concentrate the analyte in the eluate, thereby reducing band broadening; and (b) to enable otherwise impossible analytical detection when the excess reagent interferes by giving the same signal as the reaction product (e.g. with ion-pairs) or the solvent must be changed (e.g. with special detectors such as a mass spectrometer).

Sometimes, the FI manifold acts as simple interface between the chromatograph and the detector, as in the case of HPLC-MS for the analysis of major ovine milk proteins [300]. A microdialysis junction interface has also been used with this aim [301].

A dual use of the FI manifold (as such and in a post-column position) allows overall and individual determination of the target analytes. The samples can first be injected into the FI system for screening purposes and then the samples with positive response are injected in the chromatograph, the eluate from the analytical column being merged with the derivatising reagents for individual determination of the separated analytes. The application of the approach to the screening-determination of aflatoxins in foodstuffs [302] and bile acids in serum [303] demonstrate the integrating effect of the approach.

*Coupling a flow injection manifold to a gas chromatograph.* The FI-GC assembly has always aimed at developing a separation step (*viz* liquid–liquid extraction, pervaporation, etc.) prior to entering the separate phase (liquid or gas, respectively) into the chromatographic column.

It is not an easy task to couple a continuous liquid–liquid extractor to a GC. The most convenient way is to use an HPLC injection valve with a small loop volume (1–5 μL) connected to the chromatograph via a stainless steel tube with a soldered needle of the same material that is directly inserted into the injection port of the chromatograph. The carrier gas from the container is split into two lines connected to one of the ports of the injection valve and to the injection port of the chromatograph. Water traces are prevented from entering the chromatograph using a desiccating column filled with pellets of sodium aluminium silicate [304].

Pervaporation is a technique easily coupled to a GC as the separate phase is a gas which can be dragged from the acceptor chamber by the carrier gas of the chromatograph, thus constituting an excellent alternative to both static and dynamic head-space with a high capacity for total automation in a simple and cheap way. The experimental FI-pervaporation-GC assembly depends on the state of the sample as both solid and liquid samples can be handled by this approach. The different arrangements required in each case are shown in Fig. 4.14 and discussed below.

The direct use of *solid samples* requires one of the approaches in Fig. 4.14(A) or (B) depending on the concentration of the target analytes. The overall process involves the following steps: (1) pervaporation of the volatile species; (2) preconcentration/desorption—if required; (3) chromatographic separation; and (4) individual detection of the separated species. The excellent features of the methods developed with this approach (namely, determination of VOCs [174] and speciation analysis of organic and inorganic mercury [305]) demonstrate the potential of the approach. A recent contribution involving semisolid samples such as yoghurt continuously introduced into the lower chamber of the pervaporator enlarges the field of application of this coupling [175].

The coupling of an analytical pervaporator to a gas chromatograph for the determination of enological parameters in wines is an excellent example of the usefulness of this arrangement for the analysis of *liquid samples*. The FI-pervaporation-GC assembly used in this case is shown

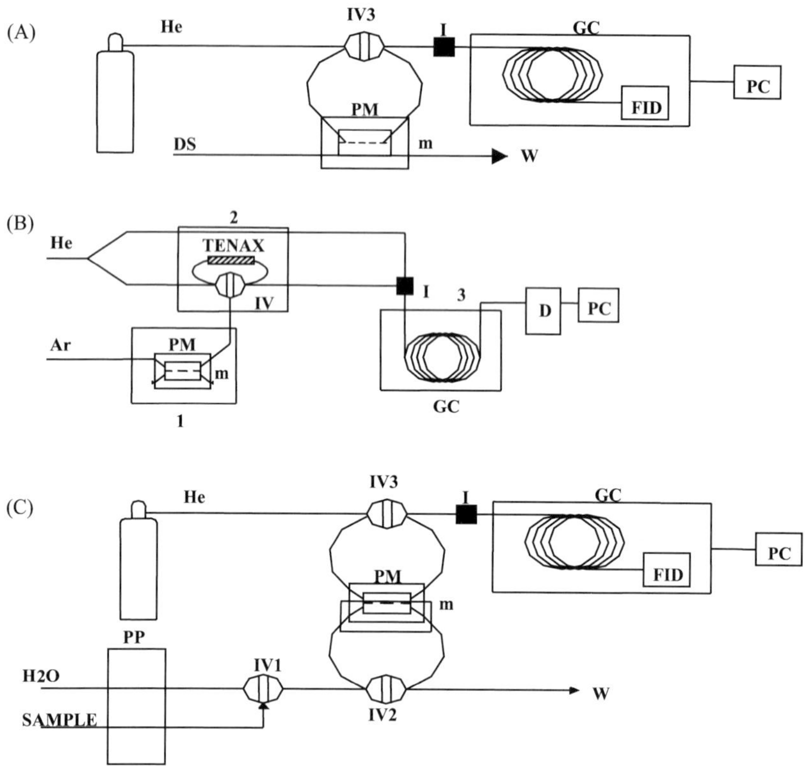


Fig. 4.14. Coupling a pervaporator to a gas chromatograph for solid and liquid samples. (A) and (B) for solid samples and different approaches for increasing the efficiency of pervaporation. (C) For liquid samples. PM, pervaporation module; I, injection; GC, gas chromatograph; FID, flame ionisation detector; DS, donor stream. Other abbreviations, as in previous figures.

in Fig. 4.14(C). It is different from that required for solid samples as a continuous method for the introduction of the liquid sample is used and the preconcentration/desorption step is absent. The auxiliary valve IV2 has the function of hindering the carrier gas from the cylinder from reaching both the inlet and outlet of the sample chamber in the pervaporator [167].

In short, the advantages of the integrating effect which results from the coupling of a pervaporator to a gas chromatograph via an FI manifold over head-space modes are as follows:

(a) the thin air gap layer above the sample requires very small amounts of analytes to establish the equilibrium with the solid or liquid sample phase and mass transfer across the membrane;
(b) continuous removal of the volatilised analytes across the membrane from the air gap displaces the mass transfer equilibrium and increases the efficiency of the separation process;
(c) in the case of solid samples, continuous removal of the pervaporated analytes to the preconcentration column allows fresh portions of acceptor gas to continuously come into contact with the diffused species, thus displacing the mass-transfer equilibrium;
(d) the separation process can readily be automated for laboratory use with minimal acquisition and maintenance costs;
(e) there is no need for a water vapour condenser, as in the case of the purge and trap mode, or a hydrophobic sorbent, since no water-crossing through the hydrophobic membrane occurs.

*Coupling flow injection to capillary electrophoresis*
Flow injection offers an elegant means for sample treatment prior to introduction into capillary electrophoresis (CE) equipment. Meanwhile FI-HPLC coupling does not cause any insurmountable problems as the volumes injected are rather large (10–100 μL) and commonly used HLPC injectors can easily be connected to almost any FI manifold. In contrast, the significantly lower injection volumes in CE (in the nL range) are definitely a limiting factor for a successful coupling of FI to CE. None of the commonly used injection techniques—pressure or electrokinetic injection—available in most commercial instruments allow a direct interfacing with FI. So far, special home-made instruments have had to be used [306].

The power of implementing a pretreatment step on-line with a CE system was first recognised by Bao and Dasgupta [307], who used a narrow-bore tubular membrane as a joint between two fused silica capillary pieces. The sample volume could be varied by changing the length and the inner diameter of the membrane joint. The sample was pumped on the outside of the membrane for a defined period of time prior to application of a high voltage (HV) across the entire capillary. This system was used for the analysis of waste water, blood and gaseous samples. The on-line coupling of *in vivo* microdialysis with CE has been proposed by implanting the microdialysis probe directly into an animal tissue. The dialysed sample portion was pumped into a microinjection valve with a transfer volume of 60 nL. After injection, the sample plug was transported via a transfer capillary into a CE interface, where a fraction of the sample was electrokinetically introduced into the separation capillary [308]. Coupling of electrodialysis to CE has also been reported and applied to the determination of adenosine triphosphate and inositol phosphates. A specially designed electrodialysis device (EDD) was used for sample introduction into the capillary [309].

All the above-mentioned approaches for on-line sample treatment prior to CE offer significant advantages over the corresponding manual procedures, which are tedious. Multiple samples can be processed in this way. However, the total analysis time is defined by the sum of the time required for the pretreatment step and the time required for the separation. A new sample cannot be processed until all the analytes in the previous sample have passed through the detector. The HV had to be interrupted during the sample pretreatment step. A more recent contribution from Kuban et al. shows that it is possible to perform consecutive, multiple runs in CE without interrupting the HV using a special FI-CE interface [310], which allows sample pretreatment (namely, dialysis, gas diffusion, etc.) to be developed simultaneously with the separation in the capillary. Other authors have also contributed to the increase of the sample throughput in FI-CE arrangements [306], thus demonstrating the integrating effect of these approaches.

## 4.7 ON-LINE COUPLING OF ROBOTICS AND FLOW INJECTION MANIFOLDS

Robotisation in the analytical laboratory can be implemented via either robotic stations or workstations. The most salient difference between both—in addition to the costs—is that a workstation can only be used for the tasks (all or some) for which it was constructed, while robotic stations can be modified by changing their software, modules or peripherals as required to undertake one or more specific tasks, or even a whole analytical process. This is not the best use of robotic stations for full automation of an analytical process as they should be used only for the tasks that cannot be developed by an other automated alternative—when this exists, it usually works faster and cheaper than the robotic station.

Users experienced in flow injection proposed this continuous alternative as the most easy-going friend of robotics in order to take advantage of both. Thus, the rapidity and simplicity of FI systems face the inability to automate several operations (some of which are basic steps such as weighing) and, on the other hand, robots, which are the most powerful systems with regards to the great variety of processes they can develop, have an inherent relative slowness as compared with FI, in addition to their high acquisition cost. The use of more than one single automated alternative to address the problems encountered in the automation of laboratory processes is desirable; however, the conservative policies of instrumentation manufacturers and the usual inertia of research groups to drastically changing investigation guidelines has so far delayed advances in the combined use of several automated approaches. The complementary features of FI and robotics can lead to a logical combined use as a mean to exploit the advantages of both approaches and also avoid the respective limitations. Surprisingly this promising hybridisation has been scarcely used so far.

Two different approaches can be accomplished in robot-FI hybridisation: (a) a simple juxtaposition of automated systems which work independently, the user acting as an active interface between both; or (b) an integrated system in which both subsystems work interrelated, thus eliminating human intervention. In the first approach the robot is devoted to automate the complex, preliminary operations prohibited to FI systems, putting the partially treated samples in a rack, which is transported by the user to the FI system. The robot acts as an automated device for introducing the samples into the instrument, with or without performing a prior separation and/or derivatisation step. In the integrated mode the FI manifold is a module of the robotics station devoted to introducing the sample into the detector (also with or without separation and/or reaction steps). In the latter approach there is no human intervention involved, which is an evident advantage as compared to the former. Some of the robot-FI integrated approaches—which are, no doubt, those providing more additional advantages—proposed so far are discussed.

When the FI configuration acts as a module of the robotics station the different operations involved in the overall process are distributed between modules searching for the highest efficiency of the overall system, but without human participation in the physical and logical communication between both systems. This type of application has been developed in the authors' laboratory.

One of the more representative examples of these applications was the method for the determination of total polyphenols in virgin olive oil [311]. The sequence of steps of the standard method is as follows: (1) weighing of the oil, (2) dilution with *n*-hexane, (3) separation of the analytes by liquid–liquid extraction, (4) derivatisation of an aliquot of the extract with the Folin Ciocalteu reagent, (5) filtration of the precipitate formed during the development of the colour, and (6) measurement of the absorbance of the product. Apparently, this sequence gives little opportunity for FI systems owing to the complexity involved in each step. Introduction into the detector seems to be the single operation that the FI system could perform easily but the help would not be significant. Nevertheless, the task of the FI manifold can be much more effective if its job starts when the analytes have been isolated. Thus, the robot directly introduces the extract into the FI manifold, which performs the derivatisation reaction and the transport at the detector. In this way, the time required for the last step is dramatically shortened (1 min *versus* 1 h) and the filtration step is unnecessary as the precipitate does not appear in the short interval elapsed between injection and detection. The border between the FI module and the robotics station is undefined in this approach as the robot, through its injection valve, introduces the extract from the test tube into the FI system, whose peristaltic pump is controlled by the robot. Moreover, the prior calibration of the FI system is performed by the robot, which prepares and injects the different standard solutions from a stock solution of a standard polyphenol, and runs the calibration graph, which is later used for analyte quantification. The treatment of the information is performed by the controller of the robotic station, thus avoiding human participation totally.

Even when the tasks in which the FI manifold is involved are simple, there is a series of advantages in coupling robots and FI. This is the case with the fully automated method for the determination of metals in soil. The use of a two-channel FI manifold for introduction of the robotic treated sample—namely, weighed, microwave-assisted digested, centrifuged and then the extract thus obtained located either in the aspiration point of the atomic absorption spectrometer or filling the loop of the injection valve of the FI system—into the instrument provides a way for performing both real and pseudo dilutions with dramatic elongation of the linear calibration range. Thus, the necessity for a prior determination to estimate the analyte concentration is avoided. In addition, an expensive peripheral such as a syringe hand is replaced by the cheap FI manifold thus reducing acquisition costs [312]. The overall operational set-up for solid sample pretreatment plus determination is shown in Fig. 4.15.

Another robotics-FI approach which shows once again the advantages of this hybridisation is that proposed for the determination of starch in food. The sequence of the time-consuming procedure for removal of sugars from the solid and the introduction of the sugar-free sample into the microwave-assisted digestor is performed by the robotic station. Then, and after hydrolysis, the liquid phase is aspirated to the injection valve of the continuous manifold through a filter. Upon injection of the solution from the digester into a carrier stream of neocuproine solution, a confluence point allows merging of the injected plug with a basic stream for development of the coloured product monitored on passage through the flow-cell [313].

## 4.8 DETECTION IN FLOW INJECTION

The usual location of a flow injection manifold is on-line and before a detector in order that proper monitoring of the species is performed after it has been subjected to the steps developed in a continuous fashion. Simple, inexpensive, conventional and easy to handle detectors capable of providing only two dimensional information (namely, those based on optical, molecular and atomic techniques such as absorptiometry and fluorimetry and electroanalytical techniques such as potentiometry, amperometry and conductimetry) have been common place in FI. In all instances, the benefits derived from a continuous monitoring, non-equilibrium conditions and non-contact with the atmosphere result in the integrated systems, so the methods from these coupling are not the subject matter of this section but special types also with special integrating effects. First, detectors providing two-dimensional information but with special features either from the detector itself or from coupling with FI, then detectors providing three-dimensional and high-discrimination detectors will be discussed, after the type of interface required in each case and the calibration modes used in these approaches have been described. Finally, the potential of FI for screening purposes is outlined.

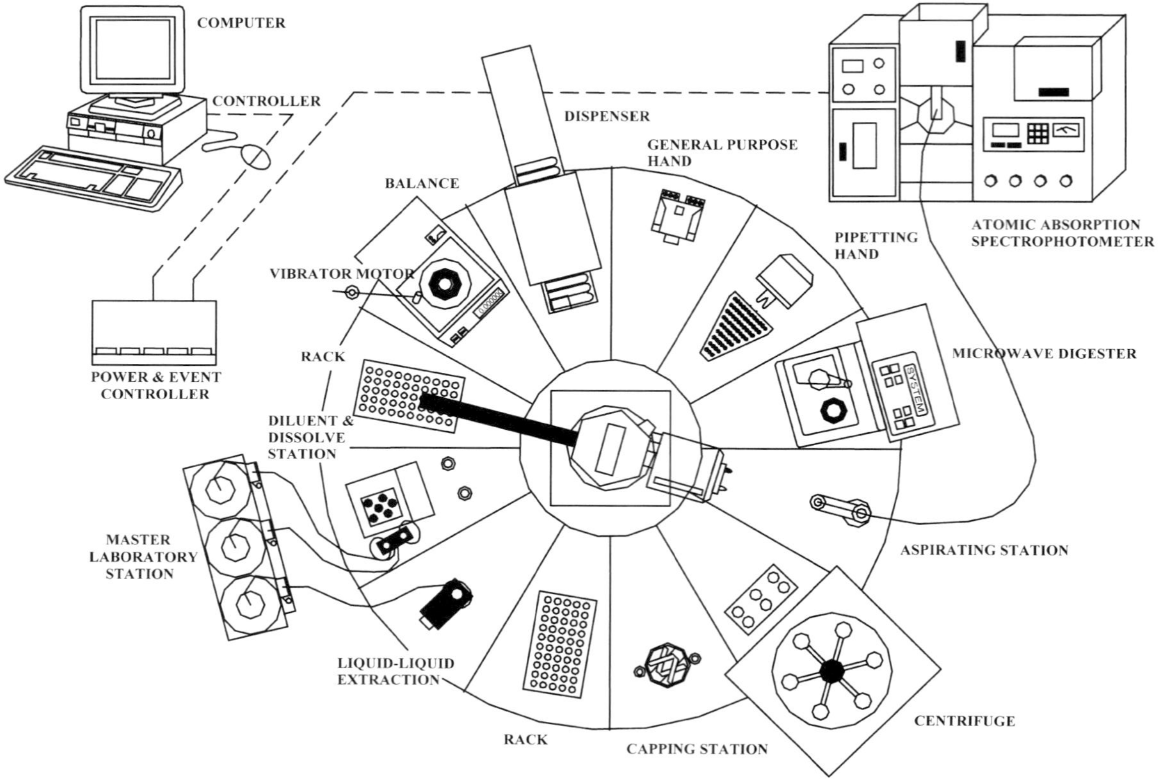


Fig. 4.15. Robotic station for the fully automated determination of metals in soil. The aspirating station can or cannot be connected with the burner through an FI manifold.

### 4.8.1 Flow injection–detector interfaces

While the interface between an FI system and a conventional detector is a commercial or laboratory-made flow cell (or a simple aspiration to the flame in the case of atomic techniques assisted by this source), the interface to a high resolution detector has a decisive influence on the performance of the hyphenated system, as analytical quality parameters such as reproducibility, accuracy, sensitivity and selectivity are highly dependent on how the coupling is accomplished. The complexity of the interface is very different depending on whether the measurement is performed in solution, plasma or vacuum.

Inexpensive flow-cells (either conventional or demountable micro-flow cells) with KBr windows and different thickness spacers (0.015–0.22 mm, 0.159 μL volume) have been used in FI-FTIR coupling when organic solvent carriers are involved [314–319]. The use of thick spacers yields a higher contribution to the blank measurement from the carrier solvent, so poorer detection limits are obtained. More sophisticated cells are used with uncommon solvents, as in the case of a 25 μL-micro-circle cell equipped with a zinc selenide crystal, used for aqueous samples in FI-FTIR [320,321], and a high-pressure flow cell (2 μL, 1 mm optical path length and 2 $mm^2$ cross-sectional area) used when supercritical $CO_2$ was the carrier [322].

The design of an NMR flow cell should provide rapid sample displacement without significant degradation of resolution. Thus, the FI injector can be connected directly to the NMR flow cell through a short and small-diameter tubing. Sufficient premagnetisation time of the sample can be accomplished by placing the injector and connectors within the magnetic field [323].

Special attention has been paid to interfacing FI systems and ICP [324]. A general interface to introduce a liquid into a plasma consists of a nebulizer, a spray chamber and a separator. The main shortcomings of using conventional interfaces in FI-ICP couplings related to continuous sample aspiration are the large dead volume and the sample loss involved as well as the band broadening. Miniaturised interfaces such as the microconcentric nebulizer [325–327] and micro glass-frit nebulizers [328] have been designed in order to overcome this drawback. A jet separator [329], a condenser [325,327] or a membrane dryer separator have been used to remove the majority of the solvent and convert the sample into a dried and desolvated aerosol in a flow of argon.

Flow injection and mass spectrometry have been interfaced by either membrane introduction or ion spray systems. The former—suitable for volatile analytes—consists of a membrane probe directly inserted into the mass spectrometer ion source: the carrier solution flows across the inside surface of the membrane while the outside surface is exposed to the vacuum of the mass spectrometer [330,331]. In ion spray systems the interface is most often a fused-silica tubing [332].

### 4.8.2 Automatic calibration

The ability for on-line development of relatively time-consuming sample pretreatment procedures such as standard additions, internal standard, isotope dilution and matrix-matching calibration in a simple way, saving both time and sample, makes FI dealing with calibration.

The standard addition method, one of the most effective in overcoming matrix effects, can be implemented in FI either by adding the standard to the sample (before, in, or after injection) or by adding the standard into the sample (reversed FI), among others, thus avoiding sample and standard contamination and the time-consuming preparation step.

In the reverse-FI mode the standard addition method is accomplished by continuous sample pumping to the detector instead of the carrier solution, the detector is zeroed for this baseline value, and the standard solutions are injected into the sample stream. The calibration curve can be run by injecting equal volumes of standards of different concentration; different volumes of the same standard solution, or using any of the FI alternatives for dilution prior to injection [318,333,334].

Sample-standard merging before injection is an easy way to implement the standard method. A third stream of internal standard can merge after the sample and standard confluence [335]. The standard addition method is used to add multielement standard solutions of variable concentrations. A complete standard addition can be performed in each sample by changing the standard solution. Unlike the reverse-FI standard addition method, continuous introduction of the sample matrix into the detector is avoided by the sample standard merging method, which has been successfully used in FI-ICP-MS couplings [336,337].

The merging-zone approach has also been used for calibration purposes in the determination of Ni, Cu and Zm in alloys [338]. The system utilised successive injections of the sample, each one accompanied by the injection of a different standard. The limitations of this system include the relatively low sampling rate and the necessity of preparing a series of standards. The latter shortcoming might be overcome by using the zone-sampling mode, which enables controlled dilution of a given standard before its injection into the final standard carrier stream [339]. Eleven additions ranging from 3 to 32% of only one standard solution can be performed in 5 min [340]. Nine toxic elements in undiluted urine have been determined in less than 5 min using an FI-ICP-MS approach in which the FI manifold, which included a splitter tee, two in-parallel injection valves with different size loops (40 and 500 μL for standard and sample, respectively) and a mixing tee, was used to inject three standard solutions sequentially into the same sample plug. Isotope dilution has also been used for the determination of lead in various matrices using a merging-zone approach with programmable time-based injections [341]. Meanwhile, matrix matching of sample and standards has been implemented in FI-ICP-AES in order to minimise the interferences due to easily ionizable elements [342]. Initially the sodium in the samples is monitored by the coupled equipment and thereafter the computer selects the appropriate sodium addition to match the saline concentration with that of the standards; after which, a suitable amount of sodium is added to the sample and the mixture introduced into the coupled system.

### 4.8.3 Special uses of conventional detectors coupled to FI

In dealing with conventional *molecular spectrometry detectors* one of the most clear integrating effects is achieved when chemiluminescence reactions are involved, as the sample–reagent mixture reaches the luminescence detector in a time as short as necessary for proper detection of the transient emission from the deactivation of the product. Thus, the method for determination of Ti(IV) using a previous minicolumn packed with Jones reductor [343], that of cyanide in environmental samples by chemiluminescence detection at low temperature [344], the simultaneous determination of nitrate and nitrite in water in gas-phase [345], the chemiluminescence sensor for the determination of vitamin $B_{12}$ [346], and the photometric biosensor for determining the activity of urease [347] are basic examples of the integrating effect achieved.

Also the FI manifold is the key for multisite detection using a simple spectrophotometer, as demonstrated by Zagatto et al. [348,349].

The use of an optical fibre as interface between the measurement point and the spectrophotometer allowed monitoring of the entire sample plug, as demonstrated by García-Mesa et al. [350].

Spectrophotometric array detectors enable more information to be collected faster from a sample injected into the FI manifold, which can be used for improving precision, for enlarging the linear range of the calibration curve [351] and for multideterminations [352], among the most important applications. The faster response of charge coupled detectors also allows the acquisition of chemiluminescence spectra profiles [353].

Lasers provide an excellent excitation source for the typical small volumes in FI. Thus, the determination of dissolved silica [354] and that of total phosphorus [355] both in water, together with that of sulphide [356], are examples of the enhanced sensitivity of fluorescence when assisted by laser excitation. A laser-based technique as thermal lens spectrometry has been used in a flowing system as FI [357,358], the method for the determination of ascorbic acid and dehydroascorbic acid being an example of this coupling [359]. The application of photoacoustic spectrometry in FI systems has also been reported [360].

*Atomic spectrometry techniques*—either those providing single or multi-information—have been widely coupled to FI and the multiple aspects of this integration have been exhaustively explained elsewhere [361], so only some examples such as: (a) the use of FI coupled to laser-enhanced ionisation detection, either for reducing electrical interference [362] or providing optimum dilution levels in matrix-interfered determinations [363] and for on-line separation concentration prior to laser assistance [364]; and (b) flow injection for arsenic and antimony hydride generation prior glow-discharge atomic-emission spectrometry detection [365] and for elimination of the matrix effect in a preconcentration step before flame atomic-absorption spectrometry [366] are needed to show both the versatility and integrated effect of these couplings.

Even more prolific in innovations are *electroanalytical techniques* due to the wider possibilities for the design of probe-sensors that can easily be converted into flow-through sensors by insertion in the appropriate flow cell. Thus, starting from the use of commercial ion selective electrodes with common potentiometric measurements [367], the two fields of research have been, (a) use of other potentiometric modes, such as bipotentiometry [368] and differential electrolytic potentiometry [369], and (b) development of new electrodes based on selective membranes [370] or ion-sensitive field-effect transistors [371].

*Potentiometric* electrode arrays have been coupled to flow injection manifolds with the aim either of improving the determination of a single analyte as is the case for the determination of ammonium [372], or simultaneous measurements, as the enzyme field-effect transistor sensor array for monitoring cultivation processes [373].

Both *voltammetric and amperometric* measurements have been implemented in FI manifolds from the very beginning of the technique. Both conventional and new excitation modes have been applied to both commercial unmodified and chemically modified electrodes. Thus, while conventional voltammetry-FI methods have been proposed for the determination of a number of analytes such as trace metals (namely, copper, lead, zinc and nickel [374]) and for that of glycerol [375] and acid phosphatase activity [376], fast-scan and dual-pulse staircase voltammetries have been used for detection in FI for the selective determination of methylmercury and ethanol in beer, respectively [377].

From conventional electrodes such as mercury-drop electrodes [378] to carbon paste electrodes modified with a manganese dioxide film [379], with copper porphyrin occluded into zeolite cavity [380], or with a phenothiazinc polymer [381], paste carbon screen printed electrodes modified with a manganese dioxide film [382], and poly(chlorotrifluoroethylene) graphite composite electrodes [383], a very wide range of these devices have been used in FI manifolds with a number of advantages additional to the intrinsic to flowing systems—such as continuous rinsing of the active surface after sample passage, as the carrier solution is always selected for optimum removal of species deposited or adsorbed on the electrode surface from the sample matrix. Arrays of gold microelectrodes modified by electrodeposition of palladium for the determination of ascorbic and uric acids in urine [384] or by electrocatalysis for carbohydrates and aminoacids [385] have also been reported. Nanoarrays of gold electrodes have also been proposed by Kauffmann et al. [386].

All *stripping modes* of electroanalysis have been coupled to FI manifolds as a way of preconcentrating the analyte(s), removing the sample matrix and performing simultaneous determinations. But the FI-stripping coupling gives the methods developed by it other big advantages such as (a) *in situ* regeneration of film-mercury electrodes by passage through the flow cell of the appropriate dissolution of Hg(II) using a switching valve for introduction; (b) ability to exchange the stripping solution in order to avoid interferences during this step; (c) ability to stop the flow during stripping for increasing reproducibility.

As regards stripping voltammetry, mercury film electrodes have been the most commonly used due to their ability to regenerate the film in a continuous way [387–389], meanwhile hanging mercury drop electrodes have been the least commonly used because of the difficulties of handling these sensors in flowing systems [390]. Potentiometric stripping has most often used *in situ* formation of the amalgam by introducing the Hg(II) with the sample [391,392]. Adsorptive stripping voltammetry with different types of electrodes depending on the analyte nature and different voltammetry modes has also been coupled to FI with significant integrating effects [393]. A calibrationless determination of mercury based on FI-stripping coulometry is a representative example of this integration [394].

*Conductimetry* and *coulometry* are other electroanalytical techniques less frequently coupled with FI. Nevertheless, the examples in the literature show the potential of these arrangements for mass distribution studies [395] and the determination of boron [396] by the former technique, and for reagent-generation in FI titrations [397] and the determination of NAD + and NADH in normal and cancer cells [398] by the latter.

*Piezoelectric measurements* of a number of analytes in samples inserted into a FI manifold also show the versatility of this combination [399–404], which has also been integrated with electrogravimetry [405].

### 4.8.4 Three-dimensional and complex detectors coupled to FI

As was the case with chromatographic techniques [406], the first attempts to obtain three-dimensional information in FI were performed using either fast scan electrochemical [407] or diode array detectors [408], thus enlarging both the integrating effect and the scope of application of FI to more complex chemical systems. In a parallel but delayed development of hyphenated systems in chromatography, the coupling of FI with high discrimination instruments showed the maturity of this dynamic technique. This step in FI started with the FI-ICP-AES coupling, the capacity of which has been reviewed by different authors [324,409–411]. More recent and less numerous have been the arrangements of FI manifolds with instruments such as MS (by direct coupling or through an ICP source), FTIR and NMR, among the most important, which have also been reviewed [324,412–417].

*Fourier transform infrared spectroscopy* has been combined with FI and all the zones of the IR spectrum have been used for measurement. Thus, the FTIR enzymic determination of glucose [418], that of oil in water [419], carbonates in water by on-line vapour generation [420], dithiocarbamate pesticides [421] and the simultaneous determination of organic acids in sugars and soft drinks with prior automated solid-phase extraction [422], near infrared spectroscopy for ethanol in liquors [423] and mid-infrared quantum cascade laser [424] are recent contributions which exploit known and new integrating aspects of FI-FTIR.

As examples of the use of *nuclear magnetic resonance* in combination with flow injection [425], readers are directed to contributions such as the studies of the interaction of catechol and ascorbic acid with the crown ether [426] and the combination with pattern recognition methods [427]; meanwhile surface enhanced Raman spectroscopy (SERS) [428] and surface-enhanced resonance Raman scattering (SERRS) [429] have been coupled to FI in a lesser extension, but with very interesting results.

Almost all possible couplings of FI to mass spectrometers have been developed, the main objectives being in most cases to increase sample throughput and/or the on-line separation/preconcentration step [430], namely: (a) direct coupling of the FI manifold to the mass spectrometer, with a wide use in combinatorial chemistry analysis [431,432]; (b) through an ICP source, as for the determination of alkyl-metal(loid) species [433], for speciation analysis of arsenic after open-focused microwave digestion to enhance the pre-reduction step and hydride generation in the FI manifold [434], and for the direct determination of impurities in high purity copper after laser ablation [435]; (c) through ICP-time-of-flight, used for the determination of trace metal in natural waters [436] and urine [437] after on-line adsorption preconcentration using a knotted reactor and hydride formation, respectively.

### 4.8.5 Screening and flow injection

Taking into account the characteristic rapidity of FI measurements, its application for screening is an immediate consequence. Thus, rapid screening of porphyrins by FI-laser assisted fluorescence [438] and that for estimation of oxidizable organic compounds in water by a solid-phase reactor [439], for polar organic micropollutants using on-column sample enrichment and separation [440], for benzodiazepines in human urine by solid-phase retention of the colour diazotised complex formed [441], and that for drugs of abuse by solid-phase extraction combined with FI-ionspray tandem mass spectrometry [442] are some of the self-explanatory examples of the potential of FI in this field.

## 4.9 INTEGRATION AND FLOW INJECTION

Flow injection is one of the techniques where the integration effect is more clearly shown. From the basic, single-channel manifold the inclusion of very simple elements such as confluence points and exchanged, modified and duplicate units endows the dynamic system with unforeseen capabilities, as does the sample–carrier exchange characteristic of reverse flow. Derivatisation reactions, either facilitated by a merging point or with a more sophisticated design—namely, solid or electrogenerated reagents, micellar media, photoinduced and catalytic reactions—provide inconceivable advantages, as does the integration of external energy sources with FI. Both in-line coupling with non-chromatographic continuous separation units and on-line arrangement with chromatographic and capillary electrophoresis equipment are excellent examples of the unforeseeable benefits from these couplings, in addition to those which are from the sum of the features of each individual technique. Maybe robotics and FI coupling is the most representative example of two very different techniques—discontinuous and continuous in nature—the coupling of which gives rise to a synergistic or integrating effect as is shown by representative pieces of research discussed in the corresponding section. The enormous number of detectors to which FI can be coupled with integrating results has also been briefly considered, as well as the capabilities for calibration and for screening purposes this dynamic technique provides.

Non-FI practitioners can ask why the utility of FI is not supported by appropriate commercial development. The reason for this imbalance between possibilities and availability of commercial devices is the low price of FI manifolds, which made them an unattractive prospect for the industry field. Despite the great number of uses of FI in the most diverse of areas, it is not an exhausted area as the only limitation of FI is the imagination of their users.

## REFERENCES


1 L.T. Skeggs, *Am. J. Pathol.*, 28 (1957) 311.
2 K.K. Stewart, G.R. Beecher and P.E. Hare, *Fed. Proc. Fed. Am. Soc. Biochem.*, 33 (1974) 1429.
3 J. Ruzicka and E.H. Hansen, *Anal. Chim. Acta*, 78 (1975) 145.
4 M. Valcárcel and M.D. Luque de Castro, *Flow Injection Analysis: Principles and Applications*, Ellis Horwood, Chichester, 1987.
5 J. Ruzicka and E.H. Hansen, *Flow Injection Analysis*, 2nd ed., Wiley, New York, 1988.
6 B. Karlberg and G.E. Pacey, *Flow Injection Analysis: A Practical Guide*, Elsevier, Amsterdam, 1989.
7 T.D. Yerian, G.D. Christian and J. Ruzicka, *Anal. Chem.*, 60 (1988) 1250.
8 A. Manz, N. Graber and H.M. Widmer, *Sens. Actuat.*, B1(1–6) (1990) 244.
9 D. Ishii, *Buseki*, 11 (2000) 690.
10 J.M. Fernández, M.E. Diaz-García and A. Sanz-Medel, *Anal. Chim. Acta*, 360 (1998) 17.
11 B.Q. Wang, J.Z. Zhang, G.J. Cheng and S.J. Dong, *Anal. Chim. Acta*, 407 (2000) 111.
12 C. Nistor, A. Oubina, M.P. Marco, D. Barceló and J. Emneus, *Anal. Chim. Acta*, 426 (2001) 185.
13 J.S. Cosano, A. Izquierdo, M.D. Luque de Castro, M. Valcárcel, C. Aguilar and G. Penelas, *Anal. Chim. Acta*, 308 (1995) 187.
14 A.G. Fogg and N.K. Bsebsu, *Analyst*, 109 (1984) 19.
15 T. Fujiwara, I.U. Mohammadzai, K. Murayama and T. Kumamaru, *Anal. Chem.*, 72 (2000) 1715.
16 H. Chen, D.B. Wei, T.C. An and Y.J. Fang, *Anal. Lett.*, 32 (1999) 787.
17 G.H. Li, L.Y. Li, C.X. Kuang and J.F. Wang, *Lihua Jianyan Huaxue Fence*, 35 (1999) 149.
18 Z.P. Wang, H. Wu, X.P. Gao and G.S. Chen, *Fenxi Shiyanshi*, 19 (2000) 13.
19 Y.X. Zhou and G.Y. Zhu, *Talanta*, 44 (1997) 2041.
20 Y.S. Li and Y. Narusawa, *J. Flow Inject. Anal.*, 9 (1992) 175.
21 D.G. Themelis and P.D. Tzanavaras, *Anal. Chim. Acta*, 429 (2001) 111.
22 M. Granda, M. de los Ángeles Serrano, A. Valdes, G. Pina and M.E. Díaz-García, *Ciencia (Maracaibo)*, 7 (1999) 277.

23 H. Tanaka, *Bunseki*, 5 (1998) 294.
24 J.M. Reijn, W.E. van der Linden and H. Poppe, *Anal. Chim. Acta*, 123 (1981) 229.
25 D.G. Themelis, P.D. Tzanavaras, A.N. Anthemidis and J.A. Stratis, *Anal. Chim. Acta*, 402 (1999) 259.
26 J.A.G. Neto, J.B.B. Silva, I.G. Souza and A.J. Curtius, *Lab. Rob. Autom.*, 11 (1999) 240.
27 C. Wyganowski, S. Motomizu and K. Toei, *Anal. Chim. Acta*, 140 (1982) 313.
28 T. Yamane and Y. Yamaguchi, *Anal. Chim. Acta*, 345 (1997) 139.
29 G. de Armas, A. Cladera, E. Becerra, J.M. Estela and V. Cerdá, *Talanta*, 52 (2000) 77.
30 A. Kazemzadeh and A.A. Ensafi, *Microchem. J.*, 69 (2001) 159.
31 K. Aikawa and R. Motohashi, *J. Flow Injection Anal.*, 6 (1989) 137.
32 Y. Narusawa and T. Hashimoto, *Chem. Lett.*, 7 (1987) 1367.
33 Y. Narusawa, *Anal. Chim. Acta*, 204 (1988) 53.
34 A. Tanaka, M. Miyazaki and T. Deguchi, *Anal. Lett.*, 18 (1985) 695.
35 N. Teshima, H. Katsumata, M. Kurihara, T. Sakai and T. Kawashima, *Talanta*, 50 (1999) 41.
36 N. Teshima, K. Ayukawa and T. Kawashima, *Talanta*, 43 (1996) 1755.
37 Z.Q. Zhang and Y.C. Jiang, *At. Spectrosc.*, 21 (2000) 100.
38 J. Karpinska and H. Puzanowska-Tarasiewicz, *Anal. Lett.*, 30 (1997) 2365.
39 J.M.T. Carneiro, R.P. Sartini and E.A.G. Zagatto, *Anal. Chim. Acta*, 416 (2000) 185.
40 C.J. Mason, G. Coe, M. Edwards and P.G. Riby, *Analyst*, 124 (1999) 1719.
41 J.A. Infantes, M.D. Luque de Castro and M. Valcárcel, *J. Pharm. Biomed. Anal.*, 10 (1992) 95.
42 J.S. Cosano, A. Izquierdo, M.D. Luque de Castro, M. Valcárcel, C. Aguilar and G. Penelas, *Anal. Chim. Acta*, 308 (1995) 185.
43 I. López-García, P. Viñas, N. Campillo and M. Hernández-Córdoba, *Anal. Chim. Acta*, 308 (1995) 85.
44 W.E. Gan, Y. Shi and Y.Z. He, *Fenxi-Shiyanshi*, 19 (2000) 40.
45 L.W. Forman, B.D. Thomas and F.W. Jacobson, *Anal. Chim. Acta*, 249 (1991) 101.
46 M. Carlsen, C. Johansen, R.W. Min, J. Nielsen and F. Lantreibecq, *Anal. Chim. Acta*, 279 (1993) 51.
47 M. Rank, J. Gram and B. Danielsson, *Anal. Chim. Acta*, 281 (1993) 521.
48 C. Tollnick and M. Beuse, *Lab. News*, April (2000) 8.
49 K. Schuegerl, L. Brandes, X.A. Wu, J. Bode, J.I. Ree, J. Brandt and B. Hitzmann, *Anal. Chim. Acta*, 279 (1993) 3.
50 E. Lamparter and C. Lunkenheimer, *J. Pharm. Biomed. Anal.*, 10 (1992) 727.
51 M.A.Z. Arruda, M. Gallego and M. Valcárcel, *Anal. Chem.*, 65 (1993) 3331.

52 R. Martínez-Ávila, V. Carbonell, A. Salvador and M. de la Guardia, *Talanta*, 40 (1993) 107.
53 M.A. Bautista, C. Pérez-Sirvent, I. López-García and M. Hernández-Córdoba Fresenius, *J. Anal. Chem.*, 350 (1994) 359.
54 I. López-García, J. Arroyo-Cortez and M. Hernández-Córdoba, *Talanta*, 40 (1993) 1677.
55 G.D. Christian and J. Ruzicka, *Anal. Chim. Acta*, 261 (1992) 11.
56 T. Tsukatani and K. Matsumoto, *Anal. Sci.*, 16 (2000) 265.
57 S.H. Wang, A.M. Zhang, L.Y. Du, C.L. Ma and D.J. Liu, *Fenxi Huaxue*, 24 (1996) 832.
58 J.H. Wang and R.H. He, *Anal. Chim. Acta*, 276 (1993) 419.
59 D. Nacapricha, S. Muangkaew, N. Ratanawimarnwong, J. Shiowatana and K. Grudpan, *Analyst*, 126 (2001) 121.
60 T. Pérez-Ruiz, C. Martínez-Lozano, V. Tomás and J. Fenoll, *Anal. Lett.*, 34 (2001) 295.
61 M.J.A. Marra, A.A. Mateos, M.M. Mata and C. García de María, *Talanta*, 47 (1998) 121.
62 T. Tsukatani and J. Matsumoto, *Anal. Sci.*, 16 (2000) 265.
63 F. Lázaro, M.D. Luque de Castro and M. Valcárcel, *Anal. Chem.*, 59 (1987) 950.
64 J.A. Infantes, M.D. Luque de Castro and M. Valcárcel, *Anal. Chim. Acta*, 242 (1991) 179.
65 R.A.S. Lapa, J.L.F.C. Lima and J.L.M. Santos, *Anal. Chim. Acta*, 407 (2000) 225.
66 J. Ruzicka and E.H. Hansen, *Anal. Chim. Acta*, 145 (1983) 1.
67 J.S. Cosano, M.D. Luque de Castro and M. Valcárcel, *Anal. Chim. Acta*, 302 (1995) 269.
68 S. Olsen, J. Ruzicka and E.H. Hansen, *Anal. Chim. Acta*, 136 (1982) 101.
69 I. López-García, P. Vinas, N. Campillo and M. Hernández-Córdoba, *Anal. Chim. Acta*, 308 (1995) 67.
70 A. Ríos, M.D. Luque de Castro and M. Valcárcel, *Talanta*, 32 (1985) 845.
71 J. Marcos, A. Ríos and M. Valcárcel, *Anal. Chim. Acta*, 261 (1992) 489.
72 J. Marcos, A. Ríos and M. Valcárcel, *Anal. Chim. Acta*, 261 (1992) 495.
73 C.N. Yarnitzky, N. Klein and O. Cohen, *Talanta*, 40 (1993) 1937.
74 M. Agudo, J. Marcos, A. Ríos and M. Valcárcel, *Anal. Chim. Acta*, 239 (1990) 211.
75 C.H. Fishcer and J.G. Rabe, *Anal. Chim. Acta*, 284 (1993) 257.
76 J. Marcos, A. Ríos and M. Valcárcel, *Trends Anal. Chem.*, 11 (1992) 373.
77 A. Ríos, M.D. Luque de Castro and M. Valcárcel, *Anal. Chem.*, 60 (1988) 1540.
78 M. Romero-Saldaña, A. Ríos and M. Valcárcel, *Fresenius J. Anal. Chem.*, 342 (1992) 547.

79 S.W. Kang, T. Sakai, N. Ohno and K. Ida, *Anal. Chim. Acta*, 261 (1992) 197.
80 G.D. Clark, J. Zable, J. Ruzicka and G.D. Christian, *Talanta*, 38 (1991) 119.
81 J. Edwards, A. Patel and D.M. Spence, *Anal. Chim. Acta*, 417 (2000) 185.
82 J. Toei, *Analyst*, 113 (1988) 475.
83 J. Wang, H. Huiliang and W. Kubiak, *Electroanalysis*, 2 (1990) 127.
84 N.S. Thomaidis and C.A. Georgiou, *Anal. Chim. Acta*, 405 (2000) 239.
85 J.W.B. Stewart and J. Ruzicka, *Anal. Chim. Acta*, 82 (1976) 137.
86 J.M. Fernández-Romero, M.D. Luque de Castro and M. Valcárcel, *Anal. Chim. Acta*, 283 (1993) 1447.
87 M.S. Jiménez and J.R. Castillo, *J. Anal. At. Spectrom.*, 12 (1997) 1397.
88 J.M. López-Fernández, M.D. Luque de Castro and M. Valcárcel, *J. Autom. Chem.*, 12 (1990) 263.
89 J. Ruz, A. Torres, A. Ríos, M.D. Luque de Castro and M. Valcárcel, *J. Autom. Chem.*, 8 (1986) 70.
90 A. Izquierdo, P. Linares, M.D. Luque de Castro and M. Valcárcel, *Fresenius J. Anal. Chem.*, 336 (1990) 490.
91 J.L. Pérez, B. Moreno, J. Hernández and R.M. Isidro, *Anal. Chem.*, 61 (1989) 1789.
92 A.A. Specht and D. Beauchemin, *Anal. Chem.*, 70 (1998) 1036.
93 B. Bermúdez, A. Ríos, M.D. Luque de Castro and M. Valcárcel, *Talanta*, 35 (1988) 810.
94 A. Ríos, M.D. Luque de Castro and M. Valcárcel, *Anal. Chem.*, 58 (1986) 663.
95 J. Ruz, A. Ríos, M.D. Luque de Castro and M. Valcárcel, *Anal. Chim. Acta*, 211 (1988) 281.
96 J.F. Van Staden and L.G. Kleuver, *Anal. Chim. Acta*, 369 (1998) 157.
97 J. Stripeikis, M. Tudino, O. Trocolli, R. Wuilloud, R. Olsina and L. Martínez, *Spectrochim. Acta*, 56B (2001) 93.
98 P.H. Petsul, G.M. Greenway and S.J. Haswell, *Anal. Chim. Acta*, 428 (2001) 155.
99 J.L. Pérez-Pavón and B. Moreno-Cordero, *Analyst*, 117 (1992) 215.
100 E. Capella-Peiro, L. Monferrer-Pons, C. García-Alvarez-Coque and J. Esteve-Romero, *Anal. Chim. Acta*, 427 (2001) 93.
101 A. Safavi and M. Mirzaee, *Talanta*, 51 (2000) 225.
102 T. Kyaw, T. Fujiwara, H. Inoue, Y. Okamoto and T. Kumamaru, *Anal. Sci.*, 14 (1998) 203.
103 M.A. Ruíz, A.J. Reviejo, C. Parrado and J.M. Pingarrón, *Electroanalysis*, 8 (1996) 529.
104 B.F. Jonson, R.E. Malick, B. Ghearing and J.G. Dorsey, *Analyst*, 117 (1992) 1833.
105 A. Alonso, M.J. Almendral, F. Merino and S.P. Palmero, *Analyst*, 120 (1995) 2401.

106 S.M.V. Fernándes, J.L.F.C. Lima and A.O.S.S. Rangel, *Fresenius J. Anal. Chem.*, 366 (2000) 112.
107 J.A. García-Bautista, J.V. García-Mateo and J. Martínez-Calatayud, *Anal. Chim. Acta*, 404 (2000) 141.
108 S. Motomizu and M. Sanada, *Anal. Chim. Acta*, 308 (1995) 406.
109 R.H. Atallah and D.A. Kalman, *Talanta*, 38 (1991) 167.
110 T. Pérez-Rutz, C. Martínez-Lozano, V. Tomás and J. Martín, *Analyst*, 124 (1999) 197.
111 H.W. Chen and Q.H. He, *Talanta*, 53 (2000) 463.
112 A.W. Knight and G.M. Greenway, *Analyst*, 120 (1995) 1077.
113 B. Li, Z. Zhang and M. Wu, *Talanta*, 51 (2000) 515.
114 J.E. Atwater, J.R. Akse, J. DeHart and R.R. Wheeler Jr., *Anal. Lett.*, 30 (1997) 1445.
115 X.N. Chen, C.X. Zhang and J.R. Lu, *Fénix Shiyanshi*, 19 (2000) 62.
116 J. Li and P.K. Dasgupta, *Anal. Chim. Acta*, 398 (1999) 33.
117 B.X. Li, Z.J. Zhang and M.L. Wu, *Anal. Chim. Acta*, 432 (2001) 311.
118 N. Maniasso and E.A.G. Zagatto, *Anal. Chim. Acta*, 366 (1998) 87.
119 A. Segura-Carretero, J. Rodríguez-Fernández, A.R. Bowie and P.J. Worsfold, *Analyst*, 125 (2000) 387.
120 N. Yonehara, S. Chaen, T. Tomiyasu and H. Sakamoto, *Anal. Sci.*, 15 (1999) 277.
121 Y.P. Zhang, D.X. Yuan, J.X. Chen, T.S. Lan and H.Q. Chen, *Clin. Chem.*, 42 (1996) 2021.
122 J.M. Fernández-Romero and M.D. Luque de Castro, *Chim. Oggi.*, 11 (1988) 17.
123 K. Matsumoto and H. Ukeda, *Bunseki*, 4 (2000) 211.
124 G. Guebitz and C. Shellum, *Anal. Chim. Acta*, 283 (1993) 421.
125 A.N. Araujo, J.A.M. Catita, J.L.F.C. Lima and E.A.G. Zagatto, *Fresenius J. Anal. Chem.*, 360 (1998) 100.
126 I. Da Cruz Vieira and O. Fatibello Filho, *Anal. Chim. Acta*, 366 (1998) 111.
127 M.D. Luque de Castro and J.L. Luque-García, *Acceleration and Automation of Solid Sample Pretreatment*, Elsevier, Amsterdam, 2002.
128 F. Lázaro, M.D. Luque de Castro and M. Valcárcel, *Anal. Chim. Acta*, 242 (1991) 283.
129 D. Chen, F. Lázaro, M.D. Luque de Castro and M. Valcárcel, *Anal. Chim. Acta*, 226 (1989) 221.
130 J.L. Luque-García and M.D. Luque de Castro, *Analyst*, 127 (2002) 115.
131 E.N. Andrade, *Phil. Trans.*, 230 (1932) 413.
132 R.Y. Wang, J.A. Jarratt, P.J. Keay, J.J. Hawkes and W.T. Coakley, *Talanta*, 52 (2000) 129.
133 H. Bergamin, F.J. Krug, E.A.G. Zagatto, E.C. Arruda and C.A. Coutinho, *Anal. Chim. Acta*, 190 (1986) 177.

134 A.P.G. Gervasio, G.C. Luca, A.M. Menegario, B.F. Reis and H.B. Filho, *Anal. Chim. Acta*, 405 (2000) 213.
135 T.J. Cardwel, R.W. Cattrall, G.N. Chen, G.R. Scollary and I.C. Hamilton, *J. AOAC Int.*, 76 (1993) 1389.
136 M. Burguera, J.L. Burguera and O.M. Alarcón, *Anal. Chim. Acta*, 214 (1988) 421.
137 M. Domínguez de Almeida, K.C. Leandro, C. Vidal da Costa, R.E. Santelli and M. de la Guardia, *J. Anal. At. Spectrom.*, 12 (1997) 1235.
138 J. Flock, F. Michael and K.D. Ohls, *Fresenius J. Anal. Chem.*, 363 (1999) 306.
139 M.E. Moreno, C. Pérez-Conde and C. Cámara, *J. Anal. At. Spectrom.*, 15 (2000) 681.
140 C. Cabrera, Y. Madrid and C. Cámara, *J. Anal. At. Spectrom.*, 9 (1994) 1423.
141 E.R. Pereira-Filho, M.A.Z. Arruda, 124 (1999) 1873.
142 E.S. Beary, P.J. Paulsen, L.B. Jassie and J.D. Fassett, *Anal. Chem.*, 69 (1997) 758.
143 E.R. Pereira, F. Jarbas, J.R. Rohwedder and M.A.Z. Arruda, *Analyst*, 123 (1998) 1023.
144 L. Bordera, V. Hernandis and A. Canals, *Fresenius J. Anal. Chem.*, 355 (1996) 112.
145 A. Morales-Rubio, M.L. Mena and C.W. McLeod, *Anal. Chim. Acta*, 308 (1995) 364.
146 M.A. Crespín, M. Gallego and M. Valcárcel, *J. Chromatogr.*, 897 (2000) 279.
147 J.L. Luque-García, S. Morales-Muñoz and M.D. Luque de Castro, *Chromatographia*, 55 (2002) 117.
148 E. Luque-Pérez and S.J. Haswell, *Anal. Proc.*, 32 (1995) 85.
149 M. de la Guardia, A. Salvador, J.L. Burguera and M. Burguera, *J. Flow Injection Anal.*, 5 (1988) 121.
150 V. Carbonell, M. de la Guardia, A. Salvador, J.L. Burguera and M. Burguera, *Anal. Chim. Acta*, 238 (1990) 417.
151 M.A.Z. Arruda, M. Gallego and M. Valcárcel, *J. Anal. At. Spectrom.*, 10 (1995) 501.
152 V. Fernández-Pérez, L.E. García-Ayuso and M.D. Luque de Castro, *Analyst*, 125 (2000) 317.
153 J.F. Tyson, S.G. Offley, N.J. Seare, H.A.B. Kibble and C. Fellows, *J. Anal. At. Spectrom.*, 7 (1992) 315.
154 B. Welz, Y. He and M. Sperling, *Talanta*, 40 (1993) 1917.
155 M. Van Son, R.C. Schothorst and G. Den Boef, *Anal. Chim. Acta*, 153 (1983) 271.
156 J. Vlcek and V. Kuban, *Collect. Czech. Chem. Commun.*, 64 (1999) 1966.
157 R. Liu, B. Sun and I. Johns, *Analyst*, 120 (1995) 2845.

158 T.J. Cardwell, R.W. Cattrall and M. Mitri, *Talanta*, 41 (1994) 115.
159 H. Chen, Y.J. Fang, L. Yuan and T.C. An, *Lab. Rob. Autom.*, 11 (1999) 157.
160 S.Z. Yao and X.L. Su, *J. AOAC Int.*, 82 (1999) 1479.
161 P. Linares, M.D. Luque de Castro and M. Valcárcel, *Anal. Chim. Acta*, 225 (1989) 443.
162 E. Miralles, R. Compano, M. Granados and M.D. Prat, *Fresenius J. Anal. Chem.*, 365 (1999) 516.
163 M.D. Luque de Castro and I. Papaefstathiou, In: R.A. Meyers (Ed.), *Analytical Pervaporation in Encyclopedia of Environmental Analysis and Remediation*, Wiley, New York, 1998, pp. 3462–3467.
164 M.D. Luque de Castro and L. Gámiz-Gracia, In: R.A. Meyers (Ed.), *Analytical Pervaporation in Encyclopedia of Analytical Chemistry*, Wiley, New York, 2000, pp. 3084–3101.
165 H. Sulistyarti, T.J. Cardwell, M.D. Luque de Castro and S.D. Kolev, *Anal. Chim. Acta*, 390 (1999) 133.
166 F. Delgado Reyes, J.M. Fernández-Romero and F. Luque de Castro, *Talanta*, 53 (2001) 961.
167 E. Mataix and M.D. Luque de Castro, *Chromatographia*, 52 (2000) 205.
168 S.Y. Sheikheldin, T.J. Cardwell, R.W. Cattrall, M.D. Luque de Castro and S.D. Kolev, *Anal. Chim. Acta*, 419 (2000) 9.
169 I. Papaefstathiou, U. Bilitewski and M.D. Luque de Castro, *Anal. Chim. Acta*, 330 (1996) 265.
170 F. Delgado-Reyes, I. Papaefstathiou, J.M. Fernández-Romero and M.D. Luque de Castro, *Analyst*, 123 (1998) 2367.
171 L. Gámiz-Gracia and M.D. Luque de Castro, *Analyst*, 124 (1999) 1119.
172 I. Papaefstathiou and M.D. Luque de Castro, *Anal. Chim. Acta*, 354 (1997) 135.
173 L. Gámiz-Gracia and M.D. Luque de Castro, *J. Pharm. Biomed. Anal.*, 22 (2000) 909.
174 I. Papaefstathiou and M.D. Luque de Castro, *J. Chromatogr.*, 779 (1997) 352.
175 E. Priego-López, M.D. Luque de Castro, *J. Chromatogr*, 976 (2002) 399.
176 A. Caballo-López, M.D. Luque de Castro, *Analyst*, submitted for publication.
177 M.D. Luque de Castro and M.M. Jiménez-Carmona, *Quím. Anal.*, 18 (1999) 263.
178 E. Mataix and M.D. Luque de Castro, *Analyst*, 123 (1998) 1547.
179 P. Cañizares and M.D. Luque de Castro, *Quím. Anal.*, 15 (1996) 239.
180 I. Papaesftathiou and M.D. Luque de Castro, *Int. J. Environ. Anal. Chem.*, 66 (1997) 107.
181 E. Mataix and M.D. Luque de Castro, *Anal. Chim. Acta*, 381 (1999) 23.
182 E. Mataix and M.D. Luque de Castro, *Talanta*, 51 (2000) 489.

183 M.D. Luque de Castro and I. Papaesftathiou, *Spectrochim. Acta*, 53B (1998) 311.
184 D.W. Bryce, A. Izquierdo and M.D. Luque de Castro, *Anal. Chim. Acta*, 324 (1996) 69.
185 M. Filippelli, F. Baldi, F.E. Brinckman and G. Olson, *J. Environ. Sci. Technol.*, 26 (1992) 1457.
186 J.L. Burguera, M. Burguera, C. Rivas and P. Carrero, *Talanta*, 45 (1998) 531.
187 J.M. González LaFuente, M.L. Fernández-Sánchez, J.M. Marchante-Gayon, J.E. Sánchez-Uria and A. Sanz-Medel, *Spectrochim. Acta*, 51B (1996) 1849.
188 R. Muñoz-Olivas and O.F.X. Donard, *Talanta*, 45 (1998) 1023.
189 B. Welz, D.L. Tsalev and M. Sperling, *Anal. Chim. Acta*, 261 (1992) 91.
190 T. Guo and J. Baasner, *Talanta*, 40 (1993) 1927.
191 M. Gallignani, H. Bahsas, M.R. Brunetto, M. Burguera, J.L. Burguera and Y. Petit de Pena, *Anal. Chim. Acta*, 369 (1998) 57.
192 R. Lane, C.W.K. Chow, D.E. Davey, D.E. Mulcahy and S. McLeod, *Analyst*, 122 (1997) 1549.
193 C.W.K. Chow, R. Lane, T.C.W. Yeow, D.E. Davey and D.E. Mulcahy, *Lab. Autom. Inf. Manage*, 33 (1998) 199.
194 R. Lane, C.W.K. Chow, D.E. Davey, D.E. Mulcahy and S. McLeod, *Anal. Lett.*, 32 (1999) 1847.
195 P.C. do Nascimento, D. Bohrer and L.M. de Carvalho, *Analyst*, 123 (1998) 1151.
196 A.M.R. Ferreira, A.O.S.S. Rangel and J.L.F.C. Lima, *Commun. Soil Sci. Plant Anal.*, 26 (1995) 183.
197 J.L.F.C. Lima, A.O.S.S. Rangel and M.R.S. Souto, *Anal. Sci.*, 12 (1996) 81.
198 J.F. Van Staden, *Talanta*, 38 (1991) 1033.
199 E. Mataix and M.D. Luque de Castro, *Anal. Chim. Acta*, 428 (2001) 7.
200 J.F. Van Staden and C.J. Hatting, *Fresenius J. Anal. Chem.*, 367 (2000) 79.
201 J.C. Cooper, J. Danzer and H.L. Schmidt, *Anal. Chim. Acta*, 282 (1993) 369.
202 N. Kasthurikrishnan and J.A. Koropchak, *Anal. Chem.*, 65 (1993) 857.
203 D. Berggren, *Int. J. Environ. Anal. Chem.*, 41 (1990) 133.
204 Y.M. Huang, Z.J. Zhang, D.J. Zhang and J.G. Lv, *Talanta*, 53 (2001) 835.
205 G.Y. Shi, F. Xu, H.G. Zhou, L.Q. Mao and L.T. Jin, *Anal. Chim. Acta*, 386 (1999) 123.
206 J.N. Wang, M.Q. Lu, F.Z. Yang, X.R. Zhang, W.R.G. Baeyens and A.M. García-Campaña, *Anal. Chim. Acta*, 428 (2001) 173.
207 Q. Fang, X.T. Shi, Y.Q. Sun and Z.L. Fang, *Anal. Chem.*, 69 (1997) 3570.
208 Y.M. Huang, Z.J. Zhang, J.G. Lv and H. Cheng, *Anal. Chim. Acta*, 419 (2000) 175.
209 L.J. Deterding, K. Dix, L.T. Burka and K.B. Tomer, *Anal. Chem.*, 64 (1992) 2636.

210 N. Kobayashi, M. Kazui and T. Ikeda, *J. Pharm. Biomed. Anal.*, 21 (2000) 1233.
211 T. Yao, S. Suzuki, H. Nishino and T. Nakahara, *Electroanalysis*, 7 (1995) 1114.
212 F. Palmisano, D. Centonze, M. Quinto and P.G. Zambonin, *Biosens. Bioelectron.*, 11 (1996) 419.
213 Q. Fang, S.S. Liu, J.F. Wu, Y.Q. Sun and Z.L. Fang, *Talanta*, 49 (1999) 403.
214 L. Nord and B. Karlberg, *Anal. Chim. Acta*, 125 (1981) 199.
215 L. Nord and B. Karlberg, *Anal. Chim. Acta*, 145 (1983) 151.
216 L. Nord and B. Karlberg, *Anal. Chim. Acta*, 164 (1984) 233.
217 M. Valcárcel and M.D. Luque de Castro, *Non-Chromatographic Continuous Separation Techniques*, Royal Society of Chemistry, Cambridge, 1991.
218 N. Maleki, B. Haghighi and A. Safavi, *Microchem. J.*, 53 (1996) 147.
219 S.L. Lin, Q. Shuai, H. Qiu and Z.Y. Tang, *Spectrochim. Acta*, 51B (1996) 1769.
220 T. Blanco, N. Maniasso, M. Fernanda Gine and A.O. Jacinto, *Analyst*, 123 (1998) 191.
221 A. Alonso, M.J. Almendral, M.D. Báez, M.J. Porras, F.L. Lavin and C. García de Marías, *Anal. Chim. Acta*, 408 (2000) 129.
222 J.F. Liu and G.B. Jiang, *Microchem. J.*, 68 (2001) 29.
223 T. Sakai, Y.S. Chung and N. Ohno, *Anal. Chim. Acta*, 276 (1993) 127.
224 T. Fujiwara, I.U. Mohammadzai, K. Murayama and T. Kumamaru, *Anal. Chem.*, 72 (2000) 1715.
225 A.R.K. Dapaah and A. Ayame, *Anal. Sci.*, 13 (1997) 405.
226 Q. Fang, Y.Q. Sun and Z.L. Fang, *Fresenius J. Anal. Chem.*, 364 (1999) 347.
227 Y. Sahlestrom and B. Karlberg, *Anal. Chim. Acta*, 179 (1986) 315.
228 P.R. Danesi, In: S. Alegret (Ed.), *Developments in Solvent Extraction*, Ellis Horwood, Chichester, 1988, Chapter 9.
229 M. Agudo, A. Ríos and M. Valcárcel, *Analyst*, 119 (1994) 2097.
230 F. Cañete, A. Ríos, M.D. Luque de Castro and M. Valcárcel, *Anal. Chem.*, 60 (1988) 2354.
231 F. Cañete, A. Ríos, M.D. Luque de Castro and M. Valcárcel, *Anal. Chim. Acta*, 224 (1989) 169.
232 J.A. García-Mesa, P. Linares, M.D. Luque de Castro and M. Valcárcel, *Anal. Chim. Acta*, 235 (1990) 441.
233 F. Ortiz-Boyer, J.A. García Mesa and M.D. Luque de Castroía Mesa, *Anal. Chem.*, 66 (1994) 2794.
234 M. Agudo, A. Ríos and M. Valcárcel, *Anal. Chem.*, 65 (1993) 2941.
235 U. Ornemak and A. Olin, *Talanta*, 41 (1994) 67.
236 P. Sooksamiti, H. Geckeis and K. Grudpan, *Analyst*, 121 (1996) 1413.
237 F. Shan and T.Y. Chen, Lihua Jianyan, *Huaxue Fence*, 36 (2000) 457.
238 T. Seki, H. Takigawa, Y. Hirano, Y. Ishibashi and K. Oguma, *Anal. Sci.*, 16 (2000) 513.

239 J.R. Baena, M. Gallego and M. Valcárcel, *Spectrochim. Acta*, 54B (1999) 1869.
240 I. Gacs and R. Ferraroli, *Anal. Chim. Acta*, 269 (1992) 177.
241 M.M. González, M. Gallego and M. Valcárcel, *J. Anal. At. Spectrom.*, 14 (1999) 711.
242 J.R. Baena, M. Gallego and M. Valcárcel, *Analyst*, 125 (2000) 1495.
243 Y. Petit de Pena, M. Gallego and M. Valcárcel, *J. Anal. At. Spectrom.*, 12 (1997) 453.
244 J.B. Borba da Silva, M.B.O. Giacomelli and A.J. Curtius, *Analyst*, 124 (1999) 1249.
245 H. Emteborg, D.C. Baxter and W. French, *Analyst*, 118 (1993) 1007.
246 S. Motlagh and J. Pawliszyn, *Anal. Chim. Acta*, 284 (1993) 265.
247 J.L. Luque-García and M.D. Luque de Castro, *J. Chromatogr*, 959 (2002) 25.
248 M.C. Herrera, R.C. Prados-Rosales, J.L. Luque-García, M.D. Luque de Castro, *Anal. Chim. Acta*, 463 (2002) 189.
249 V. Kuban, *Fresenius J. Anal. Chem.*, 346 (1993) 873.
250 P. Martínez, M. Gallego and M. Valcárcel, *Analyst*, 112 (1987) 1233.
251 R.E. Santelli, M. Gallego and M. Valcárcel, *Anal. Chem.*, 61 (1989) 1427.
252 R.E. Santelli, M. Gallego and M. Valcárcel, *J. Anal. At. Spectrom.*, 4 (1989) 547.
253 C. Dittfurth, E. Ballesteros, M. Gallego and M. Valcárcel, *Spectrochim. Acta*, 51 (1996) 1935.
254 M.C. Yebra, M.F. Enriquez and R.M. Cespón, *Talanta*, 52 (2000) 631.
255 J.L. Burguera, M. Burguera and C.E. Rondón, *Anal. Chim. Acta*, 366 (1998) 295.
256 J.L. Burguera, M. Burguera, C. Rivas, C. Rondon, P. Carrero and M. Gallignani, *Talanta*, 48 (1999) 885.
257 S. Pei and Z.L. Fang, *Anal. Chim. Acta*, 294 (1994) 185.
258 Z.L. Fang, M. Sperling and B. Welz, *J. Anal. At. Spectrom.*, 6 (1991) 301.
259 S. Sella, G.E. Sturgeon, S.N. Willie and R.C. Campos, *J. Anal. At. Spectrom.*, 12 (1997) 1281.
260 P. Martínez, M. Gallego and M. Valcárcel, *J. Anal. At. Spectrom.*, 28 (1987) 211.
261 P. Martínez, M. Gallego and M. Valcárcel, *Anal. Chim. Acta*, 193 (1987) 127.
262 J. Zorro, M. Gallego and M. Valcárcel, *Microchem. J.*, 39 (1989) 71.
263 V.V. Kuznetsov, Y.V. Ermolenko, D.V. Novikov and M.L. Bykhovskii, *J. Anal. Chem.*, 55 (2000) 630.
264 M.C. Yebram and R.M. Cespón, *Fresenius J. Anal. Chem.*, 365 (1999) 370.
265 S. Motomizu, J.P. Susanto, M. Oshima, H. Mikasa and Y. Hori, *Anal. Sci.*, 11 (1995) 155.
266 R. Montero, M. Gallego and M. Valcárcel, *J. Anal. At. Spectrom.*, 3 (1988) 725.

267 J. Martínez-Calatayud and V. García-Mateo, *J. Pharm. Biomed. Anal.*, 7 (1989) 1441.
268 M.C. Yebra, M. Gallego and M. Valcárcel, *Anal. Chim. Acta*, 308 (1995) 275.
269 C.E. Addeyinwo and J.F. Tyson, *Anal. Proc.*, 26 (1989) 375.
270 M.B. Glauert, *J. Fluid Mech.*, 1 (1956) 625.
271 D.W. Bryce, A. Izquierdo and M.D. Luque de Castro, *Anal. Chim. Acta*, 308 (1995) 96.
272 D.W. Bryce, A. Izquierdo and M.D. Luque de Castro, *Fresenius J. Anal. Chem.*, 351 (1995) 433.
273 A. Izquierdo, M.D. Luque de Castro and M. Valcárcel, *Electroanalysis*, 6 (1994) 894.
274 L. Almestrand, D. Jagner and L. Renman, *Anal. Chim. Acta*, 193 (1987) 71.
275 S. Mannino, *Analyst*, 109 (1984) 905.
276 X. Ruan and H. Chang, *Talanta*, 35 (1988) 861.
277 L. Renman, D. Jagner and R. Berglund, *Anal. Chim. Acta*, 188 (1986) 137.
278 H. Eskilsson and C. Haraldsson, *Anal. Chim. Acta*, 198 (1987) 231.
279 G. Schulze, M. Husch and W. Frenzel, *Mikrochim. Acta*, 1 (1984) 191.
280 H. Eskilsson, C. Haraldsson and D. Jagner, *Anal. Chim. Acta*, 175 (1985) 105.
281 J. Wang and S. Mannino, *Analyst*, 114 (1989) 643.
282 J.A. Polta and D.C. Johnson, *Anal. Chem.*, 57 (1985) 1373.
283 J. Wang and B.A. Frelha, *Anal. Chem.*, 55 (1983) 1285.
284 E.N. Chaney Jr. and R.P. Baldwing, *Anal. Chim. Acta*, 176 (1985) 105.
285 M.M. Jiménez-Carmona and M.D. Luque de Castro, *Anal. Chim. Acta*, 342 (1997) 215.
286 V. Fernández-Pérez, M.M. Jiménez-Carmona and M.D. Luque de Castro, *J. Anal. At. Spectrom.*, 14 (1999) 1761.
287 C.M. Rico-Varadé and M.D. Luque de Castro, *J. Anal. At. Spectrom.*, 13 (1998) 787.
288 S. Morales-Muñoz, J.L. Luque-García and M.D. Luque de Castro, *Anal. Chem*, 74 (2002) 4213.
289 J. Amador-Hernández and M.D. Luque de Castro, *J. Biochem. Biophys. Methods*, 43 (2000) 329.
290 J.D. Brewster, R.J. Maxwell and J.W. Hampson, *Anal. Chem.*, 65 (1993) 2137.
291 M.T. Tena, M.D. Luque de Castro and M. Valcárcel, *Anal. Chem.*, 67 (1995) 1054.
292 L. Manganiello, A. Ríos and M. Valcárcel, *J. Chromatogr.*, 874 (2000) 265.
293 P. Richter, J.M. Fernández-Romero, M.D. Luque de Castro and M. Valcárcel, *Chromatographia*, 34 (1992) 445.
294 C. Mardones, A. Ríos and M. Valcárcel, *Electrophoresis*, 20 (1999) 2922.

295 J.N. Dolan, J.R. Grant, N. Tanaka, R.W. Gise and B.L. Karger, *J. Chromat. Sci.*, 16 (1989) 616.
296 R. Carabias, E. Rodríguez-Gonzalo, E. Hernández-Fernández and J. Hernández-Méndez, *Anal. Chim. Acta*, 304 (1995) 323.
297 K. Grudpan, J. Jakmunee and P. Sooksamiti, *Talanta*, 49 (1999) 215.
298 M. Krachler and H. Emons, *J. Anal. At. Spectrom.*, 15 (2000) 281.
299 R. Sur, J. Begerow and L. Dunemann, *Fresenius J. Anal. Chem.*, 363 (1999) 526.
300 A.J. Trujillo, I. Casals and B. Guamis, *J. Chromatogr.*, 870 (2000) 371.
301 C.L. Liu and S.S. Verma, *J. Chromatogr.*, 835 (1999) 93.
302 F. Lázaro, M.D. Luque de Castro and M. Valcárcel, *J. Chromatogr.*, 448 (1998) 173.
303 A. Membiela, F. Lázaro, M.D. Luque de Castro and M. Valcárcel, *Anal. Chim. Acta*, 12 (1991) 461.
304 E. Ballesteros, M. Gallego and M. Valcárcel, *Anal. Chem.*, 62 (1990) 1587.
305 D.W. Bryce, A. Izquierdo and M.D. Luque de Castro, *Anal. Chem.*, 69 (1997) 844.
306 P. Kuban and B. Karlberg, *Trends Anal. Chem.*, 17 (1998) 34.
307 L. Bao and P.K. Dasgupta, *Anal. Chem.*, 64 (1992) 991.
308 B.L. Hogan, S.M. Lunte, J.F. Stobaugh and C.E. Lunte, *Anal. Chem.*, 66 (1994) 596.
309 B.A.P. Bushcer, U.R. Tjaden and J. Van der Greef, *J. Chromatogr.*, 764 (1997) 135.
310 P. Kuban, A. Engström, J.C. Olsson, G. Thorsen, R. Tryzell and B. Karlberg, *Anal. Chim. Acta*, 337 (1997) 117.
311 J.A. García-Mesa, M.D. Luque de Castro and M. Valcárcel, *Anal. Chem.*, 65 (1993) 3540.
312 P. Torres, E. Ballesteros and M.D. Luque de Castro, *Anal. Chim. Acta*, 308 (1995) 371.
313 A. Velasco-Arjona and M.D. Luque de Castro, *Anal. Chim. Acta.*, 333 (1996) 205.
314 M.B. Shabani and A. Masuda, *Anal. Chim. Acta*, 261 (1992) 315.
315 H. Mukai, Y. Ambe and M. Morita, *J. Anal. At. Spectrom.*, 5 (1990) 75.
316 L. Ebdon, A. Fisher, H. Handley and P. Jones, *J. Anal. At. Spectrom.*, 8 (1993) 979.
317 S. Garrigues, M. Gallignani and M. de la Guardia, *Analyst*, 117 (1992) 1849.
318 M. Gallignani, S. Garrigues and M. de la Guardia, *Anal. Chim. Acta*, 274 (1993) 267.
319 M. de la Guardia, M. Gallignani and S. Garrigues, *Anal. Chim. Acta*, 282 (1993) 543.
320 D.K. Morgan, N.D. Danielson and J.E. Katon, *Anal. Lett.*, 18 (1985) 1979.

321 B.E. Miller, N.D. Danielson and J.E. Katon, *Appl. Spectrosc.*, 42 (1988) 401.
322 S.V. Olesik, S.B. French and M. Novotny, *Anal. Chem.*, 58 (1986) 2256.
323 S.A. Curran and D.E. Williams, *Appl. Spectrosc.*, 41 (1987) 1450.
324 M.D. Luque de Castro and M.T. Tena, *Talanta*, 42 (1995) 151.
325 J.A. Koropchak and D.H. Winn, *Anal. Chem.*, 58 (1986) 2561.
326 J.W. Elgersma and F.J.M.J. Maessen, *Spectrochim. Acta*, 41B (1986) 1217.
327 J.W. Elgersma, J. Balke and F.J.M.J. Maessen, *Spectrochim. Acta*, 46B (191) 1073.
328 M. Ibrahim, W. Nisamaneepong and J. Caruso, *J. Chromatogr. Sci.*, 23 (1985) 144.
329 A. Gustavsson, *Spectrochim. Acta*, 42B (1987) 111.
330 M.J. Hayward, T. Kotiaho, A.K. Lister, R.G. Cooks, G.D. Austin, R. Narayan and G.T. Tsao, *Anal. Chem.*, 62 (1990) 1798.
331 G.J. Tsai, G.D. Austin, M.J. Syu, G.T. Tsao, M.J. Hayward, T. Kotiaho and R.G. Cooks, *Anal. Chem.*, 63 (1991) 2460.
332 K.W. Siu, G.J. Gadner and S.S. Berman, *Anal. Chem.*, 61 (1989) 2320.
333 Y. Israel and R.M. Barnes, *Anal. Chem.*, 568 (1984) 1188.
334 Q. Shen, Z. Jiang and Z. Liao, *Fenxi Shiyanshi*, 10 (1991) 456.
335 Z. Peng, J. Klingenberg, T. Beeren and W. Van Borm, *Spectrochim. Acta*, 46B (191) 1051.
336 P. Richner, *J. Anal. At. Spectrom.*, 8 (1993) 927.
337 A. Miyazaki and K. Bansho, *Kogai*, 24 (1989) 87.
338 E.A.G. Zagatto, A.O. Jacintho, F.J. Krug, B.F. Reis, R.E. Bruns and M.C.U. Araújo, *Anal. Chim. Acta*, 145 (1983) 169.
339 E.C. Silva, V.L. Martins, A.F. Araujo and M.C.U. Araujo, *Anal. Sci.*, 15 (1999) 1235.
340 B.F. Reis, M.F. Giné, F.J. Krug and H. Bergamin Fo, *J. Anal. At. Spectrom.*, 7 (1992) 865.
341 A. Lasztity, M. Viczian, X. Wang and R.M. Barnes, *J. Anal. At. Spectrom.*, 4 (1989) 671.
342 M.F. Giné, H. Bergamin Fo, B.F. Reis and R.L. Tuon, *Anal. Chim. Acta*, 234 (1990) 207.
343 A.A. Alwarthan, A. Almuaibed and A. Townshend, *Anal. Sci.*, 7 (1991) 623.
344 E. Yamada, C. Hamamura, K. Fukuda, M. Sato and J. Flow, *Injection Anal.*, 10 (1993) 48.
345 T. Aoki and M. Wakabayashi, *Anal. Chim. Acta*, 308 (1995) 308.
346 W. Qin, Z.J. Zhang and H.J. Liu, *Anal. Chim. Acta*, 357 (1997) 127.
347 H. Li and O.S. Wolfbeis, *Anal. Chim. Acta*, 276 (1993) 115.
348 E.A.G. Zagatto, F. Bergamin, S.M.B. Brienza, M.A.Z. Arruda, A.R.A. Nogueira and J.L.F.C. Lima, *Anal. Chim. Acta*, 261 (1992) 59.

349 A.R.A. Nogueira, S.M.B. Brienza, E.A.G. Zagatto, J.L.F. Lima and A.N. Araujo, *Anal. Chim. Acta*, 276 (1993) 121.
350 J.A. García-Mesa, M.D. Luque de Castro and M. Valcárcel, *Anal. Chim. Acta*, 297 (1994) 313.
351 F. Lázaro, A. Ríos, M.D. Luque de Castro and M. Valcárcel, *Anal. Chim. Acta*, 179 (1986) 279.
352 B. Fernández-Band, F. Lázaro, M.D. Luque de Castro and M. Valcárcel, *Anal. Chim. Acta*, 229 (1990) 177.
353 A. Segura-Carretero, J. Rodríguez-Fernández, A.R. Bowie and J.P. Worsfold, *Analyst*, 125 (2000) 387.
354 T. Korenaga and F. Sun, *Anal. Chim. Acta*, 318 (1996) 195.
355 T. Korenaga and F. Sun, *Talanta*, 43 (1996) 1471.
356 M.A. Spaziani, J.L. Davis, M. Tinani and M.K. Carroll, *Analyst*, 122 (1997) 1555.
357 A. Chartier, C.G. Fox and J. Georges, *Analyst*, 118 (1993) 157.
358 E. Simó-Alfonso, Y. Martín-Biosca, M.C. García-Alvarez-Coque and G. Ramis-Ramos, *Thermochim. Acta*, 64 (1993) 2153.
359 H.M. Sorouraddin, A. Hibara, M.A. Proskurnin and T. Kitamori, *Anal. Sci.*, 16 (2000) 1033.
360 I. Carrer, P. Cusmai, E. Zanzottora, W. Martinotti and F. Realini, *Anal. Chim. Acta*, 308 (1995) 20.
361 M.D. Luque de Castro and L. Gámiz-Gracia, In: J. Sneddon (Ed.), Recent Developments in Flow-injection Atomic Spectroscopy in 'Advances in Atomic Spectroscopy', Vol. 4, JAI Press Ltd, Stamford, 1998.
362 S.C. Wang and K.C. Lin, *Anal. Chem.*, 66 (1994) 2180.
363 S.C. Wang and K.C. Lin, *Analyst*, 120 (1995) 2593.
364 C.B. Ke and K.C. Lin, *Anal. Chem.*, 71 (1999) 1561.
365 N.G. Orellana-Velado, M. Fernández, R. Pereiro and A. Sanz-Medel, *Spectrochim. Acta*, 15 (2001) 113.
366 S. Walas, E. Borowska, M. Herda, M. Herman and H. Mrowiec, *Int. J. Environ. Anal. Chem.*, 72 (1998) 217.
367 M. Bralic and N. Radic, *Analusis*, 27 (1999) 57.
368 L.M.B.C. Alvarez-Ribeiro and A.A.S.C. Machado, *Analyst*, 123 (1998) 653.
369 A.M.S. Abdennabi and M.E. Koken, *Talanta*, 46 (1998) 639.
370 A. Florido, I. Casas, J. García Raurich, R. Arad-Yellin and A. Warshawsky, *Anal. Chem.*, 72 (2000) 1604.
371 P.L.H.M. Cobben, R.J.M. Sgberink, J.G. Bomer, E.J.R. Sudholter, P. Bergveld and D.N. Reinhoudt, *Anal. Chim. Acta*, 248 (1991) 307.
372 F.J. Sáez de Viteri and D. Diamond, *Electroanalysis*, 6 (1994) 9.
373 T. Kullick, U. Bock, J. Schubert, T. Schper and K. Schuegeri, *Anal. Chim. Acta*, 300 (1995) 25.
374 C. Colombo and C.M.G. van der Berg, *Int. J. Environ. Anal. Chem.*, 71 (1998) 1.

375 I.L. de Mattos, D. Melo and E.A.G. Zagatto, *Anal. Sci.*, 15 (1999) 537.
376 J. Castañón-Fernández, M.T. Fernández-Abedul and A. Costa-García, *Anal. Chim. Acta*, 406 (2000) 225.
377 R. Lai, E.L. Huang, F. Zhou and D.O. Wipf, *Electroanalysis*, 10 (1998) 926. Y.S. Fund and S.Y. Mo, *Analyst*, 121 (1996) 369.
378 W.W. Kubiak and Z. Kowalsky, *Talanta*, 41 (1994) 1319.
379 K. Schachl, H. Alemu, K. Kalcher, J. Jezkova, I. Svancara and K. Vytras, *Anal. Lett.*, 30 (1997) 2655.
380 S.V. Guerra, L.T. Kubota, C.R. Xavier and S. Nakagaki, *Anal. Sci.*, 15 (1999) 1231.
381 E. Domínguez, H.L. Lan, Y. Okamoto, P.D. Hale, T.A. Skotheim and L. Gorton, *Biosens. Bioelectron.*, 8 (1993) 167.
382 K. Schachl, H. Alemu, K. Kalcher, H. Moderegger, I. Svancara and K. Vytras, *Fresenius J. Anal. Chem.*, 362 (1998) 194.
383 T.Y. Ou and J.L. Anderson, *Anal. Chem.*, 63 (1991) 1651.
384 R.C. Matos, M.A. Augelli, C.L. Lago and L. Agnes, *Anal. Chim. Acta*, 404 (2000) 151.
385 Q. Chen, J. Wang, G. Rayson, B. Tian and Y. Lin, *Anal. Chem.*, 65 (1993) 251.
386 Z.P. Yang, J.M. Kauffmann, M.I. Acedo-Valenzuela and S. Ozkan, *Mikrochim. Acta*, 131 (1999) 85.
387 Z.I. Denchev, N. Elenkova and L. Ilcheva, *Anal. Lab.*, 6 (1997) 49.
388 F. Zhou, J.T. Aronson and M.W. Ruegnitz, *Anal. Chem.*, 69 (1997) 728.
389 M. Korolczuk and M. Grabarczyk, *Anal. Chim. Acta*, 387 (1999) 97.
390 C. Fernández-Bobes, M.T. Fernández-Abedul and A. Costa-García, *Electroanalysis*, 10 (1998) 701.
391 G. Schulze, E. Han and W. Frenzel, *Fresenius J. Anal. Chem.*, 332 (1989) 844.
392 L.E. Locascio and J. Janata, *Anal. Chim. Acta*, 194 (1987) 99.
393 C.M.A. Brett, M.B.Q. García and J.L.F.C. Lima, *Electroanalysis*, 8 (1996) 1169.
394 E. Beinrohr, M. Cakrt, J. Dzurov, P. Kottas and E. Kozakova, *Fresenius J. Anal. Chem.*, 356 (1996) 253.
395 F.J. Andrade, F.A. Inon, M.B. Tudino and O.E. Troccoli, *Anal. Chim. Acta*, 379 (1999) 99.
396 S.D. Kumar, B. Maiti and P.K. Mathur, *Anal. Chem.*, 71 (1999) 2551.
397 Z.K. He, B. Fuhrmann and U. Spohn, *Anal. Chim. Acta*, 407 (2000) 203.
398 F. Torabi, K. Ramanathan, P.O. Larsson, L. Gorton, K. Svanberg, Y. Okamoto, B. Danielsson and M. Khayyami, 50 (1999) 787.
399 A. Magna, A. Fernando de Oliveira and O. Fatibello Filho, *Anal. Lett.*, 29 (1996) 2411.
400 X.L. Su, H.W. Tan, W.F. Li, W.Z. Wei and S.Z. Yao, *Anal. Sci.*, 14 (1998) 553.

401 X.L. Su, W.Z. Wei, L.H. Nie and S.Z. Yao, *Analyst*, 123 (1998) 221.
402 X.L. Su, L.H. Nie and S.Z. Yao, *Fresenius J. Anal. Chem.*, 360 (1998) 272.
403 M.T.S.R. Gomes, E.V. Morgado and J.A.B.P. Oliveira, *Anal. Lett.*, 32 (1999) 2715.
404 L. Manganiello, A. Ríos, M. Valcárcel, A. Ligero and T. Tena, *Anal. Chim. Acta*, 406 (2000) 309.
405 Y. Nimoiya, S. Oda and N. Oyama, *Bunseki Kagaku*, 42 (1993) T93.
406 L. Huber and S.A. George, *Diode Array Detection in HPLC*, Marcel Dekker, New York, 1993.
407 J. Janata and J. Ruzicka, *Anal. Chim. Acta*, 139 (1982) 105.
408 F. Lázaro, A. Ríos, M.D. Luque de Castro and M. Valcárcel, *Analysis*, 14 (1986) 378.
409 J. Ruzicka, *Fresenius J. Anal. Chem.*, 324 (1986) 745.
410 Z. Fang, S. Xu, X. Wang and S. Zhang, *Anal. Chim. Acta*, 179 (1986) 325.
411 Z. Fang, S. Xu and Z. Zhang, *Anal. Chim. Acta*, 200 (1987) 35.
412 G.D. Christian and J. Ruzicka, *Spectrochim. Acta*, 42B (1987) 157.
413 M. Linscheid, *Chem. Labor. Bert.*, 41 (1990) 125.
414 T. Shimamura, *Shitsuryo Bunseki*, 36 (1988) 273.
415 R.M. Barnes, *Spectroscopy*, 1 (1986) 24.
416 E.G. Bartick and R.G. Messerschmidt, *Int. Lab.*, 58-64 (1985) 66.
417 O.F.X. Donard and F.M. Martin, *Trends Anal. Chem.*, 11 (1992) 17.
418 H. Abderrazak, B. Lendl and R. Kellner, *Quim. Anal.*, 16 (1997) 27.
419 K. Hayakawa, Y. Yoneda, Y. Okamoto, T. Kumamaru and M. Ikeda, *Anal. Sci.*, 15 (1999) 803.
420 A. Pérez-Ponce, S. Garrigues and M. de la Guardia, *Vib. Spectrosc.*, 16 (1998) 61.
421 A.R. Casella, R. Jorgensen-Cassella, S. Garrigues, R.E. Santelli, R.C. de Campos and M. de la Guardia, *Analyst*, 125 (2000) 1829.
422 H. LeThanh and B. Lendl, *Anal. Chim. Acta*, 422 (2000) 63.
423 P. Tipparat, S. Lapanantnoppakhun, J. Jakmunee and K. Grudpan, *Talanta*, 53 (2001) 1199.
424 B. Lendl, J. Frank, R. Schindler, A. Muller, M. Beck and J. Faist, *Anal. Chem.*, 72 (2000) 1645.
425 M. Spraul, M. Hofmann and P. Neidig, *Bruker. Rep.*, 147 (1999) 14.
426 S.K. Lunsford, C.A. Striley, Y.L. Ma, H. Zimmer, G. Kreisham and H.B. Mark Jr., *Anal. Lett.*, 29 (1996) 1309.
427 M. Spraul, M. Hofmann, M. Ackermann, A.W. Nicholls, S.J.P. Damment, J.N. Haselden, J.P. Shockcor, J.K. Nicholson and J.C. Lindon, *Anal. Commun.*, 34 (1997) 339.
428 N. Weissenbacher, B. Lendl, J. Frank, H.D. Wanzenboeck and R. Kellner, *Analyst*, 123 (1998) 1057.
429 V.J.P. Gouveia, I.G. Gutz and J.C. Rubim, *J. Electroanal. Chem.*, 371 (1994) 37.

430 K.L. Morand, T.M. Burt, B.T. Regg and T.L. Chester, *Anal. Chem.*, 73 (2001) 247.
431 T. Wang, L. Zeng, T. Strader, L. Burton and D.B. Kassel, *Rapid Commun. Mass Spectrom.*, 12 (1998) 1123.
432 E. Gorlach, R. Richmond and I. Lewis, *Anal. Chem.*, 70 (1998) 3227.
433 C.M. Tseng, D. Amouroux, I.D. Brindle and O.F.X. Donard, *J. Environ. Monit.*, 2 (2000) 603.
434 D. Beauchemin, *J. Anal. At. Spectrom.*, 13 (1998) 1.
435 S. Pattberg and R. Matschat, *Fresenius J. Anal. Chem.*, 364 (1999) 410.
436 K. Benkheda, H.G. Infante, E. Ivanova and F.C. Adams, *J. Anal. At. Spectrom.*, 15 (2000) 1349.
437 G. Centineo, M.M. Bayon and A. Sanz-Medel, *J. Anal. At. Spectrom.*, 15 (2000) 1357.
438 C.W. Huie, J.H. Aiken and W.R. Williams, *Anal. Chim. Acta*, 254 (1991) 189.
439 B. Ruchti, C. Schramm, S. Kubitschko and B. Neidhart, *Fresenius J. Anal. Chem.*, 342 (1992) 822.
440 K.P. Hupe, M. Riedmann and G. Rozing, *Chromatographia*, 40 (1995) 631.
441 D. Gambart, S. Cárdenas, M. Gallego and M. Valcárcel, *Anal. Chim. Acta*, 366 (1998) 93.
442 W. Weinmann and M. Svoboda, *J. Anal. Toxicol.*, 22 (1998) 319.

*Chapter 5*

# *Distributed analytical instrumentation systems*

***D. Ramírez,* A.E. Navarro and S. Casans***

## 5.1 THE REMOTE CONCEPT

The main objective of this chapter is to present the so-called *distributed analytical instrumentation systems*. From a conceptual point of view, a system is considered a group of interrelated elements or subsystems which perform a specific function. In our case, an instrumentation system has the function of taking measurements by applying analytical techniques in order to obtain a certain degree of information of a chemical nature. Physical separation of the subsystems from each other through one or more different types of connection results in a distributed measurement system. Digital data transmission networks allow the distance of the distributed measurement system from the place where the information is required to be greater.

### 5.1.1 Elements in a measurement system

A measurement system, regardless of its chemical, physical or electrical nature, and whether or not it is distributed, possesses functions or subsystems with different degrees of complexity that are necessary

* Corresponding author. Tel.: +34-963160455; fax: +34-963160466. *E-mail address:* diego.ramirez@uv.es.



ISSN: 0166-526X

for the fulfilment of those measurements of interest (Sydenham, [1]). These functions are sensing, conditioning, transmission, acquisition, processing, storage, displaying, distribution and actuation.

Additionally, the energy supply function indirectly affects the process of taking measurements, but it is necessary in order to provide energy to the above-mentioned subsystems. These are listed with more detail below.

*Sensing*: This consists of the extraction of information contained within the chemical system. It provides a signal which has a transfer function with the property to be measured.

*Conditioning*: The signal provided by the sensing system is transformed to another signal, which is generally in the electronic domain (e.g. voltage, current or frequency) so that it can be more easily treated in the acquisition stages.

*Transmission*: This function in a measurement consists of the interconnection of different stages in order to allow the information to flow through them.

*Acquisition*: In this subsystem the analog information is held and digitally codified for future access and manipulation.

*Processing*: In this subsystem the information of interest is recovered from the codified signal.

*Storage*: The information of interest can be used in real time for its own interpretation or for eventual use off line.

*Displaying*: With this function the information is shown in an adequate format to produce a complete human perception.

*Distribution*: The processing subsystem can make decisions with the information obtained. It can control another chemical system or even its own system on which measurements have been taken. The distribution subsystem sends these decisions, mainly on an electrical format signal, to the actuation system.

*Actuation*: This is the subsystem that executes the decision through any (e.g. electrical, mechanical, luminous or audible) mechanism.

The instrumentation systems, regardless of their physical, chemical or electric nature, can be divided into two main groups. If the ultimate objective of the measurements is to increase knowledge, then the system is in itself an instrumentation and measurement system; but if the measurements are used to control a particular process, then the instrumentation system is a part of a whole control system. The instrumentation and measurement systems are the basic tools in the scientific research field, while control systems are dedicated to the industry field. The content of this chapter will be explained from the scientific point of view rather than the industrial one.

Intelligence in a measurement system is related to the capacity to process the acquired information and to make distribution and control decisions as a result of this processing (Sydenham and Thorn, [2]). In this way, smart instrumentation systems are characterized by a certain degree of autonomy at the time but controlled by other systems of higher hierarchy. These characteristics involve some degree of complexity in the instrumentation system as it includes not only analog transducing blocks but also digital processing blocks, memories, registers, communication interfaces and microprocessors. The degree of autonomy will increase at the same time as its complexity.

An instrumentation system designed to be distributed must have a strong component of intelligence (smart system). At present, the distributed systems physically exceed the dimensions of an analytical laboratory or an industrial plant.

The presence of the internet in science and technology allows the configuration of instrumentation systems located in physically remote places. With this new situation, digital data transmission requirements (communication protocols, web pages support, data and email services, etc.) increase the degree of intelligence and generate an added complexity to the traditional measurement functions.

### 5.1.2 Distributed system topologies

A distributed instrumentation system is composed of sites or nodes provided by a higher or lower degree of intelligence, configured as a local or remote type, connected through a link or communication interface. At the same time, the nodes can act either as sites where information is generated (remote measurement nodes) or as sites where the information is served (measurement publishing nodes).

If the generation of information (measurement of interest) is performed on nodes physically separated from the node where such information is processed and performed, then the instrumentation system is in a *remote position*. In this configuration it is possible to locate a number of measurement systems in a remote situation which then pass the information to a central processing node, which has the role of passing it to other users or nodes of interest. These nodes of interest would not have access to the remote measurement nodes.

Nevertheless, an instrumentation system in a *local position* generates measurements, before processing and supplying them to several nodes or remote users in order to be read. In the first case, it is understood to be a remote measurement system; in the second case, a measurement publishing system is created.

From a systematic point of view, it is possible to classify distributed instrumentation systems into five categories.

*Primitive distributed measurement systems*

The simplest distributed measurement system is composed of production, processing and actuation subsystems. In the information production section a group of sensors with specific sensitivity to the parameters of chemical value of interest is obtained (Fig. 5.1). If the sensors are very different, then each of them has its own electronic

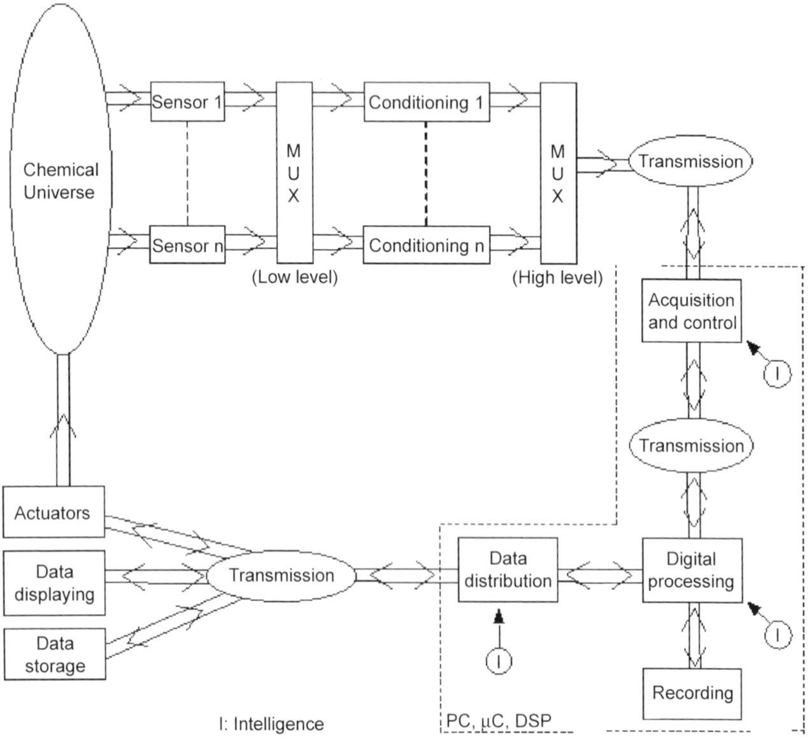


Fig. 5.1. Primitive distributed measurement system.

conditioning subsystem. If the sensors are of the same type, then they could share one conditioning subsystem. In both cases if there are several input information channels of interest a multiplexed signal block is used within the conditioning stage. The function of this block is to help the measurement system to share certain common resources, avoiding subsystems redundancy and an increase in total cost. In the case where they have very different sensors, signal multiplexing is done once they have been conditioned. This results in *high level multiplexing* when the shared resources are the eventual transmission and acquisition stages. But if the sensors are of the same type, *low level multiplexing* is used and the resources are shared from the conditioning stage (Franco, [3]). It is possible but optional that a transmission stage exists in these types of systems, acting when the conditioning blocks are separated from the acquisition stages. Distances can range from several metres to a kilometre if there is physical support of the transmission stage (i.e. voltage or current mode transmission by cable or by optic fibre) or they can be longer in the case of transmission via a radio-system.

The processing section is represented by the personal computer. It is the stage where the degree of intelligence in this distributed system category is located, because it contains the hardware and software resources needed to carry out the acquisition, digital processing and signal distribution tasks. In particular, the acquisition and distribution systems would be represented by a card (input/output board) inserted in the same backplane of the computer. The digital processing functions would be carried out by software running on the computer. This software would be composed of: data production and acquisition control routines; digital processing routines, distribution and decision control routines and information storage routines (with external or internal devices). The decisions resulting from the processing software adopt the format of output signals and are distributed in an appropriate format, either to activate an actuator device (engine, electric valve, pump, micro syringe, etc.), or to be saved (external memory devices).

In practical situations, the application requirements are less strict and instead of using a computer as an intelligent subsystem, other devices such as microcomputers could be the right choice. These subsystems are based on microprocessors and digital circuits of different characteristics (memories, registers, communications interfaces, etc.), and support specific software. With all this hardware, it is possible to perform the required intelligent functions with additional capabilities, like the possibility to display on digital screens or transmit to higher hierarchy intelligent subsystems (computers) the desired information.

In this case, only the sensing, conditioning and actuation stages could be distributed. The acquisition and processing stage is common to all, and therefore the distributed property is in the hardware but not in the software. On the other hand, the degree of intelligence of this topology is assigned to the computer. It supports the acquisition, processing and control functions and, as a consequence, the autonomy of the distributed stages is rather non existent.

*Distributed intelligence in local mode*

In this category the instrumentation system is characterized by the presence of different equipment with autonomous measurement capabilities (Fig. 5.2). Each piece of equipment is a complete measurement system as it internally includes all the stages required to perform measurements. Chromatographs, spectrometer, pH-meters and climatic chambers are some examples of these types of equipment. However, they have an additional property, which is connectivity: there is a digital communication interface that puts the equipment capabilities in the service of a more complex instrumentation system requirements. The communication interface is, in these cases, of the RS-232 type and its derivatives (RS-485, $I^2C$, etc.) are even more complete, like the instrumentation interface IEEE-488 (*intelligence of communications*). The computer is still the central element in this organization. All the processing, acquisition, displaying and distribution functions (*general intelligence*) are centralized in the computer. In this type of organization a *virtual instrumentation* concept appears as a new way to perform measurements, designing measurement functions through software. These functions were previously reserved for the hardware of measurement equipment. The computer is connected with the rest of the equipment either through the communication interface or by the use of a designed card to acquire or distribute electrical signals located in the backplane. In this way it is possible to access the information of interest in two ways: through high service equipment connected between them and with the central computer by an instrumentation bus; or by simple measurement systems (electronic modules) which enter the measurements through the acquisition cards (I/O board).

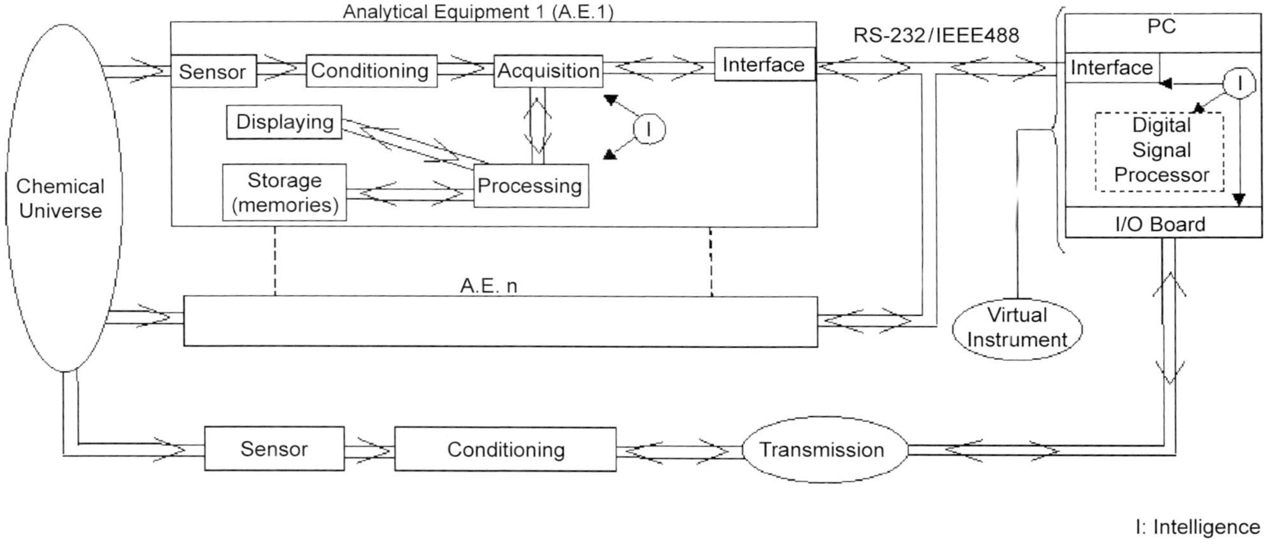


Fig. 5.2. Distributed intelligence in local mode.

This system categories have a more advanced degree of distribution than in the previous case. The equipment connected with a digital communication interface is able to carry out, in the equipment itself, certain type of functions like processing, data displaying and acquisition, which is used by the computer later (*local intelligence*). The system is distributed in a local mode due to the fact that the distance between equipment and the computer is reduced to that observed in a laboratory.

*Distributed intelligence in remote mode (instrumentation remote system)*
The presence of the internet has completely changed the design capabilities of the instrumentation and measurement systems. Its application allows us to talk of instrumentation remote systems. In Fig. 5.3, on top of the measurement stages of the previous organization, a microprocessor based measurement system is connected to the computer through the RS-232 interface. This new connection is also possible in the previous case; however, the new feature is that we can incorporate into the central computer a digital communication hardware interface and communication protocols based on software routines in order to have access to the internet. With this, other computers (web browsers) will be able to have access to the information which the remote computer choses to show (measurement server). In this organization, the measurement and instrumentation system is totally remote, including the central computer. The level of intelligence is distributed as in the previous case (measurement equipment, central computer, microprocessor), but in this case in a remote mode, far from the place where information is processed.

*Distributed intelligence in remote mode (equipment remote system)*
In this category (Fig. 5.4) the measurements equipment sends information to a module or communication controller which has access to the internet. In this way, the digital communication interfaces present in the measurement equipment (based on the standards RS-232 or IEEE-488) can send the information over the internet through the controller by networks (Ethernet) using standard protocols (TCP/IP). Processing and interpretation of measurements carried out by the computer are not performed in the same place as where they are taken. In this type of system, the intelligence functions are located in a place

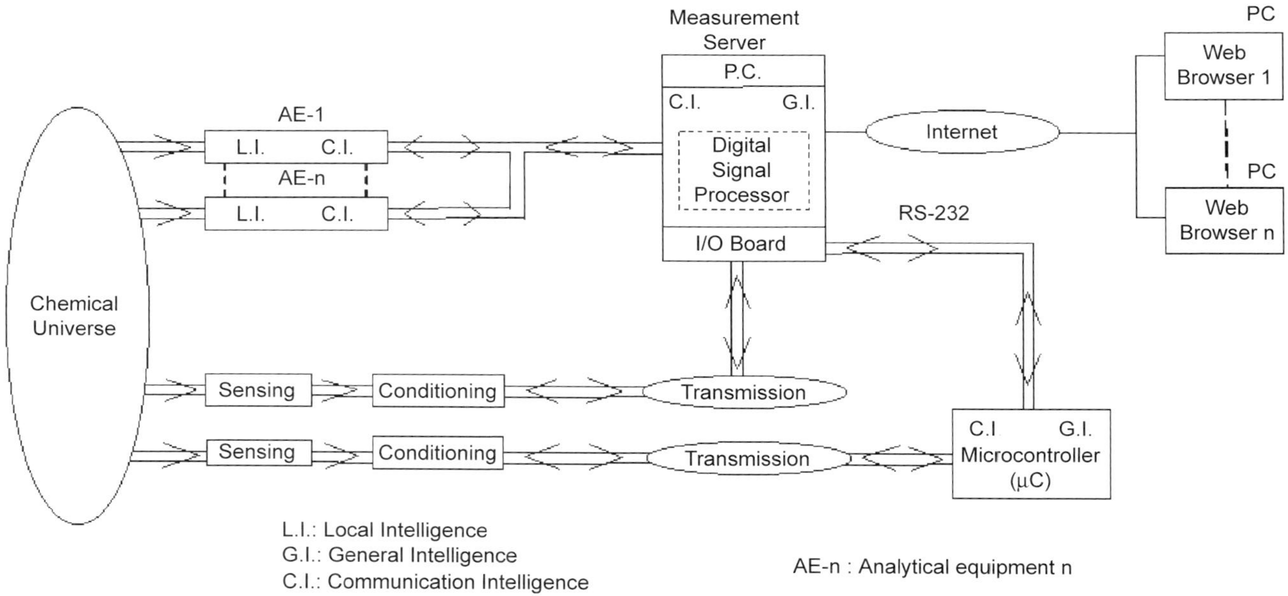


Fig. 5.3. Distributed intelligence in remote mode (instrumentation remote system).

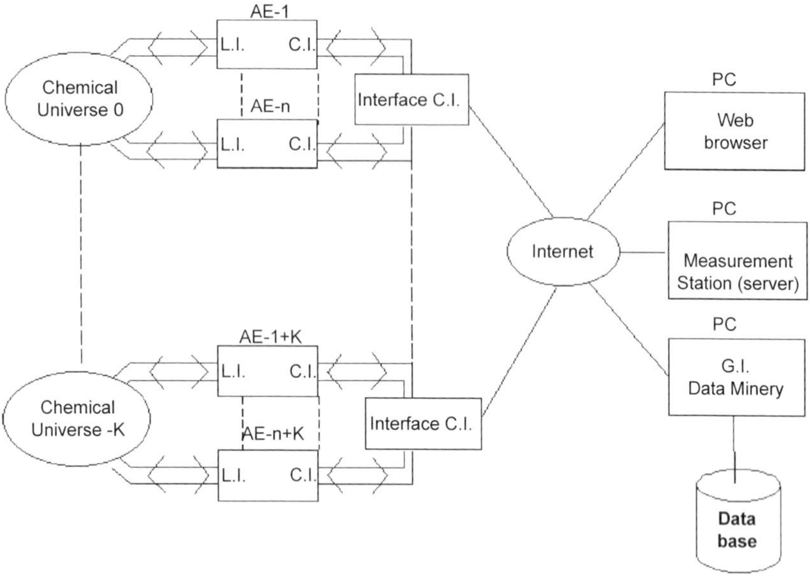


Fig. 5.4. Distributed intelligence in remote mode (equipment remote system).

that is accessible via the internet and different from the instrumentation system. At the other end, processing units acquire the remote measurement system information with various objectives: data display (web browser), data distribution (measurement server) and data storage (data base system).

In this organization mode, the intelligence of the system shows a high degree of distribution. The measurement equipment has its own processor dedicated to the production of information and its communication interface (local intelligence). The central computer is now located away from the measurement process but it is still accomplishing its functions (processing, controlling, data displaying) in the process (general intelligence) with full capacity to make a virtual instrument. The system could handle specific databases not located in the central computer. The instrumentation system is now a remote equipment system accessible through a communication controller (communication intelligence).

*Fully distributed instrumentation system (hardware and software)*
This is the most complex degree of distribution (Fig. 5.5). In this category remote measurement equipment not only concentrates

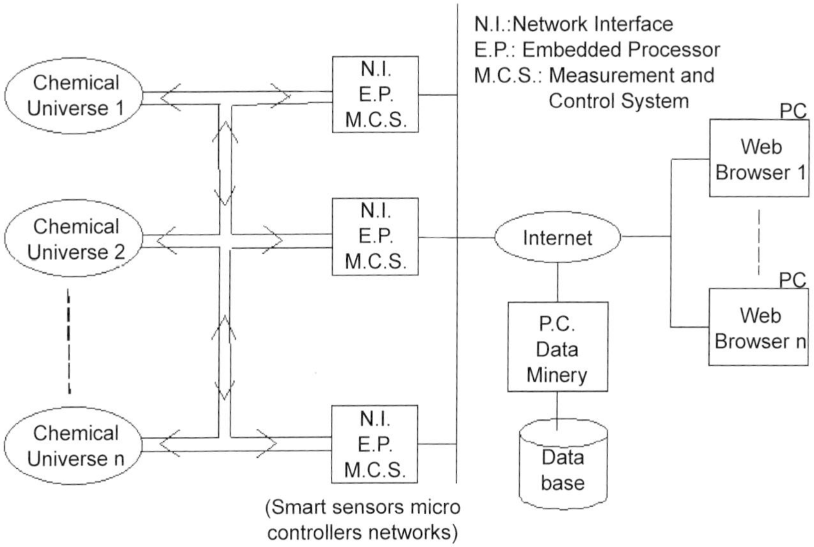


Fig. 5.5. Fully distributed instrumentation system.

the total measurement functions but also contains part of the processing software, which till now was a function assigned to the personal computer. There exists, therefore, a distribution system, not only to a hardware level (measurement equipments), but also to a level of its own execution and control capabilities (software). Each specific piece of measurement equipment has its own processor (embedded processor) which can relieve functions of the central computer. In this way, the embedded processor runs part of the main software, a measurement and control subsystem takes measurements and a network interface allows it to be connected with the rest of the equipment, with the internet access capabilities. With this type of organization, in the case of failure or malfunction of any instrument, the whole system would not break down. Software distribution would enable the defective subsystem to be excluded from the global process, allowing the rest of subsystems to run properly. The reliability and robustness of this organization are increased with respect to previous cases. In this configuration the central computer would perform coordination and supervision functions over the rest of equipments.

## 5.2 THE REMOTE PLACE

The remote place is understood as the physical area in which the initial measurement stages are located. In this remote place the sensing functions as well as the electronic, conditioning and acquisition functions are performed. The processing and distribution functions may not be found in a remote position, depending on whether the central processing unit is present.

### 5.2.1 The benefits of distributed intelligence

In the first place, the sensing and conditioning functions are located in the remote place. The sensor gathers energy (information) from the chemical universe or cell, transforming it into a signal which can then be processed. Moreover, only a small amount of energy has to be extracted, otherwise it would alter the initial conditions around the area being studied.

In the chemical analysis field, for instance, liquid electrolyte sensors, solid state electrochemical sensors, field effect chemical sensors, calorimetric chemical sensors and optochemical sensors could be found. The signal emitted by all of them is usually of low level and has a low energy content. For this reason, it is necessary to condition it by using specific electronic circuits, so that it gets the format (current or voltage), level (amplification) and frequency content (filtering) to be acquired later. In analytical instrumentation, these two functions are performed either by specific sensing and conditioning modules, or by measurement equipment. The first case is represented by hardware designed circuits to which the sensors of interest are connected. Its output is an electrical voltage signal, which must be acquired. Internally, the electronic conditioning functions are implemented and also in some cases an analog transmission stage (4–20 mA loop). It is a simple and low cost solution which covers the needs of sensing and conditioning in a great variety of measurement situations. In practice, the sensor manufacturer itself provides the necessary conditioning module. However, they do not have a digital transmission interface, and because of this they cannot be controlled and are unable to perform tests/diagnosis routines. Input/output boards and microcomputers based measurement systems are used to get the information from these modules.

A second level related to signals acquisition and distribution is represented by already manufactured equipment which has the function to control measurements in practical situations. Initially, measurements were carried out once the equipment capabilities were known, and performing measurements through the panel controls in a manual way. In many cases, the measurement equipment was connected with a central computer with the role of acquiring the measurements. These were processed by applying software designed specifically for the equipment (propietary software). Possibilities of controlling the equipment were practically non-existent. However, in the 1980s and with the invention of the General Purpose Instrument Bus (GPIB) communication interface, also known as standard IEEE 488.1, new possibilities of having automatic controllability of the measurement equipment were opened. A communication intelligence was added to the processing and displaying intelligence capabilities, which were within the equipment. This communication intelligence was represented according to the particular case through the RS-232 or IEEE-488 interfaces. Both types of interface allowed complete access to the whole equipment capabilities, taking measurements automatically. At the same time, the software programming languages were performed in structure and capacity (Basic, Pascal, etc.), so that it was possible to design an acquisition programme suitable for the experiment being considered. In this situation, the hardware was distributed (although not physically remote) to the dimensions of a laboratory. However, the software is centralized in a central computer where programme routines are carried out.

Recently, certain types of module have been designed to perform measurements which are connected to intelligent controllers capable of processing information and making control and signals distribution decisions. These modules are really autonomous processing units, as they possess the necessary elements (processor, memories, communications interfaces, etc.). In this type of organization a number of routines, which were previously present in a central computer, are now set up and distributed in the modules executing them. The use of distributed intelligence has some advantages:

- Complete modularity of the instrumentation system, which reaches software and hardware.
- It allows the central computer to unload the mundane and repetitive tasks to the distributed modules, so that it can perform

other processing functions, such as manipulation and analysis of data in real time.

- Calibration, configuration and diagnosis routines of the hardware are no longer performed by the central processor. The presence of distributed intelligence makes it possible for each module to perform these routines.
- The distributed module can reconfigurate itself automatically just by downloading the new software routine from the central processor.
- The presence of distributed intelligence makes it possible to lower the cost and complexity of the central computer.

Iotech [4] and National Instruments [5] are two industries that manufacture hardware modules which can be connected through communication interfaces such as RS-232 and derivatives, or Ethernet through protocol TCP/IP. These modules are configurated in a distributed way, including the execution software, to measure variables like temperature, voltage and current.

*The acquisition function*

Acquisition, processing and distribution are the specific functions carried out by a measurement system related to intelligence.

The signals of interest (luminous radiation, ion concentration, dissolved gases, etc.) are analog signals which, after crossing, sensing and conditioning stages, result in electrical signals (voltage or current). However, in a distributed system, the subsystem that processes these signals (computer, microcomputer, digital signal processor) is digital. The acquisition subsystem converts the analog signal into its digital representation without losing or degrading the initial information. The acquisition function is done by the electronic component called an analog-to-digital converter. This function is carried out in two phases: quantification and codification. Quantification consists of assigning a digital state, taken from a finite number, to the analog input signal. Codification implies translating the digital state to a digital code or format, which can be easily represented or processed. The analog-to-digital converter is located in the data input/output boards, which can be plugged either into the computer backplane or in the microprocessor based acquisition systems. On the other hand, manufactured equipment already includes this component in the internal modules of its circuit systems. In this way, acquisition is totally transparent to the user.

The processing function is related to the presence of software algorithms in the processor. In a distributed system there are several processors located in remote and/or local positions. The remote processor is physically in close contact with the measurements of interest, it can be placed within autonomous equipment, a microprocessor based system, or a computer. In this last case, the central processor is connected to the measurement equipment by instrumentation interfaces of the type RS-232 or IEEE-488 (Fig. 5.2). In other cases, the main processor is physically separated from the measurement process, keeping an additional communication interface with established protocols for transmitting information (Figs. 5.3–5.5).

### 5.2.2 The computer-controlling function

The basic purpose for designing measurement systems controlled by computer is to obtain measurement capabilities which otherwise would not be possible with autonomous equipment. In most instrumentation and measurement systems the computer carries out measurement, information processing and signal distribution functions. Microcomputers and digital signal processors can also perform these functions although to a lower degree and in a small number of applications. The use of a computer has several advantages; it is important to mention the large quantity of opened software (see Section 5.4) which can be used to easily design the applications of interest. On the other hand, the computer can easily support communication interfaces with the measurement equipment, and also offers the best man–machine interface. The basic objective of this part is to present the requirements to satisfy—at a hardware and software level—computer controlled measurement systems. In addition, the concept of virtual instrumentation will be presented as a tool that will focus on the measurement systems designing the software for instrumentation.

*Hardware elements*
There are four hardware elements involved in a computer-controlled measurement system: the equipment or instruments of measurement, the interface between equipment and computer, the switching system and the computer. *Equipment/instruments of measurement*: These are

the tools which capture signals of interest and generate the measurement. They can be equipment produced either by standard off-the-shelf manufacturers or designed for a specific purpose. Measurement equipment is divided into three categories: that which actually perform measurements (produces an output signal in response to an external stimulus), that which generates stimulus over a chemical system, and that which routes either measurements or stimulus from the computer to the chemical system or vice versa (switching and routing systems). Measurement equipment controlled by computer ranges from the simplest (pH electrodes, conductivity or thermometers) to advanced equipment for measurement and processing, such as equipment for protein analysis or chromatographs. *Computer/equipment interface*: This is a group of physical connections which link the computer to the equipment or instrument. These connections circulate digital signals corresponding to control and address instructions representing the measurements performed. On a local position, the buses or interfaces highly extended at the time to connect measurement equipment with a computer are the interface RS-232 and the bus IEEE-488.

The RS-232 interface is likely to be the simplest way to connect measurement equipment to a computer, but both incorporate this option in their standard connection features. It only allows connection to a device and at a distance no greater than 15 m. The standard interface located in the computer has a maximum transmission speed of 36,000 bits/s (bauds), although there are interface cards with higher transmission speeds (115 kb/s). A previous protocol between computer and equipment is not required. In this way, a dialogue between them is established through a chain of standard characters (ASCII codes). Interface RS-485 is an enhanced version of the former, in which it is possible to connect up to 32 pieces of equipment to the same computer and at a maximum distance of 1.2 km. This specific connection allows actual distributed systems to be designed. The typical speed of the RS-485 interface cards is 460 kb/s.

The standard IEEE-488 is a communication interface which was developed by Hewlett Packard in 1965 and admitted as a standard IEEE in 1975. It is the communication bus most used for connecting a computer to measurement equipment. Typical data transmission speed is 1 mb/s and up to 14 pieces of measurement equipment can be connected to a computer which already has an interface. Connection between equipment and/or a computer is done through a special IEEE-488 cable of a maximum length of 4 m. This connection is of a parallel type allowing connection of a specific piece of equipment to a terminal of another piece of equipment. In this way, it is possible to extend the physical length of the system [7]. Because the computer does not commonly incorporate this communication interface, an electronic circuit which allows correct communication between computer and instruments must be added. There are in the computer market a large number of manufacturers which produce IEEE-488 computer interface electronic cards [4,5]. On the other hand, manufacturers of high and medium capacity equipment are already incorporating this communication interface to their systems.

*Switching system*: This is a hardware component which establishes controlled connections between the measurement equipment, stimulus-generating equipment and the device or system under test. The system is formed by a group of switches controlled by logic signals and it is extensively used in the so-called testing and characterization systems.

*The computer*: When chosing a computer to control an instrumentation system, several factors must be considered. As standard as possible a computer has to be selected, in order to have access to the maximum number of software tools, and as much technical assistance and upgrade as possible. In addition, in the long run, if the computer is to be replaced, the software used for instrumentation system must be portable or admit minimal changes.

*Software elements*

There are different types of software running in the central computer related to the development of applications for testing and measurement systems [6]. The first one to keep in mind is the so-called *operating system*. This is the software that controls its own hardware. It has to be stable, robust and as safe as possible, since stability of the measurement process depends mainly on this software. Secondly, the *instrument bus driver* must be kept in mind between the computer and the measurement equipment. This software allows the reading and writing of messages between the computer and equipment, establishing specific rules to make bus connections between them. Also, it enables the operating system to use a new hardware feature, such as the controller interface. The extensive use of the IEEE-488 interface has made possible the development of printed circuit boards designed specifically to control this interface. Each manufacturer supplies the controller, and adequate software in order to make a precise set-up. Once the computer has recognized and accepted the communication interface through the controlling circuit, it is necessary to establish specific communication routines between the computer and each piece of measurement equipment. These routines constitute the third type of software present in the computer and are called *instrument control drivers*. Each piece of measurement equipment has its own control driver, although there is a command language called Standard Commands for Programmable Instruments (SCPI), which is admitted by a large number of manufacturers. This language permits easy and fast creation of control drivers using a common set of control commands. Furthermore, a *test and diagnostics program* must be considered. This is the program that controls the equipment connected to the computer making it possible for the equipment to produce and serve the desired measurements to the computer. The most extended types of test and diagnostics software are: the traditional programming languages (Pascal, C, etc.), the objects oriented programming languages (C++, Visual Basic, etc.) and the iconic programming language (LabView, VEE, etc.). It is recommended that a user with little or no background in programming techniques should use the iconic programming software. These programmes allow the user to develop control and measurements applications in a friendly and graphical way by manipulating icons (small pictures) without accessing any specific programming language. This type of software is adequate for the development of small and medium-sized projects, although they take up a large memory resource from the computer (disk space, memory, speed processor) [7,8]. Finally, the *application programs* are the highest level software used. They consist of a series of routines oriented to the instrumentation system management. Some of the functions assigned to this software are the user log-in control, test routines control, data-logging control, diagnostics control or report generation.

### 5.2.3 Virtual instruments

*The concept*

The development of computers and object oriented software has allowed the measurement procedures to be performed by the computer instead of being carried out by the equipment. The computer is then transformed into a place where complete access and total control of equipment capabilities take place. The computer screen is the equivalent to the physical front panel which is found by the user when working with the measurement equipment. Using this type of object-oriented software even makes it possible to perform some type of processing tasks over the measurements not initially available in the equipment. In general, a virtual instrument is a software routine running in a computer. It allows processing and graphical representation of the results acquired by communication interfaces flowing from measurement equipment. The virtual term is related to the capacity the user has to create the software routine with the controls and processing types which are of interest for its application. In this way, there is no longer physical equipment which meets these capabilities.

Iconic programming techniques have permitted virtual instruments to be developed easily, without excessive knowledge of textual programming languages (C, Visual Basic, Pascal, etc.). With this graphic programming it is possible to control not only the measurement capabilities but also the main sequence of the program and the flow of instructions. The general aspect produced by this programming technique shows a group of icons connected with arrows or cables with a logical/sequential connection. As for development time, the iconic programming technique implies a reduction of 1/10. The applications developed are modular and structured in such a way that a complete program can be organized as a group of icons where the other routines are more or less extended. This technique allows the development of more and more sophisticated applications.

*A virtual instrument case*

A good development example of virtual instrument is the ISFETmeter. This is an instrumentation system with the purpose of characterizing the ability of an ISFET sensor to measure pH in fluids [9]. ISFETs sensors characterization demands a variation of working conditions (voltage and current polarization) under which the sensor must work. To demonstrate the ISFET capability as a sensor device it must obtain its transconductance characteristic curve. In order to do this a bias current sweep must be done at a constant bias voltage. The instrumentation system is shown in Fig. 5.6 and is implemented by:

- A computer. This is the central component of the instrumentation system. It acquires the information of interest, and governs the

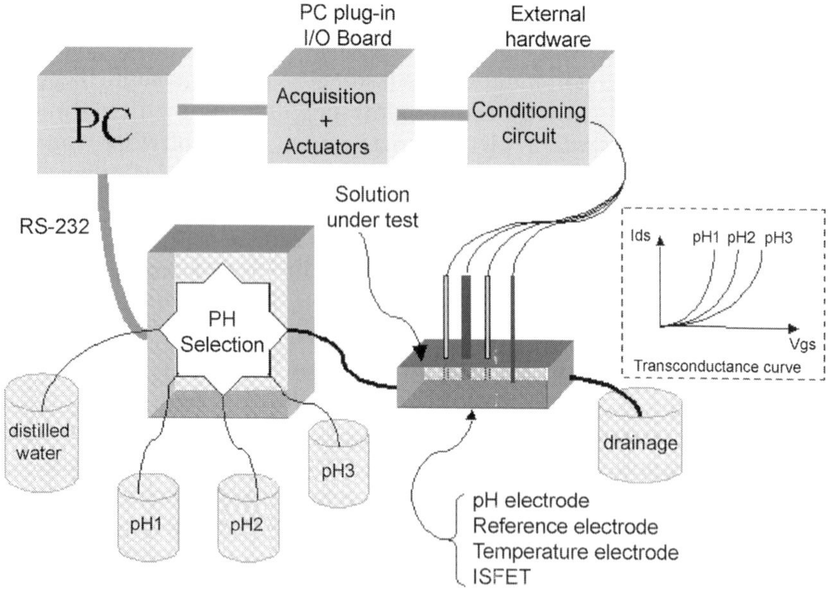


Fig. 5.6. A virtual instrument case [9].

controllable equipment through the digital communication interfaces and the acquisition cards. It also performs adequate digital processing.

- Sensors, which permit the system to obtain information on the environment where the ISFET is immersed (pH electrode and temperature sensor).
- An electronic circuit designed to condition the signals provided by this circuit, including the ISFET signals.
- The acquisition subsystem and actuators are implemented by an input/output board inserted in the computer slots. This card obtains information from the sensors through the conditioning circuits. Once it is processed by the software, it receives the appropriate instructions to act over the ISFET conditioning circuit.
- A special micro-syringe controlled by the RS-232 interface which selects and injects specific samples of selected pH concentrations. In this way, it is possible to carefully change the sample concentration of the solution where the sensor is located. At the same time, it is likely to know the answer and to perform its characterization.

This is the simplest system. The most complex part is found in the conditioning circuit design. However, other simpler instrumentation systems can be used although the global cost of the system is increased. In these simpler systems the conditioning circuit is replaced by commercial equipment (pH-meter, thermometer, source and measure units, ammeters, etc.). The only condition is that the instruments could be controlled either by the RS-232 interface or via the IEEE-488 bus in order to guarantee reliability and reproducibility of the characterization tests.

### 5.2.4 Smart/intelligent sensors

*Concept and architectures*

The newly-created intelligent sensors have permitted the design of measurement systems with high reliability and accuracy. An intelligent sensor is an electronic subsystem which can perform not only the sensing function but also calibration functions (compensation of zero, gain, non-linearity and drift errors), diagnosis (autotest), decision taking (digital processing), communication (presence of digital interfaces for internal configuration, data transmission, calibration, control, etc.), or even a combination of them [10,11]. The concept of the intelligent sensor is extended to a general sense when microcontrollers or digital signal processors are considered as a processing subsystem. Figure 5.7 shows the diagram block of an intelligent sensor with all the above-mentioned functions available. From a practical point of view, it is possible to arrange several of these blocks in the same physical package, so that several types of sensors levels are considered.

Type I: The simplest type of intelligent sensor is the one which includes its own sensor and the signal conditioning block. With this structure, the intelligent sensor is able to perform analog processing functions (amplification, gain selection, analog filtering, multiplexing, etc.).

Type II: In this case, the intelligent sensor is formed by the conditioning block and the analog-to-digital conversion stage. This level does not have great complexity, but it is a more flexible configuration as it allows an external sensor connection. On the other hand, as data are arranged in digital format, the interface with microcontrollers or digital signal processors is immediate [12].

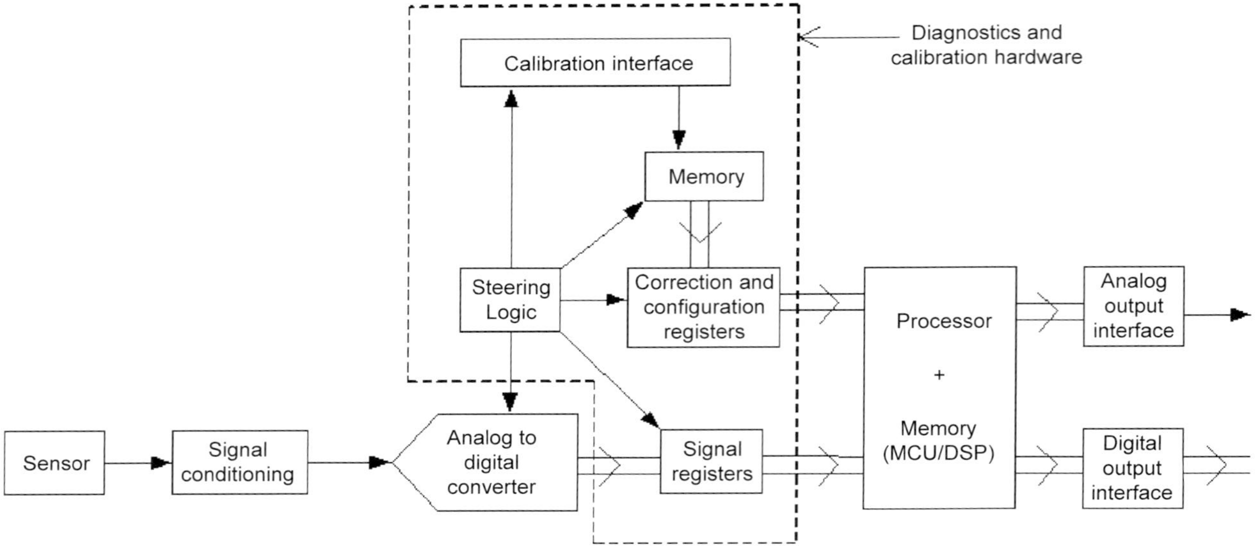


Fig. 5.7. General smart/intelligent sensor model.

Type III: In this case within the same package are integrated the analog-to-digital conversion block and the processing unit (microcontroller or digital signal processor). This last stage allows the sensor to perform calibration, self-test and digital processing functions. This is because it is now possible to implement software routines for these purposes using a specific programming language. The digital communication system is guaranteed since the microcontroller or digital signal processor incorporates some type of digital communication interface (output bits in serial or parallel format, RS-232, SPI or $I^2C$ interfaces, etc.). They also contain at least one analog communication interface (4–20 mA loop, digital-to-analog converter, pulse width modulated output pulses, among others).

Type IV: In this level the intelligent sensor is able to perform conditioning functions, self-test, calibration and digital signal processing. It is the most complete level as all the necessary hardware subsystems are contained in the same package, except the sensor itself. Moreover the signal conditioning, analog-to-digital conversion and processing subsystems, with other features like the presence of memories, storage registers and control logic, are incorporated to the hardware. This in fact allows the simplicity of the processor to be enhanced, and also the self-test and calibration routines are simplified.

In the same package there are also interface blocks used for self-calibration as well as for digital communication. In some cases, an analog communication interface is also incorporated [13].

Intelligent sensors are interesting from the point of view of instrumentation systems design because only an appropiate sensor selection is needed to measure the physical or chemical variable. After that, depending on the measured properties the intelligent system must be software and hardware configured. Once the configuration technique is known, the developing time measuring new variables of interest is notably short.

The presence of digital communication interfaces in a great number of intelligent sensors in the market makes possible the configuration of intelligent analytical systems at the level indicated in Fig. 3. In this case, access to the information given by the intelligent system is possible only through a measurement server. In order for an intelligent sensor to have its own capacity to supply information to other remote users through the internet, it is necessary to have a more complex communication interface. In this way, it is possible to establish advanced communication protocols. In the next section the basic principles of the IEEE-1451 standard are explained as a new method for designing intelligent sensors networks.

*Sensor networks: the standard IEEE 1451*

At present there are many ways to interconnect intelligent sensors, and to create device control networks or communication protocols between systems. In [12] are registered 31 different protocols to establish intelligent sensor networks, and in [14] there are 21 already registered. Each protocol performs its functions in a specific field of industry. However, the generalized use of intelligent sensors has been obstructed due to the nonexistence of a standard interface generally accepted by industry and research centres.

The objective of initiative IEEE-1451 is to propose a standard interface for designing distributed sensors networks [14,15]. It explains the connection interface between a sensor or actuator and microprocessors based systems, instrumentation systems and the so-called industrial networks. The aim of the standard is that the intelligent sensors follow the same connection procedures with data communication networks (transducers-to-network interchangeability). In addition, different data communication networks are expected to have a standard interface to be able to connect specific intelligent sensor (transducer-to-networks interoperability).

First, the IEEE-1451 standard determines the hardware interface connection from a sensor to a microprocessor based or instrumentation system and secondly, it establishes the software interface to satisfy the resulting intelligent sensor in order to be connected to the existing networks. At the same time, this norm is composed of four standards so that it can be correctly interpreted.

1. IEEE 1451.1—Network Capable Application Processor (NCAP) Information Model. (A published standard.) This standard deals with the software object model of the smart sensor and its connectivity to a network [16].
2. IEEE 1451.2—Transducer to Microprocessor Communications Protocols and Transducer Electronic Data Sheet (TEDS) Formats. (A published Standard.) In this standard a physical connection interface is determined between the sensor and the processing subsystem (microprocessor, instrumentation system) as well as the digital format of its characteristics [17,18].

3. IEEE 1451.3—Digital Communication and Transducer Electronic Data Sheet (TEDS) Formats for Distributed Multidrop Systems. (Draft specification being developed.) Its objective is to determine the communication protocol of several intelligent sensors from the same node with a digital-mode network.
4. IEEE 1451.4—Mixed-mode Communication Protocols and Transducer Electronic Data Sheet (TEDS) Formats. (Draft specification being developed.) Its objective is to determine the communication protocol of the intelligent sensor using a mixed-mode network (able to communicate in both analog and digital modes, but not simultaneously) [19].

In recent years, several reports have been issued in reference to the design of intelligent sensors connected to a network and following the IEEE-1451 standard. Conway et al. [20] show an intelligent sensor connected to an Ethernet and an application using bus CAN is developed Cámara et al. [21]. While all the time there are more and more reports related to networks design intelligent sensors the meeting IEEE-1451 standard, in the analytical chemical field there are no reports of measurement systems based on this norm.

## 5.3 THE LINK

In any given case of analytical instrumentation system there are different places (remote) where signals of interest (information) are produced, or where particular signals are sent (actuation). In addition, there could be a place (controller) where data are interpreted, actuation signals are generated or an analytical process is monitored. Distances between both places varies according to each case, from several metres to hundreds of kilometres, or even more. A physical support (link) which works as a communication route for data transmission in both directions is needed in all of them. Remote transmission of electric signals leads to the presence of other signals generated around the physical transmission support, which contributes to the degradation of the information being transmitted (noise pickup, signal conditioning complications, measurement speed restrictions, length of cabling, etc.). There are two techniques specially designed to preserve the integrity of the signal being transmitted between the remote place and the place where the signal is processed. The first one conditions the sensor signal in the same place where the sensor is located, and then transmits it to the remote location where it will be centrally processed. This method requires a point-to-point connection with the central processor. The second method shifts the acquisition functions, and if possible some processing functions, to each remote location where the data signals are produced. Transmission to a central controller, in this case, would be performed by a supporting device called a field bus in digital format. This second method completely minimizes the cabling design in the distributed system. In this way modularity, diagnostics and detection of failures, and system strength are enhanced. However, the ability to have a certain degree of intelligence on each measurement node will increase the cost of equipment at a primary level. Nevertheless, the initial investment is eventually justified by the many advantages obtained when using an intelligent system. Data of interest (performed measurements, control signals) codified in a digital format permit the distributed nodes to self-diagnose, to self-calibrate, to disconnect from the system when detecting malfunction, to transfer information to the central controller whenever it is required, to send or to receive messages, etc.

### 5.3.1 Industrial networks

The so-called field buses are the actual digital routes where signals between the remote locations and the processing or controlling nodes circulate. Industry in its different sectors (automotive, semiconductors, chemical, etc.) has designed specific field buses according to its needs. However, the necessary physical hardware to connect the digital bus, and the communication protocol to be known by the instrumentation and measurement equipments, as well as the necessary controllers (computers, microprocessors, etc.), have been specific to each bus. This fact makes it impossible to interchange equipment and controllers adapted to different field buses. In the research and development areas there are several field buses already designed [10], between them the most developed are the CAN bus, Profibus and Field Bus. A special mention must be made of Ethernet data transmission because considerable effort is being dedicated to standardize the interconnection networks between equipment by using Ethernet support [14,15,22].

*CAN Bus*: In 1980, Robert Bosch GmbH developed a bus called a Control Area Network (CAN) in order to control, in a more efficient way, the production process in an automobile industry, in particular the multiplexing cabling systems. Since then, the CAN Bus has been applied and extended to other industrial fields, including the chemical sector [23]. In this case, distance from the controller to the remote location can be between 30 m and 5 km, supporting up to 128 devices connected to a network. The CAN Bus accepts distributed real-time control with a high level of security and message integrity. It has also become attractive for use in lower speed and other distributed control applications [24]. There is a web contact with the international organization for the CAN Bus (www.can-cal.de).

*Profibus*: 'Process Field Bus', the leading open Field Bus system in Europe, was developed in Germany in 1989 and extensively used by Siemens. It was designed for factory automatization and for control processes. Up to 127 control nodes can be connected to a controller with a separation distance that can range from 100 (with cables and without repeaters) to 24 km (with optic fibre transmission and repeaters). Designed to meet a variety of application requirements, Profibus could be used for both high-speed, time critical-data transmission between controllers and input/output devices and complex communication between programmable controllers. There is a committee maintaining and updating everything related to this field bus (www.profibus.com).

*Foundation Field Bus*: This is an industrial network designed specifically for distributed process control applications. It was designed originally in 1998 by the International Society for Measurement and Control (ISA). The field bus supports up to 240 nodes per segment and up to 65,000 possible segments with a maximum distance of 1900 m.

### 5.3.2 Ethernet

The ethernet provides the possibility of implementing a fast, simple and easy network at a low cost. In fact, it is becoming a standard network. With this type of network it is possible to link equipment from 100 m away from the original connection up to 50 km away by using fibre optics, with speed ranging from 10 Mbps to 1 Gbps. The ethernet is the 'de facto' physical support for the internet, through which a great number of personal computers are able to access data coming from remote devices. Internet access through the ethernet has allowed a remote connection to equipment, instruments and measurement systems, as well as development of new applications of instrumentation and measurement. On the other hand, the protocol associated to the ethernet is the Transaction Control Protocol/Internet Protocol (TCP/IP), which is becoming a standard transmission protocol. Operating systems like Windows and Unix already support this protocol, so that even the lowest experts in data transmission networks can design their own network in a minimum period of time. The initial disadvantages related to its non-determinist feature and its low robustness to adopt the ethernet as an industrial field bus have been recently overcome by designing the so-called Switched Ethernet or Ethernet 2000 [19,22,25, 26]. On the other hand, the most developed industrial buses have already included TCP/IP protocol within their specifications in order to adapt their features to the ethernet [22]. At the hardware level, convertibility between field networks and the ethernet is performed by the so-called gateways. These devices are network nodes equipped to communicate with another network which uses different protocols, although these devices are expensive. The IEEE-1451 standard has taken the ethernet as a field network reference. From this point of view, different standard connections between intelligent sensors and instruments are developed [15]. However, there are two aspects that the ethernet has not solved yet. The first aspect refers to the impossibility transmitting remote energy and the second aspect, in a more circumstantial way, makes reference to the nonexistence of a great amount of equipment and instrumentation with ethernet interfaces incorporated to their framework. This fact differs with the easy incorporation of an ethernet interface in systems equipped with intelligence (personal computers, microcontrollers, controllers, etc.). Most of the manufacturers of measurement and analysis equipment incorporate RS-232 communication interfaces or their derivatives, or IEEE-488. In order to facilitate connectivity between this equipment and the ethernet network, many manufacturers supply the so-called ethernet adapters, which allow an address accessible via the ethernet to be assigned to programmable instruments. In this way, they are able to communicate through the ethernet network [4,27–29]. With all these possibilities with the ethernet it is possible to put data acquisition and input/output devices into the internet in a fast, reliable and simple manner with the possibility of having data available anywhere, anytime. Figure 5.8 shows a topology of a network in which there are different possibilities for using networks and interfaces to configure a distributed system.

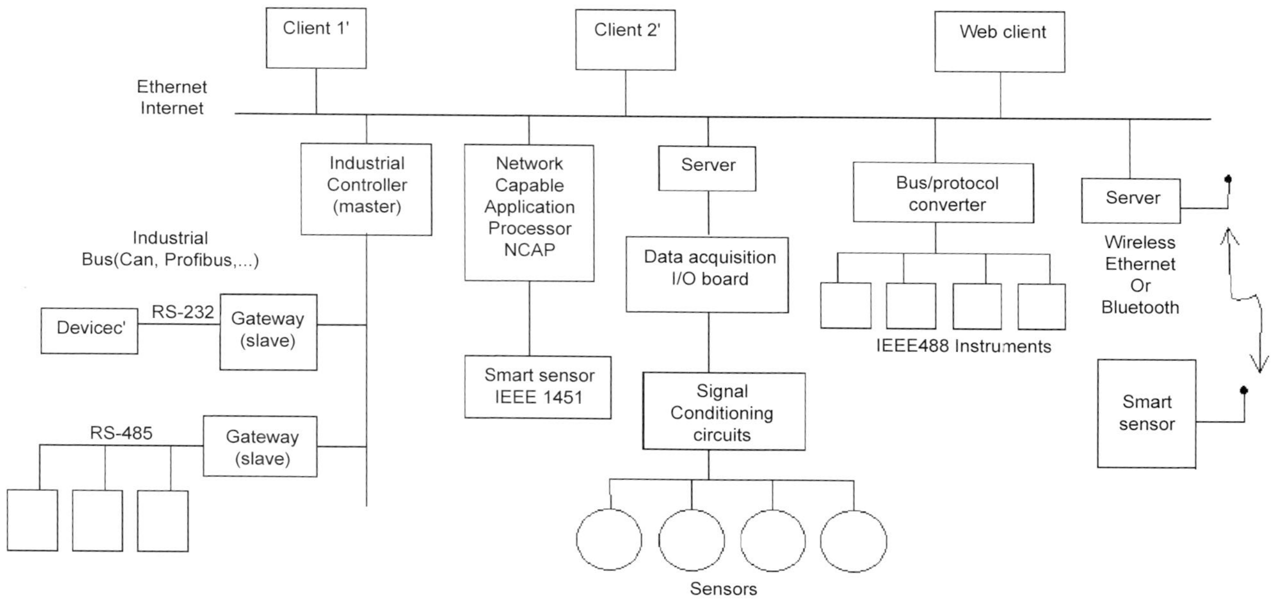


Fig. 5.8. Networks and interfaces to configure a distributed measurement system.

### 5.3.3 Wireless links

The advances in mobile communications technologies and the development of portable computers during the past years have made possible the design of measurement systems based on sensors and instruments with wireless transmission interfaces. Controlling these processes, measuring and sensing in dangerous environments, or designing acquisition/portable systems, have been, amongst others, the most immediate applications of this type of communication transmission.

Wireless information technology uses two types of signal to perform transmission. The transmission technique initially used is based on the so-called narrow-band signals. At the present time, the wide-band signals have generated the most recent transmission technologies. There is a potential application of this technology for measurement systems design, which makes it more interesting. Narrow-band technologies limit the emitted signals to a reduced interval (several tens of kHz), require a license in order to be used and have a limited bandwidth (9600 bits per second—bps-typical). Transmission based on wide-band signals, on the contrary, uses a wide interval of frequency (several tens of MHz), does not need an official license and allows a higher speed of data transmission. Specifically, a modem based on wide band with an RS-232 interface can transmit at a speed of 115 kbps and at a distance of up to 30 km. On the other hand, narrow-band signals are used for transmission over long distances by manipulating high levels of radiation, while wide-band signals work at low levels of radiation (1 W; in Europe 0.1 W) and are used for transmission at medium-range distances. The purpose of this section is to present some of the wireless transmission systems which have the most recent application for instrumentation and measurement based on wide-band technologies, particularly the wireless ethernet and Bluetooth.

*IEEE 802.11 wireless Ethernet*

This is based on the IEEE-802.11 standard for the design of an ethernet wireless local area network (ethernet wireless WLAN). When using wide-band signals (83.5 MHz), it enjoys the benefits of this technology, particularly with an ISM band (industrial-scientific-medical) of 2.4 GHz. At present, it has the capacity to transmit information at a speed of 1 or 2 Mbps, although standard 802.1 lb, for 11 Mbp, is about to be completed and in the future it is expected to reach 54 Mbps with standard 802.11a. The basic architecture of a WLAN that meets standard 802.11 consists of a group of 'stations' formed by desktop computers or notebooks and an access point (AP). Standard 802.11 allows two configuration modes: an infrastructure mode, in which the AP is connected either to a wired LAN network allowing the stations to have access to wired LAN services, or acts as the central transmitter of information between stations and the independent basic service. In this second case, the WLAN network is formed by stations without an AP [30]. At present, there are manufacturers with products (plug-in computer boards) to configure a the WLAN network according to the IEEE 802.11 standard [31]. Nevertheless, most of the products and applications are oriented towards the designing of WLAN networks for office and domestic use (small-office/home-office, or SOHO, environments). Developed applications forming WLAN networks with standard 802.11 are recent in instrumentation and measurement systems. In [32] it is possible to find practical cases related to vibration or temperature measurement, meteorology monitoring, etc. It also offers specific products that can be used with standard hardware and software and are able to configure their own WLAN with a length of approximately 300 m, offering additional possibilities to configure wireless wide area networks (WWAN). Designing a WLAN network applied to analytical chemistry would not create further problems, only the ones related to the choice of measurement equipment and chemical analysis that are needed with the appropriate interface to connect them to the computer. Acquisition and processing hardware and software are completely transparent to the IEEE 802.11 wireless link.

*Bluetooth*

Bluetooth is an open system of wireless technology for low cost and short range radio links (from 1 to 100 m) which, by transmitting signals in an ISM band of 2.4 GHz, has been designed for communication between computers, mobile telephones and other portable devices. In an asymmetric transmission mode, and for a range of 10 m, Bluetooth sends data at a speed of 721 kbps and receives them at a speed of 57.6 kbps. In symmetric transmission mode, speed reaches 432.6 kbps. It is possible to configure networks of up to 200 Bluetooth devices, although in typical applications for which it is designed (domestic or personal network designs), no more than seven devices can be connected (desktop computer, PDA, printer, video equipment, DVD, telephone, etc.). Architectural designs of networks that can be generated are organized in 'piconets', in which the nodes are distributed in conventional way, and 'scatternets', in which a node can act as a master for one piconet and at the same time can be the slave of another piconet [33].

Comparatively, Bluetooth has a transmission speed lower than IEEE-802.11 technology, although the cost of implementing a WLAN with Bluetooth is lower. Both technologies are susceptible to interference with one another, because they are operating in an ISM band of 2.4 GHz. This interference can also be with other devices which use the same band (microwave ovens).

In data acquisition/distribution, Bluetooth technology has just started its application to the design of instrumentation and measurement systems. Dunbar [34] describes a node to which four sensing links can be connected. The node has the capacity to transmit information or actuation commands through a Bluetooth link, which has an initial range of 10 or 100 m, by using an optional amplifier. The Bluetooth link allows the node to communicate with other receiver units located in desktop, laptop or handheld computers and to transfer specific technical data of a sensor configuration according to standard IEEE 1451. Information received by the computer can be processed and analysed through Window-based data-acquisition and control software packages, such as LabView from National Instruments or Excel from Microsoft. Once this is done, the information obtained is distributed through a wide area network using a TCP/IP protocol and the internet [35]. Like the IEEE-802.11 standard, Bluetooth technology offers a new and innovative possibility of designing systems of analytical distributed instrumentation based on wireless data transmission networks. In any case, both technologies offer the designer the possibility of reducing installation costs, and easily reconfiguring data acquisition systems, as well as internet connectivity for measuring and control data.

## 5.4 THE LOCAL PLACE

In a distributed instrumentation system, the location known as the working place in which measured information is obtained or appropriate control commands are generated must have, as an essential requirement, both an adequate software and hardware interface in order to fully support the system requirements being supervised.

*Hardware*: The most important aspect is the connectivity capabilities that the processor or computer have to meet. Depending on the instrumentation and measurement system topology (Section 5.1.2), the local processing unit must be provided with digital interfaces for acquisition and local control (RS-232 interfaces, IEEE-488, SPI, $I^2C$, etc.) or with analog interfaces (analog acquisition cards, 4–20 mA loops). If the position is remote, then the interfaces to be supported must be of the ethernet type for applications in wide area networks or field buses interfaces (CAN, Field bus, Profibus) in industrial areas. If a wireless local area network (WLAN) is needed, the local processing system will control an adequate interface with this technology (transceiver module and an antenna). Each of the hardware interfaces can be associated with a software package of higher or lower complexity so that the system and the measurement, control and acquisition software have adequate access to the communication hardware.

*Software*: Software for acquisition, measurement and control must be able to work in hostile industrial areas, should be portable and have few requirements to be supported by the system. Additionally, remote control features and non-critical starting functions must be satisfied.

A critical aspect when choosing a software is to select the operating system on which the application is going to be performed. At present, although there are other operating systems other than Windows, this is the system which includes most applications for acquisition and measurement. Moreover, this operating system is based on icons with identification, characteristics and functions that can be executed graphically. This fact makes it possible for the user to be able to develop applications without having an in depth knowledge of programming languages. The DOS operating system is less in use now and not recommended for developing new control and measurement applications.

Based on the Windows system, the designer of remote instrumentation and measurement system gives two choices for the selection of adequate software: either a closed or an open programming environment [36]. The open programming environment consists of a previously known programming language on which it will be possible to develop a concrete application, by applying specific programming techniques. The most representative examples of this system are the visual languages C/C++, C++ Builder and Visual Basic. These open types of software allow:

- Build applications with complete access to programming code. In this way it is possible to freely distribute or modify them to design new applications, thus reducing development time.
- Development of software dedicated exclusively to the application of particular requirements. With this feature, the processing system must support only the necessary routines and resources, so that memory and execution time are optimised.
- Many hardware manufacturers supply tools to allow easy control of equipment through an open system.
- Open programming systems generate text-based applications. This aspect would not be the most attractive to a designer with little or no experience on programming language.
- As regards software efficiency, open applications use only the necessary resources and are autonomous in respect to the central processor.

The second option for designing measurement and control software is to use closed programming environments. These are software packages already prepared with industrial characteristics, which the user can obtain and use for designing the desired application without having advanced knowledge of programming language. Such an application is possible through the development of a graphical interface. The most well-developed representatives of graphical programming environments are LabView™, from National Instruments, Agilent VEE®, from Agilent Technologies, and Test Point™, from Keythley [37].

The closed programming techniques allow the following.

- Build applications on a graphical and visual environment. This feature makes these type of system the favourite for the non-expert textual programming designers.
- Software manufacturers now offer the software driver to control, from the graphical interface, most of the measurement equipment and instruments in the market.
- As the proprietary software is designed for a large number of potential users, it must support a great number of functions and because of this uses excessive memory at the time to run. Therefore, execution time is then not as efficient as the applications derived from open systems.

- Closed systems allow the user to develop software applications oriented to instruments/equipment control and to carry out remote data acquisition more easily than the opened systems.

The designer of distributed analytical instrumentation should value the programming environment according to its own complexity and the available developing time. In addition, the designer should consider whether it is possible to use a textual or graphical programming knowledge background.

## 5.5 REMOTE ANALYTICAL INSTRUMENTS/SYSTEMS: APPLICATION EXAMPLES

The concepts related to instrumentation systems, virtual instrumentation and remote measurement and control systems have been applied to the analytical chemistry field in many different and complex cases. Applications have been developed from the simplest ones, like the close control of one or two pieces of equipment, to the highest levels of complexity oriented to acquisition and remote control of equipment located in industrial production plants. Remote analytical instruments are present in environments that include academic, pharmaceutical, industrial and environmental. In particular are the following.

- Smith [38] describes remote control of a pharmaceutical production plant through the LabView graphical programming software and a commercial distributed control system (proprietary). In order to link the equipment and the control centre, a protocol of communication Dynamic Data Exchange (DDE) has been used, under the Windows system.
- Conquergood [39] presents a system in which an oil production process is controlled in a remote way. Signals coming from the process are obtained through controlled measurement equipment using IEEE-488 and RS-232 interfaces. Additionally, certain variables are monitored through a data acquisition card. The information obtained is transmitted through fibre optic in order to have enough galvanic isolation. Communication between hardware equipment and computers is performed through the standard Ethernet under TCP/IP protocol.
- Reports by Bonastre et al. [23] present an analytical distributed system focussed on monitoring and control of an alcoholic

fermentation process. A chemical analysis of variables is performed, such as the presence of ethanol, dry extract, total acidity, carbon dioxide, sulfur dioxide, or sugars, and a measurement of physical–chemical variables like temperature and pH. For this purpose, a set of measurement and chemical analysis equipment is prepared, particularly electropneumatic injection valves, deviation valves, pH readers and an UV–vis spectrophotometer. This equipment physically connected to a CAN field bus controller allow information interchange in two directions: from the supervising node to the equipment and vice versa. The designed distributed system is running in a brewery manufacturing plant.

- Doherty et al. [40] present a networked measurement system for specific chromatography applications. This system controls analytical equipment such as chromatographs, spectrometers and other required analytical equipment. Using the Windows NT operating system it is possible to build internal measurement networks (intranets) or to have remote access to them through the internet. The system has the capacity to obtain analog signals, process them, generate reports, manage data and analytical samples and store the information on a standard database. A networked data system for a pharmaceutical plant is presented.
- The work of Torán et al. [41] shows the design of a distributed and multiplatform system for water quality monitoring, enhanced with internet capabilities. The measured variables are temperature, turbidity, pH, dissolved oxygen and electrical conductivity. The software designed permits various features like e-mail alarm notifications, automatic storage of measured data in a remote machine via the FTP protocol, dynamic generation of HTML reports and real time graphs and indicators visible from a remote web browser. The whole system supports several types of computers, operating systems and communication hardware.
- Walters [42] shows instrumental analysis teaching using educational techniques in small groups. To achieve this goal, LabView software is used as a graphical programming tool. The RS-232 interface is used for controlling the equipment and obtaining information coming from it. Other data of interest are obtained through a data acquisition board. With these elements it was possible to design a potentiostat, to develop an instrumentation system for the characterization of a spectrometer of atomic adsorption and to simulate an analytical control of lead production.

- Haight and Schwartz [43] present an instrumentation system designed to monitor based line-imaging Raman spectroscopy electrode-position processes. Manufacturing of composition modulated alloys applied to the design of hard-disks heads or material for solar cells fabrication are some examples. The designed spectro-electrochemical instrument uses LabView as graphical software tool. The software developed controls an imaging spectrograph, acquires data coming from a CCD array detector and controls the acquisition signals coming from a potentiostat/galvanostat system. The interface used was IEEE-488. The designed systems were applied to monitor the CuSCN deposition in an electrode surface and allowed to have a modular and flexible instrument which could adapt to particular measurement needs.
- Collins [44] presents a report in which a virtual instrument is developed to control a spectrophotometer of 64 channels—which is applied to endotoxine analysis—by using an RS-232 interface. The instrumentation system is applied to chemical and pharmaceutical industries. The virtual instrument has been designed using LabView graphical programming software.

Table 5.1 shows some of the manufacturers of analytical instrumentation equipment with communication interfaces prepared to be controlled by a processing unit. A large majority of them have RS-232 interfaces, while only some incorporate a more complete interface, such as the IEEE-488 interface.

## 5.6 CONCLUSION

Computer and electronic technologies joined to digital communications progress have made possible the development of measurement systems able to control and monitor chemical processes of interest. In an industrial environment, setting up instrumentation networks is related to field buses, while in a research and scientific field, instrumentation buses are the most suitable. The internet, through protocol TCP/IP and wireless standards IEEE 802.11 and Bluetooth, make the design of short range wireless instrument networks possible. The IEEE-1451 standard develops the elements to satisfy an intelligent sensor forming a sensor network.

TABLE 5.1

Communication interface capabilities of some analytical equipment

| Manufacturer | Contact | Product | Communication Interface |
|---|---|---|---|
| XYZ Scientific | www.xyz-scientific.com | Peristaltic pump | RS232; RS485 |
| Watson Marlow Ltd | www.watson-marlow.com | Peristaltic pump | RS232 |
| Desing Environment Ltd | www.designenvironmental.co.uk | Climatic chambers | RS233;RS422;IEEE488 |
| Memmert Gmbh + Co. KG | www.memmert.com | Incubators | RS232 |
| | | Heaters | RS232; RS485 |
| wtb Binder Labortechnik Gmbh | www.wtbbinder.com | Climatic chambers | RS232 |
| Riedel-de Haën | www.riedeldehaen.com | Photometers | RS232 |
| Naber | | Ovens | RS485 |
| Büchi Labortechnik AG | www.buchi.com | Evaporation | RS232 |
| Cecil Instruments Ltd. | www.cecil-instrument.co.uk | Spectrophotometers | RS232 |
| Anton Para | www.anton-paar.com | Density meters | RS232 |
| Froilabo | www.firfaxsystems.co.uk | Temperature forcing system | RS232;IEEE488 |
| Dr. Lange | www.drlange.com | Thermostats | RS232 |
| Polyscience | www.polyscience | Heating/refrigerated circulators | RS232 |
| A&D Instruments Ltd. | www.andinstruments.co.uk | Balances | RS232 |
| US Filter | www.usfilterenvirex.com | Water treatment | RS232 |
| Crison Instruments S.A. | crison@mail.cinet.es | pH-meters conductivity | RS232 |
| | | | RS232 |
| Siemens | www.Siemens.com | Gas/liquid analysis | Propietary software |
| Spectro Analytical Australia | www.Spectro.com | Spectrometer | Propietary software |
| Jasco | www.Jascoinc.com | Chromatography spectroscopy | Propietary software |
| Dionex | www.dionex.com | Shromatography spectroscopy | Propietary software |
| Perkin-Elmer Instruments | www.PerkinElmer.com | Gas/liquid chromatography spectrometer | Propietary software |
| Beckman Coulter | www.Beckman.com | Gas/liquid chromatography pH-meters spectroscopy | Propietary software |

The above-mentioned technologies are accessible to the analytical chemistry R&D community, they just need to be accessed for the development of new applications.

## REFERENCES


1 P.H. Sydenham, (Ed.), *Handbook of Measurement Science, Theoretical Fundamentals*, vol. 1, Wiley, Chichester, 1996.
2 P.H. Sydenham and R. Thorn (Eds.), *Handbook of Measurement Science*, Elements of Change, vol. 3, Wiley, Chichester, 1992.
3 S. Franco, *Design with Operational Amplifiers and Analog Integrated Circuits*, McGraw-Hill, New York, 1988.
4 www.iotech.com/scanware_3.
5 National Instruments, *Building Intelligent Ethernet-Based Distributed I/O Systems with National Instruments LabView*, Application Note (www.ni.com).
6 C.F. Coombs, *Electronic Instrument Handbook*, 2nd ed., McGraw-Hill, New York, 1995.
7 www.ni.com.
8 www.get.agilent.com/gpinstruments/products/vee.
9 S. Casans, A.E. Navarro, D. Ramírez, J. Pelegrí, A. Baldi and N. Abramova, *Sens. Actuat. B*, 76 (2001) 629.
10 B. Travis, Smart Sensors, *EDN*, 9 (1996) 57–65.
11 R. Pallás and J.G. Webster, *Sensors & Signal Conditioning*, Wiley, New York.
12 R. Frank, *Understanding Smart-sensors*, Artech House Pub, Boston, 1996.
13 A. Makdessian and M. Parsons, DSSP-Based Pressure Sensors, *Sensors*, (2001) 38–46.
14 K. Lee, Sensor Networking and Interface Standardization, *IEEE Instrumentation and Measurement Technology Conference*, Budapest, 2001, pp. 147–152.
15 K. Lee and R.D. Schneeman, *IEEE Trans. Instrum. Measur.*, 49 (2000) 621.
16 J. Warrior, IEEE-P1451 Network Capable Application Processor Information Model, *Proceedings of Sensors Expo*, Helmers Publishing, Anaheim, April 1996, pp. 15–21.
17 R.N. Johnson, IEEE-1451.2 Update, *Sensors*, January (2000) 17–27.
18 R.N. Johnson and S.P. Woods, Proposed Enhancement to the IEEE 1451.2, *Sensors*, September (2001) 74–84.
19 T.R. Licht, The IEEE 1451.4 Proposed Standard, *IEEE Instrum. Meas. Magazine*, March (2001) 12–18.

20 P. Conway, D. Heffernan, B. O'Mara, P. Burton, T. Miao, IEEE 1451.2: An Interpretation and Example Implementation, *Proc. of the Instrumentation and Measurement Technology Conference IMTC'2000*, Baltimore, May 2000, pp. 535–540.
21 L. Cámara, O. Ruiz, J. Samitier, Complete IEEE 1451 Node, STIM and NCAP, implemented for a CAN Network, *Proc. of the Instrumentation and Measurement Technology Conference IMTC'2000*, Baltimore, May 2000, pp. 541–545.
22 P. Sink, A Comprehensive Guide to Industrial Networks, *Sensors*, June (2001) 28–43.
23 A. Bonastre, R. Orts and M. Peris, *Trends Anal. Chem.*, 20 (2001) 263.
24 B. Negley, Getting control Through CAN, *Sensors*, October (2000) 18–34.
25 M. Babb, Ethernet moves to the factory floor, *Global Design News*, February (2000) 19–20.
26 P. Arpaia, F. Baccigalupi, F. Cennamo, P. Daponte, Ethernet application perspectives for distributed measurement systems, *Proceedings of XIV IMEKO World Congress*, Tampere, Finland, June, 1997, pp. 13–18.
27 J. Humphrey, V. Malhotra, V. Trujillo, Developing Distributed GPIB Test Systems Using GPIB-ENET/100 and Existing Ethernet Networks, *National Instruments*, Application Note 103, June 2000. www.ni.com.
28 *National Instruments*, Application Note 180, October 2001. www.ni.com.
29 M. Ghercioiu and T. Howard, How to Network To PCs for Remote Data Acquisition, *National Instruments* Application Note 118, April 1998. www.ni.com.
30 G. Vrana, Wireless Ethernet: Serving the Public, *EDN*, May 24, 2001, pp. 37–49, www.ednmag.com.
31 www.wavelan.com, www.intersil.com/design/prism, www.semiconductors.philips.com/wicon/80211b, www.wlana.org.
32 www.viasat.com/products/wireless/minidat.htm.
33 D. Strassberg, On the Cusp, *EDN*, 7 (2001) 59–68.
34 M. Dunbar, Plug-and-play sensors in wireless networks, *IEEE Instrum. Meas. Magazine*, March (2001) 19–23.
35 Crossbow Technology: www.xbow.com or www.crossnetwireless.com.
36 www.ni.com, www.agilent.com, www.keithley.com.
37 Keithley, *Data Acquisition and Control Handbook*, Keithley Instruments, Inc. Cleveland, Ohio, 2001.
38 A.C. Smith, Remote Monitoring of Pharmaceutical Plant Using LabView, DDE, www.ni.com.
39 S. Conquergood: High-Speed Remote Process Control with LabView, http://digital.ni.com/csma.nsf/IntAllCSDocs/667456484BC845AE8625692D00789BFD?OpenDocument&node = 124400_US.
40 L. Doherty, J. Welsh and W. Winter, *Am. Lab.*, 32 (2000) 50.

41 F. Torán, D. Ramírez, S. Casans, A.E. Navarro, J. Pelegrí, Distributed Virtual Instrument for Water Quality Monitoring Across the Internet, *Proc. of the IEEE Instrumentation and Measurement Technology Conference*, Baltimore, USA, May, 2000.
42 J.P. Walters, Learning Analytical Chemistry Using LabView and the Macintosh, http://digital.ni.com/csma.nsf/IntAllCSDocs/2FDA786120BCB3048625692D00789D42?OpenDocument&node = 124300_US.
43 S.M. Haight, T.S. Schwartz: Design of an Imaging Raman Spectroelectrochemistry System Using LabView, http://digital.ni.com/csma.nsf/IntAllCSDocs/25B52369EE5305618625692D00789C5F?OpenDocument&node = 124300_US.
44 M.P. Collins, LabView Interface Tests Pharmaceutical Products, http://digital.ni.com/csma.nsf/IntAllCSDocs/E9C7D199D5F56DD08625692D00789A9D?OpenDocument&node = 124200_US.

*Chapter 6*

# *Laboratory information management systems*

***R.D. McDowall****

## 6.1 THE ANALYTICAL LABORATORY

### 6.1.1 Role of an analytical laboratory

The role of an analytical laboratory within most organisations is to provide information upon which decisions can be made. A laboratory can be responsible for:

- process analysis of intermediates and bulk products
- quality control of raw materials and/or finished products and
- research and development.

The laboratory can be located within an industrial, medical or academic environment, but regardless of the location, it still has the same overall responsibility. Therefore, as well as the need for information, it is crucial that the information must be produced in a timely manner with appropriate quality to allow effective decision-making.

While samples enter a laboratory where they are analysed, it is rare for the information to be used there; the decision makers usually reside outside of the laboratory. Therefore, a laboratory should integrate with its client departments to become an efficient part of an organisation or

* Tel.: +44-2083-13-0934. *E-mail address:* r_d_mcdowall@compuserve.com

*Comprehensive Analytical Chemistry XXXIX, pages 287–325* 

ISSN: 0166-526X

enterprise. One element of this integration is to take the laboratory to the sample or process rather than the sample to the laboratory. When taking the laboratory to sample, the analytical chemist can use conventional equipment or use some of the processes and techniques outlined in this book such as electronic noses and tongues, membranes, sensors of all descriptions and continuous flow systems.

Regardless of the analytical system used, all it will do is generate data. Other software applications will be needed to reduce the data to information and then to disseminate it to enable decisions to be taken.

### 6.1.2 Need to increase productivity

Many laboratories are situated in industries that operate under regulatory guidelines, such as Good Manufacturing Practice in the pharmaceutical industry, or voluntary quality schemes such as ISO 17025. This can result in a growing number of samples for analysis with stricter control of administrative procedures as a means of ensuring quality. Other laboratories operate in companies and industries where there is a need to contain costs whilst maintaining analytical information to make decisions. To overcome these issues, laboratories need to increase productivity as a means of curtailing rising staff numbers and/or labour costs.

If productivity increases are sought, then there are two possible approaches. The first approach is to change procedures so that some tasks are simplified or even eliminated; as such this may not involve the use of automation. However, the viability of this approach is subject to question unless labour costs are reduced dramatically. The second approach is to use laboratory automation.

### 6.1.3 The aims of laboratory automation

The use of automation within an organisation is an opportunity to create strategic advantage [1], which is manifested in one or more of the following factors:

- Better decisions based on valid data. This can lead to research that is more effective, to faster time to market for new products or to better

products that due to their quality can increase customer satisfaction and market share.
- Improved organisational productivity.
- Reduced risk to employees.

Bringing these aims down to an analytical laboratory level, the automation of a task should produce one or more of the following advantages that will contribute to the aims above:

- Greater increases in productivity (either in numbers of samples assayed per unit time or speedier turnaround time). This results in overall cost reduction and increased productivity.
- Automated systems that should be capable of the same or better precision and accuracy as existing manual methods with improved quality and reliability of measurement.
- The freeing up of trained laboratory staff to do more creative or productive work than tedious tasks, thus improving morale.
- Reduced human contact with biological or chemical hazards.
- Lower consumption of sample and/or reagents used in an automated analysis.

### 6.1.4 Problems with laboratory automation

How well have laboratories fared in implementing systems in practice? Not very well, as from a historical perspective, Betteridge [1] has stated:

> "the development of automation in analytical chemistry can be viewed as a series of steps in which the solution of today's bottleneck creates another problem for the future to solve".

This was shown succinctly by the development of automation in analytical chemistry [1].

- Before the 1960s, the barrier to automation was actually generating the data; the solution was the development of electronic control of apparatus. When apparatus was controlled, the large amounts of chart paper produced became the new problem.
- Therefore, in the 1960s the problem became data acquisition; this was solved by digitisers, which acquired the data but produced a new problem of manipulation.

- This was solved in the 1970s with the introduction of minicomputers to reduce the data; but many computers created the problem of data transfer.
- Data management was solved in the 1980s by the use of workstations and laboratory information management systems (LIMS) but the problem changed to become the management of technical data.
- The interpretation of these data either by expert systems or artificial intelligence is the current problem for the 1990s to solve [1].

However this article was written in 1987 and was looking forward to the 1990s. The question for the first decade of the 21st Century is integration of analytical data. We have systems that can generate mountains of data and reduce it to information, but when trying to collate information between systems paper and manual data entry rule supreme. The laboratory that has integrated all its systems together is very rare.

Part of the historical problem was the lack of technology or instrumentation at the various times discussed to allow an all-encompassing solution. The other part of the problem was, and still is, the lack of vision of many chemists to see a laboratory as an integral part of an organisation.

### 6.1.5 Solutions for laboratory automation

In the past we have automated single applications and this creates islands of automation; the business needs of an organisation are rarely considered. We need to ensure that business objectives are actively considered and we automate the whole analytical process with the result that an automated laboratory is integrated into its parent organisation.

Thus, an understanding of laboratory automation must include changing the culture of analytical chemists from a 'purchase off-the-shelf and open the box' one to a strategically thinking and business oriented one. Therefore, to progress, laboratory automation should be reassessed based on the following considerations:

- development of a strategic laboratory automation plan driven by business needs
- justification of individual projects from a business perspective and
- prioritisation of individual projects according to potential payback and business and scientific justification.

However, before this can begin, laboratory automation must be understood more fully. What are the constituents of laboratory automation?

## 6.2 WHAT IS LABORATORY AUTOMATION?

### 6.2.1 A definition of laboratory automation

A practical and encompassing definition of laboratory automation has been presented by McDowall [6] as:

> "Apparatus, instrumentation, communications or computer applications designed to mechanise or automate the whole or specific portions of the analytical process in order for a laboratory to provide timely and quality information to an organisation".

This definition combined portions of earlier literature and the IUPAC definitions but presented with the aim of automating the whole of the analytical process.

For full laboratory automation, automated systems should be implemented whenever feasible and cost-effective. Automated systems have feedback mechanisms to ensure that problems can be investigated or the progress of an analysis monitored, in contrast to mechanised instruments or equipment that tend to operate in a blind manner.

Note the inclusion of the phrase 'in order for a laboratory to provide timely and quality information to an organisation': this is intended to show that laboratory automation is not an end in itself but merely a means to an end.

The objective of this definition was to alight laboratory automation with the business aims and role of the laboratory mentioned earlier: the provision of quality and timely information to an organisation.

### 6.2.2 Laboratory automation constituent groups

Using the definition above four groups were defined for laboratory automation [2], these are:

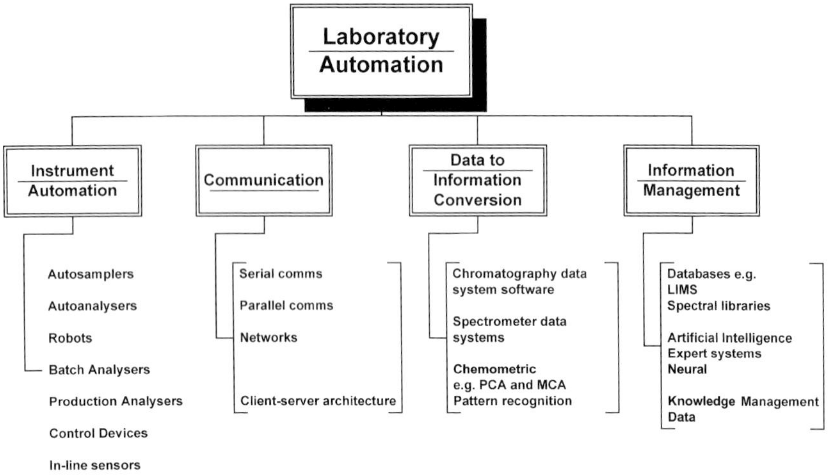


Fig. 6.1. Laboratory automation elements and constituent groups.

- instrument automation
- communications
- data to information conversion and
- information management.

The overall scheme of laboratory automation is shown in Fig. 6.1. The purpose and examples of each group of automation is outlined below; the manner by which each group interacts with the others in the form of a strategy is discussed in Section 6.3.

This discussion is important to understand as it underpins the whole discussion of an integrated approach to laboratory automation. It links the analytical systems that generate the data and the computerised systems such as LIMS that disseminate the information to the customers of the laboratory.

### 6.2.3 Instrument automation

The purpose of the equipment in this group is to speed up tasks such as sample preparation and analysis of samples. It contains most automatic or mechanical devices or instruments that are usually thought of as laboratory automation, typical examples are:

- autosamplers
- autoanalysers
- robotic systems
- continuous flow analysers
- batch analysers
- electronic noses and tongues and
- sensors,

together with the embedded chips or computer systems to control these various instruments.

Instrument automation is very task oriented and has a relatively rapid payback when implemented correctly. More importantly, it is the most easily understood of the four laboratory automation groups. However, it is regrettable that the view of laboratory automation by many chemists starts and finishes here.

Normally, the constraints on automation within a laboratory begin with the purchase of the first item of analytical instrumentation. Therefore, every item of analytical equipment purchased should be bought with automation, and communication with other equipment and computers, in mind. A further consideration may be the standardisation on a single instrument model for a particular task, e.g. balance, chromatographic detector or data capture system. This approach has much to recommend it, including the advantages of easier training and lower interfacing costs, but may have the disadvantage of less flexibility to undertake other tasks and eventually improvements in technology will make the chosen model obsolete.

It must always be possible to identify the sample so that the analytical result can be unequivocally linked with the original sample when the final report is made. Positive sample identification is the way to achieve this requirement if the reading of the identifying information is implicitly linked with the specimen during the act of sampling. Attaching machine-readable identification, e.g. a barcode, directly to the specimen in a position such that it can be read during the sampling process does this. The main advantage of positive sample identification is that a successful sampling sequence and frequency need not be assumed. Indeed, it is not even necessary for the operator to put the specimens onto the sampler in a particular sequence because their identification is always carried with them.

### 6.2.4 Communications

Communications are not usually considered as laboratory automation. However they provide the means by which data, acquired by instruments, are transmitted from the bench to an analyst and information derived from these observations is sent to a decision maker and finally to a corporate database.

The elements in this group provide the means to connect the so-called islands of automation contained in the first group and integrate the laboratory with the rest of the organisation. In addition to connecting analytical instruments together, communications include computer-to-computer links for data storage and further processing.

Communications, usually in the form of a network, allow the electronic linkage of a laboratory with its client departments, of which the latter can be either on site or at remote sites to build both an integrated organisation and an integrated information solution.

One method of integration is to take the information from a quality control LIMS into a manufacturing resources planning (MPR) system to speed up raw material or finished product acceptance or rejection.

Transmission of data and information is via basic data communication tools such as:

- binary coded decimal (BCD)
- serial communications (e.g. RS232C and RS423 formats) or
- parallel communications (e.g. HP-IB and IEEE488 protocols).

These modes of communication are used to transmit data from an automated instrument or system to a data system.

Increasing communications sophistication and capacity are supplied by:

- local area networks (LANs) such as Ethernet and Token Rings, and
- wide area networks (WANs) and virtual private networks (VPNs).

Ideally, a network should hide the complexity of the system from a user who is only interested in accessing the services that it can provide. A network allows the benefits of a distributed processing philosophy to be implemented within a laboratory. The flexibility of instruments can be integrated in a client–server environment to utilise information resources efficiently. A network is also essential to provide on-line access to databases either within an organisation (contained locally on CD-ROM) or accessed via modem to commercial databases for scientific information to allow staff to be more efficient.

Communications must play a significant part in a laboratory's strategic laboratory automation plans. When planning a new laboratory, one of the mandatory pre-requisites in the design process is to include a LAN throughout the building with sufficient capacity to attach the present instruments and terminals, and also to allow for future expansion both in instruments and in data capacity.

### 6.2.5 Data to information conversion

The purpose of this group is to convert the data produced by the analytical instruments into information upon which decisions can be made. It is sufficient for the purposes of this module to summarise that information is equivalent to data plus context, for example, the interrelationships between observations and expected results or between two or more sets of observations rather than the original observations such as the raw data file from an analytical instrument.

The constituents of this group are mainly software programs, their purpose is either to reduce raw data to information or to analyse reduced data further. Some examples are:

- chromatographic data acquisition and analysis software
- data reduction and analysis of spectroscopic data
- fast Fourier transformation analysis of spectrometric data
- pattern recognition software
- digital filtering
- statistical analysis and
- chemometric techniques such as principal component analysis (PCA) and multivariate component analysis (MCA).

Again, like data communications, this group is not normally considered under the general banner of laboratory automation. However, when considering a holistic approach to automation, the techniques and tools either form a bridge between the data acquisition and the information management techniques to be outlined in the next group or act as a means of further analysis of the information managed by the final group.

### 6.2.6 Information management

This automation group stores the laboratory information during its generation and before its onward transmission within an organisation. It is composed of database management systems (DBMS) typified by either spectral databases or LIMS which tend to be the best developed tools of this group.

Other examples are:

- expert systems and
- neural networks.

This is the least understood of the four groups as there are a few commonly applied applications in analytical chemistry apart from the database applications and the success rates can vary with the expertise used to implement them.

We will discuss LIMS in more detail, such as what it is, its role in the laboratory and the system development life cycle, in a later section of this chapter.

### 6.2.7 A laboratory automation strategy in practice

The product of the laboratory, as stated in the introduction, is information. What the new definition and its breakdown have demonstrated is that laboratory automation is not a single entity but composed of four main groups (see Sections 6.2.3–6.2.6). These groups do not exist in isolation but depend heavily on inter-relationships for overall success. Data are acquired by the automated instruments in Group 1, using the interfaces and communications from Group 2 transmitted around the laboratory to the management tools of Group 4. Conversion of data to information uses the applications in Groups 3 and 4 and finally via the networks in Group 2 it is communicated to the organisation.

An overall implementation of a strategy for laboratory automation is shown in Fig. 6.2; individual systems have been installed in achievable stages as shown by the numbered sequence.

In this illustration, the strategy starts at a laboratory bench with the purchase of an item of analytical equipment to analyse a sample and acquire data (1). This instrument can be automated or manual but

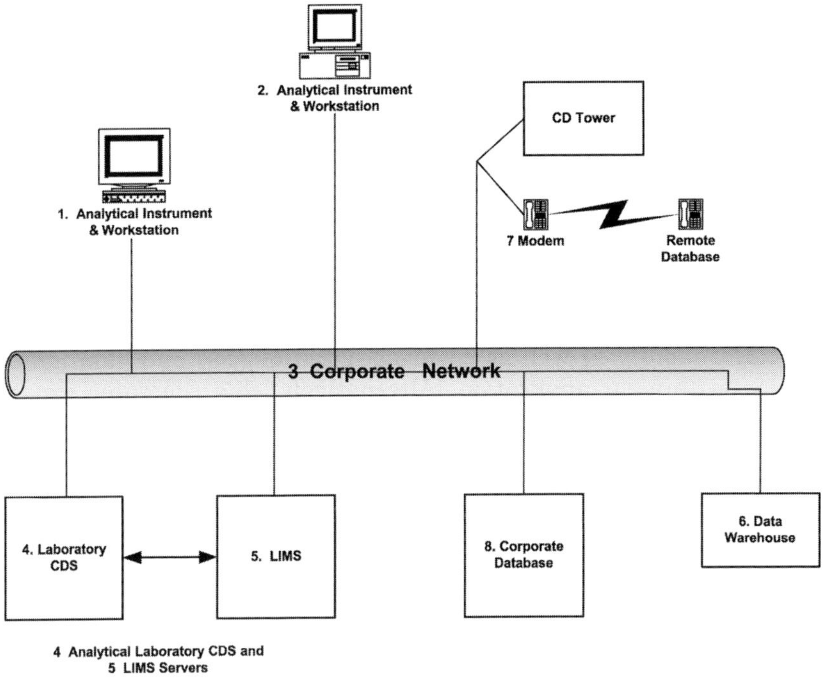


Fig. 6.2. A laboratory automation strategy in practice.

inevitably has a workstation for local control of the system as well as data capture and reduction. Such instruments are inevitably successful and cost effective and lead to the purchase of a second instrument (2). There now begins the problem which an automation strategy should avoid: the rapid generation of data which have nowhere to go, leading to backlogs and bottlenecks involving paper transfer.

The key to any automation strategy is the installation of a network (3). This is the conduit along which data and information will flow within the laboratory and which will integrate the laboratory with the parent organisation. Once a network is installed, it is relatively easy to add additional instruments (shown here with individual microcomputers for network connection but soon to have built in connectivity), computers, software, etc. to integrate all laboratory functions and provide an expansion route in the future. Moreover, it is essential that the network comply with the corporate computing standards. No network, no laboratory integration.

Once the network is installed, a data capture computer (4) and a LIMS (5) can be installed to centralise the functions of data reduction, sample tracking, report production and information generation. A network also provides the ability to add additional resources such as an additional computer (6). This computer can have data to information conversion programs running on it that integrate or interact with the LIMS shell such as chemometric programs, or it can act as a file server to the PCs within the laboratory to serve microcomputer programs across the network. The advantages of this latter approach mean that just a single copy of an application program is maintained within the laboratory for control and ease of maintenance and upgradability as well as ensuring data integrity and information sharing within an application. Alternatively, this computer can be situated in a different part of the organisation and run, say, an MRP program that controls production or a chemical structure database.

Further extensions of the network can access the library (7) with on-line optical disks, internal and external databases for retrieval of information and rapid access to current literature. Finally, for true integration of a laboratory, connection to the corporate database (8) to allow the accumulation of information and knowledge of a process or product is essential.

## 6.3 LABORATORY INFORMATION MANAGEMENT SYSTEMS

### 6.3.1 What is a LIMS?

The major function of most analytical laboratories is the creation and presentation of information quickly to allow decisions to be made. A LIMS is one of the major laboratory automation tools at the disposal of analytical chemists to help achieve this aim. Although a LIMS does not undertake analysis, a LIMS can be pivotal in integrating both the laboratory operations and the laboratory itself within an efficient organisation. A LIMS can provide a laboratory with the means to automate the processes of information creation and presentation, as well as being the platform for information dissemination to clients and senior management.

However, this is not always the case. A large number of systems fail to meet initial expectations, which implies that LIMS are not fully understood. To help overcome this, a LIMS model was proposed [4] that enabled the requirements of a system to be visualised conceptually. It has been adopted, modified and used as the basis of the LIMS concept model in the ASTM LIMS guide [5]. However, the LIMS model focuses on identifying and visualising the user functions within the laboratory environment and less on the strategic siting of a system.

To overcome this problem, a paper by McDowall proposed a matrix for the development of a LIMS with a strategic focus [6,7]. It introduced the concept of an operational, logistic and strategic LIMS. The matrix is formed by plotting these three types of LIMS against the scope of laboratory and organisational tasks that can be undertaken by such a system.

### 6.3.2 A LIMS has two targets

A LIMS is unlike any other piece of laboratory automation equipment available to the analytical chemist. It can provide benefits both within the laboratory and outside it. Thus a LIMS has two targets:

- the laboratory: the information generator; and
- the organisation: the information user.

The problem is how to site and implement a system so that it hits both targets effectively.

Figure 6.3 shows an outline of the functions that a LIMS should undertake. The diagram shows a LIMS sited at the interface between a laboratory and an organisation. Samples are generated in the organisation and logged into the LIMS; the samples are analysed within the laboratory and data are produced and reduced within the LIMS

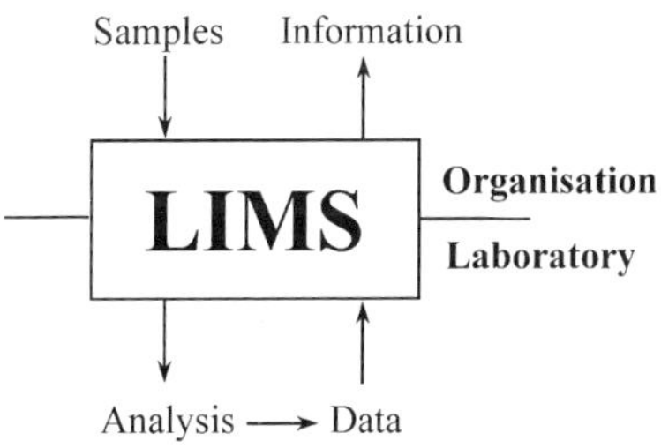


Fig. 6.3. Ideal implementation of a LIMS. The interface between the laboratory and the organisation shows that the LIMS benefits both.

environment to information that is transmitted back into the organisation. Figure 6.3 represents the ideal siting of the LIMS: both the organisation and the laboratory benefit. The line dividing the organisation and the laboratory show the system is of equal benefit to both.

However, there are two other implementations that are possible with a LIMS, which result in a different position of the interface between the laboratory and organisation. Figure 6.4 shows the more common implementation, which is probably typical of the majority of early implementations of LIMS in the 1980s. The main functions carried out by the system are the same as previously but the emphasis of the implementation is different. The boundary between the organisation and the laboratory has been moved up and the benefit of the LIMS is almost exclusively that of the laboratory with little payback for the organisation. Here the LIMS is a toy for the laboratory that few others are allowed to play with. The system is built from the bottom up but with no consideration for anyone outside of the laboratory.

The rarest alternative implementation is presented in Fig. 6.5. This is the top-down approach, where senior management or, worse, the IT group has decided that a LIMS will be implemented. There has been little consideration for the laboratory, only the organisation. The analysis and data gathering functions of a LIMS have been ignored, which allows the staff the latitude and the excuse to develop their own alternative local processing solutions. This system requires additional work by the staff to ensure its success in addition to the normal analytical function. The likelihood of failure with such a system is much higher than with the other two forms of implementation.

As can be seen when comparing the three alternatives, there is a balance to be found between the needs of the organisation and the laboratory. The interface between the two must be carefully defined;

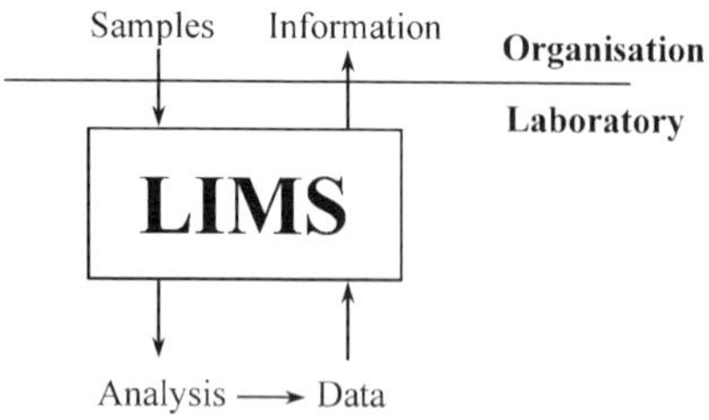


Fig. 6.4. A bottom-up implementation of a LIMS. The laboratory benefits but the organisation does not.

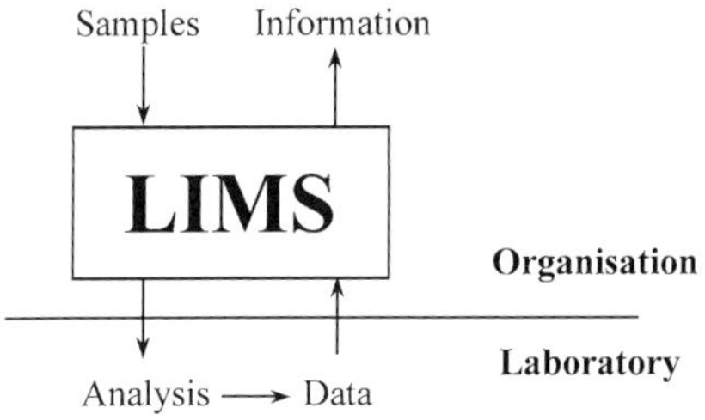


Fig. 6.5. A top-down implementation of a LIMS. The emphasis is on the overall system benefiting the organisation but the implementation allows the laboratory to develop data analysis solutions independently of the LIMS.

however, the initial implementation should be towards the analytical laboratory, the information generator. Automating the information generator is the key to success for the whole LIMS.

The importance of the LIMS matrix, described in Section 6.3.3, allows that dividing line to be drawn accurately before the project is started and the implementation of a system to be managed.

### 6.3.3 Construction of the LIMS matrix

The matrix axes consist of LIMS type plotted against function in the laboratory or organisation. Each will be outlined below.

*Three types of LIMS*

There are three types of LIMS: operational LIMS, logistic LIMS and strategic LIMS [6]. The functions and aims of these systems are described in more detail.

*Operational LIMS.* This is a basic system that automates analytical processes in the laboratory. This system increases the efficiency of a laboratory but the impact is only local. A computer system in this category would normally be concerned with the operation control of a laboratory and would automate functions such as sample entry, work-list generation and report preparation. Another way of looking at this system is usually the automation of the status quo. Such a system is probably typified in Fig. 6.4.

*Logistic LIMS*. In addition to the functions of an operational LIMS, a logistic LIMS provides users, especially managers, with the information necessary to undertake their work. Such systems generate control information. Ultimately, this information improves the effectiveness of a laboratory.

*Strategic LIMS.* The aims of a strategic LIMS are to integrate information and applications from different functional areas. From this information it may be possible to reshape operations. A strategic LIMS has the greatest impact upon the business by increasing the competitiveness of the laboratory.

Properly designed, a logistic or strategic LIMS will achieve the correct balance shown in Fig. 6.3. However, too great an emphasis at the start of the project is likely to produce the system shown in Fig. 6.5.

*The system scope*

Where does a laboratory begin and end? Clear definition of these two points help to determine the scope of a LIMS. The six areas of system scope that define the matrix are laboratory operations, monitor and control of operations, laboratory management, reporting and communications, analytical decision making and organisational integration [6]. This is the maximum scope of a LIMS. These comprise the horizontal axis of the matrix. The function carried out by each of these areas is outlined in Table 6.1.

*Matrix construction*

The matrix can now be formed as a three by six block and is depicted in Fig. 6.6. There is a division between the first three items of the system scope. Laboratory operation, monitor and control and laboratory management are concerned with functions inside the laboratory. Organisational functions are represented as reporting and communications, analytical decision making and organisational integration.

The individual components of the matrix are shown in Table 6.2. This is an outline of the functions of a LIMS within each cell.

Considering all of the functions in Table 6.2, one can see that the main functions of the laboratory are outlined in the lowest three cells of the matrix (Operational LIMS: laboratory operations, monitor and control and laboratory management). This represents the basic laboratory operation and is indicative of the data domain discussed previously. This is shown in Fig. 6.7 as the box labelled 'Laboratory'.

TABLE 6.1

Functional scope of a LIMS

| Functional area | Scope |
|---|---|
| Laboratory operations (LO) | To automate and structure work<br>The automation of the basis laboratory operations such as sample entry, work-list generation and results entry<br>Work rationalisation |
| Monitor and control (MC) | To evaluate performance<br>The monitor and control of the laboratory operations by such processes as approving results, the use of quality control schemes and the checking of transcription errors<br>Provide standards, measures, and information for performance evaluation and feedback |
| Laboratory management (LM) | Support intellectual processes<br>Organising and managing the laboratory functions and operations<br>Project and work planning |
| Reporting and communications (RC) | To augment human communication<br>Here are the means to transmit results or reports and communicate with the clients of the laboratory |
| Analytical decision making (AD) | To aid and speed decision making<br>Providing quality information in a timely manner and the right format to make decisions<br>Supporting processes in production, development or research |
| Organisational integration (OI) | Facilitate intra- and inter-organisation transactions<br>Integrating with other functional groups in the corporation and between organisations |

Ideally, the major impact of any computer system, including a LIMS, should be strategic. The organisation uses the information generated by the analytical laboratory and this can be considered as the information domain. The maximum impact of a LIMS is in the top right-hand side of the matrix. It consists of the strategic LIMS cells of results and communications, analytical decision making and organisational integration, and is marked 'Business Objectives' in Fig. 6.7. The ideal is for every LIMS to aim for the top-right hand area of the matrix.

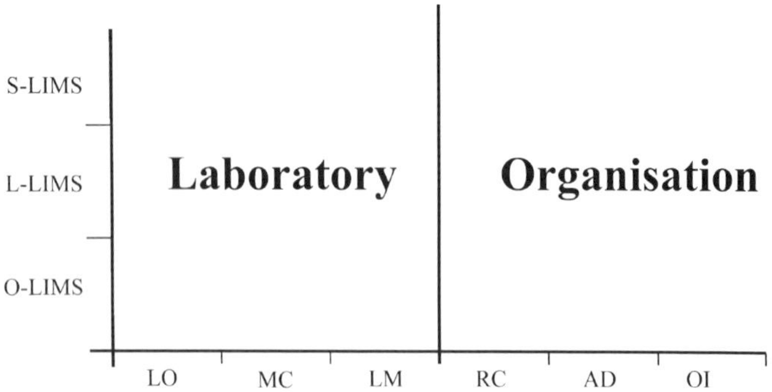


Fig. 6.6. An outline of the LIMS matrix. S-LIMS, L-LIMS and O-LIMS are strategic, logistic and operational versions of LIMS, respectively. The horizontal axis functions are laboratory operations (LO), monitor and control (MC), laboratory management (LM), reporting and communications (RC), analytical decision making (AD) and organisational integration (OI).

However, it should be stressed that not every LIMS will be developed to the full extent of the matrix. There will, inevitably, be some duplication with other computer applications that have been implemented or are being implemented within an organisation. The aim, therefore, is to develop the matrix around business objectives and a good financial justification to make the system pay for itself in a reasonable time period. One way to do this, and make economical use of resources, is to link the implementation of the LIMS with existing applications to provide the functionality described in the matrix. The development of a LIMS within a pharmaceutical environment has been shown in an earlier publication [6,7]. Here the LIMS was linked through a communications network to a document management software package that can be itself developed to produce computer-assisted new drug applications (CANDAs) which will fulfil the requirements of the matrix.

### 6.3.4 LIMS matrix views

Similar to the ability to use views to gain an insight into the contents of a database, the same concept can be used to gain an insight into the matrix and its impact. We will consider two main views in this section. The first is the vertical view through laboratory operations and a horizontal view across the logistic LIMS cells (see Table 6.2).

TABLE 6.2

The LIMS matrix

| | Laboratory operations | Monitor and control | Laboratory management | Reporting and communications | Analytical decision making | Organisational integration |
|---|---|---|---|---|---|---|
| Strategic LIMS | Integrate with client operations | Compare requests with capacity | Resources based on workload | Integrate with document management software | Decision support system integration | EDI & CANDA rapid commercialisation |
| Logistic LIMS | Integrate laboratory operations | Automate regulatory compliance | Monitor test use | Electronic reports to clients | Highlight out of specification results | Remote on-line inquiry |
| Operational LIMS | Automate existing operations | Monitor and approve results | Sample status billing | Paper reports to clients | Display results versus specification | Remote printing of results client/lab e-mail |

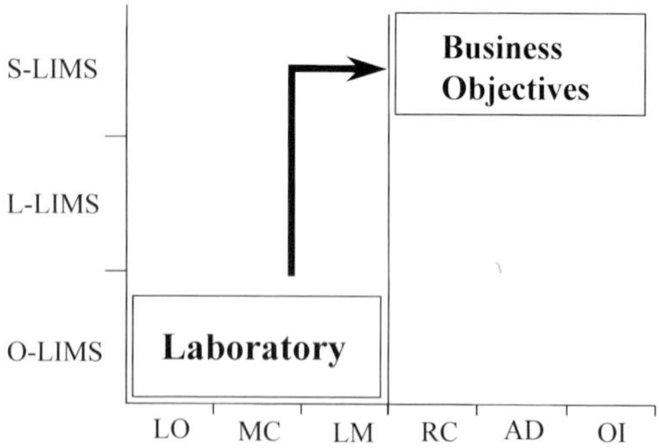


Fig. 6.7. The LIMS matrix illustrating how some LIMS are implemented with the emphasis on laboratories and the overall business objectives where the best payback on the system investment would be obtained.

*Vertical view of laboratory operations*

The vertical view through the laboratory operations cells of the matrix will show how this changes from the operational LIMS to the strategic LIMS. The lowest left-hand cell of the matrix represents the core of the analytical laboratory: the inventory of assay methods and instrumental techniques. The operational LIMS would generally automate these processes and techniques. Business rules would be incorporated into the system such as who was allowed to create, modify or delete methods within the LIMS and how many replicates were needed for each analysis. Samples would be logged in and work sheets would be produced for each analytical run. All functions are the usual LIMS functions within the analytical laboratory.

At the logistic LIMS level, the matrix is looking for integration of laboratory operations. This would be the linking of analytical instruments to the LIMS for electronic transfer of results; samples would be barcoded and would have their location monitored as they move through the laboratory. The various applications used for analysing data and producing results would be integrated into the LIMS environment and would not be standalone applications requiring re-entry of data.

At the strategic LIMS, the system should be integrated with client operations. Possible examples of this could be the clients logging onto the system to initiate sampling within their areas. The system would print barcoded labels for the samples from printers within the clients area.

The samples would be submitted to the laboratory for analysis and logged into the system via the barcode. An alternative scenario, in a production environment, could be the analytical laboratory developing robust methods for the production staff to operate and use the LIMS as a means of collating and disseminating the results. The LIMS would be the medium for the laboratory staff to monitor the performance of the analytical methods and alert them if a method were getting out of control.

*Horizontal view of a logistic LIMS*

Looking along the matrix cells at the logistic LIMS level would give us a view of a system that has integrated laboratory operations as described above. The monitor and control cell requires automated regulatory compliance, this would also include compliance with a quality scheme such as ISO 9000. There would be audit trail facilities as standard and the establishment of chain of custody from the time the sample arrived into the laboratory to when it was disposed of or returned to the client. Automated regulatory compliance is a term which means that the documentation required to meet the quality standards that the organisation is working to is generated by the system automatically. This would include receipt of samples and documentation of exceptions.

Laboratory management is concerned at this level with monitoring test use. Many laboratories accept requests for sample assays without questioning the rationale for the analysis or finding out if the information generated is ever used. The purpose of this cell is to question the analyses requested and monitor how the results are used. Is the laboratory generating the right information and offering the methods that the clients want? It is about managing the client as much as managing the future direction of the laboratory and anticipating the services it will offer. The information from the LIMS, via a retrospective search of the database, will allow managers to follow up work that was done and how the results were used. Depending upon the implementation of the whole system, this could be done electronically via feedback loops or manually.

The cells of reporting and communications, analytical decision making and organisational integration cover the functions whereby reports generated by the system are transmitted via the network, out of specification results are highlighted and the clients can have remote access for on-line query of the database. The aim of these first two cells is to transmit the report to the client effectively and highlight aberrant results; this allows the client to focus immediately on problem areas. Some of the options available to implement this section are covered in more detail under the financial justification and risk assessment in Section 6.3.4.

These views of the matrix are based on straight lines. However, not all systems will be implemented in straight lines and the views of individual systems will be based on functions required by an organisation and laboratory.

*Using the matrix for financial justification and risk assessment*

The matrix can be used to justify the spend on a LIMS on a cell by cell basis. The costs for each cell in the matrix can be calculated and the benefits quantified. This cost justification can be used to justify each cell of the matrix to senior management. It will enhance the efficient use of resources and avoid duplication of functions within an organisation. However, this idea can be taken further. Figure 6.8 shows as the LIMS is developed out of the operational level to the logistic and strategic levels and also into the organisation there is increasing benefit but there is also a corresponding increase in the risk of the project. Therefore, a combined cost justification and risk assessment can be made for each matrix cell.

Let us look at two examples of this approach. In the logistic LIMS level of the matrix corresponding to the organisation, communication is electronic, there is the ability to highlight out of specification results for

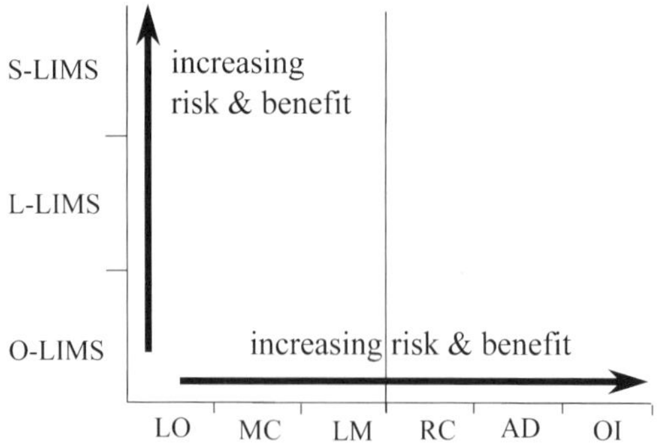


Fig. 6.8. The LIMS matrix showing increasing risk and benefits the further the system is implemented outside of the laboratory.

the client for decision making and, to help organisational integration, clients can access their results remotely. Ideally, these first two functions will help communicate results to the clients and enable them to make effective decisions. The third will enable the client to access the LIMS to see the progress of their samples and avoid the need to phone the laboratory to enquire about progress. All functions are desirable to integrate the laboratory and the organisation. However, senior management is looking for a cost benefit justification and an outline of the risks involved. Imagine two companies, one where information technology is well established and the second where use is not as successful as management would wish.

Scenario 1 would envisage an organisation with an existing network linking all functional groups. This is a key component for organisational integration and provides the medium to move information. This organisation is relatively sophisticated in its use of IT with e-mail and on-line diaries available and used regularly. The main communications component, the network, is already installed. Therefore, the cost benefit justification will only require expenditure for the customisation of reports and the linkage of the LIMS with the e-mail system to transmit a report to the client, a relatively minor cost component. To highlight out of specification results, more programming would be required to place an asterisk by the out of specification result. The final cell in this level of the matrix requires external read-only access via the network to the LIMS database. Security levels would be set up within the LIMS to prevent unauthorised users from changing results and roaming around the database; this may require a customised database view for each client or client group. With a network in place and working, the approach to implementation might be to authorise the communication cell and the organisational cell to proceed, with the ability to highlight out of specification results being implemented in the first place with the asterisk option. The risk assessment would be based upon a number of factors but would be low. The reason is that there should be little impact on the ways of working as the staff already tend to work using the network facilities. The use of the matrix is built on the use of information technology and requires relatively little capital investment to be made.

In contrast, scenario 2 envisages an organisation where information technology implementation is not always successful. A network is not in place throughout the organisation and requires investment. Here the cost benefit justification and risk assessment are changed dramatically to implement exactly the same functions. A higher capital investment is required to put the network in place. There may be the need to retrain or hire staff to run the network. Users would need to change their ways of working to use the system effectively, thus increasing the risk to the success of the whole project. Here the prudent approach would be to implement an operational level of the LIMS matrix: hard copy reporting first and concentration on getting the LIMS working within the laboratory. As a future phase of the project, a network investment to the client groups could be justified on a piece-by-piece basis.

This illustrates the combined use of cost justification and risk assessment that should be applied to the use of the matrix. It also requires an assessment to be made of the use of information technology and the maturity of staff within an organisation.

### 6.3.5 Organisational integration and LIMS

The use of information technology crosses functional boundaries: robots and other automation equipment are justified and used locally, however the impact of a LIMS is far greater as shown by the matrix. A LIMS has the ability to influence groups outside of the laboratory.

An operational LIMS has its impact mainly within the laboratory with little influence on the rest of the organisation. As discussed earlier in this article, this is the typical implementation seen in the 1980s. The logistic LIMS has its main impact on the functional groups that work closely with the laboratory. They are linked via a corporate network that is fully realised with the strategic LIMS. The strategic LIMS has its greatest impact on the organisation as well as providing benefits to the laboratory and the client groups served by the laboratory.

The implementation of a LIMS with increasing logistic and strategic elements means that there will be greater impact on the organisational structure. In instances where this occurs, a question arises: which processes should be automated?

Most LIMS automate the current ways of working. These have evolved over time, but may not be efficient or cost effective. Examples may be the way samples are received by the laboratory via multiple logbooks or several signatures are required to authorise a report. It may be prudent when considering such a system to consider re-engineering or redesign of the laboratory operations prior to automating them.

This is a higher risk approach, but it can improve the overall efficiency of the laboratory and its client groups and may be better than simply automating the status quo. We shall explore this approach in Section 6.4 of this module.

The other element in the equation is the impact of the LIMS on the existing organisational structures. It may be that organisational changes are required when implementing either a strategic or logistic LIMS. If this has not been anticipated or thought out, this requirement of organisational change is a high risk factor. To avoid project failure due to this cause, an assessment of the impact of the system on the existing organisational structures is essential.

## 6.4 LIMS AND THE SYSTEM DEVELOPMENT LIFE CYCLE

### 6.4.1 System development life cycle

The development and implementation of an information system such as a LIMS follows the traditional system development life cycle (SDLC) shown in Fig. 6.9. Normally equipment is purchased off the shelf and used with little or no modification, however a different approach is necessary with complex laboratory automation such as a robotic system

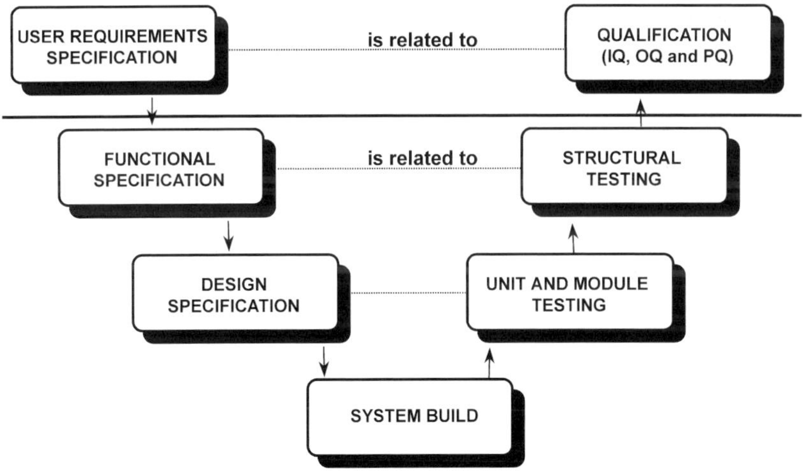


Fig. 6.9. System development life cycle.

or a LIMS. In this instance, the users must state what tasks they want the system to undertake. Often this places the chemist in a difficult position as he or she is not certain what the system should be, or is, capable of doing.

*Quality and the SDLC*

Quality is designed into a computer application by defining the requirements and functions of the system. Quality can never be tested in a LIMS.

Quality of a system is manifested in different ways such as user friendliness, undertaking the task it was designed to do, robustness of operation and tolerance of error. During the development of the system by the supplier, quality is incorporated by ensuring that design and programming standards are adhered to. Furthermore, there are testing procedures that can be undertaken by the supplier to test the system prior to its release.

*Life cycle approach to LIMS implementation*

As can be seen, there are many stages to go through before the system is fully operational, thus the acquisition of a LIMS must not be undertaken lightly. The stages outlined in this module are guidelines, which are based on the author's experience, and which can be recommended to any individual or organisation contemplating an information system. These guidelines can be tailored or modified, as appropriate, to an individual situation.

As the life cycle unfolds the reasons for following this route will become apparent. The requirements and functions of a system must be fully defined, the responsibilities clearly stated and good communications set up between all relevant parties. This is particularly important when the timescales of the SDLC are considered: the minimum overall time of a project is approximately 12–18 months, but the maximum should be 24–36 months. However, timescales can be longer if the project is poorly managed. The timescales obviously depend on the complexity of the project and how much configuration (changes of pre-set parameters in the software) or customisation (programming) is involved as well how much time and resource can be committed by the organisation. Configuration is preferable to customisation for speed of implementation and long-term support issues.

The challenge for any organisation is to implement a LIMS that supports the effective automation of information, production and distribution from the laboratory. Here the LIMS matrix can provide a high level outline of what functions the LIMS should automate. Once the outline of the LIMS and the functions in the organisation that it should automate has been discussed and agreed, the project proposal should be prepared.

### 6.4.2 Project proposal

The first stage in the life cycle is usually the preparation of a project proposal that will be reviewed by either senior management or a central computing resources committee. The information gathered during the laboratory review for the development of the laboratory strategy should be sufficient to state if computerisation is a viable and cost-effective solution. Note the use of word 'computerisation' and not LIMS, as the project proposal should be kept deliberately vague and not recommend definite hardware or software solutions.

The proposal document should contain an overview of the department and the problems that the system should overcome. It is important that the justification must be based on the business objectives of the department concerned and any solution is not to be installed at the whim of management or an enthusiast. The outline justification must include the areas and estimates of cost savings that will be made when the new system is operational compared with an estimated cost of the package.

Outline approval by senior management of the often considerable financial outlay will be dependent on the ability of the laboratory to demonstrate, and where possible quantify, a return on the investment arising from benefits in:

- increased efficiency and productivity
- reduced or limited increase in staff levels in relation to increased workload
- increased scientific standards
- production of more reliable data
- improved speed of report production and
- provision of management analysis functions.

The justification of a LIMS must be balanced between the 'tangible' benefits (i.e. those where a return on investment can be calculated) and the 'intangible' benefits (i.e. where this cannot be done such as quality of data) [3].

There is a balance between over-justifying the system to obtain project approval and being realistic about what the system will actually deliver in practice. This is very difficult to delineate, however the review of the laboratory should give management the information to decide if the LIMS is worthwhile; if it is not, the project should not proceed.

A post-implementation review of the project, about 6–9 months after the LIMS is operational, should temper overoptimistic justification of the system. The setting of criteria for success early in the project's life will help to make this review more objective than subjective.

Approval of the project proposal will normally allow the formation of a project team and allow the project to proceed to defining the requirements of the system, inviting tenders and evaluating alternative systems.

### 6.4.3 The LIMS project team

To solve the challenges posed by the design, implementation and operation of a LIMS a multidisciplinary approach is required. The formation of a LIMS project team to oversee all of the remaining stages of the SDLC up to the operation of the system is the normal solution. The minimum membership of the team will be users representatives (a vital component and often overlooked in some projects), laboratory management and computing professionals.

There may be other members on the team such as personnel from client departments. When the system is to be operated under regulatory agency or government guidelines the involvement of a member of the quality assurance unit (QAU) in all stages will be of direct benefit to the project. The project environment will entail computing and analytical experts working closely together, therefore it is vital that each group has a full understanding of each other's areas of expertise and working practices. The computer scientist must know how the analyst works and vice versa, one area in particular must be the development of a common vocabulary for efficient and unambiguous communication between both parties.

A major problem in the implementation of any computer application, including LIMS, has been the crossing of disciplinary boundaries.

The LIMS matrix can be used to help attain this goal, by giving both groups the common concepts and vision of a LIMS to help each group communicate with each other. Each discipline's own perspectives and strengths will be fused together in a project team with the remit to implement a system successfully, which will enhance this communication.

The leader of the project team must be a member of the laboratory management, this clearly demonstrates to the users that it will be their system and not something that the computer department has decided may be suitable.

The project team must operate in a matrix management style, its members being responsible for what tasks should be undertaken and when they should be started and completed. The members liase with their respective line managers who are responsible for providing the resources for the tasks and deciding how they should be carried out. This means that the project extends across department and discipline boundaries, therefore communication is essential to ensure that members of the team and their respective line managers are aware of the progress and problems of the project.

### 6.4.4 User requirements specification and system selection

This is the most critical part of the system life cycle as it is the basis of all decisions and actions taken in the later phases. If this stage is wrong, then the rest of the life cycle can be jeopardised.

The assumption made in this section is that a commercial system will be selected rather than the organisations change their ways of working. A commercial system is a viable and cost-effective option as many systems are now available. Applications are networked and are scaleable from small to large laboratories and can run from a single server across several sites and even continents.

The commercial approach has the advantage that development costs are shared over the whole user base and because of user feedback and competition there is continuous development of the software. A potential drawback may be that the vendor's conception of a laboratory may not be the way that yours works, and some configuration or customisation of the package may be necessary.

*Understand the overall process*
To obtain the best from the implementation of a LIMS, the project team will need to map and understand the whole process from sample generation to decision making using the information generated. Inevitably, this will involve crossing functional boundaries and defining the hand-offs between the different departments involved in the process.

The team will need to identify where the process can be improved or redesigned to overcome bottlenecks in the existing process. As change is involved with a LIMS project, the project team needs to give serious thought to changing the laboratory process to fit with the selected LIMS system. This will make the whole implementation faster than changing the LIMS to fit the existing laboratory processes.

*Write the user requirements specification*
The new process map of the workflow gives the project team members the information needed to write the next document in the life cycle of a LIMS. This is the user requirements specification (URS). The URS is important as it gives the project team the outline tasks that the system should be capable of undertaking. In some respects, it can be regarded as a 'wish list' of functions that the system should be capable of performing.

The document should be written in a non-technical style and follow a logical sequence starting from the entry of the sample into the laboratory, via analysis, through to the compilation of the analytical report and its distribution.

Each of the requirements needs to be uniquely referenced and be worded so that it can be tested. Guidance for writing a URS is presented in the paper by McDowall [8]. The URS is important as it defines what the laboratory wants from the system and stops the users being seduced by technology: the URS is the basis for the system selection.

*Write the invitation to tender*
The URS is then converted into an invitation to tender or a request for proposal by appending to the document:

- *A laboratory overview* describing the laboratory environment and the work undertaken. The analytical instrumentation and level of automation, the sources of work and how results are reported. The present and past workload of the laboratory.

- *Future of the laboratory.* The future developments in automation, instrumentation and techniques anticipated to be used by the laboratory. Estimated trends in workload and staffing. How is the laboratory changing?
- *Integration with existing packages.* Data input from and output to other computers, software packages and e-mail systems.
- *Computing environment* describes the preferred hardware and software systems used at the local and corporate levels.

This information enables a prospective supplier to respond effectively to the invitation to tender. At this stage, for the project team to have as much choice as possible, the document should be sent to as many vendors as can be found.

To elicit similarly structured replies from all the vendors, either a section can be inserted in the tender document instructing the vendor how to respond or a checklist containing all the relevant questions that should be answered can be enclosed. It is important to know whether a function is available as standard in the package, needs to be configured within the standard package or customised, or is not available in the current released version of the system.

*Evaluation of prospective systems*

The evaluation process is vital as it matches the laboratory requirements with those on offer from the suppliers. Unlike many other types of laboratory automation the full benefit from a system should not be expected until several months after the system has been introduced into the laboratory. This time lag means the impact of a wrong decision is much greater because much time, effort and money have been expended to date. Therefore, the need to make the right decision the first time is very important.

The importance of requirements specification is now realised as it provides the written definition for the evaluation exercise. The key approach to evaluation is to develop an evaluation checklist for comparison of systems: the users should grade each requirement as either 'essential' or 'desirable'. All those features falling into the 'essential' category form the minimum requirements of the LIMS and the 'desirable' category features are those that are useful to have but the users do not require in order to perform their normal role. To aid objectivity, essential items can be given a numeric value; this will give a 'total score' for each system on offer.

The replies by suppliers to the tender document can be compared on paper to reject those obviously unsuitable, this should reduce the list to about four or five. These remaining should be visited and their replies discussed in depth and the systems evaluated by use. At the end of this phase, the systems may be reduced further to one or two for in-depth evaluation.

It should be noted that the use of a checklist by itself should be discouraged as the subjective comments of the project team and users should be taken into consideration. This may take the form of comments on the interaction with the software or performance of the system.

Once a single system has been chosen, it should be tested in house for up to a month (dependent on the complexity of the tasks it should undertake) to confirm that it is a suitable choice. Here the opportunity for all users to have hands-on access to the system must be encouraged. This in-house testing will usually entail the purchase of the licence for a specified time from the LIMS vendor. If the evaluation is successful, then the cost of the licence should be deducted from the total cost of the system. If, however, the evaluation shows that the selected system is unsuitable then little time and money has been lost, this now means that the project team must re-evaluate their requirements and consider other vendors and approaches and decide how to proceed with the project.

A final point for the evaluation: the project team must consider the laboratory environment and whether it is changing or static. The reason is that software handles routine tasks well, it is the unusual request that will prevent easy operation. Therefore, in a changing laboratory will the software be able to cope up with changes in working practices?

This formal approach to the definition and evaluation of a LIMS will enable the project team to proceed to the purchase of a system with the confidence that it will undertake the tasks required of it. The in-house evaluation of the system will highlight areas that require tailoring to the way the laboratory wishes to operate as well as finding the deficiencies where custom software will be needed for full functionality.

*Vendor audit*

If you look at Fig. 6.9, you will see a line drawn under the URS and the qualification stages of the V model. The reason is that for a commercial system, all the stages under the line are the responsibility of the vendor and you are seeing if the system they offer is suitable to your laboratory. In regulated laboratories such as in the pharmaceutical industry, there will be a requirement to see if the application software was developed in a quality manner. Hence the need for a vendor audit; for more information see the two articles by McDowall [9,10].

*Purchase of the selected LIMS*

The purchase of a LIMS will usually follow a company's procedures for capital equipment authorisation, as the cost of a system will usually exceed the minimum limit for this process. The exception may be the instance when a company is purchasing the software alone to run on an existing processor and the cost may be considered as a revenue item.

Before placing an order, it is vital to check:

- Sufficient peripheral devices such as terminals, plotters and printers need to be available to use the LIMS. This should ensure that there would be no complaints by users being unable to access the system.
- The small print of the contract should be examined to ensure that there are no escape clauses that can allow the vendor to avoid doing what they have promised to do.
- It should be ensured that commitments to enhance or modify the system are put in writing to protect both the purchaser and the vendor from any misunderstandings that might occur.

The purchase procedure will generally entail writing a further document, the basis for which is the project proposal. For most systems this will entail abstracting and updating the information already gathered. Include in the justification:

- business objectives of the laboratory
- problems with the existing method of working
- objectives of the system and how it should overcome the existing problems
- design of the proposed system
- detailed costing of the system including support, consultancy and internal resources
- competitive quotes from other suppliers, if required, and
- costs for laying IT infrastructure and other equipments, if required.

Once the purchase has been authorised and the order placed with the vendor, planning of the implementation should begin in conjunction with the vendor.

### 6.4.5 Functional specification

Once the final system has been chosen, the URS is used as a basis for the next document, the functional specification, which is the record of the configuration or customisation of the application.

What is the difference between the requirements and functional specifications? An example of this could be a program to update records within the database of the LIMS. In the URS all that would be mentioned would be the need for a program to update records in the database. In contrast, the functional specification requires more detail:

- Functions that the program should undertake: create, copy, modify, view, and print records.
- Security considerations: who should have access to which functions e.g. new or basic users may only use the view and print functions.
- Screen layouts: how the program should appear to the user.
- Program logic: operation of the software if the module is to be customised or programmed.

As can be seen there is much more information required in a functional specification, it is for that reason that configuration of a LIMS is faster and far more preferable than customisation. It is also easier to omit details, often essential ones such as a 'delete record' function, from the specification. This is the cause of many problems and one main reason for enhancement of the system functions later in the SDLC. An apocryphal definition of enhancement is 'what we forgot the first-time around'. At first reading, it is funny, unfortunately it is also true.

Essentially, the functional specification is a document that can be read by the users and therefore is written with the minimum use of computer terminology. The functions to be undertaken by the system must be stated in a clear and concise manner. A major problem, if the wording is ambiguous or does not adequately describe the function, can be misinterpretation by programmers resulting in the program not working as required (see above). Close liaison between the chemists and the supplier is essential, as the main problem is the crossing of disciplinary boundaries. The systems analyst who is drawing up the specification document must be aware of the intimate workings of the environment where the LIMS is targeted. To this end, spend one or more days walking him or her through the laboratory and showing the operation. Great care is needed to include all working practices into the document to ensure that flexible software is written. Remember to define words to ensure common understanding.

Mistakes are expensive to correct and will cost time, effort and credibility. The best advice is to consult the users, the laboratory management and the client departments concerning the content of this document.

### 6.4.6 Qualification of the system

There are three stages to this process:

- Installation qualification (IQ): is the system installed correctly.
- Operational qualification (OQ): does the system operate as the vendor says it should.
- Performance qualification (PQ): does the system work the way the users want it to.

The IQ and OQ can be performed in a relatively short time scale but there is usually a delay in undertaking the PQ as the system needs to be configured or customised before the testing can be started.

*Installation qualification (IQ)*
Prior to purchase of the LIMS, consideration should be given to the siting of the various pieces of hardware and to any special requirements for those areas. The LIMS computer could be sited in a central computer room that is air conditioned to control humidity and temperature and with a false floor to allow for the large number of cables needed to connect the analytical instruments and peripheral devices to the processor. To allow controlled shut down of the system in case of a power failure an uninterruptible power supply (UPS) should be considered. The room housing the computer may need to be secure with

access limited to authorised personnel only. The siting of the peripheral devices and the associated cables requires detailed discussion and planning with the computer department.

The system software and components will be delivered and installed, usually by the vendor's staff, and any tests undertaken documented by them.

*Operational qualification (OQ)*

The OQ may be follow on immediately from the IQ and can be performed by the vendor's staff or trained laboratory staff. This phase of the life cycle can be very short if there is extensive alteration of the system or relatively long if the system is used with little or no configuration.

*Performance qualification (PQ)*

Performance qualification, also known as acceptance testing, has the objectives of confirming that the system operates as expected and to find errors in the LIMS.

It must be stated at the outset that the end user is responsible for PQ of the LIMS. However, it is impractical and virtually impossible to find all the errors in a system because of the multitude of different pathways through the software and the time it would take to test them all. Therefore, the PQ tends to be a compromise as software bugs will be found after the system becomes operational.

The size of most commercial LIMS is such that the PQ testing will be focussed on the key system functions that acquire data, store it, transform it and report it. If the system has been customised then the customised functions must also be tested as these are unique.

### 6.4.7 User training and roll-out strategies

Once the LIMS has been validated and the documentation is available, the users can be exposed to the new system. There are three main ways this can be accomplished:

- total immersion
- parallel operation or
- selected use.

One approach to consider is a trial operation of the system, before the validation and the documentation of the system are finished, to find the working practices that will be used after the system becomes operational. Regardless of which method is chosen, the learning curve is usually long, requiring much training and understanding. However, the benefits of a well-designed system should always outweigh the time spent on training.

*Total immersion*

An implementation using total immersion uses the LIMS from a set date and abandons the old method of operation. This is a bold form of implementation and requires total confidence in the system that has been gained during both the acceptance testing and the validation stages as well as total commitment from the users and management. This style of implementation is best suited when the LIMS is being used to implement change in the laboratory and its operation differs radically to the existing ways of working.

*Parallel operation*

Parallel operation involves the running of both the old and new systems together to allow time to cross-check the data and information produced by the two methods and to build confidence in the LIMS. This approach will be applicable where tasks undertaken by both systems are similar, however if the data produced by the two methods are not the same then the reason must be found and the problem resolved. If there are problems with the LIMS then the old system can still be used to produce results whilst a remedy is sought. This approach to implementation is resource intensive and a deadline should be given after which only the LIMS will be used.

*Selective use*

Selective use, as the name implies, is the teaching of a small number of users on how to use the LIMS. This is a cautious approach but has the advantage of allowing prompt attention to problems almost as they arise; moreover, in cases of slow learning then individual tuition can be given. This has an advantage over the other two styles in that if the first group of users are enthusiastic they can be used to 'sell' the benefits of the LIMS to other potential users by word of mouth. Care in the selection of the first group, therefore, is essential.

### 6.4.8 Project close-out

Once the system is operational, the project team should formally close the project and hand the system over to the system manager. Normally the project team will disband, however in the author's view, it should undergo a metamorphosis to become a departmental LIMS user group which would be responsible for the future development of the system, monitoring its use, highlighting and solving problems, etc.

### 6.4.9 Post-implementation review

Usually after 6 months of operation the system should be reviewed. This should be a formal process where the actual performance of the LIMS in relation to the claims made at the project proposal or the purchase justification will be critically evaluated. Successes, failures, advantages and problems should be documented: it is only now that the consequences of the decisions and actions that the project team took during the SDLC will be seen. It is essential for the organisation that the successes and the failures be recorded and applied to other computing projects either that are ongoing or that will be initiated in the future.

### 6.4.10 Enhancement of the system and controlling change

Once the initial configuration of the LIMS has been established, it will inevitably undergo change. This change comprises two areas.

The first area concerns the repair of hardware where a part is replaced due to malfunction, this should be logged in an instrument book to show the repair and maintenance records similar to those of an analytical instrument.

The second area of change involves the software: when an error is detected and remedied (by changing the code of the program), the software changes from one unique state to another. What is unknown is how many errors have been created by fixing the original fault.

Hence, there is the need to revalidate either the software module or the whole system when a fault is remedied. Therefore, any changes to the system should also be authorised, logged and validated to demonstrate that the system is under control.

If a LIMS is successful then the users will request change, this should be encouraged as it is a natural outgrowth of the system and a normal part of the SDLC. The changes requested by the users should be discussed at a meeting of the users, where they will be authorised, modified or rejected as appropriate. The enhancement, if approved, can be communicated to a vendor to implement as part of a general improvement to the standard package, they can be contracted to undertake the work as a separate task, or it can be carried out in-house with the appropriate staff.

## REFERENCES


1 D.J. Betteridge, *Anal. Proc.*, 24 (1987) 106–108.
2 R.D. McDowall, *Chemometrics Intell. Lab. Syst.: Lab. Inform. Mgmt*, 17 (1992) 265.
3 R.R. Stein, *Chemometrics Intell. Lab. Syst.: Lab. Inform. Mgmt*, 13 (1991) 15.
4 R.D. McDowall and D.C. Mattes, *Anal. Chem.*, 62 (1990) 1069.
5 American Society for Testing and Materials, *Standard guide for laboratory information management systems (LIMS)*, Standard E5178, ASTM, Philadelphia, PA, 1993.
6 R.D. McDowall, *Anal. Chem.*, 65 (1993) 896A.
7 R.D. McDowall, *Chemometrics Intell. Lab. Syst.: Lab. Inform. Mgmt*, 31 (1995) 57.
8 R.D. McDowall, What do you mean it doesn't work?, *Sci. Data Mgmt*, March (1998) 7–12.
9 R.D. McDowall, Who wrote this £$%! Program? Part 1, *Sci. Data Mgmt*, June (1998) 8–14.
10 R.D. McDowall, Who wrote this £$%! Program? Part 2, *Sci. Data Mgmt*, September (1998) 7–11.

Chapter 7

# *Chemically modified electrodes with integrated biomolecules and molecular wires*

***E.J. Calvo*, C. Danilowicz, E. Forzani, A. Wolosiuk and M. Otero***

## 7.1 INTRODUCTION

In this chapter, we describe recent advances of electrodes that are chemically modified with integrated biomolecules and molecular wires. Both hydrogels and self-assembled organized multilayer platforms will be treated in the following sections.

Electrode reactions provide a switch between two phases of different electrical conductivity, i.e. a metal electrode with electronic conductivity and an electrolyte with ionic conductivity. Therefore electrochemical reactions at electrodes are ideal transducers of molecular information into electrical signals with large amplification provided by the Faraday constant, ca 96,500 coulomb per mole. However, electrode reactions lack the needed molecular specificity and thus, development of chemically modified electrodes during the last 25 years has attempted to solve this limitation [1]. Chemically modified electrodes are electrode surfaces modified by suitable molecules or polymers with known specific interaction towards a target analyte molecule that enhance the specificity

* Corresponding author. Tel.: +54-11-4576-3378x120; fax: +54-11-4576-3341. *E-mail address:* calvo@q1.fcen.uba.ar.

*Comprehensive Analytical Chemistry XXXIX, pages 327–376*


ISSN: 0166-526X

and at the same time provide mechanisms for electron transfer to or from the underlying electrode.

In this sense, protein films on electrode surfaces can be regarded as model systems of electrochemical biosensors based on molecular recognition and electrical signal generation. Through electrocatalysis of 'wired' or electrically connected enzymes to electrodes via molecular 'wires' which can shuttle electrons to or from the enzyme prosthetic group to the electrode conveying an electrical signal through the external circuit, and yet keeping the biocatalytic activity.

Hydrogels on surfaces, obtained either by random electropolymerization or sol–gel transformation, provide a convenient aqueous environment for the biomolecules to attain stability and preserve their catalytic activity at the electrode. Self-assembled protein thin films on electrode surfaces, on the other hand, have the advantage over hydrogels in that the same biological components and molecular wires are employed with better control over the molecular orientation and organization at the nanoscale level.

Electrochemical transduction with chemically modified enzyme electrodes has advantages over radiochemical and fluorescent methods in that direct recognition and transduction can be accomplished in complex turbid samples. For instance, commercial systems [2] are available that monitor glucose in a tiny drop of whole blood taken from a Diabetes Mellitus patient [3,4]. Glucose levels in the range 20–600 mg $dL^{-1}$ can now be measured within 20 s from a drop of whole blood of volume 0.3–3.5 μL using integrated chemical sensors. Also, implantable glucose sensors have been recently developed.

Importantly, a number of electrochemical biosensors [5], based on molecular recognition systems utilizing redox enzymes and enzyme-labeled antibodies or oligonucleotides, allow the detection of biomolecules of clinical or environmental relevance through enzyme–substrate, antigen (hapten)–antibody or ss-DNA hybridization interaction and transduction of this recognition by an integrated signal generation process. These strategies lead potentially to intelligent enzyme-labeled immuno or genomic biochips where the analyte can be supplied by a microfluidic field in the same chip. Applications of lab-on-chip to genomics and proteomics, microarray technologies for drug discovery, etc. are already under development with microfluidics dynamics, microelectrophoretic separations or microdialysis for 'in vivo' monitoring. These applications are described in other chapters of this book.

## 7.2 ENZYME REDOX CATALYSIS

In order to model the kinetics of immobilized enzymes and redox mediators shuttling the electrons to and from electrodes, the equations describing diffusion and enzyme reaction must be solved if a complete description of the system is to be obtained. Since the enzyme Michaelis–Menten kinetics is non linear, no simple solution equations in a closed form are possible in the general case to describe the catalytic current response to substrate and co-substrate.

The flavoprotein glucose oxidase (GOx) has certainly been one of the most widely used enzymes in enzyme electrodes. The interest in glucose electrodes as biosensors derives from the large number of blood glucose assays (in excess of 1 billion in the USA) that are performed daily. The requirement then, is a robust and easy method to handle a health care devices for self-monitoring diabetes [6].

The mechanism of enzyme catalysis [7] in the oxidation of β-D-glucose, S, to glucono-δ-lactone, P, and the concomitant reduction of the natural or artificial co-substrate, redox mediator, Med, is depicted in Fig. 7.1.

The redox co-substrates can be regenerated over an electrode surface poised at the proper electrode potential:

$$\mathrm{Med} \rightarrow \mathrm{Med}^{+} + \mathrm{e}^{-} \tag{7.1}$$

The contribution of Cass et al. [8] represents the first report of glucose biosensors using ferrocene to mediate the electron transfer between the redox site of GOx and the electrode surface. In general two types of

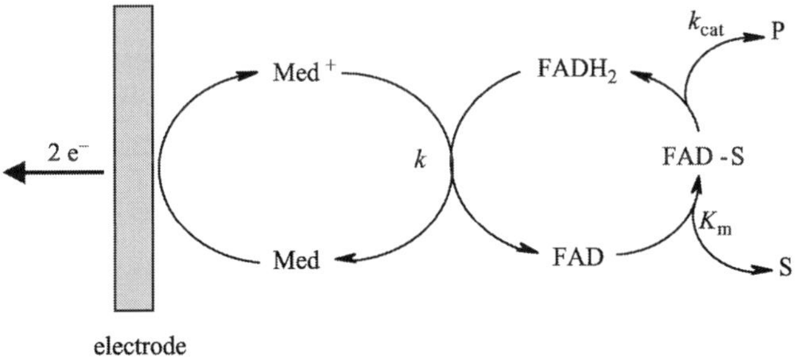


Fig. 7.1. Catalytic cycle of enzymatic oxidation of β-D-glucose in the presence of a redox mediator, Med.

artificial redox co-substrates or mediators have been employed: two-electron and two-proton redox couples, such as quinone/hydroquinone [7] and one-electron redox couples [8]. The electrical communication between buried electroactive sites in redox enzymes and electrodes has been achieved by using organometallic compounds like ferrocene and its derivatives [8–10], and osmium [11, 12] or ruthenium complexes [13].

The determination of biocatalytic kinetic parameters from electrochemical experiments is based on the catalytic currents obtained in the presence of the redox mediator, enzyme and enzyme–substrate, namely glucose. Poising or scanning the electrode potential results in the conversion of the reduced mediator into the oxidized form. It is important that the redox mediator is stable in its different oxidation states to achieve reversible heterogeneous kinetics and is able to react rapidly with the enzyme [14]. The addition of glucose to the solution increases the anodic current due to the regeneration of the mediator through the catalytic cycle as shown in Fig. 7.1.

The first electrochemical studies were done with enzyme and mediator in solution employing cyclic voltammetry [15, 16]. Assuming a steady-state approximation for FAD, FAD-S and $FADH_2$ concentrations one can obtain:

$$\frac{\partial[\mathrm{Med}^+]}{\partial t} = D\frac{\partial^2[\mathrm{Med}^+]}{\partial x^2} - \frac{2kC_{\mathrm{E}}^0[\mathrm{Med}^+]}{1+k\left(\dfrac{1}{k_{\mathrm{cat}}}+\dfrac{K_{\mathrm{M}}}{[\mathrm{S}]}\right)[\mathrm{Med}^+]} \tag{7.2}$$

where $D$ is the diffusion coefficient of the mediator; $C_{\mathrm{E}}^0$, the total concentration of enzyme; $k_{\mathrm{cat}}$, the enzyme turnover; $k$, the bimolecular re-oxidation rate constant; $K_{\mathrm{M}}$, the Michaelis constant; $t$, time and $x$, the distance from the electrode.

When large glucose concentrations, $S$, are employed, one can consider the limiting pseudo-first-order kinetics where the Michaelis–Menten kinetics term is approximated by a second-order term:

$$\frac{\partial[\mathrm{Med}^+]}{\partial t} = D\frac{\partial^2[\mathrm{Med}^+]}{\partial x^2} - 2kC_{\mathrm{E}}^0[\mathrm{Med}^+] \tag{7.3}$$

Bourdillon and co-workers [17] realized the difficulties with Eq. (7.3) to approximate Eq. (7.2): a large concentration of substrate is not a sufficient condition for the pseudo-first-order approximation; in addition, $k[\mathrm{Med}^+]/k_{\mathrm{cat}} \ll 1$, should hold. In this way they worked out a closed-form expression for the plateau current in cyclic voltammetry when a one-electron mediator re-oxidizes GOx. When the concentration of glucose is at least 50 times the mediator concentration and the enzyme is saturated in the substrate, then:

$$\frac{I_{\text{cat}}}{I_{\text{dif}}} = \frac{1}{0.446}\left(\frac{2kC_{\text{E}}^{0}RT}{\nu F}\right)^{1/2}\left(\frac{2}{\sigma}\left(1 - \frac{1}{\sigma}\ln(1+\sigma)\right)\right)^{1/2} \tag{7.4}$$

where $I_{\text{cat}}$ is the catalytic plateau current and $I_{\text{dif}}$ is the peak current of the reversible wave of the mediator in the absence of glucose, $\nu$, is the sweep rate and $R$, $T$ and $F$ have the usual meaning and $\sigma$ is given by:

$$\sigma = \frac{k[\text{Med}^{+}]}{k_{\text{cat}}}\left(1 + \frac{K_{\text{M}}}{[\text{S}]}\right) \tag{7.5}$$

The slope of $I_{\text{cat}}/I_{\text{dif}}$ versus $\nu^{-1/2}$ allows one to obtain $k$, by an iterative procedure. However, if $k_{\text{cat}}$ is considerably higher than $k[\text{Med}^{+}]$, a condition that can be achieved using small concentrations of mediator, Eq. (7.5) can be transformed by expanding in a Taylor series the term $\ln(1+\sigma)$ to yield:

$$\frac{I_{\text{cat}}}{I_{\text{dif}}} = \frac{1}{0.446}\left(\frac{2kC_{\text{E}}^{0}RT}{\nu F}\right)^{1/2} \tag{7.6}$$

This is the same expression derived by Nicholson and Shain [18] for an EC' mechanism applied to the re-oxidation of GOx with a one-electron mediator.

Numerical techniques have also been employed to solve transient and steady-state concentration profiles at electrodes based on immobilized glucose oxidase [19]. It is possible, however, to obtain approximate analytical solutions for limiting cases and to linearize the diffusion-kinetic equations.

By studying a monolayer of glucose oxidase immobilized by an antigen–antibody procedure [20] with ferrocene methanol in solution as a redox mediator, Bourdillon and Saveant derived the expression:

$$i = -FD\left(\frac{\partial[\text{Med}^{+}]}{\partial x}\right)_{0} + \frac{2Fk[\text{Med}^{+}]_{0}\Gamma_{\text{enz}}}{1 + k[\text{Med}^{+}]_{0}\left(\dfrac{1}{k_{\text{cat}}} + \dfrac{K_{\text{M}}}{[\text{S}]}\right)} \tag{7.7}$$

where $[\text{Med}]_{\text{TOT}}$ represents the analytical concentration of mediator in solution and $\Gamma_{\text{enz}}$ is the enzyme surface concentration. The first term in Eq. (7.7) describes the current observed in the absence of glucose if the $Med^+$/Med couple obeys the Nernst law:

$$[\mathrm{Med}^+] = [\mathrm{Med}]_{\mathrm{TOT}} \frac{\mathrm{e}^{(F/RT)(E-E'\mathrm{o})}}{1+\mathrm{e}^{(F/RT)(E-E'\mathrm{o})}} \tag{7.8}$$

The second term represents the current increase due to the catalytic reaction:

$$i_{\mathrm{cat}} = \frac{2Fk[\mathrm{Med}^+]\Gamma_{\mathrm{enz}}}{1+\exp\left[-\dfrac{F}{RT}(E-E^{\mathrm{o}\prime})\right] + k[\mathrm{Med}^+]\left(\dfrac{1}{k_{\mathrm{cat}}}+\dfrac{K_{\mathrm{M}}}{[\mathrm{S}]}\right)} \tag{7.9}$$

Rearranging Eq. (7.9), Bourdillon et al. [20] suggested the possibility of working with 'primary' and 'secondary' linear plots:

$$\frac{1}{i_{\mathrm{cat}}} = \frac{1}{2Fk\Gamma_{\mathrm{enz}}}\frac{1}{[\mathrm{Med}^+]} + \frac{1}{2F\Gamma_{\mathrm{enz}}}\left(\frac{1}{k_{\mathrm{cat}}}+\frac{K_{\mathrm{M}}}{[\mathrm{S}]}\right) \tag{7.10}$$

For each glucose concentration, a plot of $i_{\mathrm{cat}}^{-1}$ versus $[\mathrm{Med}^+]^{-1}$ should give a straight line. A primary plot can be derived from a cyclic voltammogram recalling the Nernst equation, since it is possible to vary the mediator concentration by means of the electrode potential and $k$, the bimolecular rate constant, can be calculated if $\Gamma_{\mathrm{enz}}$ is known (e.g. from radio-labeling experiments).

From Eq. (7.10) the intercept values of primary plots when plotted against $1/[S]$ results in the secondary plot, which yields the characteristic rate constants of the immobilized enzymes. In this chapter, the authors have demonstrated that all enzyme and mediator constants were the same when determined in homogeneous solution. The methodology has been extended to other systems: self-assembed biotinylated GOx over immobilized avidin [21] and multilayers of antigen–GOx modified antibody [22].

Conducting the experiment described requires reversible fast mediator kinetics at the electrode. However, when the catalytic current reaches a plateau far beyond the formal potential of the redox couple, the reversibility of the mediator is no longer a prerequisite in Eq. (7.10). Note that for the soluble mediator case this strategy requires changing the mediator concentration in each experiment.

For thin films where the redox mediator is not soluble but integrated with the surface enzyme layer, such as an osmium complex attached to a polymer, and in the absence of glucose mass transport hindrance, the expression for the current density is given by [15, 22]:

$$i_{cat} = \frac{2Fk_{cat}\Gamma_{enz}}{1 + \dfrac{k_{cat}}{k[\mathrm{Med}]} + \dfrac{K_m}{[\mathrm{S}]}} \tag{7.11}$$

The Michaelis–Menten constant for β-D-glucose is $K_{MS} = (k_{-1} + k_{cat})/k_1$; $k_{cat}$ and $k$ are the enzyme turnover (glucose oxidation) and the bimolecular $FADH_2$ re-oxidation rate constant, respectively, and $\Gamma_{enz}$ is the surface concentration of the total active enzyme wired by the integrated redox mediator, [Med], while [S] is the concentration of soluble β-D-glucose.

## 7.3 REDOX HYDROGELS

The integration of biomolecules on electronic transducers requires the ability to engineer a device where the biological component is immobilized and highly efficient communication between both components is achieved. The development of integrated devices has been of significant relevance in the field of biosensors during recent years. Different strategies have combined a biomolecule involved in a recognition event (such as antibody, enzyme, and receptor) with electronic transducers. It has been demonstrated for many biologically active compounds, mainly enzymes, that direct electron transfer from a conducting surface to the catalytic site is not possible. Therefore, small molecules shuttling electrons to and from the enzyme active site were used as redox mediators with enhancement of the electrical communication and increase of the signal output.

The chemistry required for biomolecules immobilization can lead to monolayers or multilayer structures and has been described in several review articles [23–26]. The integrated components can be confined in a three-dimensional structure with a random distribution. However, more ordered structures can be achieved using layer-by-layer assembling strategies.

Initial attempts to integrate a redox enzyme such as GOx were focused on covalent attachment of redox molecules acting as electron shuttles to reactive sites within the protein structure. Different mediators were bound to amine groups [27–30]; alternatively, carbohydrate residues were also modified [31]. The effective modification of the enzyme was demonstrated by the quenching of tryptophan fluorescence and it was suggested that enhanced electron transfer rates should be expected if acid residues were modified [32].

Electrochemical studies showed evidence of both intramolecular and intermolecular redox mediation for the electrochemical re-oxidation of the modified glucose oxidases in the presence of glucose. The adsorption of modified enzymes on electrodes exhibited different behavior with the different surfaces suggesting that orientation of the adsorbed enzyme is induced due to the interaction with the mediator. Construction of enzyme switches based on these modified enzymes was also carried out, and their responses were compared with those obtained using native glucose oxidase and a soluble redox mediator.

More effective redox mediation was achieved upon integration of soluble redox polymers in three-dimensional structures. Heller et al. designed redox hydrogels (Fig. 7.2) where an electrostatic complexation between the enzyme and the polymer initially occurred [33].

Finally a three-dimensional gel was formed on the electrode after addition of a cross-linking agent such as di-epoxides [34–36]. This strategy proved to be successful with large current densities in the presence of freely diffusing substrate and anaerobic conditions. It was possible to eliminate the current resulting from interfering species such as ascorbate, urate and acetaminophen when immobilizing peroxidase

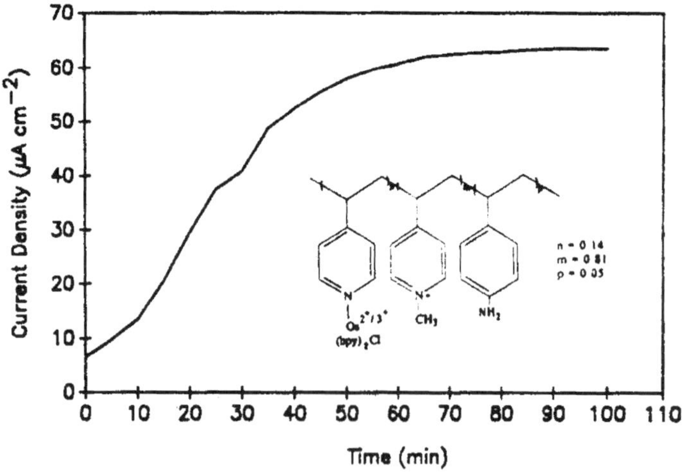


Fig. 7.2. Redox hydrogel designed by Heller et al. [33] to 'electrically' wire enzymes and catalytic amperometric response.

in the external layer [37, 38]. Furthermore, in some cases a poly(vinylimidazole) polymer was used in order to obtain redox hydrogels presenting less positive oxidizing potentials which avoided these interfering effects [39, 40]. This redox polymer proved to be useful in the design of a subcutaneously implantable glucose electrode [41]. A one-point calibration method was proposed for implanted electrodes even for rapidly changing glucose concentration [42].

Most of the work was focused on the GOx enzyme, but other enzymes were also electrically 'wired' [43–49]. Heller [50] and Anzai [51] have also described wired lactate oxidase (LOx) enzyme electrodes for the determination of lactate.

The electrical communication between acrylic copolymers and GOx was also reported [52]. Theoretical aspects of these amperometric modified electrodes were evaluated and compared with the experimental results [53]. The catalytic cycle corresponding to glucose oxidase was represented as:

$$R \rightarrow O + e \tag{7.12}$$

$$GOx(FAD) + S \underset{k_{-1}}{\overset{k_1}{\rightleftarrows}} GOx(FAD - S) \xrightarrow{k_{cat}} GOx(FADH_2) + P \tag{7.13}$$

$$GOx(FADH_2) + O \xrightarrow{k_1} GOx(FADH) + R + H^+ \tag{7.14}$$

$$GOx(FADH) + O \xrightarrow{k_2} GOx(FAD) + R + H^+ \tag{7.15}$$

The concentrations of oxidized and reduced redox sites in the polymer, soluble substrate and the different forms of enzyme entrapped in the gel layer are described by $c_O$, $c_R$, $c_S$, $c_{EO}$, $c_{ER}$ and $c_{ET}$, respectively. The electrode reaction given by Eq. (7.12) takes place at the electrode–gel interface with fast kinetics.

The electron transfer to heme-porphyrin in peroxidase was also demonstrated according to the reactions [54]:

$$H_2O_2 + HRP_{(red)} \rightarrow H_2O + HRP_{(ox)} \tag{7.16}$$

$$HRP_{(ox)} + 2R \rightarrow HRP_{(red)} + 2O \tag{7.17}$$

$$2O + 2e \rightarrow 2R \tag{7.18}$$

Heller et al. obtained larger currents by copolymerizing acrylamide and vinylimidazole with osmium redox sites and different oxidases were electrically wired [53].

Poly(allylamine) was also modified with redox compounds and further cross-linked to redox enzymes showing three-dimensional charge propagation. However, the charge of the redox species proved to be critical when obtaining catalytic currents [55–56].

Fast measurements in non-steady-state conditions were performed with a quartz crystal microbalance and the viscoelastic changes for ferrocene redox hydrogels were followed with simultaneous cyclic voltammetry responses [55–56].

The effect of ionic strength on the electron-transfer behavior of enzyme electrodes connected through ferrocene hydrogel was reported. It was shown that larger electrolyte ionic strength increased the concentration of redox sites in the film while the diffusion coefficient decreased slightly. The results were explained in terms of swelling and shrinking effects due to water and ion exchange inside the hydrogel [57].

The modification of poly(allylamine) with osmium redox complexes (Fig. 7.3) was also achieved [58] and higher catalytic currents as well as electrode stability were attained after wiring several redox enzymes [58].

The integration of osmium polymers also proved to be effective in the development of miniature sensors, particularly for the detection of pyruvate in biological fluids [59]. The fabrication of a bienzymatic microelectrode for glutamate detection was reported recently.

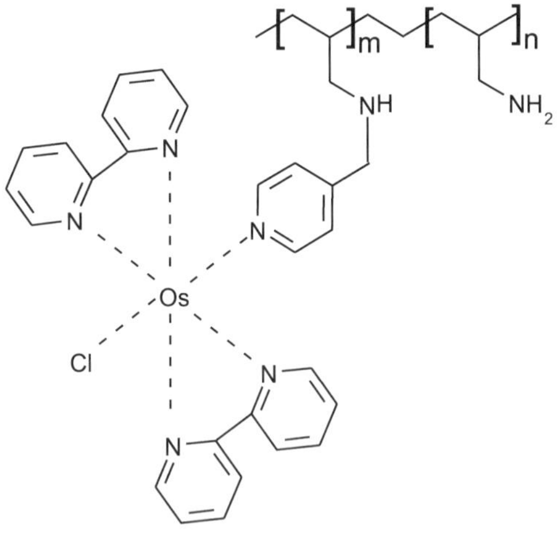


Fig. 7.3. Chemical structure of the redox polymer of poly(allylamine) modified with the osmium complex [$Os(bpy)_2ClPyCOH^+$] (PAH-Os).

This sensor was able to operate at considerably lower potentials and the possibility of monitoring in situ glutamate secretion was demonstrated.

The formation of electrostatic complexes between enzymes and redox polymers was not only limited to biosensor design but also to other relevant applications such as biofuel cells. The wiring of laccase enzyme on a cathode for oxygen electroreduction to water proved to be successful showing high current density output [60]. Finally, the combination of a microanode catalyzing the electroxidation of glucose and the laccase microcathode for oxygen reduction was reported as a high power density biofuel cell [61].

Another relevant approach involving redox enzymes integrated on modified electrodes was presented for RNA and DNA detection [62]. An initial deposition of acrylic-vinylimidazole redox copolymer and avidin was made on the electrodes. These electrodes were further employed for biotinylated oligonucleotide immobilization and sequential hybridization to target DNA or RNA analyte. Finally, a second HRP-labeled complementary oligonucleotide was added for enzyme-amplified amperometric detection [62].

A different approach for integration of biomolecules on modified electrodes is possible through electropolymerization of conducting polymers [63–65]. The properties of the resulting films can be adapted according to the polymerization parameters. The combination of electroactive monomers and redox molecules covalently attached to them has allowed enzyme entrapment in a film presenting improved electron transfer kinetics. The resulting electrodes showed less oxygen interference and efficient substrate transport through the film [66].

Electrochemically mediated deposition of GOx and non-conducting monomers was also reported and exhibited improved enzyme stability and interference rejection. These microbiosensors worked at high potentials oxidizing hydrogen peroxide produced in the catalytic reaction and regenerated oxygen within the film. The sensor also showed high sensitivity and short response times.

Reagentless amperometric biosensors were also reported using sol–gel technology. TTF initially adsorbed on the electrode surface was used as mediator and finally a sol–gel/copolymer film was applied to the electrode to immobilize HRP. The resulting film combined the rigidity of the sol–gel and the biocompatibility of hydrophilic polymers [67].

Bioencapsulation in sol–gel-built materials has proved to be successful for entrapment of biomolecules such as proteins, antigens,

DNA, and RNA, while retaining their full biological function and enabling the fabrication of bioactive nanocomposites.

Quinn and Heller have also reported a biocompatible, glucose-permeable hydrogel for in situ coating of implantable biosensors formed by cross-linking an eight-armed (amine-terminated polyethylene glycol) PEG, derivatized with a di-succinimidyl ester of an $\alpha,\omega$-di-propionic acid derivative of PEG [68].

Another interesting application of a redox hydrogel biosensor has been reported by Corton and Battaglini [69] to measure lactate in milk. The authors cross-linked osmium derivatized poly(allylamine), PAH-Os, redox polyelectrolyte with lactate oxidase (LOx) using poly(ethylene-diglicidyl) ether, PEG, as bifunctional cross-linking agent to build a bienzymatic electrode (Fig. 7.4) responsive to lactate (Fig. 7.5) [70].

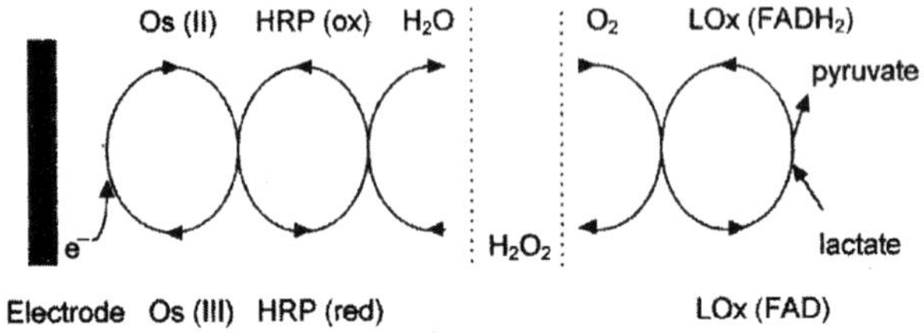


Fig. 7.4. Bienzymatic electrode for lactate measurement with lactate oxidase (LOx), peroxidase, and PAH-Os redox polymer wire [69].

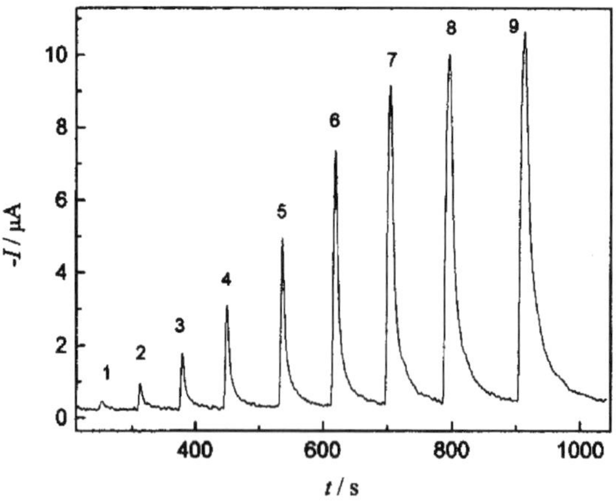


Fig. 7.5. Flow injection analysis response of a LOx/HRP bienzymatic electrode after successive additions of lactate in milk samples [70].

## 7.4 SELF-ASSEMBLED POLYELECTROLYTE AND PROTEIN FILMS

The buildup of organized protein films on surfaces represents one of the major growth areas overlapping the boundary between materials science, biology, biomedicine and microelectronics. Among other applications, protein films have been used in biosensors, medical diagnostics, bioseparations, catalysis [71–76], bio-compatibility to produce surfaces that prevent adsorption of proteins [77], to induce time-delayed specific responses at the cellular level, to modify the surface of human repair devices, or to be in contact with biological fluids or tissues (catheters, prostheses, storage vessels, etc.).

Proteins can be immobilized on solid surfaces using different techniques such as physical adsorption, entrapment, chemical binding, and electropolymerization [72–76]. In surface protein films inhomogeneities, protein denaturation and restricted film permeability to enzyme substrates are limitations to be overcome. Ordered protein multilayer films with a high protein density have been constructed by: (a) Langmuir–Blodgett deposition [73–76]; (b) biospecific interactions [78, 79] such as avidin-biotinylated protein and antigen–antibody; and (c) electrostatic layer-by-layer deposition of proteins and polyelectrolytes of different electrical charge [72].

The alternating adsorption of charged macromolecules or layer-by-layer adsorption (LBL) has recently emerged as a promising simple method to fabricate controlled and highly ordered molecular assemblies in a pre-designed architecture [80–82]. This strategy is based on the stepwise adsorption of charged species onto a charged substrate, primarily by electrostatic interactions with charge overcompensation in each deposition cycle, thereby inducing the adsorption of the next layer of oppositely charged macromolecules.

A variety of polyelectrolyte multilayer films has been reported with the LBL technique, in particular the self-assembly of soluble proteins [72,73,83–90]. There are also examples of protein self-assembled films [81,85,91–93] such as immunoglobulin G (IgG), anti-immunoglobulin G (a-IgG), cytochrome C, lysozyme, myoglobin, hemoglobin, peroxidase, and glucose oxidase. It is generally accepted that when biomolecules such as enzymes or antibodies are inserted between different polyelectrolyte layers the biological activity results additive. In some examples two oppositely charged proteins separated by at least one polyelectrolyte layer pair were inserted in the film [86,91] while in the case of antibodies inserted in the polyelectrolyte film, it was shown that the embedded proteins kept their reactivity with respect to their antigenic reaction when not more than four polyelectrolyte layers were deposited over them [87].

Decher et al. [82,94] studied the adsorption of negatively charged human serum albumin (HSA) onto a positively charged poly(ethylenimine) (PEI)-$(PSS\text{-}PAH)_3$ film or a negatively charged PEI-$(PSS\text{-}PAH)_3$-PSS polyelectrolyte film. They observed that HSA adsorbs on both types of surfaces and that on PSS-terminating films, which are similarly charged as albumin, only monolayer adsorption of HSA was found, whereas on the oppositely charged PAH-terminating films the adsorbed protein layers extended over thicknesses larger than four times the largest dimension of the HSA molecule. They also demonstrated that the intermolecular interactions involved in these processes are mostly of electrostatic origin.

## 7.5 SELF-ASSEMBLED ENZYME FILMS

It is worthwhile noting that in hydrogels there is a random distribution of components with little control over the molecular orientation while spatially ordered enzyme assemblies built with the same active components of the hydrogels offer several advantages over random polymers [95,96]. The step-by-step electrostatic adsorption between a charged surface and oppositely charged molecules in solution introduced by Decher with molecular-level control [97–100] offers the possibility of regulating the adsorption and restricting the deposition to a monolayer. The layer-by-layer electrostatic adsorption between a charged surface and oppositely charged macromolecules in solution with surface charge reversal brings the possibility of: (i) adsorption regulation by repulsion of soluble molecules of equal charge; and (ii) adsorption of oppositely charged soluble molecules on top of the surface layer [100].

As explained for redox hydrogels, electrically 'wired' enzymes have attracted attention recently because of their potential applications in biosensors and molecular devices [27]. Molecular recognition with enzymes and enzyme-labeled immuno and genomic electrodes based on antigen–antibody interaction and single stranded DNA (ss-DNA) hybridization with integrated redox relays at the surface to generate an electrical signal can be integrated in electronic circuits or biochips [101].

In large proteins like GOx (186,000 g $mol^{-1}$), direct electron transfer from the electrode surface to the prosthetic group, $FADH_2$, buried inside the protein structure is hindered. Heller demonstrated that electrical communication between the $FADH_2$ in glucose oxidase and electrodes can be facilitated by electrostatic complexing the negatively charged enzyme in a solution of pH above the isoelectric point (4.2) with a cationic quaternized poly(vinylpyridine) and poly(vinylpyridine)$Os(bpy)_2Cl_2$ redox mediator polyelectrolyte copolymer [102]. Heller then introduced a two-component epoxy technique combining GOx and other oxidases with the polycationic redox mediator cross-linked with a bifunctional reagent in a three-dimensional random polymer [34–36].

Random hydrogel segregation of hydrophobic phases may occur at high protein concentrations. Self-assembled enzyme films, on the other hand, are better suited for the study of the mechanisms involved in molecular recognition by the enzyme, redox mediation by the polymer and signal generation with a more homogeneous distribution of the components. Also, the enzyme concentration can be better quantified during the deposition process and the redox mediator concentration can be obtained by integration of the voltammetric surface waves in the absence of substrate [21]. In addition, well-defined spatial distribution of enzyme with respect to the mediator can be achieved and substrate mass transport limitations minimized in ultrathin films. Multienzyme systems can be assembled with precise control of position and distance achieving directional electron transfer to the sequential enzyme reactions and molecular wires.

Among chemically modified electrodes [1], integrated chemical systems with a high degree of organization in supramolecular architectures [103] have been developed with enzymes in ordered monolayers and multilayer systems using different assembling techniques: Langmuir–Blodgett [104], self-assembled monolayers [105–107], step-by-step electrostatic adsorption of alternate multilayers [86–88], antigen–antibody interaction [20,108–113], avidin–biotin interaction [21,114–117], surfactant films [118, 119], electrostatic adsorption of hyperbranched polyelectrolytes [120], redox-modified dendrimers [121], etc.

For glucose oxidase monolayers assembled on carbon surfaces by the antigen–antibody binding, three kinetic barriers for mediated glucose oxidation have been identified: (i) diffusion of a soluble mediator to the enzyme surface; (ii) positioning with respect to the $FADH_2$ moiety; and (iii) electron transfer from $FADH_2$ to the redox molecule [20].

Furthermore, since the diffusion of the soluble mediator towards the enzyme is fast, $k_D \gg 5 \times 10^8\ M^{-1}\ s^{-1}$, Saveant et al. concluded that the positioning of the mediator with respect to the prosthetic site ($FADH_2$) should be the rate-determining step [122].

A remarkable open structure was obtained after multilayer deposition of several antigen–antibody layers for enzyme conjugate immobilization. Furthermore the resulting film was fully active upon substrate addition [22]. The characterization of these assembled structures containing active and inactivated enzyme layers also showed the low compactness of the enzyme films. In this case, the film was enlarged and therefore, it was possible to evaluate the competition between the enzymatic reaction and mass transport within the multilayer structure [112]. The glucose oxidase conjugate was also applied to the study of biorecognition dynamics in antibody–antigen binding [123].

The design of spatially ordered assemblies was also possible using the avidin–biotin binding reaction [21]. A monolayer-assembled film using this interaction allowed the simultaneous immobilization of glucose oxidase and an electron mediator. In this work, glucose oxidase conjugated avidin was initially attached to covalently bound biotin and further interaction of a long chain biotinylated ferrocene to the remaining vacant sites determined mediator co-immobilization. The catalytic currents were at least 20 times larger when soluble mediator was used as compared to a self-contained monolayer of ferrocene and GOx. For the long chain biotinylated ferrocene in solution, a decrease in $k$ by a factor of 17 was reported and attributed to its larger viscosity with respect to ferrocene methanol. Furthermore, a decrease by a factor of 60 in $k$ was found in the GOx monolayer with co-immobilized ferrocene. In this work, the authors assumed that the same concentration of GOx that could be oxidized by the ferrocene methanol was oxidized by the immobilized long chain ferrocene.

The dynamics of the biotin–avidin multilayer film was evaluated in terms of catalytic responses in order to prove the spatial order and to estimate the average distance between layers [116].

Avidin–biotin technology also proved to be useful for a step-by-step construction of bienzyme electrodes where two monolayers of biotinylated hexokinase on top of five monolayers of biotinylated glucose oxidase were built up, using avidin as molecular cement [124].

Moiroux et al. integrated mediator and enzyme in multilayer films with active and inactive enzyme [125].

Molecular monolayers of immunoglobulins bearing terminally attached ferrocene poly(ethyleneglycol) chains, IgG-PEG-Fc, were self-assembled LBL at an electrode surface through antigen–antibody recognition with molecular spacers [125, 126]. Further immobilization of glucose oxidase conjugate yielded catalytic currents in the presence of β-D-glucose solutions. By using molecular spacers, Moiroux et al. concluded that a catalytic current is obtained only when the redox heads can reach the enzyme active sites and that the rate of enzyme re-oxidation in the integrated mediator system cannot compete kinetically with freely diffusing dioxygen or ferrocene methanol. The quantitative analysis was carried out with an elastic bound diffusion and electron propagation model previously described by the authors [126].

Willner and co-workers have reported several monolayer enzyme self-assembled systems [127–129]. The authors have also reviewed the surface functionalization of electrodes with covalently linked enzymes in integrated layered redox proteins and conductive supports for bioelectronic applications [26,130].

The reconstitution of a monolayer of GOx electrically connected to the electrode by the pyrroloquinoline quinone (PQQ) redox relay has been reported [131]. In the presence of glucose the electro-biocatalyzed oxidation of the substrate takes place as indicated by the anodic current, whereas the PQQ/FAD monolayer alone does not catalyze the oxidation of glucose. The novel method to assemble a GOx monolayer electrode via reconstitution of the apo-protein with the PQQ/FAD monolayer, yields a functionalized electrode for the electroxidation of glucose at an unprecedentedly high rate, reaching the theoretical limit for a maximally packed enzyme monolayer [131].

A novel bioelectrode based on self-assembled multilayers of poly-phenol oxidase intercalated with cationic polyallylamine built up on a thiol-modified gold surface has been reported by Solis and co-workers [132]. Hodak et al. described the redox mediation of glucose oxidase in self-assembled structures of cationic poly(allylamine) modified by ferrocene (PAH-Fc) [133] or an osmium complex (PAH-Os) [134] and anionic GOx deposited stepwise in alternate polymer enzyme multi-layers. Figure 7.6 depicts schematically an organized multilayer composed of redox polymer (PAH-Os) and GOx on a flat gold surface (Fig. 7.6a) and the amperometric response to glucose (Fig. 7.6b).

It was found that the redox charge and the amount of enzyme increases in step with the number of multilayers deposited. However, only a small fraction of the active assembled GOx molecules observed

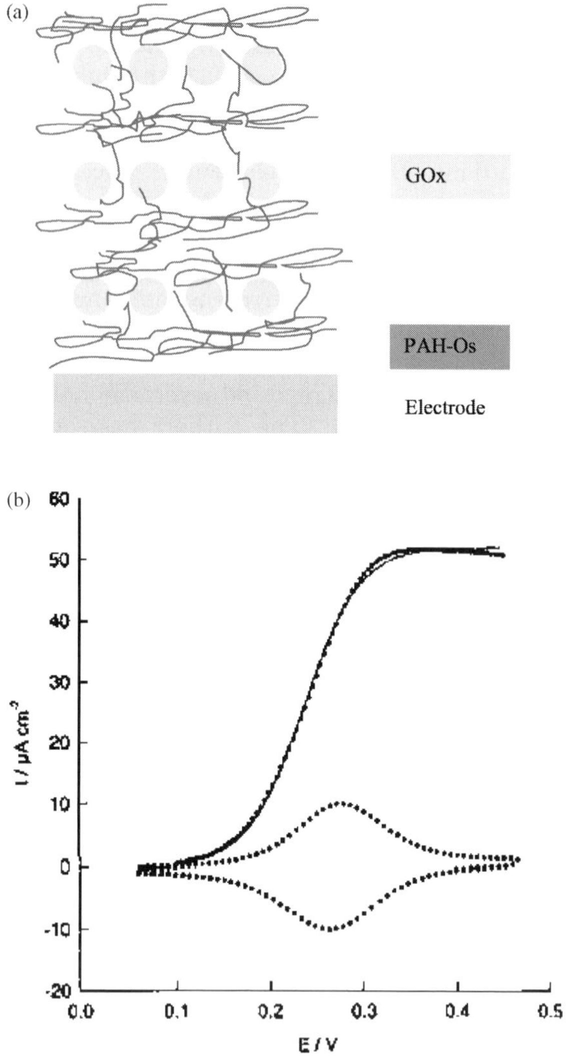


Fig. 7.6. (a) Schematic representation of a layer-by-layer self-assembled multilayer structure composed of redox polymer (PAH-Os) and GOx. (b) Voltamperometric catalytic response of a $(\text{PAH-Os})_X(\text{GOx})_X$ self-assembled film in presence of β-D-Glucose [134] and cyclic voltammogram in the absence of glucose.

were 'electrically wired' by the ferrocene polymer in the integrated chemical structure [133].

The behavior of single enzymes such as glucose oxidase (GOx), lactate oxidase (LOx) and soy bean peroxidase (SBP) and multienzyme systems (GOx + SBP) or (GOx + GDH, glucose dehydrogenase) in electrostatically self-assembled multilayers with PAH-Os has been reported [96]. These films were formed by adsorption through sequential immersion in the respective redox polycation and enzyme solutions of pH above the isoelectric point, where the enzymes carry a net negative charge [96]. The method allows the construction of well-ordered systems and quantitative determination of the enzyme electrical 'wiring' efficiency, kinetic coefficients of the reactions involved and the effect of the number of adsorbed layers. These studies led to the conclusion that although most of the enzyme molecules in the surface film were active upon adding soluble redox mediator, only a small fraction of these active assembled GOx molecules were 'electrically wired' by the ferrocene [133] or osmium polymer mediator in the integrated chemical structure [134].

## 7.6 ELECTROCATALYSIS

Different supramolecular [135, 136] multilayer structures comprised of monolayers of glucose oxidase (GOx), FAD-free glucose oxidase (Apo-GOx) and GOx with an Os complex derivatized poly(allylamine) (PAH-Os) have been built by alternate layer-by-layer (LBL) electrostatic adsorption in a self-assembly process. The resulting modified electrodes were tested as reagentless glucose biosensors and the enzyme kinetic parameters, $k$[Os], and surface concentration of 'wired' enzyme $\Gamma_{enz}$ have been obtained by analysis of the catalytic current dependence on glucose concentrations: assuming $k_{cat}$ and $K_M$ values for the ping-pong mechanism of glucose oxidase. The osmium volume concentration was estimated by integration of the redox charge density of cyclic voltammetry in the absence of glucose and the ellipsometric thickness of the multilayers, assuming homogeneous distribution of redox sites in the enzyme layer volume. The total enzyme surface concentration was measured with a quartz crystal microbalance (QCM) in each adsorption step (see below). The amperometric response of the reagentless biosensors was analyzed in terms of the fraction of 'wired' enzyme and the bimolecular rate constant for $FADH_2$ oxidation by the redox polymer for the different multilayers.

A catalytic current increase with the number of LBL layers due to the increase in the enzyme loading was found (Fig. 7.7). The efficiency of enzyme re-oxidation by the Os redox polymer, however, was found to decrease as more assembled layers were stacked because of electron-hopping diffusion limitations [136].

### 7.6.1 Electron hopping

Heller has estimated the current that flows from a monolayer of GOx covering one square centimeter as follows [6]: each enzyme molecule oxidizes 700 glucose molecules per second yielding 2 electrons to the external circuit. From the X-ray crystallographic enzyme dimensions [137] there is an estimated $1.3 \times 10^{12}$ molecules per square centimeter of densely packed GOx on the surface. The resulting maximum catalytic current density can be estimated at the level of 400 $\mu A\ cm^{-2}$, much more than anyone has measured in a mediated amperometric glucose oxidase electrode for a GOx monolayer. Therefore, we should conclude that the wiring efficiency is less than that calculated or that some enzyme deactivation has taken place.

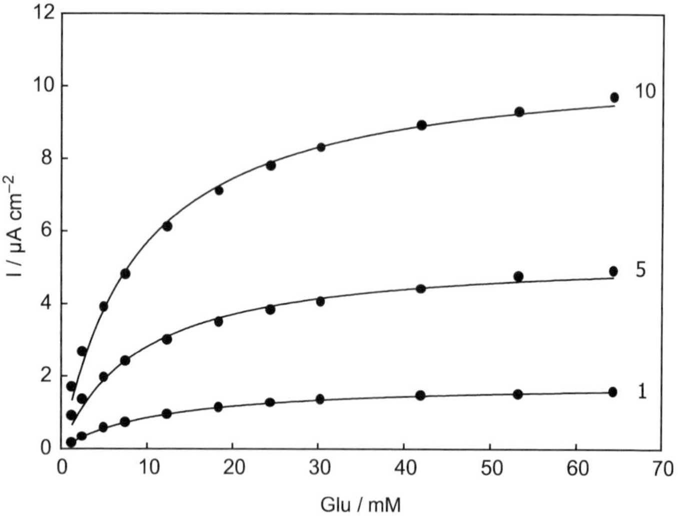


Fig. 7.7. Catalytic current vs. glucose for 1, 5, 10 PAH-Os/GOx layers.

Inspection of Eq. (7.11) shows that the catalytic current alone is not a good indication of 'wiring efficiency' since it depends on the $FADH_2$ oxidation rate, $k$; the redox polymer concentration [Os]; and the number of 'wired' enzyme molecules per unit area, $\Gamma_{enz}$. The double redox–enzyme catalytic cycle for the oxidation of β-D-glucose catalyzed by GOx and mediated by the self-contained molecular 'wire polymer' PAH-Os in the organized self-assembled multilayer is considered in Fig. 7.1.

Figure 7.8 shows typical plots of the electrochemical response in β-D-glucose solution and a nonlinear fit to Eq. (7.11) from which the quantities $k$[Os] and $E^\circ$ have been derived assuming $k_{cat} = 700\ s^{-1}$ and $K_{MS} = 25$ mM [136].

The importance of the relative position for a single active enzyme layer in the multilayer has been investigated by stepwise electrostatic adsorption of one active GOx monolayer and three inactive FAD-free apoenzymes intercalated between PAH-Os layers. One should expect the same 'electrical wiring' efficiency irrespective of the relative enzyme

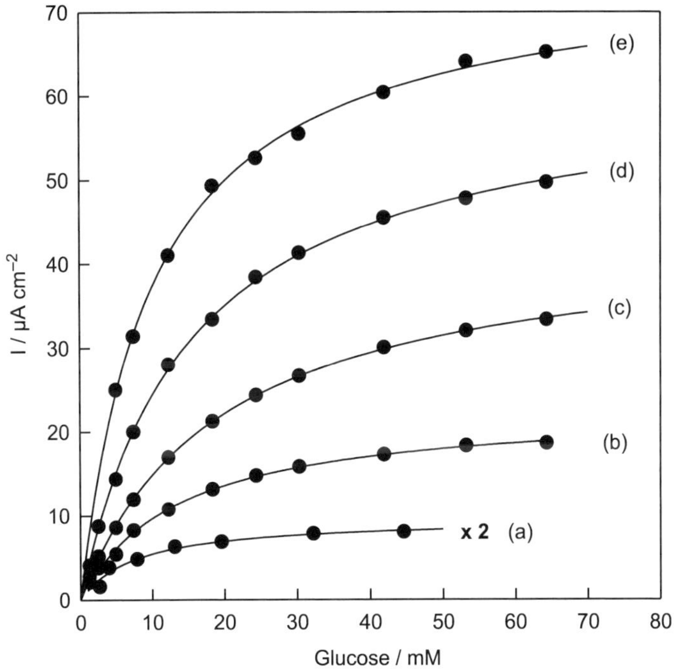


Fig. 7.8. Catalytic current vs. glucose relative position.

to osmium polymer position in the nanostructure if the bimolecular collisional rate were isotropic in the film and all sites are equally accessible. However, Wolosiuk et al. reported [138] that the catalytic response for the two limiting cases where the active GOx layer is progressively moved from the bottom layer adjacent to the thiolated gold electrode toward the top layer in contact with the electrolyte is quite different as it is shown in Fig. 7.8. A progressive decrease in the value of $k$[Os] from 2665 $s^{-1}$ for the bottom layer adjacent to the underlying electrode to 523 $s^{-1}$ for the top layer in contact with the external electrolyte has been reported demonstrating that the relative position of the active GOx in the multilayer determines the 'wiring efficiency', although the redox charge was approximately the same in all four experiments.

Cyclic voltammetry experiments have shown that all redox sites are equally accessible for electron transfer by hopping between adjacent osmium sites in the timescale of the experiment, ca $\nu F/RT \approx 20$ to $40 \times s^{-1}$ ($\nu$ is the electrode potential scan rate). However, the steady-state oxidation of a GOx at a given distance from the electrode surface demands a much higher flux of electrons (i.e. depending on the rate-determining step either $k_{cat} \approx 700\ s^{-1}$ or $k \leq 5000\ s^{-1}$).

As depicted in Fig. 7.1, the generation of an electrical signal from the molecular recognition by GOx involves the following sequence of events: $FADH_2$ re-oxidation, $k_M = k$[Os], and propagation of electrons from the active layer to the underlying electrode through the multilayer. The observed rate of $FADH_2$ re-oxidation, $k_M$(obs), can be expressed as the sum of the reciprocals of the re-oxidation rate ($k_M$) and the electron-hopping diffusion constant, $k_D$, which can be expressed as the reciprocal of the characteristic diffusion time given by the Einstein–Smoluchowski equation $\tau = \Delta x^2/2D_e$, with $\Delta x = nd_{layer}$, where $n$ is the layer number and $d_{layer}$ the interlayer average distance, and $D_e$ is the electron-hopping diffusion coefficient. A linear dependence of $1/k_M$(obs) versus $n^2$ is shown in Fig. 7.9 as predicted by the equation [138]:

$$\frac{1}{k_M(\text{obs})} = \frac{1}{k_M} + \frac{1}{k_D} \tag{7.19}$$

From the intercept, the maximum re-oxidation constant $k_M = 4500\ s^{-1}$ was obtained and from the slope and $d_{layer} \approx 5$ nm, the value of $D_e = 1.2 \times 10^{-9}\ cm^2\ s^{-1}$ is in excellent agreement with the electron-hopping diffusion constant measured in random hydrogels of the same

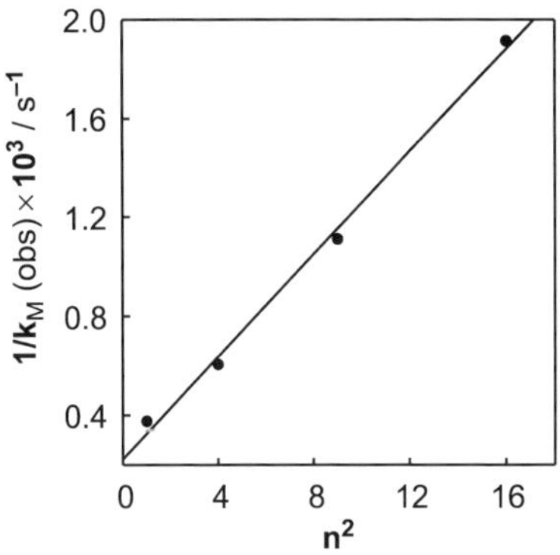


Fig. 7.9. $1/k_M$(Obs) vs. the square of the number of layer plot (From Ref. [138]).

composition by ultramicroelectrode chronoamperometry. The re-oxidation efficiency resulted in $k = 4.5 \times 10^4\ M^{-1}\ s^{-1}$ from $k_M$ and [Os] ≈ 0.1 M in the films.

These results proved that the flux of electrons that can re-oxidize the enzyme $FADH_2$ in layers farther away from the underlying electrode surface is limited by the propagation of electronic charge in the multilayers.

### 7.6.2 Different molecular architectures

The role of spatial distribution and molecular organization of enzymes in LBL self-assembled films has been highlighted by Gooding et al. who studied two enzymes in an amplified enzyme electrode using different geometries of glucose oxidase (GOx) and glucose dehydrogenase (GDH) monolayers self-assembled onto alkanethiol-modified gold [139].

The microenvironment of glucose has also been investigated by using mixed self-assembled alkanethiol monolayers to study the effect of microenvironment on the structure and enzyme activity [107]. The authors concluded from FTIR spectroscopy that conformational changes of the enzyme took place upon surface immobilization and that the hydrophobic environment resulted in higher specific activity of GOx.

Several molecular architectures, with the active enzyme layer in contact with either discrete positively charged molecules on a thiolated gold surface or in contact with positively charged redox polyelectrolyte, have been investigated [135]. A schematic representation of these structures is depicted in Fig. 7.10.

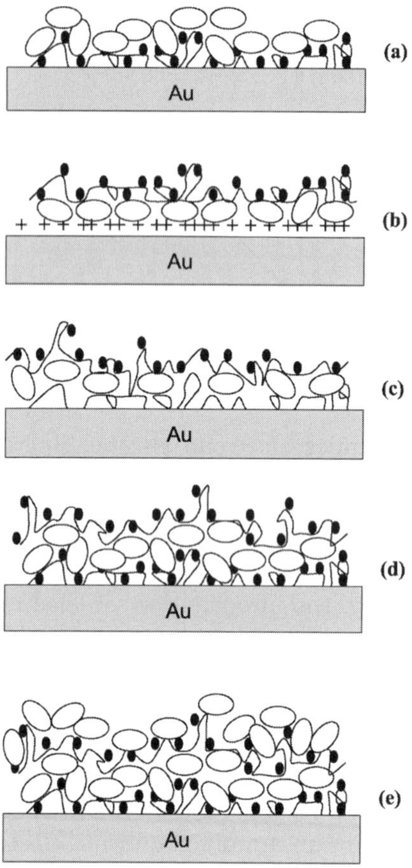


Fig. 7.10. Different molecular architectures in multilayer enzyme structures.

Table 7.1 compares the results for the different molecular architectures examined: (a) PAH-Os/GOx; (b) cystamine/GOx/PAH-Os; (c) PAH/GOx/PAH-Os; (d) PAH-Os/GOx/PAH-Os; and (e) $(PAH\text{-}Os)_2(GOx)_2$. The relevant difference is the highest fraction of 'wired' enzyme for the system labeled as (c) where almost 30% of the enzyme molecules present at the surface are 'electrically wired' unlike system (a) where only 1% of the GOx molecules contribute to the catalytic current by collision with the mediator. Structures labeled as b, d and e show a similarly low fraction of 'wired' enzyme molecules of around 6%.

TABLE 7.1

Electrocatalytic data for different molecular architectures of self-assembled enzyme electrodes

| Experiment | $k$[Os]/$s^{-1}$ | [Os]/mM | $k/M^{-1}\,s^{-1}$ | $\Gamma_{enz}/10^{-12}$ mol $cm^{-2}$ | [Os]/[GOx] |
|---|---|---|---|---|---|
| (a) MPS/PAH-Os/GOx | 322 | 68 | 4700 | 0.12 (1.1%) | 3.2 |
| (b) Cyst/GOx/PAH-Os | 846 | 96 | 8800 | 0.31 (6.6%) | 10.2 |
| (c) MPS/PAH/GOx/PAH-Os | 2481 | 278 | 8900 | 0.41 (29.3%) | 99.3 |
| (d) MPS/$(PAH\text{-}Os)_2/(GOx)_1$ | 1142 | 183 | 6300 | 0.74 (6.6%) | 8.7 |
| (e) MPS/$(PAH\text{-}Os)_2/(GOx)_2$ | 490 | 77 | 6400 | 1.35 (5.8%) | 4.8 |

*Note:* MPS is mercapto propane sulfonate; PAH is high molecular weight poly(allylamine); PAH-Os is $[Os](bpy)_2Cl(PyCH_2NH$-derivatized poly(allylamine)); GOx is glucose oxidase and Cyst is cystamine or 2,2′-diaminoethyldisulfide. [Os]/[GOx] is the ratio of the volume concentration of PAH-Os to the total enzyme concentration obtained from QCM determinations. The larger concentration [Os] for electrode (c) is due to the smaller thickness of PAH films adsorbed from pH 5 solutions than PAH-Os films adsorbed from pH 7–8 solutions.

It should be noticed that the role of the osmium poly(allylamine) polycation in the multilayer is two-fold: (i) it acts as the glue to hold the GOx building blocks together in the nanostructure; and (ii) also functions as a 'molecular wire' connecting the enzymatic $FADH_2$ to the underlying electrode. Close examination of the last column in Table 7.1 shows that the larger the ratio of the osmium to the total enzyme concentration, the larger the fraction of 'wired' enzyme. Therefore, the larger the ratio of Os to total GOx concentration, the greater the excess of flexible osmium polymer segments that can collide with the enzyme and hence the better the wiring efficiency. Conversely, if there is a shortage of polymer relative to the enzyme (cases a, b, d and e) then most of the polymer segments will be electrostatically bound to the enzyme and they have less chance they have to interact with another enzyme molecule.

The bimolecular $FADH_2$ oxidation rate constants ($k$ in Table 7.1) are very similar in all cases being only slightly higher for cases (b) and (c) where the glucose oxidase was adsorbed on osmium-free cystamine or poly(allylamine). It is interesting that the values of $k$ in Table 7.1 are comparable to that reported for the intermolecular oxidation of glucose oxidase modified with the same osmium complex in solution, namely $k = 8.3 \times 10^3\ M^{-1}\,s^{-1}$ [12], but in all cases are much less than the value for the bimolecular oxidation rate of the complex in solution, ca $k = 3.7 \times 10^5\ M^{-1}\,s^{-1}$ [12], where free diffusional collision of the osmium complex with the enzyme can take place. On the other hand, intramolecular oxidation of GOx molecules with tethered Os complexes has been shown to be less efficient, with $k = 0.2\text{–}0.4 \times 10^3\ M^{-1}\ s^{-1}$ [12].

A useful molecular architecture has been the integration of antibodies or antigens on the electrode surfaces leading to immunosensors where the recognition event is transduced into an electronic signal [140–142]. Such devices have been of great interest in the last decades due to the potential applications in environmental, food and drug monitoring as well as in clinical diagnostics, mainly point of care testing. Many strategies have been used in order to immobilize these biomolecules and the optimization of such procedures is focused on effective orientation and binding activity.

Several methods have been proposed to achieve biospecific activity by adsorption on clean surfaces [143,144] and polymers [145,146]. In either case the modified surface was stable and allows efficient antigen–antibody binding. A simple way of designing a well-organized surface involved self-assembled layers for protein deposition on substrates. A successful method required the initial adsorption of thiotic acid and further covalent immobilization of antibody molecules onto a micro-porous gold membrane [144]. Similar self-assembled layers of sulfur-containing molecules were also used for immunosensor design [109, 110]. The direct assembling of an antigen monolayer is possible and has been applied to antibody detection in a reusable sensor [111].

Another procedure to develop immunosensing surfaces using the alternate adsorption of polyelectrolyte films has been reported [16]. It was possible to obtain an increased response by deposition of several antibody layers [92] and therefore the sensitivity was controlled by increasing the protein adsorption. A detailed characterization of polyelectrolyte-protein (IgG) multilayers has been achieved employing several techniques [143]. The number of spacer layers between the biomolecule layers determined the degree of order within the film. A disordered structure was observed for the films assembled with one polymer layer separating each protein layer. On the other hand, films including five polyelectrolyte layers between each IgG layer showed that a dense layered structure was formed. The different biosensing applications of these structures, such as the recognition event involved in antigen–antibody binding, were also evaluated.

An attractive approach is the use of a functionalized dendrimer monolayer to design an affinity-sensing surface with reusability. Bound antibody molecules were dissociated from the surface via a displacement reaction by the addition of free biotin in solution, enabling the affinity surface to be renewed and repeatedly utilized.

The development of modified surfaces where the recognition event occurs efficiently is of major concern [147–150]. One of the critical issues to achieve this goal is molecule orientation. The immobilization of biomolecules onto surfaces by simple adsorption from solution results in random orientation on the surface leading to inefficient binding of target molecules. Several publications have focused on the possibility of improving orientation during the deposition of antibodies onto solid supports [151–156]. Several strategies such as simple adsorption, antibody thiolation and Protein A binding were evaluated considering orientational aspects of immobilization and immunological activity [152]. It was observed that Protein A coupling improves antibody binding. However, it was not possible to separate the contributions from non-specific binding for an unambiguous conclusion.

The preparation of Fab′ fragments covalently coupled to liposomes via linker lipids demonstrated that it was possible to control the orientation of antibodies and therefore increase the overall binding efficiency. More significantly it was shown that non-specific binding depended critically on the monolayer matrix [153]. An interesting example of site-oriented deposition of antibodies showed that the oxidation of carbohydrate residues also allowed a controlled deposition upon covalent coupling [154].

A different strategy for oriented immobilization involved using native sulfide groups of immunoglobulins in order to attach them to gold surfaces. The immobilized antibodies exhibited high antigen binding constants and no distribution of the affinity constant, providing site-oriented adsorption on the surfaces [155]. It was possible to assemble an IgG-binding protein on gold for highly oriented antibody immobilization by use of the strong affinity between cysteine and the surface. The molecules were not only highly oriented but also showed a significantly increased binding activity compared to a physically adsorbed antibody [156].

Anzai and co-workers [157, 158] reported the LBL construction of multilayer films composed of avidin and biotin-labeled poly(amines) (a) or polyanions (b). They followed the layer buildup by the frequency shift of a quartz crystal resonator.

The redox mediation of biotinylated glucose oxidase (GOx-b) assembled LBL on avidin in a self-assembled structure where the enzyme was electrically wired by poly(allylamine) with $[Os(bpy)_2ClPyCOH^+]$

covalently attached (PAH-Os) has been studied by Otero et al. [159]. Two successive PAH-Os, avidin and GOx-b layers were deposited by alternate immersion of a thiol-modified gold surface in the respective solutions. The uptake of the avidin and the GOx-b was monitored by a quartz crystal microbalance as shown in Fig. 7.11. The surface concentration of redox sites obtained from cyclic voltammetry integration increases in step with the number of PAH-Os layers deposited. From the frequency shift with the QCM the surface concentration of avidin for the two layers was 2.3 and $4.0 \times 10^{-12}$ mol cm$^{-2}$ and the surface concentration of biotinylated GOx bound to the avidin was 6.5 and $9.0 \times 10^{-12}$ mol cm$^{-2}$, respectively. The ratio of $\Gamma_{\mathrm{GOx(QCM)}}$ to $\Gamma_{\mathrm{avidin(QCM)}}$ in each avidin–GOx layer is close to 2.5, which is consistent with the fact that each molecule of avidin can bind in solution a maximum of four molecules of biotin. It should be noted that in the present case not all the binding sites of avidin are accessible to biotin because avidin is immobilized on the electrode surface.

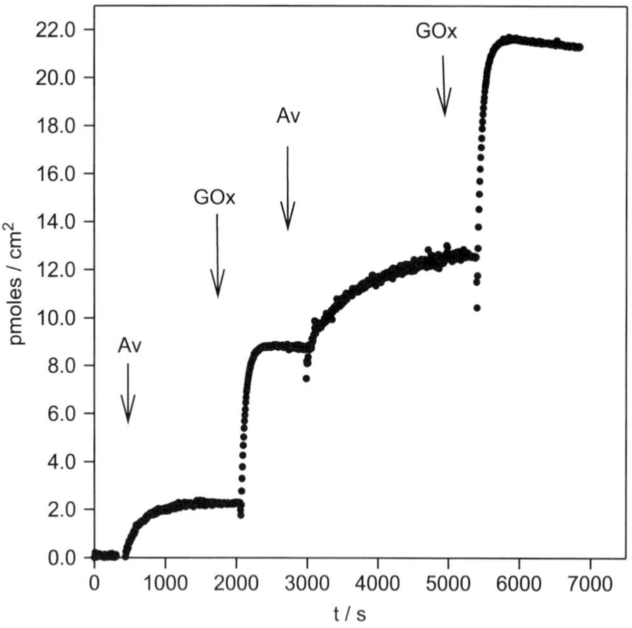


Fig. 7.11. QCM transient response during the uptake of avidin and GOx-b.

The uptake of biomolecules follows single exponential transients with time constants of 216 s for avidin on PAH-Os and 62 s for biotinylated GOx-b on the avidin-modified surface, respectively, as observed in other studies of these molecules [159].

A typical result for the electrocatalysis of glucose oxidation by GOx-b and mediated with PAH-Os is shown in Fig. 7.12. Reversible cyclic voltammetry shows a fast electrode process. The charge involved in the surface process, estimated by integration of the voltammetric peaks, increases steadily for each succesive (PAH-Os)(avidin)(GOx-b) layer with a value of 8 $\mu C\ cm^{-2}$ for the first layer and 10 $\mu C\ cm^{-2}$ for the second. The surface redox concentration of the first layer is $8.3 \times 10^{-11}$ mol $cm^{-2}$ and for the second layer is $1.03 \times 10^{-10}$ mol $cm^{-2}$. The dependence of the glucose catalytic oxidation current density on electrode potential for a $(PAH\text{-}Os)_2(avidin\text{-}GOx\text{-}b)_2$ multilayer in 10 mM HEPES buffer of pH 7.5, 0.2 M $KNO_3$ at 0.005 V $s^{-1}$ for solutions of different glucose concentration was analyzed. As the potential increases the Os(II) within the film is oxidized to Os(III) and a steady state catalytic current due to the oxidation of glucose is observed [159].

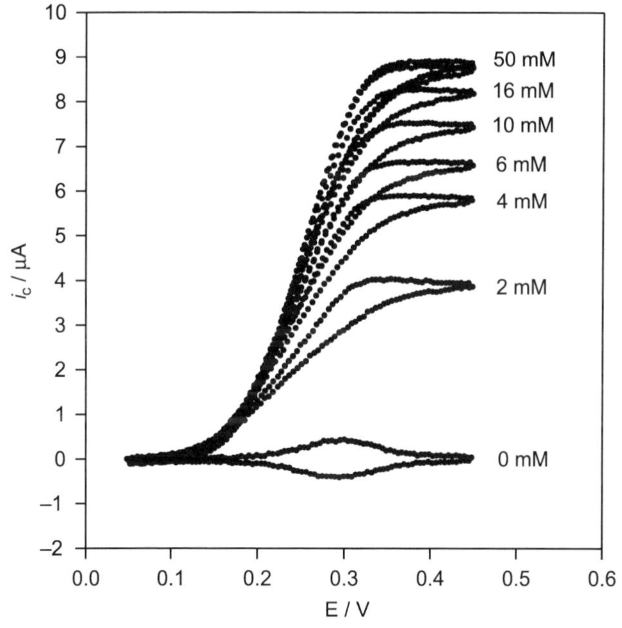


Fig. 7.12. $i_{cat}$ vs. $E$ for avidin and GOx-b and PAH-Os in different glucose concentrations.

It is interesting to compare two molecular architectures with the same surface concentration of GOx: (a) PAH-Os/avidin/GOx-b and (b) PAH-Os/avidin/ GOx-b/PAH-Os but with the osmium polymer mediator approaching the enzyme either from below or from below and above the enzyme layer (Fig. 7.13). In the second case more redox-integrated mediator is accessible to the GOx layer in spite of the fact that the concentration of redox centers in the polymer is always in large excess with respect to the enzyme molecules present. Similar results were reported for electrostatically self-assembled structures of GOx and PAH-Os [96].

Figure 7.13 shows a remarkable increase of the catalytic current wave when the enzyme 'wiring' occurs from below and above the GOx layer (Fig. 7.13b). Table 7.2 summarizes the results corresponding to the best fit to Eq. (7.11) of curves assuming $k_{cat} = 750\ s^{-1}$ and that $K_M = 25$ mM. The fraction of 'wired' enzyme increases from 0.13 to 0.44 pmol $cm^{-2}$ with the same total amount of GOx, 6.5 pmol $cm^{-2}$, as determined by the

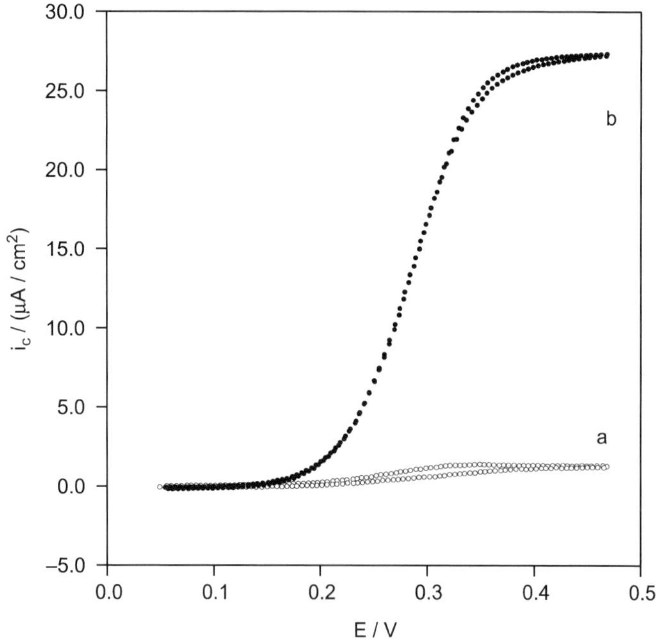


Fig. 7.13. Voltammetry of avidin and GOx-b modified electrodes with one (a) and two PAH-Os layers (b).

TABLE 7.2

Electrocatalytic data for avidin-biotinylated GOx self-assembled enzyme electrodes

| Film | $\Gamma_{enz}$/mol cm$^{-2}$ | $k$[Os]/s$^{-1}$ | Thickness (nm) | [Os]/M | $k$/M$^{-1}$ s$^{-1}$ |
|---|---|---|---|---|---|
| PAH-Os/Avidin/GOx-b | $1.3 \times 10^{-13}$ | 166 | 9.3 | 0.089 | 1860 |
| PAH-Os/Avidin/GOx-b/PAH-Os | $4.4 \times 10^{-13}$ | 880 | 13.5 | 0.112 | 7860 |

QCM. The rate of $FADH_2$ oxidation by the PAH-Os mediator, $k$[Os], and the bimolecular rate constant $k$ are also enhanced approximately by a factor of 4–5 for an osmium concentration increase of only 25%. We therefore conclude that the relative position of the PAH-Os and GOx in supramolecular architectures that favour more flexible Os polymer to approach the enzyme active center lead to the highest catalytic activity.

## 7.7 STRUCTURE OF SELF-ASSEMBLED ENZYME FILMS

From the preceding analysis it is clear that the structure of the enzyme polyelectrolyte multilayer films is relevant to the 'wiring efficiency' and catalytic activity. We therefore present in the following sections some results combining electrochemistry quartz crystal microbalance (EQCM), atomic force microscopy (AFM), ellipsometry and Fourier transform infrared studies of layer-by-layer deposited PAH-Os and GOx.

### 7.7.1 Atomic force microscopy

Some attempts to image glucose oxidase on surfaces by STM have been reported [160–162]. Atomic force microscopy (AFM) studies of enzyme monolayers adsorbed on PAH-Os [134] have shown the existence of large two-dimensional enzyme aggregates of GOx adsorbed on the osmium polymer from 1 μM solutions and less aggregation for 50 nM solutions as depicted in Fig. 7.14.

However, when the short alkanethiol, mercaptoethylamine (MEA), was pre-adsorbed on gold, a monoexponential adsorption law was observed and single GOx enzyme molecules could be detected on the surface (Fig. 7.15). It can be concluded from this evidence that aggregation of the enzyme was due to the underlying polymer

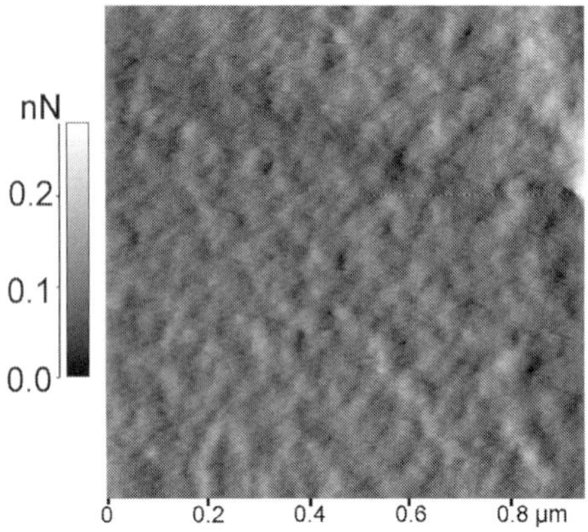


Fig. 7.14. AFM error signal mode image of adsorbed GOx/PAH-Os surface structure adsorbed from 50 nM GOx solution in water (constant force 0.25 nN) (From Ref. [134]).

interacting electrostatically. Nanodozing experiments were further carried out in order to determine whether the enzyme aggregates in two or three dimensions onto the redox polyelectrolyte at the surface. Murray et al. developed the 'nanodozing' or 'nanoindenting' technique

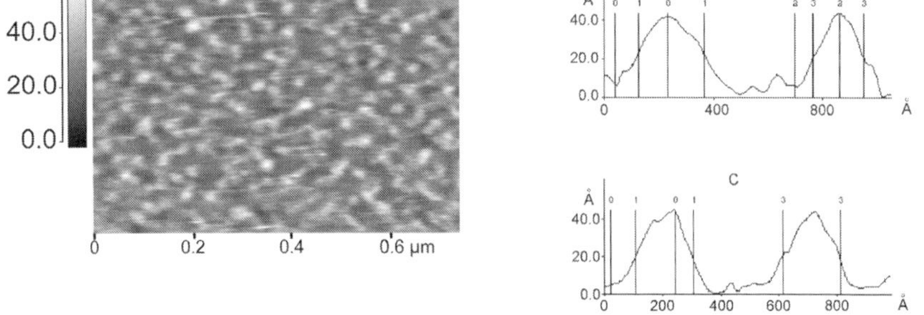


Fig. 7.15. Topography mode (unfiltered) image of GOx layer electrostatically adsorbed onto mercaptoethylamine-modified gold surface from a 50 nM GOx solution in water (constant force 0.25 nN). Lines A, B and C include selected individual enzyme molecules (From Ref. [134]).

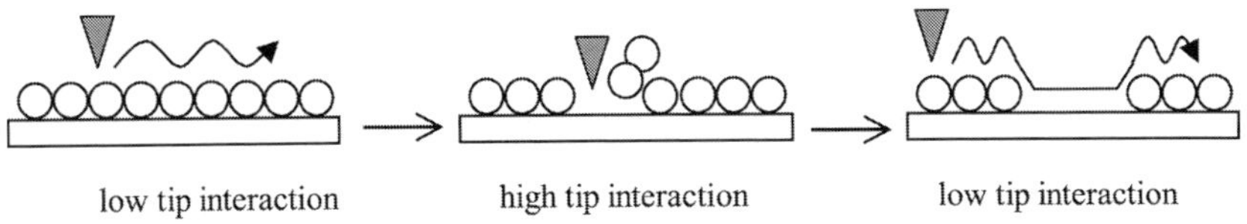


Fig. 7.16. Schematic representation of the nanodozing technique (From Ref. [134]).

with AFM, the principle of which is shown schematically in Fig. 7.16 and consists of a tip perturbation effect in AFM to pattern thin films.

The result is depicted in Fig. 7.17: a rectangular window was patterned by increasing the tip-to-surface interaction force, which can be seen as a darker or deeper zone at the MPS/PAH-Os/GOx modified surface. Two depth profiles across perpendicular lines A and B show an average depth of 4–4.5 nm, which correspond to the magnitude of the crystallographic [137] and ellipsometric [163] dimensions of glucose oxidase, ca $5 \times 5 \times 7$ nm$^3$. Therefore it is concluded that a monolayer of GOx is present on the self-assembled structure with lateral aggregation of the protein.

For a monolayer of GOx adsorbed on a Au modified with mercaptoethylamine, a different morphology was observed with the AFM as seen in Fig. 7.15. The AFM image shows a surface structure with a regular pattern of evenly distributed individual enzyme molecules on the surface. It is expected that the mercaptoethylamine produces discrete positive charges of protonated $NH_2$ groups on the Au surface unlike the poly(allylamine) with a charge distribution along the polymer trailing into solution. Cross-sectional scans of selected enzyme molecules along the lines A, B and C indicate heights that are in the 4–5 nm range, coinciding with previous STM reports for GOx molecules

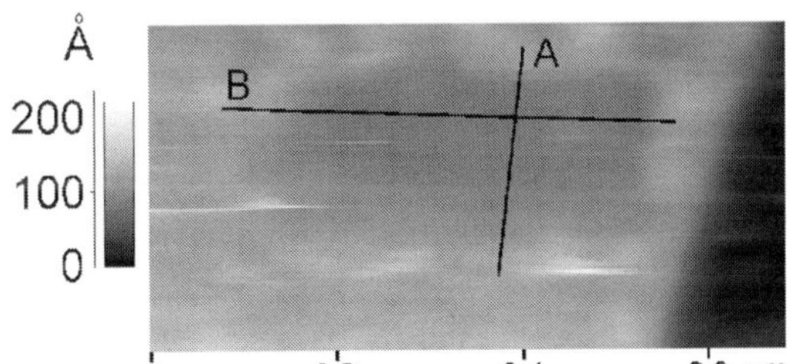


Fig. 7.17. AFM topography mode (unfiltered) image (at minimum constant force of 0.1 nN tip-to-surface force) for a layer-by-layer self-assembled PAH-Os/GOx layer on thiolated Au with nanodozing in a restricted rastered area at 5 nN tip-to-sample force (From Ref. [134]).

adsorbed on gold [161], X-ray crystallographic data [137] and ellipsometric measurements [163].

The protein width appears larger than the crystallographic radius, which may be due to AFM tip-particle convolution [164]. According to Bustamante the apparent width of the molecule, $W$, for a parabolic tip with a radius of curvature $R_c$ and a molecular radius $R_m$ can be obtained by convolution of GOx molecules and cantilever tip radius [164]:

$$W = \frac{4(R_c + R_m)\sqrt{R_m(R_c - R_m)}}{R_c}, \qquad R_c > R_m$$

For the Ultralevers™, when $R_c = 10$ nm and the crystallographic diameter of GOx as $R_m = 5$ nm, the apparent diameter results in $W = 30$ nm, which is in good agreement with the observation in Fig. 7.18.

### 7.7.2 Ellipsometry

Ellipsometry is a powerful tool to characterize thin films: thickness, refractive index, and extinction coefficient can be assessed from Fresnel equations that relate the changes of elliptically polarized light parameters measured after incidence on the sample [165]. Changes in

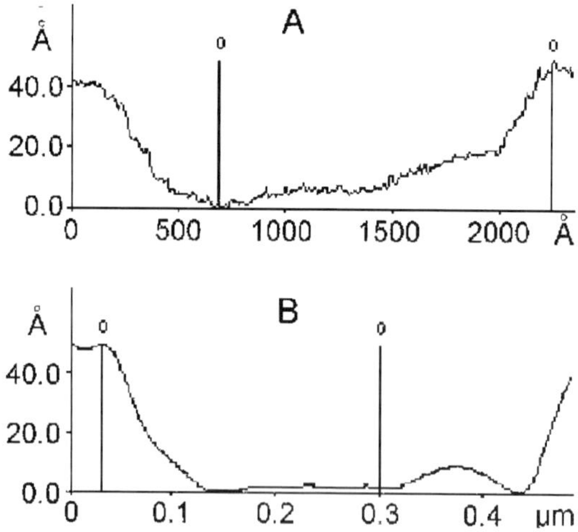


Fig. 7.18. Cross-sectional line scans along selected lines A and B show the depth of the nanodozed selected region (From Ref. [134]).

the state of polarization of linearly polarized light due to reflection at a planar surface are expressed as ellipsometric angles $\Psi$ and $\Delta$ and then related to the ellipsometric ratio $\rho$ according to:

$$\rho = \tan \Psi \exp(i\Delta) = r_p/r_s \qquad (7.21)$$

where tan $\Psi$ represents the amplitude change in $\rho$, $\Delta$ the phase difference change and $r_p$ and $r_s$ are the complex reflection coefficients for light polarized parallel and perpendicular to the plane of incidence [166] which depend on the optical properties and the structure of the film. For multiple layers, the ellipsometric ratio, $\rho$, is a function of the complex refractive index and the thickness of each layer, the angle of incidence, $\phi$, and the wavelength, $\lambda$, of the incident light.

The complex refractive index, $\hat{n}$, is usually expressed in terms of its real, $n$ (refractive index), and imaginary, $k$ (extinction coefficient), components: $\hat{n} = n - ik$ [165].

Ellipsometry has been widely employed when following the growth of multilayer self-assembled films of linear [93,166–174] and hyperbranched [120] polycations and polyanions. Ellipsometric studies of self-assembled multilayers with proteins have been reported [120,166, 175,176] including spectroscopic ellipsometric studies of avidin–biotin self-assembled multilayers and thickness estimation for GOx self-assembled LBL with a redox polymer and in Langmuir–Blodgett films.

Forzani et al. [177, 178] described ellipsometric measurements for the buildup of LBL electrostatically adsorbed multilayers of anionic enzymes (GOx and polyphenol oxidase, PPO) and polycations (polyallylamine, osmium modified polyallylamine and polydiallyl-dimethylamine, PDDA) on thiolated gold surfaces.

### 7.7.3 Combination of QCM and ellipsometric measurements

Gottesfeld et al. have shown that the combined information of ellipsometric film thickness and QCM film mass makes it possible to assess the apparent mass density of GOx films during their growth in polypyrrole [165]. For $(\text{PAH-Os})_n(\text{GOx})_n$ self-assembled systems, Forzani et al. [177] found that the amount of GOx during each adsorption step ranges from one to two-fold the value expected for a monolayer in the region adjacent to the Au substrate and as much as 10-fold after the fourth deposited bilayer.

A plot of the QCM mass versus the ellipsometric thickness for different polycation–enzyme self-assembled multilayers after sequential adsorption cycles (shown in Fig. 7.19) indicates that several enzyme multilayer structures show similar apparent density values, i.e. 2.4 ± 0.1, 2.1 ± 0.1 and 2.2 ± 0.1 g $cm^{-3}$ for PDDA/PPO, PAH/PPO and PAH/GOx films, respectively. For the first adsorption cycles a significant variation in the slope for PAH-Os/GOx self-assembled films is shown in the inset of Fig. 7.19 with an apparent density that decreases with thickness and reaches a constant value of 1.7 ± 0.2 g $cm^{-3}$ beyond the fourth bilayer. The dry film density is 2 g $cm^{-3}$ and therefore a less hydrated film prevails close to the thiolated gold surface and a more hydrated film occurs in the outermost layers [177].

Göpel et al. have reported that the QCM operated in liquids overestimates the mass by sensing water entrapped in the film [179]. The areal mass density, $\Delta m_e$, can also be obtained from the ellipsometric

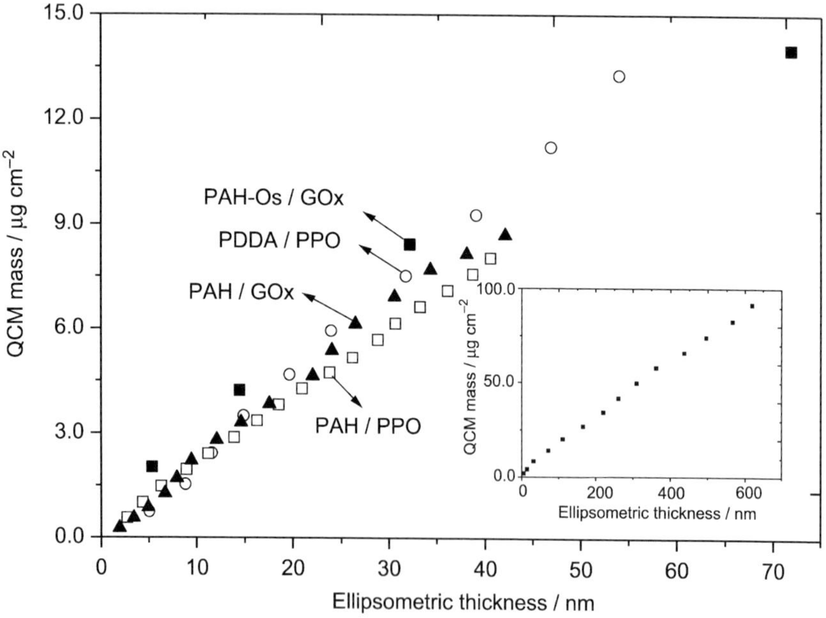


Fig. 7.19. QCM mass vs. ellipsometric thickness for self-assembled multilayer films: PAH-Os/GOx, PAH/GOx, PAH/PPO and PDDA/PPO.

film thickness ($d_f$) and film refractive index ($n_f$) through the equation proposed by de Feijter et al. [180]:

$$\Delta m_e = d_f \frac{n_f - n_o}{\partial n/\partial c} \tag{7.22}$$

where $n_o$ is the electrolyte solution refractive index, $n_f$ is the film refractive index, and $\partial n/\partial c$ is the refractive index increment of the adsorbed substance with the concentration. For GOx solutions of concentrations ranged between 61 and 321 mg mL$^{-1}$, a value of $\partial n/\partial c = 0.177$ cm$^3$ g$^{-1}$ was obtained [177], being similar to previous values found for proteolytic enzymes and other proteins (0.19 cm$^3$ g$^{-1}$) [181].

For several polycation–enzyme multilayers the calculated ellipsometric mass, $\Delta m_e$, is compared to the QCM mass, $\Delta m_{QCM}$, in Fig. 7.20 with larger values for the acoustic mass. Both acoustic and optical measurements are not in agreement since water molecules are not polarized in the visible spectral region while the shear wave senses water in the film as coupled mass. The acoustic contrast between film and liquid due to the difference in shear modulus (i.e. $10^6$ Pa for the film

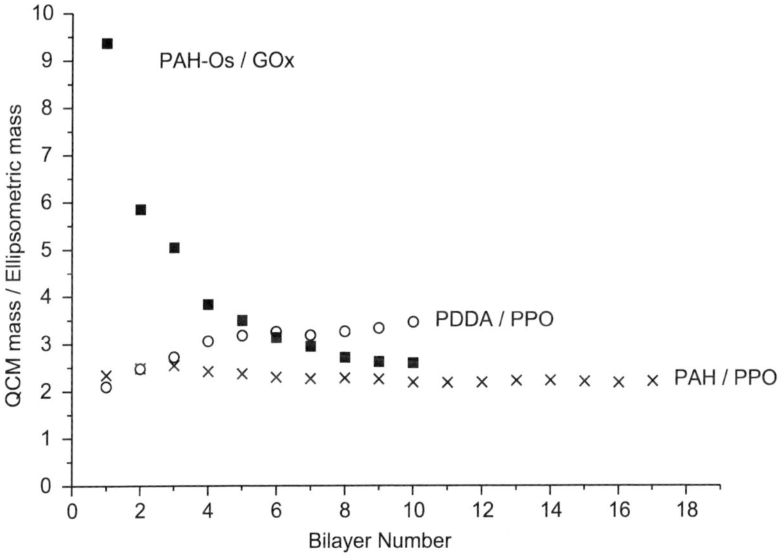


Fig. 7.20. Quartz crystal microbalance mass ($\Delta m_{QCM}$)/ellipsometric mass ($\Delta m_e$) as a function of bilayer number ($N_b$) for self-assembled multilayer films of: PAH-Os/GOx, PAH/GOx, PAH/PPO and PDDA/PPO.

and $10^4$ Pa for water) is much more pronounced than the optical contrast given by the difference in refractive index of the hydrated film (ca. 1.43 for the protein film and 1.33 for water).

Figure 7.21 is a schematic representation of the LBL self-assembled enzyme films that arise from combined QCM and ellipsometric measurements [177].

### 7.7.4 Infrared spectroscopy (FTIR)

Infrared spectroscopy allows proteins to be studied in a variety of environments. This renders the technique uniquely well-suited to probe the structure of membrane protein and protein films on surfaces [182].

The amide linkages between amino acid residues in polypeptides and proteins give rise to well-known signatures in the infrared region of the electromagnetic spectrum. For instance, the position of the amide A, I and II bands in the FTIR spectra of proteins is a sensitive indicator of conformational changes in the secondary structure [183] and provides direct evidence whether or not immobilized proteins are denatured [92]. In this context, the different variants of FTIR infrared spectroscopy of

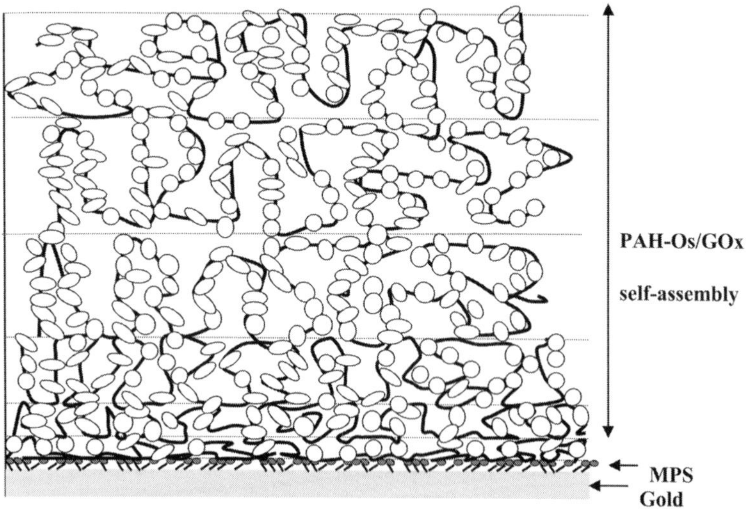


Fig. 7.21. Schematic representation of GOx spatial distribution within the multilayer structure self-assembled with PAH-Os assessed from ellipsometric, quartz crystal microbalance and electrochemical studies [177].

surfaces, namely external reflectance spectroscopy (FTIR-RAS) [92,107, 120,183], attenuated total internal reflection (FTIR-ATR) [184] and subtractively normalized interfacial (SNIF-TIRS) are very useful.

For native proteins, the exact position of the amide A, I and II bands in IR spectra usually depends on the type of protein. However, they do not have significant differences. The amide A band at $\sim 3300\ cm^{-1}$ is usually attributed to N–H asymmetric and symmetric stretching. Amide I band, assigned to the stretching mode of the carbonyl group coupled to the amide linkage, occurs at $\sim 1650\ cm^{-1}$, while the amide II band, from the C–N stretching modes and N–H in-plane bending of the polypeptide chains, occurs at $\sim 1540\ cm^{-1}$ [92,120]. It is well-known that the ratio between the absorbance of these bands brings information on the protein microenvironment [107, 184, 185].

Using external reflection FTIR-RAS, Forzani et al. [186] have studied the nature of different layer-by-layer self-assembled films, derived from enzymes such as GOx and poly(phenol-oxidase), PPO with polycations like PAH, and PAH-Os and PDDA. Figure 7.22 shows typical FTIR-RAS spectra of $(PAH\text{-}Os)_{10}/(GOx)_{10}$ and $(PAH)_{16}/(GOx)_{16}$ films self-

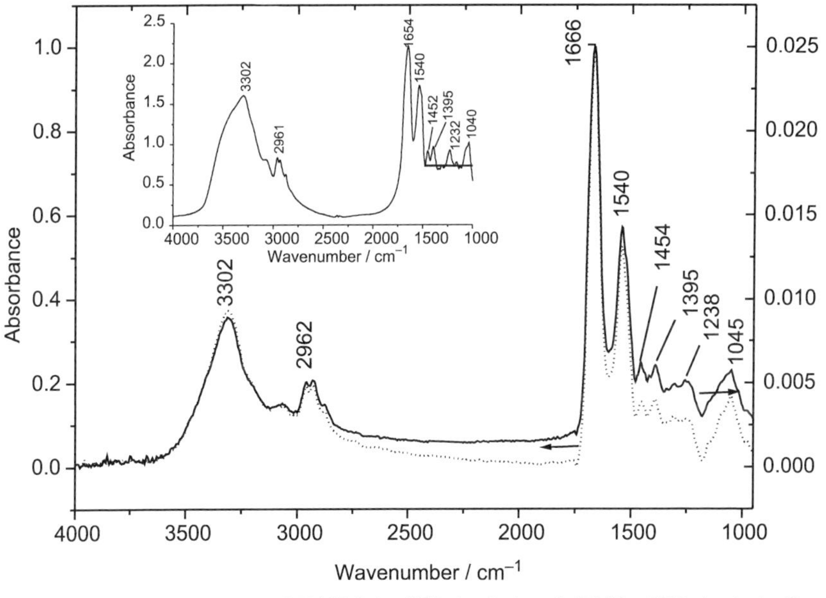


Fig. 7.22. FTIR-RAS spectra of $(PAH\text{-}Os)_{10}/(GOx)_{10}$ (—) and $(PAH)_{16}/(GOx)_{16}$ (...) films self-assembled on MPS-thiolated gold. Inset: Transmission spectra of GOx in KBr.

assembled on an MPS thiolated surface obtained with 'p' polarized light and a clean gold single beam as background.

No significant absorbance changes have been observed after polycation adsorption on GOx layers, but large absorbance increments at 3302, 1666 and 1540 $cm^{-1}$ are apparent after each GOx adsorption on PAH-Os (full line) for amide A, I and II bands, respectively. This led to the conclusion that the proteic amide linkage contributions are much more important than the corresponding bands due to the polycations in PAH-Os/GOx films.

Additional information about the enzyme microenvironment in the self-assembled structures have been obtained by comparing the spectra of enzyme solutions and pellets. Figure 7.23 shows the amide I and II infrared regions for GOx immobilized in a self-assembled structure with PAH-Os and the best fit line with a set of Gaussian peaks according to literature data. The ratio of amide I and II bands for the enzyme in solution (less than one) resulted in a value very different from the ratio in the solid state (greater than one; see inset Fig. 7.22).

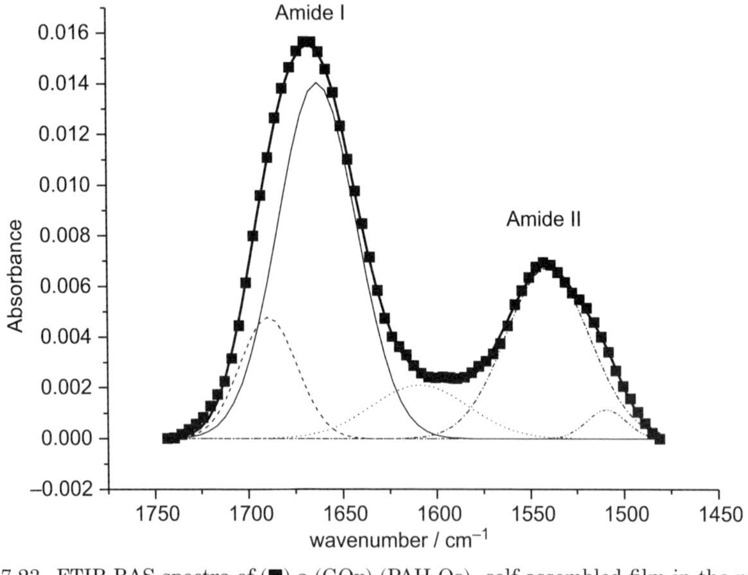


Fig. 7.23. FTIR-RAS spectra of (■) a $(GOx)_6(PAH\text{-}Os)_6$ self-assembled film in the region of amide I and II bands centered at 1660 and 1550 $cm^{-1}$, respectively. (–) Integrated deconvolution of (- - - - - -) random coils, (______) α-helix structures, (.......) β-sheet structures, (-..-..-) $NH_2$ deformations in primary alkyl amides and (-.-.-.-) NH deformations in secondary amides.

Table 7.3 shows the results of amide I and II peak deconvolution for the $(GOx)_6$ $(PAH\text{-}Os)_6$ self-assembled film in air, the enzyme in a KBr pellet and in buffer solution. It is rather surprising that the percentage of different secondary structures for GOx immobilized in self-assembled structures and for GOx in the crystalline state are very similar, in particular for α-helix, β-sheets and random coil signatures. Also, $NH_2$ deformations in primary alkyl amides, NH deformations in secondary amides and random coils retain their frequencies. However, the bands ascribed to α-helix and β-sheets for self-assembled enzyme films present 11 and 35 $cm^{-1}$ peak shifts with respect to crystalline GOx. The increase in vibration frequency may be associated with stronger interactions from the poly-cation. The structure of self-assembled GOx multilayers is very different from that in solution.

## 7.8 CONCLUSIONS

We have described enzyme-modified electrodes with the enzyme entrapped in hydrogels or self-assembled in organized multilayers. In both cases an integrated redox polymer mediator was used as a 'molecular wire' to generate an electrical signal in the external circuit as a consequence of molecular recognition by the enzyme and transduction by the electrode reaction. Hydrogels with larger content of water provide a stable environment for biomolecules such as enzymes, antibodies and oligonucleotides; however, phase segregation and random distribution of components is a limitation since there is no control of the molecular architecture as there is with layer-by-layer self-assembled biomolecular films.

These modified electrodes are finding increasing applications in clinical management (i.e. diabetes mellitus, remote medicine), genomics, proteomics and drug screening, food analysis, and environmental control. For these applications the future trend is integration with microelectronics (i.e. chip microarrays) in intelligent biochips, which take advantage of microfluidics.

For a rational design of miniaturized biosensor chips it is essential to gain control of the redox-enzyme kinetics and molecular wiring with vectorial electron transfer to and from the enzyme prosthetic group to and from the electrode. Electrocatalysis and electron hopping in the sensing surface layers have been described as well as 'in situ' methods to

TABLE 7.3

Deconvolution data for amide I and II bands for GOx in different forms[a]

| Structures | GOx in crystalline state | | Self-assembled GOx | | GOx in buffer solution | |
|---|---|---|---|---|---|---|
| | Frequency[b]/$cm^{-1}$ | %[c] | Frequency[b]/$cm^{-1}$ | %[c] | Frequency[b]/$cm^{-1}$ | %[c] |
| NH deformations in secondary amides | 1511 | 7.7 | 1509 | 2.1 | 1510 | 5.9 |
| $NH_2$ deformations in primary alkyl amides | 1541 | 20.0 | 1541 | 25.9 | 1544 | 58.9 |
| β-sheets | 1573 | 7.7 | 1609 | 9.4 | 1631 | 13.1 |
| α-helix | 1652 | 52.0 | 1663 | 50.2 | 1650 | 19.4 |
| Random coil (β or simple turns) | 1693 | 12.8 | 1689 | 12.4 | 1683 | 3.1 |

[a]Obtained from gauss function parameters.
[b]Corresponding to the peak maximum.
[c]Calculated from the ratio of peak and total area.

characterize the structure of such organized films, e.g. STM and AFM, QCM, ellipsometry and FTIR.

There is still a number of intriguing questions in connection with electrical communication and optimal enzyme response that open new avenues of research as well as a wide range of applications demanded by from the society, from the medicine and environment and antiterrorism.

## ACKNOWLEDGEMENTS

Financial support from the University of Buenos Aires, University of Córdoba, CONICET, ANPCyT, ACC and Motorola Semiconductor Sector, Arizona (US) is gratefully received. E. Forzani also thanks CONICET for a postdoctoral fellowship. E.J.C acknowledges a Guggenheim Fellowship, 2001.

## REFERENCES

1 R. Murray, In: R.W. Murray (Ed.), *Molecular Design of Electrode Surfaces*, Willey-Interscience, New York, 1992.
2 G. Ramsay, (Eds.), Commercial Biosensors, Applications to Clinical, Bioprocesss, and Environmental Samples, Vol. 148., Wiley, New York, 1998.
3 www.medisense.com.
4 www.therasense.com.
5 D.R. Thévenot, K. Toth, R.A. Durst and G.S. Wilson, *Biosensors and Bioelectronics*, 16 (2001) 121.
6 A. Heller, *J. Phys. Chem.*, 96 (1992) 3579.
7 C. Bourdillon, J.M. Laval and D. Thomas, *J. Electrochem. Soc.*, 133 (1986) 706.
8 A.E.G. Cass, G. Davis, G.D. Francis, H.A.O. Hill, W.J. Aston, I.J. Higgins, V. Plotkin, L.D. Scott and A.P.F. Turner, *Anal. Chem.*, 56 (1984) 667.
9 M.J. Green and H.A.O. Hill, *J. Chem. Soc. Faraday Trans.*, 82 (1986) 1237.
10 E. Liaudet, F. Battaglini and E.J. Calvo, *J. Electroanal., Chem.*, 293 (1990) 55.
11 R. Rajagopalan, A. Aoki and A. Heller, *J. Phys. Chem.*, 100 (1996) 3719.
12 C. Danilowicz, E. Cortón and F. Battaglini, *J. Electroanal. Chem.*, 89 (1998) 445.
13 A.L. Crumbliss, H.A.O. Hill and D.J. Page, *J. Electroanal. Chem.*, 206 (1986) 327.

14 A. Chaubey and B.D. Malhotra, *Biosensors and Bioelectronics*, 17 (2002) 441.
15 P.N. Bartlett and K.F.E. Pratt, *J. Electroanal. Chem.*, 397 (1995) 53.
16 C. Bourdillon, B. Limoges, J. Moiroux and J.M. Savéant, *J. Electroanal. Chem.*, 501 (2002) 1 and 8.
17 C. Bourdillon, C. Demaille, J. Moiroux and J.M. Savéant, *J. Am. Chem. Soc.*, 115 (1993) 2.
18 R.S. Nicholson and I. Shain, *Anal. Chem.*, 36 (1964) 706.
19 F. Battaglini and E.J. Calvo, *Anal. Chim. Acta*, 258 (1992) 151.
20 C. Bourdillon, C. Demaille, J. Gueris, J. Moiroux and J.M. Savéant, *J. Am. Chem. Soc.*, 115 (1993) 12264.
21 N. Anicet, A. Anne, J. Moiroux and J.M. Savéant, *J. Am. Chem. Soc.*, 120 (1998) 7115.
22 C. Bourdillon, C. Demaille, J. Moiroux and J.M. Savéant, *J. Am. Chem. Soc.*, 116 (1994) 10328.
23 P.N. Bartlett, D. Pletcher and J. Zeng, *J. Electrochem. Soc.*, 144 (1997) 3705.
24 W. Schuhmann, *Rev. Molec. Biotechnol.*, 82 (2002) 425.
25 B. Wang, J. Zhang, G. Cheng and S. Dong, *Anal. Chim. Acta*, 407 (2000) 111.
26 I. Willner, B. Willner and E. Katz, *Rev. Molec. Biotechnol.*, 82 (2002) 325.
27 A. Heller, *Acc. Chem. Res.*, 23 (1990) 128.
28 Y. Degani and A. Heller, *J. Am. Chem. Soc.*, 110 (1988) 2615.
29 P.N. Bartlett, S. Booth, D.J. Caruana, J.D. Kilburn and C. Santamaria, *Anal. Chem.*, 69 (1997) 734.
30 F. Battaglini, P.N. Bartlett and J.H. Wang, *Anal. Chem.*, 72 (2000) 502.
31 W. Schuhmann, T.J. Ohara, H.L. Schmidt and A. Heller, *J. Am. Chem. Soc.*, 113 (1991) 1394.
32 A. Badia, R. Carlini, A. Fernandez, F. Battaglini, S.R. Mikkelsen and A.M. English, *J. Am. Chem. Soc.*, 115 (1993) 7053.
33 Y. Katakis, Y. Ling and A. Heller, *J. Am. Chem. Soc.*, 116 (1994) 3617.
34 B.A. Gregg and A. Heller, *Anal. Chem.*, 62 (1990) 258.
35 B.A. Gregg and A. Heller, *J. Phys. Chem.*, 95 (1991) 5970.
36 B.A. Gregg and A. Heller, *J. Phys. Chem.*, 95 (1991) 5975.
37 R. Maidan and A. Heller, *J. Am. Chem. Soc.*, 113 (1991) 9003.
38 R. Maidan and A. Heller, *Anal. Chem.*, 64 (1992) 2889.
39 T.J. Ohara, R. Rajagopalan and A. Heller, *Anal. Chem.*, 65 (1993) 3512.
40 T.J. Ohara, R. Rajagopalan and A. Heller, *Anal. Chem.*, 66 (1994) 2451.
41 E. Csoregi, D.W. Schmidtke and A. Heller, *Anal. Chem.*, 67 (1995) 1240.
42 D.W. Schmidkte and A. Heller, *Anal. Chem.*, 70 (1998) 2149.
43 I. Katakis and A. Heller, *Anal. Chem.*, 64 (1992) 1008.
44 L. Ye, M. Hammerle, A. Olsthoorn, W. Schuhmann, H. Schmidt, J. Duine and A. Heller, *Anal. Chem.*, 65 (1993) 238.

45 G. Kenausis, C. Taylor, Y. Katakis and A. Heller, *J. Chem. Soc. Faraday Trans.*, 92 (1996) 4131.
46 T.J. Ohara, M.S. Vreeke, F. Battaglini and A. Heller, *Electroanalysis*, 5 (1993) 825.
47 M. Niculescu, I. Frebort, P. Pec, P. Galuszka, B. Mattiason and E. Csoregi, *Electroanalysis*, 12 (2000) 369.
48 M. Niculescu, C. Nistor, I. Frebort, P. Pec, B. Mattiason and E. Csoregi, *Anal. Chem.*, 72 (2000) 1591.
49 N. Gajovic, K. Habermuller, A. Warsinke, W. Schuhmann and F.W. Scheller, *Electroanalysis*, 11 (1999) 1377.
50 G. Kenausis, Q. Chen and A. Heller, *Anal. Chem.*, 69 (1997) 1054.
51 J. Anzai, H. Takeshita, Y. Kobayashi, T. Osa and T. Hoshi, *Anal. Chem.*, 70 (1998) 811.
52 E.J. Calvo, C. Danilowicz and L. Diaz, *J. Chem. Faraday Soc.*, 89 (1993) 377.
53 T. Woodyear, P. Rocca, J. Lindsay, Y. Dror, A. Freeman and A. Heller, *Anal. Chem.*, 67 (1995) 1332.
54 E.J. Calvo, C. Danilowicz and L. Diaz, *J. Electroanal. Chem.*, 369 (1994) 279.
55 E.J. Calvo, C. Danilowicz, L. Diaz and R. Etchenique, *Anal. Chem.*, 68 (1996) 4186.
56 E.J. Calvo, C. Danilowicz and R. Etchenique, *J. Chem. Faraday Soc.*, 91 (1995) 4083.
57 E.J. Calvo, F. Battaglini, C. Danilowicz and A. Wolosiuk, *Anal. Chem.*, 71 (1999) 1062.
58 C. Danilowicz, E. Corton, F. Battaglini and E.J. Calvo, *Electrochim. Acta*, 43 (1998) 3225.
59 N. Gajovic, G. Binyamin, A. Warsinke, F. Scheller and A. Heller, *Anal. Chem.*, 72 (2000) 2963.
60 S.C. Barton, H.H. Kim, G. Binyamin, Y. Zhang and A. Heller, *J. Am. Chem. Soc.*, 123 (2001) 5802.
61 T. Chen, S.C. Barton, G. Binyamin, Z. Gao, Y. Zhang, H.H. Kim and A. Heller, *J. Am. Chem. Soc.*, 123 (2001) 8630.
62 C.N. Campbell, D. Gal, N. Cristler, C. Banditrat and A. Heller, *Anal. Chem.*, 74 (2002) 158.
63 E. Mikeladze, A. Schulte, M. Mosbach, A. Blochl, E. Csoregi, R. Solomonia and W. Schuhmann, *Electroanalysis*, 14 (2002) 393.
64 K. Habermuller, A. Ramanavicius, V. Laurinavicius and W. Schuhmann, *Electroanalysis*, 12 (2000) 1383.
65 S. Reiter, K. Habermuller and W. Schuhmann, *Sensors and Actuators B*, 79 (2001) 150.
66 X. Chen, N. Matsumoto, Y. Hu and G.S. Wilson, *Anal. Chem.*, 74 (2002) 368.

67 X. Chen, B. Wang and S. Dong, *Electroanalysis*, 13 (2001) 1149.
68 C.A.P. Quinn, R.E. Connor and A. Heller, *Biomaterials*, 18 (1997) 1665.
69 E. Corton and F. Battaglini, *J. Electroanal. Chem.*, 511 (2001) 1.
70 E. Corton and F. Battaglini, *J. Electroanal. Chem.*, 511 (2001) 8.
71 I. Gill, *Chem. Mater.*, 13 (2001) 3404.
72 Y. Lvov and H. Möhwald (Eds.), *Protein Architecture: Interfacing Molecular Assemblies and Immobilization Biotechnology*, Marcel Dekker, New York, 2000.
73 T.A. Horbett and J.L. Brash (Eds.), *Proteins at Interfaces: Fundamentals and Applications*, ACS Symposium Series 602, American Chemical Society, Washington, DC, 1995.
74 D. Catty and C. Raykundalla (Eds.), Antibodies: A Practical Approach, Vol. II., IRL Press, Oxford, 1989, p. 97.
75 A.P.F. Turner, I. Karube and G.S. Wilson (Eds.), *Biosensors: Fundamentals and Applications*, Oxford University Press, Oxford, 1987.
76 J.L. Guesdon, S. Avrameas, In: L.B. Wingard, E. Katchalski-Katzir and L. Goldstein (Eds.), Applied Biochemistry and Bioengineering, Vol. 3., Academic Press, New York, 1981, p. 207.
77 P. Ros Seigel, R. Harder, M. Dahint, F. Grunze, M. Josse, G.M. Mrksich and Whitesides, *Anal. Chem.*, 69 (1997) 3321.
78 I.V. Turko, I.S. Yurkevich and V.L. Chashchin, *Thin Solid Films*, 210/211 (1992) 710.
79 M. Ahlers, W. Müller, A. Reichert, H. Ringsdorf and J. Venzmer, *Angew. Chem., Int. Ed. Engl.*, 29 (1990) 1269.
80 W. Müller, H. Ringsdorf, E. Rump, G. Wildburg, X. Zhang, L. Angermaier, W. Knoll, M. Liley and J. Spinke, *Science*, 262 (1993) 1706.
81 T. Cassier, K. Lowack and G. Decher, *Supramol. Sci.*, 5 (1998) 309.
82 G. Ladam, P. Schaaf, F.J.G. Cuisinier, G. Decher and J.C. Voegel, *Langmuir*, 17 (2001) 878.
83 C. Tedeschi, F. Caruso, H. Möhwald and S. Kirstein, *J. Am. Chem. Soc.*, 122 (2000) 5841.
84 M. Onda, K. Ariga and T. Kunitake, *J. Biosci. Bioeng.*, 87 (1999) 69.
85 F. Caruso, K. Niikura, D.N. Furlong and Y. Okahata, *Langmuir*, 13 (1997) 3427.
86 Y. Lvov, K. Ariga, I. Ichinose and T. Kunitake, *J. Am. Chem. Soc.*, 117 (1995) 6117.
87 M. Onda, Y. Lvov, K. Ariga and T. Kunitake, *Biotechnol. Bioeng.*, 51 (1996) 163.
88 M. Onda, Y. Lvov, K. Ariga and T. Kunitake, *J. Ferment. Bioeng.*, 82 (1996) 502.
89 G. Ladam, P. Schad, J.C. Voegel, P. Schaaf, G. Decher and E. Cuisinier, *Langmuir*, 16 (2000) 1249.
90 G. Decher, *Macromolecules*, 31 (1998) 8893.

91 Y. Lvov, K. Ariga and T. Kunitake, *Chem. Lett.*, (1994) 6117.
92 F. Caruso, D.N. Furlong, K. Ariga, I. Ichinoze and T. Kunitake, *Langmuir*, 14 (1998) 4559.
93 S.W. Keller, H.N. Kim and T.E. Mallouk, *J. Am. Chem. Soc.*, 116 (1994) 8817.
94 G. Ladam, C. Gergely, B. Senger, G. Decher, J. Voegel, P. Schaaf and F.J. Cuisinier, *Biomacromolecules*, 1 (2000) 674.
95 G. Decher, *Science*, 277 (1997) 1232.
96 E.J. Calvo, F. Battaglini, C. Danilowicz, A. Wolosiuk and M. Otero, *Faraday Discussion Chem. Soc.*, 116 (2000) 47.
97 Ph. Lavalle, C. Gergely, F.J.G. Cuisinier, G. Decher, P. Schaaf, J.C. Voegel and C. Picart, *Macromolecules*, 35 (2002) 4458.
98 G. Decher and J.D. Hong, *Ber. Bunsenges. Phys. Chem.*, 95 (1991) 1430.
99 Y. Lvov, G. Decher and G. Sukhorukov, *Macromolecules*, 26 (1993) 5396.
100 Y. Lvov, G. Decher and H. Mohwald, *Langmuir*, 9 (1993) 481.
101 R. Rajagopalan, A. Heller, In: J. Jortner and M. Ratner (Eds.), *Molecular Electronics*, Blackwell Science, Oxford, 1997, p. 241.
102 Y. Degani and A. Heller, *J. Am. Chem. Soc.*, 111 (1989) 2357.
103 A.J. Bard, *Integrated Chemical Systems, A Chemical Approach to Nanotechnology*, Wiley, New York, 1994.
104 S. Sun, P.H. Ho-Si and D.J. Harrison, *Langmuir*, 7 (1991) 727.
105 F. Mizutani, Y. Sato, S. Yabuki and Y. Hirata, *Chem. Lett.*, (1996) 251.
106 K.T. Kinnear and H.G. Monbouquette, *Langmuir*, 9 (1993) 2255.
107 A.J. Guiomar, J.T. Guthrie and S.D. Evans, *Langmuir*, 15 (1999) 1198.
108 R. Blonder, E. Katz, Y. Cohen, N. Itzhak, A. Riklin and I. Wilner, *Anal. Chem.*, 68 (1996) 3151.
109 C. Danilowicz and J.M. Manrique, *Electrochem. Commun.*, 1 (1999) 22.
110 C. Duan and M.E. Meyerhoff, *Anal. Chem.*, 66 (1994) 1369.
111 C. Berggren, B. Bjarnason and G. Johansson, *Biosensor. Bioelectron.*, 13 (1998) 1061.
112 C. Bourdillon, C. Demaille, J. Gueris, J. Moiroux and J.M. Savéant, *J. Am. Chem. Soc.*, 115 (1993) 1264.
113 C. Bourdillon, C. Demaille, J. Moiroux and J.M. Savéant, *Acc. Chem. Res.*, 29 (1996) 529.
114 M. Dijksma, B. Kamp, J.C. Hooguliet and W.P. Bennekom, *Anal. Chem.*, 73 (2001) 901.
115 R. Blonder, S. Levi, G. Tao, I. Ben-Dov and I. Willner, *J. Am. Chem. Soc.*, 119 (1997) 10467.
116 N. Anicet, C. Bourdillon, J. Moiroux and J.M. Savéant, *J. Phys. Chem. B.*, 102 (1998) 9844.
117 C. Bourdillon, C. Demaille, J. Moiroux and J.M. Savéant, *J. Phys. Chem. B*, 103 (1999) 8532.

118 J.M. Rusling, In: Y. Lvov and H. Mohwald (Eds.), *Protein Architecture*, Ch. 13,337. Marcel Dekker, New York, 2000, pp. 337.
119 J.F. Rusling, *Acc. Chem. Res.*, 31 (1998) 363.
120 J.G. Franchina, W.M. Lackowski, D.L. Dermody, R.M. Crooks, D.E. Bergbreiter, K. Sirkar, R.J. Russell and M.V. Pishko, *Anal. Chem.*, 71 (1999) 3133.
121 H.C. Yoon, M.Y. Hong and H.S. Kim, *Anal. Chem.*, 72 (2000) 4420.
122 A'.P. Alzari, N. Anicet, C. Bourdillon, J. Moiroux and J.M. Saveant, *J. Am. Chem. Soc.*, 118 (1996) 6788.
123 C. Bourdillon, C. Demaille, J. Moiroux and J.M. Saveant, *J. Am. Chem. Soc.*, 121 (1999) 2401.
124 N. Anicet, C. Bourdillon, J. Moiroux and J.M. Saveant, *Langmuir*, 15 (1999) 6527.
125 A. Anne, C. Demaille and J. Moiroux, *J. Am. Chem. Soc.*, 123 (2001) 4817.
126 A. Anne, C. Demaille and J. Moiroux, *J. Am. Chem. Soc.*, 121 (1999) 10379.
127 I. Willner, A. Riklin, B. Shoham, D. Rivenzon and E. Katz, *Adv. Mater.*, 5 (1993) 912.
128 I. Willner and A. Riklin, *Anal. Chem.*, 66 (1994) 1535.
129 I. Willner, N. Lapidot, A. Riklin, R. Kasher, E. Zahavy and E. Katz, *J. Amer. Chem. Soc.*, 116 (1994) 1428.
130 I. Willner and E. Katz, *Angew. Chem. Int. Ed.*, 39 (2000) 1180.
131 I. Willner, V. Heleg-Shabtai, R. Blonder, E. Katz, G. Tao and A. Heller, *J. Am. Chem. Soc.*, 118 (1996) 10321.
132 E.S. Forzani, V.M. Solis and E.J. Calvo, *Anal. Chem.*, 72 (2000) 5300.
133 J. Hodak, R. Etchenique, E.J. Calvo, K. Singhal and P.N. Bartlett, *Langmuir*, 13 (1997) 2708.
134 E.J. Calvo, R. Etchenique, L. Pietrasanta, A. Wolosiuk and C. Danilowicz, *Anal. Chem.*, 73 (2001) 1161.
135 E.J. Calvo, C. Danilowicz, A. Wolosiuk, M. Otero, E. Forzani, M. Lopez-Teijelo, Joint ECE/ISE Meeting of the Electrochemical Society and the International Society of Electrochemistry, San Francisco, CA, Sept. 2–7, 2001. Proceedings Symp. G.
136 A. Wolosiuk, C. Danilowicz, E.J. Calvo, in preparation.
137 H.J. Hecht, H.M. Kalisz, J. Hendle, R.D. Schmid and D.J. Schomburg, *Mol. Biol.*, 229 (1993) 153.
138 E.J. Calvo, C. Danilowicz and A. Wolosiuk, *J. Am. Chem. Soc.*, 124 (2002) 2452.
139 J.J. Gooding, L. Pugliano, D.B. Hibbert and P. Erokhin, *Electrochem. Commu.*, 2 (2002) 217.
140 L. Alfonta, A.K. Singh and I. Willner, *Anal. Chem.*, 723 (2001) 91.
141 J. Wang, P.A.V. Pamidi and K.R. Rogers, *Anal. Chem.*, 70 (1998) 1171.
142 A.J. Killard, S. Zhang, H. Zhao, R. Hohn, E.I. Iwuoha and M.R. Smyth, *Anal. Chim. Acta*, 400 (1999) 109.

143 P. Treloar, J. Kane, P. Vadgama, In: C.P. Price and D.J. Newman (Eds.), *Principles and Practice of Immunoassay*, Ch. 19. MacMillan Reference Ltd, 1997.
144 M. Liu, Q.X. Li and G.A. Rechnitz, *Anal. Chim. Acta*, 387 (1999) 29.
145 B. Köning and M. Grätzel, *Anal. Chem.*, 66 (1994) 341.
146 E. Uttenthaler, C. Köẞlinger and S. Drost, *Anal. Chim. Acta*, 362 (1998) 91.
147 M. Minunni, M. Mascini, R.M. Carter, M.B. Jacobs, G.J. Lubrano and G.G. Guilbault, *Anal. Chim. Acta*, 325 (1996) 169.
148 Y.M. Lvov, Z. Lu, J.B. Schenkman, X. Zu and J.F. Rusling, *J. Am. Chem. Soc.*, 120 (1998) 4073.
149 R. Porter, P. Van Logt, S. Howell, M. Kyrolainen-Reay and A. Badley, *Biosensors. Bioelec.*, 16 (2001) 875.
150 H.C. Yoon, D. Lee and H. Kim, *Anal. Chim. Acta*, 456 (2002) 209.
151 K. Nakanishi, H. Mugurama and I. Karube, *Anal. Chem.*, 68 (1996) 1695.
152 F. Caruso, E. Rodda and D. Neil Furlong, *J. Colloid Interface Sci.*, 178 (1996) 104.
153 I. Vikholm and W.M. Albers, *Langmuir*, 14 (1998) 3865.
154 W. Qian, D. Yao, B. Yu, F. Yu, Z. Lu and W. Knoll, *Chem. Mater.*, 11 (1999) 1399.
155 A.A. Karyakin, G.V. Presnova, M.Y. Robtsova and A.M. Egorov, *Anal. Chem.*, 72 (2000) 3805.
156 S. Kanno, Y. Yanagida, T. Haruyama, E. Kobatake and M. Aizawa, *J. Biotechnol.*, 76 (2000) 207.
157 J. Anzai, Y. Kobayashi, N. Nakamura, M. Nishimura and T. Hoshi, *Langmuir*, 15 (1999) 175.
158 J. Anzai, T. Hoshi and N. Nakamura, *Langmuir*, 16 (2000) 6306.
159 M. Otero, E.J. Calvo, in preparation.
160 A.R. Czajka, C.G.J. Koopal, M.C. Feiters, J.W. Gerritsen, R.J.M. Nolte and H. Van Kempen, *Bioelectrochem. Bioenergetics*, 29 (1992) 47.
161 B.Q. Chi, J. Zhang, S. Dong and E. Wang, *J. Chem. Soc. Faraday Trans.*, 90 (1994) 2057.
162 D. Losic, J.G. Shapter and J.J. Gooding, *Langmuir*, 18 (2002) 5422.
163 A. Szucs, G.D. Hitchens and J.O'.M. Bockris, *J. Electrochem. Soc.*, 136 (1989) 3748.
164 E. Bustamante, C. Vesenka, C.L. Tang, W. Rees, M. Guthold and R. Keller, *Biochemistry*, 31 (1992) 22.
165 S. Gottesfeld, Y.-T. Kim and A. Redondo, Recent Applications of Ellipsometry and Spectroellipsometry in Electrochemical Systems, In: I. Rubinstein (Ed.), *Physical Electrochemistry: Principles, methods and applications*, Ch. 9., Marcel Dekker, New York, 1995.
166 K. Spaeth, A. Brecht and G. Gauglitz, *J. Colloid Interface Sci.*, 196 (1997) 128.

167 H. Lee, L.J. Kepley, H.-G. Hong and T.E. Mallouk, *J. Am. Chem. Soc.*, 92 (1988) 2597.
168 H. Lee, L.J. Kepley, H.-G. Hong and T.E. Mallouk, *J. Am. Chem. Soc.*, 110 (1988) 618.
169 H.C. Yang, K. Aoki, H.-G. Hong, D.D. Sackett, M.F. Arendt, S.-L. Yau, C.M. Bell and T.E. Mallouk, *J. Am. Chem. Soc.*, 115 (1993) 11855.
170 J.J. Harris and M.L. Bruening, *Langmuir*, 16 (2000) 2006.
171 D. Li, M. Lütt, M.R. Fitzsimmons, R. Synowicki, M.E. Hawley and G.W. Brown, *J. Am. Chem. Soc.*, 120 (1998) 8797.
172 D.L. Elbert, C.B. Herbert and J.A. Hubbell, *Langmuir*, 15 (1999) 5355.
173 J. Ruths, F. Essler, G. Decher and H. Riegler, *Langmuir*, 16 (2000) 8871.
174 M. Ginzburg, J. Galloro, F. Jäkle, K.N. Power-Billard, S. Yang, I. Sokolov, C.N. Catherine Lam, A. Wilhelm Neumann, I. Manners and G.A. Ozin, *Langmuir*, 16 (2000) 9609.
175 H. Arwin, *Thin Solid Films*, 48 (2000) 377.
176 S. Sun, P.H. Ho-Si and D. Jed Harrison, *Langmuir*, 7 (1991) 727.
177 E.S. Forzani, M. Otero, M.A. Pérez, M. López Teijelo and E.J. Calvo, *Langmuir*, 18 (2002) 4020.
178 E.S. Forzani, M. López Teijelo, F. Nart, E.J. Calvo, V.M. Solis, in preparation.
179 J. Rickert, A. Brencht and W. Göpel, *Anal. Chem.*, 69 (1997) 1441.
180 F. Tiberg and M. Landgren, *Langmuir*, 9 (1993) 927.
181 I. Cecilia Hahn Berg, D. Muller, T. Arnebrant and M. Halmsten, *Langmuir*, 17 (2001) 1641.
182 W.K. Surewicz, H.H. Mantsch and D. Chapman, *Biochemistry*, 32 (1993) 389.
183 A. Gole, C. Dash, A.B. Mandale, M. Rao and M. Sastry, *Anal. Chem.*, 72 (2000) 4301.
184 P.A. Suci and G.G. Geesey, *Langmuir*, 17 (2001) 2538.
185 N. Higashi, T. Koga, Y. Fujii and M. Niwa, *Langmuir*, 17 (2001) 4061.
186 E.S. Forzani, C. Bonazzola, T. Iwasita, F. Nart, E., J. Calvo, in preparation.

*Chapter 8*

# Composite and biocomposite materials for electrochemical sensing

***Salvador Alegret* and Arben Merkoçi***

## 8.1 INTRODUCTION

Chemically and biologically modified electrodes, also known as electrochemical chemosensors and biosensors, represent one of the most interesting research areas within the field of electroanalysis and comprise a clear example of integrated analytical systems. Three aspects need to be considered prior to conceptualising an electrochemical sensor: (i) selection of a chemical or biological molecular recognition element known as a receptor; (ii) selection of an electrochemical transduction element (usually potentiometric or amperometric) known as a transducer; and finally (iii) the integration of both elements (normally via receptor immobilisation onto the transducer through surface-modification procedures as described in the previous chapter).

Traditionally, the chemical or biological receptor has been immobilised onto the surface of the transducer either directly (adsorption, cross-linking, covalent bonding, etc.) or through a gel or membrane. From the point of view of construction, this immobilisation technique requires meticulous control of the variables affecting the immobilisation

* Corresponding author. Tel.: +34-93-5812118; fax: +34-93-5812379. *E-mail address:* salvador.alegret@uab.es.

*Comprehensive Analytical Chemistry XXXIX, pages 377–411*


ISSN: 0166-526X

procedure. These variables include temperature, pH, ionic strength, reagent purity, concentration, time, etc. While it is possible to follow this technique for construction of electrochemical sensors in a laboratory environment, on a large scale of production it proves to be both difficult and costly. Also, most available methods of surface modifications are not durable and the sensor is degraded with use—requiring a new modification of the surface or the replacement of the whole device.

Among electrochemical analytical devices, biosensors suffer most from deleterious effects on their surface. Fouling effects can occur on the transducer, as well as denaturation or leaking of the biological recognition material. Additionally, some biosensing designs are based on irreversible binding reactions. Interactions such as enzyme-inhibitor, antibody–antigen, and DNA hybridisation can be considered irreversible from a practical point of view, and devices based on these concepts can be considered probes rather than sensors.

Conducting composites and biocomposites are interesting alternatives for the construction of electrochemical sensors. The possibility of integrating various materials is one main advantage. Some materials that are incorporated within the composite result in enhanced sensitivity and selectivity. This incorporation is possible either through a previous modification of one of the components of the composite before its preparation or through physical incorporation into the composite matrix. Composite and biocomposite electrodes offer many potential advantages [1–4] compared to more traditional electrodes consisting of a surface-modified single conducting phase (see Fig. 8.1). Composite electrodes can often be fabricated with great

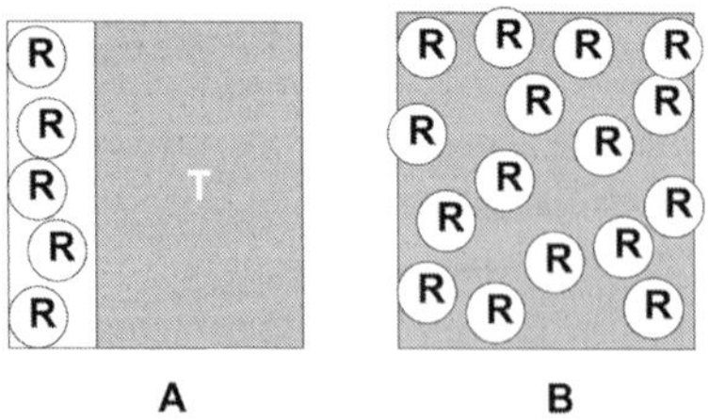


Fig. 8.1. Two basic designs for the construction of amperometric (bio)sensors, known also as chemically or biologically modified electrodes: surface-modification (A) and bulk-modification (B) of the amperometric transducer. Receptors (R) can be either chemical or biological recognition materials and transducers (T) are conducting materials such as carbon, gold or platinum. Both designs represent different approaches to integrating R and T.

flexibility in size and shape of material, permitting easy adaptation to a variety of electrode configurations.

Composite surfaces can be smoothed or polished to provide fresh active material ready to be used in a new assay. Each new surface yields reproducible results because all individual compounds are homogeneously dispersed or compressed in the bulk of the composite.

Another advantage of composite electrodes is the higher signal-to-noise (S/N) ratio (when compared with the corresponding pure conductors) that accompanies an improved (lower) detection limit. And also there is the possibility of obtaining composite electrodes from precious metals in a form with less weight and lower cost compared to their pure conductor counterparts.

Recent developments in the field of conducting composites applied to electrochemistry have opened a new range of possibilities for the construction of electrochemical sensors. The main features of these materials have been described elsewhere [1–4].

## 8.2 COMPOSITE ELECTRODE MATERIALS

The performance of electrochemical methods of analysis is strongly affected by the working electrode material. An ideal working electrode, for use in voltammetric techniques, should possess low ohmic resistance, chemical and electrochemical inertness over a broad range of potentials, high hydrogen and oxygen overvoltages resulting in a wide potential window, low residual current, ease of reproduction of the electrode surface, maximum versatility, low cost and no toxicity.

To date a range of materials has been used as working electrodes in electroanalytical techniques, the most popular consisting of a single conducting phase, such as mercury, platinum, gold or glassy carbon. In contrast to amperometric techniques is the design of potentiometric electrodes (e.g. ion selective electrodes, ISE) and consequently the choice of transducer and membrane materials will demand other characteristics to achieve enhanced analytical performance. However, the development of composite-based electrodes may lead to important advances in analytical electrochemistry, particularly in sensor devices.

### 8.2.1 Conducting composites

A composite results from the combination or integration of two or more dissimilar materials. Each individual component maintains its original characteristics while giving the composite distinctive chemical, mechanical and physical qualities. These qualities are different from those shown by the individual elements of the composite [5].

In general, composites are classified by the nature of the conducting material (platinum, gold, carbon, etc.) and the arrangement of its particles (i.e. whether the conducting particles are dispersed in the polymer matrix or if they are grouped randomly in clearly defined conducting zones and insulating zones). Furthermore, using a given conducting material, we may use different types of polymers thus establishing a new typification: epoxy composites, methacrylate composites, silicone composites, etc. Moreover, composites can be classified according to their rigidity as rigid composites or soft composites, also known as pastes.

*Conducting materials*

Carbon-based materials are used frequently as the conductive phase in composite electrodes. These materials are available in different morphologies such as graphite, carbon black, carbon fibres, etc. Carbon is an ideal choice due to its high chemical inertness, wide range of working potentials, low electrical resistance and a crystal structure responsible for low residual currents.

The inherent electrical properties of the composite depend on the nature of each of the components, their relative quantities and their distribution. The electrical resistance is determined by the connectivity of the conducting particles inside the polymeric matrix. Therefore, the relative amount of each composite component has to be assessed to achieve optimal composition. By constructing a percolation curve [6] it is possible to determine the minimum conductor content required to achieve a certain conductivity. This point is known as the percolation threshold (see Fig. 8.2).

If the surface area of a macroelectrode is reduced, the signal and the associated noise also diminish. In microelectrodes, according to Oldham [7,8], the perimeter of the sensing surface has a greater influence on the signal. By means of this 'edge effect', non-linear diffusion is established and the quality of the signal is enhanced. This enhancement is shown by

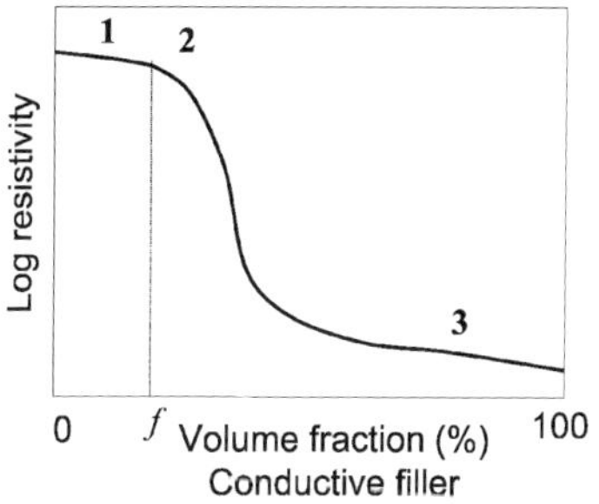


Fig. 8.2. Percolation curve of a conducting composite. Theoretical dependence of composite resistivity on conductive filler content. In zone 1, the electrical resistance of the composite is similar to that of the polymer. In zone 2, the percolation fraction *f* represents a critical conductive filler content that permits the formation of the first conducting filament consisting of particle-to-particle contacts. In zone 3, electrical resistance of the composite is similar to that of pure conductive filler.

a higher signal-to-noise ratio and lower detection limits. These features and their inherent small size have raised interest in microelectrodes. However, microelectrodes are scarcely used in sensor designs because their low currents demand complex and expensive instrumentation. If miniaturised sensors are not required, an alternative is to build an array of carbon fibres separated by an insulating matrix and connected in parallel [9,10]. The signal produced by this macroelectrode formed by a carbon fibre ensemble is the sum of the signals of the individual microelectrodes. The size of the resulting signal is equivalent to the signal produced by a carbon rod of the same active surface area, but showing the signal-to-noise ratio of a microelectrode since the conducting microdisks exposed on the surface are capacitively, resistively and diffusively independent. The construction of these ensembles is difficult. However, an equivalent device can be constructed very easily when a composite is made of small conductive particles distributed in a supporting matrix. This is supported by the recently coined phrase RAM™—a random assembly of carbon microdisks embedded in epoxy resin [10].

Composites may be also classified depending on the arrangement of the conducting phase in the polymer matrix. A composite formed by carbon fibres and an adhesive polymer matrix illustrates a composite where the carbon has an orderly distribution [11].

If the carbon is distributed randomly, the composite is easier to prepare. A composite material is said to be a *dispersion composite* if the conductive material is distributed randomly and its particles have the same probability of occupying any given space in the matrix. A graphite powder can be dispersed in a soft matrix, producing a 'carbon paste' [12]. Alternatively, a rigid matrix can be produced with a monomer that can be polymerized chemically, thermally, etc. or using a fused polymer that hardens upon cooling (see Fig. 8.3(A)).

On the other hand, *consolidation composites* exist when the conductor is dominant in clearly defined areas and areas where the insulator is the primary material (see Fig. 8.3B). There are two ways of building these composites. Firstly, *impregnation composites* are constructed from a porous conductor which is soaked with an insulating material. In this case, the consolidation of the conductive phase is determined by the porous structure of the conductor. Secondly, *segregation composites* are constructed by mixing a powder of conductive particles with an insulating material. The mixture is then subjected to a pressure. Examples of these substances are mixtures of graphite–polyethylene, graphite–Teflon, graphite–Kel F, etc. (see Tables 8.1 and 8.2).

### *Polymeric materials*

Adams proposed the use of soft carbon pastes [12] to build amperometric transducers. These pastes are prepared by mixing an inert conductor (e.g. graphite powder) with an insulating liquid (e.g. paraffin oil, silicone, Nujol). The insulating liquid has a specific viscosity and the paste has a certain consistency. The resulting materials are easy to prepare and inexpensive.

However, these pastes have limited mechanical and physical stability, especially in flow systems. Additionally, the pastes are dissolved by some non-polar solvents used in non-aqueous electrochemical determinations, leading to a deterioration of the signal. The general

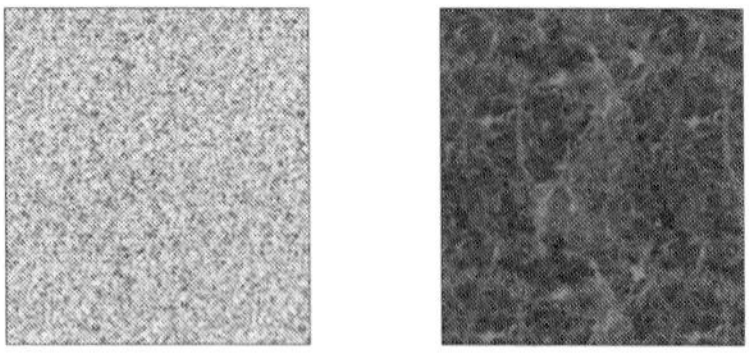

A B

Fig. 8.3. Morphology of the conducting graphite–polymer composites formed by (A) dispersion or (B) consolidation of the conducting particles.

TABLE 8.1

Carbon–polymer composites used in amperometric sensors

| Composite preparation | | | Analyte | Working medium | Potential (V) (Reference electrode) | Reference |
|---|---|---|---|---|---|---|
| Polymer (%) | Conductor (%) | Curing/Consolidation | | | | |
| *Dispersed composites* | | | | | | |
| Co-polymer of ethylenene vinyl acetate with 9% vinyl acetate (80–70) | Carbon black (20–30) | 12 min, 150°C | Ascorbic acid | $H_2SO_4$ 0.1 M | 0.65 (Ag/AgCl) | [17] |
| Polyethylene (50) | Carbon black (50) | Heated to the melting point | Phenol and chlorophenols | $H_2SO_4$ 0.025 M 28% acetonitrile | 1.2 (Ag/AgCl) | [102] |
| Cerasin wax (43.5) | Graphite (56.5) | Hot plate | Catecholamines | Acetate 1.0 M pH 4 | 0.17–0.42 (SCE) | [103] |
| Epoxy (55–50) | Graphite (45–50) | 2 h, 80°C | Ferrocyanide | KCL 0.1 M | 0.6 (SCE) | [104] |
| Epoxy (50) | Graphite (50) | Overnight, 70°C | Cd, Pb, Cu | $KNO_3$ 0.05 M pH 3 | 0.7 (Cd) 0.5 (Pb) 0.1 (Cu) (SCE) | [105] |
| Epoxy (44) | Graphite (56) | 5 min, 50–60°C | Iodide | KCl 0.1 M | 0.5 (SCE) | [106] |
| Epoxy (50) | Graphite (50) | 24 h, room temperature | Hydrogen peroxide | Phosphate buffer 0.1 M KCl 0.1 M, pH 7 | 1.15 (Ag/AgCl) | [23] |
| Epoxy (80) | Graphite (20) | 1 week, 35–40°C | Thiocholine | Phosphate buffer 0.1 M KCl 0.1 M, pH 7 | 0.7 (Ag/AgCl) | [107] |
| Polyester (60) | Graphite (40) | 24 h, room temperature | Hydrogen peroxide | Phosphate buffer 0.1 M KCl 0.1 M, pH 7 | 1.1 (Ag/AgCl) | [23] |

(*continued on next page*)

TABLE 8.1 (*continued*)

| Composite preparation | | | Analyte | Working medium | Potential (V) (Reference electrode) | Reference |
|---|---|---|---|---|---|---|
| Polymer (%) | Conductor (%) | Curing/Consolidation | | | | |
| Polymethacrylate (50) | Graphite (50) | 24 h, room temperature | Hydrogen peroxide | Phosphate buffer 0.1 M KCl 0.1 M, pH 7 | 1.1 (Ag/AgCl) | [23] |
| Silicone (40) | Graphite (60) | 24 h, room temperature | Hydrogen peroxide | Phosphate buffer 0.1 M KCl 0.1 M, pH 7 | 1.15 (Ag/AgCl) | [23] |
| *Consolidated composites* | | | | | | |
| Polyethylene (9–5) | Carbon black (91–95) | Pressure (10–15 atm) | Phenol | Phosphoric acid 0.01 M, Acetonitrile (1:1) | 1.2 (Ag/AgCl) | [108] |
| Kel-F (95–75) | Graphite (5–25) | High pressure 250–325°C | Ferrocyanide | KCl 1.0 M | 0.35–0.85 (SCE) | [109] |
| Kel-F (40) | Precious metal (60) | High pressure | $Ru(NH_3)_6^{3+}$ | $KNO_3$ 1.0 M, phosphate buffer 0.02 M, pH 6.9 | | [110] |
| Teflon (95–70) | Graphite (5–30) | Pressure (350 kg $cm^{-2}$) | Ferrocyanide | Phosphate buffer 0.1 M KCl 0.1 M, pH 6.8 | 0.16 (SCE) | [111] |

TABLE 8.2

Modified carbon–polymer composites used in amperometric sensors

| Composite preparation | | | | Analyte | Working medium | Potential (V) (Reference electrode) | Reference |
|---|---|---|---|---|---|---|---|
| Polymer (%) | Conductor (%) | Modifier (%) | Curing | | | | |
| *Dispersed composites* | | | | | | | |
| Cerasin wax (43.5) | Graphite (43.5) | Silica gel (13) | 70–80°C | Phenol | Sodium citrate 0.001 M 15% acetonitrile | 0.9 (Ag/AgCl) | [69] |
| Epoxy (46.5) | Carbon black (46.5) | CoPhC (7) | 24 h, room temperature | Thiocholine | Phosphate buffer 0.05 M, pH 7.3 | 0.3 (Ag/AgCl) | [112] |
| Epoxy (64) | Graphite (16) | Au–Pd (20) | 48 h, 60°C | Hydrogen peroxide | Phosphate buffer 0.1 M, KCl 0.1 M, pH 7 | 0.9 (Ag/AgCl) | [71] |
| Epoxy (50) | Graphite (33) | $RuO_2$ (17) | 24 h, 75°C | *N*-Nitrosamines | Phosphate buffer 0.42 M, pH 1.5 | 1.1–1.75 (SCE) | [113] |
| Epoxy-bonded (commercial) | Graphite | CoPhC (5) | 2 h, 100°C | Hydrazine, L-Cisteine, Oxalic acid | Phosphate buffer 0.05 M, pH 5.9 | 0.4 (SCE) | [19] |

CoPhC: Cobalt phthalocyanine.

degradation of these composites occurs quickly and has limited their use to the research laboratory. Reviews on chemically [13] and biologically [14] modified carbon paste electrodes have been published elsewhere.

There are also amperometric sensors and biosensors based on rigid composites that do not show the problems mentioned above. Furthermore, the fabrication of these devices can be adapted for mass production at a low cost.

Creasy and co-workers [9,15,16] reported the co-polymerisation of styrene with divinylbenzene (a cross-linking agent) and vinylferrocene (a modifier), using carbon black (semigraphitic carbon particles) or carbon fibre as a conductor. This was the basis for the construction of chemically bulk-modified electrodes. These devices showed better physical properties (the sensing surface was renewable by polishing) and enhanced chemical traits (they were stable in organic solvents) compared with carbon paste electrodes. Patriarche et al. [17] studied the effect of altering the content of carbon black in a polymeric matrix formed by a co-polymer of ethylenevinyl and vinyl acetate. They observed a sizeable shift in working potential, variations of the residual currents and changes of the settling times of the signal.

Epoxy resins comprise a family of polymers used widely for their excellent chemical properties, their good adhesion to other materials and their excellent insulating characteristics. Furthermore, they are easily prepared, inexpensive and widely available [18]. The addition of powdered fillers, metallic or carbonaceous, may increase their electrical and thermal conductivity while maintaining their ease of machination and malleability. Wang used a commercially available graphite epoxy resin (Grade RX, Dylon, Cleveland, OH, USA) to build chemically [19] and biologically [20] bulk-modified electrodes. The use of this commercial conducting composite rendered the fabrication of the sensors easier, quicker and more reproducible than the earlier proposed procedures. In this mixture the graphite content was fixed and invariant, with limited electrochemical properties.

This approach was adapted in our laboratories using a non-conducting epoxy (Epo-Tek, Epoxy Technology, Billerica, MA, USA) and graphite powder (particle size below 50 μm). Our procedure was expanded to matrices of other polymers such as silicone, polymethacrylate, polyester and polyurethane. Advantages are: polymers can be prepared in situ; they readily admit fillers (chemical and biological modifiers, catalysts, mediators, cofactors, etc.); they have a simple curing process; and are commercially available [21–23].

The polymer is chosen according to the sensor application. For example, if a biological modification is part of the construction procedure, a polymer that calls for high curing temperatures is not recommended. The same is true if the polymer contains additives harmful to the biological material. When a chemical modification is necessary (addition of redox mediators, catalysts, etc.) the modifying substance should not interfere with the curing of the polymer. Finally, considerations should be made to the environment where the sensor is going to be applied such as a living organism, an aqueous or organic solution, a biofluid, etc. This is important because biocompatibility or corrosion effects may have to be considered accordingly. Composite materials may be used in consonance with the polymer used. A few examples are epoxy, silicone, polyurethane, polystyrene, methacrylate or Teflon composites. Curing conditions such as curing time, temperature and type of atmosphere (air, nitrogen, etc.) also need to be optimised (see Tables 8.1 and 8.2).

Finally, the polymer phase of the composite defines the viscosity of the paste before curing. At this point, the material is soft and can be moulded into different shapes. If the paste is more fluid, it can be used as an ink in a screen printing process for mass production (see Section 8.3.4). In most cases this fluidity cannot be achieved since the high conductivity required calls for a high solid content, increasing its viscosity. Addition of volatile solvents, chloroform, $n$-pentanol etc. can reduce the viscosity of the paste without interfering in the curing process [24,25]. Ideally, pastes should have a certain viscosity as to be mouldable but not too dense as to require a difficult and long mixing process. Long mixing times can be a problem with fast curing polymers such as methacrylates, polyurethanes, etc. resulting in the polymer solidifying before the conducting particles are properly dispersed.

In other cases, paraffin for instance, the biocomposite is prepared by the solidification of a binder [26]. Recently, Miertus and co-workers developed composites based on a solid binding matrix (SBM) [27]. The polymer material of this matrix is an organic substance with multiple hydrophilic polar groups inserted in its carbon backbone. These substances are liquid at temperatures higher than 45°C. In this state, a biocomposite can be prepared by adding graphite, the biological component and a modifier. The mixture is homogenised and cured by cooling it at room temperature. SBMs are easy to prepare and their properties are similar to those of very rigid matrices, such as epoxies.

However, these materials cannot be used in non-aqueous solutions, as they are soluble in most organic solvents.

Finally, sol–gel matrices are being used for bulk-immobilisation of biological materials onto electrochemical transducers [28].

### 8.2.2 Conducting biocomposites

The most important component in a biocomposite is the biological receptor as it is responsible for the overall selectivity to a given analyte. However, the polymer and conductor phases are also very important as they define, respectively, the electrical conductivity needed for the conduction of the electrical signal and the mechanical consistency that guarantees the durability and resistance to the electrolyte. Biocomposites can be classified according to the nature of each phase.

*Biological materials*

As seen in Tables 8.3 and 8.4, a wide variety of lyophilised enzymes have been immobilised in a conducting matrix of the graphite–polymer type. Examples of these enzymes are oxidases, peroxidases, dehydrogenases, esterases, etc.

In the case of enzymes that show redox properties, the measurement of an analytical signal is as a result of the oxidation or reduction of a species in the electrode. These chemical species may be co-substrates, reaction products or co-factors needed for biocatalysis. At times, the analytical signal is obtained directly through the regeneration of the enzyme by the direct oxidation or reduction of its active centres on the surface of the electrode.

The applied potentials needed for the direct regeneration of the enzymes on the surface of the biosensor are very high. The same is true for the potentials required for the oxidation or reduction of products or co-substrates. These high voltages raise the number of interfering species that may contribute to a lower accuracy in the analytical data. To eliminate this problem, the biocomposite is modified with a catalyst that diminishes the activation energy and the applied potential. Electrochemical mediators may also be introduced acting as electron exchangers between the enzyme and the surface of the electrode.

Immunological materials may be immobilised also in these conducting polymer matrices. The resulting immunomaterial acts as a reservoir of the immunological component. The analytical signal will be produced

TABLE 8.3

Carbon–polymer biocomposites used in amperometric biosensors

| Composite preparation | | | | Analyte | Working medium | Potential (V) (Ag/AgCl) | Ref. |
|---|---|---|---|---|---|---|---|
| Polymer (%) | Conductor (%) | Enzyme (%) | Curing | | | | |
| *Dispersed biocomposites* | | | | | | | |
| Epoxy-bonded | Graphite (97) | Tyrosinase (3) | 48 h, room temperature | Phenol, catechol | Phosphate buffer 0.1 M, pH 6.0 | −0.1 | [114] |
| Epoxy-bonded | Graphite (95) | BO, HRP (5) | 60 h, room temperature | Bilirubin | Phosphate buffer 0.05 M, pH 7.4 $K_4[Fe(CN)_6]$ 1 mM | −0.2 | [14] |
| Paraffin (32) | Graphite (63) | Tyrosinase (5) | Mixture at 50°C, cooling at room temperature | Dopamine | Phosphate buffer 0.1 M, pH 7.0 | −0.2 | [26] |
| Epoxy (49) | Graphite (49) | GOD (2) | 24 h, room temperature | Glucose | Phosphate buffer 0.1 M, KCl 0.1 M pH 7.00 | 1.15 | [23] |
| Polyester (59) | Graphite (39) | GOD (2) | 24 h, room temperature | Glucose | Phosphate buffer 0.1 M, KCl 0.1 M pH 7.00 | 1.1 | [23] |
| Polymethacrylate (49) | Graphite (49) | GOD(2) | 24 h, room temperature | Glucose | Phosphate buffer 0.1 M, KCl 0.1 M pH 7.00 | 1.1 | [23] |
| Polyurethane (38) | Graphite (60) | GOD (2) | 24 h, room temperature | Glucose | Phosphate buffer 0.1 M, KCl 0.1 M pH 7.00 | 1.15 | [3] |
| Silicone (39) | Graphite (59) | GOD (2) | 24 h, room temperature | Glucose | Phosphate buffer 0.1 M, KCl 0.1 M pH 7.00 | 1.15 | [23] |

(continued on next page)

TABLE 8.3 (*continued*)

| Composite preparation | | | | Analyte | Working medium | Potential (V) (Ag/AgCl) | Ref. |
|---|---|---|---|---|---|---|---|
| Polymer (%) | Conductor (%) | Enzyme (%) | Curing | | | | |
| Epoxy (78) | Graphite (20) | HRP (2) | 1 week, 35–40°C | $H_2O_2$ | Phosphate buffer 0.1 M, KCl 0.1 M pH 7.00 | −0.3 | [70] |
| Epoxy (70.4) | Graphite (17.6) | AChE (12) | 1 week, 35–40°C | Glucose | Phosphate buffer 0.1 M, KCl 0.1 M pH 7.00 | 1.15 | [29] |
| 2-Hexadecanone (48) | Graphite (48) | FDH (4) | Mixture at 50°C, cooling at room temperature | Fructose | Phosphate buffer 0.2 M, pH 7.4 $K_3[Fe(CN)_6]$ 2 mM | 0.2 | [27] |
| *Consolidated bicomposites* | | | | | | | |
| Teflon (60) | Graphite (38.5) | GOD (1.5) | Pressure 7000 kg $cm^{-2}$ | Glucose | Phosphate buffer 0.05 M pH 7.4 | 0.9 | [33] |

AChE: acetylcholine esterase; BO: bilirubin oxidase; FDH: fructose dehydrogenase; GOD: glucose oxidase; HRP: horseradish peroxidase.

TABLE 8.4

Modified carbon–polymer biocomposites used in amperometric biosensors

| Composite preparation | | | | | Analyte | Working medium | Potential (V) (Ag/AgCl) | Ref. |
|---|---|---|---|---|---|---|---|---|
| Polymer (%) | Conductor (%) | Modifier (%) | Bioreceptor | Curing/ consolidation | | | | |
| *Dispersed biocomposites* | | | | | | | | |
| Epoxy-bonded (82) | Graphite | $NAD^+$ (12) | LDH (6) | 50 h, room temperature | Lactate | Phosphate buffer 0.05 M, pH 7.4 | 0.7 | [74] |
| Epoxy-bonded (94) | Graphite | $RuO_2$ (5) | Tyrosinase (1) | 48 h, room temperature | Phenol | Phosphate buffer 0.1 M, pH 6.0 | −0.1 | [72] |
| Epoxy-bonded (80) | Graphite | Au (8), Pd (12) | Tyrosinase (1) | 48 h, room temperature | Phenol | Phosphate buffer 0.1 M, pH 6.0 | −0.1 | [72] |
| Epoxy-bonded (54) | Graphite | 1,1′-dimethyl ferrocene (26) | GOD (20) | 48 h, room temperature | Glucose | Phosphate buffer 0.5 M, pH 6.0 | 0.5 | [115] |
| Epoxy (76) | Graphite (19) | Pt (3) | HRP (2) | 1 week, 35–40°C | $H_2O_2$ | Phosphate buffer 0.1 M KCl 0.1 M, pH 7.0 | −0.5 | [70] |
| Epoxy (63) | Graphite (15.8) | Au (11.8), Pd (7.9) | GOD (1.5) | 1 week, 35–40°C | Glucose | Phosphate buffer 0.1 M KCl 0.1 M, pH 7.0 | 0.9 | [71] |
| Epoxy (63) | Graphite (16) | TTF (19.5) | GOD (1.5) | 1 week, 35–40°C | Glucose | Phosphate buffer 0.1 M KCl 0.1 M, pH 7.0 | 0.15 | [73] |

(continued on next page)

TABLE 8.4 (*continued*)

| Composite preparation | | | | | Analyte | Working medium | Potential (V) (Ag/AgCl) | Ref. |
|---|---|---|---|---|---|---|---|---|
| Polymer (%) | Conductor (%) | Modifier (%) | Bioreceptor | Curing/consolidation | | | | |
| Epoxy (71) | Graphite (18) | TCNQ (9) | AChE-silica (2) | 1 week, 35–40°C | Acetyl thiocholine | Phosphate buffer 0.1 M, KCl 0.1 M, pH 7.5 | 0.3 | [116] |
| Epoxy (71) | Graphite (18) | TCNQ (9) | BChE-silica (2) | 1 week, 35–40°C | Butyryl thiocholine | Phosphate buffer 0.1 M, KCl 0.1 M, pH 7.5 | 0.3 | [116] |
| Polymethacrylate (54) | Graphite (40) | $NADP^+$ | GDH (2) | 4 days, room temperature | Glucose | Phosphate buffer 0.1 M KCl 0.1 M, pH 7.5 | 1.0 | [75] |
| *Consolidated biocomposites* | | | | | | | | |
| Teflon (70) | Graphite (27) | $K_4[Fe(CN)_6]$ (1.6) | HRP (1.4) | Pressure (10 000 kg $cm^{-2}$) 10 min | $H_2O_2$ | Ethyl acetate DOSS Phosphate buffer 0.05 M, pH 7.4 | 0.1 | [117] |
| Teflon (70) | Graphite (24) | Ferrocene (2) | HRP(2), GOD(2) | Pressure (10 000 kg $cm^{-2}$) 10 min | Glucose | Phosphate buffer 0.05 M, pH 7.4 | 0.0 | [35] |

AChE: acetylcholinesterase; BChE: butyrylcholinesterase; DOSS: dioctylsulphosuccinate; TCNQ: tetracyanoquinodimethane; TTF: tetrathiafulvalene.

by the oxidation or reduction of a product of the reaction biocatalysed by an enzyme marker of the antigen–antibody interaction. A similar approach also applies for DNA biocomposites.

*Preparation of the rigid biocomposites and construction of biosensors*
The main advantages of biocomposites are their ease of preparation, which does not require specialised personnel, and that the entire process is done in dry conditions—with the exception of biocomposites where the biological material is immobilised previously on a support (e.g. silica) using wet chemistry [29].

*Dispersed biocomposites.* The preparation of most of these composites relies on three factors: the relative content of the various phases, the homogenisation of the phases and the curing of the composite material. The mixture of the phases depends on the physical and chemical characteristics of each individual phase. As seen in Table 8.3, the proportions of the phases vary widely as one tries to achieve optimal conductivity and mechanical stability. The weight percentage of the polymer phase varies from 28% in low density polymers like silicone to 78% in epoxy resins with a high inorganic content. The nature of the polymer limits the quantity of the conducting phase. If the biocomposite calls for the presence of modifiers (catalysts or mediators), the conducting phase is reduced further as the physical and mechanical stability of the biocomposite is maximised. The proportion of the biological phase varies from 1 to 5%.

Homogenisation is a key step in the preparation of biocomposites. The composite biomaterials developed to date show high rigidity and hardness, permitting the mechanisation and the regeneration of the sensing surface by simply polishing them. The higher the degree of homogenisation of the biological component and conducting phase within the biocomposite matrix, the higher the reproducibility of the response after each polishing step. This responds to the fact that the biocatalytic activity and the conductivity of the material are uniform throughout the sensing material. The polymers with the highest fluidity favour the dispersion of all the components of the biocomposite and permit a higher filling. Good homogenisation implies a higher degree of intimate contact of the phases favouring a more sensitive and fast electrochemical interaction.

Prior to curing, the biocomposite must be confined to a vessel that forms the body of the electrode or the flow cell [30]. It may be also spread on a support material forming layers using screen-printing techniques to form a sensing surface with a desired size and form [31,32]. The next step follows where the biocomposite paste polymerises after applying a set of given conditions to the biocomposite. Generally, the curing temperature is below 40°C so that the biological component does not denaturise. If a polymer normally requires a higher temperature, the curing time has to be adjusted for total polymerisation to occur. A clear example of this is the resin known as Epo-Tek H77 (Epoxy Technology, Billerica, MA, USA) where the optimal curing temperature is 150°C during 1 h while at 40°C a week is needed.

*Consolidated biocomposites*. Segregated biocomposites are produced by the consolidation of the components under a given pressure for a given length of time. Biocomposites based on Teflon matrices [33–35] are produced in this manner. Teflon, in the form of small granules, is mixed thoroughly with graphite powder and the biological material. In some instances, the biological component is previously immobilised on the graphite particles. After the mixture is homogenised, a pressure in the range of 7000–10 000 kg $cm^{-2}$ is applied for 10 min approximately. The resulting disks are placed on electrode bodies produced previously. Segregated biocomposites based on ethylene/propylene/dieneterpolymer have been used recently in the construction of biosensors [36].

Impregnated biocomposites have been prepared using reticulated vitreous carbon (RVC) in which the insulator material used for filling the holes of the carbon electrode matrix is an epoxy resin [36].

## 8.3 COMPOSITE- AND BIOCOMPOSITE-BASED ELECTROCHEMICAL SENSORS

### 8.3.1 Conductometric sensors

Conductometric sensors based on rigid conducting filler-polymer composites have been developed. All conductive composites of the dispersion type show a characteristic percolation threshold that is dependent on the nature and size of the particles of the conductor and on the polymer used. There is a region in the percolation curve where the resistance of the composite changes abruptly as the conductor content is increased. This change is near the percolation threshold, as seen in Fig. 8.2. Composite materials that follow this characteristic composition show interesting physical properties that can be tied to this variation in conductivity.

When a conducting composite is exposed to heat, the polymer expands, the conducting particles separate and the resistivity increases. Hence, these materials may be used as thermistor materials. Conductive materials used in these composites include carbon black [37], $V_2O_3$ [38], TiO [39], NiO, $MnO_2$, $Co_2O_3$ [40], etc. Crystalline polymers such as polyethylene, poly(vinyl alcohol), etc. have been assayed [41]. This percolation disruption may be attained by means other than temperature, i.e. when pressure is applied to the composite, these composites show an electrical response.

Depending on either the conductive filler or the polymer used, a selectivity to certain organic vapours may be produced as well. These composites have been based on metal oxides such as ZnO, $SnO_2$, $Fe_2O_3$, $Co_3O_4$, $WO_3$, etc. [42] and on polymers such as those mentioned above. A large variety of devices is available in composite film chemiresistors [43] that combine the characteristics of a chemically sensitive insulating polymer with those of a conductive element such as carbon black [44,45].

### 8.3.2 Potentiometric sensors

Epoxy resins were used in the 1970s to build heterogeneous membranes used in ISE featuring an internal liquid contact. These membranes were prepared from inorganic salts mixed with Araldite epoxies [46–48] or epoxyacrylates [49,50]. All solid-state (internal solid contact) electrodes also appeared during this time, and were based on either crystalline membranes or poly(vinyl chloride) (PVC) membranes. In this type of electrode design (*coated-wire electrodes*), the sensing membrane was adhered directly to a conducting wire, without an internal solution. However, the adherence of the PVC membranes to the wire was feeble and they peeled off after a short period of use. In our laboratories, conducting composites based on an epoxy loaded with a silver or a graphite powder substituted the wire. This was an important advancement for potentiometric sensors without a liquid reference solution [51–54]. The nature of the conducting composite permitted the construction of potentiometric sensors that were easy to build at a low cost. An internal reference solution was not needed and the sensors could be machined in different shapes and sizes [55–61]. Furthermore, the adherence of the PVC membrane to the conducting composite is good and the sensors produced were robust and had an extended lifetime. An exhaustive review has been published on an ISE built with conductive composites based on epoxy resins [62].

Heterogeneous crystalline membranes based on silver halides–epoxy composites applied to conducting composites have been developed in our laboratories for iodide and bromide [63–65].

*ISFET modified with graphite–epoxy composites*

Conductive polymer composites have been used to build hybrid chemical sensors where an ISE structure is combined with ion-sensitive field-effect transistor (ISFET) technology [66]. The graphite–epoxy composite is cast on the gate area of the ISFET. This opaque layer reduces the optical sensitivity shown by some ISFETs. Additionally, it acts as a conductive support for PVC-based membranes. These membranes show good adherence to the composite, thus lengthening the lifetime of the sensor.

*Biocomposite-membrane-based potentiometric biosensors*

Consolidated composites are formed in such a manner that the conductor as well as other species (biological material etc.) are extended throughout the composite in a random, reticulated fashion with regions of pure insulator and pure conductor that do not intermix (see Section 8.2.2).

These types of consolidated composites have been used in the design of potentiometric biosensors amongst others. The coupling of an iodide-selective electrode with enzymes is one of the possibilities within a broad range of biosensor designs. Two simple examples of consolidated biocomposites are those for detection of D-amygdalin [67] and glucose [68]. The preparation procedure for these biocomposites is simple and it involves a simple compression of dry components. In a first case a mixture of electroactive salts, $Ag_2S$ and AgI, and the enzyme β-glucosidase were mixed in dried form and pressed into pellets using high pressure. The biocomposite in pellet form was used to design a potentiometric biosensor for potentiometric measurements of a D-amygdalin substrate. The specificity of enzyme toward the substrate is combined with the potentiometric detection of hydrolysis product—cyanide in this case—owing to the response of $Ag_2S$/AgI for cyanide through an iodide ion selectivity mechanism. In an another instance, glucose oxidase and peroxidase were mixed with the same electroactive salts and pressed into pellets by the same technology. The resulting consolidated biocomposite was used to produce a glucose biosensor. The pressure applied during the preparation of the biocomposite membrane determines the degree of enzyme immobilisation in the membrane. Membranes obtained at different pressures have different responses to the enzyme's substrate. Both enzymatic reaction and detection of the product occur at the same interface, resulting in no time delays associated with diffusion processes as is the case where separated layers of enzyme and transducer exist, i.e. in other classical potentiometric biosensor set-ups.

The biosensors prepared have a renewable surface that ensures the reproducibility of the transducer surface before each measurement. The electroactive salts that support the enzyme constitute a preserving medium for the enzyme providing operational lifetimes of more than three months.

### 8.3.3 Amperometric sensors

In the 1950s, Adams proposed the use of carbon paste, a soft conducting composite, as a material for the construction of electrochemical transducers [12]. These composites are easily prepared at a low cost. However, they also show some undesirable features such as low chemical and physical surface stability. The fast degradation of these materials has impeded their use outside the research laboratory. Several reviews on carbon paste electrodes (CPEs) have appeared in literature [69,13].

At the beginning of the 1990s, amperometric sensors based on rigid composites started to appear with advantageous physical, mechanical, economical and electrochemical features. Carbon composite electrodes (CCEs) do not show the same stability problems as CPEs and maintain the physical and cost advantages of the CPEs. This is why CCEs are a low-cost alternative for amperometric sensing. Additionally, the construction method can be translated to industrial mass production (see Section 8.3.4).

The mouldability of these composites permits an easy construction of amperometric sensors in a number of configurations such as planar, cylinder, tubular or microsize electrodes. When the polymer is hardened, the material can be machined, cut and polished without affecting its properties.

*Biosensors based on rigid carbon–polymer biocomposites*

These biosensors can be classified according to both the number of phases included in the composite and the nature of the modifiers used in preparing it.

*Biosensors based on carbon–polymer–enzyme biocomposites*. This type of composite design includes biosensors sensitive to substrates such as glucose, acetylthiocholine or phenol. The enzyme included in the biocomposite biocatalyses a reaction that generates an electrochemically active species. The working potential of each biosensor depends on the oxidation or reduction potential of the species generated by biocatalysis. As seen in Table 8.3, the applied potentials are generally high and the number of interfering species can be high depending on the sample under analysis. The absence of membranes reduce the size of the diffusion profiles and produces an analytical signal quickly as the electroactive species are generated close to the conducting sites on the surface of the biosensor.

Some peroxidases that possess active groups like heme c can be oxidised and reduced directly on the surface of the sensor. The physical characteristics of the biocomposites facilitates close contact between the redox centres of these enzymes and the conductive sites of the biomaterial, resulting in direct electron transfer [70]. Also, the rigid nature of the biocomposite guarantees a mechanical stability and is a reservoir of the biological material rendering the device useful for several measurements.

*Biosensors based on carbon–polymer–enzyme biocomposites modified with electrocatalysts*. The modifiers added to the biocomposite are one or several transition metals (Au,Pd) [71,72], Pt [70], ruthenium oxide [72], etc. (see Table 8.4). Although there is no electron mediation between the enzyme and the transducer, the electrocatalytic action of the modifier lowers not only the activation energy of the electrochemical reaction but also the working potential and, hence, the response time. Despite these advantages, the lowering of the working potential may have an effect on any electroactive species present in the working medium, including interferent species.

*Biosensors based on carbon–polymer–enzyme biocomposites modified with redox mediators*. Redox mediators are chemical substances with a small molecular size capable of interacting with the active redox sites of enzymes. These mediators compete with the natural co-substrates to regenerate redox enzymes and undergo regeneration on the surface of the electrode. The main advantages of using mediators is the fast electron transfer between the enzyme and the electrode and, also, the independence of the analytical signal in relation to the concentration of the natural co-substrates present in the working medium. This interaction is possible as the mediator—probably by diffusion—is close to the enzyme and to the conducting sites of the sensing biomaterial. The working potential is the regeneration potential of the mediator and should be as low as possible to screen out most of the interfering species. The first mediators were used in solution and the sample could not be used for further analysis. The technology described in this report permits the easy incorporation of these substances in the sensing material (*reagentless configuration*) (see Table 8.4) forming a mediator reservoir that is used in multiple measurements. Table 8.4, for example, shows a glucose biosensor modified with tetrathiafulvalene (TTF). TTF can interact with the active redox centres of the enzyme glucose oxidase allowing for its regeneration even in the absence of oxygen, its natural co-substrate [73]. The solubility of some forms (oxidised or reduced) of this mediator results in its leaching into solution and, hence, lowers the analytical signal upon continuous use. However, as the biocomposite is an enzyme and mediator reservoir, polishing the surface of the (sacrificial) biosensor is enough to produce a reproducible and fresh sensing (renewable) surface.

*Biosensors based on carbon–polymer–enzyme biocomposites modified with another enzyme*. Sometimes a second enzyme is added to the matrix of the biocomposite (see Table 8.4) [35]. The advantages of bi-enzyme systems is a higher and indirect analytical selectivity for the measured substrate. The linking of two enzyme systems is simple in a biocomposite as opposed to wet-chemistry immobilisation. This is true as enzyme co-immobilisation in a biocomposite does not lead to procedural incompatibilities as the immobilisation is realised under dry conditions.

*Biosensors based on carbon–polymer–enzyme biocomposites modified with cofactors*. Some dehydrogenases require the presence of co-factors in order to biocatalyse a chemical reaction. Usually this required co-factor has been used as a component of the solution resulting in its loss after the analysis cycle. Recently, the use of biocomposites based on these enzymes has added the co-factor to their matrices. Table 8.4 shows two examples where lactate dehydrogenase (LDH) [74] and glucose dehydrogenase (GDH) [75] have been immobilised together with the co-factor $NAD^+$. As with biosensors based on biocomposites modified with mediators, the signal of these devices decreases after continued use due to the co-factor leaking from the biocomposite to the solution. However, the biosensor can be regenerated by polishing the surface allowing a new analysis to be performed.

*Immunosensors based on rigid carbon–polymer immunocomposites*
Immunological assays are based on the strong specific affinity between antigen and antibody. Once this interaction is established, it is hard to break so immunosensors have to be discarded after each assay. Conducting immunocomposites [76] act as a reservoir for the immunological component. After each antigen–antibody assay, it is possible to regenerate the surface of the electrode with a simple polishing, producing a new, fresh and reproducible surface. The composite immunomaterial not only acts as a support for the specific heterogeneous interaction between the antigen and the antibody, but it also forms the electrochemical transducer. This transducer with the appropriate applied potential, oxidises or reduces the products of the reaction biocatalysed by an enzyme acting as a marker for one of the immunological components [76].

Various kinds of immunocomposite materials have been developed in our laboratories. They comprise one component of the immunological pair in the bulk of the rigid graphite–polymer composite. The composite acts as a reservoir for the immunological material.

Immunocomposites were prepared mixing graphite and an epoxy resin [77,78] or methacrylate [76,78] and then adding rabbit IgG (RIgG). They were used to analyse RIgG using GaRIgG labelled with either alkaline phosphatase (AP) [76,77] or horseradish peroxidase (HRP) [78], following the competitive technique. Two other immunosensors were also prepared using immunocomposite materials [77,79]. The immunosensors contained composites based on graphite, methacrylate and PCB (polychlorobiphenyl) or anti-β-hCG (human chorionic gonadotropin, beta subunit). Both involved HRP as the label and both were used for the determination of PCB and β-hCG, respectively, in competitive or sandwich assays.

Immunocomposite-based immunosensors are characterised by their mechanical stability and rigidity after curing. The most interesting advantage of these devices is their regeneration as their surfaces can be polished, obtaining fresh immunocomposite ready to be used in a new immunoassay. This is an alternative to the expensive and complex conventional procedures for the construction of immunosensors.

*Genosensors based on carbon–polymer composites*

An ideal material for electrochemical genosensing based on enzyme labelling allows for an effective immobilisation of the probe onto its surface, a robust hybridisation of a target with the probe, a negligible non-specific adsorption of the enzyme label and a sensitive detection of the hybridisation event [80]. An interesting application of graphite–epoxy composites (GECs) is their use as surface-modified carbon–polymer composites—a new transducing material for electrochemical DNA analysis.

GECs present several advantages. Firstly, GEC materials have a rougher surface than glassy carbon electrodes. DNA can be attached onto a GEC surface by simple 'dry-adsorption'. Single-stranded DNA binds strongly onto GEC in a way that prevents the strands from self-association, but permits the hybridisation to complementary DNA. Secondly, GEC materials present a low non-specific adsorption either for DNA probes or enzymatic labels.

The above advantages were observed in a new amperometric genosensor based on GEC in a capture format [81]. The developed genosensor relies on the dry-adsorption of a DNA capture probe onto a graphite–epoxy surface. The capture probe hybridises to a complementary sequence in the target and a biotinylated probe hybridises to a different sequence of the target DNA. The extent of the duplex formed is determined with HRP conjugated to streptavidin. The peroxidase label is detected amperometrically using a suitable substrate for the enzyme. The whole procedure takes 150 min. The detection limit is in the femtomole range. The non-specific adsorption of the enzyme is negligible. Results when compared with those obtained using other common electrochemical transducers (carbon paste, glassy carbon) as well with a genosensor based on a dot-blot format using a nylon membrane integrated onto a GEC [82,83] clearly show the previously mentioned advantages of GEC compared to the other electrochemical genosensor formats.

Another approach developed by Wang involved using an ss-DNA probe dispersed uniformly in the matrix of a polishable and renewable carbon paste electrode [84]. Here, the carbon paste acted as a reservoir for the DNA probe, such that each nucleic acid assay was performed on a new surface at the same electrode. There was, however, a need for activation of the electrode surface.

### 8.3.4 Thick-film sensors

The use of thick-film technology for the production of sensors and biosensors is an emerging discipline. Screen printing seems to be one of the most promising technologies allowing biosensors to be placed large-scale on the market in the near future because of advantages such as miniaturisation, versatility and low cost and especially the possibility of mass production.

The most critical point in manufacturing thick-film biosensors is the sensing or active layer and its adhesion to the transducer layer. The use of this technology permits either multilayer depositions or integration of various layers into a single one. In this latter configuration, enzymes and other materials such as graphite, mediators, catalysts, stabilisers, polymers etc. are mixed together forming a biocomposite ink or paste which is screen-printed onto the substrate producing biosensors in a single, one-step procedure (see Table 8.5).

Compared with multiple layer techniques, the use of biocomposite ink or pastes offers greater speed and simplicity in the manufacturing process [85]. However, biocomposite inks require more complex optimisation.

Several examples of biocomposite inks [31, 86–94] are summarised in Table 8.5. For example, in our laboratories a glucose biosensor based on a graphite–epoxy screen-printable biocomposite was designed [31]. The GOD graphite–epoxy pastes were prepared by dispersing GOD and graphite powder in an epoxy resin. The composite viscosity was adjusted with cyclohexanone. The prepared paste was printed onto a glass fibre circuit board patterned with copper tracks by a conventional photolithography process. The same procedure was implemented in a three-electrode configuration (biocomposite, graphite and silver–silver chloride tracks) [32].

This type of biocomposite paste was used to design pesticide biosensors [95,96]. It consisted of incorporating a biological material (BChE or AChE) into the graphite–epoxy paste. The enzyme remains immobilised in the bulk of the paste, which is screen-printed over

TABLE 8.5

Biocomposite pastes used in amperometric thick-film biosensors

| Biocomposite ink | | | Transducer layer/support | Analyte | Ref. |
|---|---|---|---|---|---|
| Active ingredients | Binder | Solvent | | | |
| GOD, TTF-TCNQ | PIB, NBC | Heptane | Pt/Chip | Glucose | [86] |
| LDH, graphite | UV-polymerizable paste, $SiO_2$, DEAE-dextran and lactitol | Phosphate buffer | Pt/Alumina ceramics | L-Lactate | [87] |
| LDH, NAD, activated carbon | PVP, silicone oil | | C/Polyester | L-Lactate | [88] |
| GOD-ferrocene | PVP, PVB | *n*-Pentanol | Cu/Polyamide film | Glucose | [89] |
| Tyrosinase | Carbon ink | | Alumina ceramics | Catechol (Pesticides) | [90] |
| GOD, TTF-graphite or TTF-Pt | PVP or HEC | Ethylene glycol water | Alumina ceramics | Carbohydrate | [91,92] |
| GOD, cupric hexacyanoferrate | Carbon ink | | Alumina ceramics | Glucose | [94] |
| GOD, graphite | Epoxy resin | Cyclohexanone | Cu/Glass fibre Ceramics | Glucose | [31] |
| RigG, graphite | TEOS, HPC | Water, ethanol, HCl | | Anti-RIgG | [93] |

*Biological materials:* ACh: acetylcholinesterase, GOD: glucose oxidase, LDH: lactate dehydrogenase. *Co-factor:* NAD: nicotinamide adenine dinucleotide. *Mediators:* TTF: tetratiafulvalene, TCNQ: tetracyanoquinodimethane. *Ink ingredients:* DEAE: diethanolaminoethyl, HEC: hydroxyethyl cellulose, HPC: hydroxyproyl cellulose, NBC: *N*-butylcarbital acetate, PEI: polyethylenenimine, PIB: poly(isobutylene), PVB: poly(vinylbutyral), PVP: poly(vinylpyrrolidine), TEOS: tetraethoxysilane.

copper tracks, producing the biosensor in a single-step procedure. The developed pesticide biosensor based on enzyme inhibition still lacks adequate analytical performance when compared to biosensors based on surface-immobilisation of enzymes. Nevertheless, the biosensors are sufficiently robust and are capable of maintaining the biochemical characteristics of the enzyme for long storage periods under proper conditions. Additionally, these pastes are produced at low cost—an essential feature for single-use products—solving the challenge of use in inhibition determination.

It is not only possible to produce enzymatic pastes, but also immunosensors can be made by screen printing. An example is a sol–gel thick-film amperometric immunosensor by Wang et al. [93]. The sol–gel solution was prepared by mixing tetraethoxysilane (TEOS), ethanol, water and HCl. Hydroxypropyl cellulose and RIgG (rabbit IgG) were added to the sol–gel solution and mixed thoroughly. Subsequently, this sol–gel solution was mixed with graphite and screen-printed onto ceramic substrates.

Major advantages associated with thick-film sensors are their versatility in material preparation and the simplicity in film formation, generally resulting in lowering manufacturing cost. Research in the field of thick-film biosensors is continuously improving and its evolution is focused on ways of producing biosensor strips for small portable devices, capable of monitoring different species of interest such as glucose and other clinical parameters or pollutants (pesticides etc.).

The limiting step for commercialisation is mainly the immobilisation procedure of the biological molecule, mediators and other additives, which is not suitable, in most cases, to mass production, reduction in manufacturing costs or reproducibility. However, the modification of commercially available or laboratory-made conducting inks, and also the wide range of possibilities with respect to the support material, allow the development of low-cost and high-performance biosensors which can be applied in large-scale in any field.

Currently, there is a tendency towards 'one-step deposition' configuration due to its simplicity and applicability in mass biosensor production. The use of biocomposite inks, which integrate all the necessary materials into a single one, and the absence of outer coatings greatly simplify the fabrication of screen-printed biosensors. Nevertheless, this integration process is still far from being complete. Multiple layer technology is still used with the aim of improving the analytical characteristics of the biosensor devices. Some authors began firstly with a simple configuration and then continued their work by adding additional membranes, resulting in an increase in the number of fabrication steps. Both trends make two important issues clear: simple configuration and good analytical performance. Considering the present state of knowledge, these two issues seem to be contradictory. The development of new pastes and inks that incorporate as much bioelectrocomponents and as few steps as possible is clearly the most efficient strategy in the design of new screen-printed biosensor devices.

The so-called biological inks that present the higher integration level of different materials, compatible with automatic processes, are a new and interesting field of research, directly related to the improvement of biosensor thick-film technology.

### 8.3.5 Sensors for voltammetric stripping techniques

Stripping voltammetry has proved to be a powerful electroanalytical technique for trace-metal measurements [97]. The technique offers excellent sensitivity and the possibility for multi-element determination. The performance of stripping voltammetry, as with other electroanalytical techniques, is strongly affected by the working electrode material. To date a range of materials has been used as working electrodes for detection of heavy metals, the most popular being mercury-based electrodes. The usefulness of mercury electrodes for the determination of metal ions is due to their ability to form amalgams, allowing for pre-concentration of the metal ions prior to their determination by voltammetric stripping methods. Another advantage of using mercury in working electrodes is associated with the high overpotential of hydrogen evolution on such electrodes [98].

It is the potential dangers that are associated with mercury which has led to the development of various electrodes where the manipulation of large quantities of toxic mercury solutions may be avoided. In recent years, a growing interest in the use of mercury-free electrodes has occurred [99]. Recently, the use of graphite–epoxy composite (GEC) electrodes without surface modification for the differential pulse anodic stripping voltammetry (DPASV) [100] and stripping potentiometry [101] of lead and other heavy metals of interest have been performed in our laboratories. The use of these types of electrodes is a simpler alternative to the use of mercury or other modified electrodes for analysis of trace levels of heavy metals.

The possible use of graphite–epoxy material by screen-printing technology opens the door for mass production of disposable sensors for heavy metal analysis using stripping techniques. The research into new composite materials for their application in stripping techniques seems to be an attractive and very promising field for the future.

## 8.4 CONCLUSIONS

The development of composites based on conductive materials dispersed in polymeric matrices has led to important advances in analytical electrochemistry, particularly in sensor devices. These materials integrate the electrical properties of a conducting phase (e.g. graphite particles) with the ease of processing plastics (e.g. epoxy, methacrylate, silicone, Teflon) or binders (paraffin oil, Nujol). As a consequence of this unique integration these materials show attractive mechanical, electrochemical, physical, chemical and economical features in either rigid or soft structures compared to the classic conductors (gold, platinum, graphite). The morphology, size and distribution of the conducting particles define the behaviour of the composite as a microelectrode array. Furthermore, these composites are prepared easily.

Rigid composites permit the easy construction of sensors of various shapes (cylindrical, planar, tubular). Their sensing surface can be renewed reproducibly by a simple polishing procedure. Additional outstanding features of rigid composites are their compatibility with organic media and high mechanical stability over time.

The plastic nature of composite materials makes them suitable for modification, permitting the incorporation of fillers. A great number of biological receptor materials (e.g. enzymes, antibodies, DNA) can be bulk-immobilised by blending them with these composites to form new dispersed or consolidated biocomposite materials for use in an electrochemical biosensor. The biocomposites not only act as reservoirs of biologically active materials but may also contain catalysts, mediators and co-factors that improve the response of the resulting electrochemical biosensors. The components of the sensing surface can be tailored by adjusting their content in the bulk of the biocomposite. These materials are prepared easily through ‘dry chemistry’ using inexpensive procedures, both very attractive issues for the mass fabrication of electrochemical biosensors.

## ACKNOWLEDGEMENTS

S.A. thanks the following former PhD students for valuable work on biosensor research and development involving composite materials, which is partially reported in this review: Esteve Fàbregas (UAB, Bellaterra, Spain), Antoni Florido (Universitat Politècnica de Catalunya, Spain), Josep M. Paulís (UAB, Bellaterra, Spain), Francisco Céspedes (EUPMA, Mollet del Vallès, Spain), Adela Orellana (Pharmaorion, Helsinki, Finland), Daniel Martorell (Grifols, Mollet del Vallès, Spain), Adriana Morales (UNAM, Mexico), C. A. Galán (Universidad Autónoma de Hidalgo, Mexico), Marta Santandreu (M-Química, Sabadell, Spain), Sílvia Solé (Atom, Barcelona, Spain), M. Isabel Pividori, (Universidad del Litoral, Argentina), M. Albareda (UAB, Bellaterra) and Edna Williams (Dublin City University, Ireland).

## REFERENCES

1 S. Alegret, *Analyst*, 121 (1996) 1751.
2 F. Céspedes, E. Fàbregas and S. Alegret, *Trends Anal. Chem.*, 15 (1996) 296–304.
3 F. Céspedes and S. Alegret, *Trends Anal. Chem.*, 19 (2000) 276–285.
4 S. Alegret, E. Fàbregas, F. Céspedes, A. Merkoçi, S. Solé, M. Albareda and M.I. Pividori, *Quím. Anal.*, 18 (1999) 23–29.
5 G.R. Ruschan, R.E. Newnham, J. Runt and E. Smith, *Sens. Actuat.*, 20 (1989) 269.
6 D.Y. Godovski, E.A. Koltypin, A.V. Volka and M.A. Moskvina, *Analyst*, 118 (1993) 997.
7 K.B. Oldham, *J. Electroanal. Chem.*, 122 (1981) 1.
8 D.E. Tallman and S.L. Petersen, *Electroanalysis*, 2 (1990) 499.
9 K.E. Creasy and B.R. Shaw, *Anal. Chem.*, 61 (1989) 1460.
10 S. Fletcher and M.D. Horne, *Electrochem. Commun.*, 1 (1999) 502.
11 W.L. Caudill, J.O. Howell and R.M. Wightman, *Anal. Chem.*, 54 (1982) 2532.
12 R.N. Adams, *Anal. Chem.*, 30 (1958) 1576.
13 K. Kalcher, J.-M. Kauffman, J. Wang, I. Svancara, K. Vytras, C. Neuhold and Z. Yang, *Electroanalysis*, 7 (1995) 5.
14 L. Gorton, *Electroanalysis*, 7 (1995) 23.
15 C. Wang, K.E. Creasy and B.R. Shaw, *J. Electroanal. Chem.*, 300 (1991) 365.
16 B.R. Shaw and K.E. Creasy, *Anal. Chem.*, 60 (1988) 1241.

17 J.-M. Kauffmann, C.R. Linders, G.J. Patriarche and M.R. Smyth, *Talanta*, 35 (1988) 179.
18 C.A. May, (Ed.), *Epoxy Resins, Chemistry and Technology*, 2nd ed., Marcel Dekker, New York, 1988.
19 J. Wang, T. Golden, K. Varughese and I. El-Reyes, *Anal. Chem.*, 61 (1998) 508.
20 J. Wang and K. Varughese, *Anal. Chem.*, 62 (1990) 318.
21 S. Alegret, J. Alonso, J. Bartrolí, E. Martínez-Fàbregas and F. Valdés, In: G.G. Guilbault and M. Mascini (Eds.), *Uses of Immobilized Biological Compounds*, Kluwer, Dordrecht, 1993, pp. 67–79.
22 S. Alegret, J. Alonso, J. Bartrolí, F. Céspedes, E. Martínez-Fàbregas and M. del Valle, *Sens. Mat.*, 8 (1966) 147.
23 S. Alegret, F. Céspedes, E. Martínez-Fàbregas, D. Martorell, A. Morales, E. Centelles and J. Muñoz, *Biosens. Bioelectron.*, 11 (1996) 35.
24 R. Nagata, K. Yokoyama, S.A. Clark and I. Karube, *Biosens. Bioelectron.*, 10 (1995) 261.
25 T. Vopel, A. Ladde and H. Müller, *Anal. Chim. Acta*, 251 (1991) 117.
26 C. Petit, A. González-Cortés and J.-M. Kauffmnn, *Talanta*, 42 (1995) 1783.
27 S. Miertus, J. Katrlik, A. Pizzariello, M. Stredansky, J. Svitel and J. Svorc, *Biosens. Bioelectron.*, 13 (1998) 911.
28 I. Pankratov and O. Lev, *J. Electroanal. Chem.*, 393 (1995) 35.
29 D. Martorell, F. Céspedes, E. Martínez Fàbregas and S. Alegret, *Anal. Chim. Acta*, 337 (1977) 305.
30 F. Céspedes, F. Valero, E. Fàbregas, J. Bartrolí and S. Alegret, *Analyst*, 120 (1995) 2255.
31 C.A. Galán-Vidal, J. Muñoz, C. Domínguez and S. Alegret, *Sens. Actuat. B*, 45 (1997) 55.
32 C.A. Galán-Vidal, J. Muñoz, C. Domínguez and S. Alegret, *Sens. Actuat. B*, 52 (1998) 257.
33 J. Wang, A.J. Reviejo and L. Agnes, *Electroanalysis*, 5 (1993) 575.
34 G. Ortiz, M.C. González, A.J. Reviejo and J.M. Pingarrón, *Anal. Chem.*, 69 (1997) 3521.
35 M.A. del Cerro, G. Cayuela, A.J. Reviejo, J.M. Pingarrón and J. Wang, *Electroanalysis*, 9 (1997) 1113.
36 B. Serra, S. Jiménez, M.L. Mena, A.J. Reviejo and J.M. Pingarrón, *Biosens. Bioelectron.*, 17 (2002) 217.
37 K.A. Hu, J. Runt, A. Safari and R.E. Newnham, *Mater. Sci. Res. Ser.*, 20 (1986) 475.
38 K.A. Hu, D. Moffatt, J. Runt, A. Safari and R.E. Newnham, *J. Am. Ceram. Soc.*, 70 (1987) 583.
39 K.A. Hu, J. Runt, A. Safari and R.E. Newnham, *Phase Transitions*, 7 (1986) 1.
40 P. Bergveld and B.H. van der Schoot, *Selective Electrode Rev.*, 10 (1988) 5.

41 G.R. Ruschan, R.E. Newnham, J. Runt and E. Smith, *Sens. Actuat.*, 20 (1989) 269.
42 G. Heiland, *Sens. Actuat.*, 2 (1982) 243.
43 D.M. Wilson, S. Hoyt, J. Janata, K. Booksh and L. Obando, *IEEE Sens. J.*, 1 (2001) 256.
44 N.S. Lewis, M.C. Lonergan, E.J. Severin, B.J. Doleman and R.H. Grubbs, *Proc. SPIE*, 3079 (1997) 21.
45 A. Matzger, T.P. Vaid and N.S. Lewis, *Proc. SPIE*, 3710 (1999) 315.
46 H. Hirata and K. Data, *Talanta*, 17 (1970) 883.
47 S.K. Srivastava, A.K. Jain, S. Agrawal and R.P. Singh, *J. Electroanal. Chem.*, 90 (1978) 291.
48 K. Hiiro, A. Kawahara and T. Tanaka, *Anal. Chim. Acta*, 110 (1979) 321.
49 T. Hepel, M. Hepel and M. Leszko, *Analyst*, 102 (1977) 132.
50 M. Trojanowicz, *Anal. Chim. Acta*, 114 (1980) 293.
51 S. Alegret, A. Florido, J.L.F.C. Lima and A.A.S.C. Machado, *Quím. Anal.*, 5 (1986) 36.
52 G.J. Moody, J.D.R. Thomas, J.L.F.C. Lima and A.A.S.C. Machado, *Analyst*, 113 (1988) 1023.
53 S. Alegret and A. Florido, *Analyst*, 116 (1991) 473.
54 S. Alegret and T. Escalas, *Biosens. Bioelectron.*, 6 (1991) 609.
55 S.A.H. Khalil, G.J. Moody, J.D.R. Thomas and J.L.F.C. Lima, *Analyst*, 111 (1986) 611.
56 S. Alegret, J. Alonso, J. Bartrolí, J.M. Paulís, J.L.F.C. Lima and A.A.S.C. Machado, *Anal. Chim. Acta*, 164 (1984) 147.
57 J. Alonso, J. Bartrolí, J.L.F.C. Lima and A.A.S.C. Machado, *Anal. Chim. Acta*, 179 (1986) 503.
58 S. Alegret, J. Alonso, J. Bartrolí, A.A.S.C. Machado, J.L.F.C. Lima and J.M. Paulís, *Quím. Anal.*, 6 (1987) 278.
59 S. Alegret, J. Alonso, J. Bartrolí and E. Martínez-Fàbregas, *Analyst*, 114 (1989) 1443.
60 S. Alegret, A. Florido, J.L.F.C. Lima and A.A.S.C. Machado, *Talanta*, 36 (1989) 825.
61 S. Alegret, J. Alonso, J. Bartrolí, J. García-Raurich and E. Martínez-Fàbregas, *J. Pharm. Biomed. Anal.*, 6 (1988) 749.
62 A.A.S.C. Machado, *Analyst*, 119 (1994) 2263.
63 S. Alegret, J.L.F.C. Lima, A.A.S.C. Machado, E. Martínez-Fàbregas and J.M. Paulís, *Quím. Anal.*, 6 (1987) 176.
64 S. Alegret, A. Florido, J.L.F.C. Lima and A.A.S.C. Machado, *Quím. Anal.*, 6 (1987) 418.
65 S. Alegret, A. Florido, J.L.F.C. Lima and A.A.S.C. Machado, *Talanta*, 36 (1989) 825.
66 F. Valdés-Perezgasga, S. Alegret, J. Alonso and J. Bartrolí, *Sens. Actuat. B*, 15–16 (1993) 214.

67 A. Merkoçi, S. Braga, E. Fàbregas and S. Alegret, *Anal.Chim. Acta*, 391 (1999) 65–72.
68 A. Merkoçi, E. Fàbregas and S. Alegret, *Sens. Actuat. B*, 60 (1999) 97–105.
69 V. Stara and M. Kopanika, *Electroanalysis*, 1 (1989) 251.
70 A. Morales, F. Céspedes, J. Muñoz, E. Fàbregas and S. Alegret, *Anal. Chim. Acta*, 332 (1996) 131.
71 F. Céspedes, E. Fàbregas and S. Alegret, *Anal. Chim. Acta*, 284 (1993) 21.
72 E.S. Mereike Lutz and E. Domínguez, *Electroanalysis*, 8 (1996) 117.
73 F. Céspedes, E. Fàbregas and S. Alegret, *Electroanalysis*, 6 (1994) 759.
74 U. Wollenberger, J. Wang, M. Ozsoz, E. González-Romero and F. Scheller, *Bioelectrochem. Bioenerg.*, 26 (1991) 287.
75 A. Morales, F. Céspedes and S. Alegret, *Mater. Sci. Eng. C*, 7 (2000) 99.
76 M. Santandreu, F. Céspedes, S. Alegret and E. Fàbregas, *Anal. Chem.*, 69 (1997) 2080.
77 M. Santandreu, PhD Thesis, Universitat Autònoma de Barcelona, 1998.
78 M. Santandreu, S. Solé, S. Alegret and E. Fàbregas, In: B. Hock, D. Barceló, K. Camman, P.D. Hansen and A.P.F. Turner (Eds.), *Biosensors for Environmental Diagnostics*, Teubner, Stuttgart/Leipzig, 1998, pp. 64–89.
79 M. Santandreu, S. Alegret and E. Fàbregas, *Anal. Chim. Acta*, 396 (1999) 181.
80 M.I. Pividori, A. Merkoçi and S. Alegret, *Biosens. Bioelectron.*, 15 (2000) 291.
81 E. Williams, M.I. Pividori, A. Merkoçi and S. Alegret, *Biosens. Bioelectron.*, submitted.
82 M.I. Pividori, A. Merkoçi and S. Alegret, *Biosens. Bioelectron.*, 16 (2001) 1133.
83 M.I. Pividori, A. Merkoçi and S. Alegret, *Analyst*, 126 (2001) 1551.
84 J. Wang, J.R. Fernandes and L.T. Kubota, *Anal. Chem.*, 70 (1998) 3699.
85 M. Albareda-Sirvent, A. Merkoçi and S. Alegret, *Sens. Actuat. B*, 69 (2000) 153.
86 G.F. Khan, *Electroanalysis*, 9 (1997) 325.
87 Rohm, M. Genrich, W. Collier and U. Bilitewski, *Analyst*, 121 (1996) 877.
88 H.C. Yoon and H.-S. Kim, *Anal. Chim. Acta*, 336 (1996) 57.
89 R. Nagata, K. Yokohama, S.A. Clark and I. Karube, *Biosens. Bioelectron.*, 10 (1995) 261.
90 J. Wang, V.B. Nascimento, S.A. Kane, K. Rogers, M.R. Smyth and L. Angnes, *Talanta*, 43 (1996) 1903.
91 U. Bilitewski, G.C. Chemnitius, P. Rüger and D. Schmid, *Sens. Actuat. B*, 7 (1992) 351.
92 U. Bilitewski, A. Jäger, P. Rüger and W. Weise, *Sens. Actuat. B*, 15–16 (1993) 113.
93 J. Wang and P.V.A. Pamidi, *Anal. Chem.*, 70 (1998) 1171.
94 J. Wang and X. Zhang, *Anal. Lett.*, 32 (1999) 1739.

95 M. Albareda-Sirvent, A. Merkoçi and S. Alegret, *Anal. Chim. Acta*, 442 (2001) 35.
96 M. Albareda-Sirvent, A. Merkoçi and S. Alegret, *Sens. Actuat. B*, 79 (2001) 48.
97 J. Wang, *Stripping Analysis: Principles, Instrumentation, Applications*, VCH Publishers, Deerfield Beach/Weinheim, 1985.
98 Z. Galus, In: P.T. Kissinger and W. Heineman (Eds.), *Laboratory Techniques in Electroanalytical Techniques*, 2nd ed., Marcel Dekker, New York, 1996, pp. 443–466.
99 H.H. Frey, C.J. McNeil, R.W. Keay and J.V. Bannister, *Electroanalysis*, 10 (1998) 480–485.
100 L. Moreno-Baron, A. Merkoçi, S. Alegret, Personal communication.
101 M. Serradell, S. Izquierdo, L. Moreno, A. Merkoçi, S. Alegret, *Electroanalysis*, 14 (2002) 1281.
102 D.N. Armentrout, J.D. McLem and M.W. Long, *Anal. Chem.*, 51 (1979) 1037.
103 R.J. Fens, S. Siggla and D.J. Curran, *Anal. Chem.*, 50 (1978) 1068.
104 R.C. Engstrom, M. Weber and J. Werth, *Anal. Chem.*, 57 (1985) 933.
105 J.E. Anderson and D.E. Tallman, *Anal. Chem.*, 48 (1976) 209.
106 H.P. Henriques and A.G. Fogg, *Analyst*, 109 (1984) 1195.
107 D. Martorell, F. Céspedes, E. Martínez-Fàbregas and S. Alegret, *Anal. Chim. Acta.*, 290 (1994) 343.
108 A. Liberti and C. Morgia, *Anal. Chim. Acta*, 173 (1985) 157.
109 D.E. Weisshaar and D.E. Tallman, *Anal. Chem.*, 55 (1983) 1146.
110 S.L. Petersen and D.E. Tallman, *Anal. Chem.*, 62 (1990) 459.
111 L.N. Klatt, D.R. Connell, R.E. Adams, I.L. Honigberg and J.C. Price, *Anal. Chem.*, 47 (1975) 2470.
112 P. Skladal and M. Mascini, *Biosens. Bioelectron.*, 7 (1992) 335.
113 J.A. Cox and K. Lewinski, *Electroanalysis*, 6 (1994) 976.
114 P. Onnerfjord, J. Emmeus, G. Marko-Varga, L. Gorton, F. Ortega and E. Domínguez, *Biosens. Bioelectron.*, 10 (1995) 607.
115 F. Céspedes and S. Alegret, *Food Technol. Biotechnol.*, 34 (1996) 143.
116 D. Martorell, F. Céspedes, E. Martínez-Fàbregas and S. Alegret, *Anal. Chim. Acta*, 337 (1997) 305.
117 G. Ortiz, M.C. González, A.J. Reviejo and J.M. Pingarrón, *Anal. Chem.*, 69 (1977) 3521.

*Chapter 9*

# Optical chemical sensors and biosensors

***Fortunato Sevilla III and Ramaier Narayanaswamy****

## 9.1 INTRODUCTION

Optical chemical sensors and biosensors represent integrated analytical systems employing optical transduction. In these systems, chemical measurement is performed based on the interaction of light with a chemical system and the subsequent conversion of the resulting optical signal into an electrical signal. Optical fibers are often used as the integrating medium, linking the components of an optoelectronic instrumentation with a molecular recognition element that generates a response in the presence of the analyte. These devices encode chemical information through an optical signal traveling in the optical fibers.

Integration of the constituents in an optical fiber chemical sensor has resulted in a simplification of chemical analysis. Sampling and sample treatment are eliminated, if not simplified, since the molecular recognition unit can be interfaced directly with the measurand system for the sensor to generate a response. Real time and on-site measurement can, therefore, be achieved through this sensor. The similarity of the function and operation of this optical analytical system to those of an electrode has led to the coinage of a name for this optical analytical device—optrode or optode.

* Corresponding author. *E-mail address:* ramaier.narayanaswamy@umist.ac.uk.

*Comprehensive Analytical Chemistry XXXIX, pages 413–435* 

ISSN: 0166-526X

The incorporation of an optical fiber in this analytical system has introduced attractive features that further enhanced the advantages offered by an optical method of chemical measurement. The ability of the fiber to transmit optical signals over great distances with low power loss makes possible direct measurements at sites remote from the instrumentation system. The flexibility of optical fibers enables chemical analysis to be carried out even in difficult-to-reach sites. The small radial dimensions of the fibers allow these sensors to perform measurements on micro-liter quantities of chemical samples. The absence of the need for an external reference accessory by the optical fiber sensor promotes miniaturization of the analytical system.

Optical fiber chemical sensors have found numerous applications in various areas of chemical analysis [1,2]. Optrodes have been employed for chemical measurements in the fields of agriculture, food technology, biotechnology, medicine, pharmaceutical manufacturing, automotive industry, defense industry, industrial production and environmental monitoring. The development of novel materials and configurations is continuously expanding the range of applications of these sensors.

## 9.2 SENSOR STRUCTURE

The optical fiber chemical sensor consists of the following basic components: (1) an optical fiber, (2) optoelectronic instrumentation and (3) a solid-phase molecular recognition element. A schematic diagram of the structure of an optical fiber chemical sensor is shown in Fig. 9.1.

The working principle of an optrode is essentially the same as that of the conventional optical methods. The molecular recognition unit consists of a reagent phase that interacts with the analyte, and this interaction leads to a modification of an optical property of the reagent phase. This change is detected through radiation that is supplied by the instrumentation system which travels through the optical fiber to the distal end where the molecular recognition unit is localized. The radiation is then directed via the optical fiber to a phototransducer in the instrumentation system. The phototransducer generates an electrical signal that is used as the basis for the quantitation of the analyte concentration.

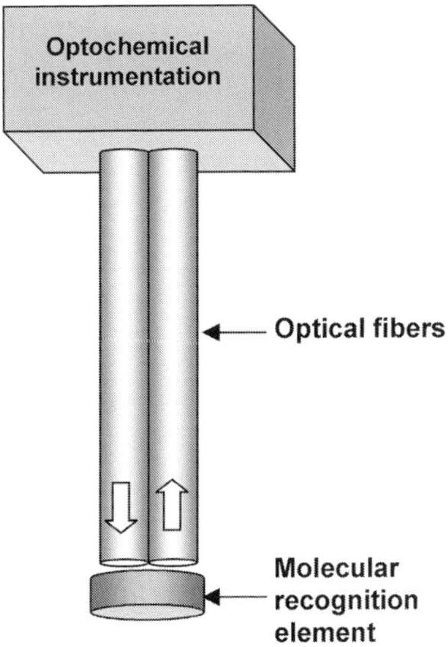


Fig. 9.1. Schematic diagram of the structure of an optical fiber chemical sensor.

### 9.2.1 Optical fibers

The optical fiber is basically a wave-guide for the radiation that interrogates the molecular recognition unit. The basic construction of an optical fiber is shown in Fig. 9.2(a). It consists of a transparent dielectric cylinder (known as the core) surrounded by a thin coating (known as cladding) of another dielectric material with a lower refractive index than the core. A jacket encloses the fiber to provide physical protection. The optical fiber can comprise glass, silica, quartz or a polymer such as polymethylmethacrylate, and have diameters as small as 3 μm or as large as 1 mm.

The propagation of light through the length of an optical fiber takes place by means of total internal reflection. Any beam that enters one end of the fiber core at an angle greater than a critical value (known as the acceptance angle, $\theta$) hits the interface of the core and the cladding, and is reflected back into the body of the core. It then strikes another portion of the core–cladding interface and is reflected once more. This process occurs repeatedly until the light reaches the other end of the fiber. Figure 9.2(b) shows the trajectory of a light beam inside an optical fiber. The light-gathering ability of a fiber is expressed through

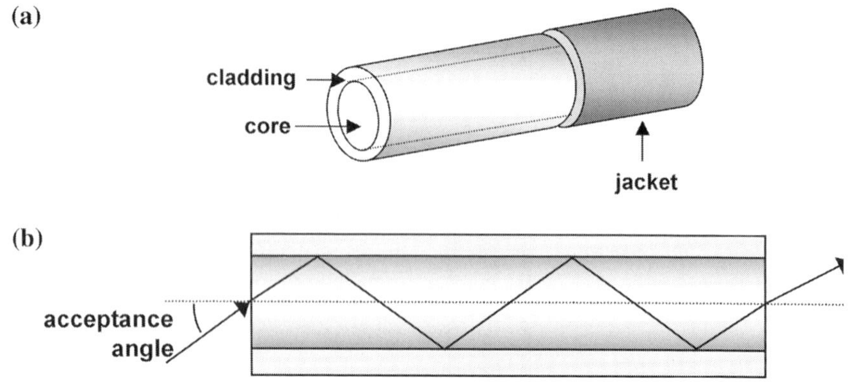


Fig. 9.2. (a) Basic construction of an optical fiber; (b) transmission of a light beam inside an optical fiber.

a parameter known as the numerical aperture (NA), which is defined in terms of the acceptance angle ($\theta$): $\mathrm{NA} = \sin\theta = (n_1^2 - n_2^2)^{1/2}$, where $n_1$ and $n_2$ are the refractive indices of the core and cladding materials, respectively. A more detailed treatment of the propagation of light in an optical fiber can be obtained from Seitz [3] and Wolfbeis [4].

It should be noted that the numerical aperture is independent of the dimensions of the fiber. It is dependent only on the refractive indices of the core and the cladding. A fiber can, therefore, be made to have a very small cross-sectional area to make it geometrically flexible without diminishing its light-gathering power. The diameter of the fiber, however, has to be of the same order as the magnitude of the wavelength of the propagated light to allow internal reflection of the electromagnetic wave.

The transmission of light through the fiber is dependent on the wavelength of the radiation and the reflection angle. Only certain combinations (modes) of wavelength and reflection angle are allowed. Each mode corresponds to a unique reflection angle for each radiation wavelength. For a particular wavelength, the number of modes of propagation decreases with the diameter of the fiber. An optical fiber with a diameter of about 3–5 μm allows only a single mode of propagation, and is, therefore, described as a single-mode fiber. Fibers with larger diameters, typically 100–200 μm, are multi-mode, and will allow the simultaneous transmission of several wavelengths through the fiber.

Not all wavelengths of light can be transmitted efficiently by an optical fiber. The limitation is determined by the fiber material, which could absorb radiation of certain wavelengths. This absorption of radiation decreases optical power, and the extent of loss increases with the length of the fiber. Glass and plastic optical fibers are useful only in the visible and near infrared region. However, quartz fibers can be used to transmit ultraviolet, visible and near-infrared radiation. A schematic diagram of the optical transmission capacity of some optical fiber materials is shown in Fig. 9.3.

During total internal reflection, interference occurs between the incident and reflected beams of light. As a consequence, there is a standing wave (known as an evanescent wave) at each point of internal reflection and the radiation can be said to penetrate the cladding (Fig. 9.4). The intensity of this wave decreases exponentially in the direction of an outward normal to the cladding. The depth of penetration for typical optical materials and visible light is of magnitude 100–1200 nm. The reflection points become closer to each other as more reflections take place within a certain length of fiber. Consequently, the evanescent waves will overlap and produce an almost continuous evanescent region under this condition. The occurrence of evanescent

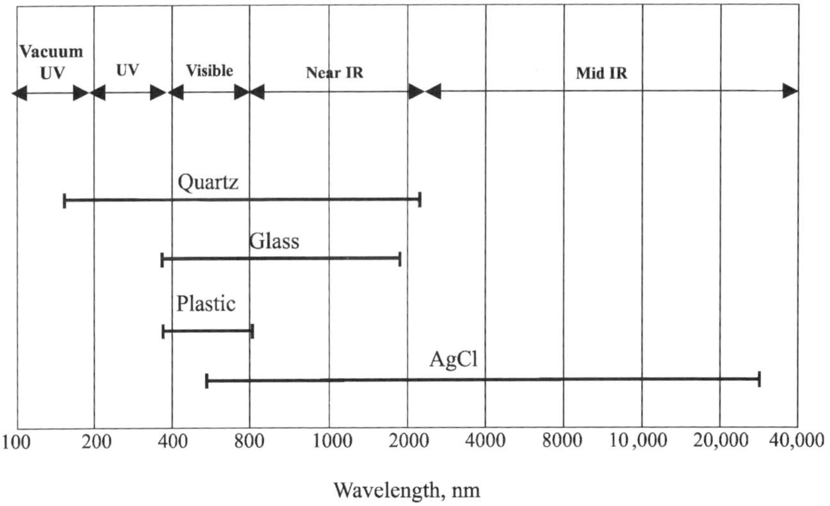


Fig. 9.3. Optical transmission capacity of some optical fiber materials.

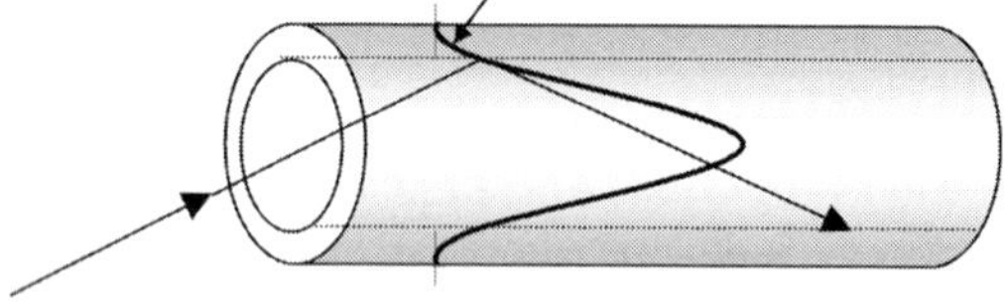

Fig. 9.4. Origin of an evanescent wave.

waves introduces another means of exploiting the optical fiber for chemical measurement.

### 9.2.2 Optoelectronic instrumentation

The instrumentation associated with optical fiber chemical sensors is similar to that involved in spectrophotometers. Its basic components include a light source, a wavelength selector, a photodetector and a read-out device. A block diagram of a typical instrumentation system employed in optical fiber chemical sensors is shown in Fig. 9.5. The radiation supplied by the light source is first made monochromatic by a wavelength selector before it enters the optical fiber sensor system. Then, the radiation emanating from the sensor is directed to the photodetector, which subsequently generates an electrical signal. The magnitude of the final signal is displayed in the read-out device.

The light source generates an intense and stable radiation signal needed to probe an optical property of the molecular recognition element in the sensor. The optical signal encodes information on analyte concentration in the sample. Widely used light sources include incandescent (filament) lamps, gas discharge lamps, light emitting diodes and lasers. Incandescent sources exhibit a broad spectral output with moderate radiance and are used in short-range optical fiber sensors. The miniature tungsten–halogen lamp is one of the most commonly used incandescent sources in optical fiber chemical sensors. Gas discharge lamps, particularly continuous arc lamps such as the xenon lamp, are characterized by a high radiance, and are ideal excitation sources in absorbance and fluorescence measurements. The light emitting diodes (LED) and laser diodes are miniature sources of relatively intense radiation of narrow bandwidth, making them highly

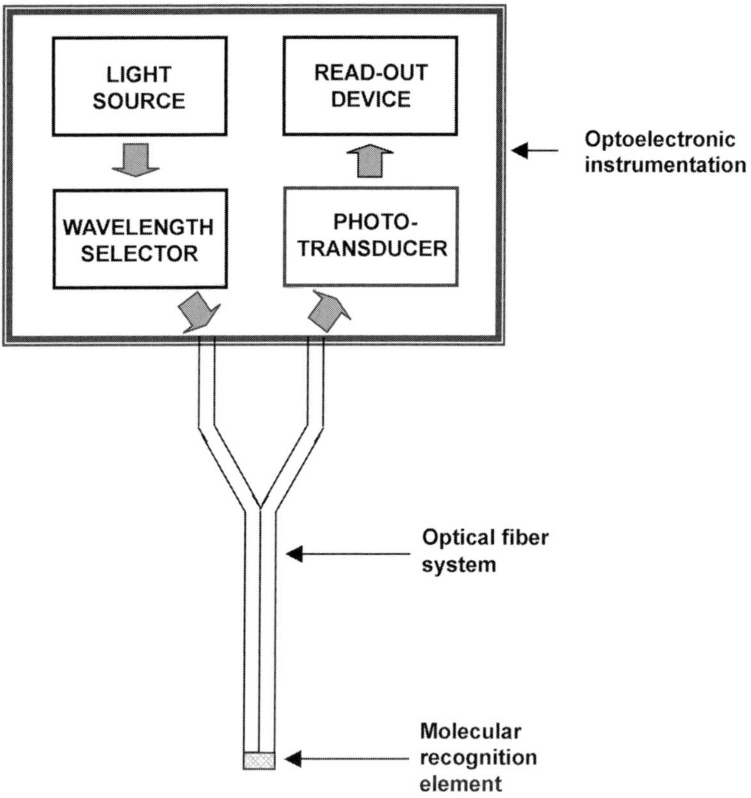


Fig. 9.5. Block diagram of a typical instrumentation system associated with optical fiber sensors.

suitable in optical fiber sensors. However, only a limited range of wavelengths can be provided by these semiconductor radiation sources. Lasers emit a highly intense and coherent radiation that is ideal for long-distance sensing systems.

The wavelength selector sorts the different optical signals according to wavelength for transmission to the photodetector. It isolates relatively narrow bands of wavelengths from a broader band to facilitate the decoding of the chemical information carried by the optical signal. It contributes to a sensitive measurement of absorbance or a selective excitation to generate fluorescence. The typical wavelength selectors employed in optical fiber chemical sensors are optical filters and grating-based monochromators. Optical filters prevent the passage of radiation of all wavelengths except that in the desired range. The removal of the unwanted wavelengths is achieved either through absorption of radiation (in absorption filters) or through the destructive interference of out-of-phase radiation (in interference filters). Monochromators facilitate geometric dispersion of the constituents of polychromatic radiation and focus the desired wavelength band onto a slit. Though compact monochromators have been based on optical gratings, integration is more easily attained through optoelectronic systems with optical filters.

The photodetector transforms the optical signal into an electrical signal which preserves the original encoded chemical information. The optical characteristics of the photodetector should match those of the light source to avoid any loss of information at critical wavelengths. Furthermore, the photodetector system must exhibit a high signal-to-noise ratio in order to ensure high sensitivity. Among common photodetection devices are photomultiplier tubes, photodiodes, photodiode arrays, phototransistors and charge-coupled devices. The photomultiplier tube is the most sensitive phototransducer and is highly preferred for low light level detection in optoelectronic systems. Quantum detectors, such as phototransistors and photodiodes, offer the advantage of compactness and miniaturization towards the achievement of a high level of integration in optical analytical systems. These solid-state photodetectors are useful in measurements conducted in the ultraviolet, visible and near-infrared regions. Multichannel detectors, such as photodiode arrays and charge-coupled devices, carry out simultaneous detection of the dispersed radiation.

The read-out device provides a visual representation of the output data in numerical form. Often, the electrical signal generated from the photodetector has to undergo simple adjustment of levels or reduction of bandwidth prior to display in the read-out device. Signal processing makes the output signal compatible with the performance characteristics of the display device. The common display systems are the analog voltmeter, digital panel meters, strip-chart recorders and computer monitors.

Optical couplers or beam splitters are sometimes needed in optical fiber instrumentation systems in order to distinguish and separate efficiently the incident and the returning radiation traveling in a single fiber. Optical splitting could be achieved either through a geometric technique, as in a small diagonal mirror with a small hole (Fig. 9.6(a)), or through interferometric means, as in a partially silvered mirror that

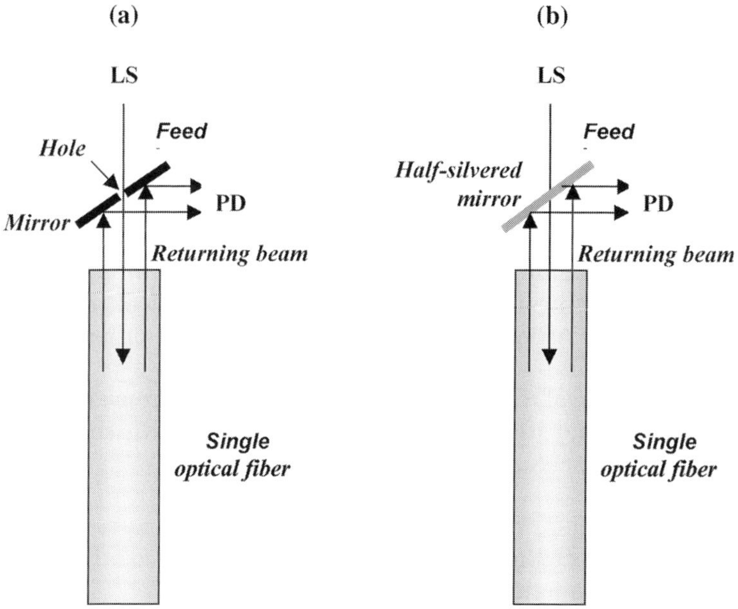


Fig. 9.6. Optical splitters based on (a) geometric technique and (b) interferometric technique.

is partially transmitting and partially reflecting (Fig. 9.6(b)). Refracting and reflecting optical components, such as lenses and mirrors, are also required in the instrumentation in order to manipulate the light and focus it more effectively on the optical fiber. A detailed discussion of the instrumentation in optical fiber chemical sensors is provided by Taib and Narayanaswamy [5] and Modlin and Milanovich [6].

### 9.2.3 Molecular recognition element

The molecular recognition element is the sensor component that comes into contact with the sample. It usually comprises a solid-phase reagent that is immobilized on the tip or the wall of an optical fiber. In the presence of the analyte, it undergoes a change of optical property that can be detected and quantified by the instrumentation system via the optical fiber. The response can be the production or modification of color or emission of light as fluorescence or chemiluminescence.

The molecular recognition element could be a single substance or a combination of several substances. For chemical sensors, the recognition unit is usually a single analytical reagent. The reagent often exhibits a chemical equilibrium that is sensitive to the presence of the analyte. The reagent R reacts with the analyte A and forms the product AR:

$$\mathrm{A} + \mathrm{R} \rightleftarrows \mathrm{AR}$$

Either R or AR is colored or luminescent and can be measured optically. The presence of the analyte species drives the equilibrium towards the product side, resulting in an enhancement of the optical property of the product species AR or the diminution of the optical property of the reagent phase R.

The concentration of the analyte, [A], is related to the concentration of the reagent, [R], and that of the reaction product, [AR], through the equilibrium constant, $K$, of the reaction:

$$K = [\mathrm{AR}]/([\mathrm{A}][\mathrm{R}]) \qquad (9.1)$$

The analyte concentration can be expressed in terms of the reagent concentration in the following manner:

$$[\mathrm{A}] = (1/K)\{(C_\mathrm{R}/[\mathrm{R}]) - 1\} \qquad (9.2)$$

It should be recognized that the concentration of AR is equal to $(C_\mathrm{R} - [\mathrm{R}])$, and that the reagent concentration is directly proportional to its absorbance or luminescence. Likewise, the analyte concentration is related to the concentration of AR as follows:

$$1/[\mathrm{A}] = K\{(C_\mathrm{R}/[\mathrm{AR}]) - 1\} \qquad (9.3)$$

wherein [AR] is related to its absorbance or luminescence. If the measured optical parameter is dependent on the concentration ratio ([AR]/[R]) the sensor response is no longer dependent on $C_\mathrm{R}$, but dependent on $K$. Thus:

$$[\mathrm{A}] = (1/K)([\mathrm{AR}]/[\mathrm{R}]) \qquad (9.4)$$

In general, reversible reactions are preferred with molecular recognition elements because they can provide continuous and unperturbed measurements. The response time (i.e. the time to reach the equilibrium) is dependent on mass transfer. Reversible reactions such as dynamic fluorescence quenching (e.g. by oxygen), can produce sensors with short response times as there is no consumption of analyte or reagent during this process. Reactions that are nonreversible may be employed in optical sensors and these can result in 'one shot' type sensor devices with high sensitivity in measurements. Indirect chemical reactions involving two or more recognition elements and/or reactions can be employed in optical chemical transducers. For example, consider the following reactions:

$$A + M \rightleftarrows AM$$

$$X + M \rightleftarrows XM$$

where M is the molecular recognition element, A is the analyte, and X is the species that competes with A for M. In this transduction process, X produces the optical signal and the binding of A to M is more favorable than of X to M. Thus, in the absence of A, all X will be bound to M and, in the presence of A, the second reaction will move towards the left releasing more X in the optical path. Such a competitive binding scheme has been applied in the development of a glucose sensor [4].

Among the widely used reagents are the acid–base indicator dyes for pH measurement, fluorophores for oxygen measurement, and chromogenic and fluorogenic reagents for the measurement of cations, anions and gases [7]. New materials, such as organic conducting polymers, molecularly imprinted polymers, supramolecular systems and organically modified sol–gel systems, have been exploited for molecular recognition in optical fiber chemical sensors [2].

For biosensors, the molecular recognition unit consists of a material of biological origin coupled with a chemical indicator usually for pH or oxygen. A wide variety of biomaterials has been employed in biosensors, ranging from isolated biomolecules, such as enzymes, antibodies, nucleic acids and receptors, to complex systems such as membranes, organelles, whole cells and intact tissues [8]. For most of these materials, a very specific interaction takes place and its occurrence is indicated by markers or labels that are detectable by the chemical reagents.

The molecular recognition element is commonly employed in the solid state in order to facilitate its integration with the optical fiber, to simplify the measurement procedure and to allow reuse of any reversible or recyclable reagents. It is conveniently localized either at one tip of the fiber or on the external surface of the fiber either by deposition or encapsulation with a polymer membrane. The solid-phase reagent is prepared by immobilizing it on an inert and solid support such as glass, silica gel, cellulose or other polymeric solids. Physical and chemical methods have been employed for immobilization [9,10].

Physical immobilization is achieved through adsorption, gel entrapment or electrostatic attraction (Fig. 9.7). In adsorption, the reagent molecules are held on the surface of the solid support through such physical forces as hydrophobic forces and hydrogen bonding. In gel entrapment, the molecules are confined in the interstices of the lattice of a polymeric gel, such as polyacrylamide and polyurethane. In electrostatic attraction, ionic sites in the solid support, usually an ion exchange resin, attract the charged portions of the reagent molecule. The physical methods for immobilization involve simple procedures and mild conditions; however, the binding strength of the reagent molecule on the solid support is weak and, as a result, reagent leaching problems may be encountered with such support materials.

Chemical methods provide a more stable means for reagent immobilization. Here, covalent bonds are formed between the reagent molecule and the polymeric support. Often, the polymeric support is activated by introducing a reactive functional group such as the aldehyde or amino group. Several types of solid supports have been used, ranging from inorganic materials (e.g. controlled pore glass) to organic polymers (e.g. cellulose and nylon). The fiber surface has also

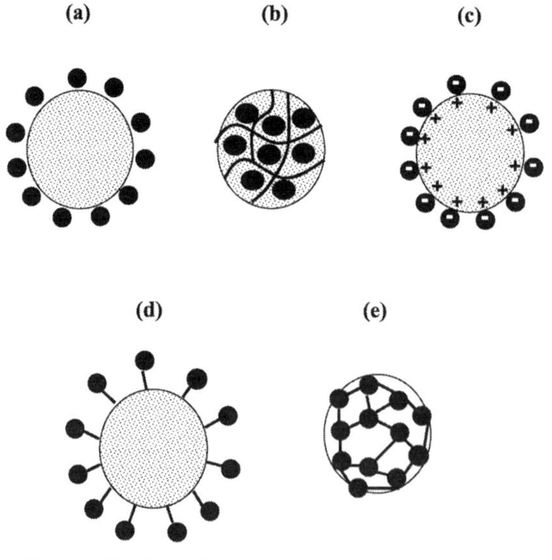


Fig. 9.7. Modes of immobilization of a reagent (●) on a substrate (◌): (a) adsorption; (b) physical entrapment; (c) electrostatic attraction; (d) covalent bonding; and (e) cross-linking.

been used as the solid support to reduce the diffusion barrier controlling the kinetics of the detection reaction.

### 9.2.4 Sensor designs

Optical fiber sensors can be designed to operate in two possible modes: (1) as an extrinsic sensor, wherein the fiber functions merely as a light guide for the probing radiation, and (2) as an intrinsic sensor, wherein the fiber acts as the transduction element. The majority of the optical fiber chemical sensors are of the extrinsic type, wherein analytical information is obtained primarily through the molecular recognition unit. It is in this type of sensor that there is an integration of the components.

There are several ways of arranging the sensor components, particularly the optical fiber and the molecular recognition unit. The configuration of the optochemical instrumentation is not as flexible, being restricted by the rigidity of the electronic circuitry. The molecular recognition unit can be set at the tip of the optical fiber, resulting in a probe configuration. Typical configurations are shown in Fig. 9.8. In the simplest arrangement, a single fiber or a fiber bundle is used to transmit the probing radiation before and after it interacts with the reagent phase. In this case, the instrumentation has a beam splitter that isolates the returning radiation and focuses it on the photodetector (Fig. 9.8(a)). Another configuration involves a bifurcated optical fiber system (actually a fiber bundle randomly divided into two groups at one end) wherein one arm feeds the probing radiation to the reagent phase and the other arm leads the radiation to the photodetector (Fig. 9.8(b)). The single-fiber configuration is miniature, and therefore, amenable to small-volume measurements. Furthermore, it has a high optical collection efficiency, since there is a maximum overlap of the incident and collected radiation. On the other hand, the bifurcated bundle arrangement requires a simpler instrumentation, not needing any beam splitter that could make the optics complicated. It also has a high optical throughput, since several fibers are used to transport the radiation.

The molecular recognition element could be deposited on the bare fiber and serve as the cladding (Fig. 9.8(c)). In this configuration, the evanescent wave interacts with the reagent phase and is attenuated according to the Beer–Lambert law. The multiple internal reflection

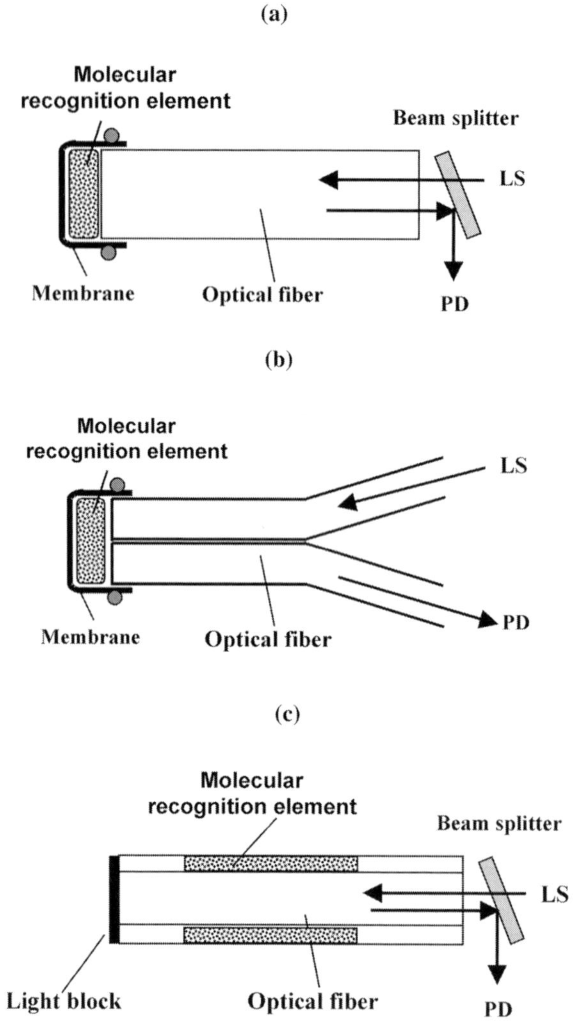


Fig. 9.8. Typical designs for chemical sensors and biosensors: (a) single fiber (or bundle) sensor; (b) bifurcated optical fiber sensor; (c) evanescent wave sensor.

that takes place as the radiation traverses the optical fiber amplifies the change in the wave intensity and contributes to an increased sensitivity. It is also possible for the evanescent wave to produce an increase in the intensity of the radiation if the cladding material is excited and emits fluorescence. This configuration generally produces a rapid response since it requires only a very thin reagent coating (about a few μm thick) and promotes an efficient mass transfer during the interaction with the analyte.

A sensor array for multi-analyte measurement can be assembled by combining several optical sensors with a probe configuration into a single bundle, with each probe containing a different reagent phase. In this instance, several light sources and photodetectors could be involved since each reagent phase would require a different wavelength for the probing radiation. However, it is possible to use a single photodetector that is temporally multiplexed. Likewise, an imaging optical fiber bundle can be employed together with a CCD camera as a detector to measure simultaneously the different reagent phases [11].

## 9.3 MODES OF OPTICAL SIGNAL MEASUREMENTS

As in the optical methods of chemical analysis, the optical fiber chemical sensors rely on the interaction of light and matter for encoding information on the concentration of the analyte. When the probing radiation from the optochemical instrumentation strikes the molecular recognition element, it interacts with the molecules in the reagent phase and produces optical signals of analytical significance. The consequence of this interaction could be an absorption, emission, scattering or reflection of light, depending on the wavelength of the radiation and on the structure of the molecules in the reagent phase. The intensity of the radiation emanating from the reagent phase carries information on the concentration of the analyte.

### 9.3.1 Absorbance measurement

The absorption of radiation is the basis of the most widely used optical method for chemical measurement, namely UV–visible spectrophotometry. It involves the reduction of the intensity of radiation of certain wavelengths as the light passes through a transparent layer containing the absorbing species. This attenuation occurs as a result of the transfer of the energy of the photon of the radiation to the absorbing species, which consequently is promoted to a higher energy or excited state.

The energy transfer is possible only if the difference in the energy states of the absorbing species exactly matches the energy of the exciting photons.

The magnitude of the attenuated optical signal carries the sought-for information on the analyte species. The extent of the decrease in the intensity of the radiation is dependent on the length of the light path and on the concentration of the absorbing species. The attenuation is usually expressed in terms of absorbance, $A$, which is defined as:

$$A = \log(I_0/I) \tag{9.5}$$

where $I_0$ is the intensity of the incident radiation and $I$ is the intensity of the transmitted radiation. Absorbance is related to the optical path length ($l$) and the concentration, $C$, of the absorbing species by the Beer–Lambert equation:

$$A = \varepsilon l C \tag{9.6}$$

where $\varepsilon$ is the molar absorptivity.

Optical fiber chemical sensors based on absorbance have been developed with the molecular recognition element immobilized as thin films on a transparent support or dispersed in a clear polymer matrix. For example, an optical sensor for ammonia was fabricated by depositing polyaniline on the inner surface of a polyethylene tube by chemical oxidation of the aniline monomer [12]. The absorbance of the reagent phase was monitored by an instrumentation system wherein the immobilized reagent phase was set between the light source and an optical fiber that was connected to a linear CCD detector (Fig. 9.9).

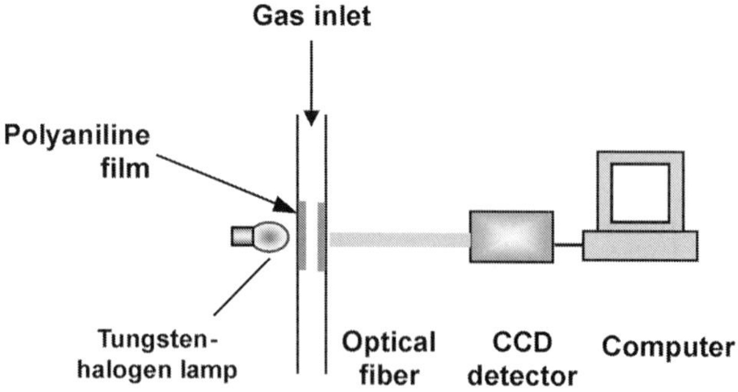


Fig. 9.9. Schematic diagram of an optical sensor for ammonia based on absorbance measurement (redrawn from Ref. [12]).

In the presence of ammonia, the polyaniline film rapidly changed its color from green to blue, and the peak in its absorption spectrum shifted from 800 to 600 nm. The sensor response, which was based on the absorbance at 600 nm, was linearly related to the ammonia concentration in the range of 180–18 000 ppm.

The absorbance of the analyte solution in the microenvironment of the optical fiber can also be monitored to provide a basis for chemical measurement. An absorbance-based biosensor for *p*-nitrophenyl phosphate was designed by positioning the enzyme reagent phase at a small distance from the tip of the optical fiber [13]. A small volume of the analyte solution seeps into this space and is hydrolyzed by alkaline phosphatase immobilized in the reagent phase according to the following reaction:

$$p\text{-nitrophenyl phosphate} + H_2O \rightarrow p\text{-nitrophenol} + \text{phosphoric acid}$$

The *p*-nitrophenol produced absorbs radiation at 404 nm, and the absorbance of the trapped solution is monitored to provide a basis for the measurement of the analyte. The sensor displayed a linear response over the concentration range of 10–380 μM *p*-nitrophenyl phosphate.

### 9.3.2 Reflectance measurement

Diffuse reflectance measurement is more often used than absorbance measurement as the basis of quantitation in optrodes particularly when the reagent phase is optically opaque. In this measurement mode, the radiation penetrates the solid phase and reappears at the surface after undergoing multiple scattering and partial absorption within the medium. Reflectance is based on the intensity of the diffusely reflected radiation relative to the incident radiation, and is often expressed in terms of the Kubelka–Munk function, $F(R)$:

$$F(R) = (I - R)^2/2R \tag{9.7}$$

where $R$ is the reflectance. This function is directly related to the concentration of the absorbing species occurring in the scattering layers, as expressed in the following equation derived from the Kubelka–Munk theory:

$$F(R) = \varepsilon C/S \tag{9.8}$$

where $\varepsilon$ is the molar absorptivity and $S$ is the scattering coefficient.

Many pH sensors have been developed based on this principle. The reagent, usually an indicator dye, is frequently immobilized on a finely powdered solid support or on a membrane that is deposited on the tip of the optical fiber and held in place by an enclosing membrane. One of the earliest reported optical fiber sensors for pH was based on the measurement of the absorbance of phenol red covalently bound to polyacrylamide mixed with polystyrene microspheres to scatter the light within the reagent phase [14]. The sensor was configured as a probe employing a bifurcated optical fiber, with one arm for the probing radiation and the other arm for the scattered radiation from the reagent phase. The sensor measures pH in the range from 7.0 to 7.4 at increments of 0.01 pH units.

The analytical application of reflectance sensors for pH is expanded by co-immobilizing an enzyme with the pH indicator used as the molecular recognition element. For example, an optical biosensor for urea was fabricated by immobilizing urease on an affinity membrane together with bromothymol blue and affixing this urea-sensitive membrane to an optical fiber in an optical instrumentation system [15]. The urease catalyzes the hydrolysis of urea:

$$NH_2CONH_2 + 3H_2O \rightarrow 2NH_4^+ + HCO_3^- + OH^-$$

wherein the products, particularly the hydroxide ion, modifies the pH of the reaction microenvironment. This pH change causes a color change, and subsequently a change in the reflectance of the reagent phase. An extensive bibliography of fiber optic biosensors based on this principle is provided by Kuswandi et al. [1].

### 9.3.3 Fluorescence measurement

Fluorescence has been exploited as the carrier of analytical information in a great number of optical fiber chemical sensors [14]. Here, the optical signal arises when certain molecules absorb radiation from an excitation source and undergo instantaneous deactivation through the emission of radiation. Under conditions of low absorbance (i.e. $A < 0.05$), the fluorescence intensity, $I_F$, is directly proportional to both the concentration, $C$, of the emitting species and the intensity of the incident radiation, $I_0$, as expressed in the following equation:

$$I_F = k_F I_0 C \tag{9.9}$$

where $k_F$ is a constant that is dependent on the absorption characteristics of the analyte, the quantum efficiency of the fluorescence and the configuration of the instrumentation system.

Fluorescence is emitted in all directions and at a longer wavelength than the incident light. It can, therefore, be easily distinguished from the excitation radiation through geometric and spectral techniques. The use of optical fibers in optrodes further facilitates the isolation of the emitted light from the incident light. Consequently, a simpler instrumentation is involved in fluorescence optrodes compared to the conventional methods of fluorescence measurement.

Fluorescence methods exhibit a greater sensitivity compared to absorption methods owing to the darker background of these measurements. Detection limits in the range of 1 pM to 1 nM can be achieved through these methods, depending on the intensity of the excitation source and efficiency of emission collection. Preconcentration through separation or immobilization of the fluorophore further lowers the detection limit.

Fluorescent reagents immobilized on a suitable solid support have been employed as the molecular recognition element in several chemical sensors. Fluorescent dyes, such as coumarin and fluorescein derivatives, were used in fluorosensors for pH. For instance, a sensor for local blood pH was developed based on seminaphthorhodamine-1 carboxylate encapsulated on a sol–gel and deposited onto the distal end of a silica optical fiber [16]. A short length of the sensing end of the fiber was decladded and tapered through etching with hydrofluoric acid in order to increase the efficiency of fluorescence collection through evanescent waves. The emission spectrum of this dye is pH sensitive, the acid form showing an emission peak at about 580 nm and its base form having an emission peak at about 640 nm. The ratio of the intensity of these peaks constituted the basis for a reproducible sensor response which was highly linear in the range of human blood pH, i.e. 6.8–8.0.

Nonfluorescent reagents capable of generating fluorescent products by their reaction with the analyte have also been exploited in a number of fluorescence optrodes. Biosensors were developed wherein a dehydrogenase enzyme was immobilized as the reagent phase and the conversion of the co-factor NAD in the reaction mixture into the fluorescent NADH was monitored for the measurement of the analyte. Among these was a glucose biosensor [17] which quantified glucose through the reaction:

glucose + $NAD^+ \rightarrow$ gluconic acid + NADH

by means of the fluorescence of the product NADH. The change in the fluorescence correlated in a linear manner with the concentration of glucose. This biosensor was demonstrated to be capable of measuring glucose concentrations as low as 0.6 mM.

Chemical information can also be conveyed through the quenching effect of certain substances on the fluorescence emitted by a fluorophore reagent. Molecules of the quenching species collide with the excited fluorophore molecules, resulting in a nonradiative energy transfer and a subsequent reduction in the fluorescence intensity. The probability of the collision is enhanced in optical fiber chemical sensors by the immobilization of the reagent. In this case, the concentration of the quenching species is inversely related to the fluorescence intensity, $I_F$, as expressed by the Stern–Volmer equation:

$$I_F^0/I_F = 1 + k_Q C \tag{9.10}$$

where $I_F^0$ is the intensity of the unquenched fluorescence and $k_Q$ is a constant that is dependent on the fluorescence quantum efficiency and on the quenching kinetics.

Fluorescence quenching has been exploited widely in oxygen optrodes. A great variety of fluorophores has been reported as the reagent phase in optical fiber sensors for oxygen, with greater preference for those with excitation spectral peaks in wavelengths close to those of the available LEDs. The widely used fluorophores in oxygen optrodes are the polycylic aromatic hydrocarbons such as perylene and decacylene, and the fluorescent ruthenium complexes. An oxygen optrode was fabricated comprising tris(4,7-diphenyl-1,10-phenanthroline)ruthenium(II) ditetrakis(4-chlorophenyl)borate adsorbed on silica gel [18]. The fluorophore was excited by blue light and emitted an intense pink fluorescence. The fluorescence was efficiently quenched by oxygen, such that the sensor showed an excellent linear Stern–Volmer behavior at oxygen levels in the range of 0.0–0.05%.

### 9.3.4 Chemiluminescence measurement

Chemiluminescence is the emission resulting from a chemical reaction. This reaction is exhibited by a small number of reagents, the most popular of which are luminol, oxalate esters and the luciferin–luciferase system. In these reactions, a significant fraction of the intermediates or products are formed in excited electronic states. The excited species then release extra energy by the emission of light or by transferring the energy to a chemical acceptor such as a fluorophore. The following equations illustrate the possible mechanisms for the generation of chemiluminescence:

$$A + B \rightarrow I + I^* \quad \hookrightarrow I + h\nu$$

$$A + B \rightarrow P + P^* \quad \hookrightarrow P + h\nu$$

$$I^* \text{ (or } P^*) + F \rightarrow I \text{ (or P)} + F^* \quad \hookrightarrow F + h\nu$$

where A and B are the reactants, I is the intermediate, P is the product, F is the fluorophore and the asterisk indicates the excited species. The majority of the common chemiluminescent reactions emit radiation in the visible region.

The instrumentation involved in these measurements is quite simple, requiring only a reaction cell and a photodetector–read-out system. In many designs, the optical fiber together with the molecular recognition element is interposed between the reaction cell and the photodetector. The measurement can be carried out either within the peak wavelength of the emission or over the entire measurable spectral range of the detector. In any of these two cases, the reaction conditions are often optimized so that the intensity of the emission is related to the analyte concentration.

In general, chemical measurements based on chemiluminescence are more sensitive than those based on absorption or fluorescence. The sensitivity is actually determined by the quantum yield of the reaction, the rate of the reaction and the efficiency of the photodetector. In solution, chemiluminescence methods can measure concentrations of analytes as low as $10^{-18}$ M.

Most of the optrodes employing chemiluminescence as the measuring source have been developed for hydrogen peroxide and for reactions producing hydrogen peroxide. A chemiluminescence optrode for hydrogen peroxide was reported as long ago as 1978 [19] wherein peroxidase was immobilized in a thin layer of polyacrylamide gel at one end of the optical fiber and the other end of the optical fiber was coupled to a photomultiplier tube. The working solution was a buffered solution of luminol, containing varying amounts of the analyte. The sensor responded instantly when the sensing end of the fiber was immersed in the working solution, a steady-state signal being obtained within 4–30 s. The detection limit was $1 \times 10^{-6}$ M $H_2O_2$, and the response was linear up to $1 \times 10^{-4}$ M $H_2O_2$.

The analytical usefulness of the chemiluminescence sensor for hydrogen peroxide can be extended by co-immobilizing an oxidase with the peroxidase in the reagent phase of the sensor. Oxidases are enzymes that catalyze the oxidation of some biomolecules with the concomitant production of hydrogen peroxide. Thus, a lysine sensor [20] was assembled wherein lysine oxidase was immobilized on glass particles and packed in a reactor that was connected to a flow cell that integrated an affinity membrane on which peroxidase was immobilized. The cell is linked to a photomultiplier tube by means of an optical fiber bundle. The buffered analyte solution upon flowing into the lysine oxidase reactor undergoes the following reaction:

$$\text{lysine} + O_2 + H_2O \rightarrow \alpha\text{-keto-}\varepsilon\text{-aminocaproate} + NH_3 + H_2O_2$$

The stream from the reactor then joins with a stream containing luminol and flows into the sensor cell where the chemiluminescence is measured. This sensing system can measure lysine concentrations in the range of 10–1000 μM and involves an analysis time of about 30–50 s per sample.

## 9.4 CONCLUSION

Optical chemical sensors and biosensors enhanced the advantages offered by the optical methods of chemical analysis. Integrating the instrumentation system and the chemical system through optical fibers introduced numerous opportunities for the development of novel approaches to chemical measurement. Progress in materials science continuously generates new materials such as electrochromic compounds and molecularly imprinted polymers that can be exploited in the molecular recognition element of the sensor. Advances in instrumentation, as well as in signal processing (e.g. artificial neural networks [21]), have further improved the characteristics and capabilities of the measurement system of the sensor. These developments likewise increase the potential for miniaturization and integration of the various components of the analytical system and also for developing multi-analyte sensor systems.

## REFERENCES


1 B. Kuswandi, R. Andres and R. Narayanaswamy, *Analyst*, 126 (2001) 1469.
2 O.S. Wolfbeis, *Anal. Chem.*, 72 (2000) 81R.
3 W.R. Seitz, *Crit. Rev. Anal. Chem.*, 19 (1988) 135.
4 O.S. Wolfbeis, In: O.S. Wolfbeis (Ed.), *Fiber Optic Chemical Sensors and Biosensors*, CRC Press, Boca Raton, 1991, pp. 1–23.
5 M.N. Taib and R. Narayanaswamy, *Analyst*, 120 (1995) 1617.
6 D.N. Modlin and F.P. Milanovich, In: O.S. Wolfbeis (Ed.), *Fiber Optic Chemical Sensors and Biosensors*, Vol. I, CRC Press, Boca Raton, 1991, pp. 237–301.
7 M. Arnold, *Anal. Chem.*, 64 (1992) 1015A.
8 D.L. Wise and L.B. Wingard Jr (Eds.), *Biosensors with Fiber Optics*, Humana Press, Clifton, NJ, 1991.
9 T. Cass and F.S. Ligler, *Immobilized Biomolecules in Analysis*, Oxford University Press, London, 1998.
10 E. Koller and O.S. Wolfbeis, In: O.S. Wolfbeis (Ed.), *Fiber Optic Chemical Sensors and Biosensors*, Vol. I, CRC Press, Boca Raton, 1991, pp. 303–357.
11 P. Pantano and D.R. Walt, *Anal. Chem.*, 67 (1995) 481A.
12 Z. Jin, Y. Su and Y. Duan, *Sens. Actuat. B*, 72 (2001) 75.
13 M.A. Arnold, *Anal. Chem.*, 57 (1985) 565.
14 J.L. Peterson, S.R. Goldstein, R.V. Fitzgerald and D.K. Buckhold, *Anal. Chem.*, 52 (1980) 864.
15 X. Xie, A.A. Suleiman and G.G. Guilbault, *Anal. Lett.*, 23 (1990) 2143.
16 S.A. Grant and R.S. Glass, *Sens. Actuat. B*, 45 (1997) 35.
17 R. Narayanaswamy and F. Sevilla III, *Anal. Lett.*, 21 (1988) 1165.
18 M.M.F. Choi and D. Xiao, *Anal. Chim. Acta*, 403 (2000) 57.
19 T.M. Freeman and W.R. Seitz, *Anal. Chem.*, 50 (1978) 1242.
20 F. Preuschoff, U. Spohn, E. Weber, K. Unverhau and K.H. Mohr, *Anal. Chim. Acta*, 280 (1993) 185.
21 I.M. Raimundo Jr and R. Narayanaswamy, *Sens. Actuat. B*, 74 (2001) 60.

*Chapter 10*

# *Electronic tongues: new analytical perspective for chemical sensors*

***Andrey Legin,* Alisa Rudnitskaya and Yuri Vlasov***

## 10.1 GENERAL APPROACH TO THE APPLICATION OF SENSOR ARRAYS

### 10.1.1 Why use sensor systems?

Traditional development of analytical instruments and methods was always aimed at obtaining the highest possible selectivity to an analyte. In the field of chemical sensors for solution analysis these efforts resulted in the development of multiple electrochemical sensors and, in particular, potentiometric sensors, widely known now as ion-selective electrodes (ISEs). Most of the currently known and advanced electronic tongues originate from ISEs [1,2]. Therefore, an historical description of the development from discrete sensors to electronic tongue systems will be limited here to the field of ISEs, though many considerations may be expanded for other types of chemical sensor. This is because electronic tongues based on different types of chemical sensor, e.g. voltammetric, have been suggested recently [3].

*Corresponding author. Tel.: +7-812-3289595; fax: +7-812-3282835. *E-mail address:* andrew@sensor.chem.lgu.spb.su.

*Comprehensive Analytical Chemistry XXXIX, pages 437–486*


ISSN: 0166-526X

The first ion-selective electrode with sensitive oxide glass membranes for the determination of hydrogen activity in aqueous solutions was suggested in 1907 by Haber and Klemensiewicz [4]. Since then significant numbers of membrane materials for ISEs, both inorganic and organic, have been implemented. The main types of ISEs are: oxide glasses for $H^+$, alkali and alkaline-earth cation determination; crystalline materials for the determination of halogenides and heavy metals; liquid or plasticized polymer compositions with ion-exchangers or neutral carriers; chalcogenide glasses for heavy metal determination; and membranes with immobilized enzymes for the detection of organic substances, etc. Some important features of ion-selective electrodes are as follows: the concentration (activity) of ionic forms, but not the total content, is being measured; the ISE response is linear; direct potentiometric measurements are often possible. Theoretically, dependence between the ISE output (electrical potential) and the logarithm of activity of the primary ion in an analyte is linear and is described by the Nernst equation, which is commonly used to build a calibration curve:

$$E = E^0 + \frac{RT}{z_i F} \ln a_i \tag{10.1}$$

where $E$ is the potential difference (e.m.f.) of the electrochemical cell comprising ion-selective and reference electrodes; $E^0$ is the standard potential; $R$ is the gas constant; $T$ is the absolute temperature; $F$ is the Faraday constant; $z_i$ the electrical charge of the primary ion; and $a_i$ is the activity of the primary ion. The term $RT/z_iF$ is known as the response slope $S$, i.e. the sensitivity of an ISE.

An extensive description of existing ion-selective electrodes can be found in books and reviews [5–10]. Features such as ease of handling, short analysis time, the possibility of automation, an option of miniaturization and low self-cost make the ISE an attractive tool for the analysis of solutions. In practice, this is particularly true for solution media, where the content of interfering species is low enough and electrode responses obey the Nernst equation. However, the lack of selectivity significantly limits practical application of many ISEs in the presence of other species in solutions besides the primary ions, and this is a common situation in real-world sample analysis, which is the most important field. Many new organic substances for ISEs with PVC-plasticized membranes are produced each year, but no significant improvements in selectivity and sensitivity have been achieved during the last two decades. Thus, in spite of the hectic efforts of numerous researchers, pH glass electrodes still remain the most selective ISE as well as the most used. Research into other types of chemical sensor has followed the same path. In spite of attempts to develop sensors with the highest possible selectivity, most sensors display insufficient selectivity in multicomponent media, which has resulted in a certain stagnation in the field. One possible way to overcome this stagnation is the application of sensor systems instead of discrete sensors.

### 10.1.2 Inspirations from chemometrics and biology

A new and emerging idea is easier to consider in relation to other contributing areas and inputs. Two fields of knowledge mainly influenced the development of electronic tongues: new achievements in biology such as a better understanding of human sensory systems and also possibilities and approaches to data handling and processing offered by chemometrics. Both have helped in generating novel ideas in the area of chemical sensors. Before describing electronic tongues in detail, it seems useful to provide a brief overview of some background knowledge.

*Chemometrics*

Application of ISEs and other chemical sensors in the analysis of multicomponent media is often hindered by their having insufficient selectivity. The influence of interfering species on ISE response can be taken into account using theory, e.g. the Nikolski–Eisenman equation:

$$E = E^0 + \frac{RT}{z_i F} \ln\left( a_i + \sum_j K_{ij}(a_j)^{z_i/z_j} \right) \tag{10.2}$$

where $K_{ij}$ is the selectivity coefficient of the ISE to the primary ion $i$ in the presence of an interfering ion $j$ and $z_i$ and $z_j$ are the charges of the primary and interfering ions, respectively.

In situations when an ISE is not highly selective the value of its output (potential) will be determined by the simultaneous presence and the content ratio of several ions or other species: the terms $a_i$ and $\sum_j K_{ij}(a_j)^{z_i/z_j}$ in Eq. (10.2) may appear comparable and the electrode response becomes non-linear. Though it is still possible to deal with a non-linear calibration curve, more than one electrode is needed to find correctly the parameters of Eq. (10.2) for multiple analytes. Evidently, the number of electrodes should not be less than the number of analytes according to simple mathematical considerations. Different theories on ISEs and other sensors have been suggested and developed but they are all based on certain sensing mechanisms and certain assumptions that make them too simplistic for real multicomponent analytes. Ultimately, this restricts the usefulness of chemical sensors as analytical instruments.

In the mid-1980s, similar reasoning led to the idea of applying an electrode array instead of a discrete ISE with the aim of improving the insufficient selectivity of ISEs in the presence of interfering ions. This approach assumes that the behaviour of each electrode of the array in multicomponent solutions can still be described by the Nikolsky–Eisenman equation. Thus, the system of such equations should be solved to find parameters such as standard potentials, selectivity coefficients and/or slopes. The parameters thus found can be used for predicting the concentration of multiple analytes. Numerous methods can be applied to calculating the parameters of Eq. (10.2) including different regression techniques and even artificial neural networks (ANNs).

The first study considering the application of ISE arrays for multicomponent analysis was that of Otto and Thomas in 1985 [9]. A sensor array comprising eight sensors was used for the simultaneous determination of sodium, potassium, calcium and magnesium at concentration levels typical for biological liquids. The main problem was insufficient selectivity of Mg- and Na-selective electrodes in the presence of calcium and potassium, respectively. Multiple linear and partial least square regressions were used for fitting two parameters of the Nikolsky equation—the standard potential and the selectivity coefficient, the slope values being preliminarily determined in the individual solutions of analytes. The array gave a certain advantage in comparison to discrete ISEs. The best results were obtained using PLS regression.

Beebe and co-workers [12,13] used a sensor array comprising five sparingly selective electrodes for the determination of Na and K in binary solutions. A non-linear regression based on a simplex algorithm and multiple linear regressions were used to determine the parameters of the Nikolsky equation including the slope values. The calibration of a sensor array in Ref. [12] was performed using a projection pursuit regression—a non-parametric multivariate technique, which required no a-priori information about the functional form of the sensor response. The best results were obtained with non-linear regression, the errors being 0.4% for Na and 5.3% for K.

A combination of three highly selective electrodes (sodium, potassium and calcium) and one sparingly selective sensor with multiple ionophores was employed for the determination of sodium, potassium and calcium in tertiary mixtures [14]. It was found that incorporation of the fourth sensor (a sparingly selective one) into the array of ISEs improved the error in the determination, i.e. 2.8% compared to 4.5% with only three highly selective sensors. The same sensor array was applied in the flow-injection set-up for Na, K and Ca determination in mineral water and human blood plasma samples [15,16].

The next step from the Nikolsky equation and theoretical description of electrode responses was the application of ANNs for sensor response processing [17–19]. ANN is a non-parametric technique that can learn relationships in the data from a set of examples, though ANN works as a 'black box' and produces mathematical models that are difficult to interpret. The feed-forward neural network was trained to detect and identify ions in samples containing sodium, potassium and calcium at different concentration levels. The sensor array included the same three ion-selective electrodes as described in Ref. [16] implemented in a flow-injection mode. Network parameters and a training algorithm were varied to investigate their influence on the neural net performance. Electrode response models were distorted in different ways including noise addition and baseline shift with the aim of investigating the ability of the network to recognize previously unseen samples.

Bos and co-workers [18,19] used a back-propagation neural net to obtain the calibration model for a sensor array, while a recurrent neural net was applied in the determination of Nikolsky equation parameters. Ion-selective sensor arrays were employed in the determination of calcium and copper and of potassium, calcium, nitrate and chloride simultaneously. Calcium- and copper-selective electrodes together with a pH electrode were used in the former situation. An average error of about 8% in $Ca^{2+}$ and $Cu^{2+}$ content determination was reported. The second array included corresponding ion-selective electrodes and also a pH electrode. The average error was found to be 6%, while the maximum was about 20%.

These analyses illustrate the first applications of an array of potentiometric sensors rather than discrete sensors for quantitative ion determination. The starting point was to use the Nikolsky equation instead of the Nernst equation to take into account interferences from other ions. This resulted in applying sparingly selective electrodes having poorer selectivity than traditional electrodes as they were more useful in the arrays than selective electrodes. This also led to increasing interest in multivariate and non-parametric data processing techniques that do not require any information about the functional form of sensor signal/analyte activity dependence.

*Biological inspiration*

Another approach that encouraged the development of chemical multisensor systems was an attempt to mimic organization and performance of biological sensory systems, particularly the olfaction of mammalians (the sense of smell) [20]. These principles were partly realized in the systems for gas analysis—'electronic noses'—later in liquid analysers—'electronic tongues'. The olfaction was earlier recognized as the most effective sensing system owing to its high sensitivity and discrimination ability. The sense of smell is capable of distinguishing thousands of different volatile molecules including some very similar ones, such as stereoisomers. The threshold of human perception for some of the odorants can be as low as a few parts per trillion. This threshold value is even lower in animals [21]. Evidently, all odorant substances are volatile but volatility (vapour pressure) and odour intensity are not proportional and some compounds with very low vapour pressure can be powerful odorants (e.g. musk) and vice versa. The relationship between compound structure and odour that it elicits is still unclear and compounds of very different chemical compositions may have similar odours. However, a relationship between fat solubility and odour intensity has been postulated: the strongest odorants are both water- and fat-soluble [22]. The impressive performance of olfactory systems is achieved through a complex set of non-specific or cross-sensitive receptors and the processing of their signals in both the neural system and the brain. The detection of odour is performed by olfactory receptor neurons through a mucus layer in the nostrils, where odorant molecules react with odorant binding proteins— the 'sensing layer' of the receptors. Since receptors are not selective, many of them respond to a given odour. This reaction results in an activity pattern, which is then transferred to the olfactory bulb, where primary signal processing is performed, and then to the higher level brain region for identification and recognition.

The sense of taste in mammalians is similarly organized to the olfaction system. The taste is perceived by non-specific taste buds, which are located on the papillae of the tongue. Conventionally, overall taste is believed to be correlated with a combination of four basic tastes: sweetness, sourness, bitterness and saltiness. Occasionally, other elementary taste characteristics are used such as umami. Umami was firstly introduced by Japanese researchers and is used to describe the delicious taste perceived in meat, cheeses and mushrooms [23]. Since taste and odour are often perceived simultaneously the term 'flavour' is widely used for describing their combination, especially when speaking about food.

The relationship between taste (flavour) and chemical composition is often not known precisely, particularly for complex media, like foodstuffs. Another interesting but highly controversial issue is the interaction between different tastes [22]. In most cases, a desensitising effect or threshold increase takes place when two substances eliciting different tastes are present simultaneously. A more curious effect is the decrease of sensitivity threshold when substances present at non-perceptible concentrations can still be detected if a contrasting taste substance is applied to the tongue. Perception thresholds of the human tongue to many taste substances are much higher than for olfaction with the exception of alkaloids, such as quinine. However, differential taste and odour thresholds are comparable [22]. Thus, the function of the mammalian sense of taste is similar to olfaction but is less developed, possibly because it is less related to the survival of living beings.

The spectacular capabilities of biological sensory systems inspired scientists to implement their organization principles in artificial sensory devices. These devices were firstly intended for gas analysis and odour recognition. According to Ref. [20] the first attempt to develop an odour detection system dates back to the early 1960s [24]. The history of intelligent multisensor systems for gas analysis starts in 1982 following the work of Persaud and Dodd [25]. Since then many different groups have tried to add to the development and application of such devices, which were named 'electronic noses' [20,26]. 'Noses' usually provide for qualitative recognition of gas mixtures and/or identification of certain individual gases, e.g. leakages of chemicals.

The same principles were applied to the development of multisensor systems for liquid analysis— so-called 'electronic tongues'. Though an electronic tongue works in liquid media like the biological tongue, the sensitivity and detection threshold of artificial 'tongues' is much better, making their performance similar to an olfactory system. Furthermore, many, if not all, volatiles will originate from either liquid or solid media, e.g. in foodstuffs. In fact, the electronic tongue can be thought of as an analogy of both olfaction and taste sense and can be used for detection of all types of dissolved compounds, including volatile ones that form odours after evaporation.

A common feature of all electronic nose and electronic tongue systems is the combination of an array of non-specific sensors together with data processing by pattern recognition methods. Unlike electronic nose systems, however, electronic tongues are much more widely employed not only for recognition and classification but also for quantitative determination of concentrations of multiple components. Multivariate analysis data processing methods are useful for these purposes and widely applied for electronic tongues.

In summarizing, we can define the electronic tongue as an analytical instrument comprising an array of non-selective chemical sensors with partial specificity to different components in liquids and an appropriate pattern recognition or multivariate calibration tool, capable of recognizing quantitative and qualitative compositions of simple and complex solutions.

### 10.1.3 Advantages of sensor systems in comparison with discrete sensors

The first study dedicated to the application of sensor arrays demonstrated that the use of chemometrics to process response-sensors allows significant compensation for the lack of selectivity and for the quantitative determination of different ions in multicomponent solutions. Experimental evidence revealed that partially selective or cross-sensitive sensors are preferable to highly selective instruments when used in a sensor array [14]. Sensor arrays including partially selective electrodes produce lower determination errors than sensor arrays made solely from highly selective sensors.

Besides overcoming insufficient selectivity, the application of cross-sensitive sensor arrays permitted lower detection limits to be obtained in comparison to the same sensors used traditionally as discrete sensors [27].

Another interesting feature of sensor arrays is the possibility of making potentiometric measurements without a reference electrode. Potential differences in this situation are measured between all sensor pairs within the array [28]. The resulting potential differences were considered during data processing and redundant values were rejected. Two PCA plots displaying the electronic tongue data measured with and without a reference electrode are shown in Fig. 10.1a and b.

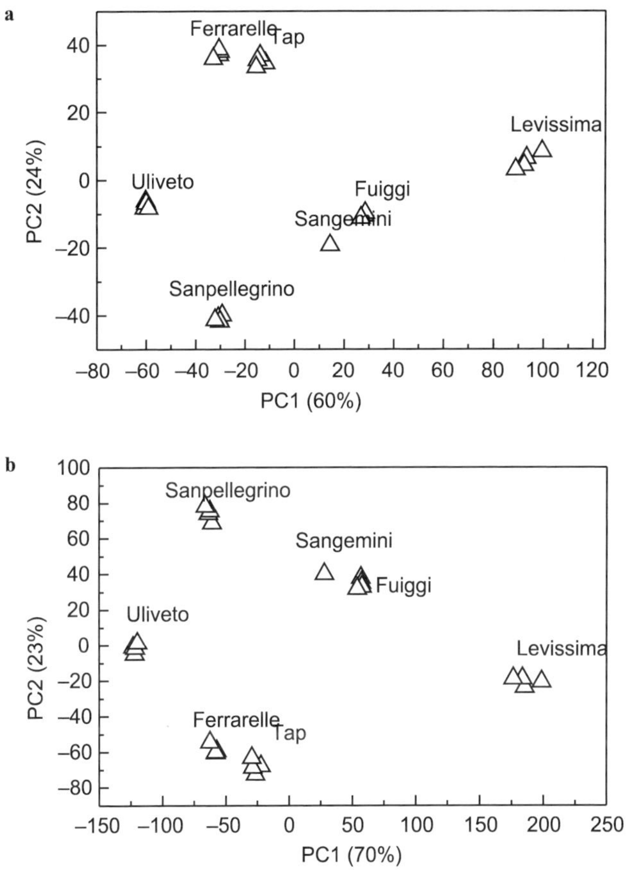


Fig. 10.1. Discrimination of Italian mineral waters using a potentiometric electronic tongue: (a) measurements made with a reference electrode; and (b) measurements made without a reference electrode.

Although the absolute positions of the points are different on these two plots, the instrument can still distinguish easily the difference between classes. Therefore, discrimination abilities of the electronic tongue do not depend on the data obtained (measurements with or without reference electrode), at least for selected applications.

The most exciting possibility provided by electronic tongue systems is the classification and recognition of complex liquids in the same way as humans. This is, however, a very untypical type of analysis in conventional analytical chemistry not only for ion-selective electrodes but also for any other traditional analytical techniques. Nevertheless, this is a new and promising analytical approach that can characterize one more important feature of the material world—the flavour of different analytes. Selected examples of different analytical applications of the electronic tongue will be discussed later.

### 10.1.4 Specific features of the sensors for the electronic tongue

The most important part of an electronic tongue is its sensor array. There is experimental evidence that cross-sensitive, non-specific or sparingly selective sensors are the most useful in multisensor systems than selective sensors [11–19]. Reproducibility of sensor response is, of course, crucial in all cases, both selective and non-specific, when serious analytical approaches and results are considered. Thus, sensors to be used in the array should display good reproducibility and maximum cross-sensitivity, which is understood as the sensitivity to many components of the analyte simultaneously. The term cross-sensitivity is commonly used when describing sensor properties in the literature devoted to multisensor systems. However, neither the theoretical description of this property nor even its unambiguous and widely accepted definition is relevant. The terms closely related to cross-sensitivity are also numerous (non-selectivity, cross-reactivity, poorer selectivity, partial specificity, global selectivity, etc.). Often these terms are used and understood interchangeably even if some of them are claimed to describe specific features. The ideas behind, e.g. 'cross-reactivity' or 'global selectivity', are very similar and very close indeed. Below we will retain the term 'cross-sensitivity' since it seems the best approximation to the phenomenon itself.

Cross-sensitivity of potentiometric sensors cannot be treated simply as a reverse value of the selectivity coefficient. The 'classical' selectivity of ion-selective electrodes has always been considered in the framework of a thermodynamic approach, on the basis of a certain sensing mechanism (ion exchange) and a situation when one primary and one interfering ion are both present. Recently, general equations describing the mixed response of sensors were suggested [29]. The approach considered the response of a polymer-based potentiometric sensor to any number of ions of different charge. It is based on the phase boundary potential model and assumes that the sensitivity mechanism remains ion exchange. Sensing mechanisms, however, may be different or even variable. For example, some sensors respond to both ionic and non-ionic species in solutions. An adequate theoretical consideration of cross-sensitivity seems to be impossible at present and more experimental evidence and theoretical considerations of the sensing mechanism of different materials with respect to different substances, including non-ionic species, are needed. For sensors other than potentiometric, theoretical considerations of cross-sensitivity are even less developed and should be one of the aims of researchers in the future. However, an empirical method of sensor cross-sensitivity assessment for potentiometric sensors, which can be used as a guide to sensor choices in practical applications, has been suggested [30,31].

The first necessary step is the determination of a set of substances, to which cross-sensitivity is to be studied, and a set of sensing materials. In most cases even 'very' non-selective or cross-sensitive sensors would not respond to any ion or substance in solution, but presumably to a certain group of substances. Calculation of cross-sensitivity parameters was based on the sensitivity study of chalcogenide glass electrodes to a set of heavy metals. Later the same parameters were successfully applied to cross-sensitivity evaluation of other types of membrane materials on a different set of analytes [32]. The experimental measurements used for cross-sensitivity estimation were simply calibrations of given sensors in individual solutions of the chosen set of compounds.

After application of different fitting procedures and consideration of literature data, three parameters were chosen for description of integral sensor response and cross-sensitivity. As parameters involved are empirical, it is possible to suggest alternative versions. However, these three appeared to be representative and successful.

The first parameter is the average sensor response slope $S$, measured in solutions of the chosen set of substances:

$$S = \frac{1}{n}\sum_i S_i$$

where $S_i$ is the sensor response slope in solutions of each individual substance, and $n$ is the number of components in the set.

The second parameter is the averaged signal-to-noise ratio of a sensor (for all components of the set):

$$K = \frac{1}{n}\sum_i K_i = \frac{1}{n}\sum_i \frac{S_i}{s_i^2}$$

where $S_i$ and $s_i$ are the response slope and its standard deviation in solutions of each substance.

The last parameter is named 'non-selectivity factor' because it describes the distribution of the sensitivity of a sensor to different components from the chosen set of substances and is calculated as follows:

$$F = \frac{S}{s^2}$$

where $S$ is the average slope (as in the first parameter) and $s$ is its standard deviation.

The average slope value is the main and the most important characteristic of the integral response and, hence, the cross-sensitivity of a sensor. The higher the parameter value the better is the overall sensitivity of the sensor to the substances from the set. An optimal range of average slope values can be estimated in each case on the basis of the following suppositions. For example, consider a study of sensor sensitivity to a set of divalent ions. In this instance, the sensor response slopes are likely to fall into a range from 0 to 29 mV/pX according to the Nernst equation. However, super Nernstian responses may be observed as well. Therefore, the average slope of a sensor close to 30 mV/pX is commonly related to a comparatively uniform distribution of sensitivity to the chosen set of divalent ions. The value of $S > 30$ mV/pX may be the result of a significantly super-Nernstian response to one of the ions. Thus, in this particular case, the range of $S$ from 25 to 30 mV/pX should be considered optimal. The sensors displaying $S > 25$ mV/pX display remarkable cross-sensitivity to all ions of the set and can be used for multisensor array analysis.

The average slope value is not the only valid response characteristic. At least two other measures appear to be useful. To characterize the distribution of sensitivity for different components the parameter, $F$, which is called the 'non-selectivity parameter', was used. $F \ll 0.1$ is typical for highly selective sensors with sharp sensitivity to the primary ion while having very poor response to the other ions. The increase of $F$ up to 0.1 is evidence of smoother distribution, but sensitivity to some components remains low. Rather uniform sensitivity distributions to most substances from the set is typical for sensors with $F > 1$. Finally, $F$ values greater than 0.5 characterize a reasonable distribution of sensitivity and significant cross-sensitivity to different species in complex solutions, and thus, may be considered as the optimal parameter for sensors designed for array applications.

A stability parameter is also important. This is because in the preliminary experiments a correlation between stability in individual ion solutions and that in complex liquids was found. The $K$ value—the average signal-to-noise ratio—is a valuable estimate of sensor stability. The higher $K$ is, the more reproducible is the sensor potential and the more stable is the electrochemical sensor behaviour both in individual solutions and in mixed liquid media. It was determined experimentally that values of $K$ greater than 2 represent a measure of reasonable sensor stability for array applications.

In conclusion, it must be noted that the exact optimal values of cross-sensitivity parameters should be determined in each case individually. In particular, the value of average slope can vary significantly. The parameters for cross-sensitivity estimation were developed and applied for potentiometric chemical sensors only. However, since no assumptions about the mechanism or theoretical description of sensor responses were considered when assessing cross-sensitivity parameters (only the experimental response value and its standard deviation was measured), the same method can be applied to assessing other types of sensors.

## 10.2 ELECTRONIC TONGUE SYSTEMS

### 10.2.1 Sensors

It was mentioned in Chapter 1 that the first studies on application of sensor arrays to the multicomponent analysis of liquids were performed using potentiometric chemical sensors (ion-selective electrodes). However, multisensor systems based on the same principles can be realized with other types of liquid sensors as well. Also, the same types of sensing material, e.g. plasticized PVC incorporating different active substances can be used in sensors having different working principles. In this chapter the main types of sensors and sensing materials that have been implemented in electronic tongue systems will be discussed.

Potentiometric sensors still remain the most widely used in electronic tongue systems. The taste sensor, which was introduced in 1990 by Toko et al. [33,34] from Kyushu University, Japan (see also Chapter 11), consisted of eight potentiometric sensors with thick film polymer membranes based on polyvinyl chloride (PVC). The membranes contained dioctyl phenylphosphonate (DOPP) as plasticizer and active substances (called 'lipids' by the authors), with tetrahydrofuran being used as a solvent [2,35,36]. Membrane compositions are shown in Table 10.1. These membranes were used for the preparation of potentiometric sensors with a liquid inner filling. Potential values of these sensors were measured against conventional Ag/AgCl electrodes. Sensors displayed sensitivity to inorganic and organic taste substances such as NaCl, KCl and KBr (salty), HCl, citric and acetic acids (sour), quinine (bitter), monosodium glutamate (umami), sucrose (sweet) [37, 38], catechin, tannic, chlorogenic and gallic acids (astringent) [39,40].

The sensors with the same polymer lipid membranes as the sensitive layer, but employing other principles of signal transduction and detection, were developed with the aim of miniaturizing the taste sensor. The most often used method of microsensor preparation is

TABLE 10.1

Lipid materials used in the multichannel electrode

| Channel | Lipid (abbreviation) |
|---|---|
| 1 | Decyl alcohol (DA) |
| 2 | Oleic acid (OA) |
| 3 | Dioctyl phosphate (DOP) |
| 4 | DOP/TOMA = 9:1 |
| 5 | DOP/TOMA = 5:5 |
| 6 | DOP/TOMA = 3:7 |
| 7 | Trioctyl methyl ammonium chloride (TOMA) |
| 8 | Oleyl amine (OAm) |

Reprinted from *Sensors and Actuators B*, 64 (2000) 205–215 (K. Toko, 'Taste sensor'), with permission from Elsevier Science.

deposition of a Langmuir–Blodgett film of the sensitive substance. Thin film membranes, which were prepared by dip-coating or by deposition of a Langmuir–Blodgett film, were used for impedance measurements [41,42]. It was found that in contrast to electric potential, the impedance of membranes changed significantly in the presence of two umami substances and sucrose. It was also found that impedance of thin film sensors increased in the presence of many bitter substances, whether or not they were both electrolytes or non-electrolytes. MOSFET technology was used to produce miniaturized versions of the taste sensor [43]. The FET taste sensor was prepared by pasting membranes having the same compositions as described described above on the gate of a FET. Another method used for FET sensitive layer preparation was the deposition of a dihexadecyl phosphate Langmuir–Blodgett film. PVC polymer film was deposited on the MOSFET gate prior to Langmuir–Blodgett layer formation to enhance adhesion of the latter to the substrate. The FET taste sensor displayed the same sensitivity to taste substances as the conventional sensor, but potential reproducibility was lower and lifetime shorter for the miniaturized device. Other types of transducers—light addressable potentiometric sensors (LAPS) and surface photovoltage (SVP) [44]—were used with polymer lipid membranes [45]. In both cases the sensitive layers were prepared by integrating polymer membranes containing active substances onto a semiconductor surface using the Langmuir–Blodgett technique.

The electronic tongue based on potentiometric sensors was developed in the Laboratory of Chemical Sensors at St Petersburg University [1,46]. These authors together with Italian colleagues suggested the term 'electronic tongue' in reference to multisensor systems comprising cross-sensitive (partially selective) sensors and pattern recognition tools for data processing.

A large number of sensing materials of different nature from inorganic compositions to organic polymers were used for sensor preparation. Membrane materials included chalcogenide glasses doped with different metals, PVC-based polymers containing various plasticizers and active substances such as ionophores, neutral carriers, metalloporphyrines, etc. and crystalline compositions [30–32]. The sensor arrays comprise from 10 to 30 sensors depending on the application. Different sensors with both liquid inner filling and solid inner contacts were manufactured and applied.

Originally, the electronic tongue was intended for the analysis of very different objects including ground and natural waters, industrial solutions, medicinal solutions, foodstuffs, etc. Thus, sensitivity of sensors to numerous inorganic and organic substances should be studied prior to their application, with special attention being paid to sensor cross-sensitivity and response reproducibility. The sensitivity of the sensors with a chalcogenide glass membrane to heavy metal cations (e.g. copper, lead, cadmium, zinc, iron and uranium at different oxidation states, etc.) has been investigated [47–51]. PVC-based sensors displayed sensitivity to alkali, alkaline-earth and heavy metal cations and inorganic anions such as chloride, sulphate, nitrate, nitrite, carbonate, etc. [32,52]. The sensitivity of about 40 different sensors to organic taste substances present in many foodstuffs and responsible for different aspects of food quality and flavour has also been investigated [53]. The taste substances were organic acids, alcohols, ethers, aldehydes, phenols, amines, alkaloids, etc. Additionally, the sensitivity to typical organic 'basic taste' substances, such as quinine, urea, and glutamate was evaluated.

Microsensors based on thin films of chalcogenide glass were prepared using a new technology—pulse laser deposition (PLD) [54,55]. The resulting thin films had the same composition as corresponding bulk glasses. Thin film microsensors showed similar electrochemical properties, such as sensitivity and selectivity, observed in conventional sensors with bulk membranes. The lifetime of microsensors was found to be about two years. Thus, PLD seems to be a promising tool for sensor miniaturization and in the preparation of microsensor arrays for the electronic tongue.

A different analytical technique, namely voltammetry, was used in electronic tongue development by Swedish scientists from Linkoping University [3,56–59]. The same general idea of combining non-specific and partly overlapping sensor signals with pattern recognition tools was realized in the voltammetric electronic tongue. In this set-up non-specific signal patterns are produced by an array of working electrodes made from different noble metals. Up to six metal electrodes, selected from platinum, gold, iridium, palladium, rhodium and rhenium, were combined and used for different tasks.

An electronic tongue developed to mimic biological sensory systems was developed at the University of Texas [60]. The sensors mimicking taste buds were prepared from poly(ethylene glycol)–polystyrene (PEG–PS) resin beads that were derivatized with a variety of indicator molecules. The sensors were: fluorescein, sensitive to pH; *o*-cresolphthalein complexone, sensitive to $Ca^{2+}$ and pH; alizarin complexone, sensitive to $Ce^{3+}$, $Ca^{2+}$ and pH; and finally a boronic ester of resorufin-derivatized galactose, sensitive to simple sugars. The sensor responses were measured against a reference sensor, which was simply a resin bead with terminal amines acetylated. The sensor sensitivity was tested in solutions containing $Ca^{2+}$, $Ce^{3+}$ and their mixture at different pH levels. The aim of this study was proof of the concept of a sensor system design mimicking a biological tongue, which explains somewhat the strange choice and limited number of analytes and the absence of any information about device application in the analysis of any real-world objects.

Four different conducting polymers were used to produce thin film sensors for an artificial taste system [61], namely, polyaniline, polypyrrole, sulfonated azobenzene polymer and the ruthenium complex mer-[$RuCl_3$ (dppb)(4-Mepy)]. Thin films of polymers were deposited by using the Langmuir–Blodgett or self-assembly techniques on integrated metallic electrodes for spectroscopic impedance measurements. Sensors were able to discriminate between basic taste substances such as NaCl, hydrochloric acid, sucrose and quinine.

One would expect the development of new sensors and even new classes of sensing materials for electronic tongues to take place in the near future.

### 10.2.2 System designs

The design of multisensor systems is defined by the measuring principle of the sensors used in these systems and is usually quite similar for the same type of sensors.

The measuring procedure when using an electronic tongue based on potentiometric chemical sensors consists generally of registering the potential value of each of the sensors in the array against a reference electrode, often Ag/AgCl. The measurements are made similarly to discrete electrodes, using high input impedance voltmeters or pH meters, which should be multichannel in the case of the sensor array. The measurements are usually guided by a PC and sensor responses are written into the computer's data file. A schematic and photograph of such a device is shown in Figs. 10.2 and 10.3. This is a laboratory device and all manipulations of sensor arrays and measuring liquids are done by hand. In some devices automatic manipulation of the sensor array and samples by the robot arm can be implemented (Fig. 10.4).

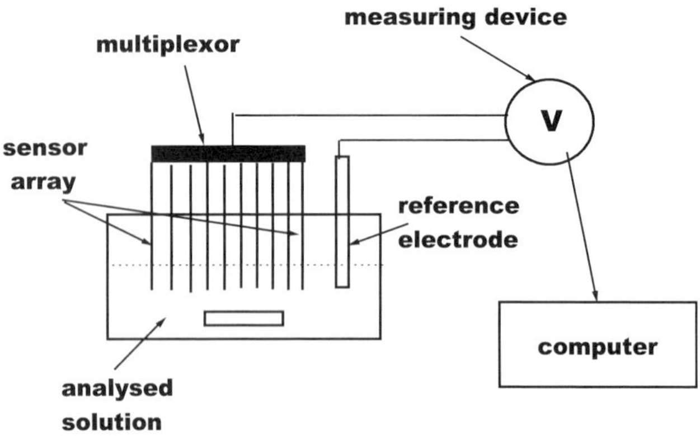

Fig. 10.2. The schematic of a potentiometric electronic tongue.

A flow-injection variant of the potentiometric electronic tongue has also been developed [50]. Measuring cells of different sizes and configurations were manufactured and the system can comprise from 3 to 11 sensors, which were made in special small bodies for this purpose. The flow-injection multisensor system allows automated sampling and calibration using a multichannel actuator. The sample volume in this case could be as small as 50 μL, usually ranging from 150 to 500 μL. The overall performance of flow sensor array systems was comparable.

Fig. 10.3. Laboratory prototype of a potentiometric electronic tongue.

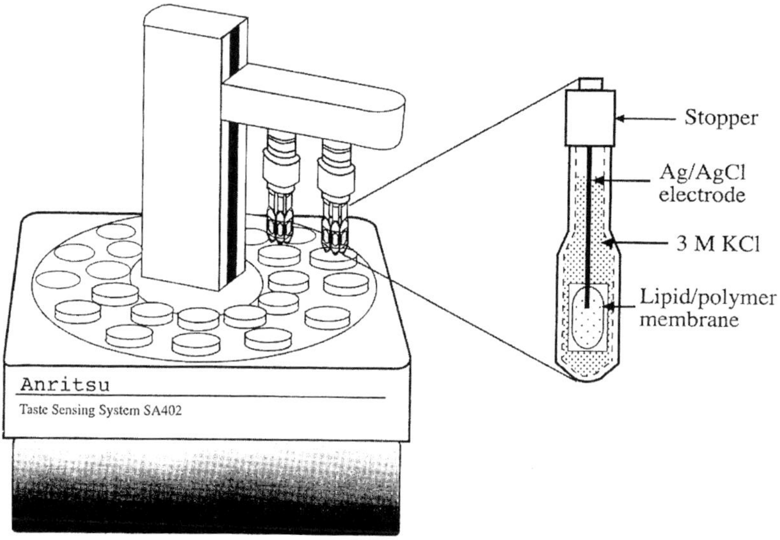


Fig. 10.4. Taste-sensing system SA402, Anritsu. Reprinted from *Sensors and Actuators B*, 64 (2000) 205–215 (K. Toko, 'Taste sensor'), with permission from Elsevier Science.

The measurements taken with a voltammetric electronic tongue are made in a conventional three-electrode scheme, which includes the Ag/AgCl reference electrode, stainless steel auxiliary electrode and an array of noble metal working electrodes instead of a single working electrode used in traditional voltammetry [3]. The schematic of the voltammetric electronic tongue with two working electrodes is shown in Fig. 10.5. The response from each of the working electrodes is recorded. Two types of voltammetric techniques were investigated: small amplitude pulse voltammetry (SAPV) and large amplitude pulse voltammetry (LAPV). LAPV was chosen for further analytical application of the electronic tongue. Different voltage steps ranging from 15 to 100 mV were used. Thus, a number of data points for each electrode were collected. The most informative data points were chosen during the data processing and used afterwards for calibration and recognition. A hybrid electronic tongue combining voltammetry with six working electrodes, three ion-selective electrodes (for pH, chloride and carbon dioxide), and conductivity and temperature sensors was also suggested [57]. In this case, each part of the system uses its particular measuring principles and the fusion of data occurs at the stage of their processing.

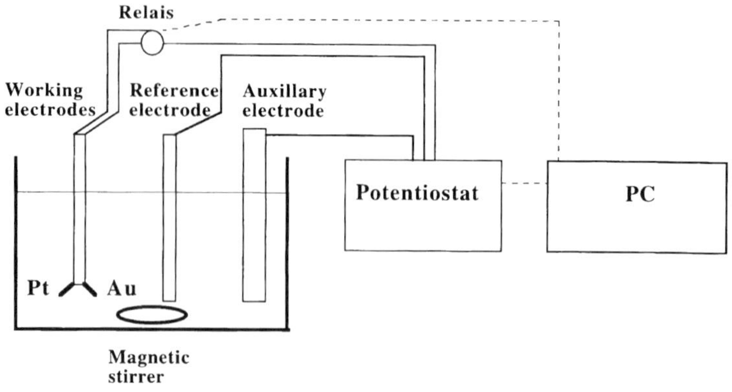


Fig. 10.5. Schematic of the voltammetric electronic tongue with two working electrodes. Reprinted from *Analytica Chimica Acta*, 357 (1997) 21–31 (F. Winquist, P. Wide and I. Lundström, 'An electronic tongue based on voltammetry'), with permission from Elsevier Science.

Optical detection of sensor responses was used in the electronic tongue based on derivatized PEG–PS resin [60]. Design of this electronic tongue was aimed at mimicking mammalian tongues. Thus, to mimic the cavities in which the taste buds of the mammalian tongue reside, the PEG–PS resin beads were positioned within micromachined wells formed inside $Si/Si_3N_4$ wafers, thus confining the beads to individually addressable positions on a multicomponent chip. The size of the wells was chosen such that they could hold the beads in swollen and unswollen states. Changes of the light absorption of the beads in different analytes were accomplished by using a charge-coupled device that was interfaced with the sensor array.

The light addressable potentiometric device (LAP) is another type of transducer used in electronic tongue systems [44,45]. The surface potential change caused by the reactions of a thin film membrane with analyte is detected by scanning a light beam across the sensor surface. Owing to a new differential measuring procedure based on a time-sharing technique, the sensitivity of the sensor system was increased by two orders of magnitude in comparison to conventional LAP system.

Application of impedance spectroscopy for sensor signal detection was based upon the fact that in the low frequency region impedance of sensors is related to the double layer, which is formed by absorption of different substances present in analysed solution [42]. AC measurements were carried out at fixed frequencies of 100 and 120 Hz, and 1, 10 and 100 kHz.

A sensing microsystem based on shear horizontal surface acoustic wave (SH-SAW) devices on 36-degrees rotated *Y*-cut *X*-propagating $LiTaO_3$ was applied to the identification of liquids [62]. The system consisted of three SH-SAW devices with three different centre frequencies of 30, 50 and 100 MHz.

More references and mentions of electronic tongue devices may be found in the literature. However, the present description is restricted to the most reliable and well-documented results regarding electronic tongue sensor systems and their analytical applications.

### 10.2.3 Hybrid systems

In humans the senses of olfaction and taste always cooperate and influence each other, especially in food flavour evaluation. Though the sensitivity of electronic and biologic noses and tongues may differ significantly in terms of the number of substances that can be determined and in their sensitivity thresholds, the idea of using an integrated sensor system comprising both an electronic nose and an electronic tongue appears logical. No wonder then that many researchers have explored the possibility of using such integrated systems, particularly for the analysis of foodstuffs [58,63,64].

The application of an integrated sensor system gives rise to a problem of combining the data derived from different origins. Different approaches have been used to combine electronic nose and tongue data. The simplest approach assumes the merging of two matrixes of normalized data. The resulting matrix should have the number of rows equal to the number of samples and the number of columns equal to the number of sensors in the electronic nose plus tongue combined system. The data fusion may also be performed at a higher level. This means that the data from the nose and the tongue are analysed separately and the results are combined after. A method described in Ref. [63] suggested applying a calculation method, e.g. PCA, to the nose and the tongue data separately and then the most significant principal components were considered together. A more sophisticated approach was put forward in Ref. [58]. A hierarchical system of neural network-based classifiers was formed. Some of them were trained with electronic nose data, others with the electronic tongue data. A gating network that combines the outputs of all classifiers produced the final results. Different analytical applications of the integrated electronic nose–electronic tongue systems are discussed below.

### 10.2.4 Data processing

Another important issue for multisensor analysis, besides the sensor arrays, is processing the sensor signals. In a multicomponent environment the sensor array produces complex signals (patterns), which, ideally, contain information about different compounds present in the media and other characteristic features, including integral ones. The sensor signals should be analysed altogether to extract valuable analytical information. Various methods of multivariate calibration and pattern recognition are now available and can be used for data processing from the sensor arrays.

In principle an electronic tongue may be applied to two main tasks: (1) for quantitative determination either of the content of components or some other formalized parameters (e.g. taste assessments); and (2) for classification (recognition, identification, discrimination) of simple and complex analytes and/or their features. The choice of the data processing technique for a particular case depends on the task to be solved and the structure of the data (non-linearity, correlations, etc.). The more detailed description of the data processing methods used in the sensor analysis is beyond the scope and format of this chapter and should be treated in a specific publication. Theoretical discussions of available methods and some case studies of different sensor applications can be found in numerous books, manuals and papers, e.g. Refs [65–67] and references therein.

## 10.3 SELECTED APPLICATIONS OF THE ELECTRONIC TONGUE

Since many of the results obtained by one of the most active research groups in the field (Toko et al.) are described in detail in a separate chapter of the present book (Chapter 11), the illustrations of the applications of the electronic tongues given below will concentrate mainly on the results produced by other researchers.

### 10.3.1 Application areas and analytes

The widest application area of electronic tongue systems is the analysis of foodstuffs, in particular beverages. The analysis of foodstuffs may include various tasks: discrimination between different individual sorts, blends or brands; recognition of products of different quality and their classification; quantitative determination of the content of multiple compounds, both inorganic and organic, and/or classes of compounds. The most exciting application area of electronic tongues, however, is taste quantification, which is understood as formalized assessment of the taste (flavour) characteristics of a product using an electronic tongue and establishing correlation of the instrument response to human sensory perception. Selected examples of analytical applications of electronic tongues will be discussed below. This chapter will be devoted to a short description of the objects that were analysed using the electronic tongue systems, and to the analytical tasks, which these systems attempted to solve.

A large number of beverages and liquid foods were analysed using electronic tongue systems. Mineral waters represent relatively simple and easily obtained samples for analysis using these sensor systems. Both classification and quantitative analyses were performed on mineral water samples. Discrimination and classification of natural and fake Georgian and Italian produced mineral waters have been performed together with a determination of the content of ions such as sodium, potassium, chloride, fluoride, sulphate, bicarbonate and nitrate [68–71]. The electronic tongue output also nicely correlates to such parameters as conductivity and dry matter residues, which are related to the total salt content in mineral water. Recognition and classification of mineral water samples according to their hardness can also be performed using the taste sensor [72].

Electronic tongue systems are used in the recognition of individual coffees of different origin and also commercial blends. This is quite an important task in the coffee industry [71,73,74]. An attempt has also been made to establish a correlation of instrument output with human perception of coffee flavour, aroma, acidity and body.

Different types of black, oolong and green teas with and without added flavour have been measured using electronic tongue systems [75, 76]. It was found that black and green teas can be reliably discriminated using both potentiometric and voltammetric sensor systems, though results for oolong teas overlapped with black ones.

Fruit juices are quite popular analytes for electronic tongue sensor systems. Measurements with an electronic tongue were made in orange, grape, apple and tomato juices with the aim of distinguishing different samples, detecting spoilage and juice dilution, correlating responses of the sensor system to juice taste and performing quantitative analysis of juice [3,44,45,74,77–80]. The electronic tongues used were capable of discriminating between different types of orange and tomato juices and fruit juice-based soft drinks [3,44,45,74,77], and juices produced from tomatoes of different sorts and ripeness grown in experimental orchards [78]. The possibility of detecting changes in the chemical composition of fruit juices induced by oxidation/ageing processes after contact with air and the subsequent spoilage process was demonstrated in many experiments [3,74,77–79]. Detection of fruit juice dilution by water and sucrose syrup, which is a common method of fruit juice adulteration, can be made with high accuracy using the electronic tongue [79]. Quantitative analysis was performed on tomato juice samples, where conventional chemical analysis methods were used for calibration. Concentrations of K, Na, Mg, Ca, phosphate, organic acids (malic and citric) and UMP were determined by the electronic tongue with an average precision of about 10–15% [78]. A comparison of multisensor system responses with those of human taste perception was performed with tomato juice. Two approaches were realized. The electronic system was calibrated using six taste estimates produced by a sensory panel (saltiness, bitterness, sweetness, sourness, sharpness and umami) and then the performance of the sensor system in test samples was evaluated [78].

Alternatively, the sensor system is calibrated in canned tomato juice, to which four taste substances (NaCl, glucose, citric acid and sodium glutamate) are added [80]. The experimental data were processed by PCA and then the measurements in several tomato varieties (without additives) were projected onto the principal axes, which represented a 'taste map'. The resulting taste assessment agreed well with that of human perception.

The electronic tongue was used for discriminating regular and diet sodas and experimental recipes [71]. The main difference between these soda samples was the type of sweeteners used, which were natural, artificial and experimental substances not approved for use in the food industry. Thus, the capability of the electronic tongue to distinguish between soda samples was related primarily to its sensitivity to different types of sweeteners and other additives in sodas.

Another widespread application area of the electronic tongue is in the analysis of fresh and fermented milk [3,56,63,81–83]. The most important practical tasks involve discrimination of milk that has undergone different heat treatments, since heat affects both milk flavour and nutrition. Recognition of milk samples that have undergone different heat treatments and the correlation of their tastes with the system output was performed in Refs. [81,82]. Also, monitoring of the milk souring process [3,82] and bacteria growth can be achieved by different electronic tongue systems [63]. The possibility of applying the electronic tongue to detection of mastitic milk has also been studied [83]. Automatic detection of mastitis is an essential part of robotic milking, though currently used technology, e.g. conductivity probes, do not provide the necessary sensitivity and discrimination.

Another large group of foodstuffs analysed by using the electronic tongue system is alcoholic beverages and spirits. The main tasks include recognition of different samples, quality control, quantitative analysis and correlation with human taste perception.

Electronic tongues have been applied to the recognition of wine samples, which were different sorts of red and white wine [64] and wine of the same denomination (Barbera) and the same vintage but from different vineyards [68,69,84]. For the electronic tongue calibration, results of standard chemical analyses and flavour quality assessments of wine samples made by the taste panel were used. Wine parameters such as total and volatile acidity pH, ethanol, contents of tartaric acid, sulphur dioxide, total polyphenols, and glycerol were determined using the electronic tongue. The taste panel produced 14 or 15 different wine flavour estimates depending on wine type, with a correlation between most of them and the electronic tongue output being established.

The electronic tongue was applied to recognizing samples of spirits such as cognac, brandy, whisky and vodka [85]. Electronic tongue reproducibly recognized different samples of cognac from each other, and from any whisky and also from brandy. Furthermore, it was possible to distinguish between samples of the same whisky in contact with different oak barrels for different periods of time. In the recognition of vodka samples it was possible to distinguish between various sorts of spirit differing in ethanol quality, water purification level and presence of taste additives. It was also possible to distinguish standard spirits from those of low quality.

Thirty-six samples of different brands of beer produced in different countries were analysed using sensor systems. All samples were distinguished from each other and the correlation of the sensor system output with beer taste characteristics such as richness/lightness and mild/ sharp taste was found [2,36].

Discrimination of four kinds of sake from the same brewery was performed using the sensor system, ordinary sake being used as a reference. An attempt was made to measure the concentration of ethanol and total acidity changes during sake fermentation [86]. Taste evaluation of sake was made where model solutions containing main sake compounds, as well as different taste substances, were used for taste sensor calibration. However, no comparison with human perception was made.

An untypical foodstuff for chemical sensor analysis, namely vegetable oil, was analysed using the potentiometric electronic tongue [87]. Since the oil is not conductive a special procedure of oil extraction by an organic solvent was employed, the resulting extract being used for the measurements. The aim was to discriminate between oils produced from different plants as well as detecting rancidity in oils. Both tasks were solved successfully using this system. A quick and easy-to-use method of rancidity determination in oils using the electronic tongue may be of considerable interest to oil producers, whereas the possibility of determining oil origin and raw material used for its production can be important for consumers and controlling agencies.

Though electronic tongues were primarily designed for the analysis of liquids, they can be applied also to measurements in suspensions, purees and other water–solid mixtures or homogenates. Sample preparation is a necessary step in these situations and it can be performed in different ways. The products with high water content such as fruits and some vegetables can be crushed or mashed and the measurements can be taken from the resulting pulp. Other products with lower water content, for example, flesh food, should be firstly minced and then mixed with water.

The applicability of the electronic tongue to flesh food analysis was demonstrated on fish [71] and pork liver [88] recognition. The measurements were made in the homogenate prepared by stirring chopped fish with distilled water. It was found that the system is capable of distinguishing between the sample of freshwater fish and two samples of seawater fish as well as detecting and monitoring fish spoilage.

Analysis of pork liver was aimed at discrimination between samples from healthy and sick animals. Also, it was found that the 'electronic tongue' could distinguish between the samples of pork liver from animals fed with an anti-stress drug before slaughter and control animals, which did not get any medication.

Another non-liquid foodstuff, miso, Japanese fermented soybean paste, was analysed using the 'taste sensor' [2]. Chemical parameters (e.g. titratable acidity) during the miso fermentation process were measured by conventional analytical methods, while humans estimated ripeness and taste quality. Thus, the electronic tongue was applied to monitoring of the miso fermentation process with the aim of replacing a part of the routine analysis by simpler measurements using a sensor system. Recognition of different sorts of miso was also performed.

The application of the electronic tongue to the analysis of foodstuffs is also popular owing to analogy with biological sensory systems. Nevertheless, the sensors used in the electronic tongues display sensitivity to many ions and many other substances and can be applied for their determination in inedible media as well. On the other hand, there is concern in society about pollution of environment by industrial discharges, which makes the detection of pollution of water an important task. Currently, this kind of analysis is usually performed in the laboratory and after sampling and, hence, is rather slow and expensive. Therefore, a simple, rapid and low-cost method for automated on-line or at-line evaluation of the content of water components or assessment of water quality would be of highly practical value. Sensor systems and the electronic tongue appear to conform well to these requirements and thus represent a prospective technique for environmental monitoring.

The potentiometric electronic tongue was applied to the analysis of modelled polluted ground waters, containing Cu, Fe, Mn, Zn, $Ca^{2+}$, $Mg^{2+}$, $Na^{+}$, $Cl^{-}$, $SO_4^{2-}$ [47–49] and modelled flood waters from abandoned uranium mines containing Fe(II), Fe(III), U(VI) and U(IV) [51]. The errors of determination were from 5 to 30%, which is acceptable for environmental monitoring especially considering the low the concentration level of components, which is as low as $10^{-7}$–$10^{-6}$ mol/L for metals.

A flow-injection multisensor system was designed and applied to determine the heavy metals (Cu, Pb, etc.) and acidic oxides ($NO_x$, $SO_2$) and HCl in the flue gases from incinerators of waste combustion plants [50]. The analysis was performed in solutions that were obtained by bubbling the flue gas through special absorbing liquids. After the absorption, heavy metals were converted in solution into ionic forms, while acidic oxides formed inorganic anions ($NO_3^-$; $NO_2^-$; $SO_4^{2-}$; $Cl^-$). The sensor array included up to 11 sensors. System calibration was performed in model solutions. Afterwards the samples from incineration plants were analysed. The same samples were analysed by ICP-MS and liquid chromatography with the aim of evaluating the multisensor flow-injection system performance and to make the calibration of the system using the independent reference data. The system was able to determine copper, lead and cadmium with an average error of 10–20% and chloride, sulphate, nitrite and nitrate with average errors of 5–15%.

Discrimination of detergents for washing machines and dishwashers was performed by a taste sensor and voltammetric electronic tongue [76]. Both systems were able to discriminate between different detergents, though to achieve discrimination between detergents for washing machines and dishwashers, the merging of data from the two devices was necessary.

Medicinal analysis is another field where rapid and low-cost analytical methods are needed. Application of the electronic tongue to quantitative analysis of dialysis solutions for an artificial kidney is described in more detail in the next chapter. Also, an attempt to use the electronic tongue as a diagnostic tool was made [63]. In this study the urine samples containing blood due to different diseases and those samples from a healthy subject were discriminated by the electronic tongue and nose.

### 10.3.2 Quantitative analysis

Determination of component concentrations is the most typical application area of ion-selective electrodes as well as of other types of chemical sensors for liquids. Although in the application of the electronic tongue systems attention is paid principally to the classification and recognition of samples, quantitative analysis remains practically important. Below, two examples of electronic tongue application to multicomponent analysis of liquids are discussed.

Nowadays, ion-selective electrodes are widely used in commercial blood analysers. However, insufficient selectivity in multicomponent media hinders further application of chemical sensors in this field. Another problem is the requirement of high accuracy of analysis, which should be usually within 1–2%, because the concentration of many components in bodily liquids change only in a very narrow range (8–10%).

The potentiometric electronic tongue was applied to component content determination in solutions with composition close to that of human blood plasma [52]. The measurements were made in complex solutions containing simultaneously $K^+$, $Na^+$, $Ca^{2+}$, $Mg^{2+}$, $Cl^-$, $HCO_3^{2-}$, $HPO_4^{2-}$, $H_2PO_4^-$ and $H^+$ in typical concentration ranges for human blood plasma. Complex solutions of 45 different components have been used for system calibration and testing.

The sensor arrays included two types of potentiometric sensors. First, conventional sensors, for example, chloride-, sodium- and potassium-selective sensors designed on the basis of solid-state and plasticized PVC organic membranes. Also, a conventional pH glass electrode was used to obtain pH values obtained as reference values during calibration and testing. Second, specially designed non-specific sensors with enhanced cross-sensitivities based on PVC membranes chosen using the cross-sensitivity estimation method. Additionally, sensors with PVC membranes sensitive to anions, i.e. hydrocarbonate and phosphate, have been included in the array. In total, ten sensors have been incorporated into the sensor arrays. A certain array composition evolved during experiments which depended on sensor stability and/or cross-sensitivity. The details of sensor preparation can be found elsewhere [14] (and references therein). PVC membranes have been dipped into solutions of primary ions between measurements. To make potential values more reproducible and comparable all sensors have been treated periodically by a conditioner solution. The sensors have been washed with water before measurements to obtain a reproducible, constant potential. The array is then exposed to the analysed solution. Repeated readings of sensor potentials were registered at 5 min intervals.

Calibration of the sensor arrays was performed using a back-propagation ANN and also by partial least squares regression (PLS). Sensor array performance was assessed using cross-validation and test validation.

Owing to a specially elaborate measuring procedure, components of multicomponent solutions, such as $Ca^{2+}$, $HCO_3^-$, $Na^+$, $K^+$, $Cl^-$ and also pH, could be determined with errors as low as 2–4%, thus complying with the requirements of clinical analysis. Results of component determination in test solutions are shown in Table 10.2. The possibility of determining precisely the concentration of magnesium in the presence

TABLE 10.2

Results of the simultaneous determination of calcium, magnesium, pH, hydrocarbonate and phosphates in solutions modelling human blood plasma using an electronic tongue

| Ion | Real (mol/L) | Found (mol/L) ($n = 3$) | $S_r$ |
|---|---|---|---|
| $Mg^{2+}$ | $1.0 \times 10^{-3}$ | $1.0 \times 10^{-3}$ | $4 \times 10^{-5}$ |
| | $7.6 \times 10^{-4}$ | $7.6 \times 10^{-4}$ | $2 \times 10^{-5}$ |
| | $5.0 \times 10^{-4}$ | $5.0 \times 10^{-4}$ | $1 \times 10^{-5}$ |
| $Ca^{2+}$ | $1.5 \times 10^{-3}$ | $1.51 \times 10^{-3}$ | $5.6 \times 10^{-5}$ |
| | $1.2 \times 10^{-3}$ | $1.2 \times 10^{-3}$ | $4 \times 10^{-5}$ |
| | $7.9 \times 10^{-4}$ | $8.2 \times 10^{-4}$ | $4 \times 10^{-6}$ |
| pH | 6.90 | 6.95 | 0.01 |
| | 7.00 | 7.02 | 0.04 |
| | 7.10 | 7.07 | 0.04 |
| $HCO_3^-$ | $4.0 \times 10^{-2}$ | $3.9 \times 10^{-2}$ | $2 \times 10^{-3}$ |
| | $3.5 \times 10^{-2}$ | $3.6 \times 10^{-2}$ | $2 \times 10^{-3}$ |
| | $3.0 \times 10^{-2}$ | $3.2 \times 10^{-2}$ | $3 \times 10^{-4}$ |
| $H_2PO_4^-$ | $1.4 \times 10^{-4}$ | $1.6 \times 10^{-4}$ | $3 \times 10^{-5}$ |
| | $1.7 \times 10^{-4}$ | $1.7 \times 10^{-4}$ | $1 \times 10^{-5}$ |
| | $2.0 \times 10^{-4}$ | $1.9 \times 10^{-4}$ | $1 \times 10^{-5}$ |

of calcium and the content of hydrophosphates using the electronic tongue has been discovered. It was shown that the application of sensor arrays allows for enhancing selectivity and decreasing determination errors. Thus, the electronic tongues could represent a new direction of chemical sensor application in medical analysis.

The next example of quantitative analysis using electronic tongue systems deals with a quite unusual analyte—bacterial colony counts. The voltammetric electronic tongue with five working electrodes (platinum, gold, iridium, palladium and rhodium) was applied to monitoring the milk souring process and bacteria growth estimation [56].

Each measurement cycle was started by cleaning the working electrodes through polishing, dipping into sulphuric acid for 1 min and rinsing with distilled water. When the sensor array was dipped into the samples, a potential of 2 V was applied for 3 s, then a potential of −2 V was applied for the same length of time as electrode cleaning. Afterwards, measurements with the sensor array using an LAPV technique were performed. Potentials applied to the working electrodes were in the range from 600 to −400 mV in steps of 100 mV.

Measurements were taken from 11 samples of milk (pasteurized, homogenized, with fat content 3%) immediately after package opening and then at 30 min intervals up to a maximum 18 h. Milk samples were stored at 21°C. A bacteriostatic agent, sodium azide, was added to two milk samples to prevent bacterial growth. Bacterial colony counts in milk were checked every 2 h using a dip-slide test and the results were used afterwards as reference values for calibration. Multivariate calibration was performed using partial least squares regression and back-propagation ANN. Similar results were obtained using both techniques. Predicted against true values for bacterial counts determined using the dip-slide test are shown in the Fig. 10.6. The results of prediction were acceptable in most cases. However, the sensitivity mechanism of the voltammetric electronic tongue to bacteria is not clear.

Electronic tongues cannot possibly substitute for multiple and widely known analytical methods of quantitative analysis. However, there are certain situations where the use of electronic tongues could be highly advantageous and desirable. Besides, the tongues may be capable of performing quantitative analysis and recognition of analytes simultaneously, which is also an unusual feature in helping to evaluate the integral quality of a product.

### 10.3.3 Qualitative analysis, recognition, identification and classification

Classification and recognition of different liquid samples is probably the largest area of electronic tongue application. We have chosen only a few examples illustrating this type of analysis. A much greater number is available in the literature.

Recognition of orange juices, apple juices and fruit beverages has been made possible by using an electronic tongue based on 29 potentiometric sensors of different types [79]. Firstly, the sensitivity and selectivity of the sensors of the array towards the six main organic acids present in fruits at the highest quantities were investigated. Measurements were made in solutions of acetic, malic, tartaric, lactic, succinic and citric acids in the concentration range from $10^{-5}$ to $10^{-1}$ mol/L. Based on this knowledge, sensors displaying sensitivity to organic acids were included in the sensor array.

The measurements were obtained from six types of 100% orange juice and three types of 100% apple juice produced from the concentrates of

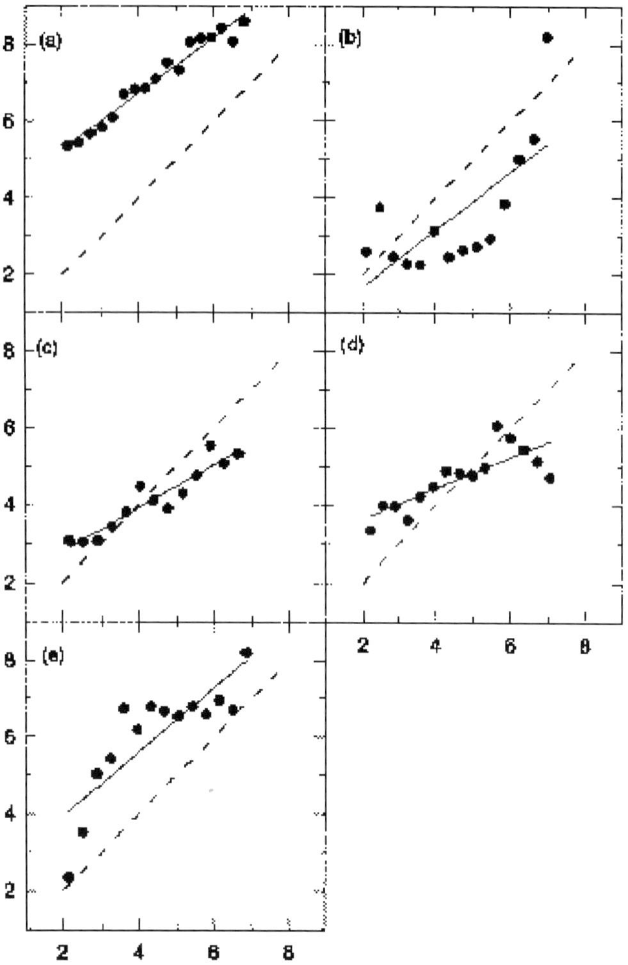


Fig. 10.6. Values of bacterial counts determined using the dip-slide test (true) and the electronic tongue (predicted). Calibration was made using PLS. True values are represented on the horizontal axis and predicted values are represented on the vertical axis: (a) sample 1; (b) sample 2; (c) sample 3; (d) sample 4; and (e) sample 7. Reprinted form *Measurement Science and Technology*, 9 (1998) 1937–1946 (F. Winquist, C. Krantz-Rülcker, P. Wide and I. Lundström, 'Monitoring of freshness of milk by an electronic tongue on the basis of voltammetry'), with permission from IoP.

different manufactures, and two types of beverages containing apple juice and freshly pressed orange juice. The sensitivity of the sensor system to juice dilution by distilled water and sugar syrup was studied. For this procedure two types of orange juice (freshly pressed and a commercial brand made from concentrate), and two types of apple juice (both commercial brands made from concentrate) were chosen. Sample descriptions are shown in Table 10.3. These samples were diluted either by distilled water or by 10% sugar syrup in such a way that the added content of water or syrup in the juice increased from 0 to 46 vol%. Three replicas of each solution were run. The aging dynamics of fruit juices was studied in three samples of orange juice including freshly pressed juice and in all the samples of apple juices and beverages. The samples were kept at room temperature (about 20°C) for a week. The measurements with the sensor array were made on each sample two times a day. The sensor responses in all samples were recorded after 5 min of stirring. The sensors were washed with distilled water between measurements.

The system was able to discriminate between orange and apple juices made from concentrates supplied by different manufactures, and also to

TABLE 10.3

Description of samples of fruit juices and beverages measured using the electronic tongue

| Name | Description |
|---|---|
| A | 100% orange juice, made from concentrate (Moscow, Russia) |
| B | Another brand of 100% juice, made by the same manufacturer as A (Moscow, Russia) |
| C | 100% orange juice, made from concentrate (St Petersburg, Russia) |
| D | 100% orange juice, made from concentrate (St Petersburg, Russia) |
| E | 100% orange juice with pulp, made from concentrate (Lipetsk, Russia) |
| Fresh | Natural orange juice, freshly pressed |
| F | 100% apple juice, made from concentrate, clarified (Moscow, Russia) |
| G | 100% apple juice, made from concentrate with sugar (St Petersburg, Russia) |
| H | 100% apple juice, made from concentrate (Schelkovo, Russia) |
| Beverage1 | Apple nectar (min 50% natural juice, sugar syrup) (Schelkovo, Russia) |
| Beverage2 | Apple beverage, containing natural juice (min 12%), sugar syrup, citric acid, vitamins and water (Moscow, Russia) |

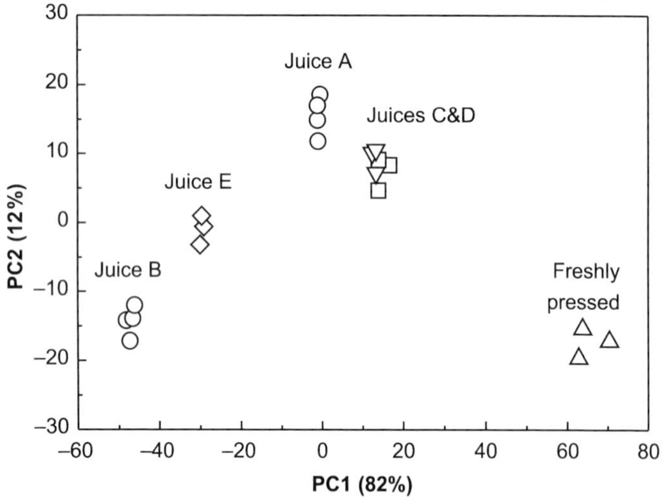


Fig. 10.7. Discrimination of orange juices made by different manufactures from concentrate and freshly pressed orange juice using an electronic tongue.

distinguish them from fresh-pressed natural juice (Figs 10.7 and 10.8). Furthermore, the electronic tongue was applied to monitor the process of the dilution of juices by water and sugar syrup. The dilution degree was determined with high accuracy and reproducibility. Even 1%

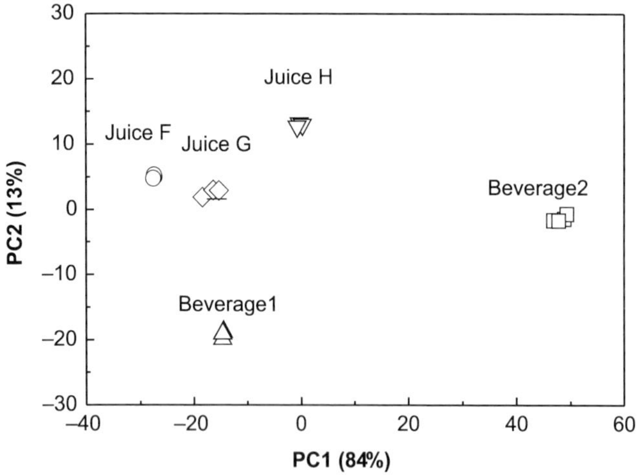


Fig. 10.8. Discrimination of apple juices produced by different manufactures from concentrate and beverages containing apple juice using an electronic tongue.

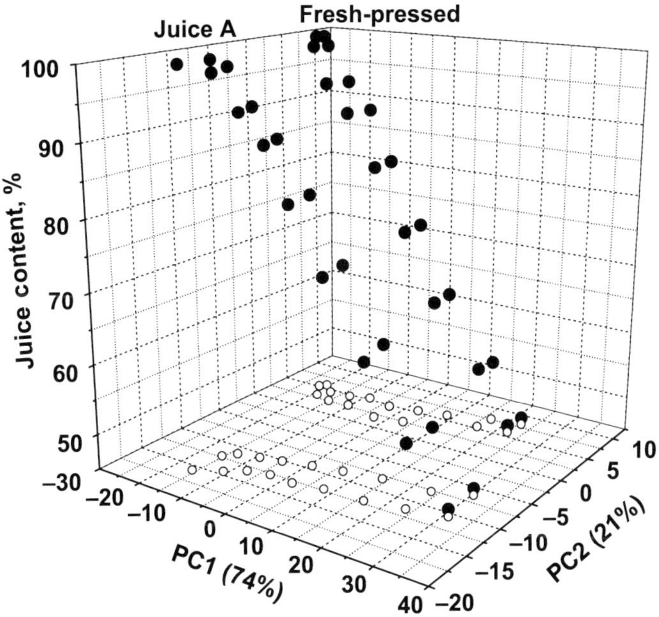


Fig. 10.9. Detection of the dilution of orange juice by water using an electronic tongue. Even minor water additives starting from 1% can be detected.

of water in the juice could have been determined (Fig. 10.9). The freshness of the juices and the dynamics of juice deterioration with time were observed in an aging experiment. Different degrees of juice spoilage were detected (Fig. 10.10). Thus, the possibility of using an electronic tongue for fruit juice quality monitoring was shown.

Another example demonstrates the application of the voltammetric electronic tongue for recognition of different types of fermented milk [57]. The voltammetric electronic tongue included six working electrodes (gold, iridium, palladium, platinum, rhenium, and rhodium). It was complemented in this experiment by other types of liquid sensors, which were conductivity and temperature probes, and three potentiometric sensors: pH, chloride-selective and carbon dioxide gas electrodes. Voltammetric measurements were made in the same way as described in the previous chapter, but the potential applied to the working electrodes was in the range from 1000 to − 1000 mV in steps of 200 mV. Measurements were made in six samples of fermented milk, which

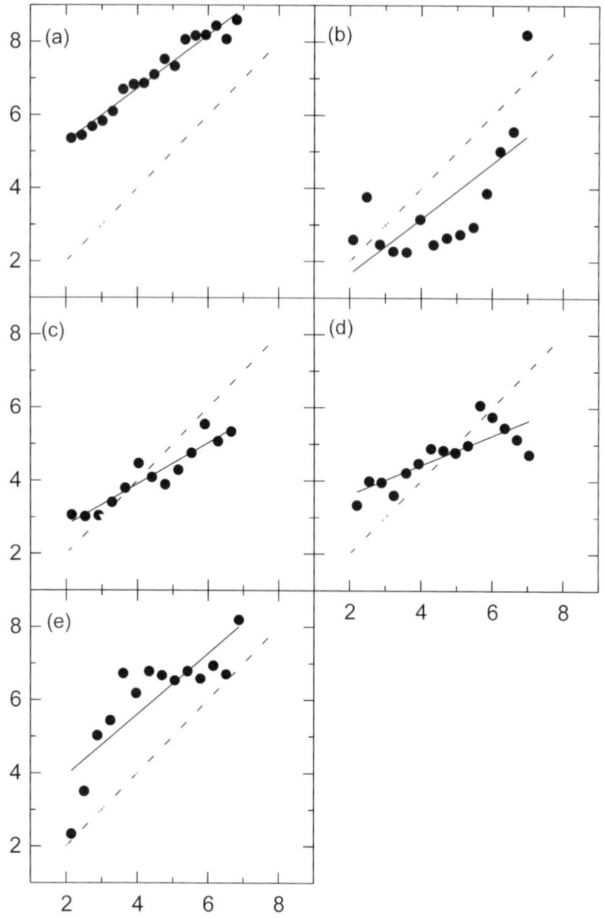


Fig. 10.10. Monitoring orange juice spoilage by an electronic tongue. The numbers denote the duration (days) of juice storage. Number 1 is the first day, when the package containing the juice was opened.

differed in microorganisms used for their preparation. Experimental data were processed by using principal component analysis and back propagation ANN. It was demonstrated that the combination of signals of different nature, e.g. voltammetric, potentiometric and conductometric, improves the discrimination abilities of the device. A mixed electronic tongue could distinguish almost all fermented milk samples (Fig. 10.11).

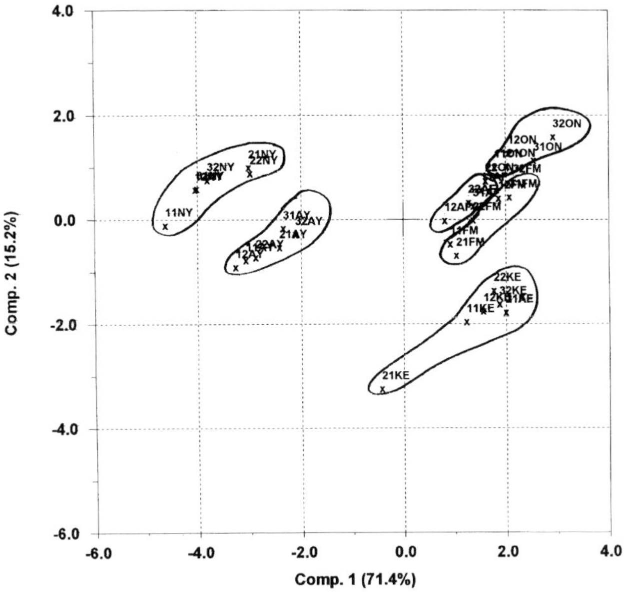


Fig. 10.11. PCA plot resulting from measurements in fermented milk samples using a voltammetric electronic tongue, ion-selective electrodes and conductivity probe, that is a hybrid electronic tongue. Points corresponding to measurements in the same fermented milk sample are encircled. Reprinted from *Analytica Chimica Acta*, 406 (2000) 147–157 (F. Winquist, S. Holmin, C. Krantz-Rülcker, P. Wide and I. Lundström, 'A hybrid electronic tongue') with permission from Elsevier Science.

### 10.3.4 Comparison with the human perception of flavours

Electronic tongues were widely applied to the recognition of different types of food, often very close in taste. This fact together with an allusion concerning the similar organization principles of artificial and biological (mammalian or human) taste and olfaction lead to the numerous attempts of looking for a correlation between human taste perception and instrument output. This approach may be illustrated by the good correlation between electronic tongue responses to different types of coffee and assessments produced by the trained sensory panel.

Almost all commercial brands of coffee are actually blends of different sorts of coffee beans. The quality of harvests differs from year to year but the same commercial coffee brand must be prepared each year with reproducible and flavour characteristics familiar to consumers. This procedure is performed now with the help of the taste panels and, hence, is time-consuming and expensive. The possibility of partly replacing humans in the coffee industry (and similar tasks) may be of significant practical impact. Attempts to use sensor systems (both noses and tongues) for coffee analysis have been carried out many times.

As an example, coffee recognition can be performed using the potentiometric electronic tongue [71]. Ten coffee samples were analysed, including seven individual beans and three commercial brands. The measurements were made in coffee brews, prepared using weighted amounts of coffee and standard water volumes and cooled to room temperature. The electronic tongue was able to distinguish all coffee samples. The experiments on different coffee concentrations and water compositions (distilled, salted distilled and tap water) showed that coffee samples were correctly distinguished in all cases (Fig. 10.12). Coffee taste assessments made by the professional sensory panel of a manufacturing company were obtained from the producer along with coffee samples.

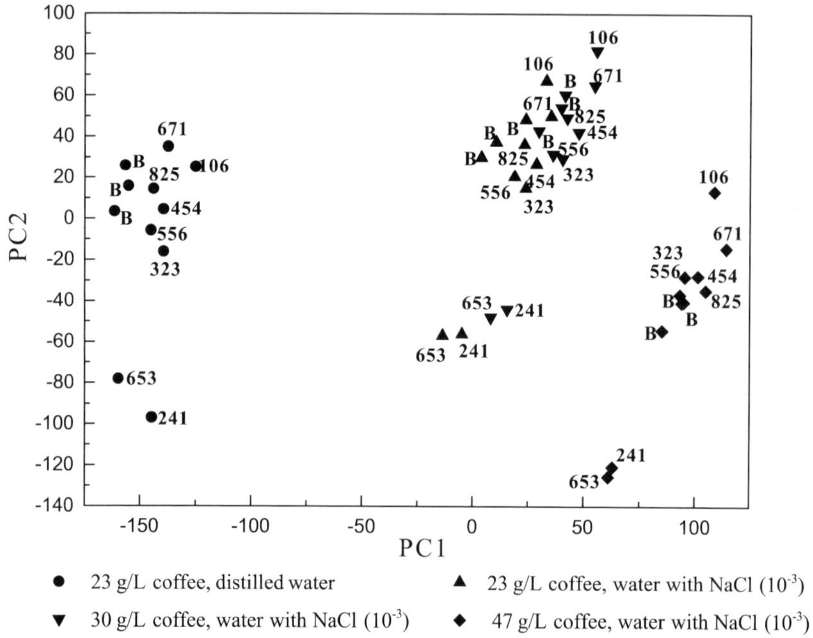


Fig. 10.12. Recognition of 11 types of coffee (eight individual coffees and three blended commercial brands) using an electronic tongue. The data were processed by PCA. The brew was prepared using different types of water (including salted water) and different coffee concentrations.

Four parameters were evaluated: flavour, acidity, body and smell, which obviously characterize taste, odour and flavour of coffee. It was found that the first PC correlates with flavour and smell of coffee. Then the multivariate calibration was performed using a back-propagation neural network, taste panel scores being used as the reference data. Afterwards, the electronic tongue could correctly predict sensory assessment values of all four flavour parameters. The results are shown in the Table 10.4.

The examples of a high correlation of perceived food flavour to the electronic tongue output are numerous but description is restricted by the size of this publication.

### 10.3.5 Taste quantification

Besides different foodstuff recognition and attempts to establish correlation between electronic tongue output and human perception of flavour, quantification of taste using multisensor systems was also performed. Taste quantification starts from the calibration of an electronic tongue when basic taste substances are used in the same way, as for sensory panel training (calibration) before food flavour assessment. Training samples of known flavour type and intensity are analysed and processed using the system. After this calibration the electronic tongue is presumed to produce taste estimates of foodstuffs that are analogous to human assessments. A further task in the taste quantification approach is the evaluation of the taste-masking effect of different substances, e.g. degree of bitterness masking by sweet substances.

The first example in this respect describes a simple experiment with taste substances used widely for human sensory panel training (calibration). Measurements with the electronic tongue were performed in solutions containing three taste substances: NaCl (salty), lactic acid (sour) and L-leucine (bitter) and their binary mixtures [53]. After electronic tongue calibration in individual solutions, the presence and intensity of each individual taste in binary mixtures was correctly predicted (Table 10.5).

An attempt to build a taste map using the taste sensor and thereby to express the tastes of foodstuffs by combinations of basic tastes was performed using the taste sensor [2]. For this purpose, 256 solutions containing four basic taste substances of NaCl, HCl, quinine and sucrose at four concentration levels were prepared and measured.

TABLE 10.4

Organoleptic scores of coffee produced by a professional panel of tasters and the values determined by the 'electronic tongue'

| Sample | Flavour | | Acidity | | Body | | Smell | |
|---|---|---|---|---|---|---|---|---|
| | e-tongue | Tasters | e-tongue | Tasters | e-tongue | Tasters | e-tongue | Tasters |
| 5 | $4.8 \pm 0.2$[a] | 3–7 | $4.8 \pm 0.1$ | 5–8 | $7.0 \pm 0.1$ | 5–9 | $5.3 \pm 0.2$ | 4–7 |
| 6 | $7.8 \pm 0.1$ | 7–9 | $3.6 \pm 0.3$ | 2–5 | $8.5 \pm 0.01$ | 8–9 | $8.4 \pm 0.01$ | 8–9 |
| 7 | $3.1 \pm 0.3$ | 1–2 | $5.7 \pm 0.3$ | 5–8 | $8.4 \pm 0.1$ | 8–9 | $3.8 \pm 0.3$ | 2–3 |
| 8 | $4.1 \pm 0.4$ | 2–4 | $5.4 \pm 0.3$ | 5–8 | $6.0 \pm 0.1$ | 5–7 | $4.7 \pm 0.3$ | 3–5 |

[a]The score and its standard deviation are shown.

TABLE 10.5

Prediction of taste intensity in binary mixtures of taste substances using an electronic tongue after its calibration in individual substances

| Sample | Sourness | | Bitterness | | Saltiness | |
|---|---|---|---|---|---|---|
| | Real | Predicted | Real | Predicted | Real | Predicted |
| B1 | Low | Low | Low | Low | 0 | 0 |
| B2 | Low | Medium | 0 | 0 | Low | Medium |
| B3 | 0 | 0 | Low | 0 | Low | Medium |
| B4 | High | High | High | High | 0 | 0 |
| B5 | High | High | 0 | 0 | High | High |
| B6 | 0 | 0 | High | Low | High | High |
| B7 | High | High | 0 | 0 | Low | Medium |
| B8 | 0 | 0 | High | Medium | Low | Low |
| B9 | 0 | 0 | Low | Low | High | High |

Dependence of each sensor potential on four component concentrations was approximated by quadratic function. Then the measurements in a commercial soft drink were obtained and the resultant pattern was used to calculate the quantity of basic taste substances that are necessary to produce the same taste. A drink made with a calculated content of basic substances appeared to have the same taste as the commercial one according to human perception. However, the possibility of modelling the taste by a combination of four basic tastes was successfully demonstrated on the single sample of a commercial soft drink only.

Suppression of certain tastes perceived by humans in the presence of other substances is well known. The suppression of a bitter taste by sweet substances, often used to mask the bitter taste of drugs, was observed in several studies.

Suppression of quinine bitterness by sucrose and artificial sweet substances such as phospholipids, was studied using a taste sensor [89,90]. The response of the taste sensor to quinine was expressed using $\tau$, a scale that is used to describe taste strength. A linear relationship between the strength of bitterness $\tau$ and first principal component calculated from the taste sensor data was established. The measurements in individual quinine solutions were used in this case. The predicted degree of bitterness dropped significantly at a sucrose concentration of 1 M (Fig. 10.13). The same experiment was repeated with phospholipids.

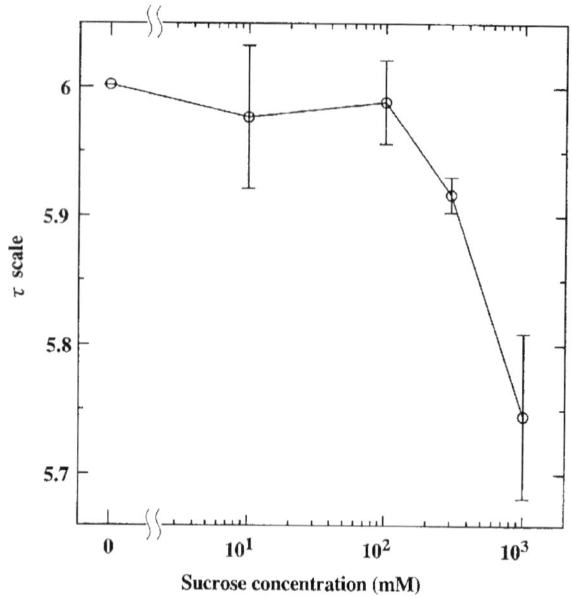


Fig. 10.13. The suppression of the bitterness of quinine by sucrose expressed by the $\tau$ scale. Reprinted from *Sensors and Actuators B*, 64 (2000) 205–215 (K. Toko, 'Taste sensor') with permission from Elsevier Science.

Recognition of different bitter drug substances was also performed by the potentiometric electronic tongue [53]. The masking effect of different flavourings and artificial and natural sweeteners added to bitter pharmaceuticals was evaluated.

The electronic tongue system appears to be useful for taste masking assessment, especially when samples of unpleasant taste have to be evaluated, or samples containing active drug substances are characterized.

### 10.3.6 Application of hybrid systems

An idea to apply combined systems consisting of liquid and gas sensors to food analysis was inspired by the fact that mammalians (and humans) use both olfaction and taste for food flavour sensing. It is worth mentioning that sensitivity and selectivity of sensors used in electronic nose and tongue systems are usually quite different from those of biological receptors due to the different physical and chemical principles artificial senses are normally based on. Nevertheless, the idea of putting together hybrid systems combining both an electronic nose and tongue and using the hybrid to discriminate substances such as wines [64], fruit juices [58,59], milk and even urine samples [63] looks logical and, hopefully, successful.

As examples the application of a hybrid system to the recognition of milk and to the classification of urine samples has been discussed [63]. Both the electronic nose and tongue were based on metalloporphyrins, which are capable of forming strong complexes with different inorganic and organic substances. Properties of metalloporphyrins may vary depending on the coordinated metal, peripheral radicals and functional groups, and conformations of the macrocyclic skeleton. The electronic nose comprised eight quartz microbalance (QMB) sensors with a thin layer of metalloporphyrin deposited onto them. Eight different metalloporphyrins were used. The electronic tongue included seven potentiometric sensors with plasticized polyvinylchloride membranes also containing different metalloporphyrins.

The measurements with both systems were made in milk samples and also in samples of urine. Six different milk samples produced by different manufacturers and thermally treated in different ways were used. Three of them were pasteurized milk and three others were ultrahigh temperature treatment (UHT) milk. Two packages of each milk were purchased at the retail store. All milk samples had the same fat content of 2.5%. Milk packages were opened simultaneously and the measurements were made in each of them. One opened package of each type of milk was stored between measurements in a fridge at a temperature of about +5°C; the other package of each milk was stored at room temperature. The measurements were made (all at room temperature (+25°C)) during the three days following package opening.

Both electronic nose and tongue were able to distinguish between fresh and sour or spoiled milk samples (Figs 10.14 and 10.15). A separation between UHT and pasteurized milk was not achieved by the electronic nose alone, but the data relating to spoiled samples are plotted at the centre of the graph, and in particular at the centre of the first principal component axis. Discrimination of milk samples that have undergone different heat treatments was possible using the electronic tongue data alone (Fig. 10.15). A clear distinction, along

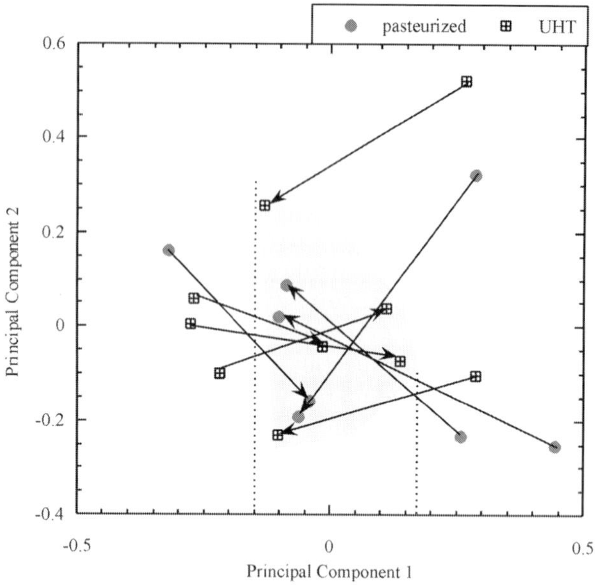


Fig. 10.14. Score plot of the PCA of the electronic nose data in a milk experiment. Reprinted from *Sensors and Actuators B*, 64 (2000) 15–21 (C. Di Natale, R. Paolesse, A. Macagnano, A. Mantini, A. D'Amico, A. Legin, L. Lvova, A. Rudnitskaya and Yu. Vlasov, 'Electronic nose and electronic tongue integration for improved classification of clinical and food samples') with permission from Elsevier Science.

the second principal component, between UHT and pasteurized milk is achieved in this case. On the other hand, the transition from fresh to spoiled samples occurs along the first and the second principal components in the case of UHT and pasteurized milk, respectively. The merging of the data from the two instruments allowed improvements in the separation of fresh and sour milk and at the same time somewhat better distinction of pasteurized and UHT milk samples (Fig. 10.16).

Urine samples were provided by the analytical laboratory of the Institute of Paediatric Clinic of hospital 'Santa Cuore' in Rome. Twenty-seven samples of human urine were measured, including three samples from healthy people and four samples containing blood due to different diseases of the patients. The data processing was performed mainly using principal component analysis. Two approaches to data fusion were used. Low-level abstraction was used for processing measurement results in urine. This means that data from the two instruments were

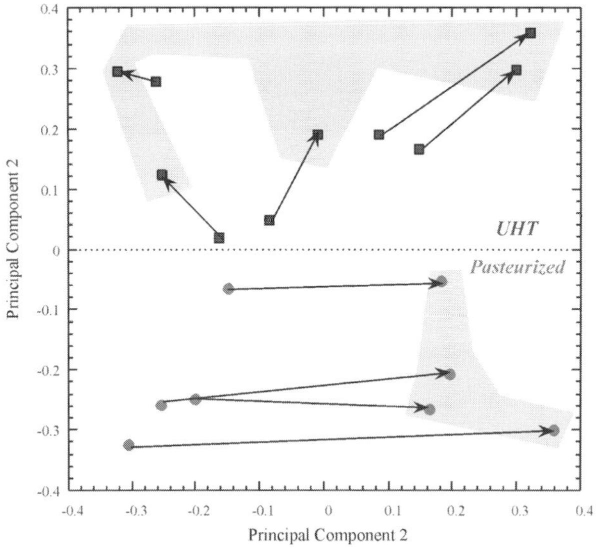


Fig. 10.15. Score plot of the PCA of the electronic tongue data in a milk experiment. Arrows connect the transition from fresh to spoiled milk for each milk sample and the shadowed region contains the spoiled samples. Reprinted from *Sensors and Actuators B*, 64 (2000) 15–21 (C. Di Natale, R. Paolesse, A. Macagnano, A. Mantini, A. D'Amico, A. Legin, L. Lvova, A. Rudnitskaya and Yu. Vlasov, 'Electronic nose and electronic tongue integration for improved classification of clinical and food samples') with permission from Elsevier Science.

merely combined into the same data set and all mathematical operations, e.g. normalization and PCA calculations, were performed with this 'supra-sensor' data set. High-level abstraction assumed that the data from different instruments were analysed separately and merged after feature extraction e.g., by PCA. This latter approach was also used for processing the results for the milk samples described above.

A correlation between the electronic tongue response in urine samples and pH and specific weight of the urine was found. In contrast, the output of the electronic nose was better correlated to blood cell count in the samples. This suggests the independence of the performance of the two systems. Reliable discrimination of urine samples from healthy and ill subjects was possible using the combined data from both instruments.

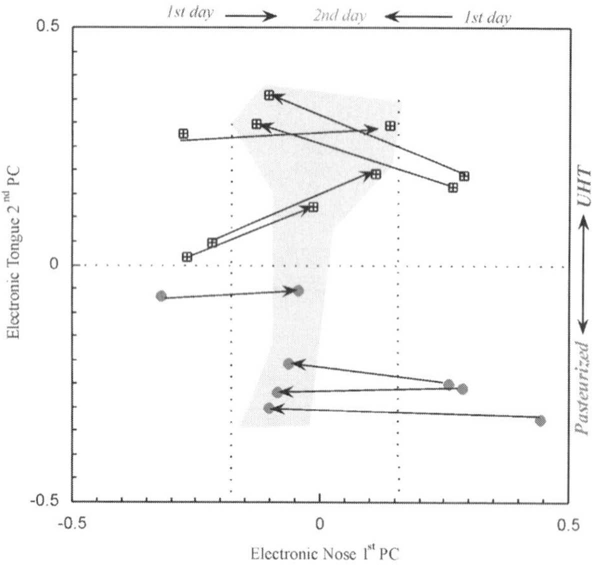


Fig. 10.16. Plot of the second principal component of the electronic tongue data vs. the first principal component of the electronic nose data. This plot represents the fusion of the two sensor systems preserving their main features. Arrows connect the transition from fresh to spoiled milk for each milk sample and the shadowed region contains the spoiled samples. Reprinted from *Sensors and Actuators B*, 64 (2000) 15–21 (C. Di Natale, R. Paolesse, A. Macagnano, A. Mantini, A. D'Amico, A. Legin, L. Lvova, A. Rudnitskaya and Yu. Vlasov, 'Electronic nose and electronic tongue integration for improved classification of clinical and food samples') with permission from Elsevier Science.

Thus, the combination of an electronic tongue with an electronic nose likely enhances recognition power of the hybrid system. This feature may be of significant practical impact when very similar samples are analysed and reliable discrimination is crucial.

## 10.4 PROBLEMS AND PERSPECTIVE

The area of electronic tongues presents an emerging and promising field of chemical sensor R&D, which should be considered as a novel and somewhat unusual branch of analytical science. Such systems may be very useful as quality control tools in the food industry, medicine, environmental monitoring, etc. Also they can be rapid and easy-to-use multicomponent quantitative analysis instruments. The correlation of the electronic tongue output with human sensory perception has been found for numerous analytes. This is an exciting feature, promising new, unconventional but practically very important and highly interesting applications. Though some electronic tongues are already offered commercially, this is still a recent scientific area and knowledge in this field remains significantly empirical. Serious efforts are demanded in sensing mechanism studies and their theoretical considerations; in the development of new sensor compositions; and in the procedures of electronic tongue applications for practical tasks. We would expect new achievements in all these fields in the near future.

## REFERENCES


1 Yu. Vlasov, A. Legin, A. Rudnitskaya, A. D'Amico and C. Di Natale, *Sens. Actuat. B*, 65 (2000) 235.

2 K. Toko, *Meas. Sci. Technol.*, 9 (1998) 1919.

3 F. Winquist, P. Wide and I. Lundstrom, *Anal. Chim. Acta*, 357 (1997) 21.

4 F. Haber and Z. Klemensiewicz, *Z. Phys. Chem. Leipzig*, 64 (1909) 385.

5 K. Camman, *Working with Ion-Selective Electrodes*, Springer, Berlin, 1979.

6 Yu. Vlasov and E. Bychkov, *Ion-Selective Electrode Rev.*, 9 (1987) 5.

7 E. Bakker, P. Buhlmann and E. Pretsch, *Chem. Rev.*, 97 (1997) 3083.

8 P. Buhlmann, E. Pretsch and E. Bakker, *Chem. Rev.*, 98 (1998) 1593.

9 D.G. Buerk, *Biosensors: Theory and Application*, Technomic, Lancaster, 1993.

10 J. Janata, M. Josowicz, P. Vanysek and D.M. DeVaney, *Anal. Chem.*, 70 (1998) 179R.

11 M. Otto and J.D.R. Thomas, *Anal. Chem.*, 57 (1985) 2647.

12 K. Beebe and B. Kowalski, *Anal. Chem.*, 60 (1988) 2273.

13 K. Beebe, D. Uerz, J. Sandifer and B. Kowalski, *Anal. Chem.*, 60 (1988) 66.

14 R.J. Forster, F. Regan and D. Diamond, *Anal. Chem.*, 63 (1991) 876.

15 R.J. Forster and D. Diamond, *Anal. Chem.*, 64 (1992) 1721.

16 D. Diamond and R.J. Forster, *Anal. Chim. Acta*, 276 (1993) 75.

17 M. Hartnett, D. Diamond and P.G. Barker, *Analyst*, 118 (1993) 347.

18 W.E. van der Linden, M. Bos and A. Bos, *Anal. Proc.*, 26 (1989) 329.

19 M. Bos, A. Bos and W.E. van der Linden, *Anal. Chim. Acta*, 233 (1995) 31.

20 J.W. Gardner and P.N. Bartlett (Eds.), *Sensors and Sensory Systems for an Electronic Nose*, NATO ASI Series E: Applied Sciences, Vol. 212. Kluwer, Dordrecht, 1992.

21 W. Gopel, Ch. Ziegler, H. Breer, D. Schild, R. Apfelbach, J. Joerges and R. Malaka, *Biosens. Bioelectron.*, 13 (1998) 479.

22 G.F. Stewart and M.A. Amerine, *Introduction to Food Science and Technology*, Academic Press, New York, 1973.
23 Y. Kawamura and M.R. Kare, *Umami: A Basic Taste*, Marcel Dekker, New York, 1987.
24 R.W. Moncrieff, *J. Appl. Physiol.*, 16 (1961) 742.
25 K. Persaud and G.H. Dodd, *Nature*, 299 (1982) 352.
26 T.A. Dickinson, J. White, J.S. Kauer and D.R. Walt, *TIBTECH*, 16 (1998) 250.
27 A. Legin, A. Rudnitskaya, Yu. Vlasov, C. Di Natale and A. D'Amico, *Sens. Actuat. B*, 58 (1999) 464.
28 A. Legin, A. Rudnitskaya, Yu. Vlasov, A. D'Amico and C. Di Natale, *Proc. Eighth Int. Conf. Electroanal.* Bonn, Germany, 2000, p. A 27.
29 M. Nagele, E. Bakker and E. Pretsch, *Anal. Chem.*, 71 (1999) 1041.
30 A. Legin, Yu. Vlasov, A. Rudnitskaya and E. Bychkov, *Sens. Actuat. B*, 34 (1996) 456.
31 Yu. Vlasov, A. Legin and A. Rudnitskaya, *Sens. Actuat. B*, 44 (1997) 532.
32 A. Legin, A. Rudnitskaya, A. Smirnova, L. Lvova and Yu. Vlasov, *J. Appl. Chem. (Russ.)*, 72 (1999) 114.
33 K. Toko, K. Hayashi, M. Yamanaka and K. Yamafuji, *Tech. Digest Ninth Sensor Symp.* 1990, pp. 193–196.
34 K. Hayashi, M. Yamanaka, K. Toko and K. Yamafuji, *Sens. Actuat. B*, 2 (1990) 205.
35 K. Toko, *Mater. Sci. Engng C*, 4 (1996) 69.
36 K. Toko, *Sens. Actuat. B*, 64 (2000) 205.
37 K. Oohira, K. Toko, H. Akiyama, H. Yoshihara and K. Yamafuji, *J. Phys. Soc. Jpn*, 64 (1995) 3554.
38 K. Hayashi, K. Toko, M. Yamanaka, H. Yoshihara, K. Yamafuji, H. Ikezaki, R. Toukubo and K. Sato, *Sens. Actuat. B*, 23 (1995) 55.
39 S. Iiyama, K. Toko, T. Matsuno and K. Yamafuji, *Chem. Senses*, 19 (1994) 87.
40 S. Iiyama, S. Ezaki, K. Toko, T. Matsuno and K. Yamafuji, *Sens. Actuat. B*, 24 (1995) 75.
41 H. Akiyama, K. Toko and K. Yamafuji, *Jpn. J. Appl. Phys. Part 1*, 35 (1996) 5516.
42 K. Toko, H. Akiyama, K. Chishaki, S. Ezaki, T. Iyota and K. Yamafuji, *Sens. Mater.*, 9 (1997) 321.
43 K. Toko, R. Yasuda, S. Ezaki and T. Fujiyishi, *Trans. IEE Jpn E*, 118 (1998) 1.
44 Y. Kanai, M. Shimizu, H. Uchida, H. Nakahara, C.G. Zhou, H. Maekawa and T. Katsube, *Sens. Actuat. B*, 20 (1994) 175.
45 Y. Sasaki, Y. Kanai, H. Uchida and T. Katsube, *Sens. Actuat. B*, 25 (1995) 819.
46 Yu. Vlasov and A. Legin, *Fresenius J. Anal. Chem.*, 361 (1998) 255.

47 C. Di Natale, F. Davide, J.A.J. Brunink, A. D'Amico, Yu. Vlasov, A. Legin and A. Rudnitskaya, *Sens. Actuat. B*, 34 (1996) 539.
48 C. Di Natale, A. Macagnano, F. Davide, A. D'Amico, A. Legin, Yu. Vlasov, A. Rudnitskaya and B. Seleznev, *Sens. Actuat. B*, 44 (1997) 423.
49 A. Rudnitskaya, A. Ehlert, A. Legin, Yu. Vlasov and S. Buttgenbach, *Talanta*, 55 (2001) 425.
50 J. Mortensen, A. Legin, A. Ipatov, A. Rudnitskaya, Yu. Vlasov and K. Hjuler, *Anal. Chim. Acta*, 403 (2000) 273.
51 A. Legin, B. Seleznev, A. Rudnitskaya, Yu. Vlasov, S. Tverdokhlebov, B. Mack, B. Abraham, T. Arnold, L. Baraniak and H. Nitsche, *Czechoslovak. J. Phys.*, 49(S1) (1999) 679.
52 A. Legin, A. Smirnova, A. Rudnitskaya, L. Lvova and Yu. Vlasov, *Anal. Chim. Acta*, 385 (1999) 131.
53 A. Legin, A. Rudnitskaya, B. Seleznev and Yu. Vlasov, *Proc. Seventh Int. Symp. Olfaction Electron. Nose ISOEN*. Brighton, UK, 2000, p. 195.
54 M. Schoning, C. Schmidt, J. Schubert, W. Zander, S. Mesters, P. Kordos, H. Luth, A. Legin, B. Seleznev and Yu. Vlasov, *Sens. Actuat. B*, 68 (2000) 254.
55 Yu. Mourzina, M. Schoning, J. Schubert, W. Zander, A. Legin, Yu. Vlasov and H. Luth, *Anal. Chim. Acta*, 433 (2001) 103.
56 F. Winquist, C. Krantz-Rulcker, P. Wide and I. Lundstrom, *Meas. Sci. Technol.*, 9 (1998) 1937.
57 F. Winquist, S. Holmin, C. Krantz-Rulcker, P. Wide and I. Lundstrom, *Anal. Chim. Acta*, 406 (2000) 147.
58 P. Wide, F. Winquist, P. Bergsten and E.M. Petriu, *IEEE Trans. Instrum. Meas.*, 47 (1998) 1072.
59 F. Winquist, I. Lundstrom and P. Wide, *Sens. Actuat. B*, 58 (1999) 512.
60 J.J. Lavigne, S. Savoy, M.B. Clevenger, J.E. Ritchie, B. McDoniel, S.J. Yoo, E.V. Anslyn, J.T. McDevitt, J.B. Shear and D. Neikirk, *J. Am. Chem. Soc.*, 120 (1998) 6429.
61 A. Riul Jr, D.S. Dos Santos Jr, K. Wohnrath, R. Di Tommazo, A.C.P.L.F. Carvalho, F.J. Fonseca, O.N. Oliveira Jr, D.M. Taylor and L.H.C. Mattoso, *Lamgmuir*, 18 (2002) 239.
62 J. Kondoh and S. Shiokawa, *Jpn. J. Appl. Phys. Part 1*, 33 (1994) 3095.
63 C. Di Natale, R. Paolesse, A. Macagnano, A. Mantini, A. D'Amico, A. Legin, L. Lvova, A. Rudnitskaya and Yu. Vlasov, *Sens. Actuat. B*, 64 (2000) 15.
64 S. Baldacci, T. Matsuno, K. Toko, R. Stella and D. De Rossi, *Sens. Mater.*, 10 (1998) 185.
65 *Neural Computing*, Neural Ware, Pittsburg, USA, 1997.
66 K. Esbensen, *Multivariate Analysis in Practice*, 5th edn, Camo ASA/CAMO Process AS, Norway, 2001.
67 H. Martens and T. Naes, *Multivariate Calibration*, Wiley, 1989.
68 A. Legin, A. Rudnitskaya, Yu. Vlasov, C. Di Natale, A. Mantini, E. Mazzone, A. Bearzotti and A. D'Amico, *Alta Frequenza (Italian)*, 10 (1999) 1.

69 A. Legin, A. Rudnitskaya, Yu. Vlasov, C. Di Natale, E. Mazzone and A. D'Amico, *Electroanalysis*, 11 (1999) 814.
70 A. Legin, A. Rudnitskaya, Yu. Vlasov, C. Di Natale, E. Mazzone and A. D'Amico, *Sens. Actuat. B*, 65 (2000) 232.
71 A. Legin, A. Rudnitskaya, B. Seleznev and Yu. Vlasov, *Food Sci. Technol.*, 37 (2002) 375.
72 S. Iiyama, M. Yahiro and K. Toko, *Sens. Mater.*, 7 (1995) 191.
73 T. Fukunaga, K. Toko, S. Mori, Y. Nakabayashi and M. Kanda, *Sens. Mater.*, 8 (1996) 47.
74 A. Legin, A. Rudnitskaya and Yu. Vlasov, *Proc. Sixth Int. Symp. Kinetics Anal. Chem. KAC'98*, Kassandra Halkidiki, Greece, 1998, p. 106.
75 P. Ivarsson, S. Holmin, N.-E. Hoejer, C. Krantz-Ruelcker and F. Winquist, *Sens. Actuat. B*, 76 (2001) 449.
76 P. Ivarsson, Yu. Kikkawa, F. Winquist, C. Krantz-Ruelcker, N.-E. Hoejer, K. Hayashi, K. Toko and I. Lundstrom, *Anal. Chim. Acta*, 449 (2001) 59.
77 A. Legin, A. Rudnitskaya, Yu. Vlasov, C. Di Natale, F. Davide and A. D'Amico, *Sens. Actuat. B*, 44 (1997) 291.
78 A. Rudnitskaya, A. Legin, C. Salles and P. Mielle, *Proc. Eighth Int. Symp. Olfaction Electron. Nose ISOEN*, Washington, DC, 2001.
79 S. Makarychev-Mikhailov, A. Legin, A. Rudnitskaya, O. Goryacheva and Yu. Vlasov, *Proc. Int. Conf. Anal. Sci*, Tokyo, Japan, 2001.
80 Y. Kikkawa, K. Toko and K. Yamafuji, *Sens. Mater.*, 5 (1993) 83.
81 A. Legin, A. Rudnitskaya, L. Lvova, Yu. Vlasov, A. D'Amico, C. Di Natale, R. Paolesse, *Proc. Fifth Ital. Conf. Sens. Microsyst.*, Lecce, Italy, World Scientific, Singapore, 2000, pp. 263–269.
82 K. Toko, T. Iyota, Y. Mizota, T. Matsuno, T. Yoshioka, T. Doi, S. Iiyama, T. Kato, K. Yamafuji and R. Watanabe, *Jpn. J. Appl. Phys. Part 1*, 34 (1995) 6287.
83 T. Mottram, A. Legin, A. Rudnitskaya, J.L. Fitzpatrick and P.D. Eckersall, *Livestock Production Science* (2003) in press.
84 A. Legin, A. Rudnitskaya, Yu. Vlasov, C. Di Natale, E. Mazzone and A. D'Amico, *Proc. Int. Conf. Eurosens. XIII*, The Hague, The Netherlands, 1999, pp. 1017–1020.
85 A. Legin, A. Rudnitskaya, B. Seleznev, Yu. Vlasov, P. Mielle, *Proc. Second Symp. In Vino Anal. Sci.*, Bordeaux, France, 2001, p. 195.
86 Y. Arikawa, K. Toko, H. Ikezaki, Y. Shinha, T. Ito, I. Oguri and S. Baba, *Sens. Mater.*, 7 (1995) 261.
87 Yu. Vlasov, B. Seleznev, A. Ivanov, A. Rudnitskaya and A. Legin, *Proc. Fifth Italian Conf. Sensors Microsyst.*, Lecce, Italy, 2000.
88 A. Legin, A. Rudnitskaya, B. Seleznev, Yu. Vlasov and V. Velikzhanin, *Proc. Eighth Int. Symp. Olfaction Electron. Nose ISOEN*, Washington, DC, 2001.
89 S. Takagi, K. Toko, K. Wada, H. Yamada and K. Toyoshima, *J. Pharm. Sci.*, 87 (1998) 552.
90 S. Tagaki, K. Toko, K. Wada and T. Ohki, *J. Pharm. Sci.*, 90 (2001) 2042.

*Chapter 11*

# A taste sensor

***Kiyoshi Toko****

## 11.1 INTRODUCTION

Taste is comprised of five basic qualities: sourness produced by hydrogen ions from HCl, acetic acid, citric acid, etc.; saltiness produced mainly by NaCl; sweetness due to sucrose, glucose, L-alanine, etc.; bitterness produced by quinine, caffeine, L-tryptophan and $MgCl_2$. The last is umami taste, which is the Japanese term for implying 'deliciousness', produced by monosodium glutamate (MSG) contained in seaweeds such as tangle, disodium inosinate (IMP) in meat and fish and disodium guanylate (GMP) in mushrooms.

In a gustatory system, substances producing taste are received by the biological membrane of gustatory cells in taste buds on the tongue. Information on the taste of substances is transduced into an electric signal, which is transmitted along the nerve fiber to the brain, where the taste is perceived. Toko and co-workers have developed a taste sensor, i.e. electronic tongue, where the transducer is composed of several kinds of lipid/polymer membranes [1–3]. The output of this system is the pattern constructed from electric potentials of eight (or seven) kinds of membranes. It does not express the amount of specific taste substances but the taste quality and intensity, because different outputs of electric patterns are obtained for chemical substances producing different taste qualities such as sourness and saltiness. In contrast, similar patterns are obtained for chemical substances

* Fax: +81-92-6423967. *E-mail address:* toko@ed.kyushu-u.ac.jp

*Comprehensive Analytical Chemistry XXXIX, pages 487–511*


ISSN: 0166-526X

producing the same taste, such as MSG, IMP and GMP, all of which have an *umami* taste, and NaCl, KCl and KBr for saltiness.

The development of this sensor is based on a concept very different from that of conventional chemical sensors, which selectively detect specific chemical substances such as glucose or urea. However, taste cannot be measured even if all the chemical substances contained in foodstuffs are quantified. Humans do not distinguish each chemical substance, but express the taste itself; the relationship between chemical substances and taste is not clear. It is also not practical to arrange as many chemical sensors as there are chemical substances, which would amount to several hundreds for one kind of foodstuff. Moreover, interactions exist between taste substances, such as the suppression effect. For example, sweet substances suppress the taste intensity of bitter substances.

Discrimination of each foodstuff is possible using the taste sensor, and also recognition of the taste itself and its quantitative expression can be made. On-line monitoring using a taste-sensing system integrated with other sensors such as pH and viscosity meters makes it possible to check and control the quality of foodstuffs. The taste sensor has a concept of global selectivity, which implies the ability to classify enormous numbers of chemical substances into several groups. It can also be applied to measurements of water pollution. The taste sensor will lead to a new era in food and environmental sciences.

## 11.2 STRUCTURE OF THE TASTE SENSOR

Different kinds of lipids were used for preparing the membranes; these were, e.g. oleic acid (OA), oleyl amine (OAm) and decyl alcohol (DA) as listed in Table 11.1. The mixed hybrid membranes composed of dioctyl phosphate (DOP) and trioctyl methyl ammonium chloride (TOMA) were also used. Each lipid was mixed with polyvinyl chloride (PVC) and plasticizer (dioctyl phenylphosphonate, DOPP) dissolved in tetrahydrofuran (THF). The mixture was then dried on a glass plate that was set on a hot plate temperature-controlled to about 30°C. Depending upon the object to be measured, different lipid materials were prepared. The lipid/polymer membrane was a transparent, soft film of approximately 200 μm thickness. The membranes can be used continuously for over one year.

TABLE 11.1

Lipid materials used in the multichannel electrode

| *Channel* | *Lipid (abbreviation)* |
|---|---|
| 1 | *n*-Decyl alcohol (DA) |
| 2 | Oleic acid (OA) |
| 3 | Dioctyl phosphate (DOP) |
| 4 | DOP:TOMA = 9:1 |
| 5 | DOP:TOMA = 5:5 |
| 6 | DOP:TOMA = 3:7 |
| 7 | Trioctylmethylammonium chloride (TOMA) |
| 8 | Oleyl amine (OAm) |

Each lipid/polymer membrane was fitted to part of a plastic tube, which has a hole, such that the inner part of the cylinder is isolated from the outside. The end of the cylinder was sealed with a stopper that holds an Ag/AgCl electrode. The tube was filled with 3 M KCl solution. Eight detecting electrodes thus prepared were separated into two groups and connected to two electrode holders, which are controlled mechanically by a robot arm. The taste sensor illustrated in Fig. 11.1 is commercially

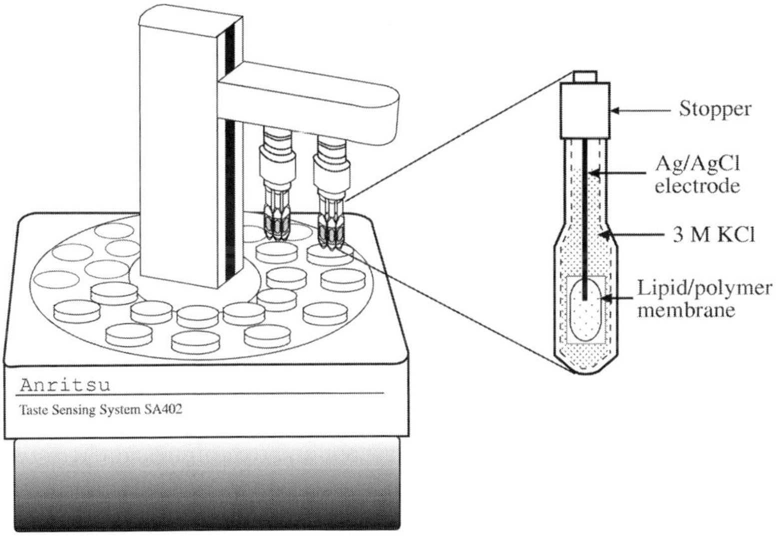


Fig. 11.1. Taste sensing system (SA402B, Insent, Inc.).

available, and used in many food and medical companies, government institutes, prefecture food research institutes and universities.

## 11.3 RESPONSE CHARACTERISTICS

Figure 11.2 shows the electric potential pattern from seven channels for two of the taste qualities (sourness and umami) measured by taking the origin to 1 mM KCl [4]. The patterns of substances producing different taste qualities are much different, and hence each taste can be easily discriminated. The reproducibility was very high, because the standard deviations were smaller than 1%. On the other hand, the taste sensor has similar response patterns to the same group of tastes, i.e. as examples of sour substances HCl, citric acid and acetic acid show similar response patterns. Umami taste substances of MSG, IMP and GMP show similar patterns, too. The same result also holds for other taste qualities such as bitterness, sweetness and saltiness. The taste sensor can respond to the taste itself.

This fact shows that the taste sensor has a global selectivity, which implies the ability to classify numerous chemical substances into several groups of taste quality. Of course, the taste sensor has an ability of molecular recognition to distinguish chemical substances even in the same taste group, e.g. the patterns for HCl, citric acid and acetic acid are slightly different. The reception mechanism in lipid/polymer membranes of the taste sensor was clarified quantitatively based on electrochemical theory, which treats the surface electric potential, surface charge density and binding of ions such as protons and hydrophobic ions [5].

## 11.4 AMINO ACIDS

### 11.4.1 Classification of taste of amino acids

Amino acids are very important, mainly because protein molecules are made up of L-amino acid units. Proteins are found in every living cell and are the principal material of skin, muscle, tendons, nerves and blood.

Each channel of the taste sensor responded to amino acids in different ways depending on their tastes [6,7]. L-Tryptophan, which elicits an almost purely bitter taste in humans, increased the potentials of channels 1–3 greatly. This tendency was observed for other amino acids,

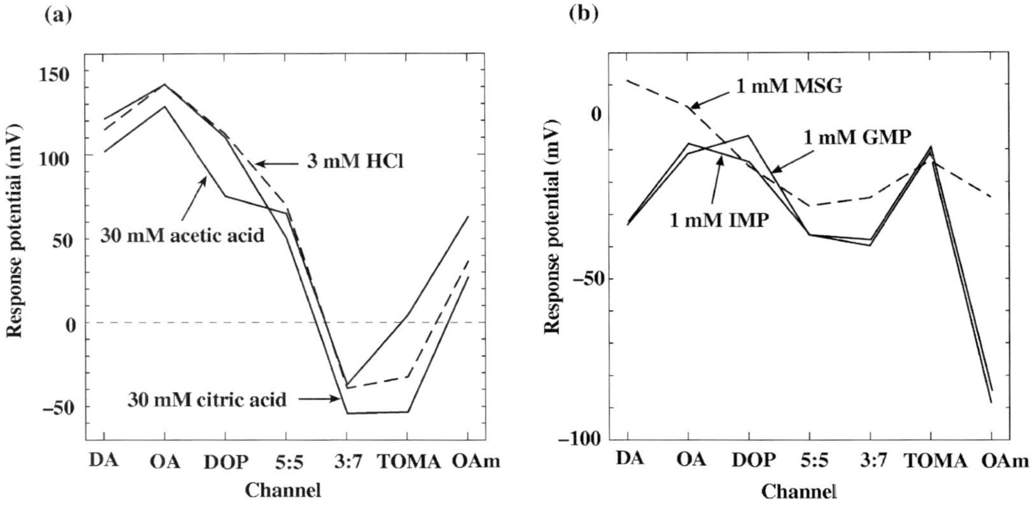


Fig. 11.2. Response electric potential patterns for (a) sour [4] and (b) umami taste substances.

which mainly exhibit bitter taste, such as L-phenylalanine and L-isoleucine. On the other hand, L-alanine, glycine and L-threonine taste mainly sweet. For these amino acids, the potentials of channels 1 and 2 decreased. L-Valine and L-methionine, which taste bitter and slightly sweet, decreased the potential of channel 5; the responses of channels 1 and 2 were small.

Next, we compared the response patterns for bitter-tasting amino acids such as L-tryptophan with that of quinine, which is a typical bitter substance. Of course, these two chemicals have very different chemical structures; nevertheless, we feel the same bitter taste quality. Figure 11.3 shows the comparison of response patterns of three amino acids (L-alanine, L-tryptophan and L-phenylalanine) with those of a bitter substance (quinine), sour substance (HCl) and umami taste substance (MSG) [6]. The response patterns were normalized using the formula:

$$v_i = \frac{V_i}{\sqrt{\sum_{i=1}^{8} |V_i|^2}} \tag{11.1}$$

where $V_i$ denotes the response electric potential of channel $i$. To make it easy to see the pattern, the response electric potentials of the positively charged membranes (channels 6–8) in Table 11.1 and the non-charged membrane (channel 5) were reversed, because they usually have a sign opposite to those of the negatively charged membranes (channels 1–4).

Figure 11.3 includes three normalized patterns for one chemical substance with three different concentrations. By the normalization procedure of Eq. (11.1), the three patterns of each chemical substance agree with each other, i.e. the pattern is independent of the concentration. This fact implies that each chemical substance has an original pattern characteristic of each taste quality. The patterns of amino acids such as L-tryptophan and L-phenylalanine are similar to that of quinine. However, they differ from those of other taste substances such as L-alanine, HCl and MSG.

The patterns for L-tryptophan and L-phenylalanine extend out to the upper-right direction (i.e. at channels 1–3), whereas the pattern for L-alanine shows the opposite tendency: its pattern bulges out to the lower-left direction. Quinine shows a bulge to the upper-right direction in a similar way to L-tryptophan.

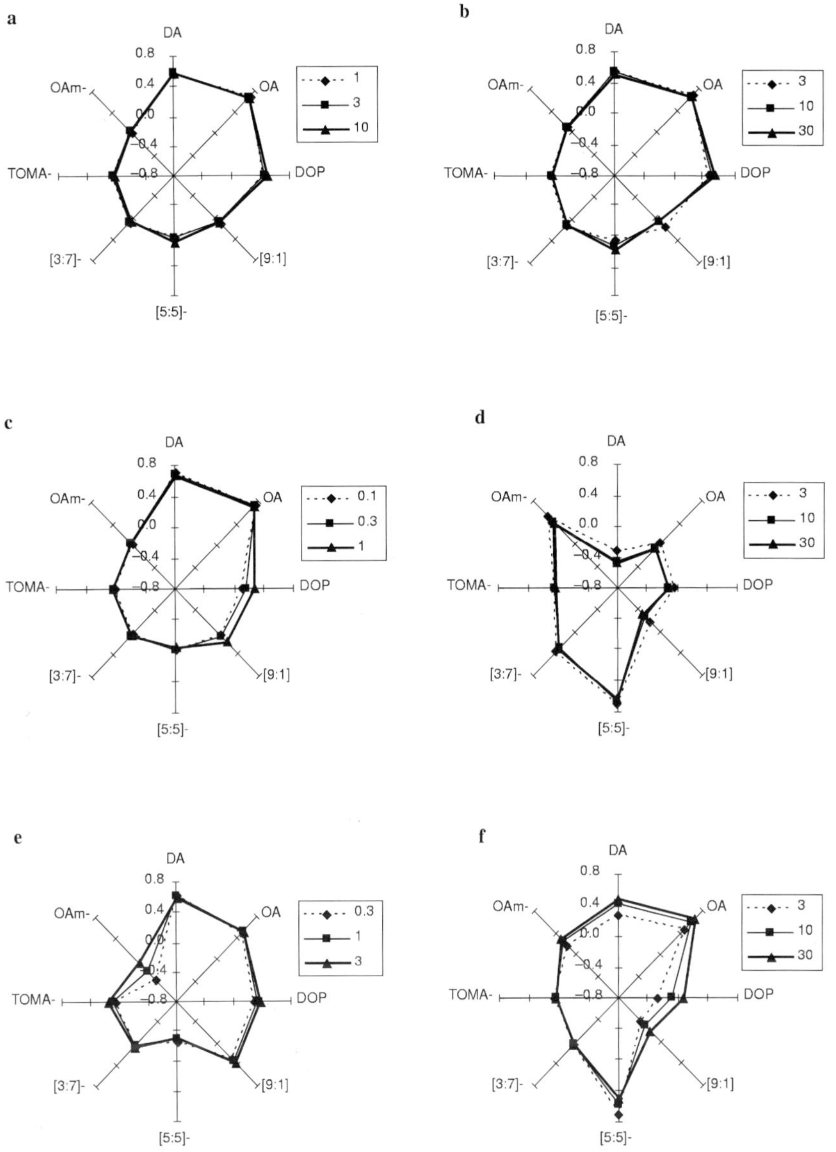


Fig. 11.3. Normalized patterns for (a) L-tryptophan, (b) L-phenylalanine, (c) quinine, (d) L-alanine, (e) HCl and (f) MSG [6]. The three numerical figures closed by a square box attached to each pattern imply the concentration (mM).

The correlation coefficients of the pattern between L-tryptophan (10 mM) and the five basic taste substances are as follows: 0.90 for 0.3 mM quinine, 0.28 for 30 mM NaCl, 0.76 for 3 mM HCl, 0.52 for 100 mM sucrose and 0.41 for 10 mM MSG. Although L-tryptophan has low correlations with salty (NaCl), sweet (sucrose) and umami (MSG) taste substances, it has a high correlation with a bitter substance, quinine. This result indicates that L-tryptophan shows the same taste as quinine, i.e. L-tryptophan tastes bitter.

Comparison of the original response pattern of L-tryptophan with that of quinine makes it possible to estimate the bitter strength of L-tryptophan in terms of the quinine concentration (see also Fig. 11.9). As a result, it was concluded that 10 mM L-tryptophan has the same bitter strength as 0.02 mM quinine. To confirm this, the author performed the sensory evaluation using his tongue. The result supported the estimation obtained using the taste sensor: humans felt the same bitter strength of 10 mM L-tryptophan by taking 0.02–0.03 mM quinine.

This result implies that the taste sensor measures the taste in itself, as humans do, irrespective of the difference of chemical structures of amino acids and alkaloids such as quinine.

Dipeptides elicit various taste qualities according to the combination of two amino acids. For bitter dipeptides such as glycyl-leucine (Gly-Leu), glycyl-phenylalanine (Gly-Phe) and leucyl-glycine (Leu-Gly), the taste sensor showed response patterns similar to L-tryptophan. In addition, the patterns characteristic of sourness, which is elicited by amino acids such as L-glutamic acid and L-histidine monohydrochloride, were obtained for dipeptides, such as glycyl-aspartic acid (Gly-Asp), seryl-glutamine (Ser-Glu), alanyl-glutamine (Ala-Glu) and glycyl-glutamine (Gly-Glu), all of which taste sour.

In contrast, dipeptides such as alanyl-glycine (Ala-Gly) and glycyl-glycine (Gly-Gly) have a little or no taste. Small response patterns were obtained for these dipeptides, as expected.

Figure 11.4 shows a taste map of five taste qualities elicited by typical chemical substances, amino acids and dipeptides [1]. The taste map was obtained by applying a principal component analysis (PCA) to the output data of the taste sensor. This analysis is one of the multivariate analyses used to reduce the dimensional space without losing information. Data points are expressed in the eight-dimensional space in the present case because of eight-channel outputs, and become expressed in the lower dimensional space by this method. The largest

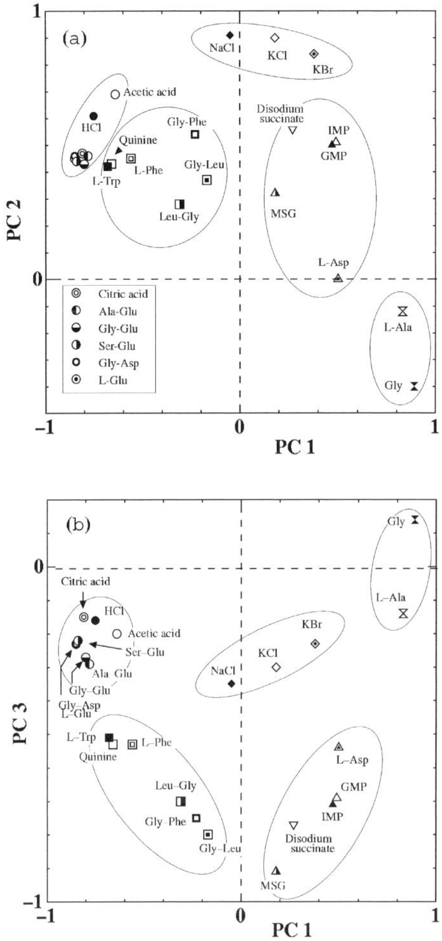


Fig. 11.4. Taste map [1] comprising typical chemical substances such as HCl, NaCl, quinine, amino acids and dipeptides: (a) PC1-PC2; (b) PC1-PC3 plane.

amount of information contained in the original data is transformed into the first principal component (PC1), and the second and third largest amounts of information are transformed into the second and third principal components (PC2 and PC3), respectively; in this way we can extract important information from the original data in order.

The mixed taste of amino acids can also be produced using the taste sensor. L-Methionine and L-valine taste both bitter and sweet simultaneously. The response pattern for L-methionine was found to be similar to that for a mixed solution composed of a sweet amino acid (L-alanine) and bitter amino acid (L-tryptophan) [7].

### 11.4.2 Discrimination of D-amino acids from L-amino acids

D-Amino acids are optical isomers of L-amino acids. These isomers are mirror images of each other and are known as enantiomers. They have identical physical properties, except for the direction of rotation of the plane of polarized light. Although they have very close physicochemical similarities, enantiomers behave differently to biological bodies; e.g. D-tryptophan and D-leucine are sweet whereas their L-forms taste bitter, and only monosodium L-glutamic acid tastes umami and enhances the flavor of food.

As mentioned above, amino acids possess different tastes and, from this point of view, isomers can be discriminated using the taste sensor [8]. Developing a sensor which can discriminate D-amino acids from L-amino acids will not only broaden their application, but also the development of such a technique, to discriminate optical isomers using a simple sensing method, will be a breakthrough in the chemical industry, medical science and pharmaceutics.

The membrane impedance change of optically active membranes due to interactions between amino acids and the membrane has been employed to discriminate D-amino acids from L-amino acids successfully [9,10]. From these results for D- and L-amino acids, it was suggested that a sufficient effect takes place in the membrane characteristics that the occurrence of an appropriate reaction such as diastereomer formation results in the membrane potential change depending on the chirality of the amino acid.

Diastereomers are also stereoisomers but are different from enantiomers because they have two or more chiral centers. They are not mirror images of each other, and two diastereomeric isomers have different physical properties.

A bimolecular combination of two chiral substances might lead to the formation of four diastereomers only if the interaction does not destroy the chiral centers. Such interactions that promote the formation of diastereomers are known as diastereomeric interactions and are most common when the interaction involves optically active organic bases and organic acids to yield diastereomeric salts.

The diastereomer formation interaction is used in the resolution of racemic modifications, i.e. the separation of racemic modifications into enantiomers. A racemic modification results in a mixture containing equal parts of the enantiomers and is optically inactive. Enantiomers making up the racemic modification have identical physical properties (e.g. boiling points, relative densities, refractive indexes, etc.) and hence cannot be separated by usual methods of fractional distillation or fractional crystallization. Addition of an optically active base, L-base, to a racemic acid (mixture of D- and L-acids) will result in a mixture of two salts, diastereomer salt D-L and diastereomer salt L-L, which are diastereomeric isomers of each other. They can be separated by usual resolution procedures, e.g. fractional crystallization and fractional distillation, since diastereomers have different physical properties.

The scheme in Fig. 11.5 shows that the membrane is prepared in such a way that its interaction with both D- and L-amino acids resembles the diastereomer formation reactions. When enantiomeric membranes reside in an optically active environment, homochiral and heterochiral diastereomeric interactions are created depending on the optical activity of the membrane and the environment. These interactions might differ sufficiently in the arrangement of the molecules of the membrane surface to permit the discrimination of optical substances due to the changes in the membrane characteristics such as membrane potential and membrane impedance.

Because D- and L-amino acids are optical isomers, we used chiral substances to manufacture the membrane. This allows both amino acids

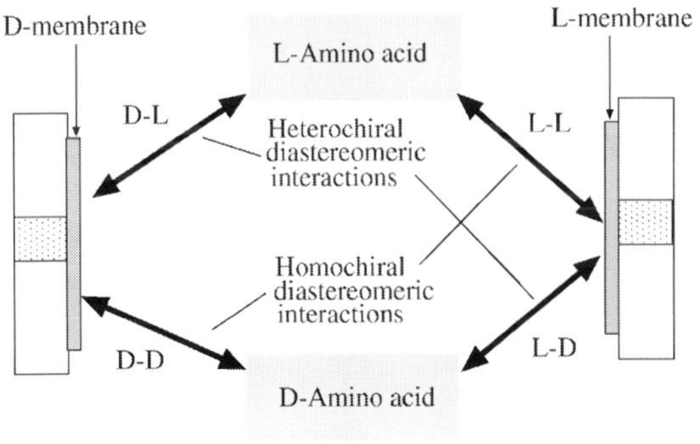


Fig. 11.5. Homochiral and heterochiral diastereomeric interactions depend on the optical relationship between the amino acid and the membrane [8].

to alter the orientation of molecules at the membrane surface, consequently resulting in the alteration of the membrane's electric potential. Two different types of optically active membranes were prepared, and each membrane comprised an optically active reagent. A chiral alkaloid, quinine (L), was used as the chiral reagent and was dissolved in PVC and THF. Quinidine (D), which is an optical isomer of quinine, was used for comparison. Finally, the plasticizers (or lipids) were mixed with the PVC–optically active reagent–THF mixture resulting in the membrane solution. The two plasticizers (or lipids) used were TOMA, which gives rise to a positively charged membrane, and DOPP, which gives a neutral membrane.

The response potential of the quinine-DOPP membrane for L- and D-aspartic acids showed sharp increases at low concentrations as shown in Fig. 11.6(a) [8]. Figure 11.6(b) shows that L-aspartic acid decreased the electric potential of the quinine-TOMA membrane at a rate higher than that of D-aspartic acid. In a similar way, L-amino acids such as L-tryptophan and L-glutamic acid were distinguished from their D-forms.

Although the quinine membrane could discriminate D-amino acids from L-amino acids, the quinidine membrane failed to distinguish

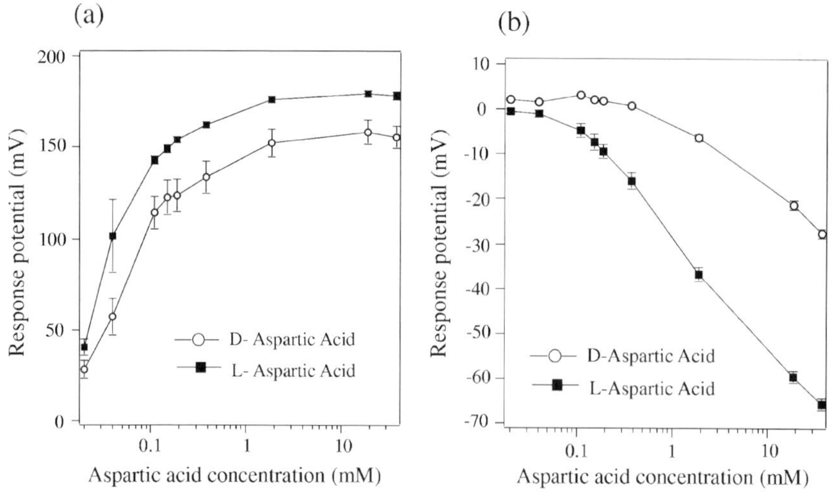


Fig. 11.6. Response potentials for aspartic acid: (a) quinine-DOPP membrane; and (b) quinine-TOMA membrane [8].

the amino acids. This confirms the fact that the potential change was attributed to the diastereomeric interactions, because quinine is used in many resolutions of racemic modifications while quinidine is rarely used.

## 11.5 QUANTIFICATION OF THE TASTE OF FOODS

The present sensor could easily discriminate several kinds of drinks such as coffee, beer and ionic drinks. The response was so fast that it occurred within 1 s, as soon as the electrode was immersed in the sample, and then the response curve of each channel was almost flat for the measuring time of 30 s [3].

Figure 11.7 shows the result of PCA applied to the response patterns for 41 brands of beer [1]. Comparison with the human taste sense indicated that PC1 corresponds to 'rich taste' and 'light taste', PC2 'sharp taste' and 'mild taste'. One of the largest merits of the taste sensor is the fact that any pretreatment of foods is unnecessary. The taste can be measured immediately after drink is poured into a cup.

Saltiness elicited by salt is one of the basic tastes. However, components within salt available on the market differ depending on the manufacturing processes and this will affect its taste. Salt manufactured by an ion-exchange membrane process is composed of more than 99% pure sodium chloride, while bay salt contains traces of coexisting components. Despite reports on sensory evaluations, the differences in taste are still uncertain because of a small amount of coexisting components. The interaction between salt and trace coexisting components has been studied [11]; the bittern (*nigari* in Japanese) was evaluated objectively and quantitatively using the taste sensor.

The model samples analyzed were composed of sodium chloride and trace coexisting components such as magnesium sulfate, magnesium chloride, calcium chloride and sodium chloride. The taste sensor clearly discriminated each sample according to the response patterns. Based on the sensor outputs, the taste by means of the combination of PCA and ionic strength was evaluated. The results showed the taste of salt with *nigari* has a correlation with a modified ionic strength, which takes into account the binding effect of $Ca^{2+}$ on lipid membranes.

As a further example, the taste of green tea was studied [12]. Green tea has many components that make up its taste. The interaction

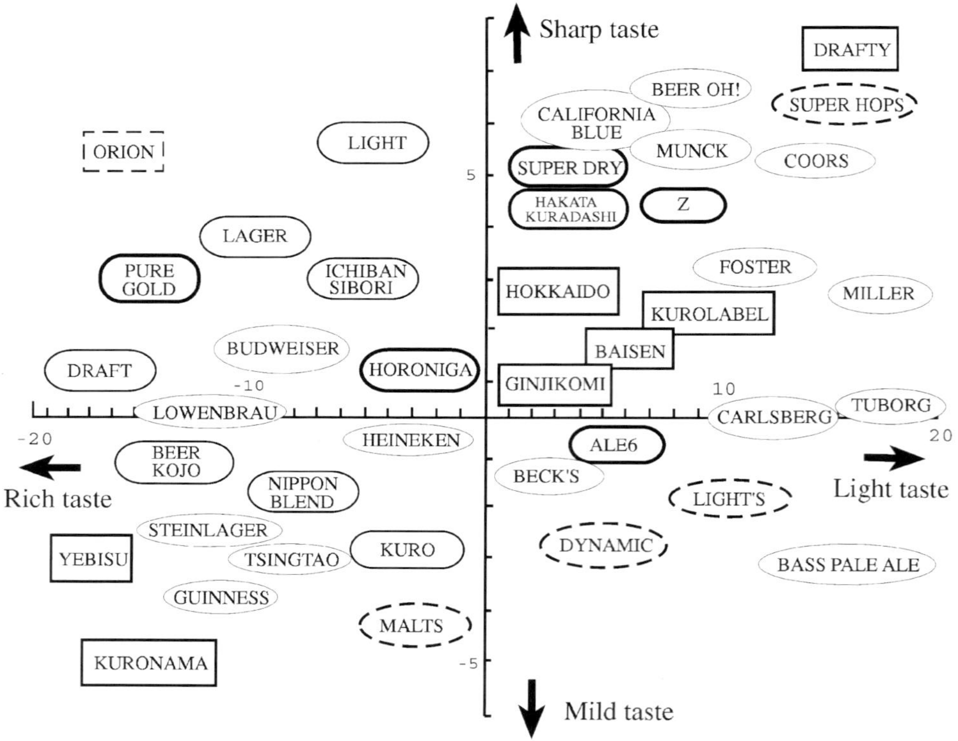


Fig. 11.7. Taste map obtained using PCA [1]. Symbols, enclosing the brand of beer, such as dashed ellipse and square indicate beer produced by different companies in Japan and the continuous ellipse indicates beer from other countries.

between the taste substances and the cells on our tongue occurs and then we recognize the taste of the green tea.

The effect of temperature on taste was studied. The samples were of the same brand and each was extracted at different temperatures: 60, 80 and 100°C. Figure 11.8 shows that the electric potentials of channels 4–8 (shown in Table 11.1) decrease with increasing temperature of extraction. Channel 4 made up of the 9:1 membrane is a little negatively charged, channel 5 neutral, channel 6 a little positively charged, with channels 7 and 8 being positively charged. The 5:5 membrane generally responds to umami taste very well, as shown in Fig. 11.2. The response pattern in Fig. 11.8 suggests that negatively charged substances are more received by the lipid membranes with increasing extraction temperature of tea. One of these substances is tannic acid, which tastes astringent. The response patterns show that astringency increases with increasing extraction temperature. In human sensory evaluations, we also obtained the same result that the green tea extracted at 100°C was more astringent than that at 60°C.

Tannic acid of 0.1, 0.3 and 1.0 mM was added to the green tea extracted at 60°C. The response pattern approached the pattern of the green tea extracted at 100°C with an increasing concentration of tannic acid. Furthermore, human sensory examinations could not distinguish between the green tea extracted at 100°C and that at 60°C to which 1.0 mM tannic acid had been added.

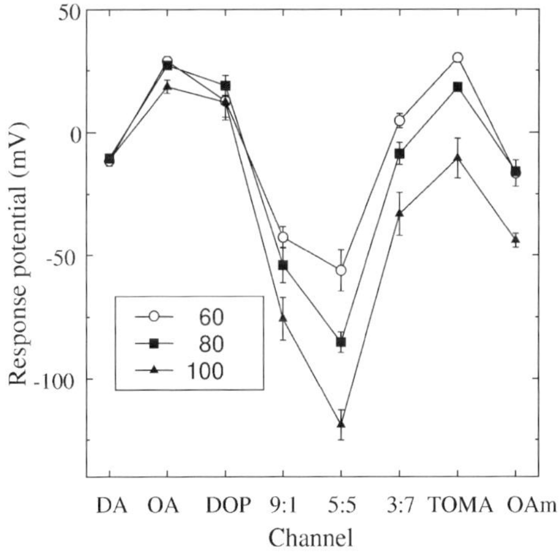


Fig. 11.8. Response to green tea by changing extraction temperature.

Both results imply that green tea extracted at high temperatures tastes astringent, and also the taste sensor is able to reproduce the human sense.

## 11.6 INTERACTION BETWEEN TASTE QUALITIES

### 11.6.1 Suppression of bitterness due to phospholipids

It is important especially for pharmaceutical and food sciences to express the extent of bitterness in their products, e.g. in the case of developing syrups. To date, the main method of measurement has been sensory evaluation by humans tasting bitterness, while conventional chemical analyses are subsidiary methods. Therefore, widely used taste-sensing devices that can detect bitterness have been desired for a long time.

A lipoprotein (PA-LG) made up of phosphatidic acid (PA) and β-lactoglobulin (LG) can completely suppress bitterness without affecting other taste qualities [13]. A similar but weaker suppression effect was found using PA only; the addition of 1% PA to solutions containing bitter substances such as quinine decreased the bitterness to the extent that humans sensed no or only a very weak bitter taste.

This suppression effect was studied [14] using the taste sensor and a commercial bitter-masking substance (BMI-60) composed of phospholipids; its ingredients are 15–20% PA, 5% phosphatidylcholine, 40% phosphatidylinositol and 10–15% phosphatidylethanolamine. We call the commercial bitter-masking substance 'phospholipid' for convenience hereafter.

### 11.6.2 Scale of bitterness

Figure 11.9 shows the responses to 3 μM–3 mM of quinine hydrochloride solutions. As the quinine concentration increased, the response potentials of channels 1–3 increased markedly, whereas those of channels 5–7 decreased slightly. The increase in channels 1–3 was brought about by the adsorption of quinine hydrochloride by the negatively charged membrane with its hydrophobic part, thereby changing the membrane electric charge from negative to positive, as previously explained quantitatively [5]. The decrease in channels 5–7,

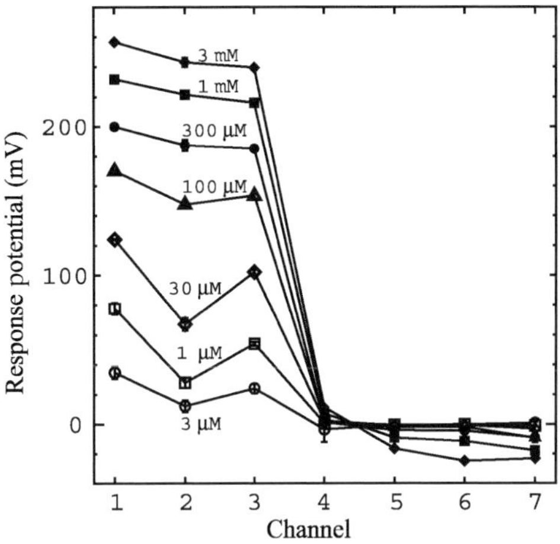


Fig. 11.9. Response to quinine hydrochloride of 3 μM–3 mM [14]. The responses in channels 1–3 markedly increase with increasing quinine concentration.

which comprise positively charged membranes, results from the electric screening effect of $Cl^-$ ions. Since most bitter substances have a hydrophobic part, like quinine, the behavior observed in channels 1–3 can be considered to reflect the characteristics of bitterness accurately. The result suggests that the response potential in channels 1–3 increases with increasing bitter intensity.

The result in Fig. 11.10 shows that the response to 1 mM quinine in channels 1–3 decreases with increasing phospholipid concentration. As shown in Fig. 11.9, the downward response in channels 1–3 means a decrease in bitter intensity. This result suggests that the increase of phospholipid concentration markedly decreases the bitter intensity of quinine hydrochloride. On the other hand, the responses to other taste substances such as 100 mM $MgCl_2$, 200 mM NaCl, 10 mM HCl, 300 mM sucrose and 5 mM MSG showed no change with increasing phospholipid concentration. This indicates that the tastes of these samples were not affected by the phospholipid. This is in agreement with the human sensory evaluations.

We applied PCA to these data in order to quantify the bitter intensity. Figure 11.11 shows the relationship between the PC1 and the quinine concentration from the data of quinine measurement (Fig. 11.9) in channels 1–3. The contribution rate, which means the relative magnitude of transformed information, of the original data to PC1 was 99.0%.

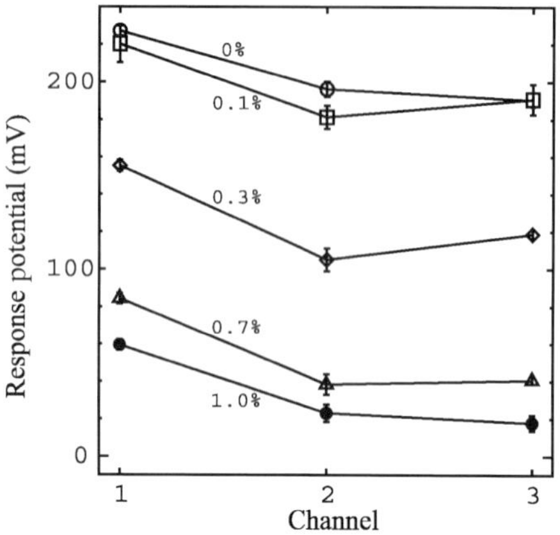


Fig. 11.10. Responses to the mixture of the phospholipid and 1 mM quinine hydrochloride [14]. It is shown that the response to quinine hydrochloride is effectively reduced by the phospholipid.

This contribution rate implies that PC1 characterizes the response to quinine in channels 1–3 and we can discuss these data using only PC1:

$$\mathrm{PC1} = \sum_{i=1}^{3} a_i (v_i - v_i') \qquad (11.2)$$

Equation (11.2) shows the expression of PC1, where $a_i$ is called the factor loading, $v_i$ is the response potential of channel $i$, and $v_i'$ is the response potential of channel $i$ averaged over the measured samples.

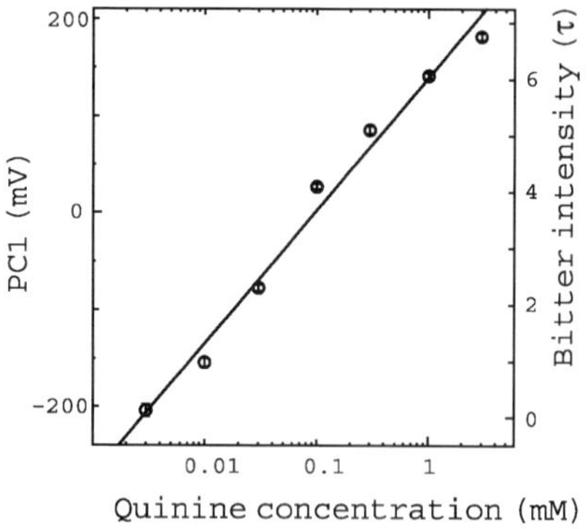


Fig. 11.11. Relationship between PC1 and quinine concentration [14]. A straight line is drawn using the method of least squares and the correlation coefficient is 0.9952.

In the present case, $a_i$ and $v_i'$ were determined as $a_i$ (0.5545, 0.6207, 0.5543) and $v_i'$ (144.8, 110.0, 108.5).

In Fig. 11.11, a straight line is drawn using the method of least squares, shown by:

$$\mathrm{PC1} = 138.9 \log C + 140.71 \tag{11.3}$$

with $C$ denoting quinine hydrochloride concentration (mM). It is noticeable that this relation agrees with the well-known Weber–Fechner's law of human sensory testing [15]. This law states that a sensation is proportional to the logarithm of stimulus intensity. The result in Fig. 11.11 suggests that PC1 can be considered to express the strength of bitterness.

Now, we use a quantitative scale in order to express the bitter intensity. Among some scales, the $\tau$ scale was employed, which expresses a relationship between the taste substance concentration and the taste intensity $\tau$ [16]:

$$\tau = 2.35 \log C + 6.01 \tag{11.4}$$

The $\tau$ scale has been produced using human sensory tests, and is a quantitative measure of the subjective feelings of humans. The $\tau$ value indicates the intensity of bitterness. For example, when the $\tau$ values of solutions are 0 and 6.01, the bitter intensity of these solutions are equivalent to 2.7 μM (human threshold) and 1 mM quinine hydrochloride solution, respectively. Since both PC1 and $\tau$ are proportional to the quinine hydrochloride concentration $C$ on a logarithmic scale, a relationship between $\tau$ and PC1 can be given from Eqs. (11.3) and (11.4) as:

$$\tau = 0.0169\mathrm{PC1} + 3.63$$

This equation shows that the bitter intensity can be estimated if PC1 is obtained. Figure 11.12 shows the $\tau$ (bitter intensity) of quinine solutions of two different concentrations of 0.1 and 1 mM as a function of phospholipid concentration. It is shown that the reduction of bitter intensity occurs at 0.1 and 1 mM quinine, whereas the reduction rate is much larger at 0.1 mM than at 1 mM. The bitter intensity of 0.1 mM quinine is effectively decreased by addition of 1% phospholipid to the level of no taste. The agreement between results of the taste sensor and the human sensory evaluations is fairly good.

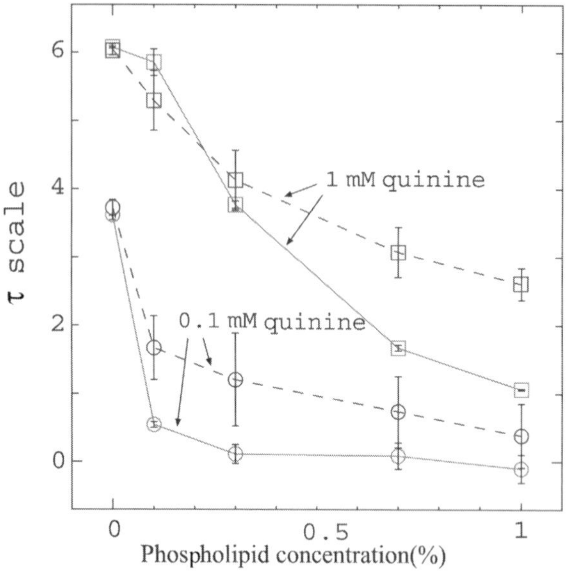


Fig. 11.12. Relationship between phospholipid concentration and bitter intensity expressed by the $\tau$ scale. The results of taste sensor and human sensory evaluations are denoted by solid and dashed lines, respectively.

### 11.6.3 Suppression of bitterness due to taste substances

It is also well known that bitterness is suppressed by coexistence of sweet and salty substances such as sucrose and NaCl, respectively. The same method as described above was applied to this situation. As a result, it was shown that bitterness is suppressed to a small extent by sugar [17] and largely also by NaCl [18].

In a similar way, the taste of commercial drug substances was studied [17]. Consequently, the decrease of the bitter strength of drug substance with increasing sucrose concentration was quantified.

We attempted also to quantify the suppression of bitterness due to saltiness by using a surface plasmon resonance (SPR) method [18,19]. The receptor membrane used was the LB membrane, which is constructed from six-layered lipid DHP (dihexadecyl phosphate). The SPR method can detect the change of adsorptive amount of taste substances to lipid membranes with the change of resonance angle. It was shown that the change of resonance angle clearly decreases with quinine plus 100 mM NaCl, compared with that of pure quinine. It can be assumed that NaCl interrupted the adsorption of quinine to the LB membrane. These facts demonstrate that the SPR measurement as well as the taste sensor can measure the suppression effect between bitter and salty substances.

In this way, the taste sensor can reproduce suppression of bitterness as felt by humans and also quantify the bitter strength. The present method using the taste sensor can be expected to provide a new automated method to measure the strength of the bitterness of a drug substance in order to replace human sensory evaluations. In addition, the study of taste interaction using the taste sensor will contribute to clarification of reception mechanisms in the gustatory system.

## 11.7 DETECTION OF WINE FLAVOR USING TASTE SENSOR AND ELECTRONIC NOSE

Wine has both taste and odor qualities due to different aromatic molecules in the liquid and vapor phases. The average wine contains about 80–85% water and over 500 different substances, some of which are very important to the flavor of wine despite their low concentrations. Wine is a suitable candidate for testing the performance of the sensory fusion of taste and odor sensors (i.e. electronic nose). Wine flavor was studied using a combined taste sensor and electronic nose [20].

The electronic-nose array is composed of four different conducting polymers. The monomer (25 mg) is dissolved in trichloroethylene (2 ml) and the oxidizing salt previously dissolved in acetonitrile is added in a dropwise manner. The polymerization process then occurs and the resulting solution is sprayed onto an alumina substrate where four interdigitated electrodes were previously evaporated. After evaporating the solvent the conducting polymer is connected with the four electrodes. Four different sensing elements were obtained by combining two different monomers and two oxidizing salts.

The electric resistance measured at the inner electrode varies when volatile molecules are adsorbed at the surface of the polymer film. The average sensitivity expressed as the ratio of the resistance change to the base resistance value was almost always less than 2% in the case of the elements used in this work for wine sensing.

Eight- and four-dimensional data arrays were obtained from measurements using the taste sensor and electronic nose, respectively.

A normalization procedure is necessary, because the output quantity is different between these two types of sensors.

After either set of data was normalized, a 12-dimensional data array was obtained for each measurement on four different wines. The combination of the two sets of data has led to a new representation of the samples in 12-dimensional space, which simultaneously contains information from the taste sensor and electronic nose concerning the sample measured.

The PCA was performed on this set of data and the results are shown in Fig. 11.13. The discrimination among the four wines is satisfactory. The relative positioning of the clusters in the principal component plane was similar to the case of the electronic nose and the relatively high distance between clusters of the red (2 and 5) and white (1 and 4) wines was successfully achieved by the contribution of the taste sensor.

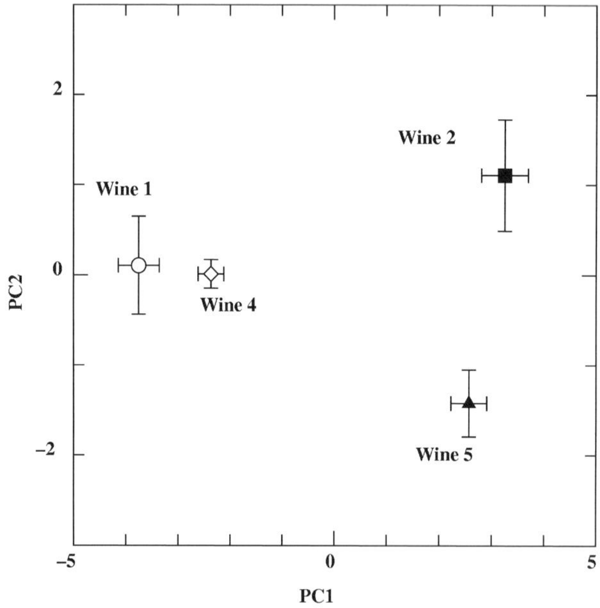


Fig. 11.13. Results of the PCA applied to the combination of the data set from the taste sensor with the data set from the electronic nose [20]. Wine 1 (white): Est! Est! Est! di Montefiascone 1995, Italy; wine 2 (red): Bon Marche' Mercian, Japan; wine 4 (white): Chablis 1994, France; and wine 5 (red): Rosso di Montalcino, Fattoria dei Barbi 1994, Italy.

The effect of the aging process after opening the bottle on the quality of wine was also studied. It was found that the system can discriminate among differently aged samples of the same red wine.

The sensor fusion is very effective because the information provided by the different arrays are to some extent independent of each other; they account for different characteristics of the wines themselves. Conventional multiple sensor arrays have several sensing elements derived from the same technology, e.g. conducting polymer sensor array, metal oxide sensor array and lipid membrane sensor array. These arrays show broad sensitivities to certain groups of substances, but on the other hand, are not sensitive to other compounds. If different sensor technologies are simultaneously applied, provided that the information from the different sources is independent, it is worthwhile combining them to obtain a broader viewpoint of the samples measured.

## 11.8 PERSPECTIVE

The taste sensor is based on a new concept of global selectivity. What is important in recognition of taste is not discrimination of minute differences in molecular structures but the transformation of molecular information contained in interactions with biological membranes into several kinds of detection groups, i.e. taste intensities and qualities. This is a high-level function, where intelligent sensing is required. In this meaning, the taste sensor is essentially an intelligent sensor able to reproduce the gustatory sense, which is a complicated, comprehensive sense of humans.

The term 'taste sensor' was coined in Japan, while 'electronic tongue' first appeared in Europe. There is some difference between these two terms. The purpose of the taste sensor is to measure the taste felt by humans, while that of the electronic tongue is aimed at discrimination of liquids. The taste sensor has a high sensitivity to discriminate drinks, which were produced in different factories on different days. The discrimination ability required is natural, because sensors should be superior to humans. Based on this requirement, the next important step is reproducing the gustatory sense, which is an integrated, complicated feeling.

Further, the concept of the taste sensor is not the same as that of the electronic nose, which uses a low selective sensor and hence utilizes multivariate signal processing. However, the receptor membranes of the taste sensor have characteristics of responding mainly to one or two taste qualities among five tastes. Thus, signal processing using a computer allows us to see the taste property at once, because there are only five basic taste qualities, such as saltiness, sourness and bitterness. This fact is very different when considering the electronic nose for the olfactory sense, where there is no basic smell. One or several kinds of gas molecules produce smell. There are many kinds of smell. In this respect, the discrimination is more important as a main analytical target than the quantification of each smell. Quantification as well as discrimination is possible using the taste sensor for the gustatory sense.

The taste sensor has so far been applied to many different foodstuffs such as beer, juice, milk, whiskey, sake, coffee, tea, mineral water, rice, vegetables and ionic drinks [1]. Another application of the taste sensor is in the environmental measurement of water quality. Detection of some toxic substances in factory drains is a time-consuming process because many different substances need to be analyzed. However, it is important to check quickly the safety of drinking water. The taste sensor is adequate for this purpose, because it can respond simultaneously with high sensitivity to many chemical substances. It is possible only to judge whether drinking water is safe or not if the safety range given by the response electric pattern of the sensor output is known.

For example, the taste sensor was applied to measurement of contamination of factory drains [21]. Many pollutants such as $CN^-$, $Fe^{3+}$ and $Cu^{2+}$ could be measured in a few minutes with detection limits lower than regulation for a drain. Cyanide was detected selectively using multiple regression analysis.

A combination of different sensors, such as the taste sensor and electronic nose, may drastically increase obtained information. Wine flavor was successfully discriminated using the taste sensor and the electronic nose that utilizes conducting polymers.

The sense of taste depends largely on subjective factors of human feelings. If we compare the standard index measured by means of the taste sensor with that of human sensory evaluation, we will be able to assess taste objectively. Moreover, this study will contribute to clarification of the mechanism of information processing of taste in the brain as well as the reception at taste cells.

# REFERENCES


1 K. Toko, *Biomimetic Sensor Technology*. Cambridge University Press, Cambridge, 2000.
2 K. Hayashi, M. Yamanaka, K. Toko and K. Yamafuji, *Sens. Actuat. B*, 2 (1990) 205.
3 K. Toko, *Sens. Actuat. B*, 64 (2000) 205.
4 K. Toko, *Meas. Sci. Technol.*, 9 (1998) 1919.
5 K. Oohira and K. Toko, *Biophys. Chem.*, 61 (1996) 29.
6 T. Nagamori and K. Toko, *Dig. Tech. Papers Transducers' 99*, 1999, p. 62.
7 K. Toko and T. Nagamori, *Trans. IEE Jpn.*, 119-E (1999) 528.
8 H. Chibvongodze, K. Hayashi and K. Toko, *Sens. Materials*, 13 (2001) 99.
9 H. Chibvongodze, K. Hayashi and K. Toko, *IEICE Trans. Electron.*, E83-C (2000) 1028.
10 H. Chibvongodze, K. Hayashi and K. Toko, *Transducers' 01 Eurosensors, The 11th. Int. Conf. Solid-State Sensors and Actuators*, 2001, p. 398.
11 M. Habara and K. Toko, *IEICE Trans. Electron.*, E83-C (2000) 1040.
12 T. Adachi, M. Yahiro, H. Nagasue and K. Toko, *Proc. 2001 Int. Conf. O-CHA Culture and Science*, 2001, p. 130.
13 Y. Katsuragi and K. Kurihara, *Nature*, 365 (1993) 211.
14 S. Takagi, K. Toko, K. Wada and T. Ohki, *J. Pharm. Sci.*, 90 (2001) 2042.
15 C. Pfaffmann, *Handbook of Physiology*. American Physiological Society, Washington, DC, 1959, p. 507.
16 T. Indow, *Percept. Psychophys.*, 5 (1969) 347.
17 S. Takagi, K. Toko, K. Wada, H. Yamada and K. Toyoshima, *J. Pharm. Sci.*, 87 (1998) 552.
18 T. Adachi and K. Toko, *Abstract Book 8th Int. Meeting Chemical Sensors*, 2001, p. 531.
19 R. Yasuda, K. Toko, H. Akiyama, T. Kaneishi, T. Matsuno, S. Ezaki and K. Yamafuji, *Trans. IEICE Jap.*, J80-CII (1997) 1 [in Japanese]; *Electron. Commun. Jap., Part 2*, 80 (1997) 1 [English translation].
20 S. Baldacci, T. Matsuno, K. Toko, R. Stella and D. De Rossi, *Sens. Materials*, 10 (1998) 185.
21 A. Taniguchi, Y. Naito, N. Maeda, Y. Sato and H. Ikezaki, *Sens. Materials*, 11 (1999) 437.

*Chapter 12*

# *Application of electronic nose technology for monitoring water and wastewater*

***Richard Stuetz****

## 12.1 INTRODUCTION

The real-time monitoring of water and wastewater quality is an unresolved problem for water companies. At present, there is infrequent continuous monitoring at most water and wastewater treatment works. This is most likely due to the high capital and operational cost associated with on-line instrumentation, and also because of the uncertainty regarding the reproducibility of current techniques used to monitor water and wastewater quality.

Conventional laboratory methods used for the analysis of water and wastewater quality are very accurate. These tests are based mainly on sample collection and retrospective analysis, which may limit their long-term application to continuous water and wastewater monitoring. However, some examples of using UV/Vis absorption [1], biosensors [2], fluorescence [3] and respiration kinetics [4] have demonstrated how chemical, biochemical and spectroscopic techniques can be integrated into a monitoring 'system' for detecting changes in water and wastewater quality. However, the failure to apply these instruments to continuous monitoring has been the poor integration and grouping of

* Tel.: +44-1234-754841; fax: +44-1234-751671. *E-mail address:* r.stuetz@cranfield.ac.uk.



ISSN: 0166-526X

these devices into an analytical system. This is because most systems either require the measurement device to have constant contact with the liquid environment or need the incorporation of a sampling handling sequence to introduce the sample, resulting in instrument and collection system fouling, respectively. This causes a loss of sensitivity and reproducibility, the need for frequent cleaning and re-calibration of the sensing instrument [5]. There is therefore considerable impetus to develop 'integrated analytical systems' for monitoring water and wastewater quality that is reproducible, rapid and non-invasive.

The recent availability of commercial 'electronic noses' for detecting and characterisation of headspace odours could offer a rapid and relatively simple technique for monitoring changes in water and wastewater quality, as odours have long been used by bioprocess operators to identify abnormalities in biological processes [6] and as a source of consumer complaint for water supply companies [7]. In this chapter we review electronic nose sensor technology and instrumentation, data processing techniques and discuss the current status of integrating this non-specific sensing technology into water and wastewater monitoring systems.

## 12.2 ELECTRONIC NOSE TECHNOLOGY

Electronic noses are analytical instruments that can characterise an odour without reference to its chemical composition. A typical system consists of a sensor array, with appropriate signal conditioning connected to a pattern recognition system for data analysis (Fig. 12.1). Several different system designs are used in today's commercial

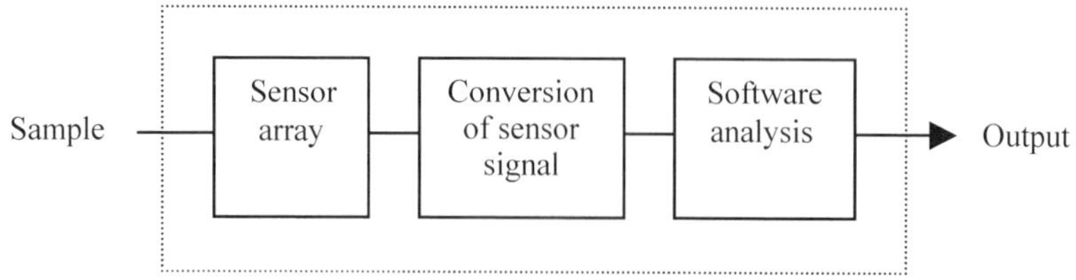


Fig. 12.1. Principal components of an electronic nose system [8].

instruments. These systems are designed to introduce the sample headspace into the sensor array chamber in a reproducible manner. An example of a commercial electronic nose is shown in Fig. 12.2.

### 12.2.1 Sensor types

A variety of sensor technologies is employed in electronic nose systems [7], the most common of these being metal oxide sensors (MOS), conducting polymers (CP), surface acoustic wave devices (SAW) and quartz crystal microbalances (QCM), all of which are shown in Fig. 12.3. The measurement principles for the different sensor types are summarised in Table 12.1. The use of an array of non-specific sensors allows for responses from many thousands of chemical species owing to the broad selectivity of the different sensor surfaces [10]. The relative responses between the sensors can be used to produce a unique odour profile that is analogous to the human olfactory system [11]. Figure 12.4 shows an example of an odour profile for wastewater from an electronic nose consisting of 12 conducting polymers [12]. The resultant odour-specific response pattern or fingerprint can then be further analysed using pattern recognition techniques and/or neural network algorithms.

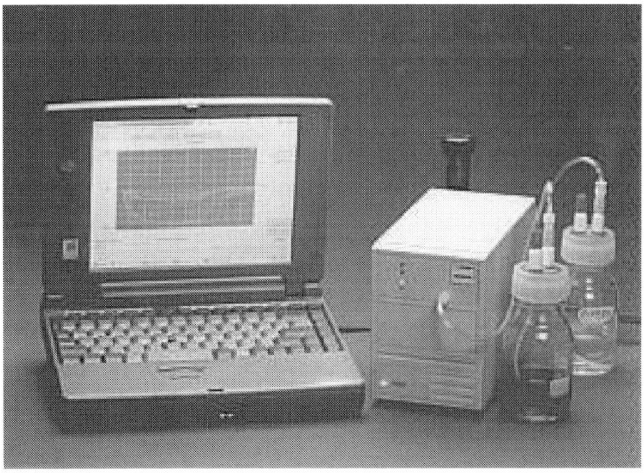

Fig. 12.2. Photograph of the BH114 electronic nose system (Courtesy of Bloodhound Sensors Ltd, UK).

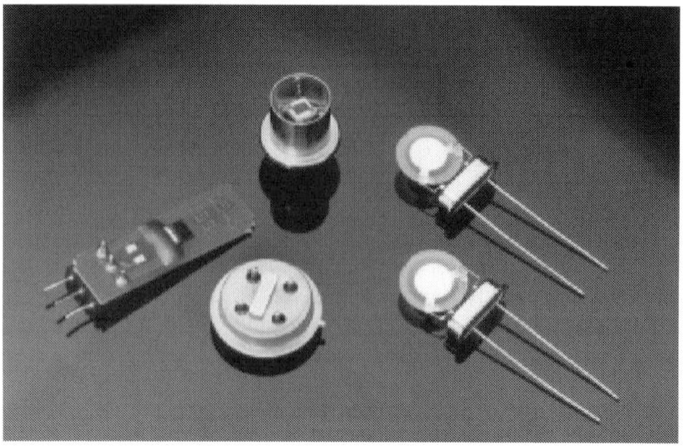

Fig. 12.3. Photograph of different sensor types used in electronic nose systems (Courtesy of Marconi Applied Technologies, UK).

### 12.2.2 Analysis of electronic nose data

The analysis of electronic nose data is dependent on the amount and nature of information available and the type of information required from the analysis (i.e. quantitative or qualitative). Output can be displayed using a variety of graphical formats that allows comparisons between samples or averaged data over a number of analysers [13]. However, to cope with large numbers of samples and of variables (i.e. number of sensors), pattern recognition techniques are often employed to process the sensor array data [14,15]. These techniques allow underlying relationships between one set of independent variables (i.e. output from an array of $n$ sensors) and another set of dependent variables (i.e. odour class or component concentrations) to be determined and can be divided into two basic methods: unsupervised and supervised [16]. In an unsupervised technique, one seeks to discriminate between unknown samples by enhancing the differences between the associated input vectors, whereas in a supervised technique unknown samples are analysed using relationships that have been learnt during an earlier calibration procedure, learning or training stage [16]. A summary of the main types of data processing techniques used for the analyses of electronic nose data is shown in Table 12.2.

TABLE 12.1

Common types of sensors used in electronic nose systems [9]

| Sensor type | Mode of activity | Comments |
| --- | --- | --- |
| Conducting polymer sensors | In the presence of a gas species a change in voltage across polymers such as polyaniline, polypyrrole and polythiophene can be measured | Selectivity is achieved by controlling surface functional groups or by varying anion chemistry during growth. Reproducibility good return to baseline resistances in short times |
| Metal oxide sensors | Metal oxide sensor passes an electrical current causing oxidation of gas molecules via electron transfer from the gas to the metal oxide leading to a change of resistance | Less selective than other sensor types. Can be subjected to poisoning. Response can be affected by presence of solvents |
| Quartz crystal microbalances | Measure change in frequency of oscillation of a quartz crystal when a gaseous species is adsorbed | Problems of reproducibility in commercial production of sensors |
| Surface acoustic wave sensors | Similar to quartz crystal microbalances but operate at much higher frequencies | Can achieve good sensitivity. Problems of reproducibility in sensor production |
| Fibre optic sensors | Use fluorescence measurements from photodeposited polymer/fluorescent dyes on bundles of fibre optics | Provide large quantities of data. Not available in commercial instruments |

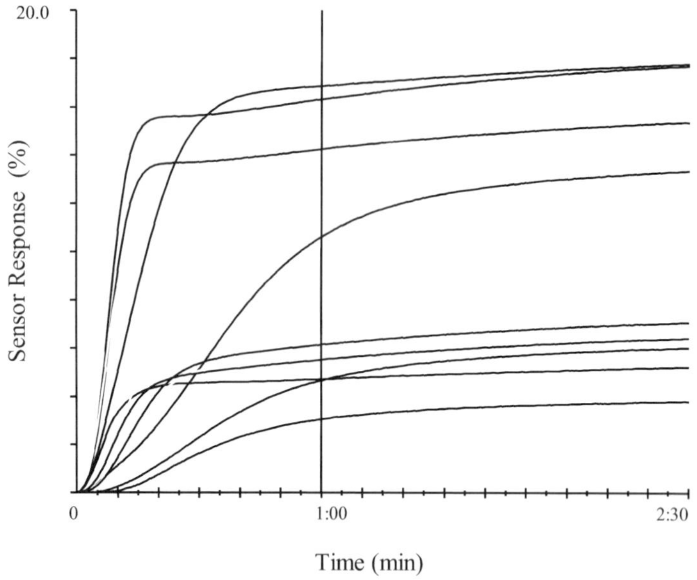


Fig. 12.4. Response pattern of an electronic nose (eNOSE 4000, Neotronics Scientific, UK) consisting of 12 conducting polymer sensors, showing the resultant odour profile over a 2.5 min acquisition [12].

*Statistical analysis*

Multivariate techniques have been used to reduce the dimensionality of the electronic nose data, so that relationships between the observations can be explored using one or two dimensions [15]. The most common *qualitative* analysis techniques are principal component analysis (PCA), cluster analysis (CA), multiple discriminant analysis (MDA) and canonical correlation analysis (CCA) (Table 12.2). More extensive reviews of the application of multivariate statistics to electronic nose data can be found in Gardner and Barlett [17,18] and Gardner and Hines [19].

*Artificial neural networks*

Artificial neural networks (ANNs) are learning architectures that consists of parallel-interconnected processing elements called *neurones*, which are based loosely upon a physical model of the biological nervous system [16]. Each processing element is weighted mathematically—these weights are determined through a process of training or

TABLE 12.2

Common techniques used for the analyses of electronic nose data

| Techniques | Supervised | Linear | Comments |
|---|---|---|---|
| Principal component analysis (PCA) | No | Yes | Used to reduce a large number of variables to a smaller number of components and at the same time extract maximum variance in the data |
| Cluster analyses | No | Yes | Used to find natural grouping or cluster of individual observation within a data-set |
| Multiple discriminant analysis | Yes | Yes | Used to understand existing groups and to create new groupings and classification based on known properties |
| Canonical correlation | Yes | Yes | Explores linear relationships between dependent sets of variables and independent sets of variables by maximising the correlation in the data |
| Artificial neural networks | Yes[a] | No | A learning architecture that consists of parallel interconnected processing elements which are based upon a physical model of the biological nervous systems |

[a]The most common technique 'back-propagation' has a supervised training stage; however, other algorithms are unsupervised

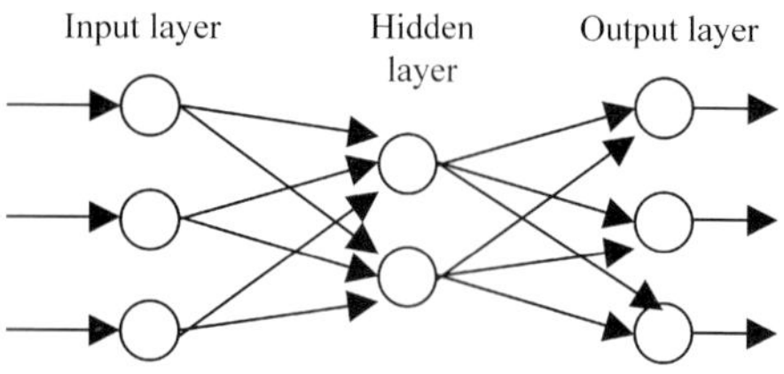


Fig. 12.5. Architecture of a typical multilayer perception.

learning [16]. The structure and learning sequence of a neural network are used to define the ANNs. The most common ANN to be applied to the analysis of electronic nose data is a multilayer network called the multilayer perceptron. The multilayer perceptron consists of three types of layer: the input layer, a number of hidden layers and the output layer (Fig. 12.5). A variety of algorithms has been used to train multilayer networks, the back-propagation technique being the most common [18]. This technique comprises a learning stage employed to train the network, followed by a recall phase or prediction during which the trained data set is used to classify unknown input data [19].

## 12.3 ELECTRONIC NOSE INSTRUMENTATION

### 12.3.1 Sensor array components

The principal components of an electronic nose instrument are its sample delivery system and the sensor array module. The delivery system is designed to transfer sample headspace from sample material to the sensor array chamber in a reproducible manner. This can be achieved by either generating a headspace above a sample and then introducing the sensor arrays to the headspace, known as *static sampling*, or pumping the sample headspace from a sample vessel to the sensor array, known as *dynamic sampling*. The different performance characteristics of these sample delivery systems have led to the development of different electronic nose instruments for different fields of application [16].

The first electronic nose systems were designed for laboratory use and employed manually transferring a sample (which had been allowed to reach constant temperature) from a sample vessel to the sensor array chamber using a glass syringe [16]. These techniques had poor reproducibility, were labour intensive and time consuming; however, the system was simple and inexpensive. To improve repeatability of this type of sampling system, commercial instruments were manufactured whereby a sensor array was mounted above a sample vessel which could be automatically raised and lowered into the sample vessel, thereby reducing any system measurement error [13]. Figure 12.6 shows a photograph of such an instrument (eNOSE 4000) previously manufactured by Neotronics Scientific (now Marconi Applied Technologies, UK).

These automated systems can reduce variation in the delivery of the sample; however, the overall sampling rate is still dependent on the slow reaction kinetics during sample equilibration [16].

The second type of sample delivery system is based on pumping or purging the sample headspace into the sensor chambers [20]. This is achieved by stripping the volatiles from the sample material using a carrier gas or gases (i.e. zero grade air or $N_2$) that is then transferred

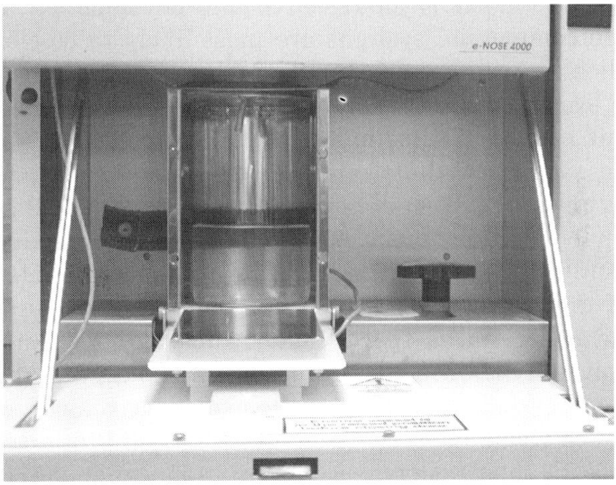


Fig. 12.6. Photograph of an eNOSE 4000 instrument (Courtesy of Marconi Applied Technologies, UK).

to the sensor chamber. Dynamic sampling systems have a number of advantages over static sampling systems: there is no dilution effect and it is possible to use different levels of humidity and different temperatures to deliver the headspace to the sensor chamber with a high level of accuracy [16]. The increased complexity in the operation of these systems does result in higher capital costs. It should be noted that a dynamic headspace may not be the same as a static headspace—either in composition or concentration—due to the mechanism of headspace generation [16].

### 12.3.2 Commercial systems

A number of electronic nose systems is available from commercial manufacturers throughout the world (Table 12.3). These systems consist of modular plug-in sensor arrays (which makes the instrument very flexible), provide precise temperature control of the sample delivery system and the sensor chamber, and include specific software for data acquisition and display and have direct links to spreadsheet packages (such as Microsoft Excel) and statistical packages (such as Statistica) to enable more detailed data analysis.

Most electronic nose instruments sold to date have been laboratory-based systems (Fig. 12.7) for research and development applications [21]. Future commercial systems are most likely to be classified into three application groups: (i) complex laboratory-based instruments; (ii) on-line systems; and (iii) portable devices for field measurements [21]. Recent commercial examples of these application-specific electronic nose systems include on-line systems for continuous process monitoring (Fig. 12.8) and portable devices for field monitoring (Fig. 12.9). In addition to the development of more flexible systems, several manufacturers have also incorporated new sensor devices or a number of different sensor types into the same instrument. These include the use of mass spectrometry and solid-state spectrometry, which involves bypassing the traditional sample preparation stages and introducing whole samples into the mass spectrometer or the solid-state sensor to give a mass spectrometry fingerprint or a spectroscopic trace [21]. With the development of more application-specific systems and complexity sensor devices, more sophisticated pre-processing and dynamic data analysis protocols (such as adaptive ANNs and fuzzy

TABLE 12.3

List of electronic nose manufacturers

| Company and location | Models |
|---|---|
| Alpha MOS, France (www.alpha-mos.com) | Fox 2000, 3000, 4000, 5000, AlphaKronos, AlphaPrometheus, AlphaCent |
| Bloodhound Sensors, UK (www.bloodhound.co.uk/bloodhound) | Bloodhound BH114 |
| Cyrano Sciences, USA (www.cyranosciences.com) | Cyranose 320 |
| Etherdata, Iceland (www.etherdata.is) | FreshSense |
| HKR Sensorsysteme, Germany (www.home.t-online.de) | QMB6/HS40XL |
| Hewlett Packard, USA (www.hp.com) | HP4440A |
| Lennartz Electronic, Germany (www.lennartz-electronic.de) | MOSES II |
| Marconi Applied Technologies, UK (www.marconitech.com) | eNOSE 5000, ProSAT |
| MoTech Sensoric, Germany (www.motech.de) | VOCmeter, VOCcheck |
| Nordic Sensor Technologies, Sweden (www.nordicsensor.se) | NST 3210, NST 3220, NST 3220A |
| Osmetech, UK (www.osmetech.co.uk) | MultiSampler-SP |
| RST Rostock, Germany (www.rst-rostock.de) | Sam |
| Smart Nose, Switzerland (www.smartnose.com) | Smart NOSE-300 |
| WMA Airsense, Germany (www.airsense.com) | PEN |

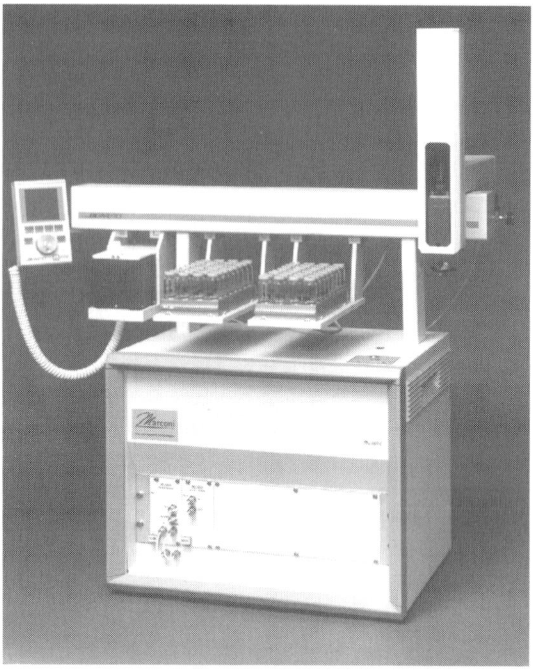

Fig. 12.7. Photograph of an eNOSE 5000 instrument with autosampler (Courtesy of Marconi Applied Technologies, UK).

logic algorithms currently under development) are being required for predictive classification, as traditional processing of electronic nose data usually involves off-line analyses.

## 12.4 APPLICATION TO WATER AND WASTEWATER MONITORING

Applications of electronic nose technology have evolved from the food and beverage industries [8]. However, more recently, studies have demonstrated its application to environmental problems. For example, some researchers have reported on the assessment of agricultural and sewage odours [14,15,22], showing relationships between the sensor

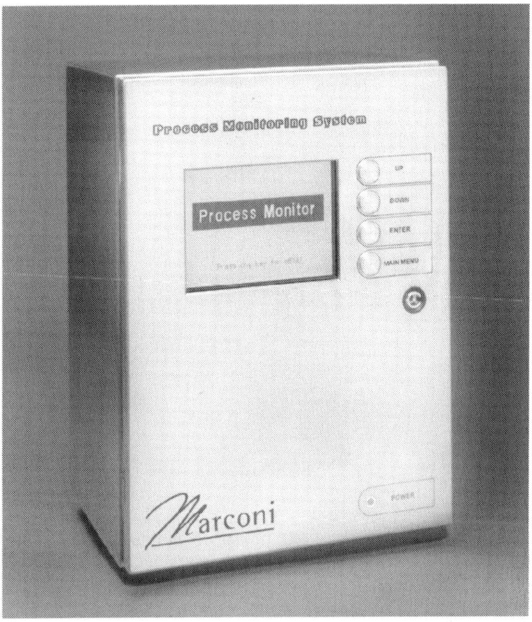

Fig. 12.8. Photograph of the ProSAT sensor array system (Courtesy of Marconi Applied Technologies, UK).

responses and threshold odour concentrations, while others have demonstrated applications in monitoring bioprocess ingredients for detecting microbial contamination [6] and predicting different bacterial types and culture growth phases [23–25]. These laboratory studies have shown the potential for this technology in terms of relationships between odour types/concentrations and sensor array data and consequently its wide application for real-time environmental monitoring [26].

Water and wastewater samples have also been used to investigate the potential application of this technology for discriminating chemical pollutants in supply water [27] and for characterising wastewater effluent [28]. Initial studies focused on the evaluation of the technology using laboratory-based systems whereas more recent research has applied these proof-of-concept results to continuous monitoring of water and wastewater.

Fig. 12.9. Photograph of a Cyranose 320 portable sensor array (Courtesy of Cyrano Sciences, USA).

### 12.4.1 Laboratory-based systems

Laboratory-based sensor array systems were initially used to discriminate between different concentrations of chemical pollutants in supply water [27] and for characterising different wastewater effluent [28]. Samples were analysed using static sampling systems such as the eNOSE 4000 (Fig. 12.6) by adding 100 ml of the samples into the sample vessel, into which the sensor head was introduced during data acquisition. The instruments were calibrated against a reference standard (such as 1,2-propanediol), with filtered nitrogen being used to purge the system during calibration and sample analysis. To ensure reproducibility between sample replicates, standard operating procedures consisting of different pre- and post-purging and sample equilibration were developed for different sample types (i.e. water and wastewater) [9].

The analysis of raw and treated water samples, tainted with trace concentrations of geosmin, methylisoborneol, 2-chlorophenol, phenol, diesel and 2-chloro-6-methylphenol has demonstrated that an electronic nose can clearly separate tainted and untainted water samples (i.e. no overlapping occurs between the sample groups). These studies also show that for some compounds the taint concentrations could be separated into distinct clusters. Figure 12.10 shows an example of the multiple discriminant analysis of raw water and water samples tainted with phenol at 10, 25, 50 and 100 ppm. The analysis of quiescent wastewater samples (from the inlet works, settlement tanks and final effluent outlet of three sewage treatment works) has also demonstrated that an electronic nose can differentiate between different wastewater types (i.e. raw sewage, settled sewage and final effluent) and samples from different treatment works [28]. The statistical analysis (using canonical correlation) of wastewater samples from a single treatment work over an extended period (up to 18 months) also suggested that a time-dependent linear relationship was evident between the sensor responses and the organic content of the wastewater as measured by the biochemical oxygen demand (BOD) [29,30]. Figure 12.11 shows an example of such a linear correlation over a three-week sample interval. Additionally, the neural network analysis of the same data set (using a three-layer back-propagation algorithm) also demonstrated that it is possible to predict BOD values from the electronic nose data [29]. Other studies using final effluent samples have also shown that by elevating the sample temperature during the electronic nose analyses the differentiation of the different sample types can be improved [31].

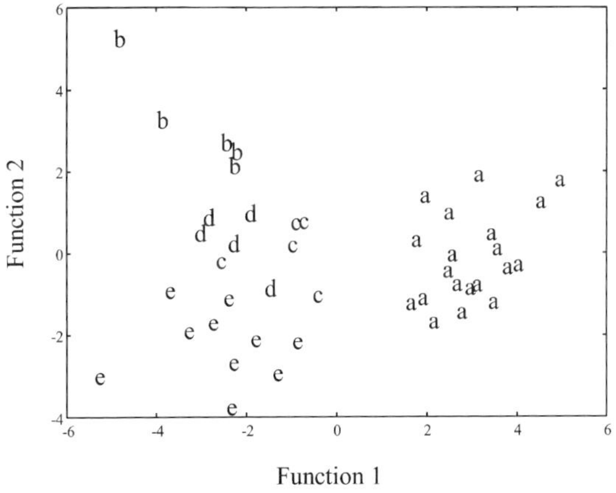


Fig. 12.10. Multiple discriminant analysis showing the differentiation of untainted (a) and phenol-tainted 100 ppm (b) 50 ppm (c) 25 ppm (d) and 10 ppm (e) water samples [27].

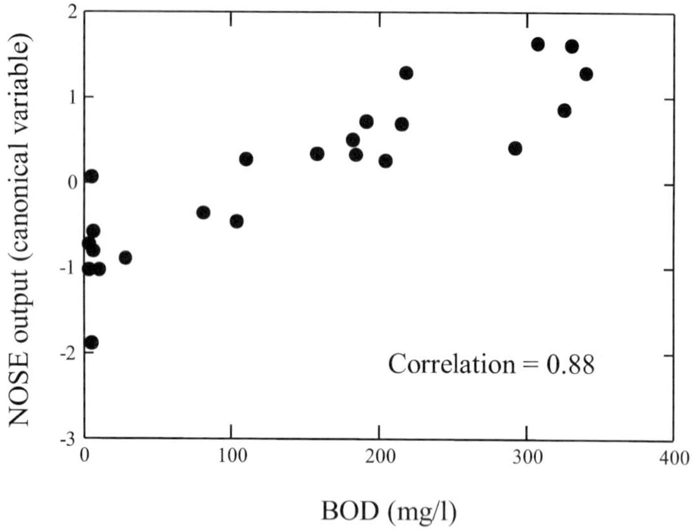


Fig. 12.11. Canonical correlation analysis showing relationship between sensor responses and BOD using raw, settled and final effluent sewage collected over a three-week interval [30].

### 12.4.2 On-line monitoring systems

In order for electronic technology to be applied to the continuous monitoring of water and wastewater, an integrated system that incorporates both the preparation of a sample headspace from a liquid stream and the analyses of this headspace gas is needed. Bourgeois and Stuetz [26] have reported on the application of using an external flow-cell for generating a headspace gas for sensor array analysis. The laboratory-based system consisted of a temperature-controlled flow-cell (25°C) into which nitrogen gas ($N_2$) was purged, and a temperature-controlled transfer line (30°C) through which the headspace gas was transferred to the sensor array module (35°C) for analysis. A schematic of the experimental apparatus is shown in Fig. 12.12.

Preliminary studies were used to determine the stability of the sensor responses as previous studies [16] have reported that conducting polymer sensors are sensitive to changes in sample temperature and humidity. A comparison between the relative humidity (RH) of the headspace and the 12 sensor responses (Fig. 12.13) shows that some of

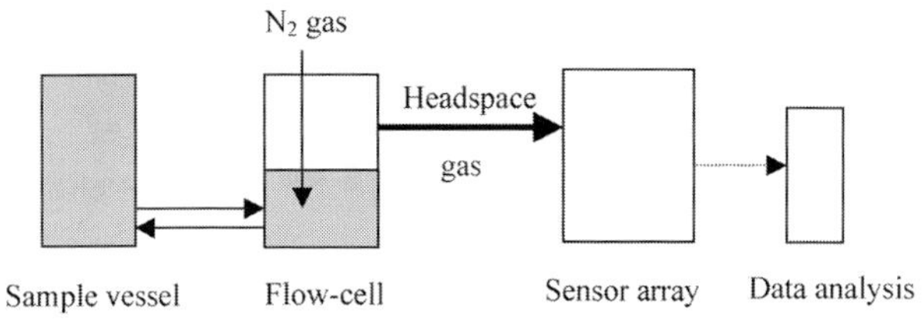


Fig. 12.12. Schematic of the experimental apparatus showing the flow-cell chamber, headspace transfer line, sensor array module and PC for process control and data analysis [26].

the individual sensors can be correlated to changes in the RH of the sample headspace. To minimise any environmental changes during headspace generation a statistical approach was used to evaluate the effect and interactions that different flow-cell parameters (flow-cell temperature, gas flow-rate and sparging porosity) had on RH generation and subsequently sensor stability [26]. Figure 12.14 shows a comparison between the sensor responses and RH (compared to

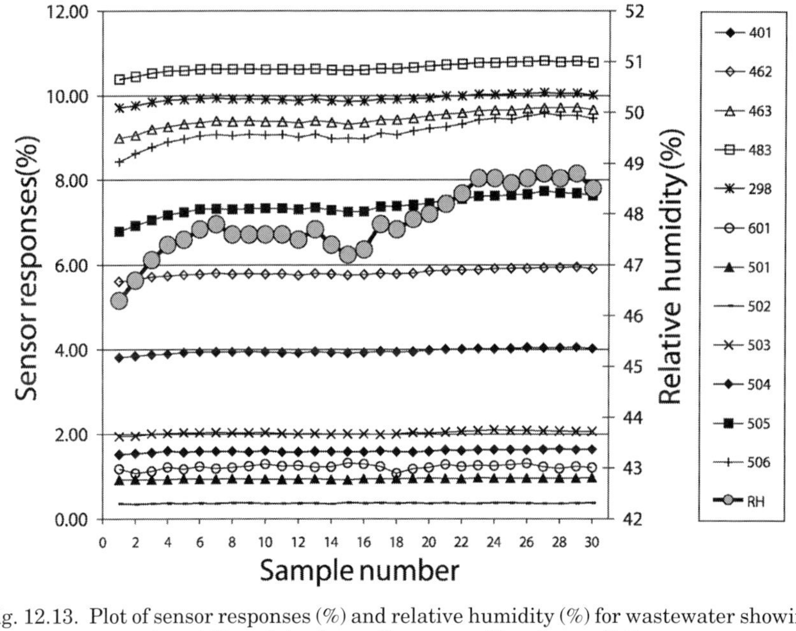


Fig. 12.13. Plot of sensor responses (%) and relative humidity (%) for wastewater showing instability of the flow-cell-generated headspace [26].

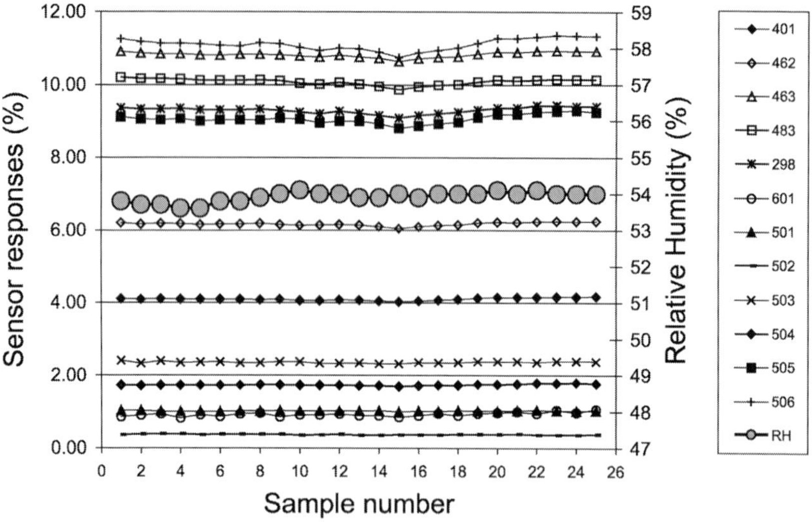


Fig. 12.14. Plot of sensor responses (%) and relative humidity (%) for wastewater once stable headspace generation had been achieved [26].

Fig. 12.13) once a stable sample headspace generation had been established. The principal component analysis of flow-cell-generated data for raw sewage, final effluent and water (Fig. 12.15) shows that different sample types can be clearly separated into a different cluster when the changes in RH generation are minimised. The spread in the data along the second principal component suggests that some variation is still present in the generation of the sample headspace. The multiple discriminant analysis (Fig. 12.16) of the same data set (as shown in Fig. 12.15) shows that when known properties are included in the statistical analysis, the different sample types can be clearly discriminated into distinct groups.

These findings have demonstrated that an external flow-cell can be used to produce a reproducible sample headspace for analysis by an electronic nose [32]. To evaluate the technique for continuous monitoring, dynamic studies whereby water and wastewater were continuously pumped through the flow-cell between headspace sampling were undertaken [33,34]. Figure 12.17 shows an example of such a study when 2-chlorophenol (equivalent to a concentration of 10 ppm) is directly injected into the flow-cell during the continuous

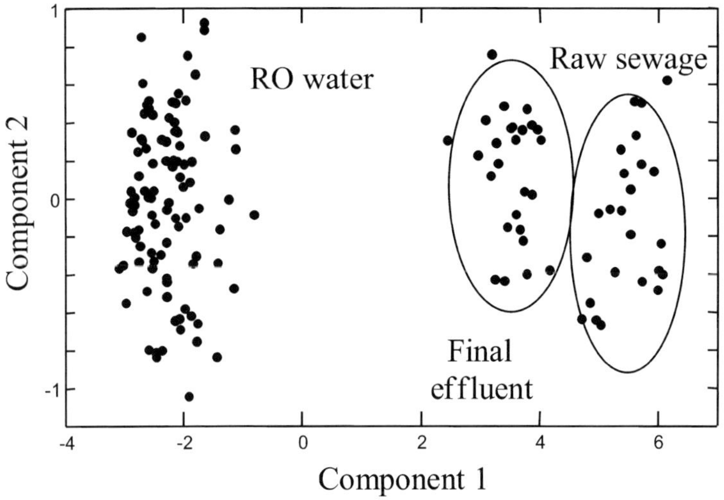


Fig. 12.15. Plot of principal components showing separation of raw sewage, final effluent and RO water from a flow-cell-generated headspace [26].

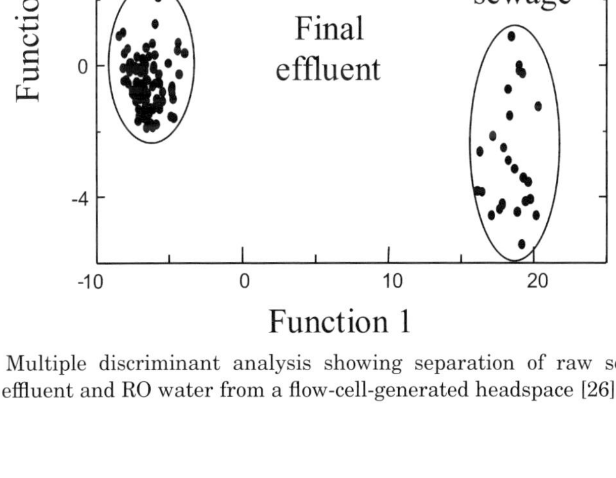


Fig. 12.16. Multiple discriminant analysis showing separation of raw sewage, final effluent and RO water from a flow-cell-generated headspace [26].

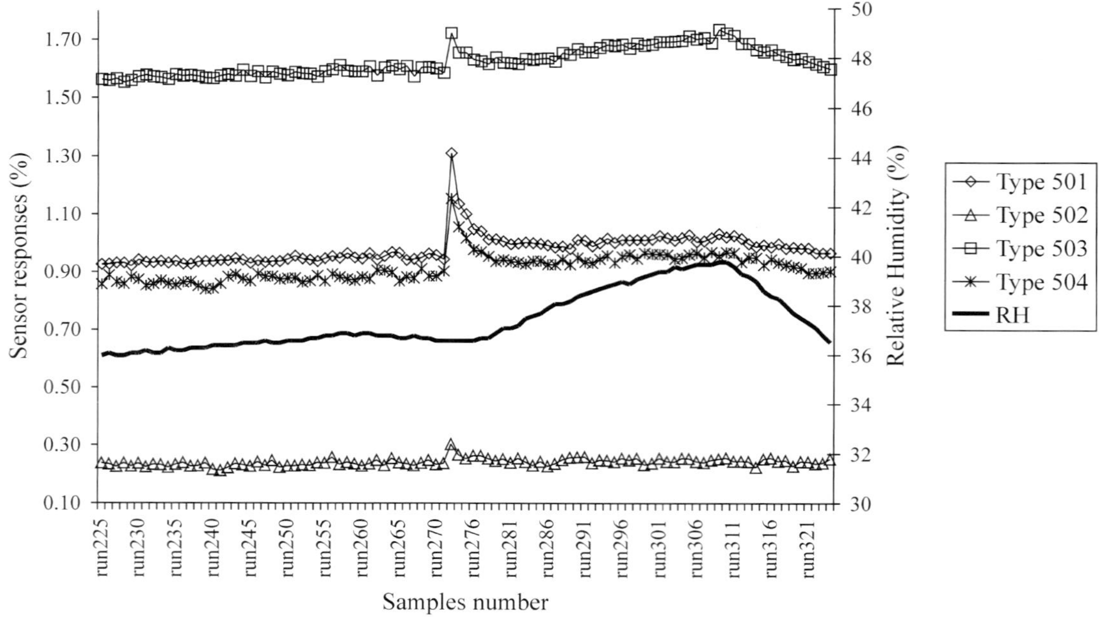


Fig. 12.17. Plot of sensor responses and relative humidity (RH) for RO water and the injection of 2-chlorophenol into the flow-cell (to an equivalent concentration of 10 ppm) [34].

monitoring of reverse osmosis (RO) water. The change in the profiles for the four plotted sensor responses shows that individual responses do differ in magnitude in the presence of 2-chlorophenol, whereas the RH remained unchanged but was observed to follow a diurnal variation, which is reflected in some of the individual sensor response profiles. The results demonstrate the advantage of using an array of non-specific sensors that can respond to many thousands of chemical species (such as 2-chlorophenol, diesel, geosmin and methylisoborneol), due to broad selectivity of the conducting polymer composition. The principal component analysis (Fig. 12.18) of the same data (from Fig. 12.17) shows that the RO water produces a distinct cluster, whereas the 2-chlorphenol-tainted samples do not, but can be clearly separated from the untainted RO water. The observed pattern mostly results from the tainted water becoming diluted after its initial injection into the flow-cell and represents the time taken to be flushed from the flow-cell.

To study the operational effects of continuous use of an electronic nose for monitoring change in water and wastewater for extended periods, two on-line monitoring systems have been constructed and installed at a water and wastewater treatment works. Figure 12.19

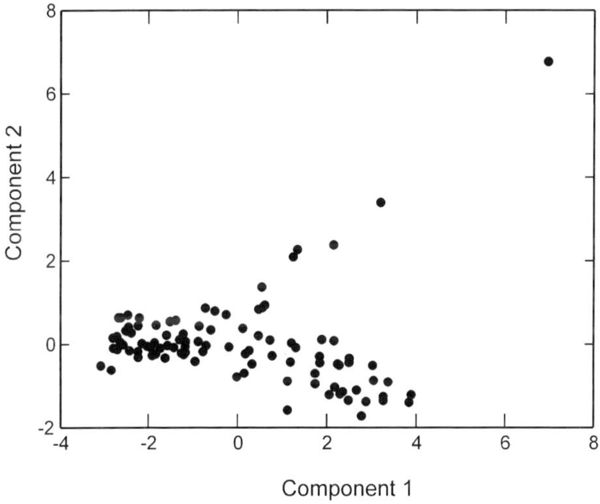


Fig. 12.18. Plot of principal components showing separation of RO water and water tainted with 2-chlorophenol at 10 ppm (using six conducting polymer sensors in the analyses) [34].

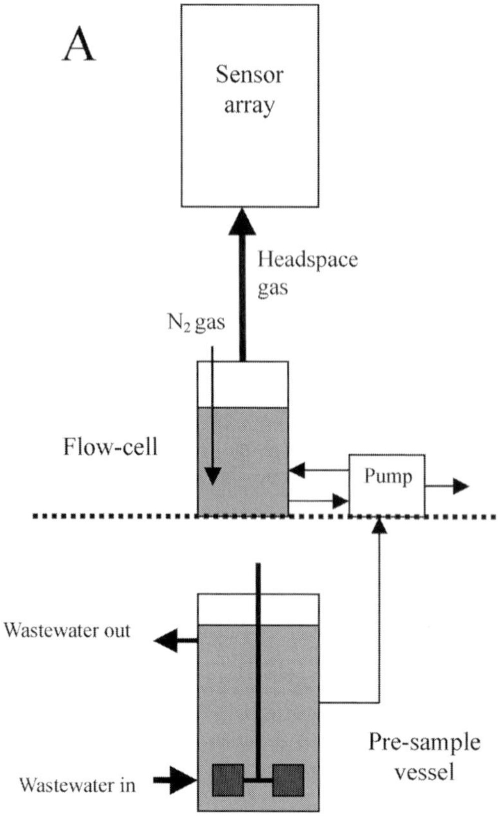


Fig. 12.19. Schematic (A) and photograph (B) of the on-line monitoring system at Cranfield University sewage works, showing pre-sample vessel, flow-cell and sensor array module [35].

shows a schematic and photography of these systems, which include a pre-sample vessel, flow-cell and sensor array module (ProSAT, Marconi Applied Technologies, UK). To simulate the presence of an intermittent pollutant in the water or wastewater, trace compounds can be dosed into the pre-sample vessel at various concentrations. An example of the detection of an unknown pollutant in the wastewater influent at Cranfield University sewage works is shown in Fig. 12.20. The rapid increase in the response profiles for three sensors corresponded with

B

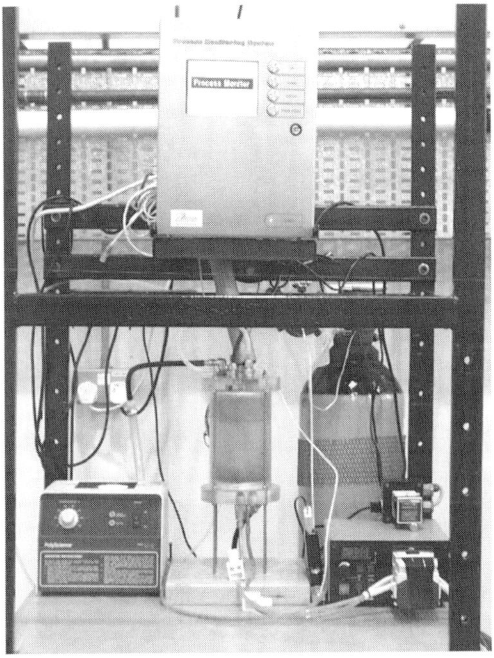


Fig. 12.19 (*continued*)

reports of a petroleum smell at the inlet of the sewage works and suggests that the unknown pollutant was in the wastewater for approximately 24 h and had reached a peak within 4 h [35]. The principal component analysis (Fig. 12.21) of the same data (from Fig. 12.20) shows that a distinct cluster is produced for the unknown pollutant along the first principal component, which represents 63% of the variance in the data set. Figure 12.21 also demonstrates that unknown samples (©) that were not included in the statistical analysis can be predicted correctly and could be applied to discriminate intermittent discharges in water and wastewater.

With further development and integration of real-time data analysis protocols into an on-line monitoring system, non-specific sensing technology could serve a number of applications in water and wastewater monitoring.

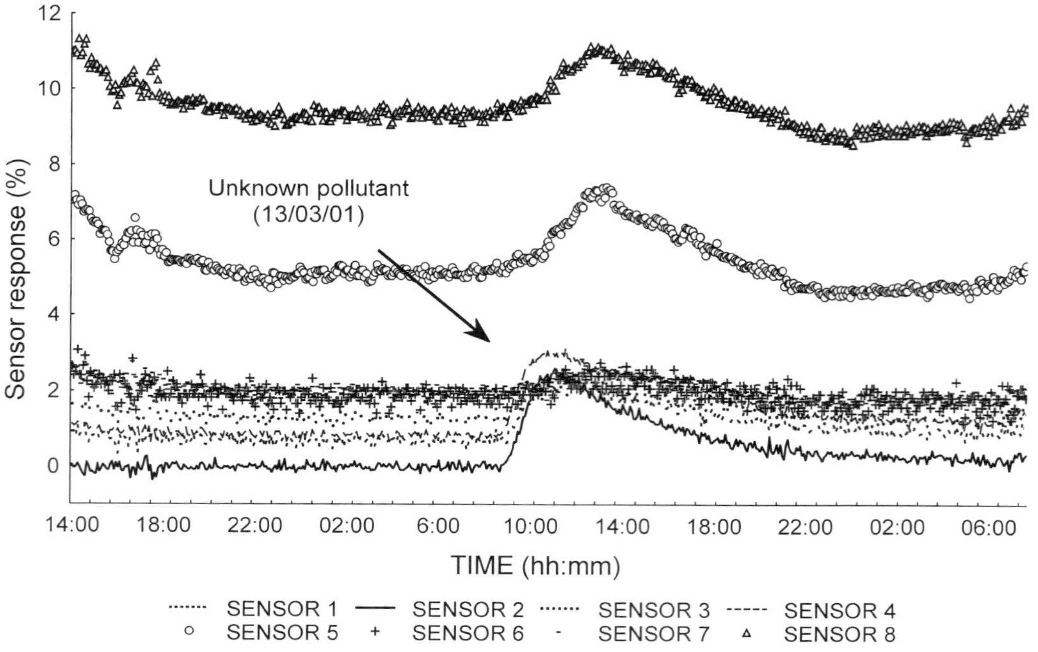


Fig. 12.20. Plot of sensor responses showing the detection of an unknown discharge into the wastewater influent at Cranfield University sewage works [35].

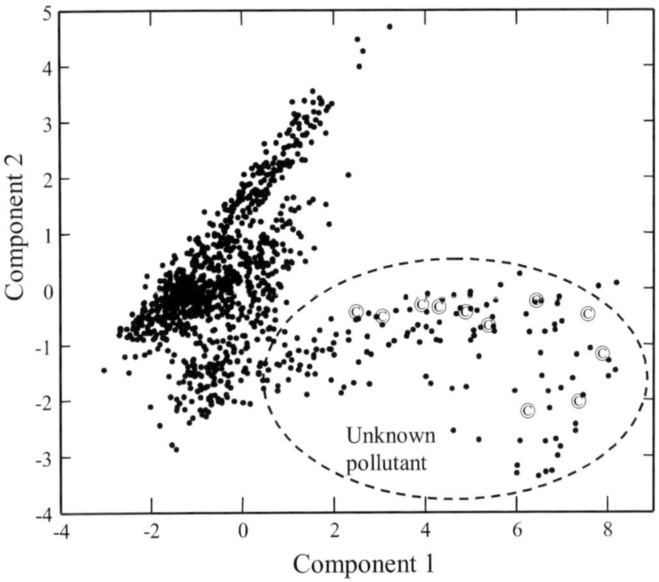


Fig. 12.21. Plot of principal components showing the separation of an unknown discharge into the wastewater influent (Fig. 12.20) and the correct prediction of 11 pollutant samples (©) within that cluster.

## REFERENCES

1 S.K.E. Brookman, *Water Res.*, 31 (1997) 372.
2 I. Karube, S. Mitsuda, T. Matsunaga and S. Suzuki, *J. Ferm. Technol.*, 55 (1977) 243–248.
3 D.M. Reynolds and S.R. Ahmad, *Water Res.*, 31 (1997) 2012.
4 L.J.S. Lukasse, K.J. Keesman and G. van Straten, *Water Res.*, 31 (1997) 2278.
5 S.R. Ahmad and D.M. Reynolds, *Water Res.*, 33 (1999) 2069.
6 P.K. Namdev, Y. Alroy and V. Singh, *Biotech. Prog.*, 14 (1998) 75.
7 W.F. Young, R. Horth, R. Crane, T. Orgen and M. Arnott, *Water Res.*, 30 (1996) 331.
8 D. Hodgins and D. Simmonds, *J. Auto. Chem.*, 17 (1995) 179.
9 R.A. Fenner and R.M. Stuetz, *Water Environ. Res.*, 31 (1999) 282.
10 K.C. Persaud and P.J. Travers, In: E. Kress-Rogers (Ed.), *Handbook of Biosensors & Electronic Noses*, CRC Press, Boca Raton, 1997, pp. 563–592.

11 J.W. Gardner and P.N. Barlett, *Sens. Actuators B*, 18 (1994) 211.
12 R.M. Stuetz and R.A. Fenner, Electronic nose technology: a new tool for odour management, *Water Quality International*, July/August (1998) 15–17.
13 D. Hodgins, *Sens. Actuators B*, 26–27 (1995) 255.
14 P.J. Hobbs, T.M. Misselbrock and B.F. Pain, *J. Agric. Eng. Res.*, 60 (1995) 137.
15 K.C. Persaud, S.M. Khaffaf, P.J. Hobbs and R.W. Sneath, *Chem. Sci.*, 21 (1996) 495.
16 J.W. Gardner and P.N. Barlett, *Electronic Nose: Principles and Applications*, Oxford University Press, New York, 1999.
17 J.W. Gardner and P.N. Barlett, In: P. Moselt, J. Norris and D. Williams (Eds.), *Techniques and Mechanisms in Gas Sensing*, Adam-Hilger, Bristol, 1991, Chapter 14.
18 J.W. Gardner and P.N. Barlett, In: J.W. Gardner and P.N. Bartlett (Eds.), *Sensors and Sensory Systems for an Electronic Nose*, Kluwer, Dordrecht, 1992, pp. 161–180.
19 J.W. Gardener and E.L. Hines, In: E. Kress-Rogers (Ed.), *Handbook of Biosensors and Electronic Noses*, CRC Press, Boca Raton, 1996, pp. 633–652.
20 P. Mielle and F. Marquis, *Sens. Actuators B*, 58 (1999) 526.
21 T. Gibson, O. Prosser and J. Hulber, Electronic noses: an inspired idea?, *Chem. Ind.*, April (2000) 287–289.
22 R.M. Stuetz, G. Engin and R. Fenner, *Water Sci. Technol.*, 38 (1998) 330.
23 T.D. Gibson, O. Prosser, J.N. Hulbert, R.W. Marshall, P. Corcoran, P. Lowery, E.A. ruck-keene and S. Heron, *Sens. Actuators B*, 44 (1997) 413.
24 J.W. Gardner, M. Graven, C. Dow and E.L. Hines, *Measure. Sci. Technol.*, 9 (1998) 120.
25 M. Holmberg, F. Gustafsson, E.G. Hornsten, F. Winquist, L.E. Nilsson, L. Ljung and I. Lundstrom, *Biotech. Technol.*, 12 (1998) 319.
26 W. Bourgeois and R.M. Stuetz, *Water Sci. Technol.*, 41 (2000) 107.
27 R.M. Stuetz, M. White and R.A. Fenner, *J. Water Supply Res. Tech.—Aqua*, 47 (1998) 223.
28 R.M. Stuetz, R.A. Fenner and G. Engin, *Water Res.*, 33 (1999) 442.
29 R.M. Stuetz, S. George, R.A. Fenner and S.J. Hall, *J. Chem. Tech. Biotech.*, 74 (1999) 1069.
30 R.M. Stuetz, R.A. Fenner, S.J. Hall, I. Stratful and D. Loke, *Water Sci. Technol.*, 41 (2000) 41.
31 T. Dewettinck, K. van Hege and W. Verstraete, *Water Res.*, 35 (2001) 2475.
32 R.M. Stuetz, Using sensor arrays for on-line monitoring of water and wastewater quality, *Am. Lab.*, January (2001) 10–13.

33 W. Bourgeois, M. Gaugler and R.M. Stuetz, On-line evaluation of a chemical sensor array for monitoring changes in effluent quality. *Proc. IWA ICA 2001*, Malmö, 2001, Vol. 1, pp. 271–278.
34 P. Hogben, R.M. Stuetz and B. Drage, Use of a chemical sensor array and an on-line flow-cell for monitoring water quality. *Proc. AWWA Water Quality Technology Conference*, Nashville, 9 pp-CD.
35 W. Bourgeois and R.M. Stuetz, *Water Res.*, 36 (2002) 4505.

Chapter 13

# Integrated optical transducers for (bio)chemical sensing

**C. Domínguez*, J.A. Rodríguez and L.M. Lechuga**

## 13.1 INTRODUCTION

Among the most promising integrated analytical systems (IAS), special attention is reserved for those designed to provide analytical information from chemical or biological environments through the concept of the chemical sensor and the use of integrated optical transducers.

In this chapter, a brief incursion into the fundamentals and advantages of these systems, with emphasis on the materials and technologies involved in their construction, and the analysis of some particular cases of integrated optical chemical sensors will be given.

The simplest waveguide (Fig. 13.1) consists of a three-layer structure in which the intermediate material has the highest refractive index. If a light beam is directed to this multilayer structure, it will remain confined to the intermediate layer by way of the well-known total internal reflection (TIR) principle, which is produced at the interface of the materials. This phenomenon is the basis of optical waveguides, the best-known example of which is the optical fiber (see Chapter 9).

At the early 1960s, the first experiments on light guidance through thin dielectric layers were performed, together with the development of semiconductor-based photodiodes and laser diodes. Owing to

* Corresponding author. Tel.: +34-935947700x1318. *E-mail address:* carlos.dominguez@cnm.es.

*Comprehensive Analytical Chemistry XXXIX, pages 541–586* 

ISSN: 0166-526X

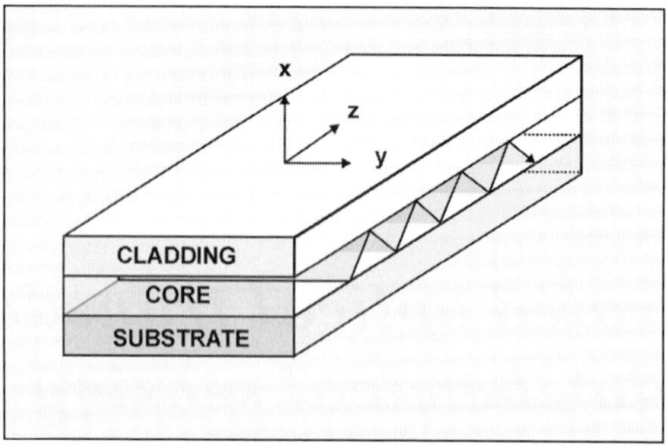

Fig. 13.1. Slab structure of a thin film dielectric waveguide.

the reduced dimensions and the low consumption required for passive and active optical components, it was thought that they could replace part of the circuitry in electronic equipments. It was before the 1970s that the effectiveness of optical components was proved, showing the feasibility of developing a sophisticated and promising technology, named *integrated optics* [1].

Integrated optics can, at least theoretically, integrate monolithically many passive and active optical devices onto the same substrate and fabricated with the same technology and can be operated at low cost. But the monolithic integration of waveguides, switches, directional couplers, polarization splitters, modulators, control circuitry, laser, photo-detectors and pre-amplifying circuitry showed a high complexity. For this reason, nowadays, integrated optics circuitry has not reached the size, the integration levels and the development of microelectronic circuits.

## 13.2 BASIC CONCEPTS

Even though the general concept of an optical sensor (optrode) was outlined above (see Chapters 1 and 9), it is appropriate to set its exact meaning as well as the advantages of using the optical sensing principle. An *intrinsic optical sensor* is defined when the interaction of the receptor (molecular recognition element) with the analyte to be detected (chemical or biological) affects, in any way, some of the properties of a light beam, properly confined by a light guiding component.

In the *extrinsic optical sensor*, the light is forced out from the guiding component, interacting directly with the sensing area, and is collected again by the guiding element for outer measurement of light property variations. Regardless of the optical sensor type, sensing is usually performed in a similar way: a change of an external magnitude modifies the optical properties of the sensing area along the guiding element and modulates the light characteristics (intensity, phase, polarization, etc.).

The optical sensing approach offers many advantages over its electrical counterpart, e.g. the absence of risk of electrical shocks or explosions, immunity to electromagnetic interferences, and in general, a higher sensitivity and a wider bandwidth. Moreover, by using optical fibers to guide light towards and from the device, remote sensing is also possible.

The advantages of optical sensing are significantly improved when this approach is used in an integration schema. Integration offers additional advantages to the optical sensing systems such as miniaturization, robustness, reliability, potential for mass production with consequent reduction of production costs, low energy consumption and simplicity in the alignment of the individual optical elements.

'Integration' is a term largely used in optics as well as in microelectronics. Since the announcement of the first optical integrated circuit [2], a rapid development of the so-called 'integrated optics' has taken place over recent decades, mainly associated with the telecommunication field. The use of microelectronic technology, with its well-known fabrication processes, has been decisive in this development.

From a technological point of view the term 'integration' is used to indicate that different components or devices are fabricated in the same substrate to make up an unique system. The final operation of this system depends on the interrelation of the functions of the individual components. In integrated optical sensors at least the sensing layer as well as the optical transducer are fabricated into the same substrate.

Indeed, an all-integrated optical system should incorporate, as a unit (i) an optical source, which provides the light at the desired wavelength and amplitude, (ii) the light guiding structure (waveguide) that directs the light to and out of the measurement area, (iii) the sensing layer, where interaction (sensing) takes place, and (iv) a light detector. This detector may measure directly the optical magnitude, which reflects the effect of the receptor–analyte interaction (intensity, phase, wavelength or polarization state), or may convert the output from the optical to the electrical domain, providing an indirect measurement. Furthermore, all these components should be fabricated in a single technological process and, preferably, with the same basic material. However, such an ideal analytical integrated system has not been obtained yet. To our best knowledge, no satisfactory integration of the source and detector with the rest of the integrated optical transducer has been achieved so far. Therefore, in the present chapter, only the development of optical systems, which integrate the optical transducer with the sensing layer, is described. In Fig. 13.2, integrated intrinsic and extrinsic optrodes based on thin film waveguides are depicted.

The fabrication process of an integrated optical sensor may be divided into three main stages. The first is the fabrication of the light guiding structure (waveguide). The so-called 'thin film technology' through different microelectronic processes is employed. A more detailed description of this technology will be given below. The second stage is the deposition of the sensitive layer or membrane, which may be composed of different materials, which confers the sensor with the ability of selective detection, either for chemical or for biological species. A wide variety of processes and technologies, ranging from standard microelectronic processing to electrodeposition techniques, is used to deposit the sensitive layers. The last stage is encapsulation or end-processing. This is a quite specific process since it should isolate the device from its surroundings, leaving only the sensitive layer in contact with the medium to be measured.

In an integrated optical sensor, special requirements are imposed on the basic materials forming the optical elements, particularly the light guiding structure. Refractive index values should be uniform and controllable, the material should be homogeneous (grain boundaries must be avoided), and also chemically and mechanically stable, with low defect, low impurity concentration and with elastic properties compatible with further thermal processing.

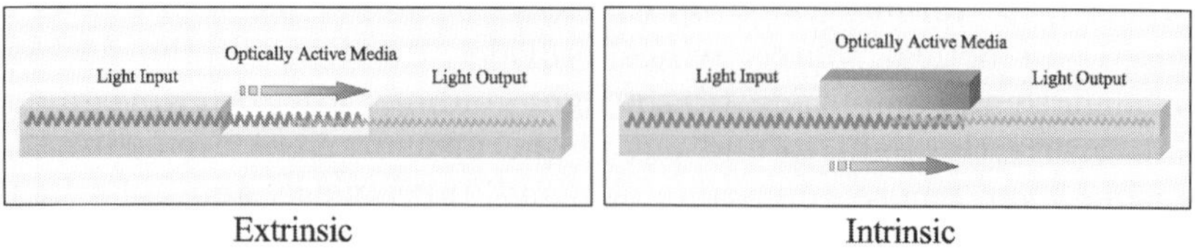


Fig. 13.2. Optrode structures based on integrated waveguides.

## 13.3 FUNDAMENTALS OF OPTICAL WAVEGUIDES [3,4]

Optical fibers are the most known, developed and used (especially in the communication field) light guiding component. However, in integrated optical sensors, the key role is played by planar waveguide structures. It is worthwhile mentioning that some of the requirements imposed on guiding elements in telecommunications are not necessary conditions in optical sensing. Thus, optical losses are not necessarily a drawback for integrated optical sensors since optical paths of short length are used. Moreover, in many applications, multimode instead of single mode structures may also be used.

A typical planar, or thin film, TIR waveguide is formed by a substrate of refractive index $n_3$, and two successively deposited layers, the guiding layer (core) and the cladding layer, with refractive indexes $n_2$ and $n_1$, respectively. It is assumed that the structure is infinite in the $y$ and $z$ directions and the condition $n_2 > n_3 > n_1$ is fulfilled (see Fig. 13.3). This shows the typical case of a dielectric layer, of thickness $t$, deposited on a substrate with a smaller refractive index, where the cladding is air. Therefore, for incidence angles higher than a critical value ($\Theta_i$), light is propagated through the core in the $z$-direction by achieving TIRs at the interfaces of core–substrate and core–cladding, alternately.

Solving the Maxwell equations, with the appropriate boundary conditions, it is possible to obtain the different electromagnetic field distributions (profiles) in the guiding structure, which are called 'guided propagation modes'. A mode could be understood as an energy

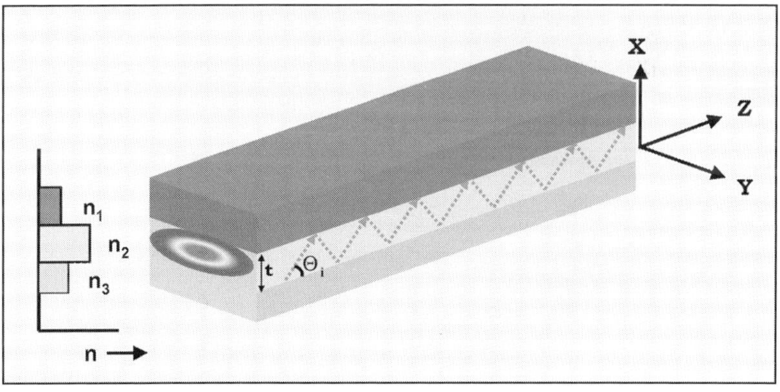


Fig. 13.3. TIR waveguide structure.

distribution inside the waveguide and the three different modal profiles, as shown in Fig. 13.4, are analytical solutions of the Maxwell equations. Firstly, (a) and (b) are radiation modes to the substrate or to both interfaces. Modes (c) are the propagating modes, the number of which depends on the wavelength and the waveguide structure. Finally, an analytical, though non-real solution, is observed (Fig. 13.4, mode (d)) due to the fact that its energy exponentially increases with $y$.

A given mode $m$ is characterized by the value of the propagation constant, $\beta_m$, and guided modes should fulfill the condition:

$$kn_3 \leq \beta_m \leq kn_2$$

where $k$ is the wavenumber. The values of $\beta_m$ are limited because there exists a maximum incidence angle at which TIR occurs. A mode is also characterized by a phase velocity ($v$) parallel to the waveguide and by an effective refractive index ($n_{\text{eff}}$). Their values are expressed as:

$$v = c(k/\beta_m) \quad (13.1)$$

$$n_{\text{eff}} = c/v = \beta_m/k \quad (13.2)$$

where $c$ is the light vacuum velocity. Since different modes move with different phase velocities, a single mode-guiding behavior is, in most applications, necessary.

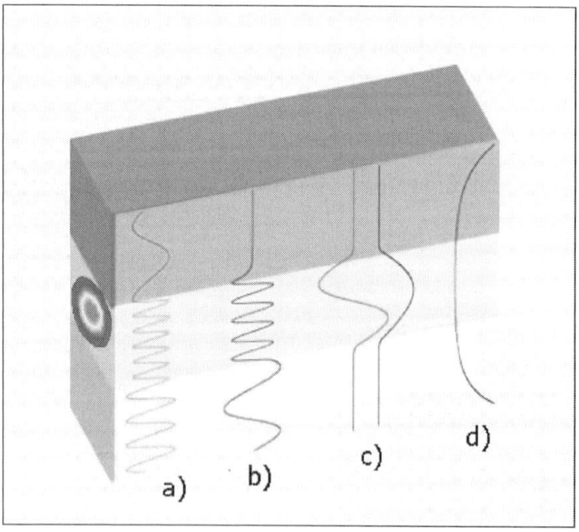


Fig. 13.4. Some mode profiles of the electromagnetic field inside a waveguide structure.

Figure 13.5 shows the electromagnetic field distribution in the $x$-direction for some guided modes traveling in the $z$-direction for a specific case. Observe that the electromagnetic field intensity does not vanish from the core, but decays exponentially in the substrate and cladding regions, i.e. it is evanescent in these regions. This *evanescent field* is fundamental to ensuring the interaction of the modes in the waveguide with the external media, which is the principle of the sensing action for intrinsic sensors.

The structure just analyzed, which confines radiation in only one dimension (the $x$-direction) is, in practice, barely employed in integrated optics sensing technology. Instead, confinement to the cross-section is frequently necessary since this allows the guiding of light from the injection point to any other point. Therefore, the strip geometry, or the so-called channel waveguide is generally used. Figure 13.6 depicts a typical example. It represents the simplest configuration, usually called a ridge or rib waveguide. Cross-section confinement is due to the dependence of the effective refractive index on the core thickness. Indeed, since the thickness of the waveguide core layer is larger in the region beneath the rib, the effective refractive index in this region is higher than in the surrounding regions of the core, thus assuring a lateral confinement of the light in the $y$-direction.

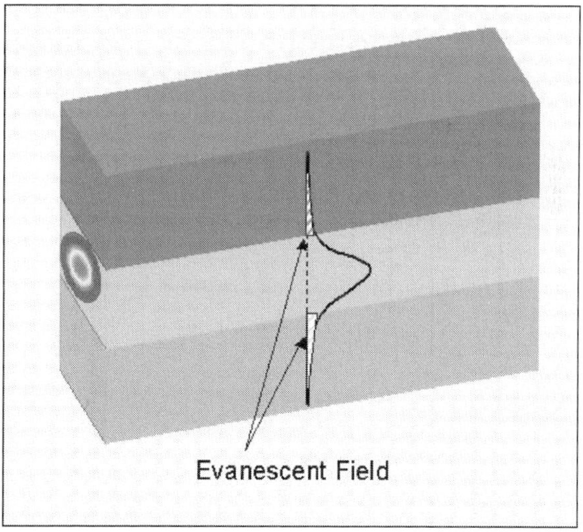


Fig. 13.5. Evanescent field due to the propagating mode.

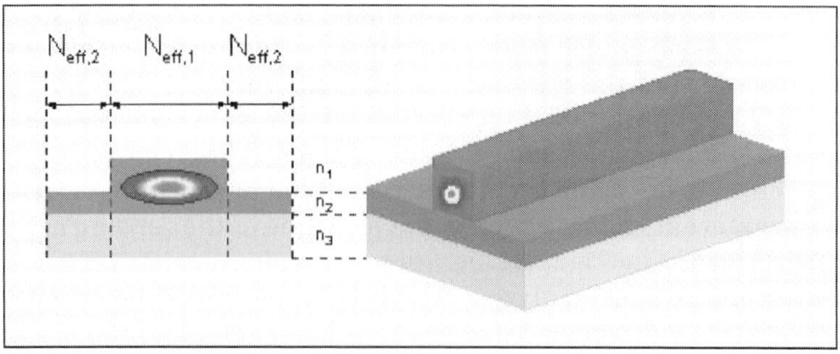


Fig. 13.6. Structure of a rib waveguide.

In addition to the TIR-type waveguide structure just described, a second variant of light guiding element that overcomes some of the most significant TIR drawbacks has been used in integrated optics. It is known as an antiresonant reflecting optical waveguide (ARROW) structure, and offers important advantages as compared to the TIR version [5]. Figure 13.7 shows a schema of the layered structure and the refractive index profile of a conventional ARROW rib structure compared with a TIR waveguide. In this ARROW structure, light confinement is achieved due to TIR at the core–air (an upper protective cover layer is sometimes employed) interface, and by a very high reflectivity (up to 99.96%) at the two interference cladding layers underneath the core. The refractive indexes and thickness of these layers are designed in such a way that they behave as a Fabry–Perot resonator operating at antiresonant conditions, for the specific working

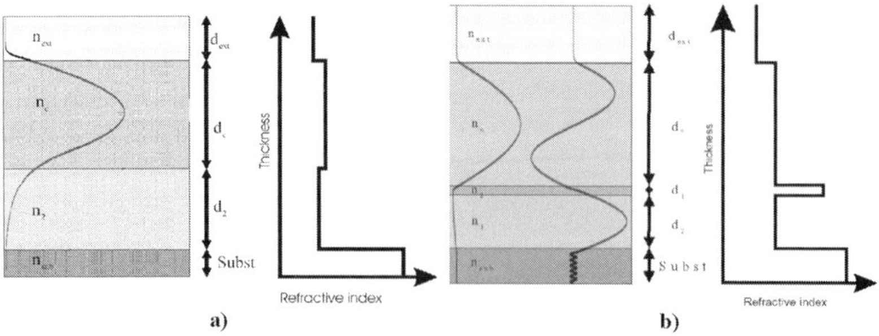


Fig. 13.7. Different waveguide single mode configurations: (a) TIR waveguide with small refractive index difference between the core and the under layer; (b) fundamental and first mode of an ARROW structure.

wavelength. ARROW guides are preferable to TIR guides in integrated optical sensors. The main reason is that, for a single-mode behavior, the core thickness of the former may be at the micrometer scale (instead of the nanometer dimensions needed for TIR waveguides with high refractive index differences between the core and the surrounding layers), making them suitable for an efficient connection to single-mode commercial optical fibers. Some other advantages of ARROW guides are: low loss single-mode behavior by loss discrimination of modes higher than the fundamental, large light confinement, high tolerance for the design with respect to refractive indexes and thickness of the cladding layers, and high surface sensitivity.

The design of the above structures is generally done using numerical methods in order to determine the values of the geometrical and optical parameters according to the requirements. Different theoretical modeling methods are used, such as spectral and effective index methods [6], multilayer analysis [7], perturbative methods [8], finite difference method [9] and finite element method [10]. The results from these methods, such as the specific thickness and refractive index of every layer, are used as technological inputs for manufacturing the chosen structure.

The designed guiding structures allow different passive optical components to be obtained by using adequate technological processes.

## 13.4 DETECTION PRINCIPLES: TYPES OF DEVICES

It is common practice to classify integrated optical sensors according to the type of effect that the receptor–analyte interaction produces in the incoming light beam.

The simplest approach is to use an extrinsic device called an *absorbance sensor*. In this device, the light coming from the source is guided by the input guiding component towards a free propagation region (the optically active medium) responsible for the detection process. After crossing the sensing region, light is collected again by an output guiding element (see Fig. 13.8). All these components are integrated in the same substrate [11]. In the simplest case, a solution containing only one analyte is analyzed. Any variation of the analyte concentration is reflected as a variation of the imaginary part, $\varepsilon$ (called the extinction or absorbance coefficient) of the complex refractive index of the medium ($N$), where $N = n + i\varepsilon$. Since the imaginary part of

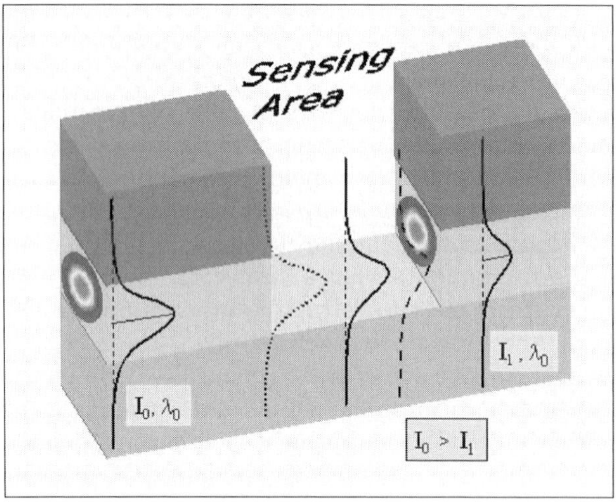


Fig. 13.8. Basic layout of an extrinsic absorbance sensor.

the refractive index is directly related to the light absorbance, a variation of the output light intensity as a function of the concentration, could be measured at the device output with a photodetector. Problems associated with this sensing configuration are that intensity changes could be produced by different factors (temperature, pressure, dust, etc.), resulting in this sensor being poorly selective if no reference can be used.

The evanescent field plays a key role in the light–matter interaction, and consequently, is the basis of the sensing principle in any intrinsic optical sensor. For the most general structure, the cladding layer is removed in a small region of the waveguide. Then, the sensitive (sensing) layer is deposited in this region, directly in contact with the core, assuring the best interaction with the evanescent field (see Fig. 13.9).

There are three fundamental types of intrinsic optical sensors according to the effect that the receptor–analyte interaction produces in the guided mode: the absorption-type sensor, the luminescent-based sensor and the real refractive index-type sensor. In the absorption-type sensor, the interaction of the analyte with the receptor (recognition) molecules in the sensing layer changes the imaginary part of the complex refractive index of this layer, thus increasing the absorbance coefficient. This change is reflected as an intensity variation of

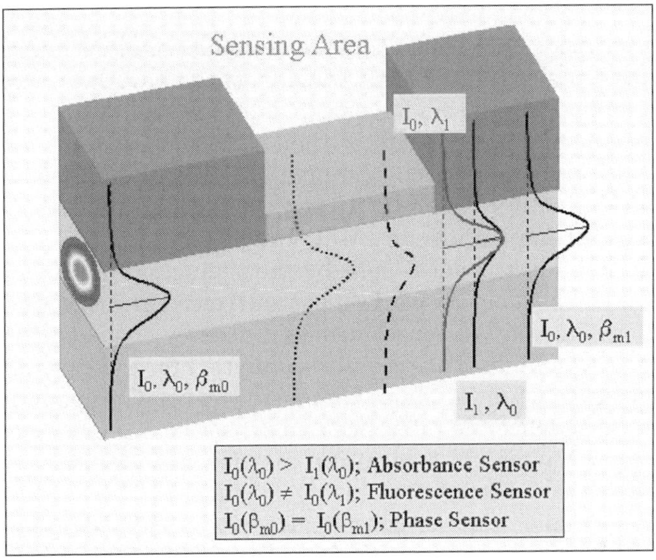


Fig. 13.9. Intrinsic absorbance sensor schema for the three detection principles.

the guided light beam at the working wavelength ($I_0(\lambda_0) > I_1(\lambda_0)$), which can be measured at the output.

The luminescent type refers to an important group of intrinsic optical sensors that are based upon the luminescent radiation produced in the sensing region. In these sensors, the sensing layer absorbs light of a concrete wavelength from the waveguide and emits light of a different wavelength, which is then collected by the same waveguide. As presented in Fig. 13.9, at the input waveguide the light intensity is $I_0(\lambda_0)$, reaching the photodetector: $I_1(\lambda_0) + I_0(\lambda_1)$, being $I_0(\lambda_0) > I_1(\lambda_0)$. The luminescent properties of the material forming the sensing layer vary as a result of its interaction with the analyte. Therefore, the characteristics of the emitted radiation are related to the nature and concentration of the analyte, and consequently the analysis of this radiation yields the desired analytical information.

The refractive index type is based upon the effect of the receptor–analyte interaction on the real part of the complex refractive index of the sensing layer. Any change of this magnitude induces a corresponding change in the propagation constant of a given waveguide mode ($\beta_{m0} \neq \beta_{m1}$), thus producing a measurable change in the guided light characteristics, due mainly to a phase shift.

Different methods have been used to measure changes in the propagation modes as a result of the receptor–analyte interaction in the sensitive region. Evanescent wave methods have been extensively used and integrated sensors have been developed as the grating coupler [12,13], resonant mirror [14,15] or the highly sensitive interferometric sensors, with the use of Mach–Zehnder [16–18], Young [19] or Fabry–Perot interferometers. Figure 13.10 depicts a schematic of an integrated Mach–Zehnder interferometer (MZI) optical sensor.

Light from a laser beam is split by the Y-junction into two identical beams that travel along the MZI sensor and reference arms. The sensing layer is deposited on the sensor arm and directly contacts the solution containing the analyte. The receptor–analyte interaction is reflected as a propagation constant or phase change in the beam propagating along the sensor arm. Once the reference and sensor beams have recombined in the second Y-junction, a resulting interference pattern is obtained, which reflects the effect of the sensing reaction.

Among intrinsic sensors, special attention should be given to the 'surface plasmon resonance (SPR)-based' evanescent wave-type sensor because of its sensibility and simplicity [20,21]. Surface plasmons are elementary excitations, which result from a collective oscillation of the free-electron plasma at a metal–dielectric film interface. Figure 13.11 shows the basic arrangement for the excitation of surface plasmons and use of the SPR principle in integrated optical sensing. A thin metal film is evaporated onto the dielectric waveguide surface. A light ray, incident at the dielectric–metal layer interface, excites a surface plasmon mode if the incidence angle is such that the momentum of the photons in the plane of this interface exactly matches that of a plasmon on the opposite surface of the metal film [22]. A drastic change of

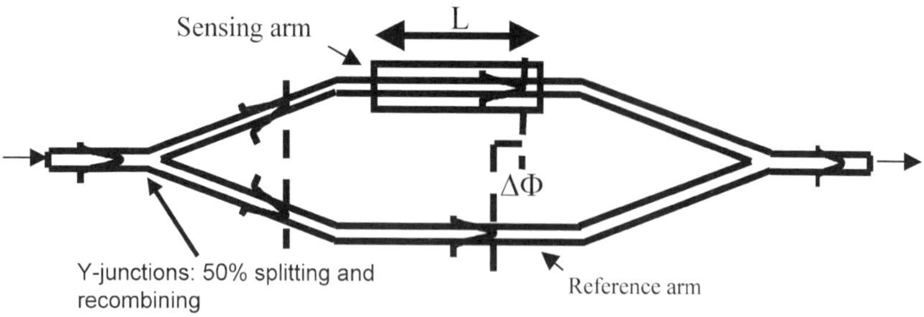


Fig. 13.10. Scheme of a Mach–Zehnder interferometer device.

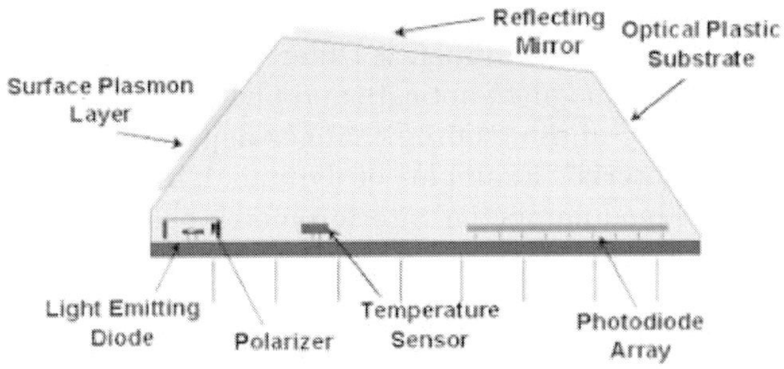


Fig. 13.11. Basic layout of a surface plasmon resonance transducer, according with http://www.ti.com/sc/docs/products/msp/control/spreeta/what.htm.

the reflection coefficient occurs at this critical incidence angle. The sensing layer is deposited on the metal layer. Any change of the real part of the refractive index of this layer, as a result of the sensing reaction, is reflected as a change of the propagation constant of the surface plasmon mode. Moreover, in this case the evanescent field intensity in the sensing region is substantially higher (possibly two orders higher) than in the case without the metal film, thus enhancing the sensor sensibility.

Some specific examples and detailed explanations of several selected integrated optical sensors will be given in a later section (Section 13.6).

## 13.5 TECHNOLOGIES FOR INTEGRATED OPTICAL TRANSDUCER FABRICATION

Chemists use and know very well the principles of optical analytical methods, but they are not so familiar with the process required in the technological definition of the devices. Hence, this section will try to present some technological aspects introducing the basic concepts of the main process steps.

Optical components may be classified according to their behavior throughout the operation [23]. The classes are given below.

*Passive*: waveguide structures that do not change their intrinsic optical properties. Within this group devices like couplers, splitters, curves, lenses or mirrors, are found. Extrinsic optrodes also belong to this group.

*Dynamic passive*: components in which guiding characteristics vary as a function of an external influence. Different physical (electro-optical, thermo-optical, and magneto-optical), and photochemical effects can be used. The efficiency of the guiding changes depends on the properties of the specific material. Examples of dynamic passive optical components are: phase modulators (i.e. interferometric optochemical sensors), amplitude and frequency (intrinsic optrodes) converters, and TE/TM splitters.

*Active*: devices that can act as light sources (laser, amplifiers) or detectors (photodiodes).

Not all materials are able to perform all optical functions. This means that for the fabrication of an optical circuit by combining several components, different materials must be involved. This causes an increasing difficulty for integration, as was mentioned before.

Hence, two types of integration schemes have been explored in order to obtain optical circuits: monolithic and hybrid integration.

The hybrid scheme [24] involves the incorporation of discrete components, individually packaged, and optically interconnected by way of optical fibers, waveguides or diffractive elements. Conventional hybrid integration has been used in the development of optoelectronic systems for telecommunications purposes, but it has not achieved the bit rate, functionality, size, and cost objectives required by the actual systems. Therefore, a new technical approach is being studied. This is multichip module (MCM) technology derived from the electronics industry, which can be considered as the most promising integration scheme [25].

To implement a monolithic integration several technologies have been developed due mainly to the different materials involved. Usually, all these technologies make use of some common basic techniques, such as layer deposition and etching to define patterns, and specific processes depending on the selected base material.

### 13.5.1 Substrate materials and specific processes

According to the performance of the optical component to fabricate, there are several materials that can be used as substrates. These can be grouped into the following categories.

*Ferroelectric crystals* [26]
Ferroelectric crystals such as lithium niobate are able to provide dynamic passive optical components due to their good electro-optic properties. The manufacturing processes tend to give crystals having a local variation of the refractive index with abrupt or gradual profile. Such processes involve conventional techniques such as thermal diffusion, ion exchange and ionic implantation. Among all of these, titanium diffusion is the best method to achieve components with good optical quality. The diffusion process takes place in an oven working at around 1000°C, in an oxygen atmosphere and for times determined by the required diffusion depth. The introduction of titanium into the $LiNbO_3$ crystalline network produces an increase in the refractive index (both real and imaginary parts) due to an increase in the polarizability and to the photo-elastic effect produced by the different size of the titanium atoms.

*Semiconductors* [27,28]
This category comprises III–V and II–VI compound semiconductors all of which have the potential for monolithic integration of sources, detectors and optical guides near electronic devices owing to their direct bandgap, and also silicon, and its derived compounds, which are able to integrate passive optical and electronic components. Due to its poor emitting properties, silicon has not yet been used to produce light sources.

Compound semiconductors provide the starting point for integrated optoelectronics, where sources and optical detectors are monolithically integrated with optical guiding devices and electronic circuitry on the same substrate. The increased performance of the optical components obtained with this type of compound runs in parallel with the development undergone by epitaxial growth techniques. These growth techniques can be grouped into beam and gaseous flow processes. In the first group, the elements of the material to be grown arrive at the substrate in the form of an atomic or molecular beam. This beam is absorbed by the surface and promotes growth through a surface reaction between the reactants. Growth is accomplished at very low pressures, when the free mean path of the reactants is higher than the distance of the beam source from the substrate, thus preventing gas phase reactions. In the second group, the constituent elements of the material arrive at the substrate in the form of a gaseous mixture, even at atmospheric or reduced pressures, and decompose or react in or very close to the surface of the substrate. To the first group belong the molecular beam epitaxy (MBE) and chemical beam epitaxy (CBE) techniques. As precursors of the material to be grown, the MBE uses gaseous or solid sources, mainly in elemental form or as hydride compounds, while the CBE uses metal-organic compounds as source materials of the integrating layer. The second group includes mainly the metal-organic vapour phase epitaxy (MOVPE) and the metal-organic chemical vapour deposition (MOCVD) techniques, where metal-organic compounds are used as source reactants. A more detailed description of these techniques follows.

Owing to its mechanical, chemical and electronic characteristics silicon is the most suitable material for integrated circuits, but this element's optical properties restrict its range of applications. However, both silicon and its compounds have been applied to obtaining passive optical components. Silicon oxide and silicon nitride are the most widely used materials in integrated optics, due to their transparency in the wavelength ranges adopted and also to their amorphousness, being structurally homogeneous. Thin film combinations of these materials obtained by similar gaseous flow techniques are used for the compound semiconductors leading to different waveguide structures resulting in simple optical components with good optical performance.

*Glasses* [29]

These are attainable and low cost materials that have been widely used for manufacture of passive components. Most of the optical components used in fiber communication networks are obtained from glasses. In a glass substrate, the refractive index variation needed to produce optical components is achieved through ionic exchange, replacing the metallic ions of the glass by other ions, such as $K^+$, $Ag^+$ or $Cs^+$. The exchange process takes place in a molten saline solution, at temperatures between 200 and 300°C, where the glass substrate is soaked. The optical component structures are defined using inert metallic masks deposited on the glass surface, which are removed after the diffusion process. The refractive index can be modified, either increased or reduced, by amounts between 0.01 and 0.1 units using appropriate ion combinations. This technique makes use of heat to produce the ion exchange, although this is a very slow process. To enhance the exchange rate an electrical field can be used as an activator.

*Organic materials* [30]
These materials are polymeric and crystalline, and present electro-optical and piezoelectric properties, with coefficients higher than those of lithium niobate. The fabrication of the optical structures is very simple and easy to implement in manufacturing lines, but only dynamically passive components can be realized. However, these materials are still appropriate for optical devices with non-linear characteristics.

Techniques able to produce layers with uniform thickness, free of bubbles and without striation can be applied to fabricating polymeric waveguide structures. To manufacture polymeric optical components, techniques not only derived from microelectronics technology but derived from different fields have been used, such as mechanical (extrusion, lamination, injection moulding), optical (photo-induced refractive index modification, photo-moulding), chemical (etching, molecular orientation) or diffusive (externally induced doping, polymerization diffusion) techniques. Each one of these techniques has been applied to a given type of polymeric material, taking into account properties such as photosensitivity, plasticity, stiffness, polymerization mechanism, transition temperature, polarizability, etc. Furthermore, its properties can be modified by varying the polymer formulation or adding new constituents, thus extending the choice of manufacturing method.

*Sol–gel materials* [31]
Sol–gel materials are porous and optically transparent glass-like materials produced at low temperature by the sol–gel technique, mainly through the hydrolysis and polycondensation of organometallic compounds. Optical structures, such as waveguides, can be fabricated using this technique with good performance. Sol–gel techniques make use of low-cost materials and equipment and is an interesting alternative for basic integrated optical circuits.

### 13.5.2 Basic technological processes

Besides the specific processes related to the selected material as substrate, there are other processes implicated in the fabrication technology. To obtain an integrated structure that is able to confine and transmit a light wave, a sequence of steps involving deposition or growth of a thin film on the substrate surface, outlining the component's pattern, and, finally, transferring the pattern into the growing layer, defines the basic processes that are used in any integrated optical technology: deposition, photolithography and etching.

*Layer fabrication* [32–34]

The most important requirements of any layer suitable for use in an integrated optics application are outlined below.

*Transparency.* At the working wavelength the materials forming the optical component must be completely transparent.

*Linearity.* In order to avoid birefringence effects, crystalline and polycrystalline materials are generally undesirable.

*Homogeneity.* A slight variation of the material's composition could mean a change in the refractive index.

*Ability to be polished* together with the substrate. Light injection into the optical devices is achieved, usually, by end-fire coupling, which requires that the cross-section of the input waveguide must be polished to acceptable optical quality. Hence, if the light guiding material has different hardness from the substrate it cannot be polished appropriately, since the process requirements for obtaining a mirror-quality surface are not the same for both materials. As a rule, materials involved in the manufacture of integrated optics devices should have comparable hardness.

*Low impurity level.* It is known that impurities generally act as scatter or absorption centres, causing an increase of light propagation losses.

*Chemical stability.* Since one of the main applications of the devices described in this chapter is in the field of chemical or biological sensors, it is extremely important that the transducer itself does not suffer from aging.

*Adherence.* Layers are expected to have good adherence both at the surface and between them in order to facilitate robust structures. Moreover, the mechanical properties of these layers, especially their thermal expansion coefficient, have to be as close as possible in order to avoid cracking during the fabrication process.

*Uniformity.* Although this is not as critical as compared to the previous requirements, variation of layer dimensions would signify a modification of all the guiding properties of the structure. If this variation is small, however, it would only mean an increase of the total losses that, although undesirable, could still be acceptable.

Although there are several techniques from which it is possible to obtain material layers, our study is restricted to those processes used mainly during the fabrication of integrated optics devices.

The most important technological step is the one that allows layer formation. The principle of its operation is based on the fact that all components forming the final structure will be delivered from an external source that can be gaseous, liquid or solid. However, deposition from a vapor phase is the preferred technique to obtain thin films of metal, dielectric and semiconductor materials, since this can be transferred easily to mass production without significant changes to its working parameters. Moreover, this technique can be used for coating irregular or even internal surfaces where it achieves good control of the properties of the final layer, such as its homogeneity, uniformity, stoichiometry and refractive index. These deposition processes can be split into two main groups according to the deposition mechanism: if the deposited layer has the same composition as the source, the technique used belongs to the physical deposition processes (physical vapor deposition, PVD); but, if the layer is obtained by a chemical reaction between different source elements, then the technique fits into chemical deposition processes (chemical vapor deposition, CVD).

Different types of thin films are deposited by thermal (or vacuum) evaporation, sputtering, or MBE. All three are examples of PVD. PVD reactors mostly use a solid as a material source. If the thermal energy is supplied to the evaporated material by means of a resistance, RF, laser or electron beam, the process is included in the thermal evaporation category. In contrast, if the source material is transferred to a gas by target ionic bombardment, the process falls into the sputtering category. Through these techniques metal overcoatings are deposited that act as masks in the diffusion processes ($LiNbO_3$, glass) or as electrodes in active and dynamical passive components.

In thermal evaporation, the source material and substrates are placed in a vacuum. The source material is heated until its vapor pressure exceeds the existing pressure in the reactor chamber. Then, the material is evaporated and, subsequently, deposited onto the substrate surface, forming the desired layer by condensation. This technique includes the direct evaporation of the material source as well as the reactive evaporation, where the evaporated material reacts with an added gaseous constituent inside the reactor chamber, before arriving at the substrate, producing a new material. An improvement of thermal evaporation processes in order to obtain crystalline semiconductor materials of higher quality, was achieved with the development of the MBE technique, described earlier.

The parameters that most influence the characteristics of the deposited layer are vacuum quality, growth rate and evaporation method. Other parameters such as substrate temperature and the surface conditions may modify the process results, but these are mostly general points and apply to all of the growth methods.

According to the vacuum quality, systems are designated as high vacuum (HV), with a base pressure of $10^{-6}$ Torr, and ultra-high vacuum (UHV), with a base pressure of $10^{-10}$ Torr. The residual composition of the vacuum chamber atmosphere affects the deposition rate since at low vacuum the growth rate and the collision rate of the residual gas molecules are comparable.

To select a particular evaporation method, the impurities produced during the source evaporation and their influence on the growing layer must be considered. Deposited layer contamination takes place during the evaporation process as a result of the high temperature reached by the crucible containing the source material (when heated by resistance or RF). Through the use of an electron beam, the source can be heated locally to avoid molten material coming into contact with the crucible walls, thus increasing layer purity. Direct sublimation of the material is an alternative method reducing the impurity content, but the source material needs to have a high vapour pressure before its fusion point, which is not feasible with all materials used for deposition.

Low-contamination layers are deposited by sputtering, which is the preferred PVD technique. Moreover, a wider choice of materials is available to work with, demonstrating better step coverage and better adhesion to the substrate. Several sputtering configurations are available. The simplest one, a diode configuration, is seen in Fig. 13.12. The target (or source material) is located in the cathode and the substrate in the anode. Plasma composed of a noble gas (e.g. argon) is generated between the electrodes. Then, a high potential is applied across this set-up. The highly ionized atoms in the plasma accelerate towards the target, physically knocking atoms from the target, which then condense onto the substrate by way of a special geometric arrangement.

In Table 13.1, the main characteristics of the electron gun evaporation and sputtering processes are compared.

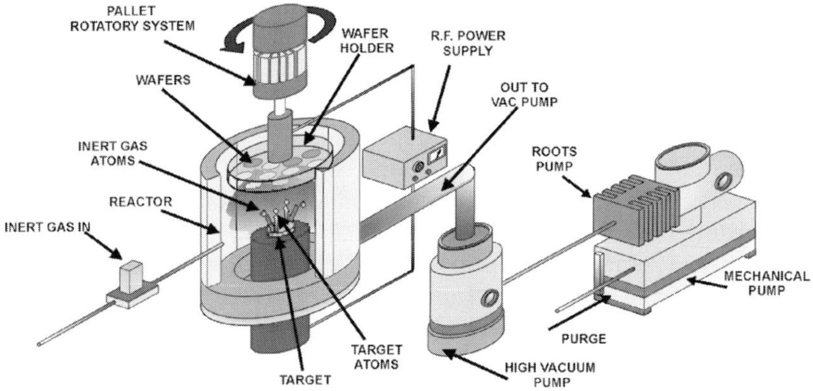


Fig. 13.12. Sputtering system with a diode configuration.

The formation of a solid layer on top of a substrate by a CVD process occurs through the chemical reaction of precursors in the gas phase near or even on the substrate surface. In all CVD techniques, temperature, pressure, precursor flow rates, reactor geometry and operating principle determine the deposition rate and the layer properties (including adherence to the substrate and/or layers underneath). It is extremely important that all the reaction products, except for those forming the expected layer, are volatile at the deposition conditions in order to assure film purity. Thus, CVD processes must proceed through heterogeneous reactions. If homogeneous reactions occur in the gas phase, particle nucleation will result, which may form

TABLE 13.1

Comparison between e-gun vacuum evaporation and sputtering methods of thin film deposition

| | E-gun | Sputtering |
|---|---|---|
| Substrate | Any | Any |
| Substrate temperature (°C) | Room temperature | 60–100 |
| Pressure (Pa) | $10^{-4}$–$10^{-2}$ | $10^{-3}$–$10^{-1}$ |
| Mean free path (m) | 1–100 | 0.1–10 |
| Kinetic energy (eV)/material | 0.3/Al | 3.0/Al |
| Stoichiometry | Poor | Good |
| Step coverage | Unsymmetrical | Symmetrical |

powdery deposits, leading to poorly adhered layers with particle contamination instead of the expected clean and uniform layers.

The most important chemical reactions used in CVD are hydrolysis, synthesis, pyrolysis, oxidation and reduction, depending on the reactants and the material to be deposited.

As compared with other deposition techniques, such as PVD, CVD has several significant advantages, among the most important are given below.

*Precursors.* Materials deposited by CVD include a wide variety of elements and compounds. Inorganic, organic and organometallic precursors are used to obtain isolators, semiconductors and conductors. With the CVD technique, it is possible to obtain layers that decompose during evaporation and then cannot be deposited on by PVD.

*Substrate.* There exists much flexibility in the choice of substrate to be used. Neither conducting nor ionized materials are required, as compared with other deposition techniques.

*Quality.* Deposition conditions are chosen to be as close as possible to equilibrium conditions. This ensures that the deposited layers will have the best possible quality.

*Temperature.* Whether amorphous, polycrystalline or single crystalline films are obtained will depend on the temperature of the process.

*Impurities.* Since CVD uses only high-purity gases, the quantity of contaminants potentially incorporated in the film is greatly reduced.

*Flexibility.* By simply varying the deposition conditions, both chemical composition and physical properties can be tailored during film growth. This allows well-defined multi-layered structures having controlled doping concentrations and stoichiometries.

*Adaptability.* CVD processes can be easily adapted from research level to mass production.

Even so, CVD does have some drawbacks. The most significant are the system complexity and the high number of variables that need controlling. This is caused mainly by the system flexibility.

There exist several CVD classifications according to the process parameters adopted: total pressure (high pressure/low pressure), substrate temperature (hot wall/cold wall), precursors (metal-organic) and method of energy transfer (resistance/RF heating/plasma enhanced/UV excitation/laser excitation). Moreover, what is commonly done is to combine several of the above-mentioned techniques so as to obtain benefit from their advantages. Table 13.2 shows the main features of the most usual CVD techniques.

TABLE 13.2

CVD process classification (with schemes of the reactors)

| Reactor scheme | CVD techniques | Process parameters | Process characteristics |
|---|---|---|---|
| 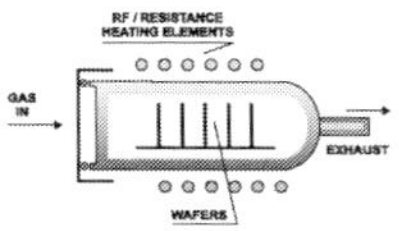  | APCVD (atmospheric pressure) | Temperature: 600–1100 °C, resistance (hot wall, isothermal process) or high frequency induction (cold wall, thermal gradient)<br>Flow rates: >3.0 slm (standard litre per minute)<br>Reactor: horizontal, vertical (pancake) | Hot wall: deposition everywhere, periodic cleaning<br>Cold wall: local deposition<br>Growth rate high (>1.0 μm/min)<br>Poor uniformity and step coverage<br>Low density of substrates<br>Particle contamination |
| 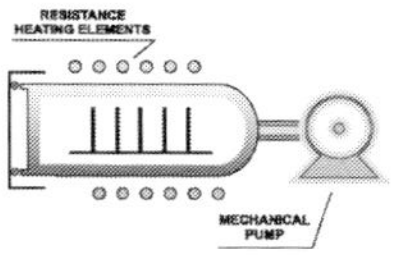  | LPCVD (low pressure) | Temperature: 600–1100 °C, resistance (hot wall, isothermal process)<br>Flow rates: <3.0 slm<br>Reactor: horizontal, vertical<br>Pressure: <1.0 Torr<br>Substrates vertical stacked | Low growth rates (<1.0 μm/min)<br>Deposition controlled by surface reaction<br>Good uniformity and step coverage<br>Low particle contamination<br>High throughput |
| 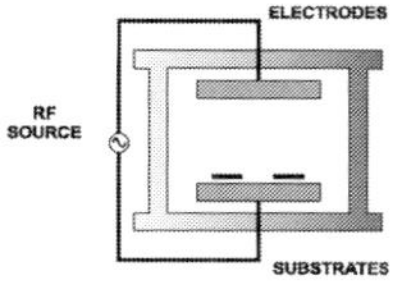  | PECVD (plasma enhanced) | Temperature: 25–400 °C<br>Flow rates: <500 sccm<br>Reactor: horizontal, vertical<br>Pressure: <1.0 Torr<br>Substrates horizontal or vertical stacked<br>Power: up to kWs<br>Frequency: 50 KHz–2.45 GHz | Low temperatures involved<br>High growth rates<br>Low melting point substrates<br>Mechanical stress reduction<br>High uniformity and step coverage<br>Contamination by by-products<br>Ion bombardment damage |

(continued on next page)

**Table 13.2** *(continued)*

| Reactor scheme | CVD techniques | Process parameters | Process characteristics |
|---|---|---|---|
| 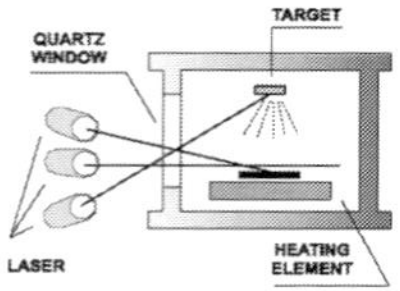  | LCVD (laser assisted) | Variable temperature<br>Pressure: 1.0 Torr–1.0 atm<br>Low flow rates | Three mechanisms: pyrolysis: continuous lasers; photolysis: pulsed and continuous lasers with specific $\lambda$; evaporation: high power pulsed lasers<br>Low growth rates |
| 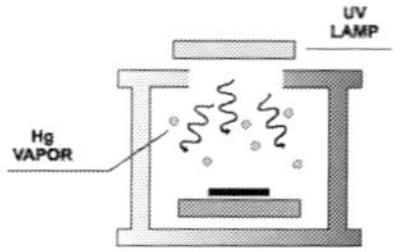  | PhCVD (photon assisted) | Room temperature<br>Variable pressure<br>Active mercury gas<br>UV lamps or lasers | Reactive photolysis (Hg)<br>Direct photolysis (UV radiation)<br>Slow deposition rate<br>Contamination<br>Poor uniformity |

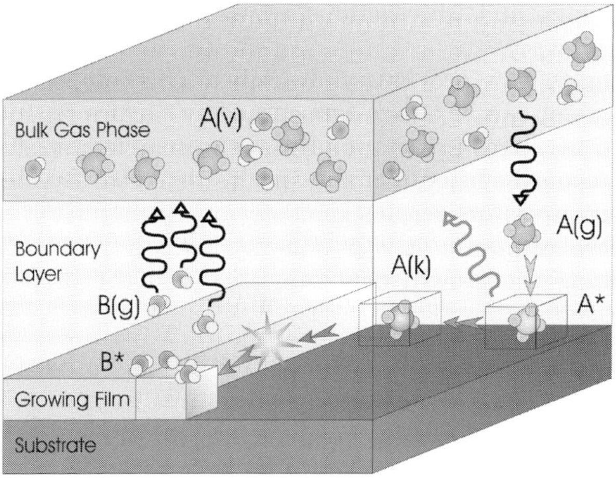


Fig. 13.13. Standard model employed for the growth mechanism description in a CVD.

Common to all the CVD systems is the basic growth mechanism. The growth in a CVD process can be illustrated with the help of Fig. 13.13. It has the following reaction sequence listed below:

- The precursor molecules, $A(v)$, diffuse due to a concentration gradient, the bulk gas zone, through the boundary layer (the thickness of which depends on the working conditions) reaching the grown surface, $A(g)$.
- Once the precursor reaches the substrate surface, it is adsorbed ($A^*$) and diffuses at substrate level until it is desorbed or until it finds a position where it can form a stable bond (i.e. a crystallographic step), $A(k)$.
- The precursor continues to diffuse along the step until it finds a stable nucleation point. A chemical reaction is then produced, providing the stable nucleus to the growing layer, $A(s)$. The residual reaction products, $B^*$, diffuse along the surface until they are finally desorbed, $B(g)$, and then diffuse across the boundary layer until they reach the bulk gas phase.

Adsorption of the precursor reactants gives rise to the first location of growth, followed by nucleation. With respect to the growth rate, it would be extremely useful to determine those factors that are the limiting steps in CVD reactions, i.e. what factors control the overall deposition process.

According to the previously described CVD deposition scheme, the limiting step can be either determined by surface reaction kinetics, i.e. the concentration of reactant gases, or by mass transport, dependent on the diffusion coefficients of the species involved. By modifying the temperature, pressure, reactor geometry or the method of transferring energy to the reactants, it is possible to change the kinetic regimen, thus changing the growth characteristics and the properties of the deposited layer.

*Photolithography* [35]

The photolithography process involves the transference of two-dimensional patterns placed on a mask to an organic polymer, which, after being developed, reveals an exact copy of the mask patterns. At this point, any technological step will be done only at the photoresist-free zones. An entire photolithographic process is schematically presented in Fig. 13.14.

In the first step of the photolithography process, a thin film of organic polymer, called the photoresist, which is sensitive to UV radiation, is deposited on the surface of the layer to be patterned. The photoresist is positive (or of positive tone) when the photochemical reaction during the exposure produces an increase of its solubility in developing solutions. Negative photoresists (negative tone), when exposed to UV radiation, undergo a photo-polymerization process which strengthens the polymer by random polymer chains cross-linking to produce an increase of its molecular weight thus becoming less soluble. A comparison between both types of photoresists is presented in Table 13.3. The choice between them depends on a wide variety of considerations, such as cost, processing time, resolution or available equipment.

The exposure to UV radiation takes place through a glass plate where two-dimensional geometric patterns are designed for each specific layer, this glass plate being designed as a photomask (or 'mask', for short). Masks are nearly flat optical glasses (transparent to near UV) or quartz plates (transparent to deep UV) with a metal (typically 800 Å thick chromium) layer where the patterns are drawn using a photolithographic process.

When obtaining any device, it is necessary to use one or more different masks, depending on its complexity. The final device is obtained

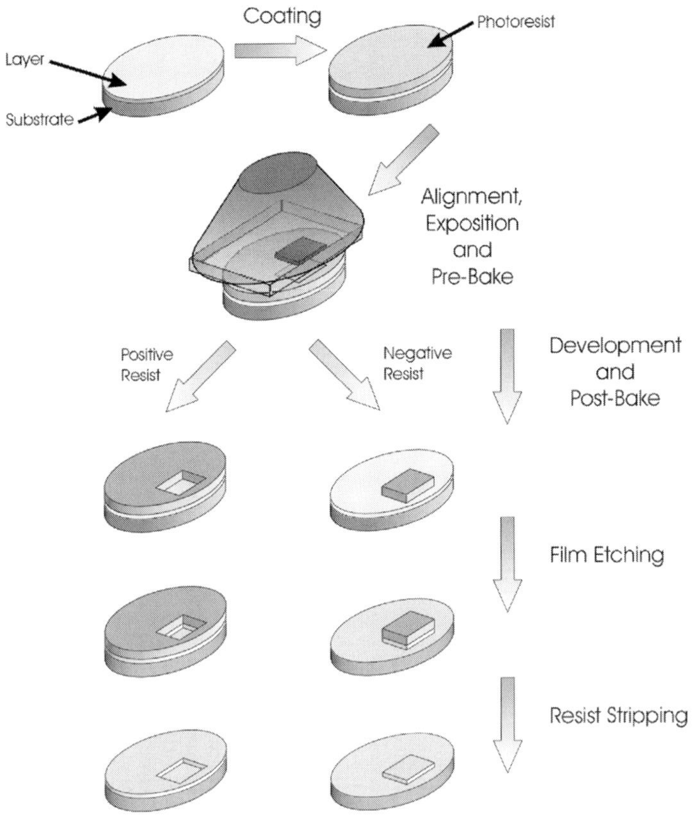


Fig. 13.14. Standard photolithographic process steps.

step-by-step, transferring sequentially the patterns of each mask to the wafer level and subsequent layer processing. Thus, the lithography step takes place as many times as the mask number needed, being one of the most critical of the manufacturing steps.

Commonly, to obtain an integrated optical device, more than one mask is needed. For that reason, it is essential that there exists a precise alignment between the used mask and the previous patterns applied to the wafer. This operation requires high positioning precision, since a small misalignment between them could cause the device to fail.

The equipment used to irradiate the wafer also performs the mechanical displacement to achieve the alignment between mask and wafer, and is therefore designated a 'mask aligner'. The aligners provide

TABLE 13.3

Comparison of positive and negative photoresists

| Properties | Positive resist | Negative resist |
|---|---|---|
| Adhesion to Si | Fair | Excellent |
| Available compositions | Many | Vast |
| Contrast/resolution | High | Low |
| Cost | Expensive | Cheap |
| Developer | Aqueous-based | Organic solvent |
| Developer process window | Small | Wide, insensitive to overdeveloping |
| Image width to resist thickness | 1:1 | 3:1 |
| Influence of oxygen | No | Yes |
| Lift-off | Yes | No |
| Minimum feature | $<0.5$ μm | 1 μm |
| Opaque dirt on clear mask effect | Not much effect | Pinhole printing |
| Pinhole count | Higher | Lower |
| Pinholes in mask | Higher | Lower |
| Proximity effect | Prints isolated trenches better | Print isolates lines better |
| Residue after development | Mostly $<1$ μm | Often a problem |
| Step coverage | Good | Fair |
| Swelling in developer | No | Yes |
| Thermal stability | Good | Fair |
| Wet chemical resistance | Fair | Excellent |
| Easy to handle | Good | Fair |

the radiation source to expose the photoresist. According to the nature of the radiation (UV, electron beam or X-rays), different lithography systems are available. Among them, photolithography, which uses UV radiation, is the most common.

The quality of a lithographic process is established through the parameters of resolution, alignment and yield. Resolution is defined as the minimal pattern size that can be repetitively transferred; the alignment gives a measure of how close patterns of successive mask levels can be; and the yield is concerned with the number of processed wafers per hour. For standard photolithography, working at a wavelength of 340 nm, resolution is approximately 1.0 μm, alignment is 0.5 μm and about 100 wafers per hour can be exposed. The electron beam systems have improved transfer performances with resolutions of less than 0.03 μm and alignments better than 0.02 μm.

*Etching* [36]

Once layers have been deposited and a photolithographic process has been completed, the most common step is the selective removal to uncover areas of the layer in the so-called etching process. This step depends on the layer history of the materials on the wafer (since their properties could have been significantly affected) and the deposition mechanism used to obtain the layer. Depending on the final geometry, etching allows the definition of the pattern that will be maintained in successive steps until the final device is obtained or until the pattern is eliminated in subsequent etching processes.

Within etching processes, two types can be distinguished: wet etching, where the active components are dissolved in aqueous or organic solutions; and dry etching, which uses ionized gases as reactive agents (plasma).

By wet etching it is understood that the elimination of a material by its dissolution takes place in an adequate etching solution. It is used mainly for cleaning, shaping and polishing. To etch different materials, a wide variety of chemical agents can be used. It should be taken into consideration, however, that solutions used in etching processes must etch exclusively the desired film, without affecting any underlying layers or lithographic masks. Thus, etching solutions must be highly selective, defining the selectivity as the ratio between the etching rates of two different films in the same etching solution. Modification of wet etchant and/or temperature can also alter its selectivity.

Two different etching types can be defined: *isotropic*, when the etching rate is equal in all directions, as occurs in amorphous materials; and *anisotropic*, when there exists a preferential etching direction that in crystalline materials is related to one of the crystallographic planes. The etching factor is defined as the relationship between etching depth and lateral etching (underetching), measured under the protection mask. Obviously, this factor will be unity for isotropic etching and greater that one for anisotropic processes. Since the mask pattern has to be transferred as accurately as possible, the etching factor of the dissolution used is extremely important. Figure 13.15 shows schematically the profiles obtained for both etching mechanisms.

For pattern dimensions below 3.0 μm, wet etching is restricted due to non-uniformity problems caused by superficial tension effects, although this effect reduces as the thickness removed increases. Moreover, isotropic etching causes a dimension reduction due to lateral underetching. By this method, a device with dimensions below 2.0 μm cannot be obtained and another technique that allows anisotropic etching of amorphous layers.

Dry etching overcomes the drawbacks of wet etching. Dry etching comprises a family of methods by which a solid-state surface is etched in the gas phase: physically by ion bombardment; chemically by a chemical reaction with reactive species at the surface; or a combined physical and chemical mechanism. The plasma dry etching techniques are categorized according to the specific set-ups as either glow discharges or ion-beam techniques. In the former technique, plasma is generated in the same vacuum chamber where the wafers are located, whereas in ion-beam techniques the plasma is generated in a chamber separate from the etching reactor. In an ion-beam etching system (in which are included sputter/ion etching and ion-beam milling) the etch is basically due to a physical effect, namely, momentum transfer between energetic ions and the substrate surface, which provides vertical etch but poor

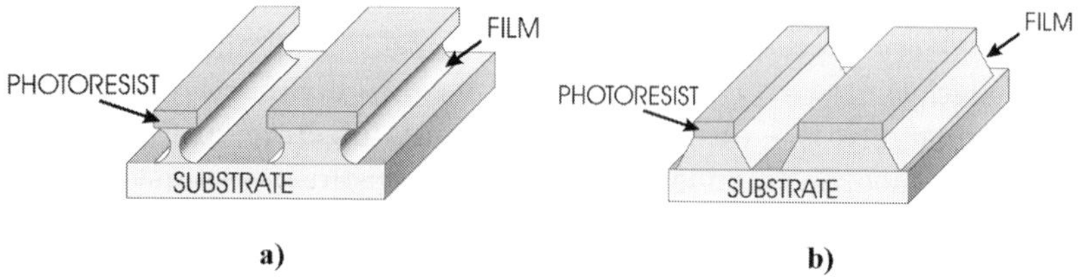


Fig. 13.15. Different etching profiles obtained: (a) isotropic; and (b) anisotropic.

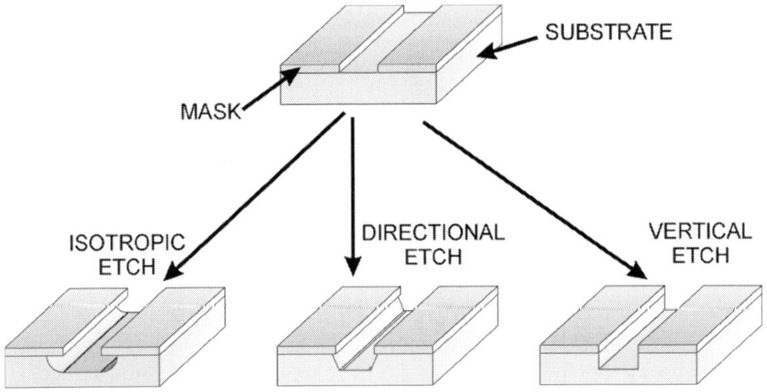


Fig. 13.16. Possible etch profiles obtained by dry etching techniques.

selectivity. A chemical reaction takes place in all other dry etching methods. In the physical/chemical case, impacting ions, electrons or photons induce chemical reactions that initiate the etching process, obtaining directional etching but with good selectivity.

When selecting a dry etching process, the desired shape of the etch profile and the etch selectivity require careful consideration. In Fig. 13.16, different possible etch profiles that can be obtained by dry etching are presented. Depending on the etching mechanism, isotropic, directional, or vertical etch profiles are obtained. The difference between wet and dry processes is that these profiles can be generated in single crystalline as well as in polycrystalline and amorphous materials. The etching anisotropy is not a result of single crystal anisotropy, as was the case in wet etching, but a result of the plasma conditions.

The most significant features of these systems can be seen in Table 13.4.

In order to obtain integrated components with well-defined optical characteristics the etching process needs to fulfill two main requirements:

- high uniformity and repetition, due to the fact that optical components are layered at the millimeter scale on the surface and variations in the layer thickness can increase intensity losses; and
- sidewalls should be very smooth and as vertical as possible to avoid acting as scattering points or prisms thus increasing losses by reflection.

TABLE 13.4

Comparison of dry etching methods

| Scheme | Etching techniques | Process parameters | Process characteristics |
|---|---|---|---|
| 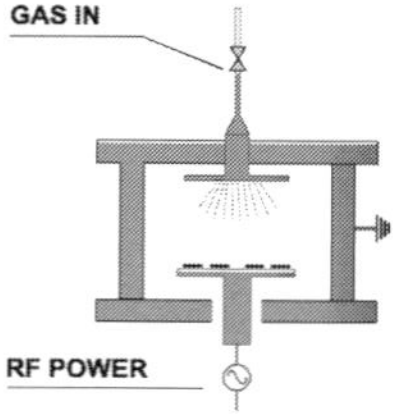  | RIE | Frequency: 13.56 MHz<br>Pressure: 5.0–100.0 mTorr<br>Ion energy: >100 eV<br>Chlorinated or fluorinated gases | Anisotropic etching<br>High uniformity and selectivity<br>High throughput<br>High ion damage<br>High etching rates<br>Physical and chemical etching |
| 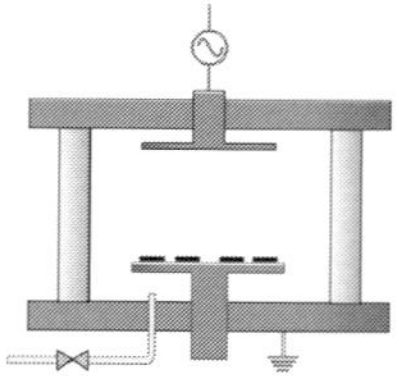 | Plasma etching | Frequency: 13.56 MHz<br>Pressure: 0.1–2.0 Torr<br>Chlorinated gases/oxygen<br>Low ion energy | Isotropic etching<br>Low ion damage<br>Good uniformity<br>Low plasma densities<br>High selectivity<br>Chemical etching |

**Table 13.4** *(continued)*

| Scheme | Etching techniques | Process parameters | Process characteristics |
|---|---|---|---|
| 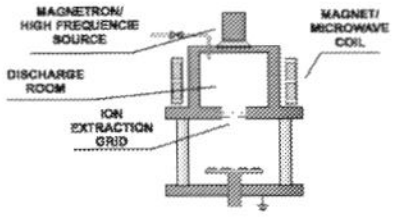  | ECR or microwave etching (electron cyclotron resonance) | Gas pressure: <2.0 mTorr<br>Ion energy: 10–20 eV<br>Remote plasma<br>Fluorinated or chlorinated gases | Poor uniformity<br>High etching rates<br>Isotropic or anisotropic etching<br>Low throughput<br>High plasma densities<br>Ion damage<br>Physical and chemical etching |
| 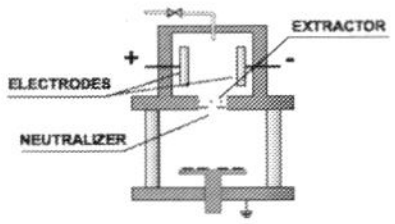  | Ion beam etching (IBE) | Gas pressure: <2.0 mTorr<br>Ion energy: >500 eV<br>Remote plasma<br>Noble gases | Poor selectivity<br>High directional etching (vertical)<br>High radiation damage<br>Physical etching |

Furthermore, in some integrated optical technologies, the etch depths must be at least several micrometers, requiring an etching process with parameters very different to those used in conventional microelectronic technologies.

Of all possible dry etching systems, the most widely used is reactive ion etching (RIE). It is based on a physical/chemical reaction that allows directional etching of high selectivity while providing precise pattern transfer. The relationship between the physical and chemical mechanism can be changed according to the process parameters.

## 13.6 INTEGRATED OPTICAL SENSORS

As was pointed out previously, integrated optical devices combine high sensitivity with compactness, simple instrumentation and a high degree of batch production. Integrated optical waveguide structures are well suited for integration of optical and electrical functions on one substrate (chip) [37,38]. Several integrated optical sensors have been reported in the literature, for example integrated absorbance sensor [39], surface plasmon resonance system [22], grating coupler [12], resonant mirror [14] and interferometer [17]. All of these sensors have advantages and drawbacks, but the common characteristic is that they all require careful waveguide design, fabrication and testing. Consequently, only a few of them have actually reached the marketplace, for example, the IAsys resonant mirror device from Affinity Sensors Ltd (Cambridge, UK) or the BIOS-1 grating coupler sensor from Artificial Sensor Instruments (Zürich, Switzerland). A detailed description of these different integrated sensors is provided below.

### 13.6.1 Absorbance sensor

Despite their simplicity, traditional absorbance/transmittance measurements are very useful owing to their broad applicability and versatility with respect to a large number of analytes. However, the discrete absorbance/transmittance optical sensors described to date have been criticized as presenting short dynamic ranges, low sensitivity and high response times. Moreover, since a light beam has to traverse the sample solution to be measured, interference effects, such as turbidity and selective sample matrix absorption, are very often present.

Technologies described in the preceding paragraph have led to the development of some specific integrated optical sensors, the simplest being an extrinsic absorbance optochemical sensor, which overcomes many of the indicated difficulties. The configuration of an integrated absorbance device designed to detect a potassium ion is schematically depicted in Fig. 13.17 [39]. It is based on a planar microoptic circuit where a chemically active membrane is deposited which constitutes a part of the guiding planar structure. In this approach, light propagates through the membrane instead of through the solution, thus avoiding sample interference. Long light paths can be used, increasing in this way the sensitivity of the measurement. Moreover, as the analyte diffusion takes place perpendicularly to the light path through a micron-thick membrane, reduced response times can be achieved.

The sensor may be divided into three main regions: the input and output light guides (termed excitation and collection waveguides, respectively) and a free propagation region between them, where the membrane is placed. Both waveguides are fabricated in a unique substrate using silica on silicon-based integrated optic technology, where they are directly aligned. The excitation waveguide comprises a single-mode 14-μm wide rib ARROW-type structure. The output waveguide collects the light that has been propagated through the membrane and it is formed by a 50-μm wide ARROW multimode structure. The active membrane, being the functional part of the planar light guiding structure, is fabricated with thickness and refractive index values similar to those of the input and output waveguide core.

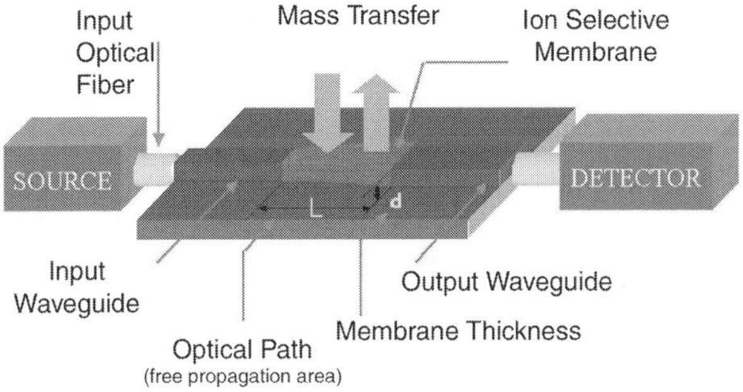


Fig. 13.17. Conceptual design of the absorbance sensor.

Therefore, in the sensing region, light remains guided in the direction transverse to the layers. However, since the rib structure is not prolonged in this region, the light beam is broadened in the lateral direction. As a consequence, a wider core waveguide is necessary at the output to guarantee low loss membrane-collection waveguide coupling.

The sensing region is also integrated in the same substrate as the excitation and collection waveguides. It is defined by photolithography followed by RIE, achieved on an ARROW waveguide multilayered structure, from which the guiding elements are formed. All the deposited layers are eliminated in this region, thus leaving the substrate free and ready for deposition of the membrane.

The membrane is formulated with a polymer support, normally poly(vinyl chloride), and an ionophore, which plays the role of chemically recognizing the analyte. When incorporated into the absorbance optical sensor described here, a second compound is included in the membrane, a so-called chromoionophore. This compound plays a double role. When selective extraction of $K^+$ from the solution toward the membrane is accomplished by the ionophore, a simultaneous expelling of $H^+$ from the membrane to the solution takes place through deprotonation of the chromoionophore, thus ensuring the electrical neutrality of the solution. At the same time, the absorbance spectrum of this compound change achieves the chemical–optical transduction.

The membrane is deposited over the free propagation region using a spin coating technique.

Figure 13.18 is an amplified real view of the sensor structure, before and after membrane deposition. Input and output waveguides as well as the sensing region are observed.

The sensor optimizes the four essential characteristics that define a high-quality optochemical sensor: enhanced and sizable sensitivity, depending on the free propagation region length; adequate selectivity

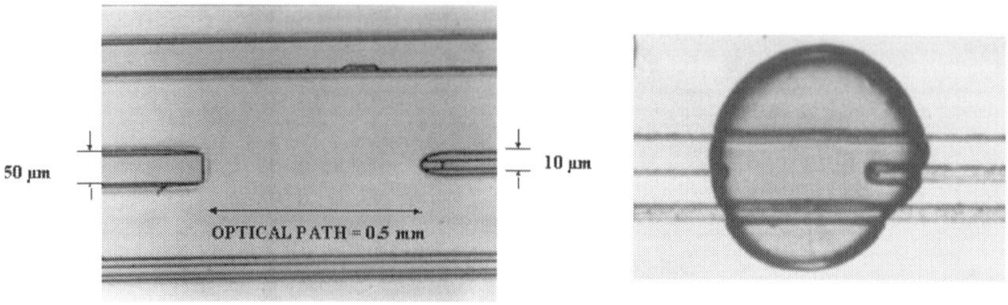


Fig. 13.18. Structure of the fabricated absorbance sensor.

toward a certain analyte, regulated by the ionophore; low response time, guaranteed by the reduced membrane thickness; and minimization of interference effects caused by the solution matrix [40].

A typical experimental recording is shown in Fig. 13.19. It displays the sensor optical power output as a function of time, corresponding to the deposition of different drops of potassium chloride standard solutions on sensing region, the compositions of which ranges from $1 \times 10^{-4}$ to $5 \times 10^{-1}$ M in 0.05 M acetic acid/magnesium acetate buffer at pH 5.5. The pH value modulates the extraction process, and, accordingly, the potassium concentration range. A dynamical range of about 30 dB in the measured optical power, which represents 3 absorbance units, is obtained for a free propagation region length of 1 mm. Moreover, response times as low as a few seconds are achieved even for a membrane thickness as large as 20 μm.

### 13.6.2 Grating coupler

This sensor, proposed by Tiefenthaler and Lukosz [12], is based on an optical grating prepared from a thin waveguide deposited over a glass substrate, as can be seen in Fig. 13.20. The grating facilitates the direct input of the laser light at an angle which excites a guide mode when

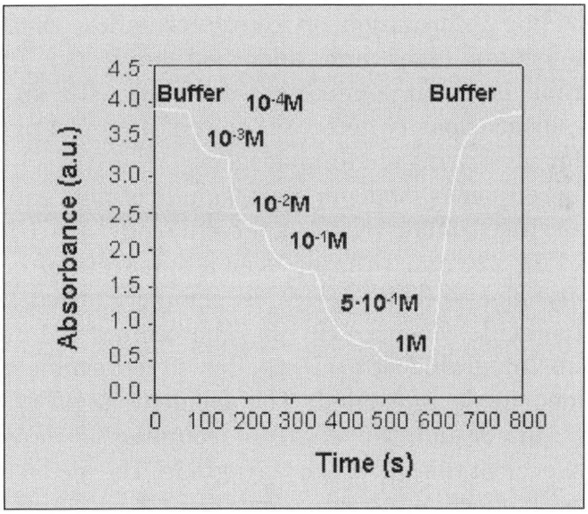


Fig. 13.19. Experimental calibration curve for a potassium optrode.

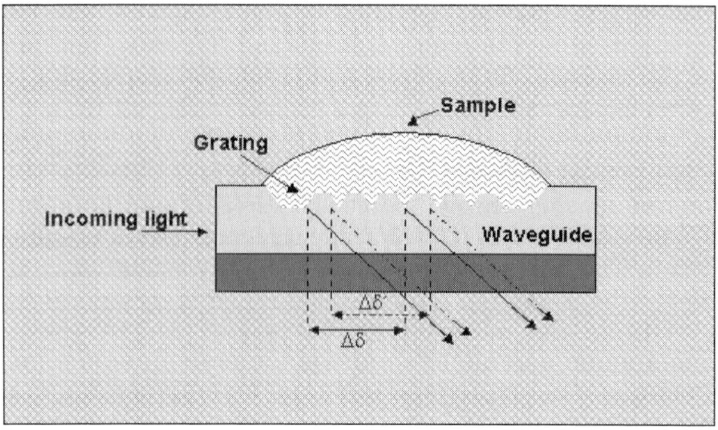


Fig. 13.20. Schematic representation of the grating coupler device and its function.

the incoupling condition is fulfilled. As the in(out)coupling angle is very sensitive to any variation in the refractive index above the grating surface, the change produced in the coupled angle as a consequence of any reaction on the grating can be used for sensing.

Varying the angle of incidence, the mode spectrum can be obtained and the effective refractive indexes calculated for both TE and TM modes. Adequate software allows real-time monitoring of any surface process, providing information on refractive index, thickness of the adsorbed molecular layer and adsorbed mass per unit surface. Measuring time is about 2 min per spectrum with an accuracy of 0.005° and a surface density detection limit of around 1 pg/mm$^2$, which is sufficient for measuring small molecules.

The grating coupler is made on a waveguide of high refractive index ($SiO_2$–$TiO_2$ or $Ta_2O_2$ made by sol–gel technology, $n = 1.75–1.82$) and thickness of 170–220 nm, situated over a substrate of low refractive index (usually glass). This device is commercialized by Artificial Sensing Instruments, Zurich ASI AG. The commercial device (BIOS-1) employs an integrated optical scanner to determine the angle at which a guided mode is excited. The incident angle of the light is continuously varied and the angle of coupling is detected by the photodiodes placed at the end of the waveguide. The model GKR 102 has a resolution with effective refractive index of $1 \times 10^{-6}$ and temperature control.

The grating coupler has been used for sensing when different sensitive layers were immobilized onto the grating. Such sensors were used for monitoring drugs [41], pesticides [42], metal ions [43], and for studies of affinity reactions [44].

More refinements to this sensor concept have been the subject of further developments [13,45], mainly more compact and miniaturized integrated optical sensors [46], as for example the replicated integrated optical sensors [47]. This further miniaturization and integration is quite important for many applications, leading to reduced cost of the sensor chips and the whole system, an increase in the density of the sensing pad on a single chip for multianalyte detection and enhanced specificity and accuracy.

### 13.6.3 Resonant mirror

The resonant mirror device (IAsys) [14] is a leaky planar waveguide sensor that uses frustrated TIR to couple light in and out of the waveguide layer. It is similar to the SPR device but in this sensor the metal film is replaced by a dielectric resonant layer of high refractive index (e.g. titania or hafnia) and thickness of approximately 100 nm, separated from a glass prism by a dielectric layer of low refractive index ($SiO_2$) and thickness of 0.5–1 μm (sandwich configuration of high–low–high $n$). This layer is thin enough to allow light to couple into the resonant layer via the evanescent field when incident light is focused onto the prism–silica interface. Efficient coupling occurs for certain angles where phase matching with the resonant guided modes in the waveguide is achieved. A schematic of the sensor is shown in Fig. 13.21.

As the waveguide layer acts as a resonant cavity (for which the device has been termed the resonant mirror), the light reflected from the RM device undergoes a full $2\pi$ phase change across the resonance through either angle (for a fixed input wavelength) or wavelength (for a fixed input angle). The phase change due to a sensing reaction can be visualized using crossed polarizers to produce a peak in intensity at the resonance angle or wavelength.

The IAsys commercial biosensor (www.affinity-sensors.com) uses this optical platform for real-time measurement of biomolecular interactions. With the fabrication technology achieved in this commercial device the sensor chips can be made in large quantities with reasonable manufacturing tolerances and cost. This device has been used for

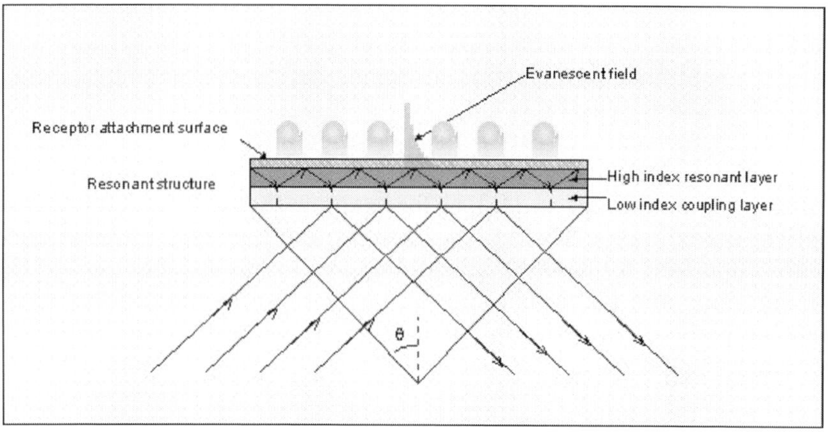


Fig. 13.21. Schematic representation of the resonant mirror device.

studies of DNA hybridization [48], protein–protein interactions and many others applications [49]. The IAsys utilizes a single channel but the model IAsys Auto+ has dual-channel flow cells for independent measurements. Kinetics software is also provided.

### 13.6.4 Mach–Zehnder interferometer

This interferometric arrangement is highly sensitive and is the only one that provides an internal reference for compensation of refractive-index fluctuations and unspecific adsorption. Several interferometric devices have been described including the difference interferometer, the Mach–Zehnder and the Young interferometer [37]. Interferometric sensors have a broader dynamic range than most other types of sensors and show higher sensitivity as compared to other integrated schemes, as is shown in Table 13.5, where a comparison of the different sensor technologies as a function of the limit of detection is presented.

The physical principle of working a Mach–Zehnder interferometer was presented in previous paragraphs. For detection purposes, when a chemical or biochemical reaction takes place in the sensor area, only the light that travels through this arm will experience a change in its effective refractive index. At the sensor output, the intensity ($I$) of the light coming from both arms will interfere, showing a sinusoidal variation that depends on the difference of the effective refractive

TABLE 13.5

Comparison between integrated transducers based on different sensing principles

| Sensing principle | Limit of detection ($pg/mm^2$) |
|---|---|
| SPR | 2–5 |
| Waveguide–SPR | 2 |
| Resonant mirror | 5 |
| Grating coupler | 1–10 |
| Mach–Zehnder interferometer | $<0.1$ |
| Differential mode interferometry | 1 |
| Reflectometry interference spectroscopy (RifS) (not EW detection) | 1–5 |

indexes of the sensor ($N_{eff,S}$) and reference arms ($N_{eff,R}$) and on the interaction length ($L$):

$$I = \frac{1}{2} > I_0\left[1 + \cos\left(\frac{2\pi}{\lambda}\left(N_{eff,S} - N_{eff,R}\right)L\right)\right] \quad (13.3)$$

where $\lambda$ is the wavelength. This sinusoidal variation can be directly related to the concentration of the analyte to be measured.

Although the Mach–Zehnder interferometer sensor seems to be one of the more promising concepts [18] for detection of low concentrations of small molecules without labels ($10^{-12}$ M or even lower), it is not commercially available yet [37]. The main problem is the complexity of the design, fabrication and optical adjustments: the overall procedure for MZI fabrication being rather laborious and single mode waveguides are requiring increasingly more complex technology.

Using a conventional TIR planar waveguide structure, a Mach–Zehnder interferometer sensor has been developed for immunological purposes with a sensitivity of $10^{-3}$ nm in the thickness of the adsorbed layer (which corresponds to an adsorbed molecular layer of 1 $pg/mm^2$). Using integrated channel waveguides for single mode behaviour where the height of the waveguide (rib) has to be less than 3 nm [16] complicates the fabrication and testing of this device [16].

An ARROW has also been used for the fabrication of an integrated MZI sensor [17,50]. The attractive characteristics of the ARROW technology was introduced before. An example of a biosensing application using the ARROW–MZI can be observed in Fig. 13.22(a) and (b), where the evaluation of the immunoreaction between the system

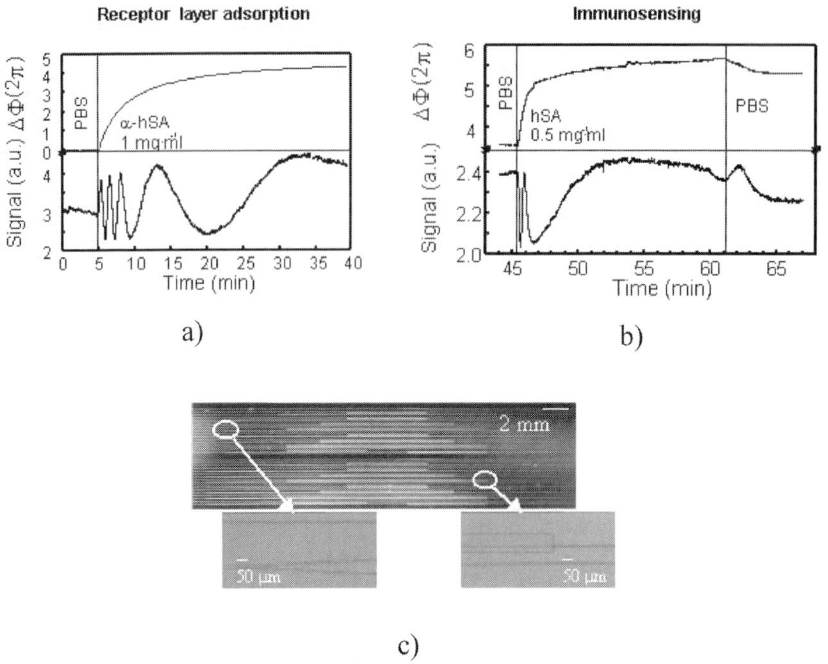


Fig. 13.22. Immunoreaction: (a) adsorption of the receptor forming the sensing layer; (b) determination of the target analyte; and (c) picture of the chip containing different MZI sensors.

antibody/antigen α-HAS/HAS has been performed. In Fig. 13.22(c) a picture of the MZI used is also shown.

## 13.7 TOWARDS A TOTAL INTEGRATED SYSTEM

Most commercial sensors are rather bulky and the prices of the instrument and/or the sensor chips are very high. There is still a need for future development of compact integrated optical sensors and some attempts have already been done [37]. The high degree of development of optical fibers and integrated optic technology, including fibers, emitters, detectors, waveguides and related devices should allow for the flexible development of miniaturized compact devices. These can be included in a remote control optical fiber network, for chemical analysis in standard analytical chemical and medical laboratories, as well as for environmental monitoring or control in manufacturing processes of the chemical industry. While size and weight are less important considerations for instruments intended for laboratory use, devices for on-site analysis or point-of-care operations are geared for portability, ease of use and low cost. IO devices have compact structures and will allow for fabricating optical sensor arrays on a single substrate for simultaneous detection of multiple analytes. Mass-production of sensors will be possible using integrated optics with the manufacture of miniaturized devices integrating the electronics and optics ('sensor-on-a-chip' in which the light source, photodiodes and sensor waveguides are combined on a single semiconductor package; see Fig. 13.23), the flow system and the reagent deposition (by ink-jet, screen-printing or other technology). A complete system fabricated with integrated optics offers low complexity, robustness, a standard device and portability.

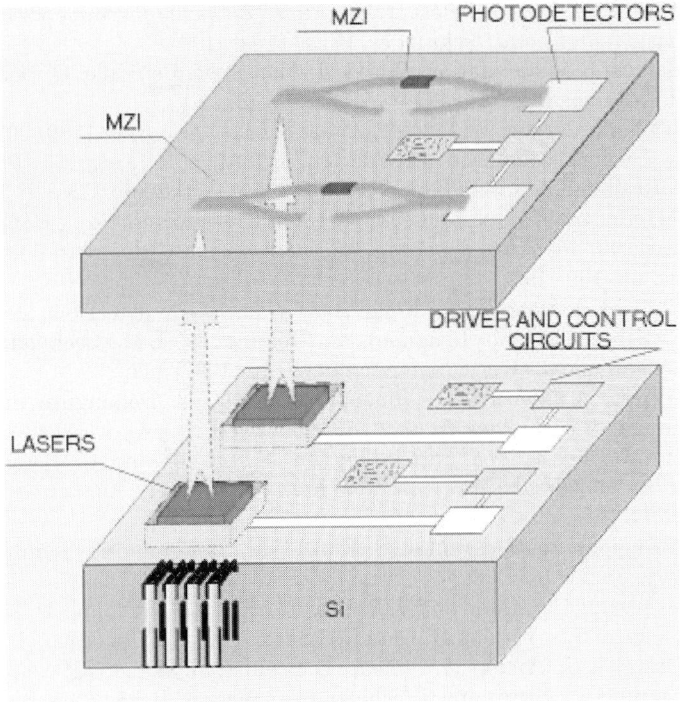


Fig. 13.23. A possible scheme for a total integration system including MZI sensor devices, laser, photo-detectors and CMOS electronics.

## REFERENCES


1 S.E. Miller, *Bell Syst. Technol. J.*, 48 (1969) 2059.
2 R.C. Alferness, H. Kogelnik and Th.H. Wood, *Bell Lab. Technol. J.*, 5 (2000) 188.
3 C. Vasallo, *Optical Waveguide Concepts*, Elsevier, Amsterdam, 1991.
4 R. März, *Integrated Optics: Design and Modeling*, Artech House, 1995.
5 M.A. Duguay, Y. Kokubun and T.L. Koch, *Appl. Phys. Lett.*, 49 (1986) 13.
6 P.N. Robson and P.C. Kendall (Eds.), *Rib waveguide theory by the spectral index method*, Research Studies Press Ltd/Wiley, Hertfordshire, UK/London, 1990.
7 D. Marcuse, *Theory of dielectric optical waveguide*, Academic Press, New York, 1991.
8 W.K. Burns and A.F. Milton, In: T. Tamir (Ed.), *Wave-guide optoelectronics*, Springer-Verlag, 1990, pp. 89–144.
9 C.M. Kim and R.V. Ramaswamy, *J. Lightwave Technol.*, 7 (1989) 1581.
10 P.P. Silvester and G. Pelosi (Eds.), *Finite Elements for Wave Electromagnetics: Methods and Techniques*, IEEE Press, 1994.
11 I. Garcés, F. Villuendas, J. Subías, J. Alonso, M. Del Valle, C. Domínguez and E. Bartolomé, *Optics Lett.*, 23 (1998) 225.
12 K. Tiefenthaler and W. Lukosz, *J. Optical Soc. Am. B*, 6 (1989) 209.
13 T. O'Brien, L.H. Johnson, J.L. Aldrich, S.G. Allen, L. Liang, A.L. Plummer, S.J. Krak and A.A. Boiarski, *Biosen. Bioelect.*, 14 (2000) 815.
14 R.P.H. Kooyman and L.M. Lechuga, In: E. Kress-Rogers (Ed.), *Handbook of Biosensors: Medicine, Food and the Environment*, CRC Press, Boca Raton, 1997, pp. 169–196.
15 P. Skládal, A. Deng and V. Kolár, *Anal. Chim. Acta*, 399 (1999) 29.
16 E.F. Schipper, A.M. Brugman, C. Domínguez, L.M. Lechuga, R.P.H. Kooyman and J. Greve, *Sens. Actuat. B*, 40 (1997) 147.
17 F. Prieto, A. Llobera, D. Jiménez, A. Calle, C. Domínguez and L.M. Lechuga, *J. Lightwave Technol.*, 18 (2000) 966.
18 D.P. Campbell and J. McCloskey, In: F. Liegler and C. Rowe (Eds.), *Optical Biosensors: Present and Future*, Elsevier, Amsterdam, 2002, pp. 277–304.
19 A. Brandenburg, R. Krauter, C. Künzel, M. Stefan and H. Schulte, *Appl. Optics*, 39 (2000) 6396.
20 L.M. Lechuga, A. Calle and F. Prieto, *Quím. Anal.*, 19(suppl. 1) (2000) 54.
21 J. Homola, S.S. Yee and G. Gauglitz, *Sens. Actuat. B*, 54 (1999) 3.
22 J. Dostálek, J. Ctyroký, J. Homola, E. Brynda, M. Skalský, P. Nekvindová, J. Spirková, J. Skvor and J. Schöfel, *Sens. Actuat. B*, 76 (2001) 8–12.
23 R.Th. Kersten, In: J. Dakin and B. Culshaw (Eds.), *Optical Fiber Sensors: Principles and Components*, Artech House, 1988, pp. 277–318.

24 H. Lockwood, P. Hangsjaa, C. Armiento and R. Boudreau, In: O. Wada (Ed.), *Optoelectronic Integration: Physics, Technology and Applications*, Kluwer, Dordrecht, 1994, pp. 375–422.
25 D.E. Smith, M.J. Murdocca and T.W. Stone, In: J. Jahns and S.H. Lee (Eds.), *Optical Computing Hardware*, Academic Press, New York, 1994.
26 Ch.-Y. Li and Y. Xu, In: M.C. Gupta (Ed.), *Handbook of Photonics*, CRC Press, Boca Raton, 1997, pp. 87–123.
27 M. Ilegems, In: O. Wada (Ed.), *Optoelectronic Integration: Physics, Technology and Applications*, Kluwer, Dordrecht, 1994, pp. 61–106.
28 B.P. Pal, In: E. Wolf (Ed.), *Progress in Optics*, Vol. 32. North-Holland, Amsterdam, 1993, pp. 1–59.
29 R.V. Ramaswamy and R. Srivastava, *J. Lightwave Technol.*, 6 (1988) 984.
30 L.A. Hornack, (Ed.), *Polymers for Lightwave and Integrated Optics: Technology and Applications*, Marcel Dekker, New York, 1992.
31 C.L. Klein, *Sol–Gel Optics: Processing and Applications*, Kluwer, Dordrecht, 1994.
32 R.G. Hunsperger, *Integrated Optics: Theory and Technology*, 4th ed, Springer, Berlin, 1995, pp. 48–72.
33 H.O. Pierson, *Handbook of Chemical Vapor Deposition*, Noyes Publishers, 1992.
34 K. Wasa and S. Hayakawa, *Handbook of Sputter Deposition Technology*, Noyes Publishers, 1992.
35 D.J. Elliott, *Microlithography: Process Technology for IC Fabrication*, McGraw-Hill, New York, 1986.
36 S. Wolf and R.N. Tauber, *Silicon Processing for the VLSI Era. Vol. I: Process Technology*, Lattice Press, 1986, pp. 514–585.
37 R.E. Kunz, In: E.J. Murphy (Ed.), *Integrated Optical Circuits and Components: Design and Applications*, Marcel Dekker, New York, 1999, pp. 335–380.
38 M. Wiki, H. Gao, M. Juvet and R.E. Kunz, *Biosens. Bioelectr.*, 16 (2001) 37.
39 M. Puyol, M. del Valle, I. Garcés, F. Villuendas, C. Domínguez and J. Alonso, *Anal. Chem.*, 71 (1999) 5037.
40 I. Garcés, F. Villuendas, I. Salinas, J. Alonso, M. Puyol, C. Domínguez and A. Llobera, *Sens. Actuat. B*, 60 (1999) 191.
41 J.J. Ramsden, *Sens. Actuat. B*, 15/16 (1993) 439.
42 J. Piehler, A. Branderburg, A. Brecht, E. Wagner and G. Gauglitz, *Appl. Optics*, 36 (1997) 6554.
43 H. Knobloch, S. Woigk, L. Brehmer, R. Yusupov and A. Rakhnyanskaya, *Thin Solid Films*, 265 (1995) 89.
44 S. Heyse, H. Vogel, M. Saenger and H. Sigrist, *Protein Sci.*, 4 (1995) 2532.
45 M. Wiki, R.E. Kunz, G. Voirin, K. Tiefenthaler and A. Bernard, *Biosens. Bioelectr.*, 13 (1998) 1181.
46 J. Dübendorfer and R.E. Kunz, *Sens. Actuat. B*, 38/39 (1997) 116.

47 P.N. Zeller, G. Voirin and R.E. Kunz, *Bios. Bioelectr.*, 15 (2000) 591.
48 H.J. Watts, D. Yeung and H. Parkes, *Anal. Chem.*, 67 (1995) 4283.
49 P.R. Edwards, A. Gill, D.V. Pollard-Knight, M. Hoare, P.E. Buckle, P.A. Lowe and R.J. Leatherbarrow, *Anal. Biochem.*, 231 (1995) 210.
50 F. Prieto, L.M. Lechuga, A. Calle, A. Llobera and C. Domínguez, *J. Lightwave Technol.*, 19 (2001) 75.

*Chapter 14*

# 'High-order' hybrid FET module for (bio)chemical and physical sensing

***Arshak Poghossian and Michael J. Schöning****

## 14.1 INTRODUCTION

### 14.1.1 Design concepts of (bio)chemical sensor arrays

The subject of miniaturised systems for (bio)chemical analysis is attracting increasing attention from researchers in the fields of analytical chemistry, biotechnology, medicine, microsensors and microsystems technology. After Manz et al. [1] introduced the concept of a 'micrototal analysis system (μTAS)' and the 'laboratory on a chip', respectively, for integrated chemical and/or biochemical analysis, significant research efforts have been concentrated on the development of such systems [2–11]. In principle, the μTAS comprises of microflow devices (e.g. micropumps, microvalves, micromixers), a microfabricated detector unit and control electronics.

Often, the detector unit represents a miniaturised multisensor module, which is intended for the multi-parameter detection of different (bio)chemical as well as physical quantities in a liquid flow. Typical parameters are the concentration of ions, gases and enzymes,

* Corresponding author. Tel.: +49-2461612973; fax: +49-2461612940. *E-mail address:* m.j.schoening@fz-juelich.de.



ISSN: 0166-526X

electrolyte conductivity, temperature, flow rate, etc. However, the detector unit also represents the most critical and important part of a microanalysis system. Therefore, how should such a multi-parameter module be realised?

Different approaches have been suggested in the literature for designing a multi-parameter sensor module for chemical sensing [12–22]. Generally, they can be presented schematically as shown in Fig. 14.1. The simplest instrumental method to obtain a large amount of chemical information is to use a large number of single-function sensors with different sensitive layers and different transducer principles. Further extensions to increase the amount of chemical information concern the utilisation of so-called 'zero-order' sensor arrays [12] that are based on different chemically sensitive layers but the same transducer principle. The combination of these two concepts is also possible. In contrast, a relatively new trend in chemical sensing is the development of the so-called 'high-order' sensor array, which implies more than one transducer principle for the same chemically sensitive layer (for example, a selective layer on a surface acoustic wave sensor can be simultaneously employed as a chemiresistor [12]). Thus, by applying this approach, the amount of chemical information obtained can be greater than the total number of sensors in the array. Generally, both the order of the sensor (or sensor array) and the number of sensing layers in the array increase the number of recording data yielding a higher information content.

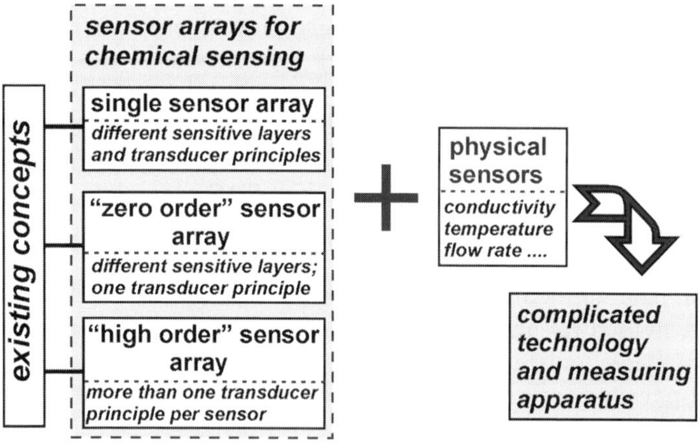


Fig. 14.1. Design concepts of multi-parameter sensor arrays (schematic).

However, a sensor module for multi-parameter detection also usually includes physical sensors for either controlling the physical parameters of the fluidic flow such as electrolyte conductivity, temperature, flow rate, flow direction, etc., or the elimination or compensation of different disturbance factors. The design, i.e. the structure, sensitive layer, transducer principle and measuring set-up of the physical sensors, mostly differs from that of (bio)chemical sensors. For example, in Ref. [21], a monolithic multisensor system, developed for the simultaneous measurement of six different physical/chemical parameters in a liquid flow, was presented. A platinum thermistor is chosen for the temperature measurement, a piezo-resistive element serves as pressure transducer, platinum electrodes are applied as the conductivity sensor, and a thermal principle is utilised for flow-rate determination. In addition, to detect the chemical parameters, UV-absorption and fluorescence methods have been followed. In Ref. [22], an electrochemical microanalysis system based on ion-selective field-effect transistors (ISFETs) is described, where in order to determine the flow rate, the amount of the solution drawn in and pumped out, is measured by weight. It is obvious that a sensor system combining different sensors, all of which are designed and optimised to detect only one physical or (bio)chemical parameter, will include a more complicated fabrication technology and sophisticated measuring apparatus.

### 14.1.2 'High-order' sensor module based on an identical transducer principle

To overcome these problems, a simple sensor structure based on interdigitated Pt electrodes is introduced, where by scheduling this structure in different modes, three sensor functions, namely, the electrolyte conductivity, temperature and hydrogen peroxide concentration and two actuator operations, the thermoresistive heating and ion generation has been achieved [23]. As an alternative approach, so-called '(bio)chemical and physical sensors based on an identical transducer principle', was proposed recently [24,25], and is currently extended by the present authors. In this approach, the same ISFET, which is well known as a (bio)chemical sensor, can also serve as a physical sensor and, consequently, the amount of (bio)chemical and physical information obtained can be significantly higher than the number of sensors present in the system ('high-order' system).

The multi-functionality in such a 'high-order' sensor module is achieved by means of different sensor arrangements and/or different operation modes. Moreover, the influence of various disturbing factors can be minimised or fully compensated by means of differential measuring set-ups. On the other hand, the modular concept, which is often favoured for the construction of μTAS or lab-on-chip devices [9–11] (a demonstrative example for the commercialisation of a microanalysis system using the modular concept, is the portable i-STAT analyser with a disposable cartridge [26]), is advantageous for the construction of a multi-parameter sensor module, too. The entire sensor system can be divided into sub-modules with different functions. Each element or sub-module combining several sensors or sensor arrays, can be fabricated on a separate chip and set as a discrete component on a printed circuit board. In this way, even a highly complicated sensor system can be constructed quite easily. Thus, such a hybrid multisensor module is flexible, easily exchangeable and can be assembled according to specific user request in an 'open loop mode' that provides a wide field of applications as well as a real chance for commercial realisation. In a sophisticated version, the control and measuring electronics can also be located on the same board. It is obvious that the proposed concept might also be extended to further semiconductor-based field-effect (bio)chemical devices, like electrolyte–insulator–semiconductor (EIS) sensors or light-addressable potentiometric sensors (LAPSs).

In this chapter, an ISFET-based 'high-order' hybrid module and multi-parameter detection system for the measurement of six (bio)-chemical and physical parameters (pH, penicillin concentration, diffusion coefficient of ions, temperature, flow velocity and flow direction) using only two sensor structures, an ion generator and a reference electrode, are presented. The basic characteristics of the tested individual sensors, including a novel ISFET-based diffusion coefficient, flow-velocity and flow-direction sensor that profits from the in situ electrochemical generation of chemical tracers, like $H^+$ or $OH^-$ ions, will be introduced. A pH-sensitive $Ta_2O_5$-gate-ISFET is applied as transducer for all suggested sensors. With regard to a further extension of the functional possibilities of the developed module to also detect gas (odour) concentration by means of integrating ISFET-based bioelectronic sensors, results of a so-called beetle/chip biosensor are discussed. Finally, some examples of application of the developed hybrid sensor module in biomedicine and environmental monitoring are demonstrated.

# 14.2 EXPERIMENTAL

## 14.2.1 Hybrid module design

Figure 14.2 shows the videomicroscopic picture of the developed hybrid module with two pH ISFETs (or one pH ISFET and one enzyme-modified FET, in this study a penicillin-sensitive field-effect transistor (PenFET)), the ion generator representing the generator electrode (Au film) and the counter electrode (Pt wire of 0.3 mm diameter). An additional Pt wire, also included in the module, can serve as a second generator electrode or as pseudoreference electrode when operating the array in a differential measurement set-up. All elements placed on a printed circuit board are electrically bonded and the whole module, except for the sensitive regions, is encapsulated using a commercially available epoxy compound (Epo-Tek 77, Epoxy Technology). The resulting sensitive area of the ISFETs amounts to about 0.5–1 $mm^2$.

## 14.2.2 ISFET fabrication

The pH ISFETs used are *n*-channel (meander- or U-shaped) devices, fabricated from a *p*-type boron-doped silicon wafer with (100)-orientation and 20 Ω cm specific resistivity. The gate composite insulator consists of an approximately 80 nm thick, thermally grown $SiO_2$ layer, covered by a $Ta_2O_5$ film (also ~80 nm thick). The $Ta_2O_5$ layer also served

Fig. 14.2. Videomicroscopic picture of the developed hybrid module with two ISFETs and Au or Pt ion generator, respectively.

as a passivation layer. The preparation of the pH-sensitive $Ta_2O_5$ films has been optimised in terms of a high pH sensitivity, high long-term stability, wide pH range, low hysteresis and small drift [27]. The $Ta_2O_5$ films are prepared by thermal oxidation of sputtered Ta in an oxygen atmosphere at 530°C for about 2 h. Aluminium was evaporated on the contact pads, providing the contacts to source and drain of the ISFET. After scribing, sensor chips having dimensions of 2 mm × 3.2 mm have been obtained. Figure 14.3 shows the videomicroscopic picture of an example of one of the chip layouts (meander gate). Here, one can recognise the meander-shaped channel of approximately 3 mm width and 25 μm length, the drain and source regions, and the Al bondpads. To guarantee an easy encapsulation of the device, the pH-sensitive gate was positioned as far as possible from the bonding pads.

### 14.2.3 Measuring system and sensor configurations

Figure 14.4 shows a photograph of the developed multi-parameter sensor system which combines a detector unit (multifunctional hybrid sensor module), a flow-through measuring cell, a pump, a 4-channel home-made ISFET/BioFET-meter (Research Centre Jülich, Germany), and a voltage source (Keithley). The hybrid module is horizontally or

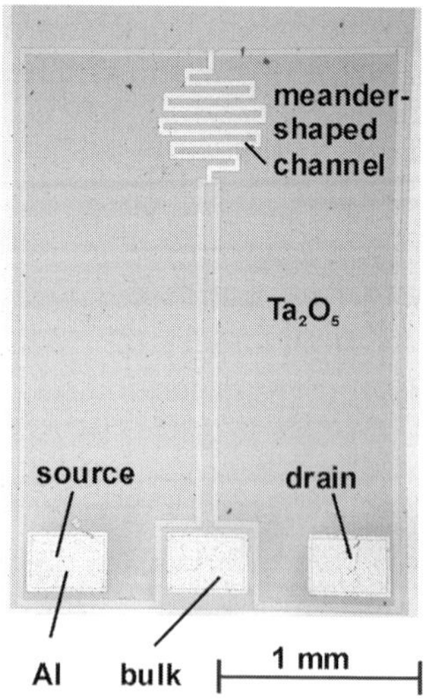


Fig. 14.3. Photograph of the ISFET layout with a meander-shaped channel.

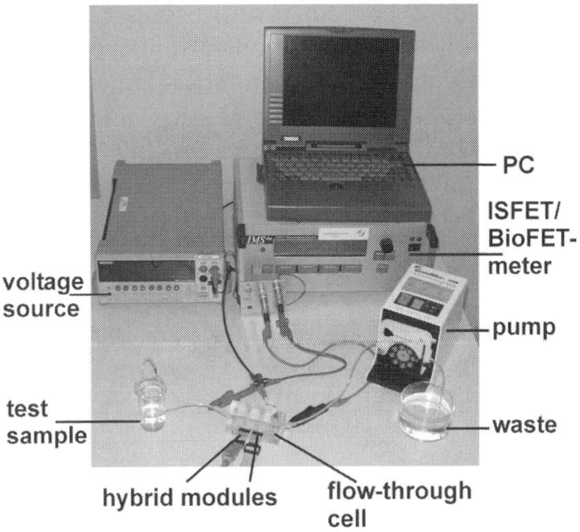


Fig. 14.4. Photograph of the developed multi-parameter sensor system consisting of hybrid module, voltage source, flow-through cell, pump, test sample and computer.

vertically placed in the middle of the rectangular flow channel of the flow-through cell. A common Ag/AgCl liquid-junction miniature reference electrode has been placed in the same flow-through cell at a certain distance from the ISFETs. The test solution is pumped through the channel.

The pH ISFET, as basic transducer, was characterised under both static and flow-through conditions, while the characterisations of the penicillin, beetle/chip, temperature and diffusion coefficient sensors were carried out in static solutions only. For the flow-velocity and flow-direction investigations, the experiments were also performed in continuous flow-operation conditions. The ISFET/BioFET-meter allows measurement of both the input and output characteristics of the sensors as well as the dynamic response using the constant charge mode (CCM) [27,28]. The CCM with a grounded reference electrode offers the possibility of carrying out simultaneous measurements of the dynamic characteristics of the multisensor by employing a single common reference electrode. All measurements were carried out at room temperature (with the exception of the temperature measurements).

The measuring system is completely driven by a personal computer, including the sequential or simultaneous scheduling of the different sensor combinations or operation modes as well as the measuring channel selection. The corresponding sensor configurations for the measurement of the six mentioned (bio)chemical and physical parameters, namely pH value, penicillin concentration, diffusion coefficient of ions, temperature, flow velocity and flow direction, are summarised in Fig. 14.5. Details of the sensors' functional principle, and the preparation and measuring procedures will be described in the following sections.

## 14.3 MULTI-PARAMETER DETECTION OF BOTH (BIO)CHEMICAL AND PHYSICAL QUANTITIES USING THE SAME TRANSDUCER PRINCIPLE

### 14.3.1 ISFET-based pH sensor

The pH sensor configuration of the developed hybrid module was tested under both static and flow-through conditions using standard pH buffer solutions. The corresponding dynamic responses of the pH ISFET at different pH values, from pH 3 to 9, are presented in Fig. 14.6. The sensor output signal was almost the same in both the upward and downward series of measurements. In contrast to the pH

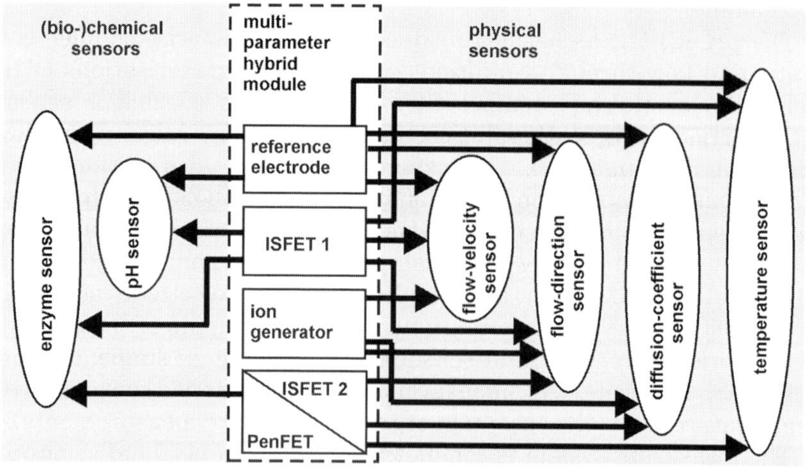


Fig. 14.5. Sensor arrangements for the measurement of six (bio)chemical and physical parameters using the developed 'high-order' hybrid sensor module.

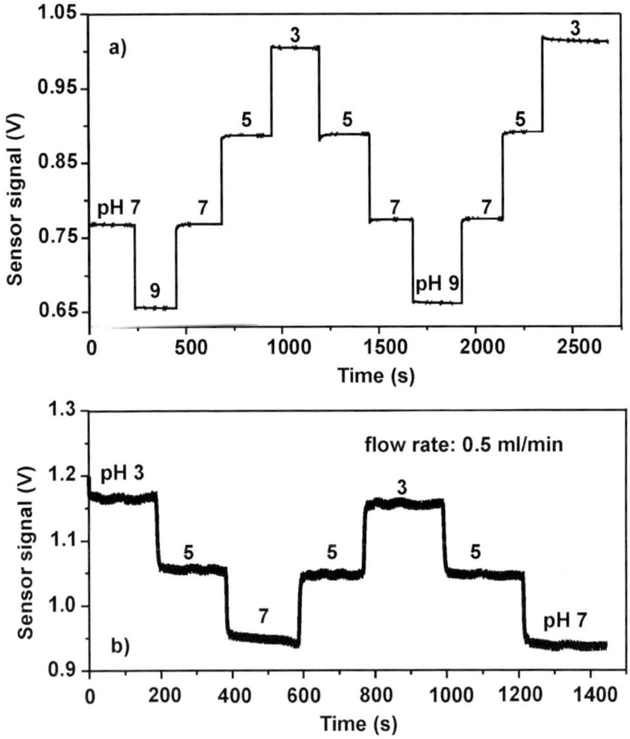


Fig. 14.6. Dynamic response of the pH ISFET at different pH values under both static (a) and flow-through (b) conditions.

response in static solution measurements (Fig. 14.6(a)), a small oscillation of the sensor signal was observed (Fig. 14.6(b)) when the ISFET operated under flow-through conditions (the flow rate was about 0.5 mL/min); this is probably due to the pump functioning and/or the possible bubble formation close to the ISFET gate region.

The basic characteristics of the developed pH-sensitive $Ta_2O_5$-gate-ISFETs were previously reported in Refs. [28,29]. A typical pH response of between pH 2 and 12 is linear with an average slope of 55–58 mV/pH. The drift of the output voltage is less than 0.5 mV/h, with the hysteresis amounting to about pH 0.02. Table 14.1 gives a survey of the ISFET parameters. ISFETs with such sensor features served as basic transducers for the development of subsequent sensor modules.

TABLE 14.1

Specifications of the developed pH-sensitive $Ta_2O_5$-gate-ISFET

| | |
|---|---|
| Sensitivity | 55–58 mV/pH |
| Linear range | pH 2–12 |
| Drift | 0.5 mV/h |
| Hysteresis | pH 0.02 |
| Response time (90%) | ~1 s |

### 14.3.2 ISFET-based penicillin sensor

The structure and the operation principle of the realised PenFET is depicted in Fig. 14.7. The PenFET is constructed by immobilising the gate of the pH ISFET with the enzyme penicillinase. The operation principle of the penicillin biosensor can be described as follows: the PenFET detects variations in the $H^+$ ion concentration resulting from the catalysed hydrolysis of penicillin by the enzyme penicillinase. A resulting local pH decrease near the gate region leads to a change in the surface charge of the pH-sensitive layer (here, $Ta_2O_5$) and consequently, to a change in the drain current of the PenFET. In this study, the enzyme penicillinase (EC 3.5.2.6., *Bacillus cereus* from Sigma, specific activity: 1650 units/mg protein) was adsorptively fixed onto the $Ta_2O_5$ gate of the pH ISFET. The main advantages of the adsorptive immobilisation technique are its simplicity and cheapness without any loss of enzyme activity and, if necessary, the possibility of a subsequent enzyme regeneration. Detailed descriptions of the immobilisation procedure as well as the measuring and storing conditions of the PenFET are given in Refs. [27,30,31]. A cross-section of the layer structure of the PenFET investigated by means of scanning electron microscopy (SEM) is shown in Fig. 14.8. In the gate region of the PenFET, one can clearly recognise the enzyme layer on the pH-sensitive $Ta_2O_5$ surface as well as the thermally grown $SiO_2$ layer on the *p*-type Si substrate.

Figure 14.9 shows a typical set of input characteristics of the developed PenFET in a working buffer (0.2 mM polymix multicomponent buffer, pH 8) and in penicillin G solutions with concentrations of 0.25 and 1 mM. With increasing concentration of penicillin in the test solution, the input curves of the PenFET are shifted towards negative voltage values, which results from the increase of the $H^+$ ion concentration at the sensor surface due to the enzymatic reaction.

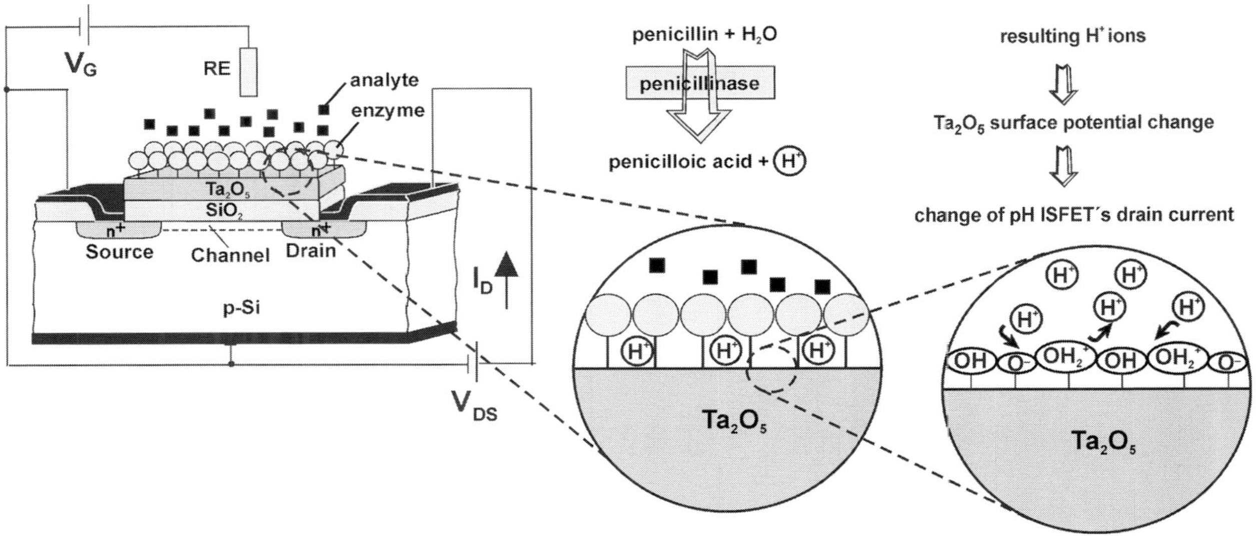


Fig. 14.7. Structure and operation principle of the penicillin-sensitive field-effect transistor (PenFET). $V_G$ : gate voltage; $V_{DS}$ : drain-source voltage; $I_D$ : drain current; and RE: reference electrode.

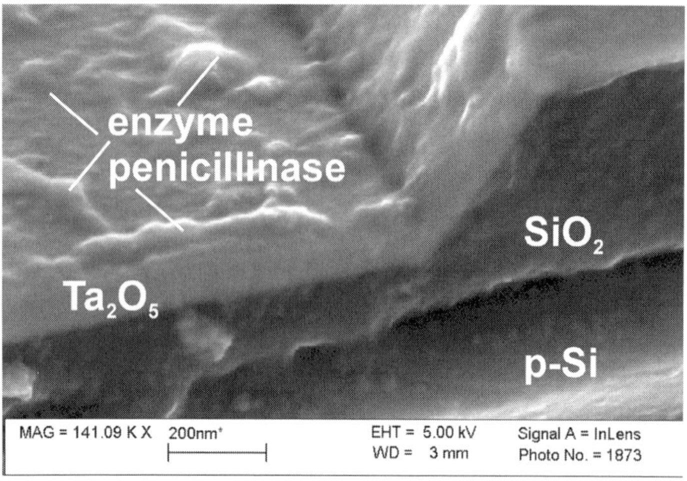


Fig. 14.8. Scanning electron microscopy (SEM) picture of the layer structure of the PenFET in the gate region.

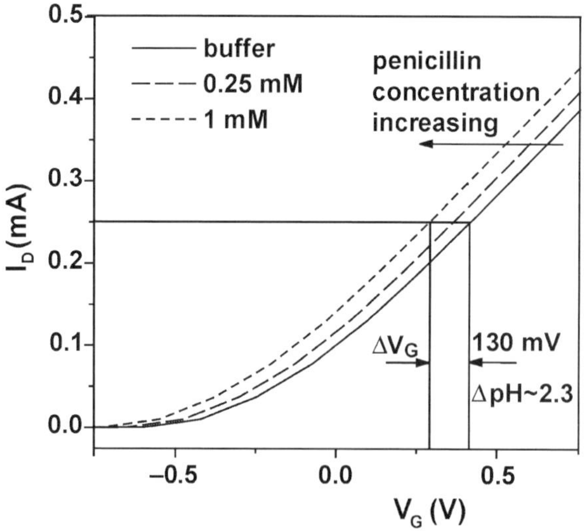


Fig. 14.9. Typical input characteristics of a PenFET ($n$-channel) in a working buffer and in penicillin G solutions with concentrations of 0.25 and 1 mM, respectively. $\Delta V_G$ : gate voltage shift.

At a penicillin concentration of 1 mM, the gate voltage shift $\Delta V_G$ was about 130 mV in comparison to buffer solutions without penicillin content. Considering an average pH sensitivity of the PenFETs of about 56.5 mV/pH, this voltage shift corresponds to a pH value at the $Ta_2O_5$ surface of approximately pH 5.7 (compared to a pH value of the bulk buffer solution of pH 8), i.e. a ΔpH of 2.3. To achieve low detection limits, the suited preferred buffer composition, buffer concentration and pH range was optimised in former experiments [32].

It is known that the pH value of freshly prepared penicillin solutions can change through spontaneous hydrolysis of penicillin in aqueous buffer solutions (0.5–1% of penicillin concentration as reported in Ref. [33]). Moreover, in a working buffer with a low buffer capacity (as used in this study), a certain change to the pH value of the bulk solution can also take place during the measurement period. Consequently, a certain blank pH response has been observed. In order to eliminate this influence of possible pH variations as well as to compensate effects that are caused by disturbing factors, like temperature and sensor drift, etc. the penicillin measurements have been performed in a differential measuring set-up [24,30] between a PenFET and a 'reference' pH ISFET with nearly identical intrinsic sensor characteristics (see Fig. 14.10). In this arrangement, disturbing values ($\Delta V_{disturb}$), such as sensor drift, drift of the reference electrode, temperature drift as well as possible pH changes of the bulk buffer solution, will affect both sensors (PenFET, pH ISFET), and thus, the resulting differential sensor output signal $\Delta V_{out}$ now depends only on the penicillin concentration ($c_{analyte}$) of the test sample.

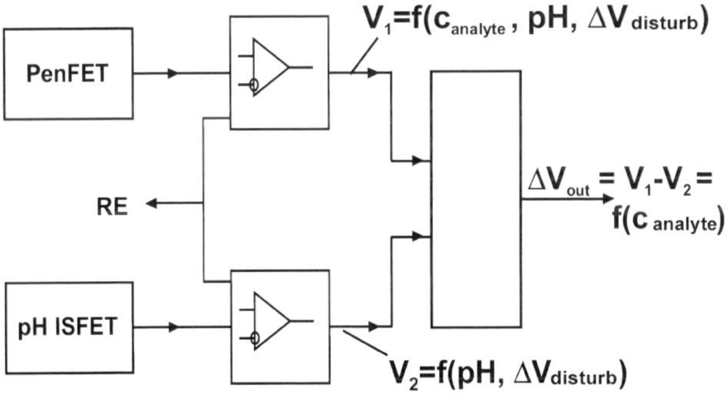


Fig. 14.10. Differential measuring set-up between a PenFET and a 'reference' pH ISFET with nearly identical intrinsic sensor characteristics.

In addition, the Au or Pt electrode can be utilised as a pseudoreference electrode in the hybrid module.

Figure 14.11 demonstrates the result of such a differential mode measurement. When increasing the penicillin concentration from 0.05 to 10 mM, the PenFET shows a normal CCM curve, while for the pH ISFET a small change in the output signal was observed that is connected to a certain change of the pH value of the bulk buffer solution. Without correcting these small deviations caused by pH variations, the

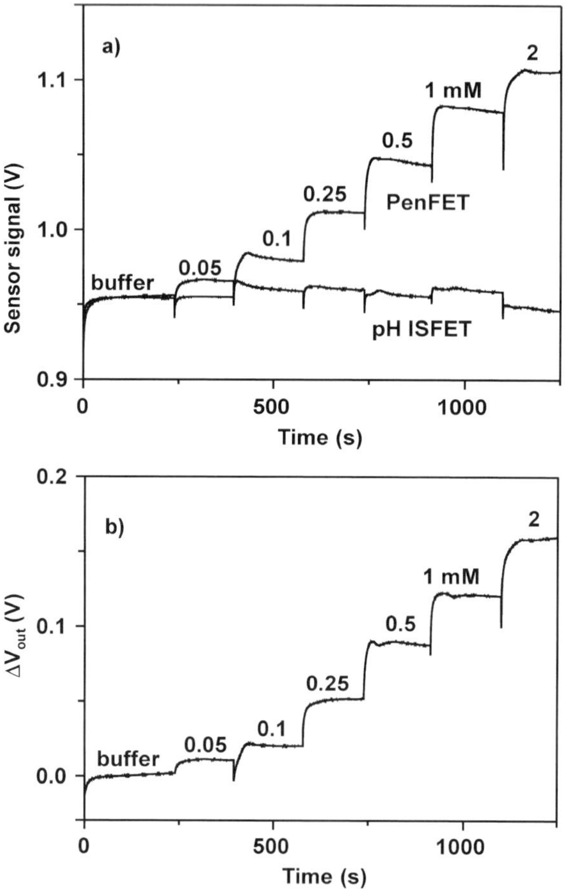


Fig. 14.11. Penicillin measurement executed in the differential measuring set-up: (a) response of the single PenFET and the single pH ISFET; and (b) net differential signal of the dual chip corrected for pH interferences of the bulk buffer solution.

sensor signal of the PenFET would be falsified. The 'true' net differential output signal corrected for the pH interferences and for the disturbing factors, is presented in Fig. 14.11(b).

PenFET characteristics were investigated over a time period of more than one year. As an example, Fig. 14.12 demonstrates the dynamic response of a PenFET measured on the 235th day after enzyme immobilisation. During this time period, the sensor has been periodically checked for about 180 measurements. Even after that operation period, the developed $Ta_2O_5$-gate-PenFET with adsorptively immobilised penicillinase shows a significant sensor response with a high sensitivity and a low detection limit. The detailed specifications of the penicillin biosensor are summarised in Table 14.2.

### 14.3.3 ISFET-based temperature sensor

The temperature characteristic of an ISFET-based sensor represents a complex function related to the reference electrode, measured electrolyte, interfacial potentials and the solid-state device itself; it depends strongly on the working point. The temperature behaviour of ISFETs

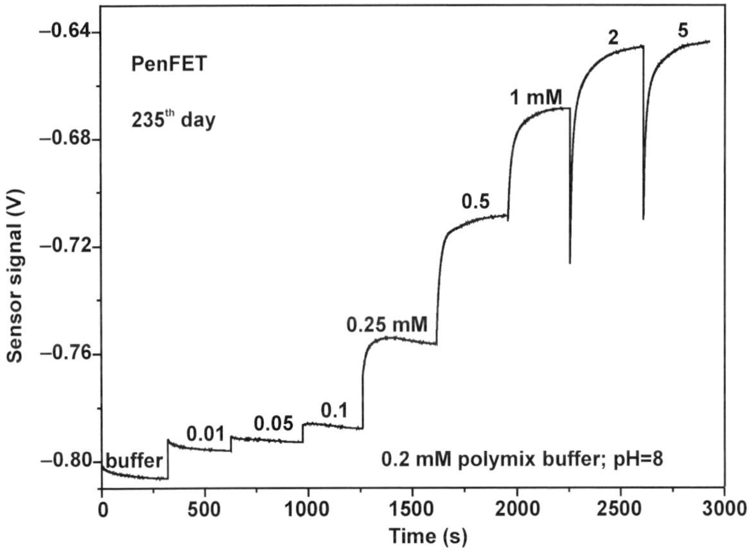


Fig. 14.12. Dynamic response of the PenFET on the 235th day after enzyme immobilisation.

TABLE 14.2

Specifications of the developed penicillin FET

| | |
|---|---|
| Sensitivity | 120 ± 10 mV/mM |
| Linear range | 0.05–1 mM |
| Lower detection limit | 5 μM |
| Upper detection limit | 20 mM |
| Hysteresis | <4 mV |
| Response time | 0.5–3 min |
| Lifetime | >1 year |

has been theoretically and experimentally studied [34–40]. The results of these investigations clarified that ISFETs possess an isothermal working point at which the drain current, $I_D$, is nearly independent of the temperature. Recently, similar behaviour for a PenFET was observed [30].

This property of ISFET-based sensors is normally used to improve their thermal stability. However, the position of the isothermal working point also depends on the pH value of the test sample, and at a fixed working point, a zero-temperature coefficient can only be achieved for one defined value of pH. At other working points, the sign and the amount of the temperature sensitivity of the ISFET depend on the respective value of the drain current. In our experiment, exactly this property of the ISFET is utilised in order to develop an ISFET-based temperature sensor.

The temperature behaviour of the developed $Ta_2O_5$-gate-ISFET was investigated in the temperature range from 22 to 37°C using a thermostatic chamber (RTE-110, Neslab). Figure 14.13 presents a typical temperature dependence of the input characteristics of the ISFET measured in two different buffer solutions of pH 5 and 8. The isothermal point in the buffer at pH 5 is reached when operating the ISFET at a working point where the drain current $I_D = 0.22$ mA and the gate voltage $V_G = 0.745$ V. For the buffer at pH 8, the corresponding isothermal working point is defined at $I_D = 0.15$ mA and $V_G = 0.76$ V.

Figure 14.14 shows the more detailed temperature characteristic of the ISFET's gate voltage in the buffer solution of pH 8 at different working points. Here, the drain current varies from $I_D = 0.05$ to 0.6 mA. The gate voltage at 22°C was normalised to zero for each working point. The temperature dependence of the gate voltage has a high linearity with a slope depending on the respective $I_D$ value. The temperature dependence spans values from −1 mV/°C at a drain current of 0.05 mA

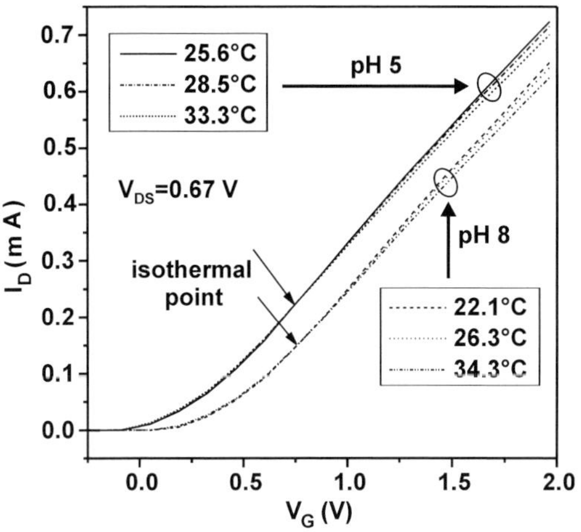


Fig. 14.13. Temperature dependence of the input curves of the pH ISFET measured in two buffer solutions of pH 5 and 8.

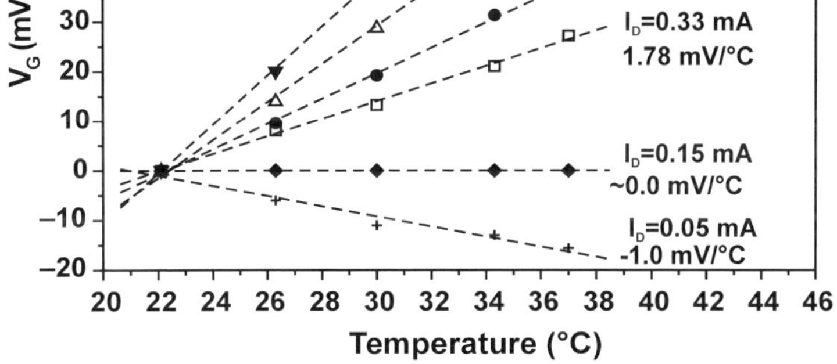


Fig. 14.14. Temperature dependence of the ISFET's gate voltage in a buffer solution of pH 8 at different working points with drain currents, $I_D$, reaching from 0.05 to 0.6 mA, evaluated from the input characteristics in Fig. 14.13.

up to about +5 mV/°C for a drain current of 0.6 mA. When the drain current is below the drain current of the isothermal point $I_{D0}$, which is equal to 0.15 mA at pH 8, the temperature coefficient is negative, whereas, positive temperature coefficients can be achieved for $I_D$ values greater than $I_{D0}$.

In spite of this distinct temperature dependence, a single pH ISFET cannot be employed as a temperature sensor because in a multisensor arrangement, the sensor signal also depends on the pH value. In this case, a distinction between the temperature and/or pH effect is not possible, since the position of the isothermal point moves with the pH value of the test solution as demonstrated in Fig. 14.13. To circumvent this drawback, a pH ISFET-based temperature sensor can be realised by also using a differential set-up consisting of two identical pH ISFETs (or a pH ISFET and a PenFET), which are operated at different working points (see Fig. 14.15). The disturbing factors, such as a possible pH change to the test solution and a drift of the sensor signal, can be compensated, and thus, the differential output signal should record only the temperature as the sensor signal, similar to the pH ISFET/PenFET dual chip. The difference in temperature sensitivity of the two ISFETs operating at different working points with different drain currents $I_{D1}$ and $I_{D2}$ allows measurement of the exact temperature. In principle, an isothermal point can also be used as a fixed working point for one of the ISFETs. However, in this case, it is necessary to carry out additional temperature measurements to determine this isothermal point.

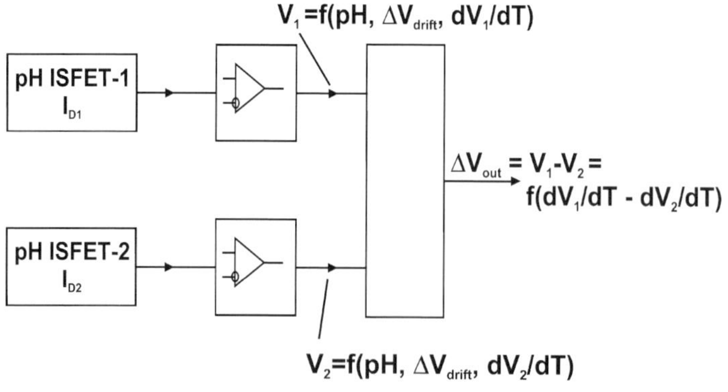


Fig. 14.15. Differential set-up for the temperature measurement consisting of two identical ISFETs operating at different working points ($I_{D1}$ and $I_{D2}$).

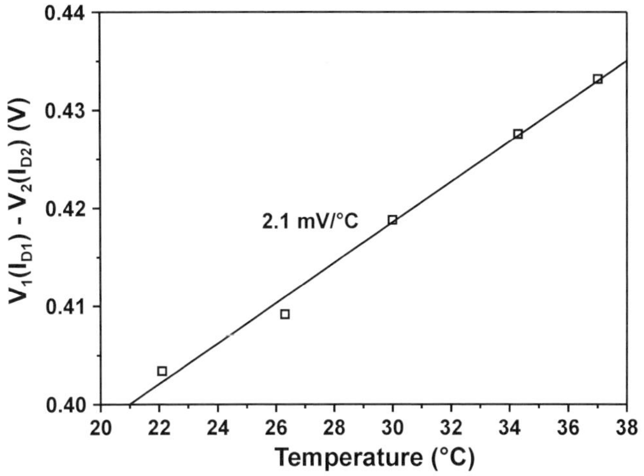


Fig. 14.16. Temperature sensitivity of the gate-voltage difference $\Delta V_G = V_1(I_{D1}) - V_2(I_{D2})$ for two varying working points $I_{D1} = 0.5$ mA and $I_{D2} = 0.33$ mA.

Figure 14.16 shows an example of the temperature dependence of the gate-voltage difference $\Delta V_G = V_1(I_{D1}) - V_2(I_{D2})$ for two working points $I_{D1} = 0.5$ mA and $I_{D2} = 0.33$ mA, which are achieved from the input characteristics of the ISFET in Fig. 14.13. Here, one can see that the temperature dependence of $\Delta V_G$ is nearly linear with a slope of 2.1 mV/°C. Thus, a direct relationship between the sensor output signal and the temperature can be established.

### 14.3.4 ISFET-based flow-velocity sensor

The most widely suggested functional principles of flow sensors are thermal dilution in a flow stream, the thermal transit time (or 'time of flight') measurement of a heat impulse injected into a flow and differential pressure detection [41–45]. The 'time-of-flight'-type thermal flow sensor is accurate and stable due to its sensing mechanism. However, it requires a relatively high heat pulse for a detectable signal downstream and additional technological processes for good thermal isolation of the heater. As an alternative to the heat impulse injection, an ion drug [46], an injection of a charge pulse carried along in the fluid as ionic species (so-called electrohydrodynamic flow sensor) [47], an injection of the electrolyte tracer from outside into the flow solution [48] and in situ electrochemical production of bubbles [49] as well as oxygen [50], were also proposed as operation principles for flow sensors. A review of recent advances in micromachined flow sensors can be found in Ref. [41].

In this chapter, a functional principle and experimental results of a novel ‘time-of-flight’-type ISFET-based flow-velocity (flow rate) and flow-direction sensor using the in situ electrochemical generation of chemical tracers, such as $H^+$ or $OH^-$ ions, are presented. The principle of the in situ coulometric production of $H^+$ or $OH^-$ ions was previously used by other groups for the development of ISFET-based microlitre-titrator devices [51,52] and pH microscopy [53].

The structure and functional principle of the ISFET-based flow-velocity and flow-direction sensor is schematically shown in Fig. 14.17. The sensor presents a chemical sensor-actuator system and consists of one ion generator (e.g. Pt, Au for $H^+$ or $OH^-$ ion generation) and one detector (e.g. ISFET) that detects the in situ electrochemically generated ions [54]. Both a generator and an ISFET detector as well as an additional reference and/or counter electrodes (not shown in the sketch) are placed in the flow channel.

The corresponding electrochemical reactions for the $H^+$ and $OH^-$ generation through the electrolysis of water are presented in Fig. 14.17(a). The direction of the current between the generator and the counter electrode placed in the same fluid channel determines which type of chemical tracer is produced. When a positive potential is applied to the generator electrode, water is dissociated and $H^+$ ions are produced (anodic electrolysis), whereas, under the condition of cathodic electrolysis, $OH^-$ ions are generated.

In contrast to ISFET-based microlitre-titrator devices [51,52], which use relatively slow processes of diffusion of generated ions, the ISFET-based flow sensor benefits from the convective mass-transport process. The in situ generated chemical tracers (in this case $H^+$ ions) flow to the downstream-placed ISFET and are detected there. The ISFET output signal (detector signal) will be determined by the $H^+$ ion concentration in the gate region. The time difference $\Delta t = t_d - t_g$ between the detection ($t_d$) and generation ($t_g$) of the ions (see Fig. 14.17(b)), i.e. the ‘time-of-flight’, is inversely proportional to the mean flow velocity of the fluid. If the geometry of the flow channel is known, the flow rate can be determined, too.

A design with both upstream and downstream ISFETs and an ion generator placed between them minimises or fully compensates the

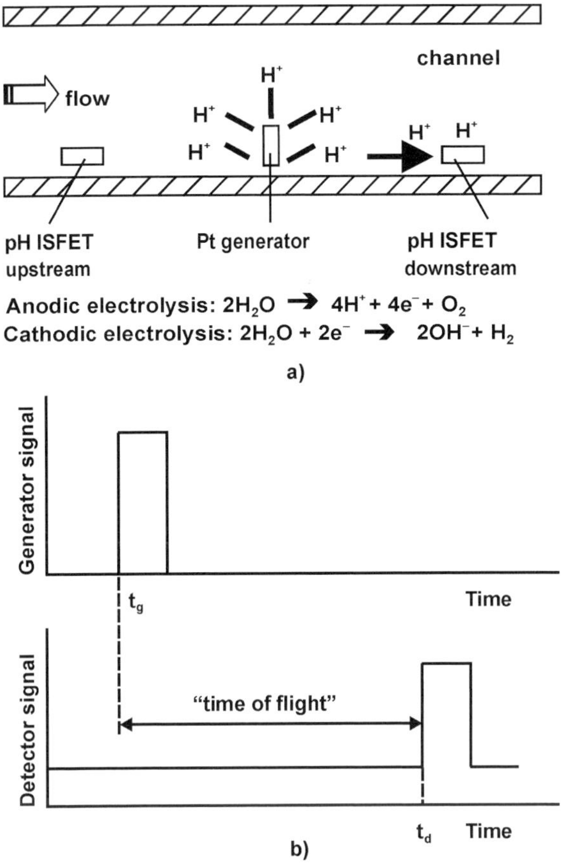


Fig. 14.17. ISFET-based flow-velocity and flow-direction sensor (schematic): (a) functional principle; and (b) 'time of flight' measurement. The corresponding electrochemical reactions for $H^+$ and $OH^-$ generation through water electrolysis are presented, too. The reference and/or counter electrodes are not shown for simplicity. At moment $t_g$, the generator produces a chemical tracer impulse, which is detected by the ISFET at moment $t_d$. The time difference, $t_d - t_g$, is defined as 'time of flight' (or transit time).

influence of various disturbing factors, such as the influence of diffusion effects. In addition, the approach based on two ISFETs might lead to a flow-direction sensor. Owing to the short response time of the ISFET in the millisecond range, it serves as an excellent detector for the measurement of the flow velocity in a wide operational range.

Moreover, since this measuring method represents a dynamic method, the detector's drift, non-linearity, temperature instability and sensitivity discrepancy between the various ISFETs are not relevant, because only relative changes in the sensor's output signal are of interest.

Flow-velocity measurements were performed with a single ISFET or an ISFET array using different positions (vertical and horizontal) in the flow-through cell as well as different modes of ion generation (constant voltage applied to the generator electrode; constant generation current; short generation impulse (10–30 s); permanent generation to reach a constant sensor output signal).

As an example, Fig. 14.18(a) demonstrates the typical output signal of two ISFETs, both placed downstream of the generator, versus the time at different flow rates from 0.1 to 0.6 mL/min. In this experiment, unbuffered 0.1 M $KNO_3$ water solution, pH ~7, was pumped through the channel by the 'MS-1 Reglo 160' pump (Ismatec, Wertheim-Mondfeld, Germany) providing a flow rate in the range from 0.005 to 45 mL/min. In this case, $H^+$ ions were generated by a constant current of 30 μA to reach a constant sensor signal. The distance between the counter electrode and the ISFET (see Fig. 14.18(b)) was sufficient to prevent any influence from the produced counter ions.

The arrows in Fig. 14.18(a) mark the time of the beginning (arrows upwards) and the end of ion generation (arrows downwards). After the ion generation, some time is taken before the induced pH change reaches the ISFET and is detected. The stepwise increase in the sensor output signal after a time period $t_i$ is derived from pH changes close to the ISFET gate, which are caused by convection of the generated $H^+$ ions. The ISFET output signal changes were about 100 mV. Considering a pH sensitivity for the ISFET of 56 mV/pH, this voltage shift corresponds to a pH change near the ISFET gate region of about $\Delta$pH ~1.8.

When increasing the flow rate, the time $t_i$ (transit time or 'time of flight') that is needed for the generated $H^+$ ions to reach the sensor gate region decreases. The 'time of flight', $t_i$, can be calculated from the time difference between the generation of the $H^+$ ions (marked by an arrow) and the intersection of the tangent at the inflection point of the response curve. The corresponding calibration curve (see Fig. 14.19), evaluated for different flow rates from 0.1 to 0.6 mL/min using the difference ($\Delta t$) of the values of the 'time of flight' for two ISFETs (ISFET1 and ISFET2), demonstrates a good linearity between the measured flow velocity of the ISFET and the delivered flow rate of the pump.

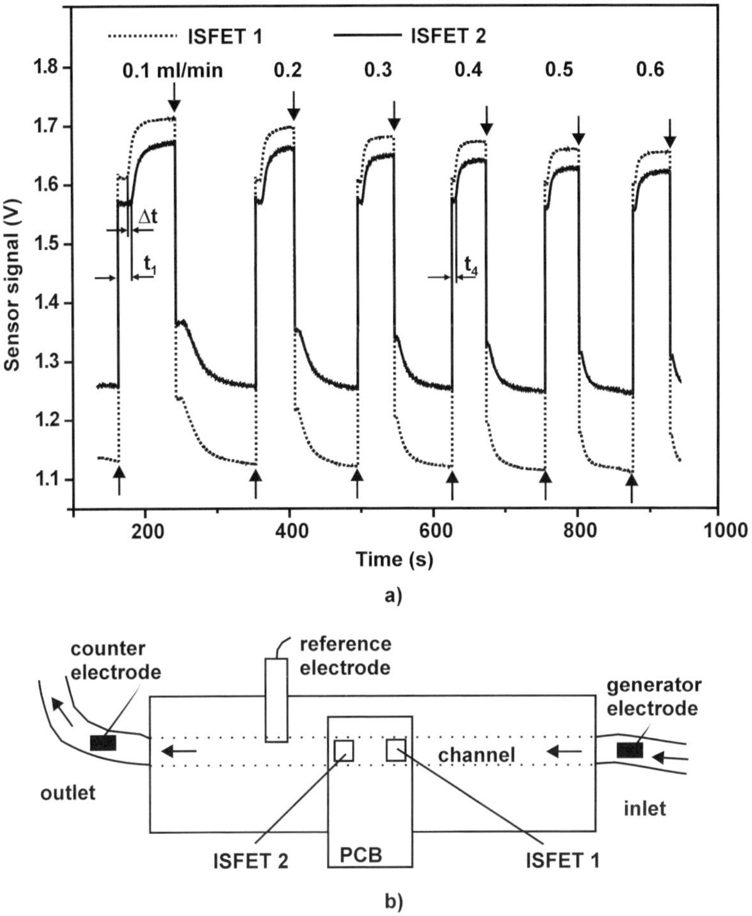


Fig. 14.18. Typical output signal of the ISFET-based flow-velocity sensor (a) and corresponding measuring set-up (b). The arrows mark the time of beginning (↑, generator on) and end (↓, generator off) of ion generation. The flow-rate values were 0.1, 0.2, 0.3, 0.4, 0.5, 0.6 mL/min, respectively; $t_i$ : 'time of flight'; and PCB: printed circuit board.

### 14.3.5 ISFET-based flow-direction sensor

As was already noted in Section 14.3.4, a flow-sensor design with both upstream and downstream ISFETs and an ion generator placed between them, also allows determination of the flow direction.

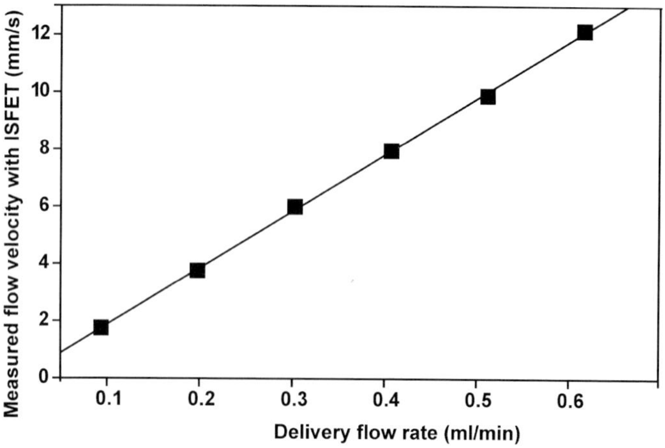


Fig. 14.19. Calibration curve of the developed ISFET-based flow-velocity sensor evaluated for the different flow rates, calculated from Fig. 14.18.

Figure 14.20 demonstrates an example of a flow-direction measurement (a) and the corresponding sensor set-up (b). The flow rate was 0.5 mL/min. The duration of the voltage impulse for the $H^+$ ion generation was about 20 s. As can be seen, the downstream ISFET shows the typical peak that is due to the change of the $H^+$ ion concentration close to the gate region, while for the upstream ISFET practically no change in the output signal is observed. Here, only a certain electrical artefact caused by the voltage drop across the solution between the generator and the counter electrode was found. It shows a small stepwise change in the output voltage at the beginning and the end of the generation. However, this artefact does not influence the flow-direction measurement itself.

### 14.3.6 ISFET-based diffusion-coefficient sensor

The ISFET-based sensor for the determination of the diffusion coefficient of ions has the same structure and set-up as the already described flow-velocity sensor (see Fig. 14.17). The sensor also presents a chemical sensor–actuator system and consists of one ion generator (e.g. Pt, Au for $H^+$ or $OH^-$ ion generation) and one detector (e.g. an ISFET) that detects the in situ electrochemically generated ions. However, the diffusion-coefficient sensor operates under static solution conditions. In contrast to the ISFET-based 'time-of-flight'-type flow-velocity sensor, which is based

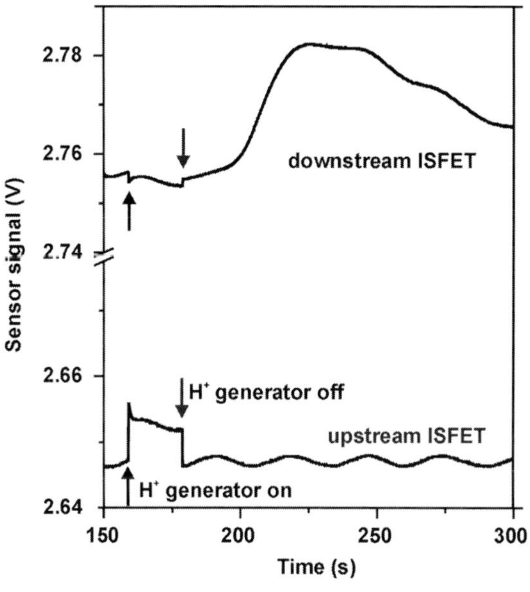


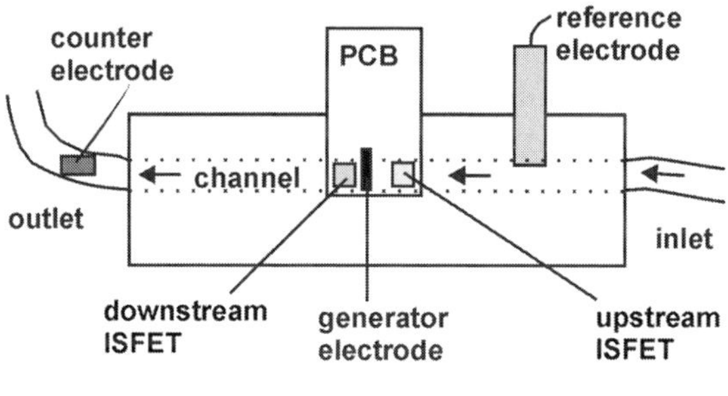


Fig. 14.20. Flow-direction measurement with an upstream and a downstream placed ISFET: (a) dynamic response; and (b) sensor set-up (schematic). The flow rate was 0.5 mL/min.

on the convective transport effect, for the diffusion-coefficient sensor the diffusion transport effect of the ions dominates. In this case, the in situ generated chemical tracers ($H^+$ or $OH^-$ ions) diffuse to the ISFET, and the ISFET output signal will be determined by the $H^+$ ion concentration in the gate region. If the time difference between the detection and generation of the ions, i.e. the diffusion time of the ions, and the distance between the generator and the ISFET are known, the diffusion coefficients can be calculated.

Preliminary experiments have shown the possible application of the developed hybrid sensor module for the determination of the diffusion coefficient of $H^+$ and $OH^-$ ions. However, since the 'open' contact area of the ISFET to the test sample was relatively large (about 1 $mm^2$), there are some problems connected with an accurate determination of the diffusion time from the experimental curves. For example, the obtained values of the diffusion coefficient for $H^+$ ions was about 1.5 times higher than the theoretical value. Hence, further investigations are necessary to optimise the design of the ISFET-based diffusion-coefficient sensor and the measuring cell.

### 14.3.7 ISFET-based bioelectronic sensor

Taking whole animals or at least complete sensory organs (e.g. an insect antenna) as a biological recognition element represents a new trend in bioelectronic sensor development. For example, insects are known to have an especially developed sense of smell. Of the one million or so insect species, only approximately 250 have so far been tested for their olfactory abilities. About 250 pheromones and about 400 host plant odours are known that can be detected with an extremely high sensitivity and selectivity using these unique features of insects.

An idea to couple an insect antenna directly to a FET and, thus, create a completely new type of biosensor, a so-called beetle/chip sensor, was first realised in Ref. [55], and then, investigated in detail and optimised in further studies [56–66]. In this approach, the voltage generated in the antenna upon detecting a certain odour was used to modify the drain current of the transitor. The responsible organ for smell, i.e. the insect antenna, usually consists of several segments, equipped with small hairs, the so-called sensilla, which contain the olfactory receptor neurons [67]. The odour recognition process is initiated at the nerve cell membranes. As a result of odour stimulation, an odour concentration-dependent voltage pulse is generated over the whole antenna yielding a dipole that leads to respective peaks in the drain current of the FET. (For details of the odour-detection mechanism of the beetle/chip sensor see, e.g. Refs [59,62].)

In order to measure the signals generated in the insect antenna, a bioelectronic interface has to be established between the insect antenna and the FET. Generally, there are two possible configurations of coupling the antenna to the FET [57]. In the first one, the whole beetle is fixed and the tip of the insect antenna is dipped into an electrolyte solution which is in direct contact with the gate insulator of the FET as shown schematically in Fig. 14.21(a). A reference electrode (e.g. platinum wire) is then placed at an appropriate site on the beetle (e.g. between the neck and the head). The experimental set-up is called 'whole-beetle' BioFET (biologically sensitive FET). In the second configuration, the antenna is resected from the beetle and contacted on both sides with the electrolyte solution (Fig. 14.21(b)). This experimental set-up is introduced as an 'isolated-antenna' BioFET. If an air current, loaded with a marker compound, is now flushed over

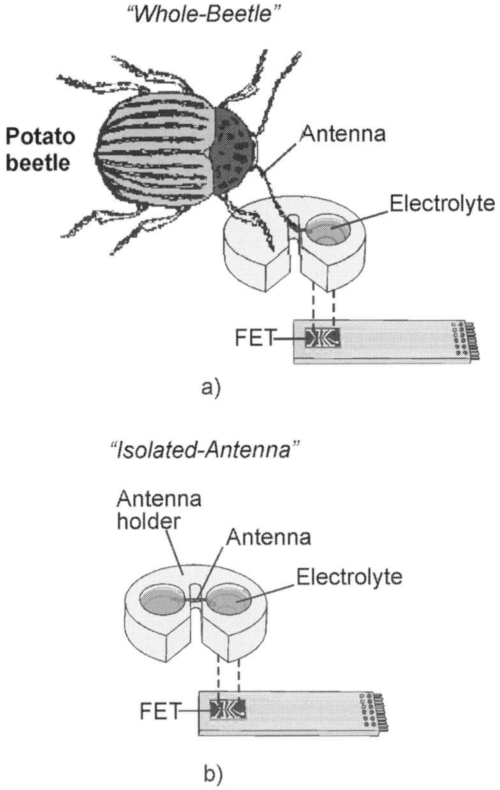


Fig. 14.21. Two general configurations of coupling the insect antenna to the FET: (a) 'whole-beetle' BioFET; and (b) 'isolated-antenna' BioFET.

the antenna (set-up (a) or (b)), the induced depolarisation response will modify the conductance of the FET channel between source and drain, yielding a variation in the drain current as the sensor signal.

As an example, Fig. 14.22(a) presents a typical dose–response curve of the beetle/chip sensor with the antenna of a Colorado potato beetle to odour pulses of different concentrations of the specific green-leaf odour *cis*-3-hexen-1-ol in air from 1 ppt to 100 ppm. As can be seen, the peak size of the sensor response clearly corresponds to the respective odour quantity. The more intensive the odour stimulus for the antenna, the higher the recorded signal. In this experiment, the measurements were performed using a specific sensor head with antenna holder

a)

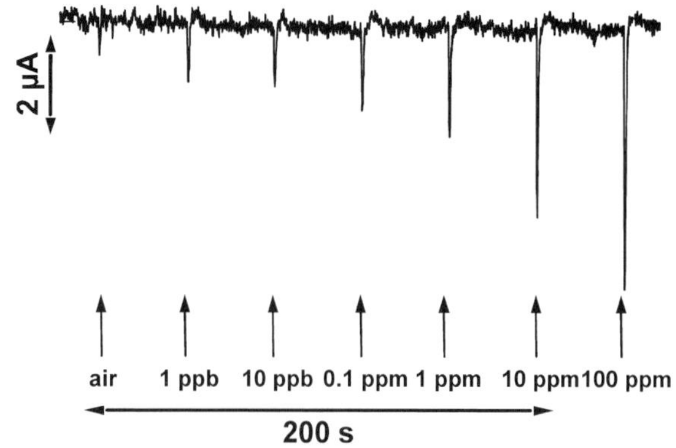


b)

| Odour \ Beetle | Potato beetle | Jewel beetle |
|---|---|---|
| cis-3-hexen-1-ol | ++ | +++ |
| guaiacol | + | ++ |
| 1-octen | + | - |

| **Substance:** | **Concentration range:** | **Possible application:** |
|---|---|---|
| cis-3-hexen-1-ol | 1 ppt ... 100 ppm | Detection of plant damages |
| Guaiacol | 50 ppt ... 500 ppb | Fire detection (wood) |
| 1-octen | 15 ppb ... 1.5 ppm | Fire detection (coal) |

Fig. 14.22. Typical dose–response curve of the beetle/chip sensor with the antenna of a Colorado potato beetle to odour pulses of different concentration of the specific green-leaf odour *cis*-3-hexen-1-ol in air (a) and the list of detectable odour substances with their respective concentration ranges and possible applications (b). The number of '+' symbols indicates the sensitivity towards the respective odour in increasing order.

and a BioFET-based sensor system, which provides a stream of air with constant temperature, humidity and velocity as well as the injection of the respective odour pulses [58,65]. With this BioFET-based beetle/chip sensor it is possible to detect odour concentrations even down to the ppt concentration range (Fig. 14.22(b)) [59].

One interesting application of the developed beetle/chip sensor is the detection of plant damage in potato fields in order to determine the optimum timing for pesticide applications. Thousands of potato fields annually fall prey to the potato beetle. Therefore, many farmers spray large quantities of pesticides that mainly save the harvest, but simultaneously contaminate the soil and groundwater and kill useful insects. By means of the developed bioelectronic device, a field's infestation with beetles can be identified at an early stage. This would alert the farmers to spray the pesticides at an exact point of time. Consequently, unnecessary insecticide applications could be avoided. Experiments performed under real conditions (i.e. in a green house) have been able to validate the detection of a single, damaged potato plant within a field of hundreds of undamaged plants [59,65].

Another area of application might be an early fire alarm system: smouldering fires, like those from burning coal, paper and wood, can be detected by means of monitoring fire-specific odours such as guaicol and 1-octen using the same biosensor only this time with the antenna of a steelblue jewel beetle [59,64,66]. The applications can be extended by employing various insect species with different sensitivity profiles. Therefore, it is necessary to build a library of different beetle species and their odour detection spectra.

## 14.4 APPLICATIONS OF THE HYBRID SENSOR MODULE

### 14.4.1 pH determination in human urine

The most promising areas of ISFET application are medicine and biology. In earlier experiments (see e.g. Refs [17,68,69]), the possibility of an application of ISFETs to the measurement of pH and concentrations of $K^+$, $Na^+$ and other ions as well as glucose and urea, could be demonstrated. These measurements have been performed in blood plasma and whole blood as well as in gastroenterological investigations for in vivo pH measurements in the stomach [70]. However, a blood-taking procedure and a process of in vivo application of the pH catheter in the stomach are rather unpleasant for patients. Therefore, in recent years, there has been an increasing demand for the development of portable, 'easy in use' urine analyser for home monitoring (e.g. diabetic patients can easily test their glucose level or pH value in urine a few times each day). In Ref. [29], an ISFET-based glucose sensor has already been utilised for determining the glucose concentration in real human urine samples.

In the above study, the possibility of applying of the developed hybrid sensor module to pH measurement in human urine was investigated. Figure 14.23 presents the results of a pH measurement in human urine samples with the ISFET versus the pH value measured by a conventional macroscopic pH glass electrode. It can be noted that there is good correlation between the measurements with the pH ISFET and the pH glass electrode over the physiologically meaningful pH range (pH 4.8–7.5) for daily human urine.

### 14.4.2 pH measurement in rain droplets

Deposition of pollutants, including acidic compounds, is one of many present global environmental problems. The analysis of the chemical components of rain, including pH, is usually carried out only in bulk

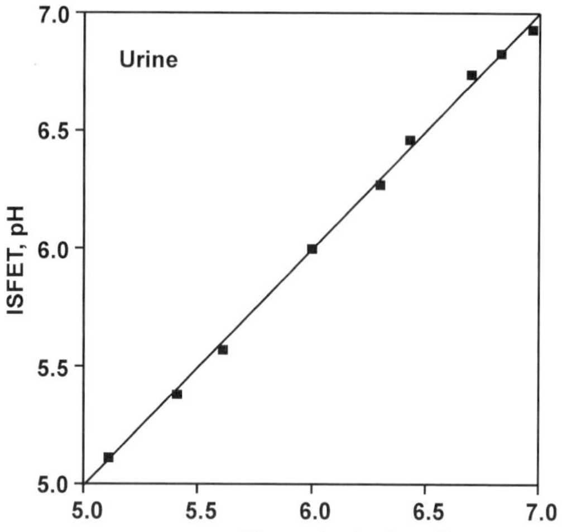


Fig. 14.23. Comparison of the pH measurements in human urine with the developed pH-sensitive $Ta_2O_5$-gate-ISFET and a conventional macroscopic pH glass electrode.

rain samples, which have been collected over a period of days, weeks or even months [71]. Since the pH can significantly change during a single rain event [72,73], a quantitative analysis on a well-defined time scale as well as the determination in small sample volumes is necessary. The analysis of chemical components in single rain drops is expected to give new and interesting information about air pollution that requires microanalytical approaches and microsensors. In this respect, a pH ISFET can serve as a useful microanalytical tool. The main advantage is the possibility of recording the pH value both in short time intervals and in small volumes (rain drop size is often $<5\ \mu l$) in order to reduce dilution effects.

The developed multi-parameter hybrid sensor module was utilised for pH measurements in bulk rain samples as well as in single rain droplets of different rain events. In addition, reference pH measurements have also been performed in bulk samples of the same rain events with a conventional pH glass electrode. The bulk rain samples and rain drops were collected from two different areas of Germany [28]. All rainwater samples were measured directly without any sample pre-treatment. Specific demands for the pH measurements in rain drops are low ionic strength of the water, the small volume and the need for sensor calibration in buffer droplets with low ionic strength.

As an example, Fig. 14.24 shows the dynamic response of the pH ISFET in rain droplets for eight different rain events (R1–R8). The pH of the bulk rain samples varied between 4.56 and 6.33. In most cases, a relatively good agreement was found between the pH values measured by the ISFET (black numbers) and the conventional glass electrode (grey numbers).

Thus far, the characterisation of rainwater has been based on the assumption that individual rain drops and bulk rain samples have the same concentration of different chemical components, in particular, the same pH value. The effect of drop size on the speciation of trace elements was almost neglected. However, the pH measurements in single rain droplets performed with the ISFET [28] shows drop-size dependence of the pH value (see Fig. 14.25). Obviously, larger drops can have a pH which is about 0.2 units higher than that of small droplets. Recently, a drop size dependence for the concentration of different rainwater components (including pH) was also observed [74,75] using alternative, but more complicated and time-consuming measuring methods.

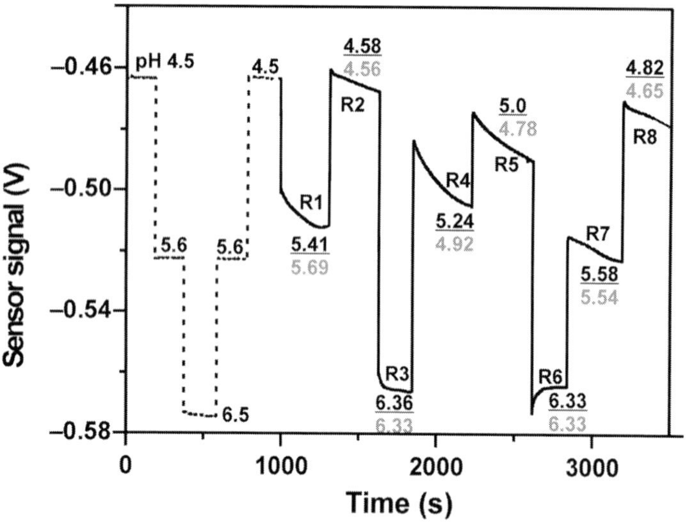


Fig. 14.24. Dynamic response of the pH ISFET in rain droplets of eight different rain events (R1–R8) and in drops of low ionic strength buffer (dotted line). The grey numbers correspond to the pH values in the bulk rain samples measured with a conventional pH glass electrode. The black numbers correspond to the pH values in the rain droplets of the same rain events measured with the pH ISFET.

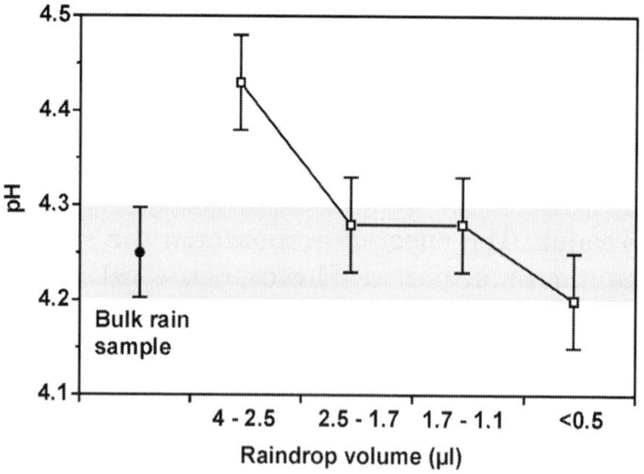


Fig. 14.25. Drop-size dependence of the pH value in single rain droplets measured with the pH ISFET.

## 14.5 SUMMARY AND CONCLUSION

A 'high-order' hybrid sensor module for (bio)chemical and physical sensing has been presented. The sensor module is based on an identical transducer principle, namely the ion-sensitive field-effect transistor. The multi-functionality of this sensor module has been achieved by means of different sensor arrangements and/or various operation modes. Thus, the number of measuring parameters can be significantly higher than the number of sensors present in the module. The developed hybrid module consists of two ISFETs only in combination with an ion generator. Several ISFET-based sensors for the determination of three (bio)chemical and four physical parameters have been realised:

- an ISFET-based pH sensor with a $Ta_2O_5$ gate for pH determination in both static and flow-through conditions;
- an ISFET-based penicillin sensor (PenFET) by adsorptively immobilising the enzyme penicillinase on top of the pH-sensitive $Ta_2O_5$ gate; the differential set-up as a pH ISFET/PenFET dual chip allows compensation for fluctuations of the test sample's pH value;
- an ISFET-based bioelectronic sensor (beetle/chip) combines both the excellent olfactory abilities of sensory organs (insect antenna) and the highly reliable electronic signal amplification of the FET;
- apart from these (bio)chemical quantities, an ISFET-based temperature sensor also operated in a differential arrangement, advantageously 'inverted' the temperature dependence of its input characteristic as a useful sensing signal;
- an ISFET-based flow-velocity sensor that electrochemically detects by 'time of flight' measurements in situ generated pH changes, produced by a sensor–actuator system;
- an ISFET-based flow-direction sensor with an upstream- and a downstream-placed sensor, to determine not only the flow velocity but also the flow direction; and
- an ISFET-based diffusion-coefficient sensor, where the diffusion mass transport of ions shows the responsible process to be in contrast to convective mass transport.

Further applications of this 'high-order' hybrid sensor module approach could potentially deal with additional ISFET-based sensors, like an ISFET-based electrolyte conductivity sensor, an ISFET-based liquid-level sensor or an ISFET-based DNA sensor. The hybrid modular concept of such a detector unit should enable its application as part of commercial microanalysis systems.

## ACKNOWLEDGEMENTS

The authors thank A. Baade, L. Berndsen, H. Emons, H. Lüth, A. Riemer, P. Schroth, S. Schütz, J. W. Schultze, A. Simonis and A. Steffen for valuable discussions and technical support. Parts of the study were supported by the Ministerium für Schule und Weiterbildung, Wissenschaft und Forschung des Landes Nordrhein–Westfalen and the BMBF (Germany).

## REFERENCES

1 A. Manz, N. Graber and H.M. Widmer, *Sens. Actuat. B*, 1 (1990) 244.
2 S. Shoji, In: H. Baltes, W. Göpel and J. Hesse (Eds.), *Sensors Update*, Wiley/VCH, Weinheim, 2000, p. 3.
3 A. van den Berg and T.S.J. Lammerink, *Topics Cur. Chem.*, 194 (1997) 21.
4 S. Shoji, *Topics Cur. Chem.*, 194 (1997) 163.
5 D.R. Reyes, D. Iossifidis, P. Auroux and A. Manz, *Anal. Chem.*, 74 (2002) 2623.
6 D. Binz, S. Keeping and A. Vogel, *Proc. Transducer-99, Sendai, Japan*, 1999, p. 684.
7 W.H. Baumann, M. Lehmann, A. Schwinde, R. Ehret, M. Brischwein and B. Wolf, *Sens. Actuat. B*, 55 (1999) 77.
8 E. Dempsey, D. Diamond, M.R. Smyth, G. Urban, G. Jobst, I. Moser, E.M.J. Verpoorte, A. Manz, H.M. Widmer, K. Rabenstein and R. Freaney, *Anal. Chim. Acta*, 346 (1997) 341.
9 W. Hoffmann and R. Rapp, *Sens. Actuat. B*, 34 (1996) 471.
10 B.H. van der Schoot, S. Jeanneret, A. van den Berg and N.F. de Rooij, *Sens. Actuat. B*, 15/16 (1993) 211.
11 G. Urban, G. Jobst, I. Moser and H. Ernst, *Proc. Eurosensors-13*, The Hague, The Netherlands, 1999, p. 343.
12 J. Janata, M. Josowicz, P. Vanysek and D.M. DeVaney, *Anal. Chem.*, 70 (1998) 179R.
13 H. Ulmer, J. Mitrovics, U. Weimar and W. Göpel, *Sens. Actuat. B*, 65 (2000) 79.

14 H. Ulmer, J. Mitrovics, G. Noetzel, U. Weimar and W. Göpel, *Sens. Actuat. B*, 43 (1997) 24.
15 A. Grisel, *Proc. Eurosensors-13*, The Hague, The Netherlands, 1999, p. 699.
16 H. Suzuki, T. Hirakawa, S. Sasaki and I. Karube, *Anal. Chim. Acta*, 405 (2000) 57.
17 F. van Steenkiste, E. Lauwers, J. Suls, D. Maes, K. Baert, W. Gumbrecht, Ph. Arquint, L. Hermans, R. Mertens, G. Gielen, W. Sansen and K. Abraham-Fuchs, *Proc. Transducer-99*, Sendai, Japan, 1999, p. 1188.
18 C. Jiménez, I. Marquès and J. Bartrolí, *Anal. Chem.*, 68 (1996) 3801.
19 W. Gumbrecht, D. Peters, W. Schelter, W. Erhardt, J. Henke, J. Steil and U. Sykora, *Sens. Actuat. B*, 18–19 (1994) 704.
20 P. Arquint, M. Koudelka-Hep, B.H. van der Schoot, P. van der Wal and N.F. de Rooij, *Clin. Chem.*, 40 (1994) 1805.
21 P. Norlin, O. Öhman, B. Ekström and L. Forssen, *Sens. Actuat. B*, 49 (1998) 34.
22 R. Rapp, W. Hoffmann, W. Süß, H.J. Ache and H. Gölz, *Electrochim. Acta*, 42 (1997) 3391.
23 G.R. Langereis, W. Olthuis and P. Bergveld, *Sens. Actuat. B*, 53 (1998) 197.
24 A. Poghossian, H. Lüth, J.W. Schultze and M.J. Schöning, *Electrochim. Acta*, 47 (2001) 243.
25 A. Poghossian, L. Berndsen, J.W. Schultze, H. Lüth and M.J. Schöning, In: M. Butler, P. Vanysek and N. Yamazoe (Eds.), *Chemical and Biological Sensors and Analytical Methods*, The Electrochemical Society Inc, Pennington, USA, 2001, p. 143.
26 K.A. Erickson and P. Wilding, *Clin. Chem.*, 39 (1993) 283.
27 A. Poghossian, M. Thust, P. Schroth, A. Steffen, H. Lüth and M.J. Schöning, *Sens. Mater.*, 13 (2001) 207.
28 A. Poghossian, A. Baade, H. Emons and M.J. Schöning, *Sens. Actuat. B*, 76 (2001) 634.
29 A. Poghossian, *Sens. Actuat. B*, 44 (1997) 361.
30 A. Poghossian, M.J. Schöning, P. Schroth, A. Simonis and H. Lüth, *Sens. Actuat. B*, 76 (2001) 519.
31 A. Poghossian, T. Yoshinobu, A. Simonis, H. Ecken, H. Lüth and M.J. Schöning, *Sens. Actuat. B*, 78 (2001) 237.
32 A. Poghossian, M. Thust, M.J. Schöning, M. Müller-Veggian, P. Kordos and H. Lüth, *Sens. Actuat. B*, 68 (2000) 260.
33 M.A.J. van Opstal, R. Wolters, J.S. Blauw, P.C. van Krimpen, W.P. van Bennekom and A. Bult, *J. Pharm. Biomed. Anal.*, 8 (1990) 49.
34 O. Leistiko, *Phys. Scripta*, 18 (1978) 445.
35 G.H. Wang, D. Yu and Y.L. Wang, *Sens. Actuat.*, 11 (1987) 221.
36 S. Martinoia, L. Lorenzelli, G. Massobrio, P. Conci and A. Lui, *Sens. Actuat. B*, 50 (1998) 60.
37 J.C. Chou, Y.F. Wang and J.S. Lin, *Sens. Actuat. B*, 62 (2000) 92.

38 J.C. Chou, Y.S. Li and J.L. Chiang, *Sens. Actuat. B*, 71 (2000) 73.
39 J.L. Chin, J.C. Chou, T.P. Sun, W.Y. Chung and S.K. Hsiung, *Sens. Actuat. B*, 76 (2001) 582.
40 J.C. Chou, C.Y. Weng and H.M. Tsai, *Sens. Actuat. B*, 81 (2002) 152.
41 N.T. Nguyen, *Flow Measur. Instrument.*, 8 (1997) 7.
42 R. Kersjes and W. Mokwa, *Sens. Actuat. A*, 46–47 (1995) 373.
43 M. Ashauer, H. Glosch, F. Hedrich, N. Hey, H. Sondmaier and W. Lang, *Sens. Actuat. A*, 73 (1999) 7.
44 S. Wu, Q. Lin, Y. Yuen and Y. Tai, *Sens. Actuat. A*, 89 (2001) 152.
45 R.E. Oosterbroek, T.S.J. Lammerink, J.W. Berenschot, G.J.M. Krijnen, M.C. Elwenspoek and A. van den Berg, *Sens. Actuat. A*, 77 (1999) 167.
46 A. Kruusing, S. Leppävuori and A. Uusimäki, *Proceedings of Eurosensors-12*, Southampton, UK, 1998, p. 783.
47 G. Fuhr, R. Hagedorn, T. Müller, W. Benecke and B. Wagner, *Proceedings of MEMS-92*, Travemunde, Germany, 1992, p. 25.
48 P. Profos and T. Pfeifer, *Handbuch der industriellen Messtechnik*, R. Oldenbourg Verlag, München, 1992.
49 A.J.N. Khalifa and M.M. Mehdi, *Energy Conv. Mgmt*, 39 (1998) 1295.
50 J. Wu and W. Sansen, *Transducers-01/Eurosensors-15, Digest of Technical Papers*, Munich, Germany, 2001, p. 1444.
51 W. Olthuis and P. Bergveld, *Mikrochim. Acta*, 121 (1995) 191.
52 S. Böhm, W. Olthuis and P. Bergveld, *Sens. Actuat. B*, 47 (1998) 48.
53 E. Klusmann and J.W. Schultze, *Electrochim. Acta*, 42 (1997) 3123.
54 A. Poghossian, L. Berndsen, H. Lüth and M.J. Schöning, *Proceedings of SPIE*, 4560, 2001, p. 19.
55 S. Schütz, B. Weissbecker, H.E. Hummel, M.J. Schöning, A. Riemer, P. Kordos and H. Lüth, *Naturwissenschaften*, 84 (1997) 86.
56 M.J. Schöning, S. Schütz, P. Schroth, B. Weissbecker, A. Steffen, P. Kordos, H.E. Hummel and H. Lüth, *Sens. Actuat. B*, 47 (1998) 235.
57 P. Schroth, M.J. Schöning, S. Schütz, Ü. Malkoc, A. Steffen, M. Marso, H.E. Hummel, P. Kordos and H. Lüth, *Electrochim. Acta*, 44 (1999) 3821.
58 P. Schroth, M.J. Schöning, P. Kordos, H. Lüth, S. Schütz, B. Weissbecker and H.E. Hummel, *Biosens. Bioelectron.*, 14 (1999) 303.
59 M.J. Schöning, P. Schroth and S. Schütz, *Electroanalysis*, 12 (2000) 212.
60 M.J. Schöning, *Am. Lab.*, (2000) 24.
61 M.J. Schöning and H. Lüth, *Phys.Stat. Sol. A*, 185 (2001) 65.
62 P. Schroth, H. Lüth, H.E. Hummel, S. Schütz and M.J. Schöning, *Electrochim. Acta*, 47 (2001) 293.
63 P. Schroth, B. Weissbecker, S. Schütz, H. Ecken, T. Yoshinobu, H. Lüth and M.J. Schöning, *Biocybernet. Biomed. Eng.*, 21 (2001) 27.
64 P. Schroth, M.J. Schöning, H. Lüth, B. Weissbecker, H.E. Hummel and S. Schütz, *Sens. Actuat. B*, 78 (2001) 1.

65 S. Schütz, M.J. Schöning, P. Schroth, Ü. Malkoc, B. Weissbecker, P. Kordos, H. Lüth and H.E. Hummel, *Sens. Actuat. B*, 65 (2000) 291.
66 M.J. Schöning, P. Schroth, H. Lüth, H.E. Hummel and S. Schütz, *Proceedings of SPIE*, 4205, 2001, p. 152.
67 M.J. Huotari, *Sens. Actuat. B*, 71 (2000) 212.
68 W. Gumbrecht, W. Schelter, B. Montag, M. Rasinski and U. Pfeiffer, *Sens. Actuat. B*, 1 (1990) 477.
69 S.V. Dzyadevich, Y.I. Korpan, V.N. Arkhipova, M.Yu. Alesina, C. Martelet, A.V. El'skaya and A.P. Soldatkin, *Biosens. Bioelectron.*, 14 (1999) 283.
70 J. Han, D. Cui, Y. Li, H. Zhang, Y. Huang, Z. Zheng, Y. Zhu and X. Li, *Sens. Actuat. B*, 66 (2000) 203.
71 R. Sequeira and M.R. Peart, *Sci. Total Environ.*, 159 (1995) 177.
72 S. Yu, Z. Cheng, X. Cai, X. Cheng and W. Shong, *Atmos. Res.*, 32 (1994) 297.
73 B. Railsback, *Sci. Total Environ.*, 198 (1997) 233.
74 K. Bächmann, P. Ebert, I. Haag, T. Prokop and K. Steigerwald, *Atmos. Environ.*, 30 (1996) 1027.
75 K. Bächmann, I. Haag and K. Steigerwald, *Atmos. Environ.*, 29 (1995) 175.

*Chapter 15*

# *Microdialysis based lab-on-a-chip, applying a generic MEMS technology*

***P. Bergveld, W. Olthuis,* A.J. Sprenkels, D.G. Pijanowska, H.J. van der Linden and S. Böhm***

## 15.1 INTRODUCTION

In reaction to the biocompatibility problems of catheter type chemical sensors, which are meant for in vivo monitoring, the microdialysis concept has been introduced. It is in fact a sampling technique, producing a protein- and cell-free sample containing an imprint of the chemical composition of blood, which can be monitored by sensors outside the body. This chapter treats the integration of a microdialysis probe with a silicon chip, which contains the sensors, as well as calibration facilities to calibrate the sensors and a pump for the liquid handling. Because the chip contains all relevant system components for the chemical analysis, it is in fact a lab-on-a-chip. Glucose as well as potassium concentrations have been measured successfully with the system.

*Corresponding author. Tel.: +31-534892688; fax: +31-534892287. *E-mail address:* w.olthuis@el.utwente.nl.

*Comprehensive Analytical Chemistry XXXIX, pages 625–663* 

ISSN: 0166-526X

### 15.1.1 The need for in vivo monitoring

About 30 years ago, silicon chip technology, developed by the integrated circuit (IC) industry, started also to be exploited in the field of sensor development, especially for biomedical purposes. Because the chips are so rigid and small, they can easily be incorporated into the tip of a catheter to be applied for in vivo monitoring. Especially for chemical blood analysis of critically ill patients in an intensive care room, a strong need for in vivo monitoring exists as an alternative to the usual laboratory tests with blood samples. Only through continuous measurement can trends in a patient's condition, for instance with respect to their blood concentration of potassium ions or lactate concentration, be observed and used as an early warning. Another example is the development of chip-based glucose sensors for diabetic patients, in order to notify hypoglycaemic situations during the night, for glycaemic stabilization during hospitalization or for the control of an implanted insulin delivery system (an ultimate goal of glucose sensor development).

Unfortunately, independent of the specific sensor concept or manufacturing technology, all attempts to make (bio)chemical sensors, which are in intimate contact with whole blood for a relatively long period of time have so far failed due to the blood incompatibility of the applied materials. Invariably proteins will adsorb to the sensor interface, ultimately followed by encapsulation with fibroblasts, resulting in interference with the actual sensor function, for instance changing its sensitivity and selectivity. Separating the actual sensor interface from the blood by means of a haemocompatible membrane has not proved to be a successful solution. Usually the response time of the sensor is increased to an unacceptable degree. Moreover, in both situations, with or without an additional membrane, the sensor should be calibrated at regular time intervals, for which up to now no adequate procedure has been developed. Furthermore, the risk of thrombosis, embolism and septicaemia is thought to be too great with the use of invasive sensors, which certainly does not enhance the application when the sensor signal is also not 100% reliable.

A solution to the problems as described above is the separation of the sensor (chip) and the sample in such a way that the ions and metabolites to be measured can reach the sensor interface, but the proteins and other large molecules cannot. In addition, the sensor should be placed extracorporeally, where it can be calibrated. Furthermore the sample should be taken from the living organism by means of a minimally invasive procedure. This approach is developed as microdialysis and was first applied in clinical practice about a decade ago.

### 15.1.2 Microdialysis

The microdialysis concept involves the use of a hollow fibre, made partly of a semi-permeable membrane material. The hollow fibre, having an outer diameter of about 300 μm, is inserted into the subcutaneous tissue, thus mimicking a blood capillary. The fibre is flushed by a buffered saline solution, the perfusate. At the site of the semi-permeable membrane, ions and water-soluble molecules of size below the molecular cut-off of the membrane will diffuse from the tissue and thus from the surrounding blood capillaries into the perfusate. The dialysate leaving the probe will thus contain an imprint of the chemical constitution of the tissue, which may be an almost 100% faithful representation in case the perfusate flow is stopped for a time that is sufficient for equilibration. For shorter times, in situations of continuous flow of the perfusate, this number will be smaller, depending on the flow rate. This percentage is called the relative recovery (see Section 15.2.2). The actual sensors are placed off-stream and can thus measure the clean dialysate and can also be calibrated by intermediate contact with calibration liquids. Biocompatibility of the sensor surface is no longer a requirement. The technique is regarded as minimally invasive when compared to the insertion of needle-type sensors of millimetre size, because microdialysis probes are much smaller in diameter.

Originally microdialysis was introduced by Ungerstedt et al. [1] for monitoring neurotransmitter release in the brain, but has now become a frequently used research tool for the investigation of chemical compositions in many other tissues or body compartments, as introduced by Lönnroth et al. [2]. Up to now no clinical applications have been launched, due to the fact that the commercially available microdialysis probes have to be connected by means of tubes to existing conventional chemical analysis equipment. This equipment is usually too bulky for application near a patient's bed and moreover the tubing will cause too large dead volumes.

The same can be said with respect to the introduction of microdialysis in non-medical applications, such as monitoring biotechnological processes, including the automatic control of these processes.

Additionally, direct contact between a sensor surface and, for instance, the fermentation solution should be prevented. This can be achieved by applying the microdialysis concept to this technological field also, which is becoming increasingly important.

Regardless of the disappointing results of the application of small and rigid chip-based sensors when brought into direct contact with a living system, the sensors as such are well established at present; potentiometric as well as amperometric sensors are commercially available. The same can be said for the microdialysis probes. The sensors are extremely suitable for measuring small volumes, such as the dialysate volumes from a microdialysis probe. Therefore it has been recognized by the first author of this chapter that a combination of the tiny sensors and the tiny microdialysis probes would be a very favorable development. In other words, if the present plastic connector of a microdialysis probe can be replaced by a silicon connector in which the sensors are placed, the best of the two systems are combined. This idea formed the start of a research program 'the comprehensive integration of the microdialysis sampling technique and silicon sensors' in 1996, subsidized by the Dutch Foundation for Fundamental Research on Matter (FOM, Utrecht, the Netherlands). The result at the present moment (2002) is a lab-on-a-chip, containing a sensor array with integrated calibration facilities, to which a microdialysis probe is connected without additional tubing.

### 15.1.3 The microdialysis lab-on-a-chip

The term 'lab-on-a-chip' is reserved especially in the USA for systems that consist of a sandwich of glass plates in which a certain configuration of channels is etched in one or another way. After insertion of a sample into an input port, the chips are placed in a benchtop apparatus, which contains electrodes to contact the various liquid input ports of the channel structure. By means of electro-osmotic forces (EOF pump) the liquids are forced to move through the channels and are separated into different components (electrophoresis), with the result usually monitored by means of an optical system (the actual sensor), which is an integral part of the relatively large equipment, as reviewed by Kopf-Sill of Caliper Technologies Corp. [3]. The chips are presently available commercially for gene analysis, drug screening, etc. Note that not all components of the analysis system have been miniaturized, but only the liquid handling system, whereas the pumping action of liquids is limited to EOF.

In Europe the term μTAS has been introduced and is more often used. The term denotes micro total analysis system, a stand-alone system containing all the components necessary for a particular chemical analysis. Such a microsystem comprises for the liquid handling micropumps and valves, and for the actual analysis the necessary micromixers, microreactors, microsensors and the means to provide calibration. Some of these systems are built up in a modular way, a so-called hybrid approach as described by Wissink et al. [4]. For a higher degree of integration, or in other words, for a complete laboratory on one and the same chip, a monolithic technology is required, as derived from IC technology. For the design of the microdialysis probe with incorporated sensors this monolithic integration is chosen as the starting point. This choice limits the design of the already known components like micropumps and microsensors, because of the different technologies used up to now for these components. For instance, it will be difficult to combine in one and the same process the technology for making piezoelectrically driven pumps for the liquid transport and chemFETs for the sensors. Instead of applying difficult-to-combine technologies for the microdialysis system, an a priori choice should be made for one generic technology. With this requirement in mind the necessary components should be (re)designed or even newly developed.

In the case of developing the microdialysis lab-on-a-chip, the chosen generic technology consists of etching the channels and cavities into a silicon substrate, depositing metal electrodes onto a glass wafer and bonding the silicon substrate to the glass cover. With this technology, the following different subsystems have to be realized:

- the microdialysis probe is directly connected to the silicon/glass sandwich in order to realize a liquid handling system with minimum dead volumes;
- the sensor array, containing potentiometric as well as amperometric sensors and a reference electrode, applies a flow-through concept;
- facilities exist to calibrate the different sensors as an integral part of the system; and
- a pump for the transport of perfusate/dialysate for continuous as well as stop-flow mode of operation.

Figure 15.1a shows how the combined subsystems will result in the complete lab-on-a-chip, as patented by FOM [5], whereas Fig. 15.1b shows how the sensors subsequently measure the calibration pulses cal.2 and cal.1 as injected on the chip by the two calibration systems, followed by the actual sensor signal.

In the following sections, the development of the first three of the four subsystems mentioned above will only be treated, because the perfusion liquid pump is not yet ready for integration. Nevertheless the fully integrated system will be described in Section 15.6, although for the

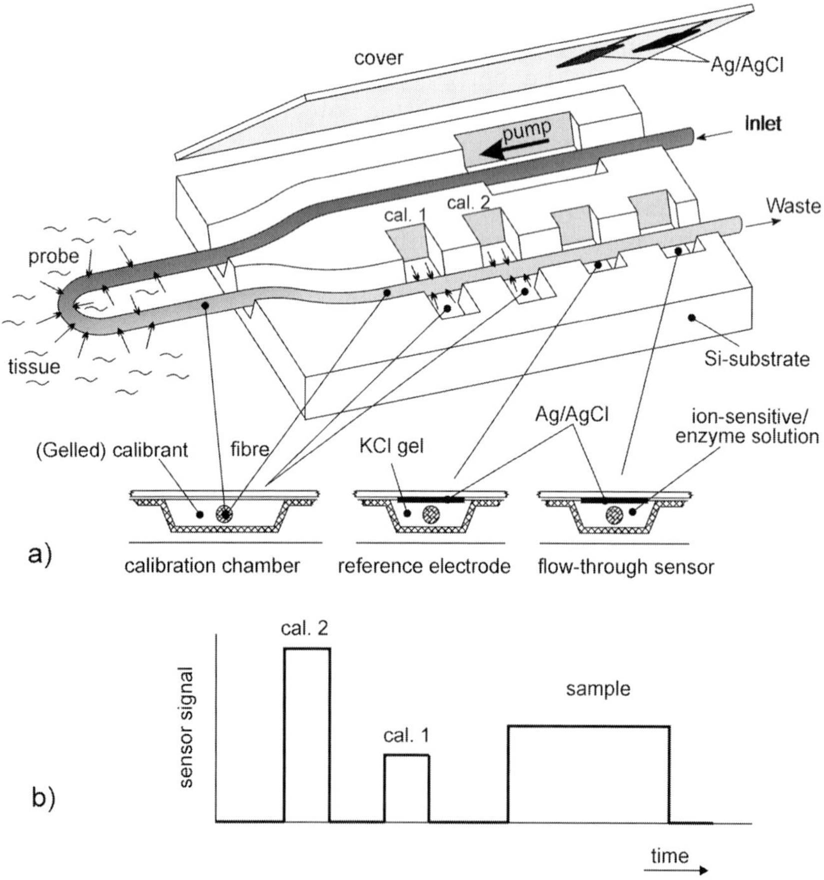


Fig. 15.1. (a) Drawing taken from the international patent application of a microdialysis probe, integrated with silicon chip; (b) two calibration pulses, followed by the sample signal.

time being applied with an external perfusion liquid pump. The integration necessitates partly redesigning the separate components as described in previous sections, as dicted by the actual integration process.

## 15.2 THE MICROMACHINED DOUBLE LUMEN MICRODIALYSIS PROBE CONNECTOR

Microdialysis probes are at present commercially available. They consist of an ultra thin shaft with, on the indwelling side, the actual semi-permeable membrane section and at the other end a plastic connector with small tubes, to which extension tubes have to be connected for introduction of a perfusion liquid and for transport of the dialysate to a remote chemical analyzer. These tube connections are in fact the weak point of the present microdialysis concept, as well as the lack of suitable microchemical analyzers. Replacement of the plastic connector by a silicon chip, which may contain the sensors, will be the first step towards the complete lab-on-a-chip system as described in this chapter.

### 15.2.1 The conventional microdialysis probe

The construction and functioning of a typical commercially available double lumen microdialysis probe is given in Fig. 15.2. Basically,

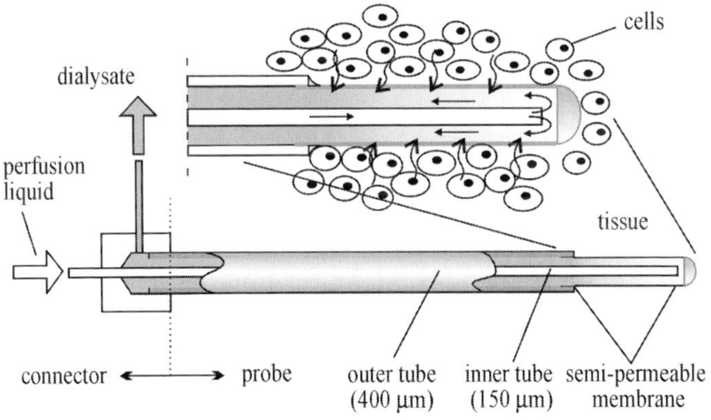


Fig. 15.2. Construction of a typical commercially available double lumen microdialysis probe.

the probe consists of a small inner tube of about 150 μm outer diameter, located within an outer tube of about 400 μm outer diameter. The inner tube is connected to a precision pump that drives the perfusion liquid (physiological saline) at a continuous flow rate in the range of 0.2–2 μL/min. The perfusion fluid leaves the inner tube at the probe tip and then flows alongside the semi-permeable membrane, via the outer lumen, back to the connector. However, in the application of on-line monitoring, the very low flow rate associated with microdialysis requires swept volumes to be minimized in order to reduce the induced lag time in measurements. The volume between the probe and the analysis system especially requires attention. In a microdialysis probe the total swept volume is made up of two parts. The first part is between the probe tip and the connector, whilst the second part is formed by the connector and its output tube, through which the dialysate is guided to the sensor array. To indicate the effect, a total swept volume of 1 μL introduces a lag time in the measurement, i.e. the time to reach the sensor, of the order of a minute.

### 15.2.2 Theory

*Probe recovery*

As the microdialysis sampling technique yields a sample stream, which reflects to some extent the actual composition at the implantation site, the dialysate could be linked to this actual sample composition. The process of sampling will be described with reference to Fig. 15.2. The probe inlet for the perfusion liquid is connected to a precision pump which drives the perfusion liquid at a continuous pump rate. This perfusion liquid can contain a certain concentration $C_0$ (mol/L) of the analyte of interest (for instance an ion like sodium or chloride). On its passage alongside the semi-permeable probe tip, the analyte present at a concentration $C_s$ in the sampling region will diffuse according to the concentration gradient. This gradient, the difference in concentration in the sampling region and the perfusion liquid, $C_s - C_0$, is the driving force for diffusional mass transport from the sample to the perfusion liquid. This mass flux through the semi-permeable membrane enriches the perfusion liquid and results in a dialysate having a concentration $C_d$. The relative recovery, RR, which describes the relation between the analyte concentration in the dialysate coming from the probe and the actual external concentration at the probing site, is generally given by

$$\mathrm{RR} = \frac{C_\mathrm{d} - C_0}{C_\mathrm{s} - C_0} \tag{15.1}$$

This relative recovery, often expressed as a percentage, is besides a number of other parameters, inversely dependent on the flow rate. A higher flow rate will for a certain probe geometry result in a dilution of the incoming diffusional mass flux by the perfusion liquid and as a consequence RR will be smaller. Other factors having a large effect on the relative recovery are the type of semi-permeable membrane used and the sample matrix at the probing site (i.e. subcutaneous tissue, blood stream, etc.). Because the relative recovery is needed to calculate the actual sample concentration $C_\mathrm{s}$ from the measured dialysate concentration $C_\mathrm{d}$, identification and understanding of this ratio is of prime importance.

*Lag time*

Lag time is defined as the time between a change in composition of the actual sample and the moment that this change is detected by the sensor system (moment of onset of the detector). It is assumed here that the response time of this sensor system is infinitely small, so that the main factor that determines the lag time results from the transport of the dialysate from the probe tip to the detector. As the diameters of the flow channels are small (around 100 μm) and flow rates are low, it is assumed that dispersion resulting from laminar flow and diffusion will balance. As a consequence the sample will travel as a 'plug' through the fluidic system (flow injection analysis conditions).

If the volumetric flow rate of the perfusion liquid is $q$ ($\mathrm{m^3/s}$) and the total swept volume is $V_\mathrm{total}$ ($\mathrm{m^3}$), the lag time $\tau$ (s) induced by the swept volume is given by

$$\tau = \frac{V_\mathrm{tip} + V_\mathrm{connector}}{q} = \frac{V_\mathrm{total}}{q} \tag{15.2}$$

where $V_\mathrm{tip}$ and $V_\mathrm{connector}$ ($\mathrm{m^3}$), respectively, represent the swept volume in the probe tip and shaft, and the swept volume within the connector. Because the construction of the tip and shaft of the microdialysis probe described in this work is exactly the same as for conventional double lumen probes, the main benefit that can be achieved from the application of microsystem technologies will be in the minimization of the swept volume $V_{connector}$ in the connecting part.

*Chloride sensitive electrode*
In the current micromachined probe, as a case study for fast on-line measurements, a chloride sensitive electrode has been incorporated in the probe connector. This sensor allows the measurement of chloride concentrations in the dialysate.

### 15.2.3 Experimental

*Device construction*
An exploded view of the proposed micromachined connector/sensor construction is given in Fig. 15.3. Basically it is a sandwich construction of two multi-step KOH etched wafers. In the silicon parts, a number of features can be identified. The actual connecting part is formed by the wide etched U-trench for receiving the outer capillary tube. This trench, which is present in both the upper and lower part, narrows via a funnel-like section down to a V-shaped channel in which the inner capillary tube is to be inserted. Also this channel is present in both parts, however, in the lower part it ends in a flow-through cell for sensor integration.

Besides channels for the fixing and aligning of the capillary tubes, a number of holes are etched. In the lower part, at the place where the funnel-like structure ends in the V-channel, an opening for the perfusion liquid is located, whilst at the end of the flow-through cell the waste outlet is located. In the upper part, two holes are etched through which, after insertion of the capillary tubes, small amounts of glue can be dosed for fixing and sealing these tubes.

The perfusion liquid enters the connector via the lower part and travels in the annular space between the inner and outer tube towards the semi-permeable tube, which forms the tip of the probe. After flowing alongside the membrane, the formed dialysate enters the inner tube and flows back to the connector. After leaving the inner tube, the dialysate flows through the flow-through cell that contains the sensor. Finally after being measured, the dialysate exits the device via the waste output.

For the experiments, two probes of different dimensions were assembled, each having identical outer capillary tube lengths

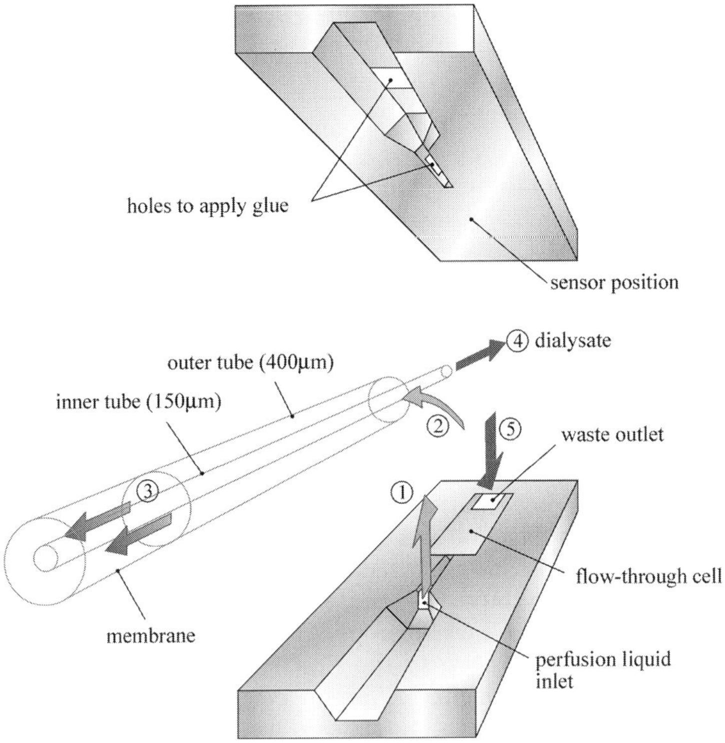


Fig. 15.3. Exploded view of the connector/sensor construction. The fluid flow from inlet to outlet is marked with numbers 1–5 (1–3: perfusion liquid, 4: dialysate entering the flow-through cell, 5: dialysate leaving the system).

(15 mm), but with semi-permeable tube lengths of 5 and 20 mm, respectively.

*Sensor incorporation*

For the incorporation of a sensor an easy to integrate Ag/AgCl electrode was chosen for evaluation of the performance of the proposed system. This electrode for the measurement of chloride ions was placed in the flow cell at the location 'sensor position' as indicated in Fig. 15.3. The mean sensitivity of the resulting electrode in the range between 10 and 100 mM $Cl^-$ was determined to be 55 mV/pCl. To compare the micromachined probe with a commercial probe, having comparable dimensions (BR-4, 4 mm membrane length, BAS Lafayette, USA), a conventionally produced flow-through sensor for chloride was made by piercing a 0.5 mm internal diameter PVC tube with a 0.2 mm chlorinated silver wire.

*Measurement set-up*

Because the target analyte is chloride, as 'sample matrix' a 100 mM KCl solution was chosen. To provide a stable baseline, a perfusion liquid containing 10 mM KCl was used.

As the main object of the experiment was to compare the lag time of measurements with conventional probes coupled to a sensor and the micromachined probe connector with integrated sensor, probes were switched between two stirred sample solutions. The first sample solution was identical to the perfusion liquid pumped through the probe whilst the actual sample solution was formed by the 100 mM KCl solution.

To evaluate the lag time of a particular probe/sensor combination, first a stable sensor baseline was obtained with the probe placed into the first (stirred) solution and perfused at a fixed rate. Then the probe was transferred swiftly to the stirred sample solution and the sensor output was recorded on a storage oscilloscope. After recording the response, the probe was transferred back to the first solution. This measurement procedure was repeated for different perfusion rates in order to investigate the difference in lag time as well as the relation between perfusion rate and the relative recovery of the probe under study (see Fig. 15.4).

### 15.2.4 Results and discussion

*Conventional probe / sensor combination*

First measurements were performed with the conventional probe connected with minimal swept volume to the wire chloride sensor. The typical microdialysis perfusion flow rate of 1 μL/min was chosen to result in a reference response with which the responses of the micromachined probe could be compared. The response (not shown) to 100 mM KCl resulted in a final amplitude of 47 mV whilst the lag time was about 60 s. The amplitude is lower than the sensors' sensitivity of 55 mV/pCl, which is the result of the relative recovery of the probe. Using the Nernst equation, a RR of about 0.68 was calculated ($C_d = 71$ mM).

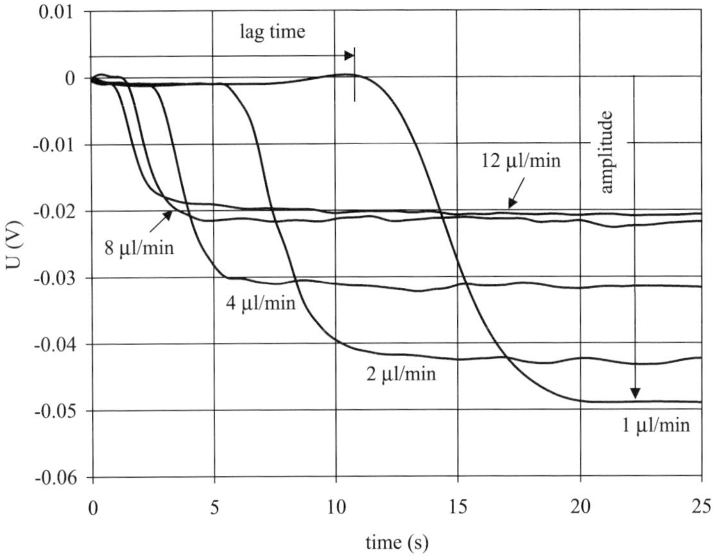


Fig. 15.4. Measured response to 100 mM KCl for a micromachined probe/sensor (membrane length 5 mm).

*Micromachined probe / sensor combination*

Next the response of the micromachined probe with 5 mm membrane length was measured for increasing perfusion flow rates. From the results plotted in Fig. 15.4, two effects can clearly be seen. The first is the decrease in lag time with increasing flow rate, which is expected from Eq. (15.2). At a flow rate of 12 μL/min, within 1.5 s the output signal starts to change and within 4 s the signal is at its final value of 20 mV, considerably smaller than 47 mV. Due to the high flow rate, the residence time of the perfusion liquid in the semi-permeable probe tip is in the order of a second, and consequently the relative recovery RR is small. These two effects are inversely dependent as can be seen from the same figure. For a flow rate of 1 μL/min, the amplitude is 49 mV, slightly larger than the amplitude obtained with the conventional probe in the previous experiment. This is probably due to a slightly larger permeability of the membrane material. However, the lag time is about 11 s, which is considerably shorter due to the fact that the swept volume in the connecting part was minimized by the application of micromachining.

The measured lag time and relative recovery for both the 5 and 20 mm probes are summarized in Fig. 15.5. It can be seen that for both membrane lengths the lag time $\tau$ is inversely proportional to the perfusion flow rate $q$, as is expected from Eq. (15.2). The relative recovery RR also shows inverse proportionality, which is in agreement with literature. The longer lag time for the 20 mm probe is caused by the larger swept tip volume $V_{\text{tip}}$ being an integral part of the total swept volume $V_{\text{total}}$. On the other hand, this longer membrane length results in a higher relative recovery RR approaching unity for flow rates $q$ smaller than about 5 μL/min. If the probe is operated in this range, the analyte concentration in the dialysate $C_{\text{d}}$ equals the analyte concentration $C_{\text{s}}$ in the sample (see Eq. (15.1)), so that no calibration for the sampling process has to be performed.

It can be concluded that a microdialysis probe/sensor combination was constructed in which the swept volume and consequently the lag time has been decreased by a factor of more than five, as compared to a conventional probe connected to a sensor located in the output tube. The minimal volume has been mainly achieved by the application of appropriate micromachining techniques for the manufacturing of a microconnector between a conventionally assembled double lumen probe shaft and tip, and the incorporation of a sensor in silicon.

Besides the reduction of lag time, another main advantage of the realized micromachined probes is the fact that they can be

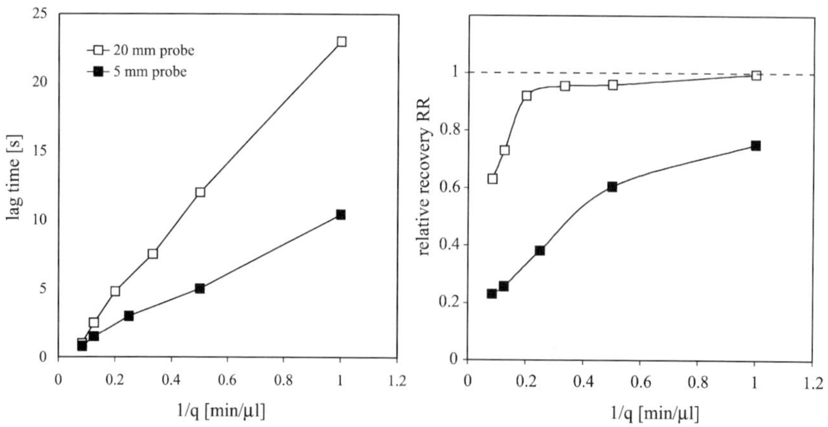


Fig. 15.5. Lag time and relative recovery as a function of flow rate for the 5 and 20 mm micromachined probes.

directly integrated with other microfluidic systems in silicon. Another advantage over the use of conventional probes in hybrid combination with microfluidic systems is that all critical fluidic connections are inside the system, from for the user. This feature ensures that no extra lag time is introduced by tubing and additional connectors. As a final advantage, it can be concluded that the resulting microconnector enables the easy (automated) assembly of double lumen microdialysis probes in a highly parallel fashion. This is mainly due to the self-aligning structure for receiving the concentrically positioned capillaries and the fact that the through holes for glue application allow the automatic distribution of the specially formulated epoxy.

## 15.3 THE PASSIVE AND THE ACTIVE CALIBRATION SYSTEM

For micrototal analysis systems to work in a reliable fashion it is important that all steps in the system, like fluid handling, separation and detection are performed in a reproducible and accurate manner. Particularly the detection requires attention as every microsensor exhibits time-varying sensitivity and offset. Therefore an integrated calibration system is a useful microcomponent.

In flow-through sensors, such as those used in FIA-based systems, calibration can be performed by formation of reference plugs in the fluid stream leading to the microsensor. When two plugs of different concentration are used, reference responses are generated in the microsensor, which can be used to correct both offset and sensitivity variations.

A microcalibration unit can be made using passive or active control.

### 15.3.1 Passive control of a calibration plug

A passive microcalibration unit can be made by using the passive diffusion of reference substances from a hydrogel into an adjacent microchannel. The hydrogel is filled with a known concentration of the reference compound and placed in direct contact with the carrier stream that leads to the sensors to be calibrated. A reference plug is formed by stopping the flow for a time long enough to allow the reference compounds to diffuse from the hydrogel into the microchannel and form a reference plug. Important parameters for the calibration unit are the time needed for reference plug formation and the number of calibrations that can be performed before the unit is depleted. A prototype of a passive microcalibration unit has been made that comprises a microfluidic channel, a hydrogel loaded with pH buffer and an ISFET downstream of the hydrogel for detection of the pH changes generated by the plug formation, as schematically shown in Fig. 15.6.

Results from measurements obtained with the prototype have shown that reference plugs can be formed in a time period of a few seconds. Also, it has been shown that sequential reference plugs show a decrease in the reference compound concentration due to depletion of the hydrogel, which is a drawback of the system. Another drawback is the continuous out-diffusion of reference compound from the hydrogel resulting in an offset in the measurement signal and an interference with the sample to be analyzed.

It can be concluded that instead of passive plug formation, a method of active fluid dispensing on demand is needed for accurate calibration of the sensors. By using an active pumping mechanism that provides undiluted calibrant only when this is needed, accurate reference plugs can be generated without generating an offset in the measurement signal and without interference to the sample.

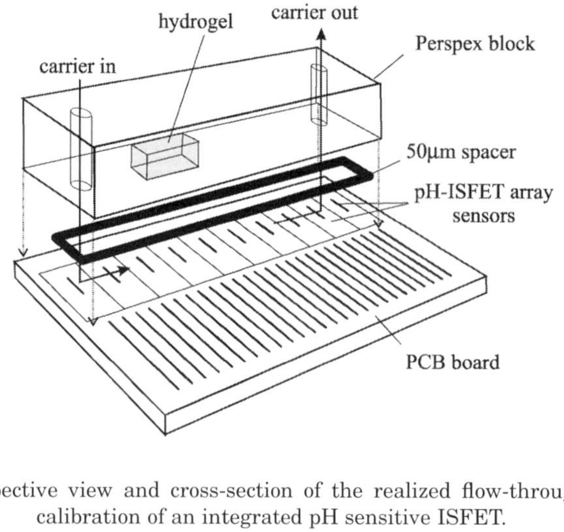


Fig. 15.6. Perspective view and cross-section of the realized flow-through system for calibration of an integrated pH sensitive ISFET.

### 15.3.2 Active control of a calibration plug

As an alternative to the passive formation of a calibration plug, an active microdosing system has been developed that can be used as a microcalibration unit. The system is based on the displacement of a fluid from a reservoir by the electrochemical production of gas bubbles. This principle was used earlier in a portable miniature pump for drug infusion by Koch et al. [6]. Because the electrochemical gas production can be accurately controlled, the pump can dose small volumes with high accuracy.

The microdosing system consists of a channel structure micromachined in silicon, closed by a Pyrex® cover with noble metal electrodes. In the silicon structure, the liquid to be dispensed is stored in a meander-shaped channel that starts in an electrolyte-filled reservoir above which two platinum electrodes are positioned. By sending an electric current through the liquid via the electrodes, gas bubbles can be produced. As these gas bubbles expand in the reservoir, the liquid stored in the meander is driven out into the channel structure leading to the microsensor to be calibrated. Figure 15.7 shows a realized double calibration unit, where the meander structures and the gas pump reservoirs can be easily distinguished.

*Theory*

If in the system shown in Fig. 15.7 the appropriate conditions are chosen (electrolyte composition, electrode material, cell configuration) then the total volume of gas bubbles produced and consequently the amount of fluid that is dispensed, can be controlled by means of the on-time of the current source and the current amplitude. If two noble metal electrodes are chosen, the following reactions take place in water at neutral pH,

$$\text{Anode}: \quad 2H_2O \rightarrow 4H^+ + 4e^- + O_2(g) \tag{15.3}$$

$$\text{Cathode}: \quad 2H_2O + 2e^- \rightarrow 2OH^- + H_2(g) \tag{15.4}$$

From both half-reactions, the rate of the gas production $dV/dt$ ($m^3/s$) and consequently the dosing rate $q$ ($m^3/s$) at atmospheric pressure can be derived if a constant current $i$ (A) is assumed

$$q = \frac{dV}{dt} = \frac{3}{4}\frac{i}{F}V_m \tag{15.5}$$

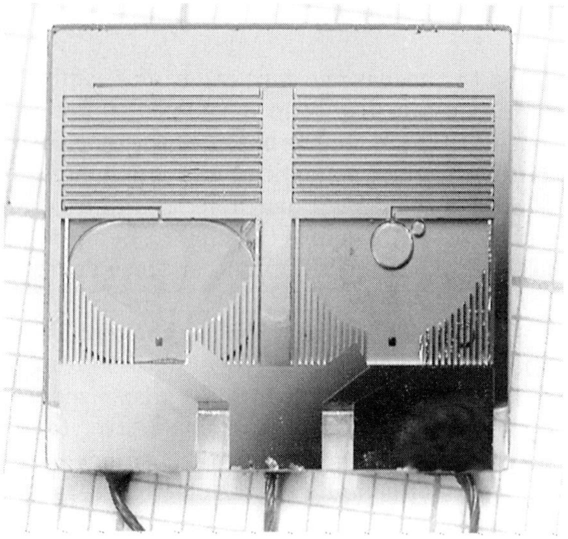

Fig. 15.7. Photograph of a realized dual-pump. Note that the left pump is almost completely filled with generated gas whilst the right pump and meander is filled with electrolyte.

where $F$ is Faraday's constant ($96.49 \times 10^3$ C/mol) and $V_m$ is the molar gas volume at 25°C and atmospheric pressure ($24.7 \times 10^{-3}$ m$^3$/mol).

If a current pulse of amplitude $i$ and duration $\Delta t$ (s) is applied, the total dosed volume $\Delta V$ is given by

$$\Delta V = q\Delta t \tag{15.6}$$

*Results*

The dosed volume as a function of the amplitude $i$ of a current pulse of 250 ms duration, applied to the electrodes, is shown in Fig. 15.8, together with the theoretical expected dosed volume. It can be seen that the dosed volume corresponds well with the theoretical expectations.

A problem with this system is the catalytic back reaction of the generated oxygen/hydrogen mixture at the noble metal electrodes (platinum). This back reaction results in a decrease in the gas volume in the reservoir and in a back-flow of dosed volumes. Also the produced gas sometimes remained dissolved in the electrolyte and did not give rise to evolving gas bubbles, which resulted in errors in the dosed volume.

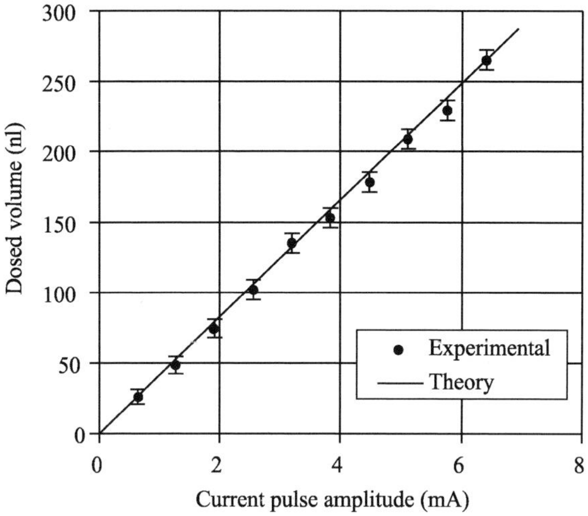


Fig. 15.8. Measured and calculated dosed volume as a function of the actuation current (250 ms actuation, $r^2 = 0.996$).

As the dosed liquid volume equals the volume of generated gas bubbles, precise dosing can be obtained by closed-loop control, with the amount of generated gas as the control parameter.

### 15.3.3 Closed-loop controlled electrochemically actuated microdosing system

The dosing accuracy of the above-mentioned active dosing system could be improved by incorporating a feedback system in which the gas bubble volume is measured and used to control the dosing action. In this set-up, a pair of interdigitated electrodes is used to electrochemically generate gases and to simultaneously measure the generated gas volume, via an impedance measurement of the gas/liquid mixture in the reservoir. As this measured gas bubble volume equals the dosed volume, active control of dosed volumes can be achieved.

Using this set-up, accurate dosing of volumes in the nanoliter range with an accuracy of 5 nL was reached.

The realized system is depicted in Fig. 15.9. The system has separate chambers for the generation of hydrogen and oxygen gas, which are in ohmic contact by a chevron that acts as a diffusion barrier for

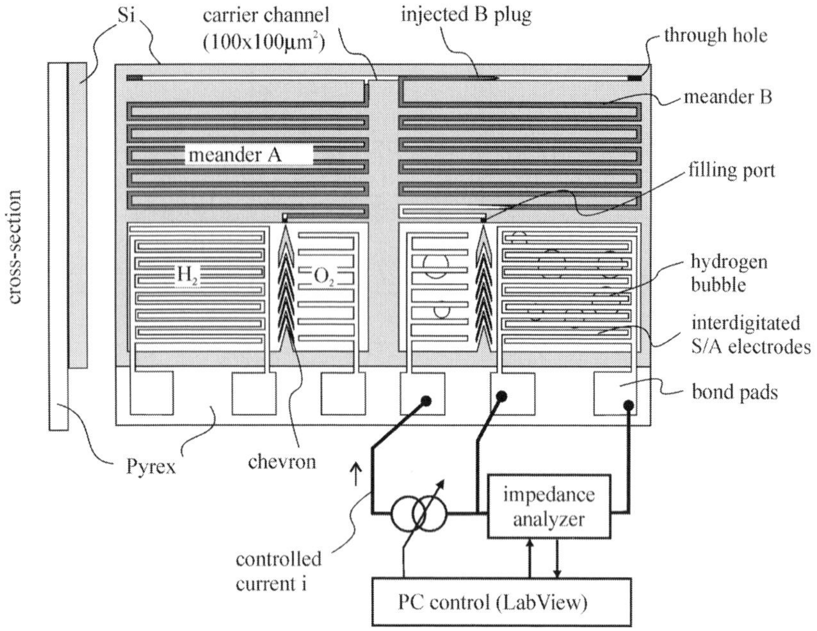


Fig. 15.9. Proposed dosing system with feedback control, geometry and electrical connections.

the generated gases. These two measures have been taken to compensate the catalytic back-reaction of the gases at the platinum electrodes.

To link the cell impedance with the total gas bubble volume a void volume fraction $\varepsilon$ has to be defined:

$$\varepsilon = \frac{V_{H_2}(\text{or } V_{O_2})}{V_{\text{reservoir}}} \tag{15.7}$$

where $V_{\text{reservoir}}$ ($m^3$) is the internal volume of the corresponding bubble reservoir. By this definition, $\varepsilon$ is zero before any gas bubbles are generated and can in theory increase to unity, which represents the case of a bubble reservoir completely filled with gas bubbles.

If the impedance is determined at a high enough frequency (200 kHz) to eliminate the influence of the double layer capacitancies, the generated gas volume can be calculated from the following equation, where $R_{\text{electrolyte}}$ is the electrolyte resistance, $R_{\text{series}}$ is the resistance of the connecting wires and the planar electrode structure and $R(\varepsilon)$ is the real part of the cell impedance:

$$V_{H_2}(\text{or } V_{O_2}) = V_{\text{reservoir}}\left(1 - \frac{R_{\text{electrolyte}}}{R(\varepsilon) - R_{\text{series}}}\right) \quad (15.8)$$

The feedback loop was controlled with a computer on which a proportional control routine was implemented for continuously controlling the amplitude of the current through the electrodes based on the measured cell impedance (i.e. the applied electric current was proportional to the difference between the preset value and the actual value of the gas volume determined via the measured resistance $R_{\text{electrolyte}}$).

Figure 15.10 shows the graph for a dosing step of 100 nL. In the graph the controlled actuation current as well as the graph representing the injected volume are shown. It can be seen that as the dosed volume goes up the current is ramped down when the dosed volume approaches the set-point of 100 nL.

By using this approach it is possible to accurately dose volumes in the nanoliter range with 5 nL accuracy and to overcome the unpredictability in bubble formation associated with the open loop active dosing system. The resulting device can, because of the remarkably simple

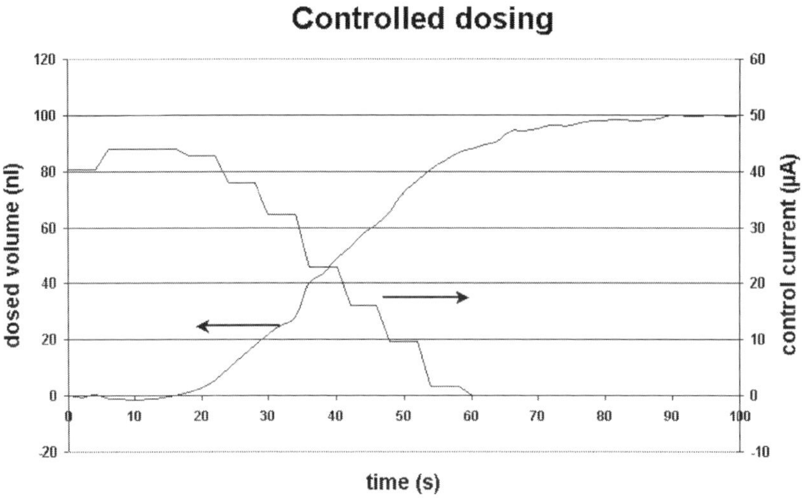


Fig. 15.10. Dosed volume of 100 nL and controlled current as a function of time.

design and implementation, be easily integrated within silicon or polymer based microfluidic systems for the accurate dispensing of liquids such as reagents or calibrants.

## 15.4 THE FLOW-THROUGH POTENTIOMETRIC AND AMPEROMETRIC SENSOR ARRAY

A number of microsensors has been developed for the purpose of being built into microchannels. However, the integration of these sensors can impose some problems if they have to be closely integrated within microfluidic systems. A specific disadvantage of these sensors is that they have mostly a planar construction. Therefore, especially for the microdialysis lab-on-a-chip system, a new sensor concept has been developed. In this concept the microchannel itself is an integral part of the sensor geometry and is formed by a length of tubular semi-permeable membrane. This semi-permeable tubing, similar to the tubing used in the microdialysis probe, forms a tubular frit for the separation of the dialysate to be analyzed and a cavity filled with an appropriate solution containing a specific receptor (ionophore, ion exchanger or enzyme). It results in a stable interface between two phases, allowing only the passage of the small molecules of interest. The major advantage of this geometry is the fact that off the shelf ion-selective cocktails or enzyme solutions can be applied, yielding sensors for a wide variety of analytes. As a consequence no specific immobilization of the receptors is required. In this case the tubing prevents leakage of enzyme out of the cavity into the dialysate, as the pores in the tube only allow passage of small molecules.

Based on potentiometric and amperometric detection, sensors for ions such as $Na^+$, $K^+$, $Li^+$, $H^+$ and for metabolites such as lactate, glucose and glutamate were implemented, respectively.

### 15.4.1 The flow-through potentiometric sensor array

The flow-through potentiometric sensor is, regarding its geometry, comparable to the liquid membrane ion selective electrode. The layout of the potentiometric flow-through ion selective electrode is shown in Fig. 15.11. The sensor is formed by a half-cell consisting of the sample

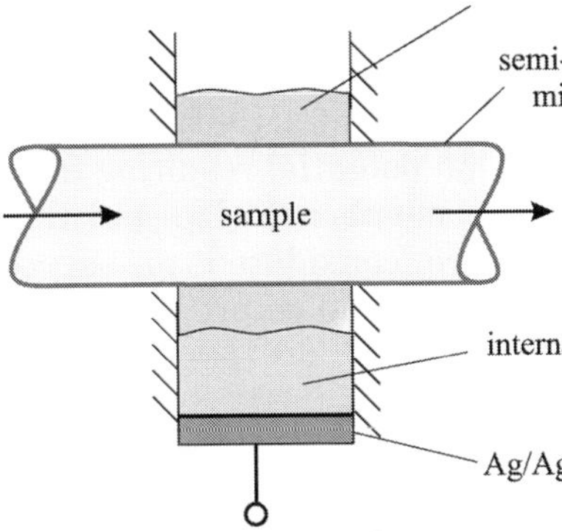


Fig. 15.11. Layout of the flow-through potentiometric sensor.

liquid in the tube and the ion selective filling, the contacting electrolyte and the Ag/AgCl electrode in a cavity around the tube. Note that the flow-through sensor is in fact identical to a conventional liquid-membrane ISE, but inside out. Therefore, the operational mechanism is also identical.

In the half-cell, two liquid/liquid interfaces are present. The first interface is formed between the sample solution and the lipophilic ion selective (liquid) membrane, whilst the second exists between this membrane and the internal reference solution. It is assumed that the semi-permeable dialysis membrane is inert and only functions to form a stable interface and to limit leakage of the (liquid) membrane to the sample liquid, comparable with the porous plug of a conventional ISE. The outer boundary potential ($E_\mathrm{b}$) is described by the well-known Nernst equation:

$$E_\mathrm{b} = E^{0'} + \frac{RT}{z_i F}\ln\left(\frac{a_{i,\mathrm{s}}}{a_{i,\mathrm{m}}}\right) \tag{15.9}$$

$$E^{0'} = \frac{\mu^0_{i,\mathrm{s}} - \mu^0_{i,\mathrm{m}}}{zF} \tag{15.10}$$

where $R$ is the universal gas constant (8.31 J mol$^{-1}$ K$^{-1}$), $F$ is Faraday's number ($9.65 \times 10^4$ C mol$^{-1}$), $z_i$ the valence of ion $i$, $T$ (K) is the absolute temperature, $a_{i,\mathrm{s}}$ and $a_{i,\mathrm{m}}$ (mol/L) are the activities of ion $i$ in, respectively, the sample and the membrane phase, whilst $\mu^0_{i,\mathrm{s}}$ and $\mu^0_{i,\mathrm{m}}$

($J\,mol^{-1}$) represent the standard chemical potential of ion $i$ in, respectively, the sample and membrane phases. The main function of the ion selective species in the membrane (ionophores or ion-exchangers) is to provide a constant activity $a_{i,\mathrm{m}}$ of the specific ion in this phase. Since the diffusion potential resulting from a concentration gradient of ions within the membrane ($E_\mathrm{d}$) and the inner reference boundary potential ($E_\mathrm{r}$) are considered to be constant and the membrane potential ($E_\mathrm{m}$) can be expressed as the sum of the contributing potentials:

$$E_\mathrm{m} = E_\mathrm{r} + E_\mathrm{d} + E_\mathrm{b} \tag{15.11}$$

it follows that

$$E_\mathrm{m} = E^0 + \frac{RT}{z_i F}\ln(a_{i,\mathrm{s}}) \tag{15.12}$$

where $E^0$ is the reference potential (i.e. $E_\mathrm{m}$ for $a_{i,\mathrm{s}} = 1$ mol/L). From this equation it can be concluded that for a monovalent ion at the standard temperature of 298 K, the half-cell potential theoretically changes about 59 mV per decade of change in ion activity (for low concentrations, the ion activity can be approximated by the ion concentration). Based on this potential $E_\mathrm{m}$, which is measured against a reference electrode also placed in the sample solution (Section 15.4.2), the actual ion activity can be determined.

As a proof of principle the sensor body was machined from a piece of Perspex® by conventional precision engineering. To yield a stable Ag/AgCl electrode, the silver surface of the rod was carefully cleaned prior to chemical chlorination by a 1% $FeCl_3$ solution for about 30 min. Next a length of semi-permeable tubing (300 μm outer diameter) adapted from an artificial kidney (regenerated cellulose, Filtral® 6, AN69 HF, Hospal, France) was inserted and glued with epoxy in a drilled horizontal hole. To form liquid type ion selective membranes, off the shelf ion selective cocktails for sodium, potassium, lithium and hydrogen ions, obtained from Fluka, were applied. The performance of various potentiometric sensors, operated in a continuous flow mode, are summarized in Table 15.1.

The sodium and lithium sensors showed excellent sensitivities, close to the theoretically expected value of about 59 mV/decade. The sensitivities of the potassium and pH sensors are somewhat lower. The response time of all sensors, except that of the pH sensor, is in the order of 10–15 s, which is sufficiently small for most applications.

TABLE 15.1

Potentiometric sensors' performance

| Sensor | Sensitivity (mV/dec) | Response time (s) | Detection limit log $a$ | Range (mM) |
|---|---|---|---|---|
| $Na^+$ | 59.2 | 15 | $<-3.5$ | 0.1–500 |
| $K^+$ | 50.5 | 10 | $<-4$ | 0.1–1000 |
| $Li^+$ | 58.0 | 10 | $<-2.5$ | 5–100 |
| $H^+$ | 52.7 | 60 | $<-4$ | pH 4–10 |

### 15.4.2 The flow-through reference electrode

Based on the same generic electrode design, a reference electrode was successfully implemented, enabling the construction of a sensor array with integrated reference electrode. In this case, the sensor cavity as shown in Fig. 15.11 was filled with a 3 M KCl solution only and the tube membrane acts as the frit of a conventional reference electrode. The stability was measured with respect to a conventional reference electrode and appeared to be within 1 mV.

Due to the relatively large contact area, provided by the tubular membrane section, the sensors and reference electrode exhibit a very low impedance. All measurements could be performed without applying shielded wires. However, to extend the lifetime of the reference electrode, it is preferable to use a gelled KCl solution and to limit the surface of the (frit)membrane.

### 15.4.3 The flow-through amperometric sensor

One of the limiting factors in biosensor fabrication is immobilization of the enzyme. This is due to the inhibition of enzyme during the immobilization procedure as well as the limited enzyme loading within a membrane or onto a sensor surface. Moreover, the immobilization methods are in some cases rather elaborate and it is also suggested by Danilich et al., Seo et al., and Almeida et al. [7–9] that enzyme immobilization affects the enzyme activity and stability.

The flow-through amperometric sensor is designed as a three-electrode electrochemical cell in the form of a microreactor, shaped around a semi-permeable membrane tubing, similar to the potentiometric sensors as described in Section 15.4.1. A platinum spiral wire working electrode is wound around the dialysis tube, whereas an Ag/AgCl reference electrode is placed at the bottom of the cell, as is the case in the potentiometric sensors. An additional Pt counter electrode is placed downstream. The layout of the amperometric sensor is shown in Fig. 15.12.

Based on this construction, sensors for glucose, lactate and glutamate determination were developed. The microreaction cell was filled with an appropriate aqueous enzyme solution to achieve an oxidation reaction of the analyte, catalyzed by the enzymes: glucose oxidase (GOD), lactate oxidase (LOD) and glutamate oxidase (GltOD), respectively, according to:

$$\text{Glucose} + O_2 + H_2O \xrightarrow{\text{GOD}} \text{Gluconolactone} + H_2O_2 \tag{15.13}$$

$$\text{L-Lactate} + O_2 + H_2O \xrightarrow{\text{LOD}} \text{Pyruvate} + H_2O_2 \tag{15.14}$$

$$\text{L-Glutamate} + O_2 \xrightarrow{\text{GltOD}} \alpha\text{-Ketoglutarate} + NH_3 + H_2O_2 \tag{15.15}$$

The electrochemical oxidation of hydrogen peroxide, being the product of all enzymatic reactions, was utilized for the actual measurement, according to:

$$H_2O_2 \rightarrow O_2 + 2H^+ + 2e^- \tag{15.16}$$

The current resulting from this reaction was measured between working and counter electrodes. The platinum working electrode was biased at about 0.65 V versus the Ag/AgCl reference electrode.

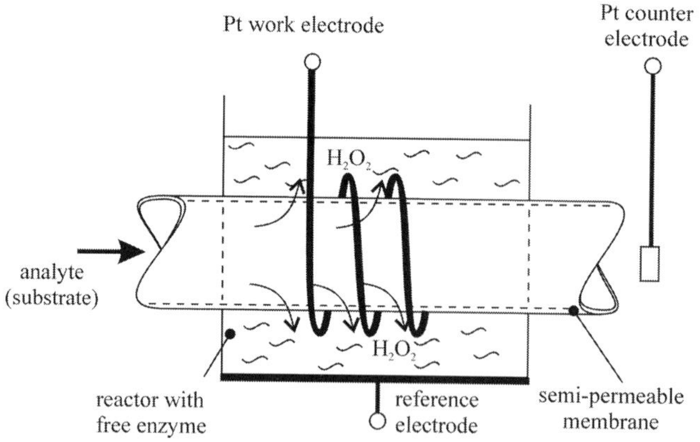


Fig. 15.12. Layout of the flow-through amperometric sensor.

The sensors were evaluated in both continuous-flow mode as well as in an FIA system (Table 15.2).

Due to the fact that the enzyme is present in its native form, the total enzyme activity can easily be changed by addition of an extra amount of the enzyme into the reaction cell. This results in changes to the sensor sensitivity, e.g. the sensitivity for the glucose sensor increased by only 35% when the glucose oxidase activity in the reaction cell was increased by a factor of 3.2.

All amperometric sensors show a high sensitivity and a large range covering all physiological ranges [10]. The stability of the sensors was remarkably good: one lactate sensor was used intermittently for about 30 days without any significant reduction in sensitivity.

The performance of the amperometric sensors operating in an FIA system is shown in Table 15.3. The experimental values of travel time ($t_A$) and peak width ($t_B$) are somewhat higher than those obtained from the semi-empirical model of Vanderslice et al. [11], in which the detector response has been assumed to be very fast. This is a result of the increased response time of the sensor caused by the fact that the analyte has to diffuse through the semi-permeable membrane into the reaction cell towards the working electrode. Taking into account that the total distance for diffusion is about 100 μm, which gives a diffusion time of about 10 s (diffusion coefficient for lactate = $1.04 \times 10^{-9}$ m$^2$/s), the experimental values for $t_A$ and $t_B$ are in agreement with theory.

Dynamic response of the lactate sensor operating in an FIA system is shown in Fig. 15.13. The reproducibility of the sensor response is expressed by a relative standard deviation (RSD) for 50 consecutive sample injections.

TABLE 15.2

Summarized amperometric sensors' performance for continuous mode

| Sensor | Enzyme activity (U/ml) | Slope (nA/mM) | Corr. coef. | Linear range (mM) | Detection limit (μM) |
|---|---|---|---|---|---|
| Glucose | 100 | 188 | 0.9990 | 0–26 | 50 |
| | 320 | 297 | 0.9988 | 0–26 | |
| Lactate | 80 | 240–250 | 0.9995 | 0–3 | 25 |
| Glutamate | 5 | 76 | 0.9984 | 0–5 | 15 |

TABLE 15.3

Amperometric sensors' performance, operating in an FIA system

| Sensor | Peak height (nA/mM) | RSD (%) | Range (mM) | $t_A$, $t_B$ experimental (s) | $t_A$, $t_B$ theoretical (s) |
|---|---|---|---|---|---|
| Glucose | 39.2 | <5 | 0–10 | 52, 130 | 22–44, 55–116 |
| Lactate | 42.9 | <2 | 0–9.8 | 50, 120 | 47–94, 51–103 |
| Glutamate | 8.35 | 1.5 | 0–5 | 62, 135 | 28–56, 70–140 |

Measurements were performed in 20 mM phosphate buffer at pH 7.3.

## 15.5 THE INTEGRATED MICRODIALYSIS-BASED LAB-ON-A-CHIP

In previous sections, the results of all separate parts of the microdialysis lab-on-a-chip have been discussed. In particular a microdialysis probe (Section 15.2), a calibration facility (Section 15.3) and a potentiometric and amperometric ion and enzyme sensor (Section 15.4) have been

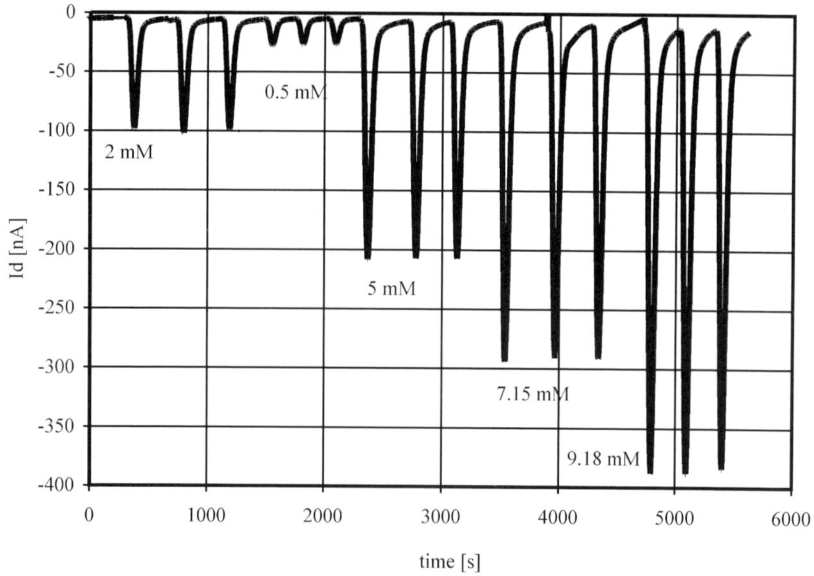


Fig. 15.13. Calibration peaks for lactate for the lactate sensor operating in an FIA system.

developed separately using different precision micromachining techniques.

From the beginning it was clear that an integrated lab-on-a-chip cannot be realized by applying the different technologies that were used for making the various parts because of conflicting process conditions. Therefore it was decided that all system components should be made with one and the same generic technology, namely a stack of a silicon wafer and a glass wafer that are anodically bonded.

In this concept, the silicon wafer contains the double lumen tube connections, meander-formed cavities that contain the calibration solutions, a dosing pump chamber, a peristaltic perfusion-liquid pump and various microcavities for potentiometric and amperometric ion and enzyme sensors as well as for the reference electrode. The glass wafer with a thin patterned Pt layer contains all the necessary electrodes and electrical connections for the sensors and the calibration facility. Both wafers are anodically bonded, yielding a hermetically sealed liquid handling system.

### 15.5.1 The complete integrated microdialysis lab-on-a-chip

The final integrated system design is shown in Fig. 15.14, where all components as described in the previous sections can clearly be distinguished, except for the peristaltic pump. At the place of this micropump a glass tube is temporarily inserted into the input channel (perfusate in) for connecting an external perfusion pump.

Perfusion liquid is pumped via this input channel to the microdialysis probe (left) at a highly continuous flow rate in the range of 0.2–2 μL/min. This fluid leaves the inner glass tube at the probe tip and flows along the semi-permeable membrane through the liquid channels to the sensor area. During the passage in the probe tip small molecular substances can diffuse through the membrane and enrich the perfusion liquid to form a continuous stream of dialysate. In the sensor area the same process takes place through the second semi-permeable membrane in the potentiometric and amperometric ion and enzyme sensor as well as in the reference electrode. Finally the dialysate leaves the chip to a waste container.

On the left side of Fig. 15.14 two calibration facilities are also connected to the main liquid channel between the probe and the sensor area, and are used for the calibration of the sensors. In addition, various

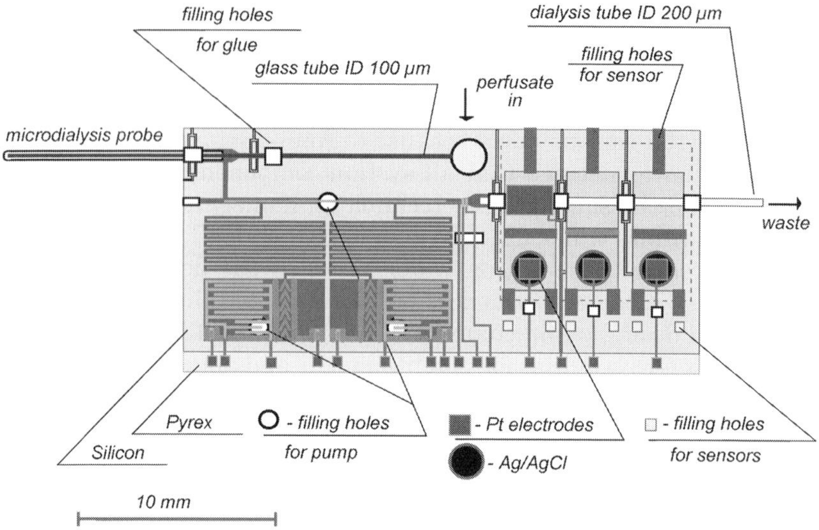


Fig. 15.14. Layout of the final chip.

filling holes for the pumps and the sensors can be distinguished. On the Pyrex glass part bonding pads are visible for the connection of the various components with appropriate measuring and controlling electronics.

*The microdialysis probe*

The actual microdialysis probe is a double lumen type and consists of an inner silica tube (150 μm OD) surrounded by a semi-permeable tube adapted from an artificial kidney (regenerated cellulose, MWCO 20 kD, Filtral® 6, AN69 HF, Hospal France, 200 μm ID). The layout of the integrated probe is slightly modified to suit the adapted generic technology of a glass–silicon wafer stack compared with the self-aligning silicon–silicon structure as described in Section 15.2.

The effective volume of the microdialysis probe tip amounts to approximately 13 nL/mm, which equals the sensor volume (100 nL) if the length of the probe is 8 mm.

*The calibration facility*

The complete design comprises a double calibration facility for two (different) calibrants. Each consists of a meander-shaped etched channel that contains the appropriate calibrant solution. The actual pump, which can dose these calibrants into the central channel between the microdialysis probe and the sensor area, consists again of an etched cavity filled with electrolyte and two gas-generating electrodes. These electrodes are shaped as a pair of interdigitated electrodes providing the possibility of closed-loop dosing by measuring the gas/electrolyte fraction in the pump chamber [12]. The volume of each meander-shaped calibrant reservoir is 1500 nL, which is sufficient for about 15 calibration cycles for a single sensor.

*The sensor area*

The sensor area (Fig. 15.15) comprises an amperometric sensor, a potentiometric sensor and a reference electrode. For all three sensors a second microdialysis tube (ID = 200 μm) is used to separate the dialysate from the internal sensor liquids. Each sensor is 2 mm wide, thus the volume of the sample inside the microdialysis tube in each sensor is approximately 100 nL.

The middle sensor in Fig. 15.15 is the potentiometric sensor filled with an ion selective cocktail that is in contact with a KCl solution and an internal Ag/AgCl electrode.

Leaving out the ion selective cocktail results in a reference electrode.

By replacing the ion selective cocktail by an enzyme solution and adding a Pt working electrode on the glass nearby the microdialysis tube, the same design has been adopted for use as an amperometric sensor.

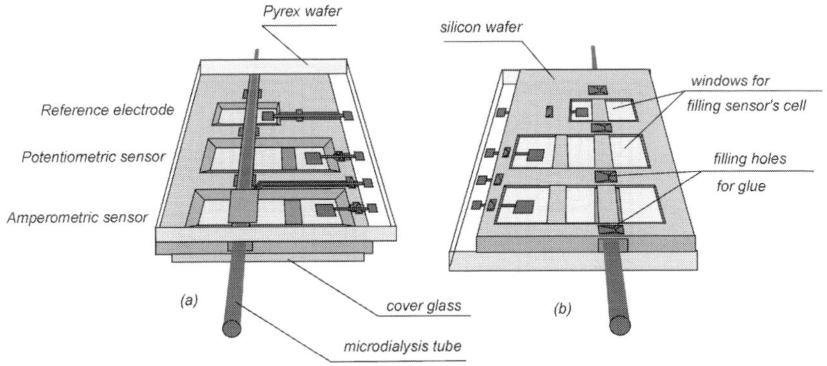


Fig. 15.15. The sensor area comprising an amperometric sensor, a potentiometric sensor and a reference electrode: (a) top side; (b) bottom side.

Advantages of this approach are the generic technology to make sensors with incorporated reference electrodes, capable of measuring a wide variety of analytes for which cocktails are commercially available, and the fact that sub-microliter samples can be measured. In addition, the electrical impedance of the proposed flow-through sensors is small due to the relatively large sensing surface formed by the tube wall.

*The perfusion liquid pump*

At present an external syringe pump is used to provide the system with perfusion liquid in a continuous flow regime. In the meantime a perfusion liquid pump has been developed that can be integrated into the system using the same generic technology: etched channels and cavities through which a tube can be led. The cavities contain a temperature- or pH-sensitive hydrogel that can swell and shrink as a function of the applied stimulus. The necessary temperature or pH actuators of Pt electrodes exist, deposited onto the cover plate. Actuation by current pulses results in a controlled swelling or shrinking of the gel in the cavities. With three cavities in a row a peristaltic pump is developed, based on pH-sensitive hydrogels. Figure 15.16 shows the design of such a pump. The tube consists of a silicone rubber fiber of which the wall thickness is etched back to about 20 μm, in order to make it more elastic and easier to close by the swelling hydrogel. First results are promising, but the pump is not yet ready for integration.

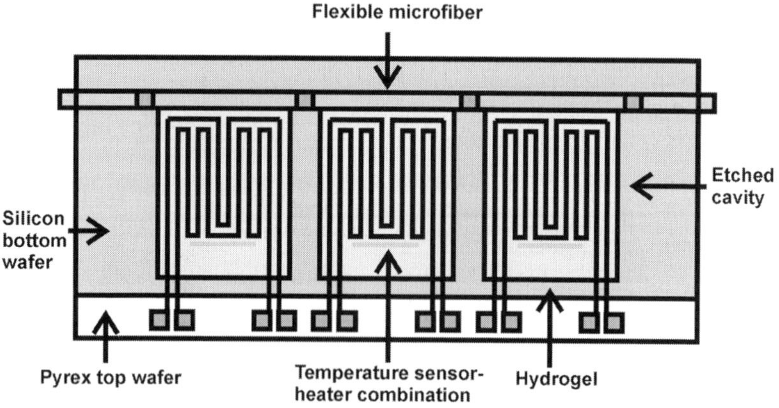


Fig. 15.16. Design of the peristaltic micropump.

### 15.5.2 Experimental

*Device preparation*
The complete lab-on-a-chip systems are produced in the cleanroom of the MESA$^+$ Research Institute of the University of Twente, Enschede, The Netherlands.

Figure 15.17 shows a realized glass wafer on top of a matching silicon wafer, prior to the anodic bonding process. The original layout as given in Fig. 15.14 can clearly be recognized.

All channels and cavities are etched using reactive ion etching (RIE) in a double-sided polished 100 mm silicon wafer. The filling holes for the internal liquids are etched through the wafer in a second step. In the same processing step, several holes are etched for the fixation glue of the microdialysis tubes connected to the anti-creep channels as will be discussed below.

All Pt electrodes are deposited on a 3-inch Pyrex glass wafer using standard sputtering, photolithography and wet chemical etching. Then both wafers are anodically bonded and subsequently diced in separate systems.

After fixing the system on a small PCB (Fig. 15.18) the silica tube and both microdialysis tubes are introduced into the appropriate channels and fixed with epoxy through the various glue holes. After application of

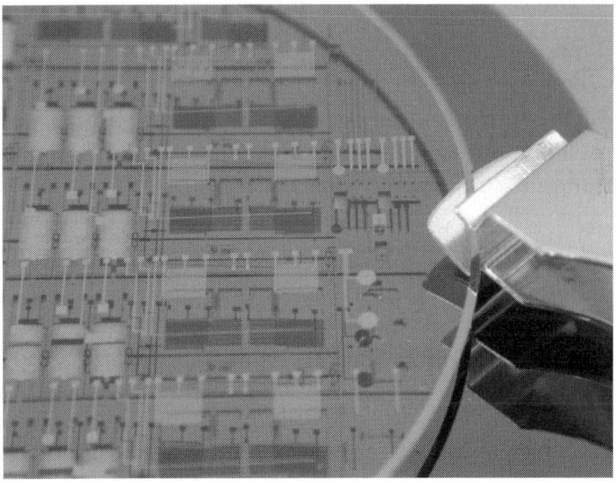

Fig. 15.17. Photograph of a silicon and a glass wafer just before anodic bonding.

Fig. 15.18. The complete lab-on-a chip, mounted on a small PCB, with a double lumen microdialysis probe on the right side and perfusion liquid input connection as well as the dialysate waste output connection on the left side.

a small drop of Ag/AgCl paste on the three Pt electrodes inside the sensor cavities through the windows in the silicon substrate, these windows are closed with a small microscopic cover glass using UV curing glue.

Finally the different liquids are injected into the appropriate cavities through the filling holes, which are subsequently closed with epoxy.

*The anti-creeping system*

For fixation of the various tube sections epoxy is applied through holes in the silicon. However, this technique worked well during assembly of separate components but appeared to be troublesome with respect to the assembly of a complete system. At some places the epoxy should close the gap between a tube and the surrounding silicon–glass channel, but should not creep into the cavities, nor cover any of the electrodes. Therefore an anti-creep design is developed, consisting of overflow capillaries in the silicon at these critical points. Figure 15.19 shows a photograph of one of the anti-creep subsystems, where the epoxy (glue) is injected through a hole into the silicon just beneath the center of the picture. It fixes the microdialysis tube near the working electrode of an amperometric sensor, whereas the surplus of glue is taken up by the capillary structure.

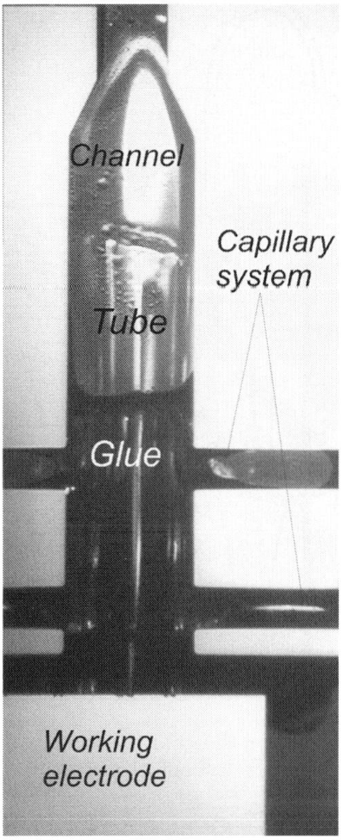


Fig. 15.19. Details of the glue anti-creep system.

*Measurement set-up*

The measurement set-up to characterize the microdialysis system is shown in Fig. 15.20. The perfusion liquid is delivered by a microdialysis pump (CMA 102, CMA Microdialysis AB, Sweden) via the inlet to the actual probe that is immersed in the sample solution. The calibration facilities P1 and P2, containing the specific calibrants, are controlled by the connected PC, while the amperometric sensor is measured with a potentiostat (EG&G Model 263A) in a three-electrode configuration. The potentiometric sensor is connected to a high-impedance instrumentation amplifier. Measurements are recorded on a standard PC.

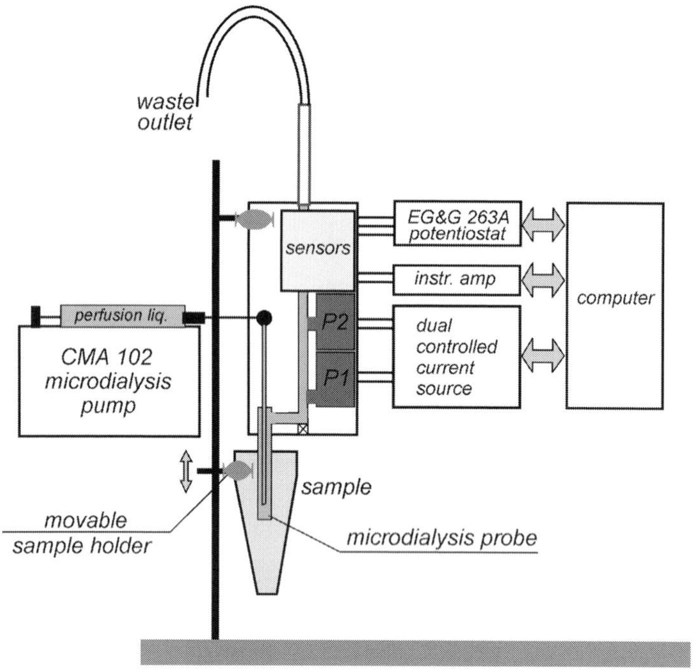


Fig. 15.20. Measurement set-up for the characterization of the microdialysis lab-on-a-chip.

### 15.5.3 Measurements

As an example, the results of experiments with an amperometric sensor and a potentiometric sensor will be discussed.

Testing the amperometric sensor was carried out by making it sensitive for glucose by filling the sensor cavity with the enzyme glucose oxidase (GOD) with an activity of 15,500 U/g solid (Sigma).

The perfusion solution was a 45 mM phosphate buffer, driven through the system by the external pump with a constant flow rate of 2 μL/min. All measurements were carried out by immersing the microdialysis probe for 25 s in small sample reservoirs (Fig. 15.20) containing solutions of different concentrations of glucose ranging from 1.25 to 25 mM (Fig. 15.21). This was repeated twice for each sample. Between consecutive samplings, the probe was placed in a 45 mM phosphate buffer solution.

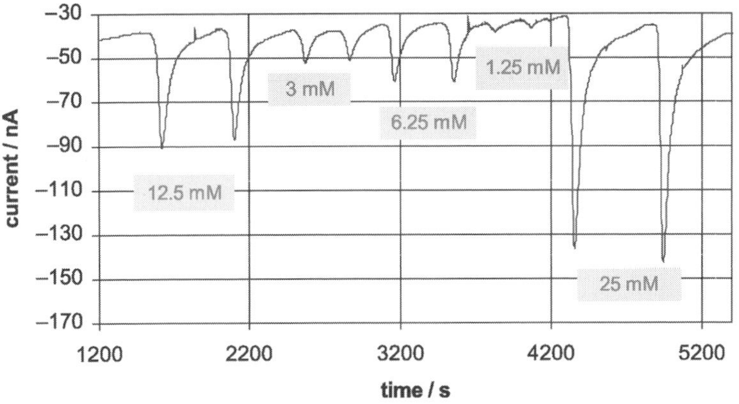


Fig. 15.21. Typical FIA mode response of the described glucose sensor.

A similar experiment has been carried out for testing the potentiometric sensor (Fig. 15.22). In this case the sensor was sensitive for potassium, after application of an ionophore cocktail based on valinomycin (60031) from Fluka. The microdialysis probe was immersed in a sample solution of 0.1 mM KCl, while again a perfusion

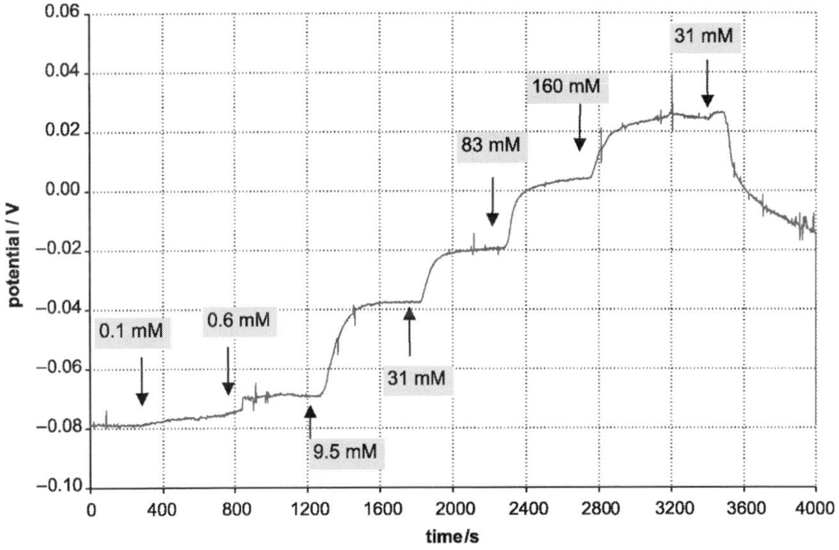


Fig. 15.22. Typical results of the potassium sensor in continuous-flow mode.

solution of a 45 mM phosphate buffer at a constant flow rate of 2 μL/min was used. The concentration of the potassium ions in the sample solution was changed by a standard addition method from 0.1 up to 160 mM KCl. The final step from 160 to 31 mM KCl corresponds to dilution of the sample.

## 15.6 GENERAL CONCLUSIONS

Based on the proof of principle of the different subsystems, as described in Sections 15.2–15.4, the ultimate integrated system, as described in Section 15.5, has been produced. It is clear that this is nearly the same as initially intended and sketched in Fig. 15.1. Only the perfusion liquid pump is missing in the final design, but the chip contains enough space to integrate a small peristaltic pump. Also more sensors can easily be integrated as well as the relevant additional calibration units, although it might also be possible to calibrate with a mixture of all components to be measured, with only the existing two calibration reservoirs.

All measurements have been carried out with standard laboratory equipment, which is not especially designed or configured for this typical lab-on-a-chip. In a final product all the necessary functions should be integrated into a single dedicated IC or ASIC.

Independent of the original aim of the project to develop a small bedside monitoring system, the different subsystems can also be applied individually, such as the microdosing system and the flow-through sensor array. In addition the microdialysis concept is not limited to biomedical applications, but can also be applied with advantage in other application fields, such as biotechnology and environmental monitoring. More detailed information is given in open literature [12–21].

## REFERENCES


1 U. Ungerstedt and C. Pycock, *Bull. Schweiz. Akad. Med. Wiss.*, 30 (1974) 44.
2 P. Lönnroth, P.A. Jansson and U. Smith, *Am. J. Physiol.*, 253 (1987) E228.
3 A.R. Kopf-Sill, *Micro Total Analysis Systems*, Kluwer, Dordrecht, 2000, ISBN 0-7923-6387-6, p. 233.
4 J. Wissink, A. Prak, M. Wehrmeijer, R. Mateman, Proc. Vol. 2 MICRO.tec 2000-VDE World Microtechnologies Congress, EXPO Hannover, 2000, p. 51.

5 Int. Patent 'Microdialysis probe integrated with Si-chip' PCT/NL99/00057, 1999.
6 M. Koch, A.G.R. Evans and A. Brunnschweiler, *J. Micromech. Microengng*, 8 (1998) 119.
7 M.J. Danilich, D. Gervasio, R.E. Marchant, *Ann. Biomed. Engng*, 21, 655.
8 H. Seo, K. Itoyama, K. Morimoto, T. Takagishi, M. Oka and T. Hayashi, *Eur. Polym. J.*, 34(7) (1998) 917.
9 N.F. Almeida, E.J. Beckman and M.M. Ataai, *Biotechnol. Bioengng*, 42 (1993) 1037.
10 H.A. Harper, *Review of Physiological Chemistry*, Lange Medical Publications, Los Altos, CA, 1975.
11 J.T. Vanderslice, K.K. Stewart, A.G. Rosenfeld, et al., *Talanta*, 28 (1981) 11.
12 S. Böhm, B. Timmer, W. Olthuis and P. Bergveld, *J. Micromech. Microengng.*, 10 (2000) 498.
13 S. Böhm, W. Olthuis, P. Bergveld, Proc. of the SPIE Conf. on Industrial and Environmental Sensing, Boston, USA, 1999, p. 24.
14 S. Böhm, D. Pijanowska, W. Olthuis and P. Bergveld, *Biosen. Bioelectron.*, 16 (2001) 391.
15 A.J. Sprenkels, W. Olthuis, P. Bergveld, First Annual International IEEE-EMBS Special Topic Conference on Microtechnologies in Medicine & Biology, Lyon, France, October 2000, p. 326.
16 S. Böhm, W. Olthuis and P. Bergveld, *Sens. Actuat. B*, 63 (2000) 201.
17 P. Bergveld, *Biomed. Microdev.*, 2 (2000) 185.
18 S. Böhm, W. Olthuis and P. Bergveld, *Mikrochimi. Acta*, 134 (2000) 237.
19 S. Böhm, W. Olthuis, P. Bergveld, Proc. of Eurosensors XIII, The Hague, the Netherlands, September 1999, p. 165.
20 S. Böhm, W. Olthuis, P. Bergveld, Proc. of the Int. Mech. Engng Congress ME'99, MEMS, Nashville, USA, November 1999, p. 391.
21 S. Böhm, W. Olthuis, P. Bergveld, Proc. of Transducers 99, Sandai, Japan, June 1999, p. 880.

Chapter 16

# Design methodology for a lab-on-a-chip for chemical analysis: the MAFIAS chip

**T.T. Veenstra* and A. van den Berg**

## 16.1 INTRODUCTION

This chapter gives insight into the design path that has to be followed to come to a complete design of a chemical analysis system based on micromachined components. This path has led to the design of a lab-on-a-chip for the measurement of ammonia concentrations in (surface-)water: a so-called micro ammonia flow injection analysis system (MAFIAS). The MAFIAS-chip is a miniaturized FIA system incorporating pumps, mixers, a reaction chamber as well as the detector all on one chip. The Berthelot reaction for the conversion of ammonia into indophenol blue is employed in this system to convert ammonia into a blue dye, which can be quantified by a light-absorption measurement.

Advances in micromechanics and microfluidics have opened up opportunities to create complete systems based on micromachined components. With respect to conventionally sized systems, micro-sized systems behave differently in a number of ways. On the micro scale, a few phenomena, which are hardly noticeable at the macro-scale, now play a very important role.

* Corresponding author. Tel.: +31-53-489-4435; fax: +31-53-489-3343. *E-mail address:* t.t.veenstra@el.utwente.nl (T.T. Veenstra).



ISSN: 0166-526X

Due to the (mostly) very low Reynolds number (*Re* typically smaller than 1), all flow is perfectly laminar. Though this is easy for modeling, it has a side effect that, since all fluid moves in sheets over each other, mass and heat transport are based on diffusion instead of turbulence [1].

At the microscale diffusion is very fast. If channel dimensions are chosen carefully, dispersion may be significantly reduced by this fast diffusion, since samples will tend to stay together in a plug instead of being smeared out over an entire channel length [2]. Due to this effect smaller samples can be used without signal loss owing to dilution of the sample.

The last noticeable effect on the microscale is the effect of surface tension. A fluid–gas interface in a microchannel results in a net force on that interface, dragging fluids into a channel or blocking the entrance to a channel [3]. The exact nature depends on the geometry of the channel, but the effect can easily add up to a degree that several millibar of pressure has to be applied to overcome the surface tension.

The mentioned phenomena have to be taken into account while designing an analysis system based on micromachined components. In this chapter, it will be shown how positive downscaling effects can be used whereas negative downscaling effects can be circumvented.

## 16.2 THE DESIGN PATH

To take full advantage of the microscale character of the involved fluidic components, care has to be taken in the design process to address these effects in a proper way.

The first thing needed is a detailed specification of the system's desired functionality, with respect to the desired concentration range, sample rate, sample size, sample pre-treatment and maybe even lifetime expectancy.

From this set of specifications, a choice can be made for the detection method (e.g. light absorption measurement, conductivity measurement, etc.). Once this has been decided, a schematic of the system can be drawn, showing how the different stages of the sample treatment are interconnected.

From the specifications for the number of samples that have to be processed per minute, the flow rate and channel dimensions have to be established. The maximum number of samples that can be processed is dependant on the dispersion of the system. The dispersion of the system is determined by the channel dimensions (radial diffusion) and the flow rate. Small channels as well as a low flow rate will enable a high sample rate since samples will tend to remain separated from each other in the channel. Of course it has to be noted that smaller channels have a higher fluidic resistance, resulting in a higher pressure drop which has to be overcome by the system's pumps. One more constraint is that there will be a lower limit for the sample size that can be used for the detector.

Knowing what the system layout looks like, the next step is to quantify specifications for the components of the system. Based on lumped element specifications all of the required micro fluidic components can be designed.

The last step of the whole system design is the physical layout of the system on the chip. All components probably can be made standalone, but integrating all components on one chip may result in process conflicts, which may lead to the need for component redesign. A way to circumvent this process compatibility problem can be found in a modular design of the system: all fluidic components are placed on a channel plate with all the connecting channels in it. Not only does this solve process problems, but it also reveals the possibility of replacing a presumably malfunctioning component with a new one or replacement of one of the components with an improved version of that component. A drawback of such a modular assembly may be the introduction of undesirable dead volume at the fluidic interconnections resulting in larger sample dispersion.

## 16.3 THE DESIGN

In the following sections, the proposed design path will be applied to the design and realization of a micro-FIA for the detection of ammonia.

### 16.3.1 Chemistry

The following section will describe how the design path in the previous section is implemented into the design of the MAFIAS system, an FIA system based on micromachined components for the detection of ammonia. With this system it should be possible to measure ammonia concentrations in water within a concentration range of 0.1–10 μM. It is envisioned that 60 samples per hour will be handled with this system. Though multiple detection mechanisms exist for measuring the ammonia concentration in water [4,5], for the MAFIAS system the reaction of Berthelot is chosen since it is suitable for the very low concentrations and has few side effects (e.g. the formation of pellets or etching of the channels) which slowly might change the system (e.g. clog a channel or etch away a valve seat). The Berthelot reaction [6] is used to convert ammonia into indophenol blue, a blue dye with its absorption peak at 600 nm. Quantification of the ammonia concentration is achieved by measuring the light absorption at that wavelength.

Though there might be a market for this kind of system, the MAFIAS system was intended as a 'proof of principle', the proof that it is actually possible to build an analysis system based on micromachined components. Therefore, the requirements for this system in terms of sample pretreatment and life expectancy are very loose.

From the use of Berthelot for the conversion of ammonia a number of constraints for the system have been deduced.

The Berthelot method is a two-step chemical reaction. In the first step, hypochlorite is added to the ammonia sample, converting the available ammonia into monochloramine. This reaction is highly pH dependent. If the pH drops below 9, the main reaction products will be dichloramine or even trichloramine (pH $<6$). In order to obtain a reasonable reaction rate as well as a pure monochloramine product, the pH has to be between 10.5 and 11.5.

The second step of the Berthelot mechanism is the formation of indophenol from monochloramine by adding phenol to the sample. This reaction is highly temperature dependent; at room temperature the reaction is at 10% of its maximum rate, whereas at 60°C it is already at 90%. A side effect of this reaction is the formation of gas bubbles when a temperature of 37°C is exceeded (presumably due to the reduced gas solubility of the carrier fluid caused by the Berthelot reaction products). This practically limits the use of this reaction to 37°C. Nitroprusside (hexacyanonitrosylferrate) can be used as a catalyst to increase the reaction rate. Using the catalyst and a temperature of 37°C, the time needed to come to full completion of the reaction is about 15 min. However, if the system treats every subsequent sample reproducibly, the absorption measurement can be completed at an earlier time even though the sample has not been converted fully.

### 16.3.2 System schematics

From the above specifications, the system schematics are drawn (Fig. 16.1). A carrier fluid (deionized water) flows continuously through the system's main channel. The sample is injected into this main stream and the first reagent (hypochlorite) is added to the sample. These two liquid plugs have to be mixed to ensure that all the available ammonia is converted into onochloramine. After completion of the first rapid reaction step, the second reagent—phenol—can be added to the sample. Again mixing will be necessary. The sample is then kept at 37°C for 30 s, after which the absorption measurement can be done. All fluids need a pump to transport them through the system. So the design of the whole system incorporates four pumps, two mixers, a reaction chamber and a light absorption cell. The first thing to be designed is the channel structure that couples all components.

### 16.3.3 Channel geometry

Though the channel connecting all components seems trivial, it certainly is not. By choosing its geometry carefully, dispersion can be minimized resulting in the need for smaller sample size as well as a smaller minimum sample-to-sample distance, resulting in a higher possible sample rate.

The mechanism that is of importance is the ratio between radial diffusion and mass transport due to the laminar flow profile. Due to the laminar flow regime, there will be a parabolic flow profile in the microchannels. In the middle of the channel, the fluid flows faster than at the walls. Therefore, a sample plug will be stretched along the channel (Fig. 16.2a).

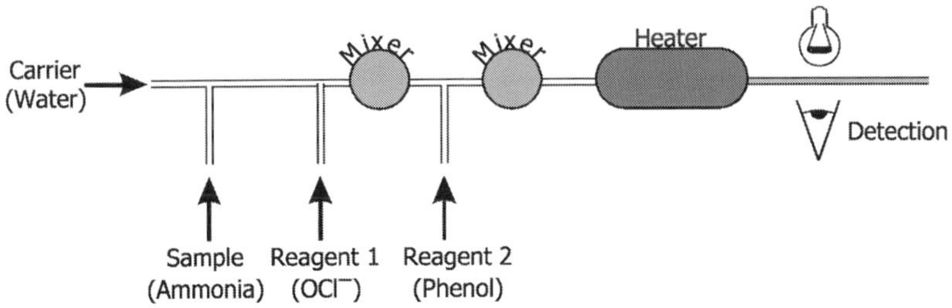


Fig. 16.1. A schematic of the system.

Fig. 16.2. (a) Sample plug distortion due to laminar flow and (b) sample plug distortion in the Aris–Taylor flow regime.

Diffusion tends to reduce the concentration gradient that is thus created across the channel. The distance that is covered by diffusion is dependent on the square root of time. Therefore, the effect of diffusion in small channels is much more important than in large channels (Fig. 16.2b).

The two mentioned phenomena work in opposite directions. The flow profile distorts the sample plug, whereas diffusion restores the plug. Models have been developed to describe the net effect of both phenomena [2]. Input for these models is the total residence time and the diffusion time. The residence time in the MAFIAS system will be about 1 min, whereas the diffusion time from the side of the channel to the middle is in the range of only half a minute (assuming a square cross-section of the channel of $200 \times 200\ \mu m^2$). From these figures it follows that the system is well into the Aris–Taylor regime [2]. This means that an injected sample plug will tend to stay together instead of being smeared out over the channel. In terms of dispersion, this is highly favourable.

Though working in the Aris–Taylor regime is favourable, a side effect of the small cross-section is the high pressure drop that accompanies it. An estimate of the channel length that is needed to couple the components is 20–25 cm. The intention is to flush the system once per minute, so the flow rate will be $10\ \mu L\ min^{-1}$. The pressure that is needed for this flow rate is approximately 7 mbar. This pressure will not be a problem for a micropump that is incorporated in the system since such a pump will have to be designed for higher pressures to overcome surface tension effects (see below).

### 16.3.4 Specifications for the components

With the use of the information in the previous paragraph, more detailed specifications for the fluidic components can be made.

Four pumps are needed. These pumps have to be capable of providing $10\ \mu L\ min^{-1}$ at a pressure of 7 mbar. Next to these obvious features, the pump should also be leak-tight (no backflow when not actuated). For ease of handling, it is also desired that the pumps are self-priming as well as bubble tolerant, meaning that the pumps cannot only fill itself with the pumping fluid starting with channels filled with gas, but they can also pump fluid containing gas bubbles. These last two features place high demands on the pump design since pumping gas requires a high compression ratio (i.e. a high stroke volume versus dead volume) [3]. A pressure of around 70 mbar (depending on exact geometry) is needed to overcome the surface tension effects involved when pumping gas/liquid mixtures. Including the pressure drop over the system's main channel with some headspace for the pressure drop involved with the other components, it seems reasonable to aim at a total pressure of 100 mbar from the pump.

Two mixers have to be incorporated. These mixers have to be capable of a complete mixing of two (aqueous) fluid streams at 10 $\mu L\ min^{-1}$. This mixing should be completed within a few seconds. To stay within the 100 mbar pressure from the pumps, the pressure drop over the mixer should be in the order of 10 mbar.

For the reaction chamber, the specifications are quite simple. The reaction chamber should be capable of maintaining the temperature of the sample at 37°C for at least 30 s.

The final component needing specification is the light absorption cell. With this light cell the absorption of light at 600 nm by the sample due to the formed indophenol has to be measured. The expected concentration range of indophenol is 0.15–15 μM. Together with an extinction coefficient of $0.15 \times 10^6\ L\ mol^{-1}\ m^{-1}$, it can be calculated that a path length of at least 4 mm is needed to ensure high sensitivity. The detection volume of the light cell should be of the order of 1 μL.

With the above-mentioned specifications for all involved components, the components can be selected or designed. The components that were chosen for the MAFIAS system are considered in Section 16.3.5.

### 16.3.5 The components

In this section, the components that will be used for the MAFIAS system will be discussed. Based on the specifications from the previous section the selection or design of the components will be justified.

*Pump*
The wish to incorporate a micropump having no backflow restricts the choice to pumps with fully closing (passive or active) valves. This excludes electro-osmotic pumps [7] and valveless pumps as presented by Bardell [8] and Stemme [9]. From a practical point of view, active valves would be advantageous since these allow for bi-directional flow, which enables easy flushing of all pumps and channels. Peristaltic pumps are also in the category of pumps with active valves. With the use of powder blasting [10] and selective anodic bonding [11] it is fairly simple to make such a peristaltic pump.

The design of the micropump is shown in Fig. 16.3. The pump consists of three identical membranes, which are interconnected with channels. The membranes are created by bonding Pyrex glass to silicon, leaving three circles open with the selective bonding technique. Piezo discs are glued to the membranes. Actuating the piezo discs will cause the membranes to bulge, creating some additional volume underneath the membranes. This is the stroke volume. Actuating the three membranes in a specific sequence will promote the transport of fluid from left to right (Fig. 16.4). The ratio of the stroke volume and the volume in the channels (dead volume) is designed such that the maximum compression corresponds to a pressure of approximately 700 mbar. Though this is much more than needed, it must be noted that this ratio is only reached for the membrane at its maximum deflection. From calibration measurements it was found that the peristaltic micropump can deliver

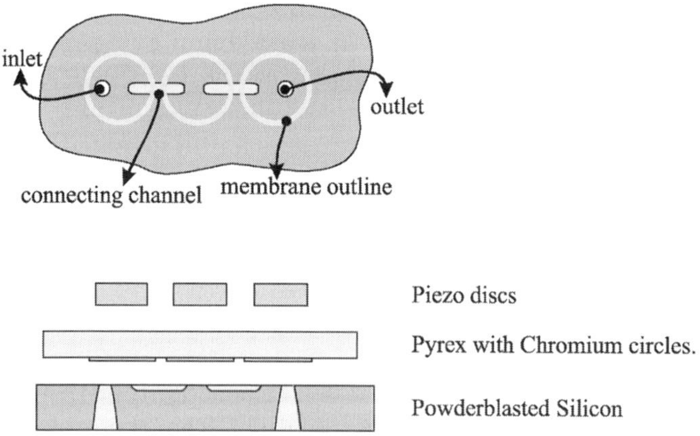


Fig. 16.3. A schematic top view and cross-section of the micropump.

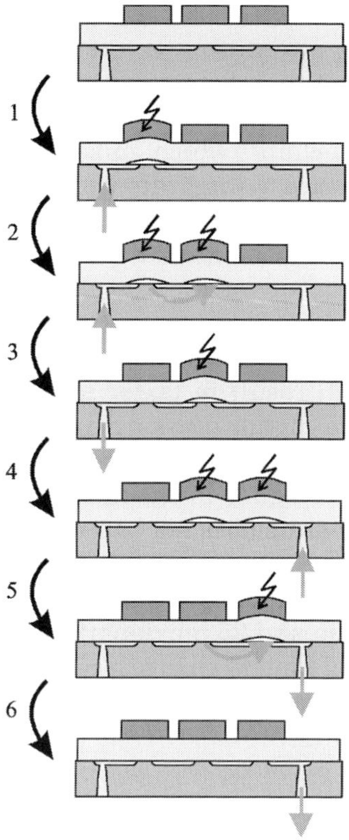


Fig. 16.4. The actuation sequence of the pump.

flow rates from 0 to 30 μL $min^{-1}$. The backflow through the pump is negligible at 100 mbar pressure at the outlet.

*Mixer*

Owing to the laminar flow regime in which the system is working, the mixing of two aqueous solutions has to be based on the process of diffusion. The average distance $\bar{x}$ travelled by a particle with diffusion constant $D$ in a given time $t$ is $\bar{x} = \sqrt{Dt}$. Diffusion over a distance of 1 mm will take more than 10 min. On the microscale, however, diffusion can be quite rapid: whereas diffusion over 50 μm will be well within 2 s, diffusion over 1 μm will take only a few hundredths of microseconds.

The key to the mixing of microfluidics therefore is to minimize the diffusion distance in order to reduce mixing time to a minimum.

Solutions presented in literature consider mostly multilamination [12,13], i.e. generating alternating layers of fluid A and fluid B. The thickness of these layers then determines the mixing time.

The complexity of the fabrication process of the mixer for the MAFIAS system is used to determine which mixer will be used. A complex process scheme will be hard to integrate with the other components. The mixer design chosen is shown in Fig. 16.5. It consists basically of a narrowing in a channel. In the narrow part, diffusion from one side to the other will be rapid.

To determine the geometry of the narrow part of the channel, two effects of reducing the channel width have to be taken into account. First, a narrower channel will result in faster mixing. A narrower channel, however, will result in a higher pressure drop over the channel. This is not only due to the fluidic resistance of the channel being higher, but also the fluid velocity will be higher in the narrow channel. Care has to be taken that the pressure drop does not become too high.

A trade-off is found for a channel of 100 μm width. In this channel, diffusion will take about 3 s. The pressure drop over a channel of 100 μm width, a depth of 200 μm and 25 mm length will be around 7 mbar. The channel needs to be that long to ensure a residence time of 3 s, in which the mixing has to be performed.

*Reaction chamber*

To ensure a complete as possible reaction, the second step of the Berthelot mechanism needs to be controlled at 37°C for 30 s. The proposed solution is to place an electric heater next to the channel (Fig. 16.6).

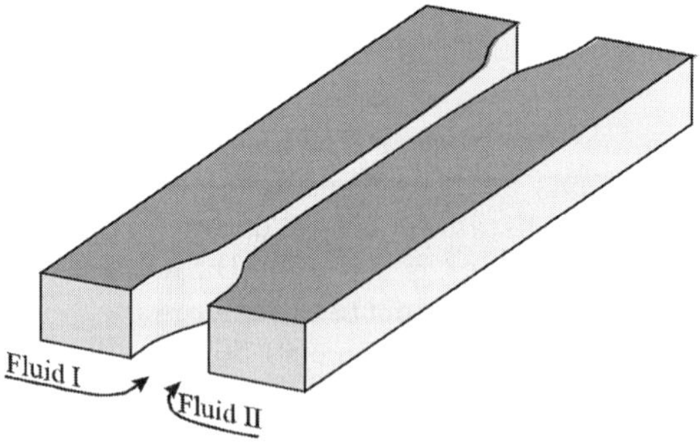


Fig. 16.5. The layout of the proposed mixer.

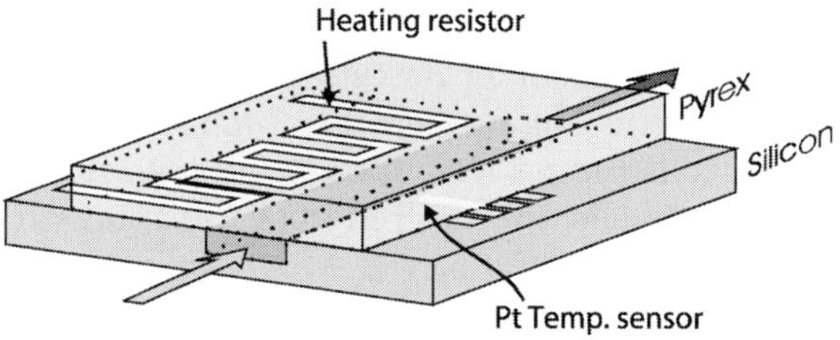


Fig. 16.6. Reaction chamber.

The temperature of the chip can be controlled if the temperature of the chip is measured. The high heat conductivity of silicon ensures that all walls of the channel will be at the same elevated temperature [13]. In this way, the temperature of the chip is regulated. The fluid that flows through the channel will be heated through the walls of the channel. Diffusion of heat through the fluid then has to bring the temperature of all of the fluid to the wall temperature. For a channel of 200 μm width, this heat diffusion takes only in the order of 100 ms.

*Detection cell*

The final component that has to be designed is the detection cell. This device needs to provide a light path of at least 4 mm length. The maximum length that can be achieved by using a hole through a silicon wafer is only 500 μm [15]. Therefore a solution is used in which the light path lies in the plane of the used silicon wafer. With the use of a wet etching technique involving KOH/IPA [16], channels can be etched in silicon with 45° sloped sidewalls. This specific angle makes it very easy to couple light in and out of the channel by using the sidewalls as mirrors [17]. The light source (LED) and detector (photodiode) can be placed directly over such a mirror. Figure 16.7 shows the working principle as

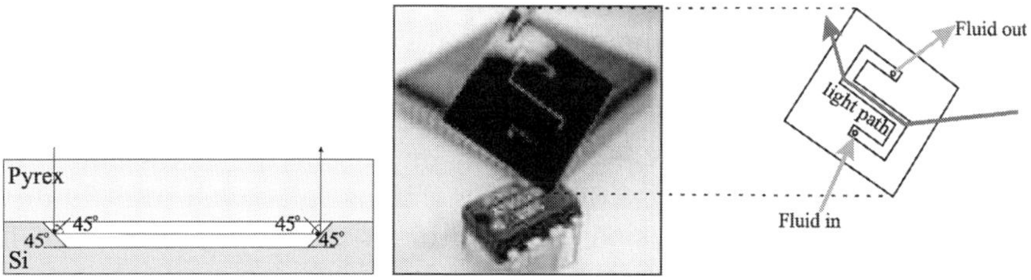


Fig. 16.7. (a) Cross-sectional view of the chip and (b) a photograph of the chip.

well as an SEM photograph of a cross-section of the channel. With this etching technique, a number of detection channels were made with their optical path length ranging from 4 to 7 mm. The detection volume of these detection cells ranges from 0.48 to 0.84 μL. It was shown that with these detection cells the desired concentration range could be measured easily.

All of the components that are necessary to build a MAFIAS system have now been designed and discussed as integrated elements. The following step is to combine the components into one design, which should result in a functional analysis system.

### 16.3.6 System

All the components from which the analysis system will be built work separately. To combine these components into a working system, all have to be connected to each other. This can be done by either integrating all components on to one chip or placing those components on a fluidic 'main board'.

Both options have been realized and will be shown in more detail here. First the system with all components on one chip is discussed, after which the system with the fluidic main board will be treated.

*Monolithic system*

The components mentioned in the previous section were all combined on one chip layout. Though bringing all components on to one chip reveals the possibility of a low-dispersion system (owing to low dead volumes in macro-interconnects), it also involves making compromises in terms of the required fabrication process steps.

In Table 16.1, it is shown which process steps were involved in fabricating the separate components. Most of those process steps can be changed to bring them in line with the processes for the other components. The KOH–IPA etching for the 45° sidewalls cannot be made compatible with the other channel etching techniques. Therefore, the detection cell will be placed on top of the main chip. As can also be seen in Fig. 16.8, all connections (electrical and fluidic) are available on the top side of the chip. A planar layout is shown in Fig. 16.9 and a photograph of the realized device is shown in Fig. 16.10.

TABLE 16.1

Process steps for the different components

| Pump | Reaction chamber | Mixer | Detection cell |
|---|---|---|---|
| 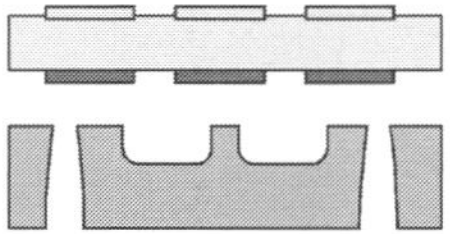 | 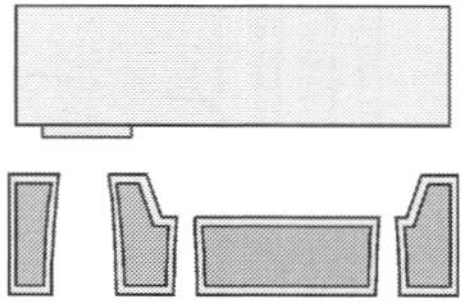 | 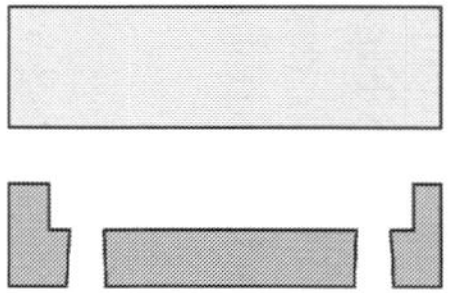 | 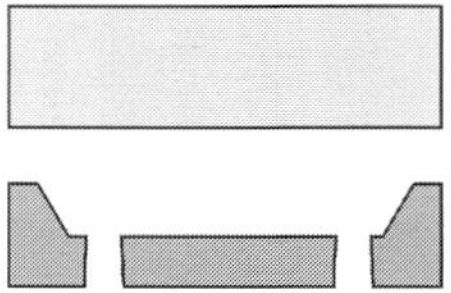 |
| Powder blasted channels | KOH-etched channels | DRIE etched channels | KOH–IPA etched channel |
| Powder blasted through holes | Powder blasted through holes | Powder blasted through holes | Powder blasted through holes |
| Chromium anti-bond layer on bottom side of Pyrex | Platinum layer on bottom side of Pyrex for heating element and temperature sensor | | |
| Platinum layer on top-side of Pyrex for electrical contacts for piezodiscs | $SiO_2$ layer on silicon wafer to prevent shortcircuiting of the Platinum layer | | |
| Pyrex wafer is thinned to match the thickness of the piezodiscs | | | |

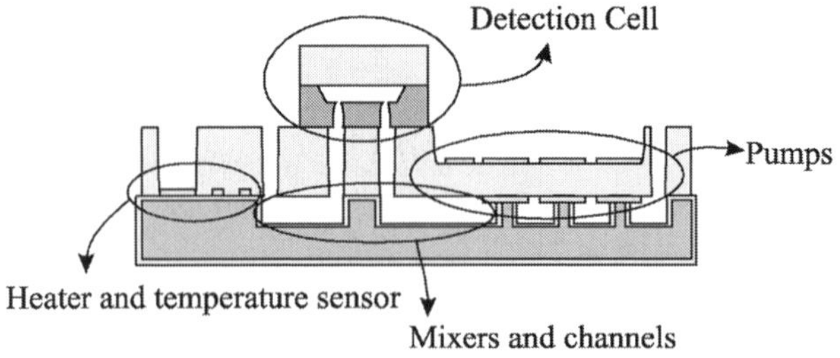


Fig. 16.8. Monolithic MAFIAS system (cross-section).

Testing the system was performed to evaluate its performance. As can be seen from the measurement results shown in Fig. 16.11, this system does not meet the desired specifications. Though reproducibility seems to be reasonably good, it can also be seen that air bubbles cause sharp

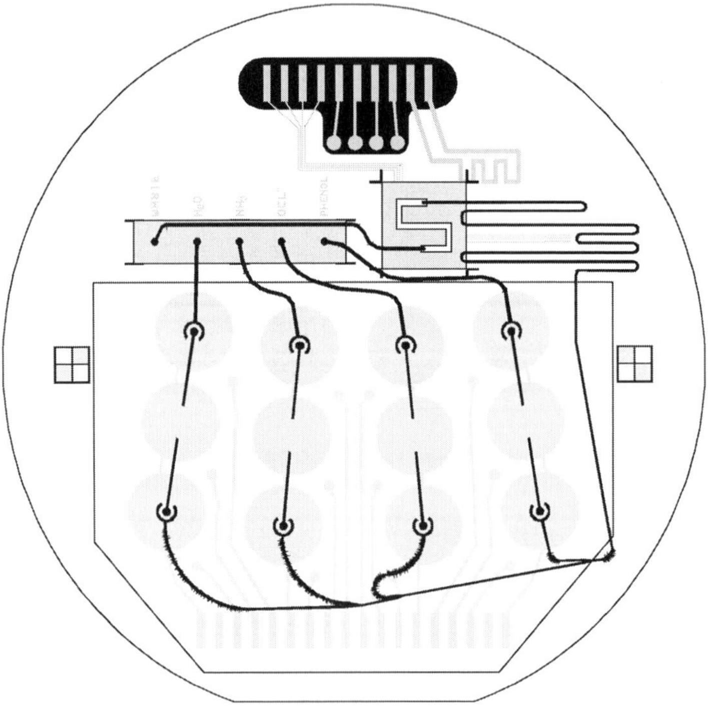


Fig. 16.9. The planar layout of the monolithic system.

Fig. 16.10. A photograph of the monolithic system.

peaks or steps in the output signal. The poor performance of this system could be tracked to the pumping performance of the second reagent (phenol). This reagent solution did not wet the interior of the pump chamber. Adding pH buffer to this reagent solved this problem.

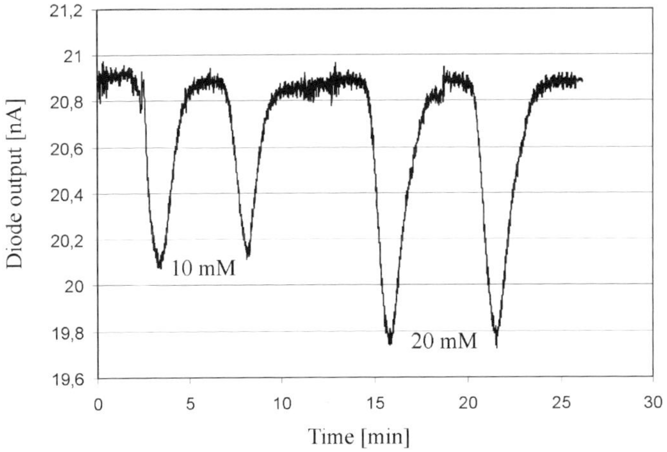


Fig. 16.11. Measurement series for the monolithic system.

After 24 h of operation, the pumps were clogged with particles from the reagent solutions. Hence, the system no longer functioned. Though more severe filtering of the reagents will partly solve this problem, the possibility of replacing a clogged component would have immediately fixed the problems. The possibility of exchanging components is shown in the next section, in which the design of a modular-built system is shown.

*Modular system*

In the previous section the design and performance of a monolithically built system were shown. This section will discuss the design and performance of a modular analysis system.

Figure 16.12 shows how the system is built. The different micromachined components are placed on a silicon-glass channel plate (fluidic circuit board). This fluidic circuit board connects the different components in the correct way to form the analysis system.

To ensure correct placement of the components, the components are placed in module housings which fit into cut-outs in a printed circuit board which in turn are glued to the fluidic circuit board. Viton O-rings are used to connect the fluidic in/outlets of the components to the fluidic

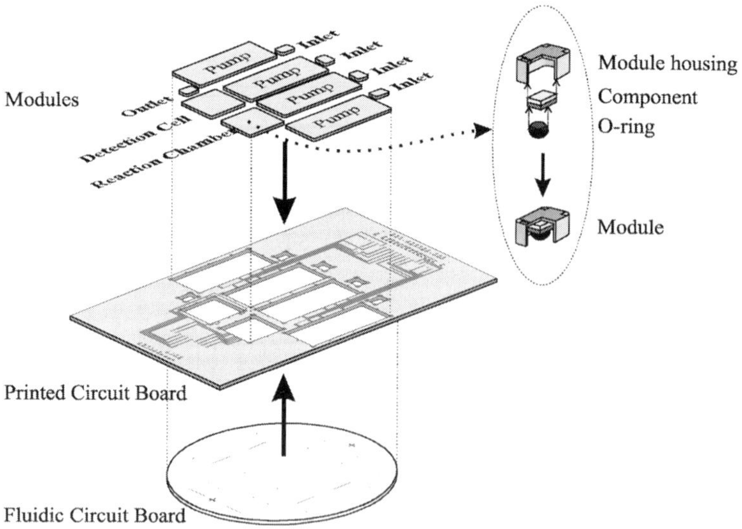


Fig. 16.12. A schematic of the MATAS technology for the modular system.

base plate. By placing the modules, the O-rings are slightly compressed, ensuring a leaktight connection. The modules are fixed by solder joints at the corners of the modules.

These solder joints are also used for electrical connections of the components in the modules to the printed circuit board, on which external electrical connections are available. For a photograph of the completed system see Fig. 16.13.

The performance of this system is shown in Fig. 16.14. The desired specifications of 5.6–560 μM are easily met. The depth of the 5.6 μM peak together with the noise level suggest a lower measurement limit of less than 1 μM. The repeatability of the measurements was calculated to be 8%.

This system may be improved by reducing the flow rate as well as the size of the injected sample. By reducing the flow rate, the total time that is available for the (slow) Berthelot reaction will increase, resulting in a higher indophenol concentration and hence a higher S/N ratio. Reducing the sample size opens the possibility of a higher sample throughput. The S/N ratio gain resulting from the longer reaction time due to the lower flow rate can be used to compensate for the signal loss from the smaller sample size.

A disadvantage of this solution is the fact that the 'dead time' between sample injection and signal readout will be longer, which might be a problem if the signal readout is used in a feedback loop for process control.

Fig. 16.13. Photograph of the modular MAFIAS system.

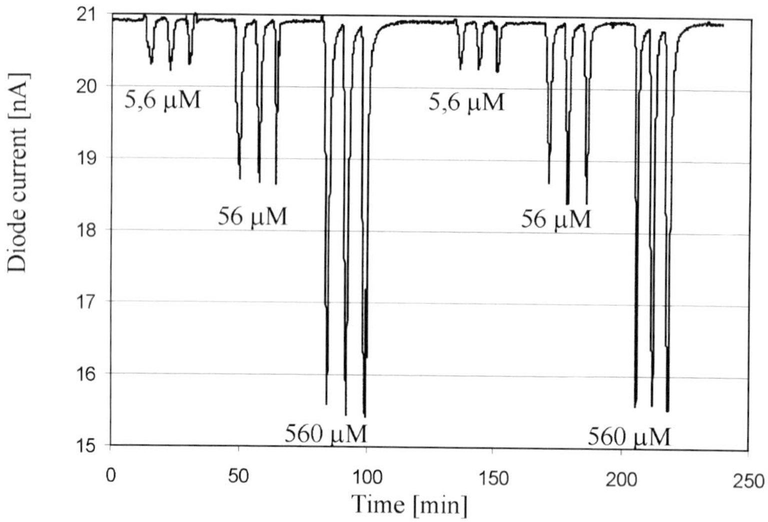


Fig. 16.14. Measurement series for the modular system set-up.

## 16.4 CONCLUSIONS

In this chapter, a method has been shown to design a chemical analysis system based on micromachined components. The main part of designing such a system consists of deriving specifications for all the components involved (including the connecting channels).

Following the path laid out in the design method, an FIA system for the detection of ammonia based upon the classic Berthelot reaction was designed. First an inventory of the required components was made, resulting in a general layout of the system. From this layout it was seen what components are needed to build the system. Following this inventory, all components were considered lumped elements, for which specifications were deduced. With these specifications, the components could then be designed. The final step of putting the components together into a system was then completed in two different ways.

The first option of creating a system from the different components was to combine all the components into one chip design. Though this leads to (theoretical) advantages in terms of dispersion and optimal connections of the different components, it also means that the malfunction of only one component renders the complete system useless.

After preliminary testing, this is exactly what happened: one of the pumps clogged, due to which no further testing could be done. The measurements that were performed showed that it is possible to make a full lab-on-a-chip. It was also realized that strict care has to be taken to prevent particles from entering the system.

Due to the system being clogged, the modified second reagent (pH 11 buffer added to the phenolic solution) could not be tested on the monolithic system.

The second system set-up was the modular system. Using the MATAS technology for coupling the different components, a highly flexible system was created. Not only was it easy to replace a malfunctioning component, but the MATAS technology also enabled changing a component for a different or improved version.

Measurement results from the modular system show that the desired concentration range was easily covered. Also, the repeatability of the measurements was good.

The results from the modular system indicate that the monolithic system could perform to the same level as the modular system if the modified reagents are used.

From all that was learned in building these two analysis systems, it can be concluded that it is possible to realize chemical analysis systems based on micromachined components that perform at the highest level. Though this is true, it must also be noted that these micromachined components require different handling procedures in order to protect them from malfunctioning.

## REFERENCES


1 J. Branebjerg, B. Fabius and P. Gravesen, Application of miniature analyzers: from microfluidic components to μ-TAS, *Proceedings Micro-TAS'94*, (1994) 141.
2 W.E. van der Linden, *Trends Anal. Chem.*, 6 (1987) 2.
3 M. Richter, R. Linneman and P. Woias, *Sens. Actuat. A*, 68 (1998) 480.
4 G. Schulze, C.Y. Liu, M. Brodowski, O. Elsholz, W. Frenzel and J. Moller, *Anal. Chim. Acta*, 214 (1988) 121.
5 F.J. Krug, B.F. Reis, M.F. Giné and E.A.G. Zagatto, *Anal. Chim. Acta*, 151 (1983) 39.
6 P.L. Searle, *Analyst*, 109 (1984) 549.
7 P.D.I. Fletcher, S.J. Haswell and X. Zhang, *Lab Chip*, 1 (2001) 115.

8 R.L. Bardell, F.K. Forster, M.A. Afromowitz, N.R. Sharma and A. Blanchard, Designing high-performance micro-pumps based on no-moving-parts valves, *MEMS'97*, 1997, pp. 47–53.
9 A. Olsson, G Stemme and E. Stemme, A numerical design study of the valveless diffuser pump using a lumped mass model, *JMM*, 9 (1999) 34–44.
10 H. Wensink, Fabrication of microstructures by powder blasting, PhD Thesis, University of Twente, Enschede, The Netherlands, 2002.
11 T.T. Veenstra, J.W. Berenschot, J.G.E. Gardeniers, R.G.P. Sanders, M.C. Elwenspoek and A. van den Berg, *J. Electrochem. Soc.*, 148 (2001) G68.
12 M. Koch, D. Chatelain, A.G.R. Evans and A. Brunnschweiler, Two simple micromixers based on silicon, *JMM*, 8 (1998) 123–126.
13 N. Schwesinger, T. Frank and H. Wurmus, A modular microfluid system with an integrated micromixer, *JMM*, 6 (1996) 99–102.
14 J.C.T. Eijkel, A. Prak, S. Cowen, D.H. Craston and A. Manz, *J. Chromatogr. A*, 815 (1998) 265.
15 A. Daridon, M. Sequeira, G. Pennarun-Thomas, J. Lichtenberg, E. Verpoorte, D. Diamond and N.F. de Rooij, Comparative Kinetic Studies of the Berthelot Reaction in Microfluidic and Spectrophotometric systems, *Proceedings Eurosensors XIV*, Copenhagen, Denmark, 2000.
16 C. Strandman, L. Rosengren, H.G.A. Elderstig and Y. Bäcklund, *J. Microelectro-Mech. Syst.*, 4 (1995) 213.
17 R.M. Tiggelaar, T.T. Veenstra, R.G.P. Sanders, J.G.E. Gardeniers, M.C. Elwenspoek and A. van den Berg, *Talanta*, 56 (2002) 331.

*Chapter 17*

# *Nanosensor and nanoprobe systems for in vivo bioanalysis*

***Tuan Vo-Dinh****

## 17.1 INTRODUCTION

Advances in nanotechnology have recently led to the development of fiberoptics-based nanosensor systems having nanoscale dimensions suitable for intracellular measurements. The possibilities to monitor in vivo processes within living cells could dramatically improve our understanding of cellular function, thereby revolutionizing cell biology. Fiberoptic sensors provide significant advantages for in situ monitoring applications due to the optical nature of the excitation and detection modalities. Fiberoptic sensors are not affected by electromagnetic interferences from static electricity, strong magnetic fields, or surface potentials. Another advantage of fiberoptic sensors is the small size of optical fibers, which allow for sensing intracellular/intercellular physiological and biological parameters in microenvironments.

Biosensors that use biological probes coupled to a transducer have been developed during the last two decades for environmental, industrial and biomedical diagnostics. Extensive research and development activities in the author's laboratory have been devoted to the development of a variety of fiberoptic chemical sensors, specifically biosensors [1–9]. The application of a submicron fiberoptic chemical sensor has been reported [10,11]. Submicron tapered optical fibers with

**E-mail address:* vodinht@ornl.gov.

*Comprehensive Analytical Chemistry XXXIX, pages 685–700*


ISSN: 0166-526X

distal diameters between 20 and 500 nm have been employed to study the submicron spatial resolution achievable using near-field scanning optical microscopy (NSOM). The combination of NSOM and surface-enhanced Raman scattering (SERS) has been demonstrated to detect chemicals on solid substrates with sub-wavelength 100-nm spatial resolution [12,13]. Submicron optical fiber probes have been developed for chemical analyses [14,15]. Nanosensors with antibody probes have been developed and used to detect biochemical targets inside single cells [16–21].

In this chapter, we provide an overview of the principle, development and applications of fiberoptics nanosensor systems using antibody-based probes. The chapter provides background information on biosensors, a description of the fabrication methods for fiberoptic nanosensors and detection systems, and applications in single-cell analysis. The usefulness and potential of fiberoptics nanosensor technology in biological research and applications are discussed.

## 17.2 BACKGROUND ON BIOSENSORS AND BIORECEPTORS

### 17.2.1 Biosensing systems

A biosensor is generally defined as a measurement system that consists of a probe with a biological recognition element, often called a bioreceptor, and a transducer. The interaction of the analyte with the bioreceptor is designed to produce an effect measured by the transducer, which converts the information into a measurable effect, e.g. an electrical signal. Figure 17.1 illustrates the operating principle of a typical biosensing system.

Biosensors have seen a wide variety of applications primarily in two major areas, biological monitoring and environmental sensing applications. The integration of biological methods, laser systems and advanced optical sensor technology promises to open up new applications in environmental and biomedical monitoring of toxic chemicals and bioindicators. Based on antibody–antigen interactions, immunoassay techniques are very powerful monitoring tools because of their excellent specificity and reasonable sensitivity. The immunological principle can be combined with laser and fiberoptics technology to develop a new generation of nanosensors for intracellular measurements.

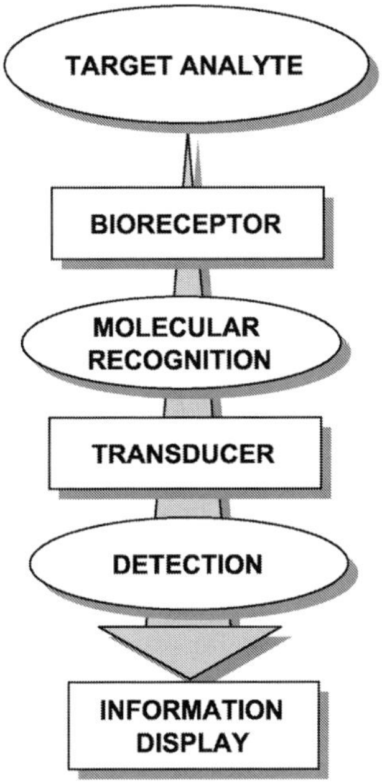


Fig. 17.1. Operating principle of biosensor systems.

### 17.2.2 Bioreceptor probes

Bioreceptors are used because they are important elements in specificity for biosensor technologies. They allow for the binding of the specific analyte of interest to the sensor for the measurement with minimum interference from other components in complex mixtures. A bioreceptor is a biological molecular species (e.g. an antibody, an enzyme, a protein, or a nucleic acid) or a living biological system (e.g. cells, tissue, or whole organisms) that utilizes a biochemical mechanism for recognition. The sampling component of a biosensor contains a biosensitive layer. The layer can either contain bioreceptors or be made of bioreceptors covalently attached to the transducer. The most common forms of bioreceptors used in biosensing are based on (a) antibody/antigen interactions, (b) nucleic acid interactions, (c) enzymatic interactions, (d) cellular interactions (i.e. microorganisms, proteins), and (e) interactions using biomimetic materials (i.e. synthetic bioreceptors). Various techniques (optical, electrochemical, and mass sensitive) can be used for detection in biosensors. This chapter deals mainly with biosensor systems using antibody probes, often called immunosensors, with optical detection.

*Antibody probes*

Due to the diversity of antibody molecules, there is a wide variety of applications for immunosensors. The antigen–antibody (Ag–Ab) binding reaction, which is a key mechanism by which the immune system detects and eliminates foreign matter, provides the basis for specificity of immunoassays. Antibodies are complex biomolecules, made up of hundreds of individual amino acids arranged in a highly ordered sequence. Antibodies are produced by immune system cells (B cells) when such cells are exposed to substances or molecules, which are called antigens. The antibodies called forth following antigen exposure have recognition/binding sites for specific molecular structures (or substructures) of the antigen. In this way an antigen and antigen-specific antibody interaction is analogous to a lock and key fit, in which specific configurations of a unique key enable it to open a lock. In the same way, an antigen-specific antibody fits its unique antigen in a highly specific manner, so that the three-dimensional structures of antigen and antibody molecules are complementary. Due to this three-dimensional shape fitting, and the diversity inherent in individual antibody make-up, it is possible to find an antibody that can recognize and bind to any one of a large variety of molecular shapes. This unique property of antibodies is the key to their usefulness in immunosensors; this ability to recognize molecular structures allows one to develop antibodies that bind specifically to chemicals, biomolecules, microorganism components, etc. One can then use such antibodies as specific probes to recognize and bind to an analyte of interest that is present, even in extremely small amounts, within a large number of other chemical substances. Another property of great importance to antibodies' analytical role in immunosensors is the strength or avidity/affinity of the Ag–Ab interaction. Because of the variety of interactions that can take place at Ag–Ab surfaces that lie in close proximity to one another, the overall strength of the interaction can be considerable, with correspondingly favorable association and equilibrium constants. What this means in practical terms is that the Ag–Ab interactions can take place very rapidly (for small antigen molecules, almost as rapidly as diffusion processes bring antigen and antibody together), and that, once formed, the Ag–Ab complex has a reasonable lifetime.

Only an immunogenic species can trigger the production of antibodies. For a substance to be immunogenic (i.e. capable of producing an immune response), a certain molecular size and complexity are necessary: proteins with molecular weights greater then 5000 Da are generally immunogenic. Radioimmunoassay (RIA), which utilizes radioactive labels, has been one of the most widely used immunoassay methods. RIA has been applied to a number of fields including pharmacology, clinical chemistry, forensic science, environmental monitoring, molecular epidemiology and agricultural science. The usefulness of RIA, however, is limited by several shortcomings, including the required use of radioactive labels, the limited shelf life of radioisotopes, and the potential deleterious biological effects inherent to radioactive materials, and the cost of radioactive waste disposal. For these reasons, there are extensive research efforts aimed at developing simpler, more practical immunochemical techniques and instrumentation, which offer comparable sensitivity and selectivity to RIA. In the 1980s, advances in spectrochemical instrumentation, laser miniaturization, biotechnology and fiberoptics research have provided opportunities for novel approaches to the development of sensors for the detection of chemicals and biological materials of environmental and biomedical interest. Since the first development of a remote fiberoptics immunosensor for in situ detection of the chemical carcinogen benzo[a]pyrene [1], antibodies have become common bioreceptors used in biosensors today.

## 17.3 FIBEROPTICS NANOSENSOR SYSTEM

### 17.3.1 Fabrication of the fiberoptic nanoprobe

This section discusses the protocols and instrumental systems involved in the fabrication of fiberoptics nanoprobes. The fabrication of near-field optical probes is a crucial prerequisite for the development of nanosensors. There are two methods for preparing the nanofiber tips. The most frequently used technique is the so-called 'heat and pull' method. It is based on local heating of a glass fiber using a laser or a filament and subsequently pulling the fiber apart. The resulting tip shapes depend strongly on the temperature and the timing of the procedure. The second method, often referred to as the 'Turner's method', is based on chemical etching of glass fibers [22]. Recently, a variation of the standard etching scheme was proposed, where the taper is formed inside the polymer cladding of the glass fibers [23].

Figure 17.2 illustrates the experimental steps involved in the fabrication of nanosensors using the heat and pull method [21]. Fabrication of nanosensors involved techniqes capable of making optical fibers with submicron-sized diameter cores. Since these nanoprobes are not commercially available, they have to be fabricated in the laboratory. One procedure consists of pulling from a larger silica optical fiber using a special fiber-pulling device (Sutter Instruments P-2000). This method yields fibers with submicron diameters. One end of a 600-μm silica/silica fiber is polished to a 0.3 μm finish with an Ultratec fiber polisher. The other end of the optical fiber is then pulled to a submicron length using a fiber puller. Figure 17.3 shows a scanning electron microscopy photograph of one of the fiber probes fabricated for our preliminary studies. The scale on the photograph of this sample indicates that the distal end of the fiber is approximately 30 nm.

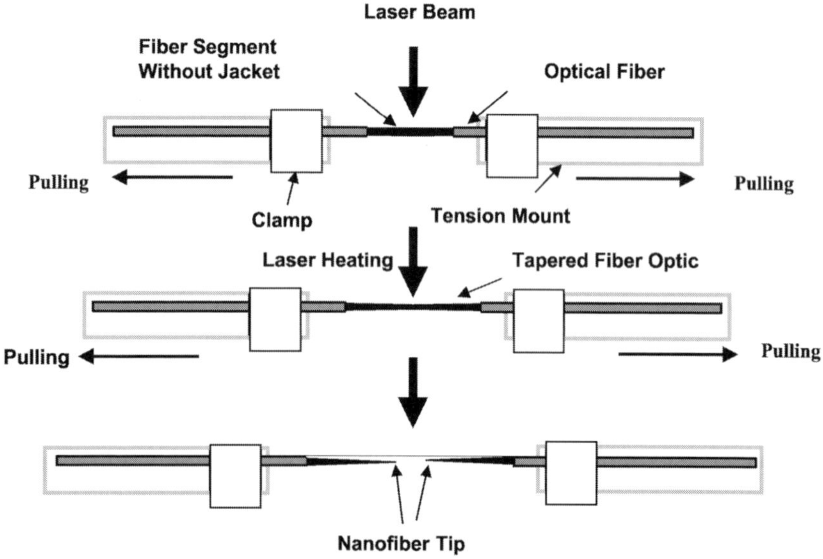


Fig. 17.2. The 'heat and pull' method for the fabrication of nanofibers.

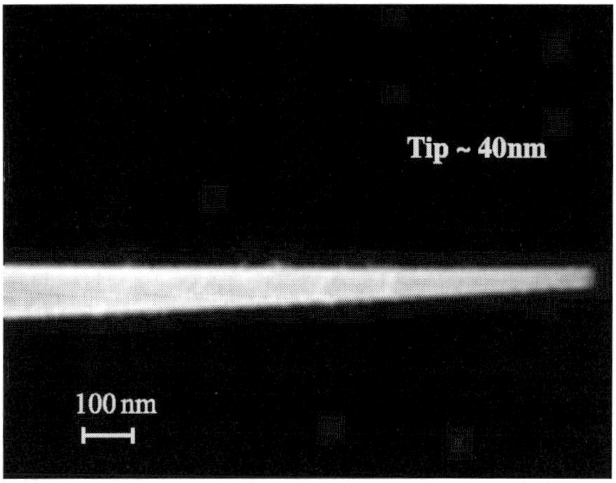


Fig. 17.3. Scanning electron photograph of a nanofiber. The fiber tip diameter is approximately 30 nm.

The side wall of the tapered end is then coated with a thin layer of silver, aluminum or gold (100–300 nm) to prevent light leakage of the excitation light on the tapered side of the fiber. The coating procedure is designed to leave the distal end of the fiber free for subsequent binding with bioreceptors. Such a coating system is illustrated in Fig. 17.4. The fiber probe is attached on a rotating plate inside a thermal evaporation chamber [3,19,21]. The fiber axis and the evaporation direction formed an angle of approximately 45°. While the probe is rotated, the metal is allowed to evaporate onto the tapered side of the fiber tip to form a thin coating. Since the fiber tip is pointed away from the metal source, it remains free from any metal coating. The tapered end is coated with 300–400 nm of silver in a Cooke Vacuum Evaporator system using a thermal source at $10^{-6}$ Torr. With the metal coating, the size of the probe tip is approximately 250–300 nm.

### 17.3.2 Immobilization of receptors onto fiberoptics nanoprobes

The next step in the preparation of the biosensor probes involves covalent immobilization of receptors onto the fiber tip. For antibody binding, several strategies can be used to retain the antibody at

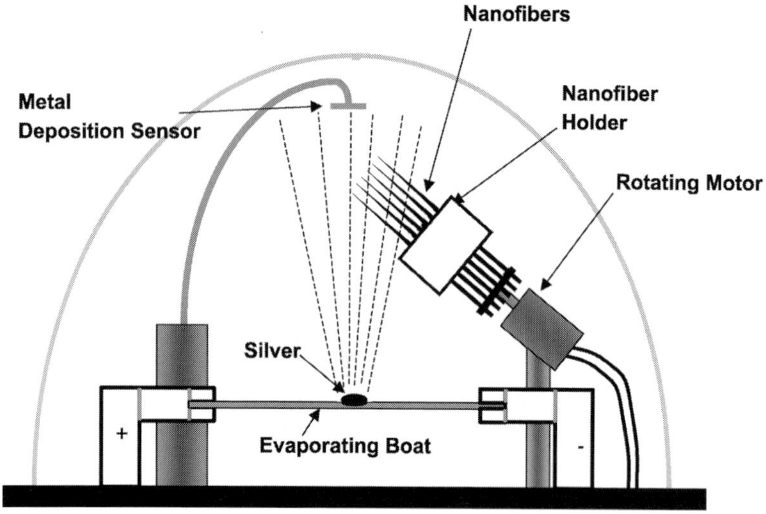


Fig. 17.4. Instrumental setup for coating the nanofiber tips with silver.

the sensing probe. Whatever procedure is involved, one requirement is that the antibody should retain its antigen-binding activity as much as possible. Perhaps the easiest procedure involves enclosure of the antibody in solution, within a semi-permeable membrane cap, which fits over the end of the sensor [2]. However, this design is more complicated and would increase the size of the tip of the nanosensor.

Antibodies can be immobilized onto the fiberoptics probes by using a method previously described. Briefly, the fiber is derivatized in 10% GOPS in $H_2O$ (v/v) at 90° for 3 h. The pH of the mixture is maintained below 3 with concentrated HCl (1 M). After derivatization, the fiber was washed in ethanol and dried overnight in a vacuum oven at 105°C. The fiber is then coated with silver as described previously. The derivatized fiber is activated in a solution of 100 mg/mL 1,1′-carbonyldiimidazole (CDI) in acetonitrile for 20 min followed by rinsing with acetonitrile and then phosphate buffered saline (PBS). The fiber tip is then incubated in a 1.2-mg/mL anti-BPT solution (PBS solvent) for 4 days at 4°C and then stored overnight in PBS to hydrolyze any unreacted sites. The fibers are then stored at 4°C with the antibody immobilized tips stored in PBS. This procedure has been shown to maintain over 95% antibody activity for BPT [21].

### 17.3.3 Experimental system and protocol for nanoprobe investigation of single cells

A unique application of the nanosensor involves monitoring single cells in vivo [18–21]. This application will be discussed later. This section discusses the procedures to grow cell cultures for analysis using the nanosensors. Cell cultures were grown in a water-jacketed cell culture incubator at 37°C in an atmosphere of 5% $CO_2$ in air. Clone 9 cells, a rat liver epithelial cell line, were grown in Ham's F-12 medium (cat #11765-054, Gibco/BRL, Grand Island, NY) supplemented with 10% fetal bovine serum and an additional 1 mM glutamine (cat #15039-027 Gibco, Grand Island, NY). In preparation for an experiment, 1 × 105 cells in 5 mL of medium were seeded into 60-mm diameter dishes (cat #25010, Corning Costar Co., Corning, NY). The growth of the cells was monitored daily by microscopic observation, and when the cells reached a state of confluence of 50–60%, BPT was added and left in contact with the cells for 18 h (i.e. overnight). The growth conditions were chosen so that the cells would be in log phase growth during the chemical treatment, but would not be so close to confluence that a confluent monolayer would form by the termination of the chemical exposure. Benzopyrene tetrol was prepared as a 1-mM stock solution in reagent grade methanol and further diluted in reagent grade ethanol (95%) prior to addition to the cells. The final concentration of BPT in the culture medium of the dish was $1 \times 10^{-7}$ M and the final alcohol concentration (combination of methanol and ethanol) was 0.1%. Following chemical treatment, the medium containing BPT was aspirated and replaced with standard growth medium, prior to the nanoprobe procedure.

Monitoring BPT molecules in single cells was then carried out using antibody nanoprobes. Investigation of BPT in single cells using the nanoprobe was carried out in the following way. A culture dish of cells was placed on the pre-warmed microscope stage, and the nanoprobe, mounted on the micropipette holder, was moved into position (i.e. in the same plane of the cells), using bright field microscopic illumination, so that the tip was outside the cell to be probed. The total magnification was usually 400×. All room light and microscope illumination light were extinguished, the laser shutter opened, and laser light allowed to illuminate the optical fiber and excitation light transmitted into the fiber tip. Usually, if the silver coating on the nanoprobe was appropriate, no light leaked out of the side wall of the tapered fiber. Only a faint glow of laser excitation at the tip could be observed on the nanoprobe. A reading was taken with the nanoprobe outside the cell and the laser shutter closed. The nanoprobe was then moved into the cell, inside the cell membrane and extending a short way into the cytoplasm, but care was taken not to penetrate the nuclear envelope. The laser was again opened, and readings were then taken and recorded as a function of the time during which the nanoprobe was inside the cell.

### 17.3.4 Optical measurement system

Figure 17.5 shows a schematic diagram of the optical measurement system used for monitoring single cells using the nanosensors. The 325-nm line of a HeCd laser (Omnichrome, 8 mW laser power) or the 488-nm line of an argon ion laser (Coherent, 10 mW) was focused onto a 600-μm delivery fiber, which terminated with an SMA connector. The antibody immobilized tapered fiber was coupled to the delivery fiber through

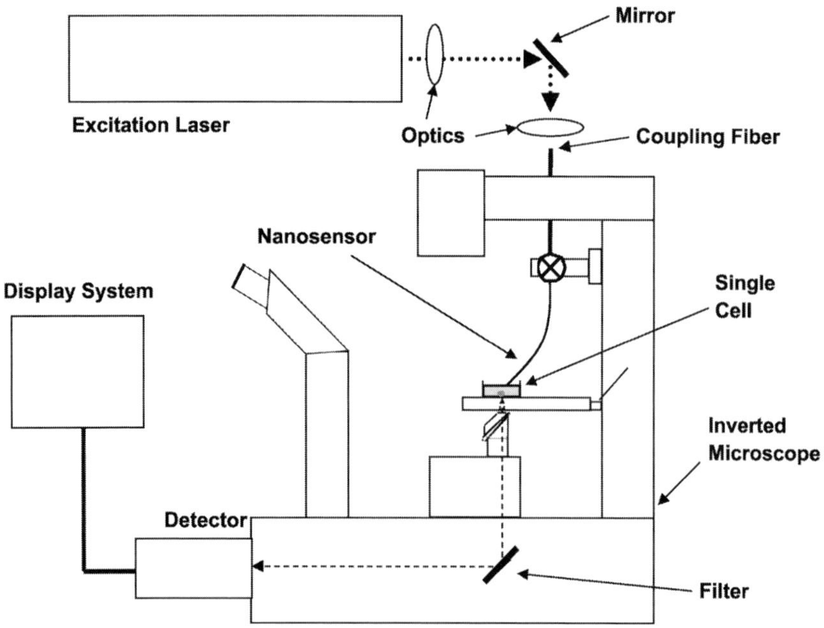


Fig. 17.5. Instrumental system for fluorescence measurements of single cells using nanosensors.

the SMA connector and was secured to the micromanipulators on the microscope. The fluorescence emitted from the cells was collected by the microscope objective and passed through a 400-nm longpass dichroic mirror and then focused onto a photomultiplier tube (PMT) for detection. The output from the PMT was passed through a picoammeter and recorded on a strip chart recorder or a personal computer (PC) for further data treatment. The experimental setup used to probe single cells was adapted to this purpose from a standard micromanipulation/microinjection apparatus. A Nikon Diaphot 300 inverted microscope (Nikon, Inc., Melville, NY) with Diaphot 300/Diaphot 200 Incubator, to maintain the cell cultures at ~37°C on the microscope stage, was used for these experiments. The micromanipulation equipment used consisted of MN-2 (Narishige Co., Ltd, Tokyo, Japan) Narishige three-dimensional manipulators for coarse adjustment, and Narishige MMW-23 three-dimensional hydraulic micromanipulators for final movements. The optical fiber nanoprobe was mounted on a micropipette holder (World Precision Instruments, Inc., Sarasota, FL). To record the fluorescence of BPT molecules binding to antibodies at the fiber tip, a Hamamatsu PMT detector assembly (HC125-2) was mounted in the front port of the Diaphot 300 microscope, and fluorescence was collected via this optical path (80% of available light at the focal plane can be collected through the front port).

## 17.4 APPLICATIONS IN BIOANALYSIS

### 17.4.1 Optical nanofiber probes for fluorescence measurements

Optical nanofiber probes have been fabricated and used to detect fluorescence emission inside the cells from species [17]. In this study, mouse epithelial cells were incubated with a fluorescent dye by incubating the cells in the dye solution and allow membrane permeabilization to take place. Another procedure for loading cells with fluorophors involved the method called 'scrape loading'. In this procedure, a portion of the cell monolayer was removed by mechanical means, and cells along the boundary of this 'scrape' were transiently permeabilized, allowing the dye to enter these cells. The dye was subsequently washed away, and only permeabilized cells retain the dye molecules, as they were not internalized by cells with intact membranes.

Following incubation, the fluorescence signal in single cells was detected using laser light excitation through the optical nanofibers fabricated using the procedures described previously. Micromanipulators were used to move the optical fiber into contact with the cell membrane. The fiber tip was then gently inserted just inside the cell membrane for fluorescence measurements. Light from an argon-ion laser (10 m; 488-nm line) was used and passed through the optical nanofiber. The dye inside the cell was excited, and fluorescence emission was collected and observed through the filter cube of a Nikon Microphot microscope, which effectively filters out the laser excitation light at 488 nm. Background measurements were performed with cells that were not loaded with the fluorophores. Fluorescence signals were successfully detected inside the fluorophore-loaded cells and not inside non-loaded cells. As another control, the optical fiber probe was moved to an area of the specimen where there were no cells, and the laser light was again passed down the fiber for excitation. No visible fluorescence (at the emission wavelength) was detected for this control measurement, thus demonstrating the successful detection of the fluorescent dye molecules inside single cells. These results demonstrated the capability of optical nanofibers for measurements in intracellular environments of luminophores incorporated into cells.

### 17.4.2 Single-cell measurements using antibody-based nanoprobes

Since cells have very small sizes (1–10 μm) the success of intracellular investigations depends largely on not only the sensitivity of the measurement system, but also on the selectivity of the probe and also on the small size of the probes. As discussed previously, biosensors can provide the required selectivity due to the Ag–Ab recognition process, which is one of the most selective molecular recognition processes. Several strategies for designing antibody-based nanosensors may be considered. For example, a membrane-based antibody sensor produced a concentration effect of 40 times greater than the non-antibody sensor [8,9]. Membrane-based probes, however, increase the size of the measurement system, making it difficult to be inserted inside a cell. For biosensors used in intracellular measurements, covalent binding of antibody molecules directly on the tip of the bare nanofiber probe is the most straightforward procedure. This method does not significantly increase the probe sizes.

Nanosensors having antibody-based probes used to measure fluorescent targets inside a single cell has been demonstrated [18–21]. The antibody probe was targeted against benzopyrene tetrol (BPT), an important biological compound, which was used as a biomarker of human exposure to the carcinogen benzo[a]pyrene (BaP), a polycyclic aromatic hydrocarbon (PAH) of great environmental and toxicological interest because of its mutagenic/carcinogenic properties and its ubiquitous presence in the environment. BaP has been identified as a chemical carcinogen in laboratory animal studies [24]. The measurements were performed on rat liver epithelial cells (Clone 9) used as the model cell system. The cells had been previously incubated with BPT molecules prior to measurements. The results demonstrated the possibility of in situ measurements of BPT inside a single cell.

Figure 17.6 shows a photograph of an antibody-based nanosensor inside a single cell. The small size of the probe allowed manipulation of the nanosensor at specific locations within the cells. To demonstrate proof of concept of single-cell measurements with antibody-based nanoprobes, experiments were performed on a rat liver epithelial Clone 9 cell line, which was used as the model cell system [17–21]. The cells were first incubated with BPT prior to measurements. Interrogation of single cells for the presence of BPT was then carried out using antibody nanoprobes for excitation and a photometric system for fluorescence signal detection.

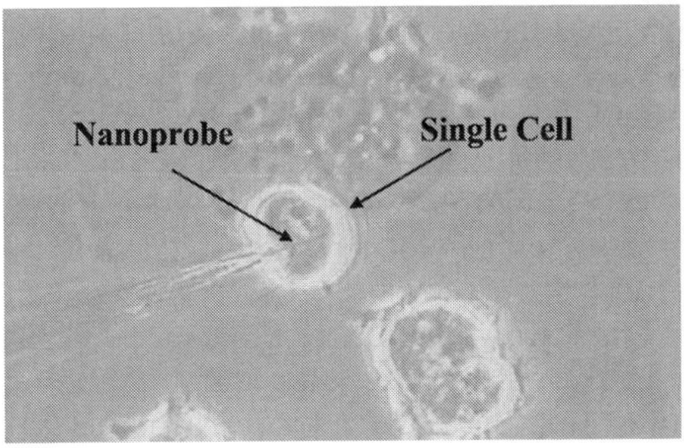


Fig. 17.6. Photograph of single-cell sensing using the nanosensor system. (The small size of the probe allowed manipulation of the nanoprobe at specific locations within a single cell.)

The nanoprobes in this work are single-use bioprobes because they were used to obtain only one measurement at a specific time and could not be reused due to the strong association constant of the Ag–Ab-binding process. Multiple (e.g. five) recordings of the fluorescence signals could be taken with each measurement using a specific nanoprobe. We have made a series of calibration measurements of solutions containing different BPT concentrations in order to obtain a quantitative estimation of the amount of BPT molecules detected. For these calibration measurements, the fibers were placed in petri dishes containing solutions of BPT with concentrations ranging from $1.56 \times 10^{-10}$ to $1.56 \times 10^{-8}$ M. By plotting the increase in fluorescence from one concentration to the next versus the concentration of BPT, and fitting these data with an exponential function in order to simulate a saturated condition, a concentration of $9.6 \pm 0.2 \times 10^{-11}$ M has been determined for BPT in the individual cell investigated [17–21].

## 17.5 CONCLUSION

Concrete knowledge of subcellular architecture and subcellular dynamic processes is important to the fundamental biological understanding of cellular processes. Many traditional microscopy techniques involve incubation of cells with fluorescent dyes and examining the interaction of these dyes with compounds of interest. However, when a dye is incubated into a cell, it is transported to certain intracellular sites that may or may not be where it is most likely to stay and not to areas where the investigator would like to monitor. The fluorescence signals which are supposed to reflect the interaction of the dyes with chemicals of interest, is generally directly related to the dye concentration as opposed to the analyte concentration. Only with optical nanosensors can excitation light be delivered to specific locations inside cells. Combined with the molecular recognition of antibody probes, such optical sensors are capable of providing selectivity in location and analyte parameters.

Other approaches involve 'fixing' of the sample before analysis; this fixing procedure often destroys cellular viability and may significantly alter intracellular architecture as compared to the living state. Nanosensors could provide the tools to investigate important biological processes at the cellular level in vivo. Not only can antibodies be developed against specific epitopes, but also an array of antibodies can be established, so as to investigate the overall structural architecture of a given protein. The integration of these advances in biotechnology and nanotechnology could lead to a new generation of nanosystems with unprecedented sensitivity and selectivity to probe cells at the molecular level.

## ACKNOWLEDGEMENTS

The author acknowledges the contribution of G.D. Griffin, J.P. Alarie, and B.M. Cullum. This research is sponsored by the LDRD Project (Advanced Nanosensors), and by the Office of Biological and Environmental Research and the National Nuclear Safeguard Agency (NN-20 Program), US Department of Energy under contract DE-AC05-00OR22725 managed by UT-Battelle, LLC.

## REFERENCES

1 T. Vo-Dinh, B.J. Tromberg, G.D. Griffin, K.R. Ambrose, M.J. Sepaniak and E.M. Gardenshire, *Appl. Spectrosc.*, 41 (1987) 735.
2 T. Vo-Dinh, G.D. Griffin and M.J. Sepaniak, Fiberoptic Immunosensors, In: O.S. Wolfbeis (Ed.), *Chemical Sensors and Biosensors*, CRC Press, Boca Raton, Florida, 1991.
3 T. Vo-Dinh, M.J. Sepaniak, G.D. Griffin and J.P. Alarie, *Immunomethods*, 3 (1993) 85.
4 J. Alarie and T. Vo-Dinh, *Polycyclic Arom. Comp.*, 8 (1996) 45.
5 J.P. Alarie and T. Vo-Dinh, *Talanta*, 38 (1991) 529.
6 J.P. Alarie, M.J. Sepaniak and T. Vo-Dinh, *Anal. Chim. Acta.*, 229 (1990) 69.
7 B.J. Tromberg, M.J. Sepaniak, J.P. Alarie, T. Vo-Dinh and R.M. Santella, *Anal. Chem.*, 60 (1998) 1901.
8 J.P. Alarie, J.R. Bowyer, M.J. Sepaniak, A.M. Hoyt and T. Vo-Dinh, *Anal. Chim. Acta*, 236 (1990) 237.
9 J.R. Bowyer, J.P. Alarie, M.J. Sepaniak, T. Vo-Dinh and R.Q. Thompson, *Analyst*, 116 (1991) 117.
10 E. Betzig, J.K. Trautman, T.D. Harris, J.S. Weiner and R.L. Kostelak, *Science*, 251 (1991) 1468.
11 E. Betzig and R.J. Chichester, *Science*, 262 (1993) 1422.
12 D. Zeisel, V. Deckert, R. Zenobi and T. Vo-Dinh, *Chem. Phys. Lett.*, 283 (1998) 38.
13 V. Deckert, D. Zeisel, R. Zenobi and T. Vo-Dinh, *Anal. Chem.*, 70 (1998) 2646.

14 W. Tan, Z-Y. Shi and R. Kopelman, *Anal. Chem.*, 64 (1992) 2985.
15 W. Tan, Z-Y. Shi, S. Smith, D. Birnbaum and R. Kopelman, *Science*, 258 (1992) 778.
16 B. Cullum, G.D. Griffin, G. Miller and T. Vo-Dinh, *Analyt. Biochem.*, 277 (2000) 25.
17 T. Vo-Dinh, G.D. Griffin, J. Alarie, B. Cullum, B. Sumpter and D. Noid, *J. Nanoparticle Res.*, 2 (2000) 17.
18 T. Vo-Dinh and B. Cullum, *Fresenius J. Anal. Chem.*, 366 (2000) 540.
19 B. Cullum and T. Vo-Dinh, *Trends Biotech.*, 18 (2000) 388.
20 T. Vo-Dinh, B.M. Cullum and D.L. Stokes, *Sens. Actuat. B*, 74 (2001) 2.
21 T. Vo-Dinh, J. Alarie, B. Cullum and G. Griffin, *Nature Biotech.*, 18 (2000) 76.
22 P. Hoffmann, B. Dutoit and R.P. Salathe, *Ultramicroscopy*, 61 (1995) 1.
23 P. Lambelet, A. Sayah, M. Pfeffer, et al., *Appl. Optics*, 37 (1998) 7289.
24 T. Vo-Dinh, (Eds.), *Chemical Analysis of Polycyclic Aromatic Compounds*, Wiley, New York, 1989.

# Index


absorbance measurements, 427
absorbance sensor, 549, 574
absorption filters, 420
absorption-type sensor, 550
acquisition, 246, 249
acquisition function, 258
active calibration system, 639
active dosing system, 643
actuation, 246
actuator device, 249
adsorptive preconcentration, 203
amperometric (bio)sensor, 17, 397
amperometric measurements, 223, 397
amperometric sensors, 397
analog-to-digital converter, 258
analytical decision making, 303
analytical instrument, 3
analytical laboratory, 287
analytical microsystem, 23, 25
analytical system, 10
antibody-based nanoprobes, 696
antibody probes, 688
antiresonant reflecting optical waveguide, 548
application programs, 262
aqua-impregnated resin, 58
array microsystem, 28
array sensor, 19, 20, 24
array system, 19
ARROW, 548, 575
artificial neural networks, 518
artificial sensory devices, 443
atomic force microscopy, 357
atomic spectrometry, 222
automated analytical system, 11
automatic calibration, 220

batch analyser, 11
beam splitters, 420
beetle/chip sensor, 612
biochip, 29
biocomposite, 17, 378
biocomposite inks, 402
biocomposite-based electrochemical sensors, 394
biocomposite-membrane based potentiometric biosensors, 396
biocomposite pastes, 403
bioelectronic sensor, 612
bioFET, 613
biologically sensitive FET, 613
biomimetic system, 22
bioreceptor probes, 687
bioreceptors, 687
biosensing systems, 686
biosensors, 16, 333, 377, 413, 686
bi-phase separation, 38
bluetooth, 275

calibration facility, 654
CAN bus, 271
capillary electrophoresis, 213
carbon composite electrodes, 397
carbon paste electrodes, 397
carbon pastes, 382
carbon–polymer–enzyme biocomposites, 398
catalytic reactions, 176
catheter type chemical sensor, 625
central processing node, 247

channel geometry, 669
chemical analysis system, 665
chemical beam epitaxy, 556
chemical image, 20
chemical sensors, 15, 437, 541
chemical timer, 145
chemical vapor deposition, 559
chemically modified electrodes, 327
chemiluminescence measurements, 432
chemosensors, 16, 377
chip, 574
chromatographic microsystems, 27
cladding, 415
closed programming environments, 278
coagulated ion-exchanger spectrophotometry, 98
cold vapour mercury generation, 189
combined ion exchange–solvent extraction, 58
combined separation techniques, 57
combined taste sensor and electronic nose, 507
communication controller, 252
communication interfaces, 282
communications, 294
composite, 17, 378
composite-based electro chemical sensors, 394
compound semiconductors, 555
computer-controlling function, 259
conditioning, 246
conductimetry, 224
conducting biocomposites, 388
conducting composites, 380
conductometric sensors, 394
connectivity, 250
consolidated biocomposites, 394
consolidation composites, 382
continuous flow analyser, 11
continuous flow analytical systems, 161
core, 415
coulometry, 224
coupling a liquid chromatograph to a flow injection, 209
coupling flow injection to a gas chromatograph, 211
coupling flow injection to capillary electrophoresis, 213
coupling sub- and supercritical fluid extractors with fluid injection, 206
couplings with liquid–liquid separators, 190
couplings with liquid–solid separators and solid phase formation, 196
couplings with techniques involving gas-separation, 184
cross-selective sensor array system, 21
cross-sensitivity, 446

data base system, 254
data processing techniques, 516
data processing, 458
data to information conversion, 295
derivatisation reactions, 174
design, 667
design path, 666
detection cell, 675
detection in flow injection, 217
detection of wine flavor, 507
dialysers, 191
dialysis, 191
diffuse reflectance spectroscopy, 103
digital communication interfaces, 267
digital processing, 249
digital signal processor, 258
dip-and-read strip, 145
discrete sensors, 444
dispersed biocomposites, 393
dispersion composite, 381

displacement chromatography, 54
displaying, 246
disposable cartridge systems, 13
distributed analytical
instrumentation systems, 245
distributed instrumentation system,
254
distributed intelligence, 250, 252, 256
distributed measurement system,
245, 248
distribution, 246
DNA microarray, 29
Donnan dialysis, 191
drop planar sensor, 84
dry etching, 570

electrical energy, 180
electrically 'wired' enzymes, 340
electrocatalysis, 345
electrochemical biosensors, 328
electrochemical sensors, 17
electrode array, 440
electrogenerated reagents, 175
electrolytic preconcentration, 203
electron hopping, 346
electronic nose instrumentation, 520
electronic nose manufacturers, 523
electronic nose technology, 514
electronic noses, 21, 514
electronic tongue systems, 449
electronic tongues, 22, 442, 509
electrophoretic microsystems, 27
ellipsometry, 360
embedded processor, 255
energy sources, 176
enzyme redox catalysis, 329
epitaxial growth techniques, 555
equipment remote system, 252, 254
etching processes, 569
etching, 569
ethernet, 252, 271
evanescent field, 547
evanescent wave-type sensor, 552
evanescent wave, 417
external energy sources integrated
with flow injection, 176
extrinsic optical sensor, 543

ferroelectric crystals, 555
fiber-impregnated system, 133
fiberoptic chemical sensors, 685
fiberoptic sensors, 685
fiberoptics nanoprobcs, 689
fiberoptics nanosensor systems, 686,
689
fiberoptics-based nanosensor systems,
685
fibre-optic chemical sensor, 19
field buses, 270
filtration, 201
fixation process, 87
flow injection analysis system, 665
flow injection analysis, 11, 161
flow injection, 161
flow injection–detector interfaces,
219
flow-through amperometric sensor,
649
flow-through potentiometric
and amperometric sensor array,
646
flow-through reference electrode, 649
flow-through sensors, 223
fluorescence measurements, 430, 695
fluorescence sensor, 551
foundation field bus, 271
Fourier transform infrared
spectroscopy, 225
frontal ion exchange chromatography,
50
fully distributed instrumentation
system, 254, 255

gas-diffusers, 184
gas-diffusion, 184
gene chip, 29

general intelligence, 250
giga scale integration, 2
glasses, 556
global selectivity, 488
grating coupler, 577
guided propagation modes, 545

hand-held dedicated device, 84
heat sources, 176
heterogeneous monolayer membrane, 134
high level multiplexing, 249
'high order' sensing microsystem, 24
'high order' sensor array, 588
high-performance liquid chromatography, 12, 209
homogenous monolayer membrane, 133
horizontal systems, 143
HPLC, 12, 209
hybrid integration., 554
hydride generation, 190
hydrogels, 327
hyphenated system, 12, 225

$I^2C$, 250
IEEE-1451 standard, 268
IEEE-488, 250
immersion solid phase spectrometric procedures, 130
immobilization, 16, 424, 691
immunosensors, 688
impregnation composites, 382
industrial networks, 270
in-line coupling of simple non-chromatographic continuous separation units and flow injection, 182
in vivo monitoring, 626
information management, 296
input/output board, 249
installation qualification, 321
instrument automation, 292
instrument bus driver, 261
instrument control driver, 262
instrumentation systems, 245
instrumentation remote system, 252, 253
integrated analytical system, 4, 413
integrated chemical and/or biochemical analysis, 587
integrated chemical system, 3, 341
integrated circuit, 1
integrated microdialysis lab-on-a-chip, 653
integrated optical chemical sensors, 541
integrated optical system, 543
integrated optical transducer, 25, 541
integrated optics, 542
integrated separation systems, 37
integrated waveguides, 544
integrating effect of conventional flow injection units, 164
integration, 1, 3, 37, 40, 48, 83, 162, 288, 333, 377, 413, 543, 625, 686
integration and flow injection, 226
intelligence, 247
intelligence of communications, 250
intelligent sensor, 265
interferant removal substances, 139
interference filters, 420
interference removal layer, 141
internet, 247
intrinsic optical sensor, 425, 542
ion exchange-crystallisation, 65
ion exchange isothermal supersaturation, 65
ion-exchange resin thin-layer spectrophotometry, 98
ion exchange supersaturation of electrolytes, 71
ion exchange supersaturation of zwitterlites, 67
ion-exchanger phase absorptiometry, 86
ion-selective electrodes, 18, 437, 438

ion-sensitive field-effect transistor, 24, 396
ISE arrays, 440
ISFET, 263, 396
ISFET-based bioelectronic sensor, 612
ISFET-based diffusion-coefficient sensor, 610
ISFET-based flow-direction sensor, 609
ISFET-based flow-velocity sensor, 605
ISFET-based 'high order' hybrid module, 590
ISFET-based microlitre-titrator, 606
ISFET-based penicillin sensor, 596
ISFET-based pH sensor, 594
ISFET-based temperature sensor, 601
ISFET fabrication, 591
ISFETmeter, 263
'isolated-antenna' BioFET, 613

lab-on-a-chip, 26, 625, 628, 665
laboratory automation, 288, 291
laboratory-based sensor array systems, 526
laboratory-based systems, 526
laboratory information management system, 14, 287, 298
laboratory management, 303
laboratory on a chip, 587
laboratory operations, 303
laboratory unit operation, 14
lag time, 633
large scale integration, 2
layer fabrication, 558
levels of integration, 41
light source, 418
LIMS, 14, 298
LIMS matrix, 301, 305
liquid–liquid extraction, 192
liquid–liquid extractors, 190
liquid membranes, 45
local area network, 14
local intelligence, 252
local position, 248
logistic LIMS, 301
low level multiplexing, 249
luminescent-based sensors, 550

Mach–Zehnder interferometer, 552
mask aligner, 567
masks, 566
MATAS technology, 683
measurement publishing nodes, 247
measurement server, 252
membrane-based separation techniques, 182
membrane filtration solid-phase spectrometric procedures, 126
membrane filtration systems, 123
membrane 'problem' deposit systems, 129
membrane 'problem' equilibration systems, 125
membrane separations, 42
membrane separation processes, 44
membrane solid-phase spectrometric assays, 122
membranes, 42, 122
MEMS technology, 625
metal organic chemical vapour deposition, 556
metal organic vapour phase epitaxy, 556
metering layer, 139
micellar media, 174
micro, 22, 665
microanalytical system, 23
microarraying, 29
microcolumn, 197
microcomputers, 249
microdialysis, 192, 627
microdialysis lab-on-a-chip, 628, 652
microdialysis probe, 631, 654
microdiffusion, 190
microdistillation, 190
microdosing system, 643

microelectrodes, 23, 380
microelectromechanical system, 25, 625
micro-FIA, 667
micropump, 672
microsensor, 23
microsensor array, 28
microstructure, 22
microsystem, 22
micro scale, 665
micro-sized systems, 665
micro total analysis systems, 26, 587, 629
microwaves, 180
milli, 13
mini, 22
miniaturised systems, 13
minicolumn, 182
ministructure, 13
mixer, 673
modified electrodes, 17, 327
modified solid phases, 83
modular analysis system, 680
modular plug-in sensor arrays, 522
molecular architectures, 349
molecular beam epitaxy, 556
molecular recognition element, 421
molecular wires, 327
monitor and control, 303
monolayer system, 132
monolithic system, 676
monolithic, 554
multi-channel analysers, 20
multilayer system, 137
multiple parameter analyser, 20
multiplexed signal block, 249
multiplexing, 249
multisensor systems, 453

nano, 22
nanodevice, 31
nanodozing, 358
nanoindenting, 358
nanoprobe, 685
nanosensors, 686
nanosystem, 30
networked system, 14
nodes, 247
nose plus tongue combined system, 457, 507, 509
nuclear magnetic resonance, 225
numerical aperture, 416

olfaction, 442
on-line coupling of robotics and flow injection, 214
on-line monitoring systems, 528
on-line separation equipment and flow injection, 205
opacifying layer, 138
open programming environment, 277
operating system, 261
operational LIMS, 301
operational qualification, 322
optical chemical sensors, 413
optical components, 553
optical couplers, 420
optical fiber chemical sensors, 414
optical fibers, 413, 415
optical nanofiber probes, 695
optical waveguides, 545
optochemical sensor, 19
optode, 19, 413, 542
optoelectronic instrumentation, 418
optrode, *see optode*
organic materials, 557
organisational integration, 303

parallel integration, 58
particulated solid phase spectrofluorimetric procedures, 115
particulated solid-phase spectrometric assays, 86

particulated solid phase spectrophotometric procedures, 113
passive calibration system, 639
performance qualification, 322
perfusion liquid pump, 656
peristaltic pump, 168
personal computer, 249
pervaporation, 186
pervaporators, 186
phase sensor, 551
photoacoustic spectrometry, 103
photodetector, 420
photoinduced reactions, 175
photolithography, 566
photomask, 566
photoresists, 568
photoresponsive layer, 142
physical vapor deposition, 559
piezoelectric measurements, 224, 361
planar sensor, 18
planar waveguide, 545
point-of-care diagnostics, 13
potentiometric sensors, 395
probe recovery, 632
processing, 246
profibus, 271
protective layer, 143
protein films, 339
pump, 672
purge and trap, 186

quantification of taste, 499, 509
quartz crystal microbalance, 363

radiation-blocking layer, 138
reaction chamber, 674
read-out device, 420
reagent film, 133
reagent layer, 139
reagentless amperometric biosensors, 337
reagentless glucose biosensors, 345
real refractive index-type sensor, 550
redox hydrogels, 333
redox mediation, 334
redox mediators, 329, 333
redox reactions, 94, 174
redundant-sensor array system, 20
reflectance measurements, 429
reflecting layer, 138
registration layer, 141
remote analytical systems, 279
remote measurement nodes, 247
remote place, 256
remote position, 247
reporting and communications, 303
resin spot tests, 86
resonant mirror, 579
reverse flow injection, 164
reverse frontal ion exchange chromatography, 52
rib waveguide, 547
rigid carbon–polymer biocomposites, 398
rigid carbon–polymer immunocomposites, 400
robotic stations, 214
RS-232, 250
RS-485, 250

scale of bitterness, 502, 509
SCPI, 262
screen printing, 402
screening and flow injection, 226
segregation composites, 382
selective-sensor array system, 20
self-assembled enzyme films, 340, 357
self-assembled organized multilayer, 327
self-assembled polyelectrolyte, 339
semiconductors, 555
sensing, 246
sensor array, 19, 24, 427, 437, 444, 514, 587
sensor networks, 268
sensor-on-a-chip, 583

sensor strip, 18
sensor systems, 444
sensor technologies, 515
sensors, 248, 449
separation column, 48
separation layer, 142
separation membranes, 9
separation techniques for volatile species coupled to FI, 189
sequential injection analysis, 12
sequential integration, 55
signal distribution, 249
single-cell measurements, 696
single-use sensor, 84
smart instrumentation systems, 247
smart system, 247
software programming languages, 257
solid-phase extraction, 197
solid-phase microextraction, 197
solid phase spectrofluorimetry, 102
solid phase spectrometry, 87
solid reagents, 174
solid-phase spectrometric assays, 81
sol–gel materials, 557
solvent impregnated resin, 59
sorption, 89, 185
spectrofluorimetry, 101
spectrophotometric array detectors, 222
spot tests, 7, 86
spreading layer, 139
standard protocols, 252
statistical analysis, 518
stopped flow, 168
storage, 246
strategic LIMS, 302
strip tests, 7, 82
stripping techniques, 204, 405
subbing layer, 143
subcritical extractors, 206
supercritical fluid extractors, 207
supported liquid membrane, 9
supported reagents, 7
supramolecular assemblies, 32
surface plasmon resonance (SPR)-based, 552
switching and routing systems, 260
system, 3,10, 245
system development life cycle, 311

tandem ion-exchange fractionation, 55
taste, 443, 487, 509
taste of amino acids, 490, 509
taste qualities, 502, 510
taste quantification, 475
taste sensor, 22, 488, 509
taste sensor and electronic nose, 507
TCP/IP, 252, 272
test and diagnostics program, 262
test paper, 131
test strip, 131, 146, 84
thick-film sensors, 402
thick-film technology, 402
thin films, 545, 559
'time-of-flight'-type thermal flow sensor, 605
TIR waveguide, 545, 548
total analysis systems, 10
total integrated system, 582
transmission, 246

ultramicroelectrodes, 23
ultrasounds, 177
unstable reagents layer, 142
UV–Vis spectrophotometric, 99

vertical systems, 132
virtual instrumentation, 250
virtual instruments, 262
voltammetric measurements, 223
voltammetric stripping techniques, 224, 405

water and wastewater monitoring, 524
wavelength selector, 419
web browsers, 252
wet etching, 569
'whole-beetle' bioFET, 613
wide area network, 14
wide-band signals, 274
'wired' enzyme, 351
wireless data transmission networks, 276
wireless Ethernet, 274
wireless local area network, 274
wireless transmission interfaces, 274
wireless wide area networks, 275
workstations, 214

'zero order' sensor arrays, 588

ISBN 0-444-51037-0

9 780444 510372